Student Study and Solutions Guide for

INTERMEDIATE ALGEBRA
GRAPHS AND FUNCTIONS

THIRD EDITION

Larson / Hostetler / Neptune

Carolyn F. Neptune
Johnson County Community College

Houghton Mifflin Company   Boston   New York

Publisher: Jack Shira
Managing Editor: Cathy Cantin
Development Manager: Maureen Ross
Development Editor: Laura Wheel
Assistant Editor: Rosalind Martin
Supervising Editor: Karen Carter
Senior Project Editor: Patty Bergin
Editorial Assistant: Meghan Lydon
Production Technology Supervisor: Gary Crespo
Executive Marketing Manager: Michael Busnach
Senior Marketing Manager: Danielle Potvin
Marketing Associate: Nicole Mollica
Senior Manufacturing Coordinator: Jane Spelman

Calculator Key font used with permission of Texas Instruments Incorporated. Copyright 1990, 1993, 1996.

Printed in the U.S.A.

ISBN: 0-618-21880-7

23456789-PO-05 04 03

Preface

This *Student Study and Solutions Guide* is a supplement to *Intermediate Algebra: Graphs and Functions,* Third Edition, by Ron Larson, Robert P. Hostetler, and Carolyn F. Neptune. This guide includes solutions for all odd-numbered exercises in the text, including the chapter reviews, chapter tests, and cumulative tests. These solutions give step-by-step details of each exercise. I have tried to see that these solutions are correct. Corrections to the solutions or suggestions for improvements are welcome.

I would like to thank the staff at Larson Texts, Inc. for their contributions to the production of this guide.

Carolyn F. Neptune
Johnson County Community College
Overland Park, Kansas

Contents

CHAPTER 1
Concepts of Elementary Algebra

C H A P T E R 1
Concepts of Elementary Algebra

Section 1.1   Algebraic Expressions

Solutions to Selected Odd-Numbered Exercises

1. $10x, 5$

3. $-3y^2, 2y, -8$

5. $4x^2, -3y^2, -5x, 2y$

7. $x^2, -2.5x, -\dfrac{1}{x}$

9. The coefficient of $5y^3$ is 5.

11. The coefficient of $-\frac{3}{4}t^2$ is $-\frac{3}{4}$.

13. Commutative Property of Addition

15. Associative Property of Multiplication

17. Multiplicative Inverse Property

19. Distributive Property

21. Multiplicative Identity Property

23. (a) Distributive Property:
$$5(x + 6) = 5 \cdot x + 5 \cdot 6 \text{ or } 5x + 30$$
(b) Commutative Property of Multiplication:
$$5(x + 6) = (x + 6)5$$

25. (a) Commutative Property of Multiplication:
$$6(xy) = (xy)6$$
(b) Associative Property of Multiplication:
$$6(xy) = (6x)y$$

27. (a) Additive Inverse Property:
$$4t^2 + (-4t^2) = 0$$
(b) Commutative Property of Addition:
$$4t^2 + (-4t^2) = (-4t^2) + 4t^2$$

29. $x^3 \cdot x^4 = (x \cdot x \cdot x) \cdot (x \cdot x \cdot x \cdot x)$

31. $(2y)^3 = (2y)(2y)(2y)$

33. $(-2x)^3 = (-2x)(-2x)(-2x)$

35. $\left(\dfrac{y}{5}\right)^4 = \left(\dfrac{y}{5}\right)\left(\dfrac{y}{5}\right)\left(\dfrac{y}{5}\right)\left(\dfrac{y}{5}\right)$

37. $\left(\dfrac{6}{x}\right)^4 = \left(\dfrac{6}{x}\right)\left(\dfrac{6}{x}\right)\left(\dfrac{6}{x}\right)\left(\dfrac{6}{x}\right)$

39. $(5x)(5x)(5x)(5x) = (5x)^4$

41. $(y \cdot y \cdot y)(y \cdot y \cdot y \cdot y) = y^3 \cdot y^4$

43. $(-z)(-z)(-z)(-z)(-z)(-z)(-z) = (-z)^7$

45. $x^5 \cdot x^2 = x^{5+2} = x^7$

47. $(a^2)^4 = a^{2 \cdot 4} = a^8$

49. $\dfrac{x^7}{x^3} = x^{7-3} = x^4$

51. $\left(\dfrac{a^2}{b}\right)^2 = \dfrac{(a^2)^2}{b^2} = \dfrac{a^{2 \cdot 2}}{b^2} = \dfrac{a^4}{b^2}$

53. $3^3 y^4 \cdot y^2 = 3^3 y^{4+2} = 27y^6$

55. $(-4x)^2 = (-4)^2 \cdot x^2 = 16x^2$

57. $(-5z^2)^3 = (-5)^3(z^2)^3 = -125z^6$

59. $(a^4 b^4)^4 = (a^4)^4(b^4)^4 = a^{4 \cdot 4} b^{4 \cdot 4} = a^{16} b^{16}$

61. $(x^3)(-x) = -x^{3+1} = -x^4$

63. $(2xy)(3x^2 y^3) = (2 \cdot 3) \cdot (x \cdot x^2) \cdot (y \cdot y^3)$
$$= 6 \cdot (x^{1+2}) \cdot (y^{1+3}) = 6x^3 y^4$$

65. $\dfrac{3^7 x^5}{3^3 x^3} = 3^{7-3} x^{5-3}$
$$= 3^4 x^2$$
$$= 81x^2$$

67. $\dfrac{(2xy)^5}{6(xy)^3} = \dfrac{2^5 x^5 y^5}{6x^3 y^3}$

$\qquad = \dfrac{32 x^5 y^5}{6 x^3 y^3}$

$\qquad = \dfrac{16(\not{2})}{3(\not{2})} x^{5-3} y^{5-3}$

$\qquad = \dfrac{16}{3} x^2 y^2 \text{ or } \dfrac{16 x^2 y^2}{3}$

69. $(5y)^2(-y^4) = 5^2 y^2 (-y^4)$

$\qquad = -5^2 y^{2+4}$

$\qquad = -25 y^6$

71. $-5z^4(-5z)^4 = -5z^4(-5)^4 z^4$

$\qquad = (-5)(-5)^4 z^4 \cdot z^4$

$\qquad = (-5)^{1+4} z^{4+4}$

$\qquad = (-5)^5 z^8$

$\qquad = -3125 z^8$

73. $(-2a)^2(-2a)^2 = (-2a)^{2+2}$

$\qquad = (-2a)^4$

$\qquad = (-2)^4 a^4$

$\qquad = 16 a^4 \qquad$ or

$(-2a)^2(-2a)^2 = (-2)^2 a^2 (-2)^2 a^2$

$\qquad = 4a^2 \cdot 4a^2$

$\qquad = 16 a^{2+2}$

$\qquad = 16 a^4$

75. $\dfrac{(2x)^4 y^2}{2x^3 y} = \dfrac{2^4 x^4 y^2}{2x^3 y}$

$\qquad = 2^{4-1} x^{4-3} y^{2-1}$

$\qquad = 2^3 xy$

$\qquad = 8xy$

77. $\dfrac{6(a^3 b)^3}{(3ab)^2} = \dfrac{6(a^3)^3 b^3}{3^2 a^2 b^2}$

$\qquad = \dfrac{6 a^9 b^3}{9 a^2 b^2}$

$\qquad = \dfrac{2(\not{3})}{3(\not{3})} a^{9-2} b^{3-2}$

$\qquad = \dfrac{2}{3} a^7 b \text{ or } \dfrac{2 a^7 b}{3}$

79. $\dfrac{-x^2 y^3}{(-xy)^2} = \dfrac{-x^2 y^3}{(-x)^2 (y)^2}$

$\qquad = \dfrac{-x^2 y^3}{x^2 y^2}$

$\qquad = -x^{2-2} y^{3-2}$

$\qquad = -1 y^1$

$\qquad = -y$

81. $-\left(\dfrac{2x^4}{5y}\right)^2 = -\dfrac{(2x^4)^2}{(5y)^2}$

$\qquad = -\dfrac{2^2 (x^4)^2}{5^2 y^2}$

$\qquad = -\dfrac{4 x^{4 \cdot 2}}{25 y^2}$

$\qquad = -\dfrac{4 x^8}{25 y^2}$

83. $\dfrac{x^{n+1}}{x^n} = x^{n+1-n} = x^1 = x$

85. $(x^n)^4 = x^{4n}$

87. $x^n \cdot x^3 = x^{n+3}$

89. $\dfrac{r^n s^m}{rs^3} = r^{n-1} s^{m-3}$

91. $3x + 4x = 7x$

93. $9y - 5y + 4y = 8y$

95. $3x - 2y + 5x + 20y = (3x + 5x) + (-2y + 20y)$

$\qquad = 8x + 18y$

97. $8z^2 + \dfrac{3}{2}z - \dfrac{5}{2}z^2 + 10 = \left(8z^2 - \dfrac{5}{2}z^2\right) + \dfrac{3}{2}z + 10$

$\qquad = \left(\dfrac{16}{2}z^2 - \dfrac{5}{2}z^2\right) + \dfrac{3}{2}z + 10$

$\qquad = \dfrac{11}{2}z^2 + \dfrac{3}{2}z + 10$

99. $2uv + 5u^2 v^2 - uv - (uv)^2 = 2uv + 5u^2 v^2 - uv - u^2 v^2$

$\qquad = (5u^2 v^2 - u^2 v^2) + (2uv - uv)$

$\qquad = 4u^2 v^2 + uv$

101. $5(ab)^2 + 2ab - 4ab = 5a^2 b^2 + (2ab - 4ab)$

$\qquad = 5a^2 b^2 - 2ab$

103. $10(x - 3) + 2x - 5 = 10x - 30 + 2x - 5$

$\qquad = (10x + 2x) + (-30 - 5)$

$\qquad = 12x - 35$

105. $(4x + 1) - (2x + 2) = 4x + 1 - 2x - 2$

$\qquad = (4x - 2x) + (1 - 2)$

$\qquad = 2x - 1$

107. $-(3z^2 - 2z + 4) + (z^2 - z - 2) = -3z^2 + 2z - 4 + z^2 - z - 2$

$$= (-3z^2 + z^2) + (2z - z) + (-4 - 2)$$

$$= -2z^2 + z - 6$$

109. $-3(y^2 + 3y - 1) + 2(y - 5) = -3y^2 - 9y + 3 + 2y - 10$

$$= (-3y^2) + (-9y + 2y) + (3 - 10)$$

$$= -3y^2 - 7y - 7$$

111. $4[5 - 3(x^2 + 10)] = 4[5 - 3x^2 - 30]$

$$= 4[-3x^2 - 25]$$

$$= -12x^2 - 100$$

113. $2[3(b - 5) - (b^2 + b + 3)] = 2[3b - 15 - b^2 - b - 3]$

$$= 2[-b^2 + 2b - 18]$$

$$= -2b^2 + 4b - 36$$

115. $y^2(y + 1) + y(y^2 + 1) = y^3 + y^2 + y^3 + y$

$$= (y^3 + y^3) + y^2 + y$$

$$= 2y^3 + y^2 + y$$

117. $x(xy^2 + y) - 2xy(xy + 1) = x^2y^2 + xy - 2x^2y^2 - 2xy$

$$= (x^2y^2 - 2x^2y^2) + (xy - 2xy)$$

$$= -x^2y^2 - xy$$

119. $-2a(3a^2)^3 + \dfrac{9a^8}{3a} = -2a(3^3)(a^2)^3 + 3a^{8-1}$

$$= -2(27)a \cdot a^6 + 3a^7$$

$$= -54a^7 + 3a^7$$

$$= -51a^7$$

121. (a) $5 - 3(5) = 5 - 15 = -10$

(b) $5 - 3\left(\dfrac{2}{3}\right) = 5 - 2 = 3$

123. (a) $10 - |3| = 10 - 3 = 7$

(b) $10 - |-3| = 10 - 3 = 7$

125. (a) $3(-1)^2 - (-1) + 7 = 3(1) + 1 + 7 = 3 + 1 + 7 = 11$

(b) $3\left(\frac{1}{3}\right)^2 - \left(\frac{1}{3}\right) + 7 = 3\left(\frac{1}{9}\right) - \frac{1}{3} + 7 = \frac{1}{3} - \frac{1}{3} + 7 = 7$

127. (a) $\dfrac{0}{0^2 + 1} = \dfrac{0}{1} = 0$

(b) $\dfrac{3}{3^2 + 1} = \dfrac{3}{9 + 1} = \dfrac{3}{10}$

129. (a) $3(1) + 2(5) = 3 + 10 = 13$

(b) $3(-6) + 2(-9) = -18 - 18 = -36$

131. (a) $(-1)^2 + 3(-1)(2) - (2)^2 = 1 - 6 - 4 = -9$

(b) $(-6)^2 + 3(-6)(-3) - (-3)^2 = 36 + 54 - 9 = 81$

133. (a) $|5 - 2| = |3| = 3$

(b) $|(-2) - (-2)| = |-2 + 2| = |0| = 0$

135. (a) $40\left(5\frac{1}{4}\right) = \left(\frac{40}{1}\right)\left(\frac{21}{4}\right) = 210$

(b) $35(4) = 140$

137. (a) Perimeter $= 2 + 2x + 4 + x + 2 + x$

$\qquad = (2x + x + x) + (2 + 2 + 4)$

$\qquad = 4x + 8$

(b) Area of upper rectangle + Area of lower rectangle = Area of region

Area of upper rectangle $= 2(x) = 2x$

Area of lower rectangle $= 4(x) = 4x$

Area of region $= 2x + 4x = 6x$

139. (a) Perimeter $= (b + 1) + \left(\frac{1}{2}b + 1\right) + b$

$\qquad = \left(b + \frac{1}{2}b + b\right) + (1 + 1)$

$\qquad = \frac{5}{2}b + 2$

(b) Area $= \frac{1}{2}$(Base)(Height)

$\qquad = \frac{1}{2}(b)\left(\frac{1}{2}b + 1\right)$

$\qquad = \frac{1}{4}b^2 + \frac{1}{2}b$

141. Area $= \frac{1}{2}$(Base)(Height)

$\qquad = \frac{1}{2}(b)(b - 3)$

$\qquad = \frac{1}{2}b(b - 3)$

$\qquad = \frac{1}{2}b^2 - \frac{3}{2}b$

When $b = 15$, Area $= \frac{1}{2}(15)^2 - \frac{3}{2}(15)$

$\qquad = \frac{1}{2}(255) - \frac{3}{2}(15)$

$\qquad = 90$ square units

143. Area $=$ (Base)(Height)

$\qquad = \left(\frac{5}{4}h + 10\right)(h)$

$\qquad = \frac{5}{4}h^2 + 10h$

When $h = 12$, Area $= \frac{5}{4}(12)^2 + 10(12)$

$\qquad = \frac{5}{4}(144) + 120$

$\qquad = 300$ square units

145. Graph: $6500 million

Model: $51.65(6)^2 + 82.8(6) + 4142 = 1859.4 + 496.8 + 4142$

$\qquad \approx \$6498$ million

$\qquad \approx \$6,498,000,000$

147. $b_1 h + \frac{1}{2}(b_2 - b_1)h = b_1 h + \left(\frac{1}{2}b_2 - \frac{1}{2}b_1\right)h$

$\qquad = b_1 h + \frac{1}{2}b_2 h - \frac{1}{2}b_1 h$

$\qquad = \left(b_1 h - \frac{1}{2}b_1 h\right) + \frac{1}{2}b_2 h$

$\qquad = \frac{1}{2}b_1 h + \frac{1}{2}b_2 h$

$\qquad = \frac{1}{2}h(b_1 + b_2)$

$\qquad = \frac{1}{2}(b_1 + b_2)$

So, the second expression for the area of the trapezoid is equivalent to the first.

Note: The second form for the area of the trapezoid comes from recognizing that the area of the trapezoid is equal to the area of the rectangle plus the area of the triangle.

Area of trapezoid $= b_1 h + \frac{1}{2}(b_2 - b_1)h$

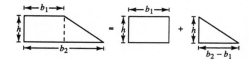

149. Area of Entire Rectangle = Area of Rectangle on Left + Area of Rectangle on Right

$a(b + c) \qquad\qquad = ab \qquad\qquad + ac$

151. Constants are real numbers and variables are letters used to represent numbers.

153. In the first expression, the base of the exponent is $2x$, but in the second expression, the base of the exponent is x. The first expression, $(2x)^3 = 2^3 x^3$ or $8x^3$.

155. True. This statement is justified by the Distributive Property.

Section 1.2   Operations with Polynomials

1. Standard form: $10x - 4$
Degree: 1
Leading coefficient: 10

3. Standard form: $-3y^4 + 5$
Degree: 4
Leading coefficient: -3

5. Standard form: $-16z^2 + 8z$
Degree: 5
Leading coefficient: -16

7. Standard form: $4t^5 - t^2 + 6t + 3$
Degree: 5
Leading coefficient: 4

9. Standard form: $x^3 - 5x^2 + x - 5$
Degree: 3
Leading coefficient: 1

11. Standard form: x
Degree: 1
Leading coefficient: 1

13. $12 - 5y^2$ is a binomial.

15. $x^3 + 2x^2 - 4$ is a trinomial.

17. $1.3x^2$ is a monomial.

19. A monomial of degree 3 is a polynomial of the form ax^3 where $a \neq 0$. Examples include $5x^3$, $-2x^3$, and $\frac{7}{4}x^3$.

21. A trinomial of degree 4 and leading coefficient -2 contains the term $-2x^4$ and two other terms of degree less than 4. Examples include $-2x^4 + x^2 + 1$ and $-2x^4 + x^3 - x$.

23. A monomial of degree 1 and leading coefficient 7 has the term with an exponent of one. Examples include $7x$ or $7y$.

25. A monomial of degree 0 is a constant. Examples include $10, -2, \frac{5}{4}$, and π.

27. (a) $(-2)^3 - 12(-2) = -8 + 24 = 16$

(b) $0^3 - 12(0) = 0 - 0 = 0$

29. (a) $(-1)^4 - 4(-1)^3 + 16(-1) - 16 = 1 + 4 - 16 - 16 = -27$

(b) $\left(\frac{5}{2}\right)^4 - 4\left(\frac{5}{2}\right)^3 + 16\left(\frac{5}{2}\right) - 16 = \frac{625}{16} - \frac{125}{2} + 40 - 16 = \frac{9}{16}$

31. $(2x^2 - 3) + (5x^2 + 6) = (2x^2 + 5x^2) + (-3 + 6)$
$$= 7x^2 + 3$$

33. $(x^2 - 3x + 8) + (2x^2 - 4x) + 3x^2 = (x^2 + 2x^2 + 3x^2) + (-3x - 4x) + (8)$
$$= 6x^2 - 7x + 8$$

35.
$$
\begin{array}{r}
5x^2 - 3x + 4 \\
-3x^2 \quad\ \ - 4 \\
\hline
2x^2 - 3x \quad\ \
\end{array}
$$

37.
$$
\begin{array}{r}
2b - 3 \\
b^2 - 2b \quad\ \ \\
-b^2 \quad\quad\ \ + 7 \\
\hline
4
\end{array}
$$

39. $(3x^2 - 2x + 1) - (2x^2 + x - 1) = 3x^2 - 2x + 1 - 2x^2 - x + 1$
$$= (3x^2 - 2x^2) + (-2x - x) + (1 + 1)$$
$$= x^2 - 3x + 2$$

41. $(10x^3 + 15) - (6x^3 - x + 11) = 10x^3 + 15 - 6x^3 + x - 11$
$$= (10x^3 - 6x^3) + x + (15 - 11)$$
$$= 4x^3 + x + 4$$

43. $\begin{aligned} x^2 - x + 3 &\Rightarrow x^2 - x + 3 \\ -(x - 2) &\Rightarrow \underline{ - x + 2} \\ & \quad x^2 - 2x + 5 \end{aligned}$

45. $\begin{aligned} -2x^3 - 15x + 25 &\Rightarrow -2x^3 - 15x + 25 \\ -(2x^3 - 13x + 12) &\Rightarrow \underline{-2x^3 + 13x - 12} \\ & \quad -4x^3 - 2x + 13 \end{aligned}$

47. $\begin{aligned} (3x^2 + 8) + (7 - 5x^2) &= (3x^2 - 5x^2) + (8 + 7) \\ &= -2x^2 + 15 \end{aligned}$

49. $\begin{aligned} (4x^2 + 5x - 6) - (2x^2 - 4x + 5) &= 4x^2 + 5x - 6 - 2x^2 + 4x - 5 \\ &= (4x^2 - 2x^2) + (5x + 4x) + (-6 - 5) \\ &= 2x^2 + 9x - 11 \end{aligned}$

51. $\begin{aligned} (10x^2 - 11) - (-7x^3 - 12x^2 - 15) &= 10x^2 - 11 + 7x^3 + 12x^2 + 15 \\ &= (7x^3) + (10x^2 + 12x^2) + (-11 + 15) \\ &= 7x^3 + 22x^2 + 4 \end{aligned}$

53. $\begin{aligned} 5s - [6s - (30s + 8)] &= 5s - [6s - 30s - 8] \\ &= 5s - [-24s - 8] = 5s + 24s + 8 \\ &= 29s + 8 \end{aligned}$

55. $\begin{aligned} (8x^3 - 4x^2 + 3x) - [(x^3 - 4x^2 + 5) + (x - 5)] &= 8x^3 - 4x^2 + 3x - [x^3 - 4x^2 + 5 + x - 5] \\ &= 8x^3 - 4x^2 + 3x - [x^3 - 4x^2 + x] \\ &= 8x^3 - 4x^2 + 3x - x^3 + 4x^2 - x \\ &= (8x^3 - x^3) + (-4x^2 + 4x^2) + (3x - x) \\ &= 7x^3 + 2x \end{aligned}$

57. $\begin{aligned} [5(2x^3 + 1) - (3x^3 - 12x^2 + 4x + 2)] + 3(x^2 + 2x - 1) &= 10x^3 + 5 - 3x^3 + 12x^2 - 4x - 2 + 3x^2 + 6x - 3 \\ &= (10x^3 - 3x^3) + (12x^2 + 3x^2) + (-4x + 6x) + (5 - 2 - 3) \\ &= 7x^3 + 15x^2 + 2x \end{aligned}$

59. $\begin{aligned} 2(t^2 + 12) - 5(t^2 + 5) + 6(t^2 + 5) &= 2t^2 + 24 - 5t^2 - 25 + 6t^2 + 30 \\ &= (2t^2 - 5t^2 + 6t^2) + (24 - 25 + 30) \\ &= 3t^2 + 29 \end{aligned}$

61. $\begin{aligned} (2z^2 + z - 11) + 3(z^2 + 4z + 5) - 2(2z^2 - 5z + 10) &= 2z^2 + z - 11 + 3z^2 + 12z + 15 - 4z^2 + 10z - 20 \\ &= (2z^2 + 3z^2 - 4z^2) + (z + 12z + 10z) + (-11 + 15 - 20) \\ &= z^2 + 23z - 16 \end{aligned}$

63. $\begin{aligned} 2(5x^3 - 13x) + (4x^3 - 9x^2 + 3x) - 3(x^3 - 2x^2 + 6x - 5) &= 10x^3 - 26x + 4x^3 - 9x^2 + 3x - 3x^3 + 6x^2 - 18x + 15 \\ &= (10x^3 + 4x^3 - 3x^3) + (-9x^2 + 6x^2) + (-26x + 3x - 18x) + 15 \\ &= 11x^3 - 3x^2 - 41x + 15 \end{aligned}$

65. $8.04x^2 - 9.37x^2 + 5.62x^2 = 4.29x^2$

67. $(4.098a^2 + 6.349a) - (11.246a^2 - 9.342a) = 4.098a^2 + 6.349a - 11.246a^2 + 9.342a$

$$= -7.148a^2 + 15.691a$$

69. $(-2a^2)(-8a) = (-2)(-8)a^2 \cdot a = 16a^{2+1} = 16a^3$ **71.** $2y(5 - y) = (2y)(5) - (2y)(y) = 10y - 2y^2$

73. $4x^3(2x^2 - 3x + 5) = (4x^3)(2x^2) - (4x^3)(3x) + (4x^3)(5)$

$$= 8x^5 - 12x^4 + 20x^3$$

75. $-2x^2(5 + 3x^2 - 7x^3) = (-2x^2)(5) + (-2x^2)(3x^2) - (-2x^2)(7x^3)$

$$= -10x^2 - 6x^4 + 14x^5$$

$$\qquad\qquad \text{F} \quad \text{O} \quad \text{I} \quad \text{L}$$

77. $(x + 7)(x - 4) = x^2 - 4x + 7x - 28$

$$= x^2 + 3x - 28$$

$$\qquad\qquad \text{F} \quad \text{O} \quad \text{I} \quad \text{L}$$

79. $(5 - x)(3 + x) = 15 + 5x - 3x - x^2$

$$= 15 + 2x - x^2$$

$$= -x^2 + 2x + 15$$

$$\qquad\qquad \text{F} \quad \text{O} \quad \text{I} \quad \text{L}$$

81. $(2t - 1)(t + 8) = 2t^2 + 16t - t - 8$

$$= 2t^2 + 15t - 8$$

$$\qquad\qquad \text{F} \quad \text{O} \quad \text{I} \quad \text{L}$$

83. $(3a^4 + 5)(2a^4 - 7) = 6a^8 - 21a^4 + 10a^4 - 35$

$$= 6a^8 - 11a^4 - 35$$

$$\qquad\qquad \text{F} \quad \text{O} \quad \text{I} \quad \text{L}$$

85. $(2x + y)(3x + 2y) = 6x^2 + 4xy + 3xy + 2y^2$

$$= 6x^2 + 7xy + 2y^2$$

$$\qquad\qquad \text{F} \quad \text{O} \quad \text{I} \quad \text{L}$$

87. $(5x^2 - 3y)(x^2 + y) = 5x^4 + 5x^2y - 3x^2y - 3y^2$

$$= 5x^4 + 2x^2y - 3y^2$$

$$\qquad\qquad \text{F} \quad \text{O} \quad \text{I} \quad \text{L}$$

89. $\left(4y - \frac{1}{3}\right)(12y + 9) = 48y^2 + 36y - 4y - 3$

$$= 48y^2 + 32y - 3$$

91. $-3x(-5x)(5x + 2) = 15x^2(5x + 2)$

$$= 75x^3 + 30x^2$$

93. $5a(a + 2) - 3a(2a - 3) = 5a^2 + 10a - 6a^2 + 9a$

$$= (5a^2 - 6a^2) + (10a + 9a)$$

$$= -a^2 + 19a$$

$$\qquad\qquad \text{F} \quad \text{O} \quad \text{I} \quad \text{L}$$

95. $(2t - 1)(t + 1) + 3(2t - 5) = 2t^2 + 2t - t - 1 + 6t - 15$

$$= 2t^2 + (2t - t + 6t) + (-1 - 15)$$

$$= 2t^2 + 7t - 16$$

97. $(x^3 - 3x + 2)(x - 2) = x^3(x - 2) - 3x(x - 2) + 2(x - 2)$

$$= x^4 - 2x^3 - 3x^2 + 6x + 2x - 4$$

$$= x^4 - 2x^3 - 3x^2 + 8x - 4$$

99. $(u + 5)(2u^2 + 3u - 4) = u(2u^2 + 3u - 4) + 5(2u^2 + 3u - 4)$

$$= 2u^3 + 3u^2 - 4u + 10u^2 + 15u - 20$$

$$= 2u^3 + 13u^2 + 11u - 20$$

101.

$$
\begin{array}{r}
7x^2 - 14x + 9 \\
\times \quad\quad x + 3 \\
\hline
21x^2 - 42x + 27 \\
7x^3 - 14x^2 + 9x \\
\hline
7x^3 + 7x^2 - 33x + 27
\end{array}
$$
$\Leftarrow 3(7x^2 - 14x + 9)$
$\Leftarrow x(7x^2 - 14x + 9)$

103.

$$
\begin{array}{r}
-x^2 + 2x - 1 \\
\times \quad\quad 2x + 1 \\
\hline
-x^2 + 2x - 1 \\
-2x^3 + 4x^2 - 2x \\
\hline
-2x^3 + 3x^2 \quad\quad - 1
\end{array}
$$
$\Leftarrow 1(-x^2 + 2x - 1)$
$\Leftarrow 2x(-x^2 + 2x - 1)$

105. Special Product: $(x - 4)(x + 4) = x^2 - 4^2 = x^2 - 16$

FOIL: $(x - 4)(x + 4) = x^2 + 4x - 4x - 16 = x^2 - 16$

107. Special Product: $(a - 6c)(a + 6c) = a^2 - (6c)^2 = a^2 - 36c^2$

FOIL: $(a - 6c)(a + 6c) = a^2 + 6ac - 6ac - 36c^2 = a^2 - 36c^2$

109. Special Product: $(2t + 9)(2t - 9) = (2t)^2 - 9^2 = 4t^2 - 81$

FOIL: $(2t + 9)(2t - 9) = 4t^2 + 18t - 18t - 81 = 4t^2 - 81$

111. Special Product: $\left(2x - \frac{1}{4}\right)\left(2x + \frac{1}{4}\right) = (2x)^2 - \left(\frac{1}{4}\right)^2 = 4x^2 - \frac{1}{16}$

FOIL: $\left(2x - \frac{1}{4}\right)\left(2x + \frac{1}{4}\right) = 4x^2 + \frac{1}{2}x - \frac{1}{2}x - \frac{1}{16} = 4x^2 - \frac{1}{16}$

113. Special Product: $(0.2t + 0.5)(0.2t - 0.5) = (0.2t)^2 - (0.5)^2 = 0.04t^2 - 0.25$

FOIL: $(0.2t + 0.5)(0.2t - 0.5) = 0.04t^2 - 0.1t + 0.1t - 0.25 = 0.04t^2 - 0.25$

115. Special Product: $(x^3 + 4)(x^3 - 4) = (x^3)^2 - 4^2 = x^6 - 16$

FOIL: $(x^3 + 4)(x^3 - 4) = x^6 - 4x^3 + 4x^3 - 16 = x^6 - 16$

117. Special Product: $(x + 5)^2 = x^2 + 2(x)(5) + 5^2 = x^2 + 10x + 25$

FOIL: $(x + 5)^2 = (x + 5)(x + 5) = x^2 + 5x + 5x + 25 = x^2 + 10x + 25$

119. Special Product: $(5x - 2)^2 = (5x)^2 - 2(5x)(2) + 2^2 = 25x^2 - 20x + 4$

FOIL: $(5x - 2)^2 = (5x - 2)(5x - 2) = 25x^2 - 10x - 10x + 4 = 25x^2 - 20x + 4$

121. Special Product: $(2a + 3b)^2 = (2a)^2 + 2(2a)(3b) + (3b)^2 = 4a^2 + 12ab + 9b^2$

FOIL: $(2a + 3b)^2 = (2a + 3b)(2a + 3b) = 4a^2 + 6ab + 6ab + 9b^2 = 4a^2 + 12ab + 9b^2$

123. Special Product: $(2x^4 - 3)^2 = (2x^4)^2 - 2(2x^4)(3) + 3^2 = 4x^8 - 12x^4 + 9$

FOIL: $(2x^4 - 3)^2 = (2x^4 - 3)(2x^4 - 3) = 4x^8 - 6x^4 - 6x^4 + 9 = 4x^8 - 12x^4 + 9$

125. $(x^2 + 4)(x^2 - 2x - 4) = x^2(x^2 - 2x - 4) + 4(x^2 - 2x - 4)$

$$= x^4 - 2x^3 - 4x^2 + 4x^2 - 8x - 16$$

$$= x^4 - 2x^3 - 8x - 16$$

127. $(t^2 + 5t - 1)(2t^2 - 5) = t^2(2t^2 - 5) + 5t(2t^2 - 5) - 1(2t^2 - 5)$

$$= 2t^4 - 5t^2 + 10t^3 - 25t - 2t^2 + 5$$

$$= 2t^4 + 10t^3 - 7t^2 - 25t + 5$$

129. $(a + 5)^3 = (a + 5)(a + 5)(a + 5)$

$\qquad = (a^2 + 5a + 5a + 25)(a + 5)$

$\qquad = (a^2 + 10a + 25)(a + 5)$

$\qquad = a^2(a + 5) + 10a(a + 5) + 25(a + 5)$

$\qquad = a^3 + 5a^2 + 10a^2 + 50a + 25a + 125$

$\qquad = a^3 + 15a^2 + 75a + 125$

131. $(2x - 3)^3 = (2x - 3)(2x - 3)(2x - 3)$

$\qquad = (4x^2 - 6x - 6x + 9)(2x - 3)$

$\qquad = (4x^2 - 12x + 9)(2x - 3)$

$\qquad = 4x^2(2x - 3) - 12x(2x - 3) + 9(2x - 3)$

$\qquad = 8x^3 - 12x^2 - 24x^2 + 36x + 18x - 27$

$\qquad = 8x^3 - 36x^2 + 54x - 27$

133. $(a^2 + 9a - 5)(a^2 - a + 3) = a^2(a^2 - a + 3) + 9a(a^2 - a + 3) - 5(a^2 - a + 3)$

$\qquad = a^4 - a^3 + 3a^2 + 9a^3 - 9a^2 + 27a - 5a^2 + 5a - 15$

$\qquad = a^4 + 8a^3 - 11a^2 + 32a - 15$

135. Special Product: $[(x + 2) - y]^2 = (x + 2)^2 - 2(x + 2)(y) + y^2$

$\qquad = x^2 + 2(x)(2) + 4 - 2y(x + 2) + y^2$

$\qquad = x^2 + 4x + 4 - 2xy - 4y + y^2$

FOIL: $[(x + 2) - y]^2 = [(x + 2) - y][(x + 2) - y]$

$\qquad = (x + 2)^2 - y(x + 2) - y(x + 2) + y^2$

$\qquad = (x + 2)(x + 2) - xy - 2y - xy - 2y + y^2$

$\qquad = x^2 + 2x + 2x + 4 - 2xy - 4y + y^2$

$\qquad = x^2 + 4x + 4 - 2xy - 4y + y^2$

137. Special Product: $[(2z + (y + 1)]^2 = (2z)^2 + 2(2z)(y + 1) + (y + 1)^2$

$\qquad = 4z^2 + 4yz + 4z + y^2 + 2(y)(1) + 1^2$

$\qquad = 4z^2 + 4yz + 4z + y^2 + 2y + 1$

FOIL: $[2z + (y + 1)]^2 = [2z + (y + 1)][2z + (y + 1)]$

$\qquad = 4z^2 + 2z(y + 1) + 2z(y + 1) + (y + 1)^2$

$\qquad = 4z^2 + 2yz + 2z + 2yz + 2z + (y + 1)(y + 1)$

$\qquad = 4z^2 + 4yz + 4z + y^2 + y + y + 1$

$\qquad = 4z^2 + 4yz + 4z + y^2 + 2y + 1$

139. $(x + 3)(x - 3) - (x^2 + 8x - 2) = x^2 - 3^2 - x^2 - 8x + 2$

$\qquad = x^2 - 9 - x^2 - 8x + 2$

$\qquad = -8x - 7$

141. $(t + 3)^2 - (t - 3)^2 = (t^2 + 2(t)(3) + 3^2) - (t^2 - 2(t)(3) + 3^2)$

$\qquad = (t^2 + 6t + 9) - (t^2 - 6t + 9)$

$\qquad = t^2 + 6t + 9 - t^2 + 6t - 9$

$\qquad = 12t$

143. Perimeter $= 2(\text{Length}) + 2(\text{Width})$

$\qquad = 2x^2 + 2x$

145. Perimeter $= 3y + (y + 4) + (4y - 5) + 2y$

$\qquad = 3y + y + 4 + 4y - 5 + 2y$

$\qquad = 10y - 1$

147. Area = (Length)(Width)

$$= (3x + 1)(2x - 5)$$

$$= 6x^2 - 15x + 2x - 5$$

$$= 6x^2 - 13x - 5$$

149. Area $= 3x(3x + 10) - x(x + 4)$

$$= 9x^2 + 30x - x^2 - 4x$$

$$= 8x^2 + 26x$$

151. Area $= \left(\frac{1}{2}\right)(6t)(7t + 4) - \left(\frac{1}{2}\right)(6t)(5t - 2)$

$$= (3t)(7t + 4) - (3t)(5t - 2)$$

$$= 21t^2 + 12t - (15t^2 - 6t)$$

$$= 21t^2 + 12t - 15t^2 + 6t$$

$$= 6t^2 + 18t$$

153. (a) $x(x + 3) = x^2 + 3x$

(b)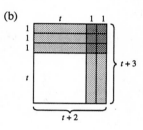

155. (a) $(t + 3)(t + 2) = t^2 + 2t + 3t + 6$

$$= t^2 + 5t + 6$$

(b)

157. Area = (Length)(Width)

$$= (x + a)(x + b)$$

Area $= A_1 + A_2 + A_3 + A_4$

$$= x(x) + (b)(x) + (a)(x) + (a)(b)$$

$$= x^2 + bx + ax + ab$$

Because both expressions represent the area, the two must be equal. Therefore,

$$(x + a)(x + b) = x^2 + bx + ax + ab.$$

This statement illustrates the pattern for the FOIL Method.

159. (a) Perimeter: 2(Length) + 2(Width) $= 2(1.5w) + 2(w)$

$$= 3w + 2w$$

$$= 5w$$

(b) Area: (Length)(Width) $= (1.5w)(w)$

$$= 1.5w^2 \text{ or } \frac{3}{2}w^2$$

161. Interest $= 1000(1 + r)^2 = 1000(1 + r)(1 + r)$

$$= 1000(1 + 2r + r^2)$$

$$= 1000 + 2000r + 1000r^2$$

163. $-16t^2 + 100 = -16t^2 + 0t + 100$

A position polynomial has the form $-16t^2 + v_0t + s_0$. In this example, the initial velocity $v_0 = 0$, so the object was dropped. At time $t = 0$, the height is $-16(0)^2 + 100 = 100$ feet. (**Note:** This height is s_0, the initial height.)

165. A position polynomial has the form $-16t^2 + v_0 t + s_0$. In this example, the initial velocity $v_0 = -24$, so the object was thrown downward ($v_0 < 0$). At time $t = 0$, the height is $-16(0)^2 - 24(0) + 50 = 50$ feet. (**Note:** This height is s_0, the initial height.)

167. Height $h = -16t^2 + 40t + 200$

$t = 1 \Longrightarrow h = -16(1)^2 + 40(1) + 200 = -16 + 40 + 200 = 224$ feet

$t = 2 \Longrightarrow h = -16(2)^2 + 40(2) + 200 = -64 + 80 + 200 = 216$ feet

$t = 3 \Longrightarrow h = -16(3)^2 + 40(3) + 200 = -144 + 120 + 200 = 176$ feet

169. The expression $x^2 - 3\sqrt{x}$ is not a polynomial because the second term cannot be expressed in the form ax^k where k is a nonnegative integer and a is a real number; $\sqrt{x}$ is not an integer power of x.

171. No. For example, the trinomial $5x^9 + x^7 + 8$ is a ninth-degree polynomial, and the trinomial $x^3 + 2x + 3$ is a third-degree polynomial.

173. (a) $(x - 1)(x + 1) = x^2 - 1^2$

$\qquad\qquad\qquad\quad = x^2 - 1$

(b) $(x - 1)(x^2 + x + 1) = x(x^2 + x + 1) - 1(x^2 + x + 1)$

$\qquad\qquad\qquad\qquad\quad = x^3 + x^2 + x - x^2 - x - 1$

$\qquad\qquad\qquad\qquad\quad = x^3 - 1$

(c) $(x - 1)(x^3 + x^2 + x + 1) = x(x^3 + x^2 + x + 1) - 1(x^3 + x^2 + x + 1)$

$\qquad\qquad\qquad\qquad\qquad\quad = x^4 + x^3 + x^2 + x - x^3 - x^2 - x - 1$

$\qquad\qquad\qquad\qquad\qquad\quad = x^4 - 1$

Following the pattern of the products in (a)–(c), $(x - 1)(x^4 + x^3 + x^2 + x + 1) = x^5 - 1$.

Section 1.3   Factoring Polynomials

1. $48 = 2^4 \cdot 3$

$90 = 5 \cdot 2 \cdot 3^2$

$GCF = 2 \cdot 3 = 6$

3. $3x^2 = 3 \cdot x \cdot x$

$12x = 2^2 \cdot 3 \cdot x$

$GCF = 3x$

5. $30z^2 = 2 \cdot 3 \cdot 5 \cdot z \cdot z$

$12z^3 = 2^2 \cdot 3 \cdot z \cdot z \cdot z$

$GCF = 2 \cdot 3 \cdot z \cdot z = 6z^2$

7. $28ab^2 = 7 \cdot 2^2 \cdot a \cdot b \cdot b$

$14a^2b^3 = 7 \cdot 2 \cdot a \cdot a \cdot b \cdot b \cdot b$

$42a^2b^5 = 7 \cdot 3 \cdot 2 \cdot a \cdot a \cdot b \cdot b \cdot b \cdot b \cdot b$

$GCF = 7 \cdot 2 \cdot a \cdot b \cdot b = 14ab^2$

9. $42(x + 8)^2 = 7 \cdot 3 \cdot 2 \cdot (x + 8)^2$

$63(x + 8)^3 = 7 \cdot 3^2 \cdot (x + 8)^3$

$GCF = 7 \cdot 3(x + 8)^2 = 21(x + 8)^2$

11. $4x(1 - z)^2 = 2^2 \cdot x \cdot (1 - z)^2$

$x^2(1 - z)^3 = x \cdot x \cdot (1 - z)^3$

$GCF = x(1 - z)^2$

13. $8z - 8 = 8(z - 1)$

15. $24x^2 - 18 = 6(4x^2 - 3)$

17. $2x^2 + x = x(2x + 1)$

19. $21u^2 - 14u = 7u(3u - 2)$

21. $11u^2 + 9$

No common factor other than 1 or -1

23. $3x^2y^2 - 15y = 3y(x^2y - 5)$

25. $28x^2 + 16x - 8 = 4(7x^2 + 4x - 2)$

27. $14x^4 + 21x^3 + 9x^2 = x^2(14x^2 + 21x + 9)$

29. $17x^5y^3 - xy^2 + 34y^2 = y^2(17x^5y - x + 34)$

31. $9x^3y + 6xy^2 = 3xy(3x^2 + 2y)$

33. $3x^3y^3 - 2x^2y^2 + 5xy = xy(3x^2y^2 - 2xy + 5)$

35. $10 - 5x = -5(-2 + x) = -5(x - 2)$

37. $7 - 14x = -7(-1 + 2x) = -7(2x - 1)$

39. $8 + 4x - 2x^2 = -2(-4 - 2x + x^2)$
$$= -2(x^2 - 2x - 4)$$

41. $2t - 15 - 4t^2 = -1(-2t + 15 + 4t^2)$
$$= -1(4t^2 - 2t + 15)$$

43. $\frac{3}{2}x + \frac{5}{4} = \frac{6}{4}x + \frac{5}{4} = \frac{1}{4}(6x + 5)$

The missing factor is $(6x + 5)$.

45. $\frac{5}{6}x + \frac{5}{9}y = \frac{15}{18}x + \frac{10}{18}y = \frac{5}{18}(3x + 2y)$

The missing factor is $(3x + 2y)$.

47. $2y(y - 3) + 5(y - 3) = (y - 3)(2y + 5)$

49. $5t(t^2 + 1) - 4(t^2 + 1) = (t^2 + 1)(5t - 4)$

51. $a(a + 6) + (2a - 5)(a + 6) = (a + 6)[a + (2a - 5)]$
$$= (a + 6)(3a - 5)$$

53. $y^3 - 6y^2 + 2y - 12 = (y^3 - 6y^2) + (2y - 12)$
$$= y^2(y - 6) + 2(y - 6)$$
$$= (y - 6)(y^2 + 2)$$

55. $14z^3 + 21z^2 - 6z - 9 = (14z^3 + 21z^2) + (-6z - 9)$
$$= 7z^2(2z + 3) - 3(2z + 3)$$
$$= (2z + 3)(7z^2 - 3)$$

57. $x^3 + 2x^2 + x + 2 = (x^3 + 2x^2) + (x + 2)$
$$= x^2(x + 2) + 1(x + 2)$$
$$= (x + 2)(x^2 + 1)$$

59. $a^3 - 4a^2 + 2a - 8 = (a^3 - 4a^2) + (2a - 8)$
$$= a^2(a - 4) + 2(a - 4)$$
$$= (a - 4)(a^2 + 2)$$

61. $z^4 + 3z^3 - 2z - 6 = (z^4 + 3z^3) + (-2z - 6)$
$$= z^3(z + 3) - 2(z + 3)$$
$$= (z + 3)(z^3 - 2)$$

63. $cd + 3c - 3d - 9 = (cd + 3c) + (-3d - 9)$
$$= c(d + 3) - 3(d + 3)$$
$$= (d + 3)(c - 3)$$

65. $x^2 - 64 = x^2 - 8^2$
$$= (x + 8)(x - 8)$$

67. $100 - 9y^2 = 10^2 - (3y)^2$
$$= (10 + 3y)(10 - 3y)$$

69. $121 - y^2 = 11^2 - y^2$
$$= (11 + y)(11 - y)$$

71. $16y^2 - 9z^2 = (4y)^2 - (3z)^2$
$$= (4y + 3z)(4y - 3z)$$

73. $x^2 - 4y^2 = x^2 - (2y)^2$
$$= (x + 2y)(x - 2y)$$

75. $a^8 - 36 = (a^4)^2 - 6^2$
$$= (a^4 + 6)(a^4 - 6)$$

77. $a^2b^2 - 16 = (ab)^2 - 4^2$
$$= (ab + 4)(ab - 4)$$

79. $(a + 4)^2 - 49 = (a + 4)^2 - 7^2$
$$= [(a + 4) + 7][(a + 4) - 7]$$
$$= (a + 11)(a - 3)$$

81. $81 - (z + 5)^2 = 9^2 - (z + 5)^2$
$$= [9 + (z + 5)][9 - (z + 5)]$$
$$= (9 + z + 5)(9 - z - 5)$$
$$= (14 + z)(4 - z)$$

83. $x^3 - 8 = x^3 - 2^3$
$$= (x - 2)(x^2 + 2x + 4)$$

85. $y^3 + 125 = y^2 + 5^3$
$$= (y + 5)(y^2 - 5y + 25)$$

87. $8t^3 - 27 = (2t)^3 - 3^3$

$\quad = (2t - 3)(4t^2 + 6t + 9)$

89. $27s^2 + 64 = (3s)^3 + 4^3$

$\quad = (3s + 4)(9s^2 - 12s + 16)$

91. $8x^3 - y^3 = (2x)^3 - y^3$

$\quad = (2x - y)(4x^2 + 2xy + y^2)$

93. $y^3 + 64z^3 = y^3 + (4z)^3$

$\quad = (y + 4z)(y^2 - 4yz + 16z^2)$

95. $50x^3 - 8 = 2(25x^2 - 4)$

$\quad = 2[(5x)^2 - 2^2]$

$\quad = 2(5x + 2)(5x - 2)$

97. $x^3 - 144x = x(x^2 - 144)$

$\quad = x(x^2 - 12^2)$

$\quad = x(x + 12)(x - 12)$

99. $b^4 - 16 = (b^2)^2 - 4^2$

$\quad = (b^2 + 4)(b^2 - 4)$

$\quad = (b^2 + 4)(b^2 - 2^2)$

$\quad = (b^2 + 4)(b + 2)(b - 2)$

101. $y^4 - 81x^4 = (y^2)^2 - (9x^2)^2$

$\quad = (y^2 + 9x^2)(y^2 - 9x^2)$

$\quad = (y^2 + 9x^2)(y^2 - (3x)^2)$

$\quad = (y^2 + 9x^2)(y + 3x)(y - 3x)$

103. $2x^3 - 54 = 2(x^3 - 27)$

$\quad = 2(x^3 - 3^3)$

$\quad = 2(x - 3)(x^2 + 3x + 9)$

105. $3a^3 + 192 = 3(a^3 + 64)$

$\quad = 3(a^3 + 4^3)$

$\quad = 3(a + 4)(a^2 - 4a + 16)$

107. $4x^{2n} - 25 = (2x^n)^2 - 5^2$

$\quad = (2x^n + 5)(2x^n - 5)$

109. $3x^3 + 4x^2 - 3x - 4 = (3x^3 + 4x^2) + (-3x - 4)$ or $3x^3 + 4x^2 - 3x - 4 = (3x^3 - 3x) + (4x^2 - 4)$

$\qquad = x^2(3x + 4) - 1(3x + 4)$

$\qquad = (3x + 4)(x^2 - 1)$

$\qquad = (3x + 4)(x + 1)(x - 1)$

$\qquad = 3x(x^2 - 1) + 4(x^2 - 1)$

$\qquad = (x^2 - 1)(3x + 4)$

$\qquad = (x - 1)(x + 1)(3x + 4)$

111. $P + Prt = P(1 + rt)$

113. Area $=$ (Length)(Width)

$\quad 32w - w^2 = $ (Length)(w)

$\quad 32w - w^2 = w(32 - w)$ or $(32 - w)w$

The length is $32 - w$.

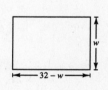

115. Area $=$ (Length)(Width)

$\quad w^2 + 8w = $ (Length)(w)

$\quad w^2 + 8w = w(w + 8)$ or $(w + 8)w$

The length is $w + 8$.

117. Area $=$ (Length)(Width)

$\quad 10x^3 + 4x^2 = $ (Length)$(2x^2)$

$\quad 10x^3 + 4x^2 = 2x^2(5x + 2)$

The length is $5x + 2$.

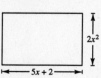

119. $S = 2\pi r^2 + 2\pi rh$

$\quad = 2\pi r(r + h)$

121. $kQx - kx^2 = kx(Q - x)$

123. A polynomial is in factored form if it is written as a product of polynomials. [**Note:** A polynomial has been *completely* factored if none of its polynomial factors can be factored further using integer coefficients.]

125. $79 \cdot 81 = (80 - 1)(80 + 1)$
$$= 80^2 - 1^2$$
$$= 6399$$

Mid-Chapter Quiz for Chapter 1

1. Distributive Property

2. Multiplicative Identity Property

3. Additive Inverse Property

4. $(5x)(5x)(5x)(5x) = (5x)^4$

5. $4x^2 \cdot x^3 = 4x^{2+3} = 4x^5$

6. $(-2x)^4 = (-2)^4 x^4 = 16x^4$

7. $\left(\dfrac{y^2}{3}\right)^3 = \dfrac{(y^2)^3}{3^3} = \dfrac{y^{2\cdot3}}{27} = \dfrac{y^6}{27}$

8. $\dfrac{18x^2y^3}{12xy} = \dfrac{\cancel{6}(3)}{\cancel{6}(2)}x^{2-1}y^{3-1} = \dfrac{3}{2}xy^2$ or $\dfrac{3xy^2}{2}$

9. $4x^2 - 3xy + 5xy - 5x^2 = (4x^2 - 5x^2) + (-3xy + 5xy)$
$$= -x^2 + 2xy$$

10. $(4x^2 - x + 7) - (x^2 - 3x - 1) = 4x^2 - x + 7 - x^2 + 3x + 1$
$$= 3x^2 + 2x + 8$$

11. $3(x - 5) + 4x = 3x - 15 + 4x$
$$= (3x + 4x) - 15$$
$$= 7x - 15$$

12. $3[x - 2x(x^2 + 1)] = 3[x - 2x^3 - 2x]$
$$= 3x - 6x^3 - 6x$$
$$= -6x^3 + (3x - 6x)$$
$$= -6x^3 - 3x$$

13. $(6r + 5s)(6r - 5s) = (6r)^2 - (5s)^2$
$$= 36r^2 - 25s^2$$

14. $(2x^2 - 4)(x + 3) = 2x^3 + 6x^2 - 4x - 12$

15. $(2x - 3y)^2 = (2x - 3y)(2x - 3y)$ or $(2x - 3y)^2 = (2x)^2 - 2(2x)(3y) + (3y)^2$
$$= 4x^2 - 6xy - 6xy + 9y^2 \qquad\qquad = 4x^2 - 12xy + 9y^2$$
$$= 4x^2 - 12xy + 9y^2$$

16. $(x + 1)(x^2 - x + 1) = x(x^2 - x + 1) + 1(x^2 - x + 1)$
$$= x^3 - x^2 + x + x^2 - x + 1$$
$$= x^3 + (-x^2 + x^2) + (x - x) + 1$$
$$= x^3 + 1$$

17. $2z(z + 5) - 7(z + 5) = 2z^2 + 10z - 7z - 35$
$$= 2z^2 + (10z - 7z) - 35$$
$$= 2z^2 + 3z - 35$$

18. $(v - 3)^2 - (v + 3)^2 = (v^2 - 6v + 9) - (v^2 + 6v + 9)$
$$= v^2 - 6v + 9 - v^2 - 6v - 9$$
$$= -12v$$

19. (a) $10(5)^2 - |5(5)| = 250 - |25| = 225$

(b) $10(-1)^2 - |5(-1)| = 10 - |-5| = 5$

20. Perimeter $= 2(2x + 5) + 2(x + 3)$
$$= 4x + 10 + 2x + 6$$
$$= (4x + 2x) + (10 + 6)$$
$$= 6x + 16$$

21. Standard form: $-16x^4 + 7x^2 + 4$

Degree: 4

Leading coefficient: -16

22. Area $= \left(\dfrac{1}{2}\right)(x + 2)(x + 2) - \left(\dfrac{1}{2}\right)(x)(x)$

$\qquad = \dfrac{(x + 2)(x + 2)}{2} - \dfrac{x^2}{2}$

$\qquad = \dfrac{x^2 + 4x + 4 - x^2}{2}$

$\qquad = \dfrac{4x + 4}{2}$

$\qquad = 2x + 2$

23. $24x^3 + 28x = 4x(6x^2 + 7)$

24. $3y(y - 5) + (y - 5) = (y - 5)(3y + 1)$

25. $x^3 - 9x^2 + 5x - 45 = x^2(x - 9) + 5(x - 9)$

$\qquad\qquad\qquad\qquad = (x - 9)(x^2 + 5)$

26. $64a^3 + b^3 = (4a)^3 + b^3$

$\qquad\qquad = (4a + b)(16a^2 - 4ab + b^2)$

27. $-16x^3 + 54 = -2(8x^3 - 27)$

$\qquad\qquad = -2[(2x)^3 - 3^3]$

$\qquad\qquad = -2(2x - 3)(4x^2 + 6x + 9)$

28. $81x^2 - 16y^2 = (9x)^2 - (4y)^2$

$\qquad\qquad = (9x + 4y)(9x - 4y)$

29. $x^3 - 9xy^2 = x(x^2 - 9y^2)$

$\qquad\qquad = x[x^2 - (3y)^2]$

$\qquad\qquad = x(x + 3y)(x - 3y)$

30. $t^3 + 2t^2 - t - 2 = (t^3 + 2t^2) + (-t - 2)$

$\qquad\qquad\qquad = t^2(t + 2) - 1(t + 2)$

$\qquad\qquad\qquad = (t + 2)(t^2 - 1)$

$\qquad\qquad\qquad = (t + 2)(t^2 - 1^2)$

$\qquad\qquad\qquad = (t + 2)(t + 1)(t - 1)$

31. Height $= -16t^2 + 45t + 350$

$\quad t = 1 \implies -16(1)^2 + 45(1) + 350 = -16 + 45 + 350 = 379$ feet

$\quad t = 2 \implies -16(2)^2 + 45(2) + 350 = -64 + 90 + 350 = 376$ feet

$\quad t = 3 \implies -16(3)^2 + 45(3) + 350 = -144 + 135 + 350 = 341$ feet

Section 1.4   Factoring Trinomials

1. $x^2 + 5x + 4 = (x + 4)(x + 1)$

The missing factor is $x + 1$.

3. $y^2 - y - 20 = (y + 4)(y - 5)$

The missing factor is $y - 5$.

5. $z^2 - 6z + 8 = (z - 4)(z - 2)$

The missing factor is $z - 2$.

7. $x^2 + 4x + 3 = (x + 3)(x + 1)$

9. $y^2 + 7y - 30 = (y + 10)(y - 3)$

11. $t^2 - 4t - 21 = (t - 7)(t + 3)$

13. $x^2 - 12x + 20 = (x - 10)(x - 2)$

15. $t^2 - 17t + 60 = (t - 12)(t - 5)$

17. $x^2 - 20x + 96 = (x - 12)(x - 8)$

19. $u^2 + 5uv + 6v^2 = (u + 3v)(u + 2v)$

21. $x^2 - 2xy - 35y^2 = (x - 7y)(x + 5y)$

23. $5x^2 + 18x + 9 = (x + 3)(5x + 3)$

The missing factor is $5x + 3$.

25. $5a^2 + 12a - 9 = (a + 3)(5a - 3)$

The missing factor is $5a - 3$.

27. $2y^2 - 3y - 27 = (y + 3)(2y - 9)$

The missing factor is $2y - 9$.

29. $3x^2 + 4x + 1 = (3x + 1)(x + 1)$

31. $8t^2 - 6t - 5 = (4t - 5)(2t + 1)$

33. $6b^2 + 19b - 7 = (3b - 1)(2b + 7)$

35. $2a^2 - 13a + 20 = (2a - 5)(a - 4)$

37. $2x^2 + 5xy + 3y^2 = (2x + 3y)(x + y)$

39. $11y^2 - 43yz - 4z^2 = (11y + z)(y - 4z)$

41. $6a^2 + ab - 2b^2 = (2a - b)(3a + 2b)$

43. $20x^2 + x - 12 = (5x + 4)(4x - 3)$

45. $2u^2 + 9uv - 35v^2 = (2u - 5v)(u + 7v)$

47. $15x^2 + 3xy - 8y^2$

This polynomial is not factorable using integer coefficients.

49. $-2x^2 - x + 6 = (-1)(2x^2 + x - 6)$

$\qquad\qquad\quad = (-1)(2x - 3)(x + 2)$

51. $1 - 11x - 60x^2 = -60x^2 - 11x + 1$

$\qquad\qquad\quad\ = (-1)(60x^2 + 11x - 1)$

$\qquad\qquad\quad\ = (-1)(15x - 1)(4x + 1)$

53. $ac = 3(8) = 24$ and $b = 10$

The two numbers with a product of 24 and a sum of 10 are 6 and 4.

$$3x^2 + 10x + 8 = 3x^2 + 6x + 4x + 8 \qquad \text{or} \qquad 3x^2 + 10x + 8 = 3x^2 + 4x + 6x + 8$$

$$= (3x^2 + 6x) + (4x + 8) \qquad\qquad\qquad = (3x^2 + 4x) + (6x + 8)$$

$$= 3x(x + 2) + 4(x + 2) \qquad\qquad\qquad = x(3x + 4) + 2(3x + 4)$$

$$= (x + 2)(3x + 4) \qquad\qquad\qquad\qquad = (3x + 4)(x + 2)$$

55. $ac = 6(-2) = -12$ and $b = 1$

The two numbers with a product of -12 and a sum of 1 are 4 and -3.

$$6x^2 + x - 2 = 6x^2 + 4x - 3x - 2 \qquad \text{or} \qquad 6x^2 + x - 2 = 6x^2 - 3x + 4x - 2$$

$$= (6x^2 + 4x) + (-3x - 2) \qquad\qquad\qquad = (6x^2 - 3x) + (4x - 2)$$

$$= 2x(3x + 2) - 1(3x + 2) \qquad\qquad\qquad = 3x(2x - 1) + 2(2x - 1)$$

$$= (3x + 2)(2x - 1) \qquad\qquad\qquad\qquad = (2x - 1)(3x + 2)$$

57. $ac = 15(2) = 30$ and $b = -11$

The two numbers with a product of 30 and a sum of -11 are -6 and -5.

$$15x^2 - 11x + 2 = 15x^2 - 6x - 5x + 2 \qquad \text{or} \qquad 15x^2 - 11x + 2 = 15x^2 - 5x - 6x + 2$$

$$= (15x^2 - 6x) + (-5x + 2) \qquad\qquad\qquad = (15x^2 - 5x) + (-6x + 2)$$

$$= 3x(5x - 2) - 1(5x - 2) \qquad\qquad\qquad = 5x(3x - 1) - 2(3x - 1)$$

$$= (5x - 2)(3x - 1) \qquad\qquad\qquad\qquad = (3x - 1)(5x - 2)$$

59. $x^2 + 4x + 4 = x^2 + 2(x)(2) + 2^2$

$\qquad\qquad\quad = (x + 2)^2$

61. $a^2 - 12a + 36 = a^2 - 2(a)(6) + 6^2$

$\qquad\qquad\qquad = (a - 6)^2$

63. $25y^2 - 10y + 1 = (5y)^2 - 2(5y)(1) + 1^2$

$\qquad\qquad\qquad\ = (5y - 1)^2$

65. $9b^2 + 12b + 4 = (3b)^2 + 2(3b)(2) + 2^2$

$\qquad\qquad\qquad = (3b + 2)^2$

67. $4x^2 - 4xy + y^2 = (2x)^2 - 2(2x)(y) + y^2$

$\qquad\qquad\qquad = (2x - y)^2$

69. $u^2 + 8uv + 16v^2 = u^2 + 2(u)(4v) + (4v)^2$

$\qquad\qquad\qquad\quad = (u + 4v)^2$

71. $3x^5 - 12x^3 = 3x^3(x^2 - 4)$

$\qquad\qquad\quad = 3x^3(x^2 - 2^2)$

$\qquad\qquad\quad = 3x^3(x + 2)(x - 2)$

73. $10t^3 + 2t^2 - 36t = 2t(5t^2 + t - 18)$

$\qquad\qquad\qquad = 2t(5t - 9)(t + 2)$

75. $5a^2 - 25a + 30 = 5(a^2 - 5a + 6)$

$\qquad\qquad\qquad = 5(a - 2)(a - 3)$

77. $6u^2 + 3u - 63 = 3(2u^2 + u - 21)$

$\qquad\qquad\qquad = 3(2u + 7)(u - 3)$

79. $9y^2 - 66y + 121 = (3y)^2 - (2)(3y)(11) + 11^2$

$\qquad\qquad\qquad\quad = (3y - 11)^2$

81. $4x(3x - 2) + (3x - 2)^2 = (3x - 2)[4x + (3x - 2)]$

$\qquad\qquad\qquad\qquad\quad = (3x - 2)(7x - 2)$

83. $x^2(x + 3) + 5x(x + 3) = [x(x + 3)](x) + [(x + 3)](5)$

$\qquad\qquad\qquad\qquad = x(x + 3)(x + 5)$

85. $36 - (z + 3)^2 = 6^2 - (z + 3)^2$

$\qquad\qquad\qquad = [6 + (z + 3)][6 - (z + 3)]$

$\qquad\qquad\qquad = (6 + z + 3)(6 - z - 3)$

$\qquad\qquad\qquad = (9 + z)(3 - z)$

87. $(2y - 1)^2 - 9 = (2y - 1)^2 - 3^2$

$\qquad\qquad\qquad = [(2y - 1) + 3][(2y - 1) - 3]$

$\qquad\qquad\qquad = (2y + 2)(2y - 4)$

$\qquad\qquad\qquad = 2(y + 1)(2)(y - 2)$

$\qquad\qquad\qquad = 4(y + 1)(y - 2)$

89. $54x^3 - 2 = 2(27x^3 - 1)$

$\qquad\qquad\quad = 2[(3x)^3 - 1^3]$

$\qquad\qquad\quad = 2(3x - 1)(9x^2 + 3x + 1)$

91. $v^3 + 3v^2 + 5v = v(v^2 + 3v + 5)$

93. $2x^3y - 2x^2y^2 - 84xy^3 = 2xy(x^2 - xy - 42y^2)$

$\qquad\qquad\qquad\qquad\quad = 2xy(x + 6y)(x - 7y)$

95. $5x^2y - 20y^3 = 5y(x^2 - 4y^2)$

$\qquad\qquad\qquad = 5y(x + 2y)(x - 2y)$

97. $x^3 + 2x^2 - 16x - 32 = (x^3 + 2x^2) + (-16x - 32)$

$\qquad\qquad\qquad\qquad = x^2(x + 2) - 16(x + 2)$

$\qquad\qquad\qquad\qquad = (x + 2)(x^2 - 16)$

$\qquad\qquad\qquad\qquad = (x + 2)(x + 4)(x - 4)$

99. $x^3 - 6x^2 - 9x + 54 = (x^3 - 6x^2) + (-9x + 54)$

$\qquad\qquad\qquad\qquad = x^2(x - 6) - 9(x - 6)$

$\qquad\qquad\qquad\qquad = (x - 6)(x^2 - 9)$

$\qquad\qquad\qquad\qquad = (x - 6)(x + 3)(x - 3)$

101. $x^2 - 10x + 25 - y^2 = (x^2 - 10x + 25) - y^2$

$\qquad\qquad\qquad\qquad = (x - 5)^2 - y^2$

$\qquad\qquad\qquad\qquad = [(x - 5) + y][(x - 5) - y]$

$\qquad\qquad\qquad\qquad = (x - 5 + y)(x - 5 - y)$

103. $a^2 - 2ab + b^2 - 16 = (a^2 - 2ab + b^2) - 16$

$\qquad\qquad\qquad\qquad = (a - b)^2 - 4^2$

$\qquad\qquad\qquad\qquad = [(a - b) + 4][(a - b) - 4]$

$\qquad\qquad\qquad\qquad = (a - b + 4)(a - b - 4)$

105. $x^8 - 1 = (x^4)^2 - 1^2$

$\qquad\qquad = (x^4 + 1)(x^4 - 1)$

$\qquad\qquad = (x^4 + 1)[(x^2)^2 - 1^2]$

$\qquad\qquad = (x^4 + 1)(x^2 + 1)(x^2 - 1)$

$\qquad\qquad = (x^4 + 1)(x^2 + 1)(x + 1)(x - 1)$

107. $b^4 - 216b = b(b^3 - 216)$

$\qquad = b(b^3 - 6^3)$

$\qquad = b(b - 6)(b^2 + 6b + 36)$

109. $x^2 + bx + 81 = x^2 + bx + 9^2$

$\quad (x + 9)^2 = x^2 + 2(x)(9) + 9^2 = x^2 + 18x + 81 \implies b = \quad 18$

$\quad (x - 9)^2 = x^2 - 2(x)(9) + 9^2 = x^2 - 18x + 81 \implies b = -18$

111. $9y^2 + by + 1 = (3y)^2 + by + 1^2$

$\quad (3y + 1)^2 = (3y)^2 + 2(3y)(1) + 1^2 = 9y^2 + 6y + 1 \implies b = \quad 6$

$\quad (3y - 1)^2 = (3y)^2 - 2(3y)(1) + 1^2 = 9y^2 - 6y + 1 \implies b = -6$

113. $4x^2 + bx + 9 = (2x)^2 + bx + 3^2$

$\quad (2x + 3)^2 = (2x)^2 + 2(2x)(3) + 3^2 = 4x^2 + 12x + 9 \implies b = \quad 12$

$\quad (2x - 3)^2 = (2x)^2 - 2(2x)(3) + 3^2 = 4x^2 - 12x + 9 \implies b = -12$

115. $x^2 + 8x + c = x^2 + 2(4x) + c$

$\quad (x + 4)^2 = x^2 + 2(x)(4) + 4^2 = x^2 + 8x + 16 \implies c = 16$

117. $y^2 - 6y + c = y^2 - 2(3y) + c$

$\quad (y - 3)^2 = y^2 - 2(y)(3) + 3^2 = y^2 - 6y + 9 \implies c = 9$

119. $16a^2 + 40a + c = (4a)^2 + 2(20a) + c = (4a)^2 + 2(4a)(5) + c$

$\quad (4a + 5)^2 = (4a)^2 + 2(4a)(5) + 5^2 = 16a^2 + 40a + 25 \implies c = 25$

121.

$\qquad (x + 6)^2 = x^2 + 2(x)(6) + 6^2$

$\qquad\qquad\quad = x^2 + 12x + 36$

$x^2 + 12x + 50 = x^2 + 12x + (36 + 14)$

$\qquad\qquad\quad = (x^2 + 12x + 36) + 14$

$\qquad\qquad\quad = (x + 6)^2 + 14$

The missing number is 14.

123. Possible factors of 18 are $(18)(1)$, $(-18)(-1)$, $(9)(2)$, $(-9)(-2)$, $(6)(3)$, and $(-6)(-3)$.

$\quad (x + 18)(x + 1) = x^2 + 19x + 18 \implies b = \quad 19$

$\quad (x - 18)(x - 1) = x^2 - 19x + 18 \implies b = -19$

$\quad (x + 9)(x + 2) = x^2 + 11x + 18 \implies b = \quad 11$

$\quad (x - 9)(x - 2) = x^2 - 11x + 18 \implies b = -11$

$\quad (x + 6)(x + 3) = x^2 + \ 9x + 18 \implies b = \quad 9$

$\quad (x - 6)(x - 3) = x^2 - \ 9x + 18 \implies b = \ -9$

125. Possible factors of -21 are $(21)(-1)$, $(-21)(1)$, $(7)(-3)$, and $(-7)(3)$.

$\quad (x + 21)(x - 1) = x^2 + 20x - 21 \implies b = \quad 20$

$\quad (x - 21)(x + 1) = x^2 - 20x - 21 \implies b = -20$

$\quad (x + 7)(x - 3) = x^2 + \ 4x - 21 \implies b = \quad 4$

$\quad (x - 7)(x + 3) = x^2 - \ 4x - 21 \implies b = -\ 4$

127. Possible factors of 8 are $(8)(1)$, $(-8)(-1)$, $(4)(2)$, and $(-4)(-2)$.

$\quad (5x + 8)(x + 1) = 5x^2 + 13x + 8 \implies b = \quad 13$

$\quad (5x + 1)(x + 8) = 5x^2 + 41x + 8 \implies b = \quad 41$

$\quad (5x - 8)(x - 1) = 5x^2 - 13x + 8 \implies b = -13$

$\quad (5x - 1)(x - 8) = 5x^2 - 41x + 8 \implies b = -41$

$\quad (5x + 4)(x + 2) = 5x^2 + 14x + 8 \implies b = \quad 14$

$\quad (5x + 2)(x + 4) = 5x^2 + 22x + 8 \implies b = \quad 22$

$\quad (5x - 4)(x - 2) = 5x^2 - 14x + 8 \implies b = -14$

$\quad (5x - 2)(x - 4) = 5x^2 - 22x + 8 \implies b = -22$

129. Some pairs of numbers with a sum of 6 are $4 + 2$, $8 + (-2)$, $3 + 3$, and $-4 + 10$.

$(x + 4)(x + 2) = x^2 + 6x + 8 \Rightarrow c = 8$

$(x + 8)(x - 2) = x^2 + 6x - 16 \Rightarrow c = -16$

$(x + 3)(x + 3) = x^2 + 6x + 9 \Rightarrow c = 9$

$(x - 4)(x + 10) = x^2 + 6x - 40 \Rightarrow c = -40$

131. Some pairs of numbers with a sum of -3 are $-2 + (-1)$, $-8 + 5$, $2 + (-5)$, and $-4 + 1$.

$(x - 2)(x - 1) = x^2 - 3x + 2 \Rightarrow c = 2$

$(x - 8)(x + 5) = x^2 - 3x - 40 \Rightarrow c = -40$

$(x + 2)(x - 5) = x^2 - 3x - 10 \Rightarrow c = -10$

$(x - 4)(x + 1) = x^2 - 3x - 4 \Rightarrow c = -4$

133. Some pairs of numbers with a sum of -4 are $-2 + (-2)$, $4 + (-8)$, $-1 + (-3)$, and $-5 + 1$.

$(t - 2)(t - 2) = t^2 - 4t + 4 \Rightarrow c = 4$

$(t + 4)(t - 8) = t^2 - 4t - 32 \Rightarrow c = -32$

$(t - 1)(t - 3) = t^2 - 4t + 3 \Rightarrow c = 3$

$(t - 5)(t + 1) = t^2 - 4t - 5 \Rightarrow c = -5$

135. $9(4) - x^2 = 36 - x^2$

$\qquad = (6 - x)(6 + x)$

$\qquad$ or $(-x + 6)(x + 6)$

137. Area $= \left(\frac{1}{2}\right)(2x)(x) - \left(\frac{1}{2}\right)(6)(3)$

$\qquad = x^2 - 9$

$\qquad = x^2 - 3^2$

$\qquad = (x + 3)(x - 3)$

139. Area = (Length)(Width)

$2x^2 + x - 15 = (2x - 5)(\text{Width})$

$2x^2 + x - 15 = (2x - 5)(x + 3)$

The width of the rectangle is $x + 3$.

141. Matches geometric factoring model (c).

143. Matches geometric factoring model (a).

145. $x^2 + 4x + 3 = (x + 3)(x + 1)$

147. $9x^2 - 9x - 54 = (3x + 6)(3x - 9)$

$\qquad = 3(x + 2)(3)(x - 3)$

$\qquad = 9(x + 2)(x - 3)$

The error was the omission of the common factor 3 which was factored out from the second binomial factor. The error could have been avoided by factoring out the greatest common factor as the first step.

$9x^2 - 9x - 54 = 9(x^2 - x - 6)$

$\qquad = 9(x + 2)(x - 3)$

149. $52^2 = (50 + 2)^2$

$\qquad = 50^2 + 2(50)(2) + 2^2$

$\qquad = 2500 + 200 + 4$

$\qquad = 2704$

Section 1.5   Solving Linear Equations

1. (a) $\qquad x = 0$

$3(0) - 7 \overset{?}{=} 2$

$-7 \neq 2$

No

(b) $\qquad x = 3$

$3(3) - 7 \overset{?}{=} 2$

$9 - 7 \overset{?}{=} 2$

$2 = 2$

Yes

3. (a) $\qquad x = -3$

$3 - 2(-3) \overset{?}{=} 21$

$3 + 6 \overset{?}{=} 21$

$9 \neq 21$

No

(b) $\qquad x = -9$

$3 - 2(-9) \overset{?}{=} 21$

$3 + 18 \overset{?}{=} 21$

$21 = 21$

Yes

5. (a)
$$x = -11$$
$$3(-11) + 3 \overset{?}{=} 2(-11 - 4)$$
$$-33 + 3 \overset{?}{=} 2(-15)$$
$$-30 = -30$$
Yes

(b)
$$x = 5$$
$$3(5) + 3 \overset{?}{=} 2(5 - 4)$$
$$15 + 3 \overset{?}{=} 2(1)$$
$$18 \neq 2$$
No

7.
$$3(x - 1) = 3x$$
$$3x - 3 = 3x$$
$$3x - 3x - 3 = 3x - 3x$$
$$-3 \neq 0$$
This equation has no solution.

9. $5(x + 3) = 2x + 3(x + 5)$
$$5x + 15 = 2x + 3x + 15$$
$$5x + 15 = 5x + 15$$
This equation is an identity.

11.
$$3x + 4 = 10$$
$$3x + 4 - 10 = 10 - 10$$
$$3x - 6 = 0$$
This *is* a linear equation because it can be written in the form $ax + b = 0$, $a \neq 0$.

13.
$$\frac{4}{x} - 3 = 5x$$
$$x\left(\frac{4}{x} - 3\right) = x(5x)$$
$$4 - 3x = 5x^2$$
This *is not* a linear equation. It cannot be written in the form $ax + b = 0$ because of the x^2 term.

15.

$3x + 15 = 0$	Original equation
$3x + 15 - 15 = 0 - 15$	Subtract 15 from each side.
$3x = -15$	Combine like terms.
$\dfrac{3x}{3} = \dfrac{-15}{3}$	Divide each side by 3.
$x = -5$	Simplify.

17.

$-2x + 5 = 12$	Original equation
$-2x + 5 - 5 = 12 - 5$	Subtract 5 from each side.
$-2x = 7$	Combine like terms.
$\dfrac{-2x}{-2} = \dfrac{7}{-2}$	Divide each side by -2.
$x = -\dfrac{7}{2}$	Simplify.

19.
$$y + 7 = 0$$
$$y + 7 - 7 = 0 - 7$$
$$y = -7$$

21. $3x = 12$
$$\frac{3x}{3} = \frac{12}{3}$$
$$x = 4$$

23.
$$23x - 4 = 42$$
$$23x - 4 + 4 = 42 + 4$$
$$23x = 46$$
$$\frac{23x}{23} = \frac{46}{23}$$
$$x = 2$$

25.
$$12y + 7 = 31$$
$$12y + 7 - 7 = 31 - 7$$
$$12y = 24$$
$$\frac{12y}{12} = \frac{24}{12}$$
$$y = 2$$

27.
$$7 - 8x = 13x$$
$$7 - 8x + 8x = 13x + 8x$$
$$7 = 21x$$
$$\frac{7}{21} = \frac{21x}{21}$$
$$\frac{1}{3} = x$$

29. $15t = 0$
$$\frac{15t}{15} = \frac{0}{15}$$
$$t = 0$$

31.
$$6a + 2 = 6a$$
$$6a + 2 - 6a = 6a - 6a$$
$$2 \neq 0$$
No solution

33.
$$4x = -12x$$
$$4x + 12x = -12x + 12x$$
$$16x = 0$$
$$\frac{16x}{16} = \frac{0}{16}$$
$$x = 0$$

35.
$$4x - 7 = x + 11$$
$$4x - 7 - x = x + 11 - x$$
$$3x - 7 = 11$$
$$3x - 7 + 7 = 11 + 7$$
$$3x = 18$$
$$\frac{3x}{3} = \frac{18}{3}$$
$$x = 6$$

37.
$$2 - 3x = 10 + x$$
$$2 - 3x - x = 10 + x - x$$
$$2 - 4x = 10$$
$$2 - 4x - 2 = 10 - 2$$
$$-4x = 8$$
$$\frac{-4x}{-4} = \frac{8}{-4}$$
$$x = -2$$

39.
$$8(x - 8) = 24$$
$$8x - 64 = 24$$
$$8x - 64 + 64 = 24 + 64$$
$$8x = 88$$
$$\frac{8x}{8} = \frac{88}{8}$$
$$x = 11$$

41.
$$5 - (2y - 4) = 15$$
$$5 - 2y + 4 = 15$$
$$-2y + 9 = 15$$
$$-2y + 9 - 9 = 15 - 9$$
$$-2y = 6$$
$$\frac{-2y}{-2} = \frac{6}{-2}$$
$$y = -3$$

43.
$$8x - 3(x - 2) = 12$$
$$8x - 3x + 6 = 12$$
$$5x + 6 = 12$$
$$5x + 6 - 6 = 12 - 6$$
$$5x = 6$$
$$\frac{5x}{5} = \frac{6}{5}$$
$$x = \frac{6}{5}$$

45.
$$2(x + 3) = 7(x + 3)$$
$$2x + 6 = 7x + 21$$
$$2x + 6 - 7x = 7x + 21 - 7x$$
$$-5x + 6 = 21$$
$$-5x + 6 - 6 = 21 - 6$$
$$-5x = 15$$
$$\frac{-5x}{-5} = \frac{15}{-5}$$
$$x = -3$$

47.
$$-3(x + 2) = 2(2x + 4)$$
$$-3x - 6 = 4x + 8$$
$$-3x - 6 - 4x = 4x + 8 - 4x$$
$$-7x - 6 = 8$$
$$-7x - 6 + 6 = 8 + 6$$
$$-7x = 14$$
$$\frac{-7x}{-7} = \frac{14}{-7}$$
$$x = -2$$

49.
$$3(x + 5) - 2x = 3 - (4x - 2)$$
$$3x + 15 - 2x = 3 - 4x + 2$$
$$x + 15 = -4x + 5$$
$$x + 15 + 4x = -4x + 5 + 4x$$
$$5x + 15 = 5$$
$$5x + 15 - 15 = 5 - 15$$
$$5x = -10$$
$$\frac{5x}{5} = \frac{-10}{5}$$
$$x = -2$$

51.
$$5(t + 2) - (3t + 4) = 0$$
$$5t + 10 - 3t - 4 = 0$$
$$2t + 6 = 0$$
$$2t + 6 - 6 = 0 - 6$$
$$2t = -6$$
$$\frac{2t}{2} = \frac{-6}{2}$$
$$t = -3$$

53.
$$\frac{u}{5} = 10$$
$$5\left(\frac{u}{5}\right) = (10)5$$
$$u = 50$$

55.
$$t - \frac{2}{5} = \frac{3}{2}$$
$$10\left(t - \frac{2}{5}\right) = \left(\frac{3}{2}\right)10$$
$$10t - 4 = 15$$
$$10t - 4 + 4 = 15 + 4$$
$$10t = 19$$
$$\frac{10t}{10} = \frac{19}{10}$$
$$t = \frac{19}{10}$$

57. $\dfrac{t}{14} + \dfrac{2}{7} = \dfrac{2}{7}$

$$14\left(\dfrac{t}{14} + \dfrac{2}{7}\right) = 14\left(\dfrac{2}{7}\right)$$

$$t + 4 = 4$$

$$t + 4 - 4 = 4 - 4$$

$$t = 0$$

59. $\dfrac{1}{9}x + \dfrac{1}{3} = \dfrac{11}{8}x$

$$72\left(\dfrac{1}{9}x + \dfrac{1}{3}\right) = 72\left(\dfrac{11}{8}x\right)$$

$$8x + 24 = 99x$$

$$8x + 24 - 8x = 99x - 8x$$

$$24 = 91x$$

$$\dfrac{24}{91} = \dfrac{91x}{91}$$

$$\dfrac{24}{91} = x$$

61. $\dfrac{11x}{6} + \dfrac{1}{3} = 2x$

$$6\left(\dfrac{11x}{6} + \dfrac{1}{3}\right) = 6(2x)$$

$$11x + 2 = 12x$$

$$11x + 2 - 11x = 12x - 11x$$

$$2 = x$$

63. $\dfrac{t}{5} - \dfrac{t}{2} = 1$

$$10\left(\dfrac{t}{5} - \dfrac{t}{2}\right) = (1)10$$

$$2t - 5t = 10$$

$$-3t = 10$$

$$\dfrac{-3t}{-3} = \dfrac{10}{-3}$$

$$t = -\dfrac{10}{3}$$

65. $\dfrac{8x}{5} - \dfrac{x}{4} = -3$

$$20\left(\dfrac{8x}{5} - \dfrac{x}{4}\right) = 20(-3)$$

$$32x - 5x = -60$$

$$27x = -60$$

$$\dfrac{27x}{27} = -\dfrac{60}{27}$$

$$x = -\dfrac{20}{9}$$

67. $\dfrac{4u}{3} = \dfrac{5u}{4} + 6$

$$12\left(\dfrac{4u}{3}\right) = 12\left(\dfrac{5u}{4} + 6\right)$$

$$16u = 15u + 72$$

$$16u - 15u = 15u + 72 - 15u$$

$$u = 72$$

69. $\dfrac{8 - 3x}{4} - 4 = \dfrac{x}{6}$

$$24\left(\dfrac{8 - 3x}{4} - 4\right) = 24\left(\dfrac{x}{6}\right)$$

$$6(8 - 3x) - 24(4) = 4x$$

$$48 - 18x - 96 = 4x$$

$$-18x - 48 = 4x$$

$$-18x - 48 + 18x = 4x + 18x$$

$$-48 = 22x$$

$$-\dfrac{48}{22} = \dfrac{22x}{22}$$

$$-\dfrac{24}{11} = x$$

71. $\dfrac{2}{3}(2x - 4) = \dfrac{1}{2}(x + 3) - 4$

$$6\left[\dfrac{2}{3}(2x - 4)\right] = 6\left[\dfrac{1}{2}(x + 3) - 4\right]$$

$$4(2x - 4) = 3(x + 3) - 24$$

$$8x - 16 = 3x + 9 - 24$$

$$8x - 16 = 3x - 15$$

$$8x - 16 - 3x = 3x - 15 - 3x$$

$$5x - 16 = -15$$

$$5x - 16 + 16 = -15 + 16$$

$$5x = 1$$

$$\dfrac{5x}{5} = \dfrac{1}{5}$$

$$x = \dfrac{1}{5}$$

73.

$$0.3x + 1.5 = 8.4 \quad \text{or}$$
$$10(0.3x + 1.5) = (8.4)10$$
$$3x + 15 = 84$$
$$3x + 15 - 15 = 84 - 15$$
$$3x = 69$$
$$\frac{3x}{3} = \frac{69}{3}$$
$$x = 23$$

$$0.3x + 1.5 = 8.4$$
$$0.3x + 1.5 - 1.5 = 8.4 - 1.5$$
$$0.3x = 6.9$$
$$\frac{0.3x}{0.3} = \frac{6.9}{0.3}$$
$$x = 23$$

75.

$$1.2(x - 3) = 10.8$$
$$1.2x - 3.6 = 10.8$$
$$1.2x - 3.6 + 3.6 = 10.8 + 3.6$$
$$1.2x = 14.4$$
$$\frac{1.2x}{1.2} = \frac{14.4}{1.2}$$
$$x = 12$$

77. (a)

$$2x - 3y = 6$$
$$2x = 6 + 3y$$
$$x = \frac{6 + 3y}{2}$$

(b)

$$2x - 3y = 6$$
$$-3y = 6 - 2x$$
$$y = \frac{6 - 2x}{-3}$$
$$y = -\frac{6 - 2x}{3} \quad \text{or} \quad \frac{2x - 6}{3}$$

79. (a)

$$7x + 4 = 10y - 7$$
$$7x = 10y - 11$$
$$x = \frac{10y - 11}{7}$$

(b)

$$7x + 4 = 10y - 7$$
$$7x + 11 = 10y$$
$$\frac{7x + 11}{10} = y$$

81. (a)

$$12(x - 2) + 7(y + 1) = 25$$
$$12x - 24 + 7y + 7 = 25$$
$$12x + 7y - 17 = 25$$
$$12x + 7y = 42$$
$$12x = 42 - 7y$$
$$x = \frac{42 - 7y}{12}$$

(b)

$$12(x - 2) + 7(y + 1) = 25$$
$$12x - 24 + 7y + 7 = 25$$
$$12x + 7y - 17 = 25$$
$$12x + 7y = 42$$
$$7y = 42 - 12x$$
$$y = \frac{42 - 12x}{7}$$

83. (a)

$$\frac{x}{2} + \frac{y}{5} = 1$$
$$10\left(\frac{x}{2} + \frac{y}{5}\right) = 10(1)$$
$$5x + 2y = 10$$
$$5x = 10 - 2y$$
$$x = \frac{10 - 2y}{5}$$

(b)

$$\frac{x}{2} + \frac{y}{5} = 1$$
$$10\left(\frac{x}{2} + \frac{y}{5}\right) = 10(1)$$
$$5x + 2y = 10$$
$$2y = 10 - 5x$$
$$y = \frac{10 - 5x}{2}$$

85.

$$\frac{y - 5}{-1} = -\frac{y - 5}{1} \quad \text{or}$$
$$= -(y - 5)$$
$$= -y + 5$$
$$= 5 - y$$

$$\frac{y - 5}{-1} = \frac{y}{-1} + \frac{-5}{-1} \quad \text{or}$$
$$= -y + 5$$
$$= 5 - y$$

$$\frac{y - 5}{-1} = \frac{(-1)(y - 5)}{(-1)(-1)}$$
$$= \frac{-y + 5}{1}$$
$$= -y + 5$$
$$= 5 - y$$

87. $\dfrac{h + 4\pi}{2} = \dfrac{h}{2} + \dfrac{4\pi}{2}$

$\qquad\quad\ = \dfrac{h}{2} + 2\pi$

89. $\dfrac{3x - 7}{6} = \dfrac{3x}{6} - \dfrac{7}{6}$

$\qquad\quad\ = \dfrac{3}{6} \cdot \dfrac{x}{1} - \dfrac{7}{6}$

$\qquad\quad\ = \dfrac{1}{2}x - \dfrac{7}{6}$

91. $E = IR$

$\quad \dfrac{E}{I} = \dfrac{IR}{I}$

$\quad \dfrac{E}{I} = R$

93. $V = lwh$

$\quad \dfrac{V}{lw} = \dfrac{lwh}{lw}$

$\quad \dfrac{V}{lw} = h$

95. $C = 2\pi r$

$\quad \dfrac{C}{2\pi} = \dfrac{2\pi r}{2\pi}$

$\quad \dfrac{C}{2\pi} = r$

97. $\qquad P = a + b + c$

$\quad P - b - c = a$

99. $A = lw$

$\quad \dfrac{A}{l} = \dfrac{lw}{l}$

$\quad \dfrac{A}{l} = w$

101. $A = \left(\dfrac{1}{2}\right)bh$

$\quad 2A = 2\left(\dfrac{1}{2}\right)bh$

$\quad 2A = bh$

$\quad \dfrac{2A}{b} = \dfrac{bh}{b}$

$\quad \dfrac{2A}{b} = h$

103. $\qquad S = \dfrac{n}{2}(a_1 + a_n)$

$\quad 2S = \dfrac{2}{1}\left(\dfrac{n}{2}\right)(a_1 + a_n)$

$\quad 2S = n(a_1 + a_n)$

$\quad \dfrac{2S}{a_1 + a_n} = \dfrac{n(a_1 + a_n)}{a_1 + a_n}$

$\quad \dfrac{2S}{a_1 + a_n} = n$

105. $\qquad S = C + rC$

$\qquad S = C(1 + r)$

$\quad \dfrac{S}{1 + r} = \dfrac{C(1 + r)}{1 + r}$

$\quad \dfrac{S}{1 + r} = C$

107. $\qquad A = P + Prt$

$\qquad A = P(1 + rt)$

$\quad \dfrac{A}{1 + rt} = P$

109. $\qquad L = a + (n - 1)d$ or

$\qquad L = a + nd - d$

$\quad L - a = nd - d$

$\ L - a + d = nd$

$\quad \dfrac{L - a + d}{d} = \dfrac{nd}{d}$

$\quad \dfrac{L - a + d}{d} = n$

$\qquad L = a + (n - 1)d$

$\quad L - a = (n - 1)d$

$\quad \dfrac{L - a}{d} = \dfrac{(n - 1)d}{d}$

$\quad \dfrac{L - a}{d} = n - 1$

$\quad \dfrac{L - a}{d} + 1 = n$

111. $F = \dfrac{km_1 m_2}{r^2}$

$\quad Fr^2 = \dfrac{km_1 m_2}{r^2} \cdot \dfrac{r^2}{1}$

$\quad Fr^2 = km_1 m_2$

$\quad \dfrac{Fr^2}{km_1} = \dfrac{km_1 m_2}{km_1}$

$\quad \dfrac{Fr^2}{km_1} = m_2$

113. (a)

t	1	1.5	2	3	4	5
Width	300	240	200	150	120	100
Length	300	360	400	450	480	500
Area	90,000	86,400	80,000	67,500	57,600	50,000

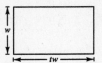

$t = 1 \Rightarrow 1200 = 2w + 2(1w)$

$1200 = 4w$

$300 = w$

$l = 1w = 1(300) = 300$

$A = lw = (300)(300) = 90,000$

$t = 1.5 \Rightarrow 1200 = 2w + 2(1.5w)$

$1200 = 5w$

$240 = w$

$l = 1.5w = 1.5(240) = 360$

$A = lw = (360)(240) = 86,400$

$t = 2 \Rightarrow 1200 = 2w + 2(2w)$

$1200 = 6w$

$200 = w$

$l = 2w = 2(200) = 400$

$A = lw = (400)(200) = 80,000$

$t = 3 \Rightarrow 1200 = 2w + 2(3w)$

$1200 = 8w$

$150 = w$

$l = 3w = 3(150) = 450$

$A = lw = (450)(150) = 67,500$

$t = 4 \Rightarrow 1200 = 2w + 2(4w)$

$1200 = 10w$

$120 = w$

$l = 4w = 4(120) = 480$

$A = lw = (480)(120) = 57,600$

$t = 5 \Rightarrow 1200 = 2w + 2(5w)$

$1200 = 12w$

$100 = w$

$l = 5w = 5(100) = 500$

$A = lw = (500)(100) = 50,000$

(b) In a rectangle of fixed perimeter with length l equal to t times width w and $t \geq 1$, as t increases, w decreases, l increases, and the area A decreases. The maximum area occurs when the length and width are equal (when $t = 1$).

115.

$v = 48 - 32t$

$0 = 48 - 32t$

$0 + 32t = 48 - 32t + 32t$

$32t = 48$

$\dfrac{32t}{32} = \dfrac{48}{32}$

$t = \dfrac{3}{2}$ or $1\dfrac{1}{2}$ seconds

It takes 1.5 seconds for a drop of water to travel from the base to the maximum height of the fountain.

117.

$\dfrac{t}{10} + \dfrac{t}{15} = 1$

$30\left(\dfrac{t}{10} + \dfrac{t}{15}\right) = (1)30$

$3t + 2t = 30$

$5t = 30$

$\dfrac{5t}{5} = \dfrac{30}{5}$

$t = 6$

The two people can complete the task in 6 hours.

119. $50,500 = 2069.46t + 48,434.71$

$2065.29 = 2069.46t$

$\dfrac{2065.29}{2069.46} = t$

$0.998 \approx t \Rightarrow t \approx 1$

The salary reached $50,500 in 1991.

121. (a) An expression is a collection of constants and variables combined using the operations of addition, subtraction, multiplication and division. An equation is a statement that two expressions are equal.

(b) An identity is an equation that has the set of all real numbers as its solution set. A conditional equation is an equation whose solution set is not the entire set of real numbers.

Section 1.6   Solving Equations by Factoring

1. $2x(x - 8) = 0$

$\quad 2x = 0 \implies x = 0$

$\quad x - 8 = 0 \implies x = 8$

3. $(y - 3)(y + 10) = 0$

$\quad y - 3 = 0 \implies y = 3$

$\quad y + 10 = 0 \implies y = -10$

5. $25(a + 4)(a - 2) = 0$

$\quad 25 \neq 0$

$\quad a + 4 = 0 \implies a = -4$

$\quad a - 2 = 0 \implies a = 2$

7. $4x(2x - 3)(2x + 25) = 0$

$\quad 4x = 0 \implies x = 0$

$\quad 2x - 3 = 0 \implies 2x = 3 \implies x = \frac{3}{2}$

$\quad 2x + 25 = 0 \implies 2x = -25 \implies x = -\frac{25}{2}$

9. $(x - 3)(2x + 1)(x + 4) = 0$

$\quad x - 3 = 0 \implies x = 3$

$\quad 2x + 1 = 0 \implies 2x = -1 \implies x = -\frac{1}{2}$

$\quad x + 4 = 0 \implies x = -4$

11. $x^2 - 3x - 10 = 0$

$\quad (x - 5)(x + 2) = 0$

$\quad x - 5 = 0 \implies x = 5$

$\quad x + 2 = 0 \implies x = -2$

13. $y^2 + 20 = 9y$

$\quad y^2 - 9y + 20 = 0$

$\quad (y - 4)(y - 5) = 0$

$\quad y - 4 = 0 \implies y = 4$

$\quad y - 5 = 0 \implies y = 5$

15. $3x^2 + 9x = 0$

$\quad 3x(x + 3) = 0$

$\quad 3x = 0 \implies x = 0$

$\quad x + 3 = 0 \implies x = -3$

17. $x^2 - 25 = 0$

$\quad (x + 5)(x - 5) = 0$

$\quad x + 5 = 0 \implies x = -5$

$\quad x - 5 = 0 \implies x = 5$

19. $3x^2 - 300 = 0$

$\quad 3(x^2 - 100) = 0$

$\quad 3(x + 10)(x - 10) = 0$

$\quad 3 \neq 0$

$\quad x + 10 = 0 \implies x = -10$

$\quad x - 10 = 0 \implies x = 10$

21. $m^2 - 8m = -16$

$\quad m^2 - 8m + 16 = 0$

$\quad (m - 4)(m - 4) = 0$

$\quad m - 4 = 0 \implies m = 4$ (Repeated solution)

23. $4z^2 + 9 = 12z$

$\quad 4z^2 - 12z + 9 = 0$

$\quad (2z - 3)(2z - 3) = 0$

$\quad 2z - 3 = 0 \implies 2z = 3 \implies z = \frac{3}{2}$ (Repeated solution)

25. $7 + 13x - 2x^2 = 0$

$\quad -2x^2 + 13x + 7 = 0$

$\quad -1(2x^2 - 13x - 7) = 0$

$\quad -1(2x + 1)(x - 7) = 0$

$\quad -1 \neq 0$

$\quad 2x + 1 = 0 \implies 2x = -1 \implies x = -\frac{1}{2}$

$\quad x - 7 = 0 \implies x = 7$

27. $x(x - 3) = 10$

$\quad x^2 - 3x = 10$

$\quad x^2 - 3x - 10 = 0$

$\quad (x - 5)(x + 2) = 0$

$\quad x - 5 = 0 \implies x = 5$

$\quad x + 2 = 0 \implies x = -2$

29.
$$y(y + 6) = 72$$
$$y^2 + 6y = 72$$
$$y^2 + 6y - 72 = 0$$
$$(y + 12)(y - 6) = 0$$
$$y + 12 = 0 \Longrightarrow y = -12$$
$$y - 6 = 0 \Longrightarrow y = 6$$

31. $x(x + 2) - 10(x + 2) = 0$
$$(x + 2)(x - 10) = 0$$
$$x + 2 = 0 \Longrightarrow x = -2$$
$$x - 10 = 0 \Longrightarrow x = 10$$

33.
$$(t - 2)^2 - 16 = 0$$
$$[(t - 2) + 4][(t - 2) - 4] = 0$$
$$(t + 2)(t - 6) = 0$$
$$t + 2 = 0 \Longrightarrow t = -2$$
$$t - 6 = 0 \Longrightarrow t = 6$$

35.
$$6t^3 - t^2 - t = 0$$
$$t(6t^2 - t - 1) = 0$$
$$t(3t + 1)(2t - 1) = 0$$
$$t = 0$$
$$3t + 1 = 0 \Longrightarrow 3t = -1 \Longrightarrow t = -\tfrac{1}{3}$$
$$2t - 1 = 0 \Longrightarrow 2t = 1 \Longrightarrow t = \tfrac{1}{2}$$

37. $x^3 - 19x^2 + 84x = 0$
$$x(x^2 - 19x + 84) = 0$$
$$x(x - 12)(x - 7) = 0$$
$$x = 0$$
$$x - 12 = 0 \Longrightarrow x = 12$$
$$x - 7 = 0 \Longrightarrow x = 7$$

39. $x^2(x - 25) - 16(x - 25) = 0$
$$(x - 25)(x^2 - 16) = 0$$
$$(x - 25)(x + 4)(x - 4) = 0$$
$$x - 25 = 0 \Longrightarrow x = 25$$
$$x + 4 = 0 \Longrightarrow x = -4$$
$$x - 4 = 0 \Longrightarrow x = 4$$

41. $z^2(z + 2) - 4(z + 2) = 0$
$$(z + 2)(z^2 - 4) = 0$$
$$(z + 2)(z + 2)(z - 2) = 0$$
$$z + 2 = 0 \Longrightarrow z = -2 \quad \textbf{(Repeated solution)}$$
$$z - 2 = 0 \Longrightarrow z = 2$$

43.
$$c^3 - 3c^2 - 9c + 27 = 0$$
$$(c^3 - 3c^2) + (-9c + 27) = 0$$
$$c^2(c - 3) - 9(c - 3) = 0$$
$$(c - 3)(c^2 - 9) = 0$$
$$(c - 3)(c + 3)(c - 3) = 0$$
$$c - 3 = 0 \Longrightarrow c = 3 \quad \textbf{(Repeated solution)}$$
$$c + 3 = 0 \Longrightarrow c = -3$$

45.
$$a^3 + 2a^2 - 9a - 18 = 0$$
$$(a^3 + 2a^2) + (-9a - 18) = 0$$
$$a^2(a + 2) - 9(a + 2) = 0$$
$$(a + 2)(a^2 - 9) = 0$$
$$(a + 2)(a + 3)(a - 3) = 0$$
$$a + 2 = 0 \Rightarrow a = -2$$
$$a + 3 = 0 \Rightarrow a = -3$$
$$a - 3 = 0 \Rightarrow a = 3$$

47.
$$x^4 - 5x^3 - 9x^2 + 45x = 0$$
$$x(x^3 - 5x^2 - 9x + 45) = 0$$
$$x[(x^3 - 5x^2) + (-9x + 45)] = 0$$
$$x[x^2(x - 5) - 9(x - 5)] = 0$$
$$x[(x - 5)(x^2 - 9)] = 0$$
$$x(x - 5)(x + 3)(x - 3) = 0$$
$$x = 0$$
$$x - 5 = 0 \Rightarrow x = 5$$
$$x + 3 = 0 \Rightarrow x = -3$$
$$x - 3 = 0 \Rightarrow x = 3$$

49. (a) $3(x + 4)^2 + (x + 4) - 2 = 0$

Let $u = x + 4$.

$$3u^2 + u - 2 = 0$$
$$(3u - 2)(u + 1) = 0$$
$$3u - 2 = 0 \quad \Rightarrow 3u = 2 \Rightarrow u = \tfrac{2}{3}$$
$$u + 1 = 0 \quad \Rightarrow u = -1$$

Substituting $x + 4$ for u: $x + 4 = \tfrac{2}{3} \quad \Rightarrow x = -4 + \tfrac{2}{3}$ or $-\tfrac{10}{3}$

$$x + 4 = -1 \Rightarrow x = -5$$

(b)
$$3(x + 4)^2 + (x + 4) - 2 = 0$$
$$3(x^2 + 8x + 16) + (x + 4) - 2 = 0$$
$$3x^2 + 24x + 48 + x + 4 - 2 = 0$$
$$3x^2 + 25x + 50 = 0$$
$$(3x + 10)(x + 5) = 0$$
$$3x + 10 = 0 \Rightarrow 3x = -10 \Rightarrow x = -\tfrac{10}{3}$$
$$x + 5 = 0 \Rightarrow x = -5$$

(c) Answers will vary.

51. First Method: $8(x + 2)^2 - 18(x + 2) + 9 = 0$

Let $u = x + 2$.

$$8u^2 - 18u + 9 = 0$$
$$(4u - 3)(2u - 3) = 0$$
$$4u - 3 = 0 \Rightarrow 4u = 3 \Rightarrow u = \tfrac{3}{4}$$
$$2u - 3 = 0 \Rightarrow 2u = 3 \Rightarrow u = \tfrac{3}{2}$$

Substituting $x + 2$ for u: $x + 2 = \tfrac{3}{4} \Rightarrow x = -2 + \tfrac{3}{4}$ or $-\tfrac{5}{4}$

$$x + 2 = \tfrac{3}{2} \Rightarrow x = -2 + \tfrac{3}{2} \text{ or } -\tfrac{1}{2}$$

—CONTINUED—

51.—CONTINUED—

Second Method: $8(x + 2)^2 - 18(x + 2) + 9 = 0$

$$8(x^2 + 4x + 4) - 18(x + 2) + 9 = 0$$

$$8x^2 + 32x + 32 - 18x - 36 + 9 = 0$$

$$8x^2 + 14x + 5 = 0$$

$$(4x + 5)(2x + 1) = 0$$

$$4x + 5 = 0 \Rightarrow 4x = -5 \Rightarrow x = -\tfrac{5}{4}$$

$$2x + 1 = 0 \Rightarrow 2x = -1 \Rightarrow x = -\tfrac{1}{2}$$

53. $ax^2 - ax = 0$

$ax(x - 1) = 0$

$ax = 0 \Rightarrow x = 0$

$x - 1 = 0 \Rightarrow x = 1$

55. $x = 1, x = 6$

$(x - 1)(x - 6) = 0$

$x^2 - 7x + 6 = 0$

57. $x = -2, x = -10$

$(x + 2)(x + 10) = 0$

$x^2 + 12x + 20 = 0$

59. $x + x^2 = 72$

$$x^2 + x - 72 = 0$$

$$(x + 9)(x - 8) = 0$$

$$x + 9 = 0 \Rightarrow x = -9$$

$$x - 8 = 0 \Rightarrow x = 8$$

(Discard the negative solution.) The number is 8.

61. $-16t^2 + 48t + 1053 = 0$

$(-4t + 39)(4t + 27) = 0$

$$-4 + 39 = 0 \Rightarrow -4t = -39 \Rightarrow t = \tfrac{39}{4}$$

$$4t + 27 = 0 \Rightarrow 4t = -27 \Rightarrow t = -\tfrac{27}{4}$$

(Discard the negative solution.) The object reaches the ground in $\tfrac{39}{4}$ seconds.

63. $(28 - 2w)(23 - 2w) = 414$

$644 - 56w - 46w + 4w^2 = 414$

$4w^2 - 102w + 644 = 414$

$4w^2 - 102w + 230 = 0$

$2(2w^2 - 51w + 115) = 0$

$2(2w - 5)(w - 23) = 0$

$2 \neq 0$

$2w - 5 = 0 \Rightarrow 2w = 5 \Rightarrow w = \tfrac{5}{2}$

$w - 23 = 0 \Rightarrow w = 23$

(Discard the solution of $w = 23$ that doesn't make sense in this application.) The width of the frame is $\tfrac{5}{2}$ centimeters.

65.   Area = $\frac{1}{2}$(Height)(Base)

$70 = \frac{1}{2}(b - 4)(b)$

$2(70) = 2 \cdot \frac{1}{2}(b - 4)(b)$

$140 = (b - 4)b$

$140 = b^2 - 4b$

$0 = b^2 - 4b - 140$

$0 = (b - 14)(b + 10)$

$b - 14 = 0 \implies b = 14$ and $b - 4 = 10$

$b + 10 = 0 \implies b = -10$

(Discard the negative solution.) The base of the triangle is 14 inches and the height is 10 inches.

67. $S = x^2 + 4xh$

$880 = x^2 + 4x(6)$

$880 = x^2 + 24x$

$0 = x^2 + 24x - 880$

$0 = (x - 20)(x + 44)$

$x - 20 = 0 \implies x = 20$

$x + 44 = 0 \implies x = -44$

(Discard the negative solution.) The base is a 20-inch square.

69.   Revenue = Cost

$x^2 - 35x = 150 + 12x$

$x^2 - 47x - 150 = 0$

$(x - 50)(x + 3) = 0$

$x - 50 = 0 \implies x = 50$

$x + 3 = 0 \implies x = -3$

(Discard the negative solution.) To break even, 50 units must be produced and sold.

71. False. The zero-factor property can be applied only to a product that is equal to zero.

73. The maximum number of solutions of an nth-degree polynomial equation is n. The third-degree equation $(x + 1)^3 = 0$ has only one real number solution: $x = -1$.

Review Exercises for Chapter 1

1. Terms: $14y^2, -9$
Coefficients: $14, -9$

3. Terms: $15t^3, -2t^2, 19t$
Coefficients: $15, -2, 19$

5. $5 + (4 - y) = (5 + 4) - y$
Associative Property of Addition

7. $(x + y) + 0 = x + y$
Additive Identity Property

9. $x^2 \cdot x^3 \cdot x = x^{2+3+1} = x^6$

11. $y^3(-2y^2) = -2 \cdot y^{3+2} = -2y^5$

13. $(-2a^2)^3(8a) = (-2)^3(a^2)^3(8a)$
$\qquad = -8a^6(8a)$
$\qquad = (-8)(8)a^{6+1}$
$\qquad = -64a^7$

15. $(xy)(-3x^2y^3) = -3x^{1+2}y^{1+3}$
$\qquad = -3x^3y^4$

17. $-(u^2v)^2(-4u^3v) = -(u^2)^2v^2(-4u^3v)$
$\qquad = -u^4v^2(-4u^3v)$
$\qquad = 4u^{4+3}v^{2+1}$
$\qquad = 4u^7v^3$

19. $\dfrac{120u^5v^3}{15u^3v} = \dfrac{120}{15}u^{5-3}v^{3-1} = 8u^2v^2$

21. $5(2x - 4) - 7x = 10x - 20 - 7x$
$\qquad = 3x - 20$

23. $y(3y - 10) + y^2 = 3y^2 - 10y + y^2$
$\qquad = 4y^2 - 10y$

25. $30 - (10x + 80) = 30x - 10x - 80$
$\qquad = 20x - 80$

27. $3x - (y - 2x) = 3x - y + 2x$
$$= 5x - y$$

29. $-2(11x - y) + 7(2x + 3y) = -22x + 2y + 14x + 21y$
$$= -8x + 23y$$

31. $3[b + 5(b - a)] = 3[b + 5b - 5a]$
$$= 3[6b - 5a]$$
$$= 18b - 15a$$

33. (a) $2(-4)^2 + (-4) - 6 = 32 - 4 - 6 = 22$
(b) $2\left(\frac{1}{2}\right)^2 + \left(\frac{1}{2}\right) - 6 = \frac{1}{2} + \frac{1}{2} - 6 = -5$

35. (a) $4(1) - (1)(5) = 4 - 5 = -1$
(b) $4(-3) - (-3)(4) = -12 + 12 = 0$

37. Standard form: $12x^6 - 4x^2 + x + 15$
Degree: 6
Leading coefficient: 12

39. $(5x + 3x^2) + (x - 4x^2) = (5x + x) + (3x^2 - 4x^2)$
$$= 6x - x^2$$

41. $(5x^2 - 2x + 7) - (3x - 10) = 5x^2 - 2x + 7 - 3x + 10$
$$= 5x^2 - 5x + 17$$

43. $(-x^3 - 3x) - 4(2x^3 - 3x + 1) = -x^3 - 3x - 8x^3 + 12x - 4$
$$= (-x^3 - 8x^3) + (-3x + 12x) + (-4)$$
$$= -9x^3 + 9x - 4$$

45. $3y^2 - [2y - 3(y^2 + 5)] = 3y^2 - [2y - 3y^2 - 15]$
$$= 3y^2 - 2y + 3y^2 + 15$$
$$= (3y^2 + 3y^2) - 2y + 15$$
$$= 6y^2 - 2y + 15$$

47. $(-2x)^3(x + 4) = (-2)^3x^3(x + 4)$
$$= -8x^3(x + 4)$$
$$= -8x^4 - 32x^3$$

F   O   I   L

49. $(2z + 3)(3z - 5) = 6z^2 - 10z + 9z - 15$
$$= 6z^2 - z - 15$$

F   O   I   L

51. $(5x + 3)(3x - 4) = 15x^2 - 20x + 9x - 12$
$$= 15x^2 - 11x - 12$$

53. $(2x^2 - 3x + 2)(2x + 3) = 2x^2(2x + 3) - 3x(2x + 3) + 2(2x + 3)$
$$= 4x^3 + 6x^2 - 6x^2 - 9x + 4x + 6$$
$$= 4x^3 - 5x + 6$$

55. $(4x - 7)^2 = (4x)^2 - 2(4x)(7) + 7^2$
$$= 16x^2 - 56x + 49$$

57. $(5u - 8)(5u + 8) = (5u)^2 - 8^2$
$$= 25u^2 - 64$$

59. $6x^2 + 15x^3 = 3x^2(2 + 5x)$

61. $28(x + 5) - 70(x + 5)^2 = [14(x + 5)][2 - 5(x + 5)]$
$$= 14(x + 5)(-5x - 23)$$
$$= -14(x + 5)(5x + 23)$$

63. $y^3 + 5y^2 + 7y + 35 = y^2(y + 5) + 7(y + 5)$
$$= (y + 5)(y^2 + 7)$$

65. $x^2 - 16 = x^4 - 4^2$
$$= (x + 4)(x - 4)$$

67. $9a^2 - 100 = (3a)^2 - 10^2$
$$= (3a + 10)(3a - 10)$$

69. $(y - 3)^2 - 16 = (y - 3)^2 - 4^2$
$$= [(y - 3) + 4][(y - 3) - 4]$$
$$= (y - 3 + 4)(y - 3 - 4)$$
$$= (y + 1)(y - 7)$$

71. $u^3 - 1 = u^3 - 1^3$
$$= (u - 1)(u^2 + u + 1)$$

73. $27x^3 + 64 = (3x)^3 + 4^3$
$$= (3x + 4)(9x^2 - 12x + 16)$$

75. $x^4 + 7x^3 - 9x^2 - 63x = x(x^3 + 7x^2 - 9x - 63)$
$$= x[(x^3 + 7x^2) + (-9x - 63)]$$
$$= x[x^2(x + 7) - 9(x + 7)]$$
$$= x[(x + 7)(x^2 - 9)]$$
$$= x(x + 7)(x + 3)(x - 3)$$

77. $x^2 + 18x + 81 - 4y^2 = (x^2 + 18x + 81) - 4y^2$
$$= (x + 9)(x + 9) - 4y^2$$
$$= (x + 9)^2 - (2y)^2$$
$$= [(x + 9) + 2y][(x + 9) - 2y]$$
$$= (x + 9 + 2y)(x + 9 - 2y)$$

79. $x^2 - 11x + 24 = (x - 8)(x - 3)$

81. $3x^2 + 23x - 8 = (3x - 1)(x + 8)$

83. $2x^2 - x - 15 = 2x^2 - 6x + 5x - 15$
$$= 2x(x - 3) + 5(x - 3)$$
$$= (x - 3)(2x + 5)$$

85. $a^2 + 4ab + 4b^2 = (a + 2b)(a + 2b)$ or $a^2 + 4ab + 4b^2 = a^2 + 2(a)(2b) + (2b)^2$
$$= (a + 2b)^2 \qquad\qquad\qquad\qquad = (a + 2b)^2$$

87. $18y^3 - 32y = 2y(9y^2 - 16)$
$$= 2y[(3y)^2 - 4^2]$$
$$= 2y(3y + 4)(3y - 4)$$

89. $6h^3 - 23h^2 - 13h = h(6h^2 - 23h - 13)$
$$= h(2h + 1)(3h - 13)$$

91. $x^4 - 4x^3 + x^2 - 4x = x(x^3 - 4x^2 + x - 4)$
$$= x[x^2(x - 4) + (x - 4)]$$
$$= x[(x - 4)(x^2 + 1)]$$
$$= x(x - 4)(x^2 + 1)$$

93. $\qquad 7x = 7(x + 5)$
$$7x = 7x + 35$$
$$7x - 7x = 7x + 35 - 7x$$
$$0 = 35 \ \text{False}$$

This is an equation with no solution.

95. (a) $10(2) - 3 \overset{?}{=} 17$
$$17 = 17$$
Yes, $x = 2$ is a solution.

   (b) $10(-1) - 3 \overset{?}{=} 17$
$$-13 \neq 17$$
No, $x = -1$ is not a solution.

97. $\qquad 5x - 2 = 13$
$$5x - 2 + 2 = 13 + 2$$
$$5x = 15$$
$$\frac{5x}{5} = \frac{15}{5}$$
$$x = 3$$

99. $\qquad 8x + 4 = 6x - 10$
$$8x + 4 - 6x = 6x - 10 - 6x$$
$$2x + 4 = -10$$
$$2x + 4 - 4 = -10 - 4$$
$$2x = -14$$
$$\frac{2x}{2} = \frac{-14}{2}$$
$$x = -7$$

101.
$$8 - 5t = 20 + t$$
$$8 - 5t - 5 = 20 + t - t$$
$$8 - 6t = 20$$
$$8 - 6t - 8 = 20 - 8$$
$$-6t = 12$$
$$\frac{-6t}{-6} = \frac{12}{-6}$$
$$t = -2$$

103.
$$4y - 6(y - 5) = 2$$
$$4y - 6y + 30 = 2$$
$$-2y + 30 = 2$$
$$-2y + 30 - 30 = 2 - 30$$
$$-2y = -28$$
$$\frac{-2y}{-2} = \frac{-28}{-2}$$
$$y = 14$$

105.
$$3(x + 1) = 5(x - 9)$$
$$3x + 3 = 5x - 45$$
$$3x + 3 - 5x = 5x - 45 - 5x$$
$$-2x + 3 = -45$$
$$-2x + 3 - 3 = -45 - 3$$
$$-2x = -48$$
$$\frac{-2x}{-2} = \frac{-48}{-2}$$
$$x = 24$$

107.
$$2[(x + 7) - 9] = 5(x - 4)$$
$$2(x + 7 - 9) = 5x - 20$$
$$2(x - 2) = 5x - 20$$
$$2x - 4 = 5x - 20$$
$$2x - 5x - 4 = 5x - 5x - 20$$
$$-3x - 4 = -20$$
$$-3x - 4 + 4 = -20 + 4$$
$$-3x = -16$$
$$\frac{-3x}{-3} = \frac{-16}{-3}$$
$$x = \frac{16}{3}$$

109.
$$\frac{4}{5}x - \frac{1}{10} = \frac{3}{2}$$
$$10\left[\frac{4}{5}x - \frac{1}{10}\right] = \left[\frac{3}{2}\right]10$$
$$8x - 1 = 15$$
$$8x - 1 + 1 = 15 + 1$$
$$8x = 16$$
$$\frac{8x}{8} = \frac{16}{8}$$
$$x = 2$$

111.
$$4.2(2x + 1.5) = 16.8$$
$$8.4x + 6.3 = 16.8$$
$$8.4x + 6.3 - 6.3 = 16.8 - 6.3$$
$$8.4x = 10.5$$
$$\frac{8.4x}{8.4} = \frac{10.5}{8.4}$$
$$x = 1.25$$

113.
$$V = \pi r^2 h$$
$$\frac{V}{\pi r^2} = \frac{\pi r^2 h}{\pi r^2}$$
$$\frac{V}{\pi r^2} = h$$

115.
$$-3x + 8 = 5y - 11$$
$$-3x + 19 = 5y$$
$$\frac{-3x + 19}{5} = y$$

117. $10x(x - 3) = 0$
$$10x = 0 \Rightarrow x = 0$$
$$x - 3 = 0 \Rightarrow x = 3$$

119. $(x - 4)(x + 5) = 0$
$$x - 4 = 0 \Rightarrow x = 4$$
$$x + 5 = 0 \Rightarrow x = -5$$

121.
$$x^2 - 5x + 6 = 0$$
$$(x - 3)(x - 2) = 0$$
$$x - 3 = 0 \Rightarrow x = 3$$
$$x - 2 = 0 \Rightarrow x = 2$$

123.
$$7 + 13x - 2x^2 = 0$$
$$-1(7 + 13x - 2x^2) = -1(0)$$
$$-7 - 13x + 2x^2 = 0$$
$$2x^2 - 13x - 7 = 0$$
$$(2x + 1)(x - 7) = 0$$
$$2x + 1 = 0 \Rightarrow 2x = -1 \Rightarrow x = -\frac{1}{2}$$
$$x - 7 = 0 \Rightarrow x = 7$$

125.
$$z(5 - z) + 36 = 0$$
$$5z - z^2 + 36 = 0$$
$$-z^2 + 5z + 36 = 0$$
$$-1(-z^2 + 5z + 36) = -1(0)$$
$$z^2 - 5z - 36 = 0$$
$$(z - 9)(z + 4) = 0$$
$$z - 9 = 0 \Rightarrow z = 9$$
$$z + 4 = 0 \Rightarrow z = -4$$

127.
$$v^2 - 100 = 0$$
$$(v + 10)(v - 10) = 0$$
$$v + 10 = 0 \Longrightarrow v = -10$$
$$v - 10 = 0 \Longrightarrow v = 10$$

129.
$$3y^2 - 48 = 0$$
$$3(y^2 - 16) = 0$$
$$3(y + 4)(y - 4) = 0$$
$$3 \neq 0$$
$$y + 4 = 0 \Longrightarrow y = -4$$
$$y - 4 = 0 \Longrightarrow y = 4$$

131.
$$2y^3 - 2y^2 - 24y = 0$$
$$2y(y^2 - y - 12) = 0$$
$$2y(y - 4)(y + 3) = 0$$
$$2y = 0 \Longrightarrow y = 0$$
$$y - 4 = 0 \Longrightarrow y = 4$$
$$y + 3 = 0 \Longrightarrow y = -3$$

133.
$$b^3 - 6b^2 - b + 6 = 0$$
$$(b^3 - 6b^2) + (-b + 6) = 0$$
$$b^2(b - 6) - 1(b - 6) = 0$$
$$(b - 6)(b^2 - 1) = 0$$
$$(b - 6)(b + 1)(b - 1) = 0$$
$$b - 6 = 0 \Longrightarrow b = 6$$
$$b + 1 = 0 \Longrightarrow b = -1$$
$$b - 1 = 0 \Longrightarrow b = 1$$

135. Perimeter $= 3x + 4 + 4x + 3 + x + 1$
$$= 8x + 8$$

Area $=$ Area of upper rectangle $+$ Area of lower rectangle
$$= 3x(1) + 4x(3)$$
$$= 3x + 12x$$
$$= 15x$$

137. (a) Perimeter $= 2(\text{Length}) + 2(\text{Width})$
$$= 2(5w) + 2(w)$$
$$= 10w + 2w$$
$$= 12w$$

(b) Area $= (\text{Length})(\text{Width})$
$$= (5w)(w)$$
$$= 5w^2$$

139. Area $= (\text{Length})(\text{Width})$
$$16l - l^2 = l(\text{Width})$$
$$16l - l = l(16 - l)$$

The width is $16 - l$.

141. $(\text{Length})(\text{Width}) = \text{Area}$
$$x\left(\tfrac{3}{4}x\right) = 432$$
$$\tfrac{3}{4}x^2 = 432$$
$$4\left(\tfrac{3}{4}x^2\right) = 4(432)$$
$$3x^2 = 1728$$
$$3x^2 - 1728 = 0$$
$$3(x^2 - 576) = 0$$
$$3(x - 24)(x + 24) = 0$$
$$3 \neq 0$$
$$x - 24 = 0 \Longrightarrow x = 24$$
$$x + 24 = 0 \Longrightarrow x = -24$$

(Discard the negative solution.) The length of the rectangle is 24 inches and the width is $24\left(\tfrac{3}{4}\right) = 18$ inches.

Chapter Test for Chapter 1

1. $(3x^2y)(-xy)^2 = (3x^2y)(x^2y^2)$

$\qquad\qquad\quad = 3x^{2+2}y^{1+2}$

$\qquad\qquad\quad = 3x^4y^3$

2. $(5x^2y^3)^2(-2xy)^3 = 5^2(x^2)^2(y^3)^2(-2)^3x^3y^3$

$\qquad\qquad\qquad\quad = 25x^4y^6(-8)x^3y^3$

$\qquad\qquad\qquad\quad = 25(-8)x^{4+3}y^{6+3}$

$\qquad\qquad\qquad\quad = -200x^7y^9$

3. $\dfrac{12x^3y^5}{4x^2y} = 3x^{3-2}y^{5-1}$

$\qquad\quad = 3xy^4$

4. $3x^2 - 2x - 5x^2 + 7x - 1 = (3x^2 - 5x^2) + (-2x + 7x) - 1$

$\qquad\qquad\qquad\qquad\qquad = -2x^2 + 5x - 1$

5. $-5x + 4x^2 - 6x + 3 + 2x^2 = (4x^2 + 2x^2) + (-5x - 6x) + 3$

$\qquad\qquad\qquad\qquad\qquad = 6x^2 - 11x + 3$

6. $(x^2 - 7x + 4) + (9x^2 + x + 1) = x^2 - 7x + 4 + 9x^2 + x + 1$

$\qquad\qquad\qquad\qquad\qquad\quad = (x^2 + 9x^2) + (-7x + x) + (4 + 1)$

$\qquad\qquad\qquad\qquad\qquad\quad = 10x^2 - 6x + 5$

7. $(16 - y^2) - (16 + 2y + y^2) = 16 - y^2 - 16 - 2y - y^2$

$\qquad\qquad\qquad\qquad\qquad = (-y^2 - y^2) + (-2y) + (16 - 16)$

$\qquad\qquad\qquad\qquad\qquad = -2y^2 - 2y$

8. $-2(2x^4 - 5) + 4x(x^3 + 2x - 1) = -4x^4 + 10 + 4x^4 + 8x^2 - 4x$

$\qquad\qquad\qquad\qquad\qquad\qquad = (-4x^4 + 4x^4) + 8x^2 - 4x + 10$

$\qquad\qquad\qquad\qquad\qquad\qquad = 8x^2 - 4x + 10$

9. $4t - [3t - (10t + 7)] = 4t - [3t - 10t - 7]$

$\qquad\qquad\qquad\quad = 4t - [-7t - 7]$

$\qquad\qquad\qquad\quad = 4t + 7t + 7$

$\qquad\qquad\qquad\quad = 11t + 7$

10. $2y\left(\dfrac{y}{4}\right)^2 = \dfrac{2y}{1} \cdot \dfrac{y^2}{4^2}$

$\qquad\qquad = \dfrac{2y^{1+2}}{16}$

$\qquad\qquad = \dfrac{2y^3}{2(8)}$

$\qquad\qquad = \dfrac{y^3}{8}$

$\qquad\qquad\qquad\quad$ F $\quad$ O $\quad$ I $\quad$ L

11. $(2x - 3y)(x + 5y) = 2x^2 + 10xy - 3xy - 15y^2$

$\qquad\qquad\qquad\qquad = 2x^2 + 7xy - 15y^2$

12. $(2s - 3)(3s^2 - 4s + 7) = 2s(3s^2 - 4s + 7) - 3(3s^2 - 4s + 7)$

$\qquad\qquad\qquad\qquad\qquad = 6s^3 - 8s^2 + 14s - 9s^2 + 12s - 21$

$\qquad\qquad\qquad\qquad\qquad = 6s^3 + (-8s^2 - 9s^2) + (14s + 12s) - 21$

$\qquad\qquad\qquad\qquad\qquad = 6s^3 - 17s^2 + 26s - 21$

13. $(4x - 3)^2 = (4x)^2 - 2(4x)(3) + 3^2$

$\qquad = 16x^2 - 24x + 9$

$\quad$ or

$(4x - 3)^2 = (4x - 3)(4x - 3)$

$\qquad = 16x^2 - 24x + 9$

14. $(6 - 4y)(6 + 4y) = 6^2 - (4y)^2$

$\qquad = 36 - 16y^2$

15. $18y^2 - 12y = 6y(3y - 2)$

16. $5x^3 - 10x^2 - 6x + 12 = (5x^3 - 10x^2) + (-6x + 12)$

$\qquad = 5x^2(x - 2) - 6(x - 2)$

$\qquad = (x - 2)(5x^2 - 6)$

17. $9u^2 - 6u + 1 = (3u - 1)(3u - 1)$

$\qquad = (3u - 1)^2$

$\quad$ or

$9u^2 - 6u + 1 = (3u)^2 - 2(3u) + 1^2$

$\qquad = (3u - 1)^2$

18. $6x^2 - 26x - 20 = 2(3x^2 - 13x - 10)$

$\qquad = 2(3x + 2)(x - 5)$

19. $b^2 - 2b - 48 = (b - 8)(b + 6)$

20. $49x^2 - 36 = (7x)^2 - 6^2$

$\qquad = (7x + 6)(7x - 6)$

21. $(a + 2)^2 - 9b^2 = (a + 2)^2 - (3b)^2$

$\qquad = [(a + 2) + 3b][(a + 2) - 3b]$

$\qquad = (a + 2 + 3b)(a + 2 - 3b)$

22. $2y^2 + 15y + 18 = (2y + 3)(y + 6)$

23. $2x^3 + 26x^2 + 24x = 2x(x^2 + 13x + 12)$

$\qquad = 2x(x + 12)(x + 1)$

24. $2x^3 - 128y^3 = 2(x^3 - 64y^3)$

$\qquad = 2[x^3 - (4y)^3]$

$\qquad = 2(x - 4y)(x^2 + 4xy + 16y^2)$

25. $\quad 6x - 5 = 19$

$6x - 5 + 5 = 19 + 5$

$\qquad 6x = 24$

$\qquad \dfrac{6x}{6} = \dfrac{24}{6}$

$\qquad x = 4$

26. $15 - 7(1 - x) = 3(x + 8)$

$15 - 7 + 7x = 3x + 24$

$8 + 7x = 3x + 24$

$8 + 7x - 3x = 3x + 24 - 3x$

$8 + 4x = 24$

$8 - 8 + 4x = 24 - 8$

$4x = 16$

$\dfrac{4x}{4} = \dfrac{16}{4}$

$x = 4$

27. $\qquad \dfrac{2x}{3} = \dfrac{x}{2} + 4$

$6\left(\dfrac{2x}{3}\right) = 6\left(\dfrac{x}{2} + 4\right)$

$4x = 3x + 24$

$4x - 3x = 3x + 24 - 3x$

$x = 24$

28.
$$3y^2 - 5y = 12$$
$$3y^2 - 5y - 12 = 0$$
$$(3y + 4)(y - 3) = 0$$
$$3y + 4 = 0 \Longrightarrow 3y = -4 \Longrightarrow y = -\tfrac{4}{3}$$
$$y - 3 = 0 \Longrightarrow y = 3$$

29.
$$(x + 5)(x - 2) = 60$$
$$x^2 + 3x - 10 = 60$$
$$x^2 + 3x - 70 = 0$$
$$(x + 10)(x - 7) = 0$$
$$x + 10 = 0 \Longrightarrow x = -10$$
$$x - 7 = 0 \Longrightarrow x = 7$$

30.
$$2x^3 + 10x^2 + 8x = 0$$
$$2x(x^2 + 5x + 4) = 0$$
$$2x(x + 4)(x + 1) = 0$$
$$2x = 0 \Longrightarrow x = 0$$
$$x + 4 = 0 \Longrightarrow x = -4$$
$$x + 1 = 0 \Longrightarrow x = -1$$

31. Area of larger rectangle $= 2x(x + 15)$

Area of smaller rectangle $= x(x + 4)$

Area of shaded region $= 2x(x + 15) - x(x + 4)$
$$= 2x^2 + 30x - x^2 - 4x$$
$$= (2x^2 - x^2) + (30x - 4x)$$
$$= x^2 + 26x$$

32.
$$5a + 2b - 10 = 8a + 7$$
$$5a + 2b - 10 - 5a = 8a + 7 - 5a$$
$$2b - 10 = 3a + 7$$
$$2b - 10 + 10 = 3a + 7 + 10$$
$$2b = 3a + 17$$
$$\frac{2b}{2} = \frac{3a + 17}{2}$$
$$b = \frac{3a + 17}{2} \quad \text{or} \quad b = \tfrac{1}{2}(3a + 17)$$

C H A P T E R 2
Introduction to Graphs and Functions

C H A P T E R   2
Introduction to Graphs and Functions

Section 2.1   Describing Data Graphically

Solutions to Odd-Numbered Exercises

1.

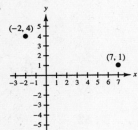

3.

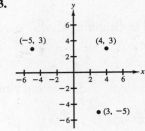

5.

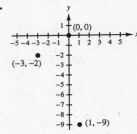

7.

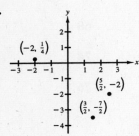

9. A: $(3, 0)$, B:$(-2, -2)$, C: $(-1, 5)$

11. A: $(4, -2)$, B: $\left(-3, -\frac{5}{2}\right)$, C: $\left(3, \frac{1}{2}\right)$

13.

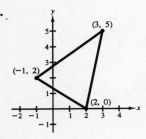

15.

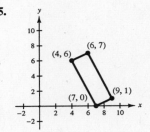

17.

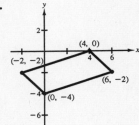

19. The point $(-3, -5)$ is in Quadrant III. The point is located to the *left* of the vertical axis and *below* the horizontal axis in Quadrant III.

21. The point $(5, 8)$ is in Quadrant I. The point is located to the *right* of the vertical axis and *above* the horizontal axis in Quadrant I.

23. The point $(15, -15)$ is in Quadrant IV. The point is located to the *right* of the vertical axis and *below* the horizontal axis in Quadrant IV.

25. $(x, 4)$ may be located in Quadrants I or II.

If $x > 0$, the point would be located to the *right* of the vertical axis and *above* the horizontal axis in the first quadrant.

If $x < 0$, the point would be located to the *left* of the vertical axis and *above* the horizontal axis in the second quadrant.

Note: If $x = 0$, the point would be located on the y-axis *between* the first and second quadrants.

27. $xy < 0 \Rightarrow$ the point is located in Quadrants II or IV.

The product xy is negative; therefore x and y have opposite signs. If x is positive and y is negative, the point would be located to the *right* of the vertical axis and *below* the horizontal axis in Quadrant IV. If x is negative and y is positive, the point would be located to the *left* of the vertical axis and *above* the horizontal axis in Quadrant II.

29. $x < 0, y > 0 \Rightarrow (x, y)$ is in Quadrant II. Because $x < 0$, the point is located to the *left* of the vertical axis. Because $y > 0$, the point is located *above* the horizontal axis. So the point is located in Quadrant II.

31. $(-5, 2)$

33. $(3, 6)$

35. $(10, 0)$

37. $(0, 12)$

39.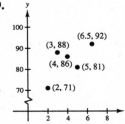

41.

43. (a) $y = 3x + 8$

$17 \overset{?}{=} 3(3) + 8$

$17 \overset{?}{=} 9 + 8$

$17 = 17$

Yes

(b) $y = 3x + 8$

$10 \overset{?}{=} 3(-1) + 8$

$10 \overset{?}{=} -3 + 8$

$10 \neq 5$

No

(c) $y = 3x + 8$

$0 \overset{?}{=} 3(0) + 8$

$0 \neq 8$

No

(d) $y = 3x + 8$

$2 \overset{?}{=} 3(-2) + 8$

$2 \overset{?}{=} -6 + 8$

$2 = 2$

Yes

45. (a) $y = \dfrac{7}{8}x$

$1 \overset{?}{=} \dfrac{7}{8} \cdot \dfrac{8}{7}$

$1 = 1$

Yes

(b) $y = \dfrac{7}{8}x$

$\dfrac{7}{2} \overset{?}{=} \dfrac{7}{8} \cdot 4$

$\dfrac{7}{2} \overset{?}{=} \dfrac{7(\cancel{4})}{\cancel{8}(2)}$

$\dfrac{7}{2} = \dfrac{7}{2}$

Yes

(c) $y = \dfrac{7}{8}x$

$0 \overset{?}{=} \dfrac{7}{8} \cdot 0$

$0 = 0$

Yes

(d) $y = \dfrac{7}{8}x$

$14 \overset{?}{=} \dfrac{7}{8}(-16)$

$14 \overset{?}{=} \dfrac{7(\cancel{8})(-2)}{\cancel{8}}$

$14 \neq -14$

No

47. (a) $4y - 2x + 1 = 0$

$4(0) - 2(0) + 1 \overset{?}{=} 0$

$1 \neq 0$

No

(b) $4y - 2x + 1 = 0$

$4(0) - 2\left(\tfrac{1}{2}\right) + 1 \overset{?}{=} 0$

$-1 + 1 \overset{?}{=} 0$

$0 = 0$

Yes

(c) $4y - 2x + 1 = 0$

$4\left(-\tfrac{7}{4}\right) - 2(-3) + 1 \overset{?}{=} 0$

$-7 + 6 + 1 \overset{?}{=} 0$

$0 = 0$

Yes

(d) $4y - 2x + 1 = 0$

$4\left(-\tfrac{3}{4}\right) - 2(1) + 1 \overset{?}{=} 0$

$-3 - 2 + 1 \overset{?}{=} 0$

$-4 \neq 0$

No

49.

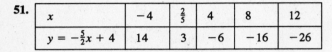

x	-2	0	2	4	6
$y = 5x - 1$	-11	-1	9	19	29

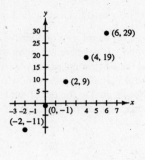

$x = -2 \Longrightarrow y = 5(-2) - 1 = -10 - 1 = -11$

$x = 0 \Longrightarrow y = 5(0) - 1 = 0 - 1 = -1$

$x = 2 \Longrightarrow y = 5(2) - 1 = 10 - 1 = 9$

$x = 4 \Longrightarrow y = 5(4) - 1 = 20 - 1 = 19$

$x = 6 \Longrightarrow y = 5(6) - 1 = 30 - 1 = 29$

51.

x	-4	$\frac{2}{5}$	4	8	12
$y = -\frac{5}{2}x + 4$	14	3	-6	-16	-26

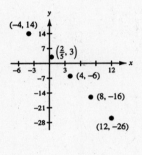

$x = -4 \Longrightarrow y = -\frac{5}{2}(-4) + 4 = \frac{20}{2} + 4 = 10 + 4 = 14$

$x = \frac{2}{5} \Longrightarrow y = -\frac{5}{2}\left(\frac{2}{5}\right) + 4 = -\frac{10}{10} + 4 = -1 + 4 = 3$

$x = 4 \Longrightarrow y = -\frac{5}{2}(4) + 4 = -\frac{20}{2} + 4 = -10 + 4 = -6$

$x = 8 \Longrightarrow y = -\frac{5}{2}(8) + 4 = -\frac{40}{2} + 4 = -20 + 4 = -16$

$x = 12 \Longrightarrow y = -\frac{5}{2}(12) + 4 = -\frac{60}{2} + 4 = -30 + 4 = -26$

53.

x	-2	0	2	4	6
$y = 4x^2 + x - 2$	12	-2	16	66	148

55. *Original Points*   *New Points*

   $(2, 1)$ $(-2, 1)$

   $(-3, 5)$ $(3, 5)$

   $(7, -3)$ $(-7, -3)$

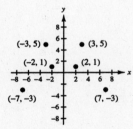

When the sign of the *x*-coordinate is changed, the location of the point is reflected about the *y*-axis.

57. $(-2, -1)$ shifted 2 units right and 5 units up is $(-2 + 2, -1 + 5) = (0, 4)$.

    $(-3, -4)$ shifted 2 units right and 5 units up is $(-3 + 2, -4 + 5) = (-1, 1)$.

    $(1, -3)$ shifted 2 units right and 5 units up is $(1 + 2, -3 + 5) = (3, 2)$.

59. $(-3, 5)$ shifted 6 units right and 3 units down is

    $(-3 + 6, 5 - 3) = (3, 2)$.

    $(-5, 2)$ shifted 6 units right and 3 units down is

    $(-5 + 6, 2 - 3) = (1, -1)$.

    $(-3, -1)$ shifted 6 units right and 3 units down is

    $(-3 + 6, -1 - 3) = (3, -4)$.

    $(-1, 2)$ shifted 6 units right and 3 units down is

    $(-1 + 6, 2 - 3) = (5, -1)$.

61.

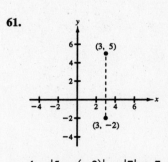

$d = |5 - (-2)| = |7| = 7$

The points lie on a vertical line.

63.

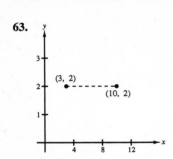

$d = |10 - 3| = |7| = 7$

The points lie on a horizontal line.

65. $d = \sqrt{(5 - 1)^2 + (6 - 3)^2}$

$= \sqrt{4^2 + 3^2}$

$= \sqrt{25}$

$= 5$

67. $d = \sqrt{(12 - 0)^2 + (-9 - 0)^2}$

$= \sqrt{12^2 + (-9)^2}$

$= \sqrt{225}$

$= 15$

69. $d = \sqrt{(-4 - (-9))^2 + (3 - (-9))^2}$

$= \sqrt{(-5)^2 + (12)^2}$

$= \sqrt{169}$

$= 13$

71. $d = \sqrt{(4 - (-2))^2 + (2 - (-3))^2}$

$= \sqrt{6^2 + 5^2}$

$= \sqrt{61}$

≈ 7.81

73. $d = \sqrt{(3 - 1)^2 + (-2 - 3)^2}$

$= \sqrt{2^2 + (-5)^2}$

$= \sqrt{29}$

≈ 5.39

75. $d = \sqrt{\left(\frac{10}{3} - \frac{1}{3}\right)^2 + (-5 - (-1))^2}$

$= \sqrt{(3)^2 + (-4)^2}$

$= \sqrt{25}$

$= 5$

77. $d = \sqrt{\left(\frac{3}{2} - \frac{1}{2}\right)^2 + (2 - 1)^2}$

$= \sqrt{1^2 + 1^2}$

$= \sqrt{2}$

≈ 1.41

79. (a) Length of horizontal side: $|10 - 2| = 8$

Length of vertical side: · $|8 - 2| = 6$

Length of hypotenuse: $\sqrt{(10 - 2)^2 + (8 - 2)^2} = \sqrt{8^2 + 6^2}$

$= \sqrt{64 + 36}$

$= \sqrt{100}$

$= 10$

(b) Is $8^2 + 6^2 = 10^2$?

Yes, $100 = 100$.

81. $d_1 = \sqrt{(2 - 2)^2 + (6 - 3)^2} = \sqrt{0^2 + 3^2} = \sqrt{9} = 3 \Rightarrow d_1{}^2 = 9$

$d_2 = \sqrt{(6 - 2)^2 + (3 - 6)^2} = \sqrt{4^2 + (-3)^2} = \sqrt{25} = 5 \Rightarrow d_2{}^2 = 25$

$d_3 = \sqrt{(6 - 2)^2 + (3 - 3)^2} = \sqrt{4^2 + 0^2} = \sqrt{16} = 4 \Rightarrow d_3{}^2 = 16$

$9 + 16 = 25 \Rightarrow d_1{}^2 + d_3{}^2 = d_2{}^2$

The points *are* vertices of a right triangle.

83. $d_1 = \sqrt{(5 - 8)^2 + (2 - 3)^2} = \sqrt{(-3)^2 + (-1)^2} = \sqrt{10} \Rightarrow d_1{}^2 = 10$

$d_2 = \sqrt{(1 - 5)^2 + (9 - 2)^2} = \sqrt{(-4)^2 + 7^2} = \sqrt{65} \Rightarrow d_2{}^2 = 65$

$d_3 = \sqrt{(1 - 8)^2 + (9 - 3)^2} = \sqrt{(-7)^2 + 6^2} = \sqrt{85} \Rightarrow d_3{}^2 = 85$

$d_1{}^2 + d_2{}^2 \neq d_3{}^2$

The points *are not* vertices of a right triangle.

85. Points $(-2, 0)$, $(0, 5)$, $(1, 0)$

$$d_1 = \sqrt{(0 + 2)^2 + (5 - 0)^2} = \sqrt{(2)^2 + (5)^2} = \sqrt{29} \approx 5.385$$

$$d_2 = \sqrt{(1 - 0)^2 + (0 - 5)^2} = \sqrt{(1)^2 + (-5)^2} = \sqrt{26} \approx 5.099$$

$$d_3 = \sqrt{(1 + 2)^2 + (0 - 0)^2} = \sqrt{(3)^2 + (0)^2} = \sqrt{9} = 3$$

Perimeter $\approx 5.385 + 5.099 + 3 \approx 13.48$

87. Midpoint: $\left(\dfrac{-2 + 4}{2}, \dfrac{0 + 8}{2}\right) = (1, 4)$

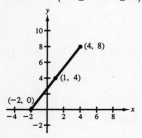

89. Midpoint: $\left(\dfrac{1 + 5}{2}, \dfrac{9 + (-3)}{2}\right) = (3, 3)$

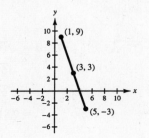

91. Midpoint: $\left(\dfrac{1 + 6}{2}, \dfrac{6 + 3}{2}\right) = \left(\dfrac{7}{2}, \dfrac{9}{2}\right)$

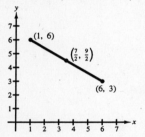

93. Midpoint: $\left(\dfrac{-2 + 5}{2}, \dfrac{-2 + 10}{2}\right) = \left(\dfrac{3}{2}, 4\right)$

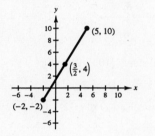

95. There appears to have been no change between the fourth and fifth hours.

97. Yes, there appears to be a strong relationship between the temperature outside and the amount of natural gas used.

About 500 cubic feet of natural gas would be used in a month with an average temperature of 45°F.

99. The "order" in an ordered pair is important because the first number describes how far to the right or left of the vertical axis the point is located and the second number describes how far above or below the horizontal axis the point is located. In other words, the first measures horizontal distance and the second measures vertical distance. Each number in the ordered pair has a particular interpretation. The ordered pairs $(5, -7)$ and $(-7, 5)$ represent different points.

101. $(x^4 y^2)^2 (-2xy^3)^2 = (x^8 y^4)(4x^2 y^6)$

$\qquad\qquad\qquad\qquad = 4x^{10} y^{10}$

103. $5(2x - 3) + x(x + 9) = 10x - 15 + x^2 + 9x$

$\qquad\qquad\qquad\qquad\qquad = x^2 + 19x - 15$

105. $2x^2 - 32 = 2(x^2 - 16)$

$\qquad\qquad\quad = 2(x + 4)(x - 4)$

107. $4x^2 - x - 5 = (4x - 5)(x + 1)$

109. $\quad x^2 - x - 6 = 0$

$\quad (x - 3)(x + 2) = 0$

$\qquad\qquad x - 3 = 0 \implies x = 3$

$\qquad\qquad x + 2 = 0 \implies x = -2$

111. $2000(1 + 0.085)^2 = \$2354.45$

Section 2.2   Graphs of Equations

1. Graph (e)   **3.** Graph (f)   **5.** Graph (d)

7. $2x + y = 3$

$\qquad y = -2x + 3$

x	-4	-2	0	2	4
y	11	7	3	-1	-5

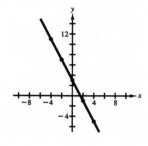

$x = -4 \Rightarrow y = -2(-4) + 3 = 11$

$y = 7 \Rightarrow 2x + 7 = 3 \Rightarrow 2x = -4 \Rightarrow x = -2$

$y = 3 \Rightarrow 2x + 3 = 3 \Rightarrow 2x = 0 \Rightarrow x = 0$

$x = 2 \Rightarrow y = -2(2) + 3 = -1$

$x = 4 \Rightarrow y = -2(4) + 3 = -5$

9.

x	± 2	-1	0	2	± 3
y	0	3	4	0	-5

$y = 0 \Rightarrow 0 = 4 - x^2 \Rightarrow 0 = (2 + x)(2 - x)$

$\qquad\qquad\qquad 2 + x = 0 \Rightarrow x = -2$

$\qquad\qquad\qquad 2 - x = 0 \Rightarrow x = 2$

$x = -1 \Rightarrow y = 4 - (-1)^2 \Rightarrow y = 3$

$y = 4 \Rightarrow 4 = 4 - x^2 \Rightarrow 0 = -x^2 \Rightarrow x = 0$

$x = 2 \Rightarrow y = 4 - 2^2 \Rightarrow y = 0$

$y = -5 \Rightarrow -5 = 4 - x^2 \Rightarrow 0 = 9 - x^2 \Rightarrow 0 = (3 + x)(3 - x)$

$\qquad\qquad\qquad\qquad 3 + x = 0 \Rightarrow x = -3$

$\qquad\qquad\qquad\qquad 3 - x = 0 \Rightarrow x = 3$

11.

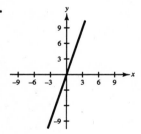

13.

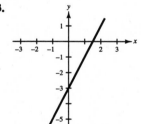

15.

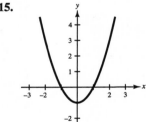

17.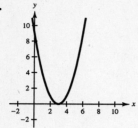

19.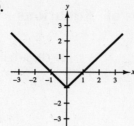

21. From the graph, it appears that the y-intercept is $(0, 3)$ and there is no x-intercept.

$$y = x^2 + 3$$

x-intercept	y-intercept
Let $y = 0$.	Let $x = 0$.
$0 = x^2 + 3$	$y = 0^2 + 3$
$-3 = x^2$	$y = 3$
No real solution	$(0, 3)$
No x-intercept	

23. From the graph, it appears that the x-intercept is $(2, 0)$ and the y-intercept is $(0, 2)$.

$$y = |x - 2|$$

x-intercept	y-intercept
Let $y = 0$.	Let $x = 0$.
$0 = |x - 2|$	$y = |0 - 2|$
$x - 2 = 0$	$y = |-2|$
$x = 2$	$y = 2$
$(2, 0)$	$(0, 2)$

25.

x-intercept	y-intercept
Let $y = 0$.	Let $x = 0$.
$x + 2(0) = 10$	$0 + 2y = 10$
$x = 10$	$2y = 10$
$(10, 0)$	$y = 5$
	$(0, 5)$

27.

x-intercept	y-intercept
Let $y = 0$.	Let $x = 0$.
$15x - 18(0) + 20 = 0$	$15(0) - 18y + 20 = 0$
$15x + 20 = 0$	$-18x + 20 = 0$
$15x = -20$	$-18y = -20$
$x = -\frac{20}{15}$	$y = \frac{-20}{-18}$
$x = -\frac{4}{3}$	$y = \frac{10}{9}$
$\left(-\frac{4}{3}, 0\right)$	$\left(0, \frac{10}{9}\right)$

29.

x-intercept	y-intercept
Let $y = 0$.	Let $x = 0$.
$0 = \frac{3}{4}x + 15$	$y = \frac{3}{4}(0) + 15$
$0 = 3x + 60$	$y = 15$
$-60 = 3x$	$(0, 15)$
$-20 = x$	
$(-20, 0)$	

31.

x-intercept	y-intercept
Let $y = 0$.	Let $x = 0$.
$0 = (x - 4)(2x + 7)$	$y = (0 - 4)[2(0) + 7]$
$x - 4 = 0 \Rightarrow x = 4$	$y = (-4)(7)$
$2x + 7 = 0 \Rightarrow 2x = -7 \Rightarrow x = -\frac{7}{2}$	$y = -28$
$(4, 0)$ and $\left(-\frac{7}{2}, 0\right)$	$(0, -28)$

33. *x*-intercept

Let $y = 0$.

$$0 = x^2 + x - 42$$
$$0 = (x + 7)(x - 6)$$
$$x + 7 = 0 \implies x = -7$$
$$x - 6 = 0 \implies x = 6$$
$$(-7, 0) \text{ and } (6, 0)$$

y-intercept

Let $x = 0$.

$$y = 0^2 + 0 - 42$$
$$y = 0 + 0 - 42$$
$$y = -42$$
$$(0, -42)$$

35. *x*-intercept

Let $y = 0$.

$$0 = x^3 - 16x$$
$$0 = x(x^2 - 16)$$
$$0 = x(x + 4)(x - 4)$$
$$x = 0$$
$$x + 4 = 0 \implies x = -4$$
$$x - 4 = 0 \implies x = 4$$
$$(0, 0), (-4, 0), \text{ and } (4, 0)$$

y-intercept

Let $x = 0$.

$$y = 0^3 - 16(0)$$
$$y = 0 - 0$$
$$y = 0$$
$$(0, 0)$$

37.

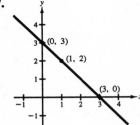

x	-2	-1	0	1	2	3
$y = 3 - x$	5	4	3	2	1	0
(x, y)	$(-2, 5)$	$(-1, 4)$	$(0, 3)$	$(1, 2)$	$(2, 1)$	$(3, 0)$

39.

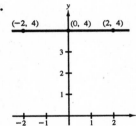

x	-4	-2	0	2	4
$y = 4$	4	4	4	4	4
(x, y)	$(-4, 4)$	$(-2, 4)$	$(0, 4)$	$(2, 4)$	$(4, 4)$

41.

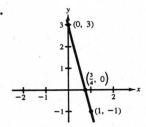

$4x + y = 3$

$y = -4x + 3$

x	-1	0	$\frac{3}{4}$	1	2
$y = -4x + 3$	7	3	0	-1	-5
(x, y)	$(-1, 7)$	$(0, 3)$	$\left(\frac{3}{4}, 0\right)$	$(1, -1)$	$(2, -5)$

43.

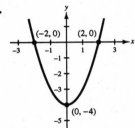

x	-3	-2	-1	0	1	2	3
$y = x^2 - 4$	5	0	-3	-4	-3	0	5
(x, y)	$(-3, 5)$	$(-2, 0)$	$(-1, -3)$	$(0, -4)$	$(1, -3)$	$(2, 0)$	$(3, 5)$

45.

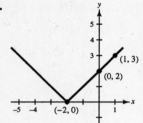

x	-5	-4	-3	-2	-1	0	1
$y = \lvert x + 2 \rvert$	3	2	1	0	1	2	3
(x, y)	$(-5, 3)$	$(-4, 2)$	$(-3, 1)$	$(-2, 0)$	$(-1, 1)$	$(0, 2)$	$(1, 3)$

47.

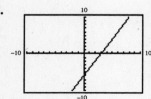

The y-intercept appears to be located at $(0, -6)$.
For $y = 2x - 6$, $x = 0 \Rightarrow y = 2(0) - 6$
$$y = -6$$
$(0, -6)$

49.

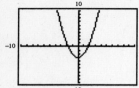

The y-intercept appears to be located at $(0, -3)$.
For $y = x^2 - 3$, $x = 0 \Rightarrow y = 0^2 - 3$
$$y = -3$$
$(0, -3)$

51.

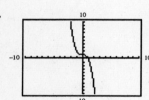

The y-intercept appears to be located at $(0, 1)$.
For $y = 1 - x^3$, $x = 0 \Rightarrow y = 1 - 0^3$
$$y = 1$$
$(0, 1)$

53.

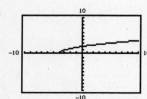

The y-intercept appears to be located at $(0, 2)$.
For $y = \sqrt{x + 4}$, $x = 0 \Rightarrow y = \sqrt{0 + 4}$
$$y = \sqrt{4}$$
$$y = 2$$
$(0, 2)$

55.

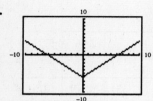

The y-intercept appears to be located at $(0, -6)$.
For $y = \lvert x \rvert - 6$, $x = 0 \Rightarrow y = \lvert 0 \rvert - 6$
$$y = -6$$
$(0, -6)$

57. From the graph it appears that the x-intercept is $(4, 0)$.

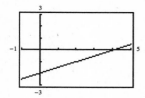

$y = \frac{1}{2}x - 2$

x-intercept: Let $y = 0$.

$$0 = \frac{1}{2}x - 2$$
$$2(0) = 2\left(\frac{1}{2}x - 2\right)$$
$$0 = x - 4$$
$$4 = x$$

The x-intercept is $(4, 0)$.

59. From the graph it appears that the x-intercepts are $(0, 0)$ and $(6, 0)$.

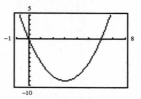

$y = x^2 - 6x$

x-intercept: Let $y = 0$.

$$0 = x^2 - 6x$$
$$0 = x(x - 6)$$
$$x = 0$$
$$x - 6 = 0 \Rightarrow x = 6$$

The x-intercepts are $(0, 0)$ and $(6, 0)$.

61.

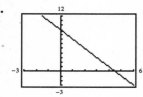

The solution is $\frac{9}{2}$.

Verify: $7 - 2\left(\frac{9}{2} - 1\right) \overset{?}{=} 0$

$$7 - 2\left(\frac{7}{2}\right) \overset{?}{=} 0$$
$$7 - 7 = 0 \checkmark$$

63.

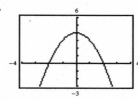

The solutions are 2 and -2.

Verify: $4 - 2^2 \overset{?}{=} 0$

$$4 - 4 = 0 \checkmark$$
$$4 - (-2)^2 \overset{?}{=} 0$$
$$4 - 4 = 0 \checkmark$$

65.

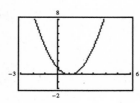

The solution is 1.

Verify: $1^2 - 2(1) + 1 \overset{?}{=} 0$

$$1 - 2 + 1 \overset{?}{=} 0$$
$$0 = 0 \checkmark$$

67.

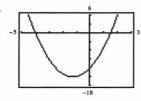

The solutions are -4 and $\frac{3}{2}$.

Verify: $2(-4)^2 + 5(-4) - 12 \overset{?}{=} 0$

$$2(16) - 20 - 12 \overset{?}{=} 0$$
$$32 - 20 - 12 \overset{?}{=} 0$$
$$0 = 0 \checkmark$$

$2\left(\frac{3}{2}\right)^2 + 5\left(\frac{3}{2}\right) - 12 \overset{?}{=} 0$

$$2\left(\frac{9}{4}\right) + \frac{15}{2} - 12 \overset{?}{=} 0$$
$$\frac{9}{2} + \frac{15}{2} - 12 \overset{?}{=} 0$$
$$\frac{24}{2} - 12 \overset{?}{=} 0$$
$$12 - 12 = 0 \checkmark$$

69.

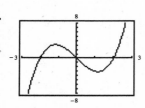

The solutions are 0, 2, and -2.

Verify:

$0^3 - 4(0) \overset{?}{=} 0$ $2^3 - 4(2) \overset{?}{=} 0$ $(-2)^3 - 4(-2) \overset{?}{=} 0$

$0 - 0 = 0 \checkmark$ $8 - 8 = 0 \checkmark$ $-8 + 8 = 0 \checkmark$

71.

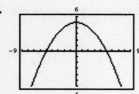

The x-intercepts of the graph of $y = 5 - 0.2x^2$ are $(5, 0)$ and $(-5, 0)$, so the solutions of the equation are $x = 5$ and $x = -5$.

73.

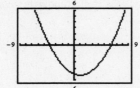

The x-intercepts of the graph of $y = 0.2x^2 - 0.4x - 5$ are approximately $(6.1, 0)$ and $(-4.1, 0)$, so the solutions of the equation are $x \approx 6.1$ and $x \approx -4.1$.

75. (a) $y = \frac{4}{3}x$

x	0	3	6	9	12
F	0	4	8	12	16

(c) When the force is doubled, the length of the spring is also doubled.

(b)

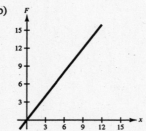

77. (a)

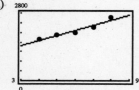

(b) The model fits the data well.

(c) S = 210(15) + 820 thousands of dollars

= 3970 thousands of dollars

= \$3,970,000

(d) As t increases, the values of S get infinitely large.

79.

t	0	5	8
y	225,000	125,000	65,000

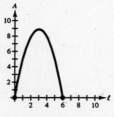

81. (a)

$$P = 2l + 2w$$
$$12 = 2l + 2w$$
$$12 - 2l = 2w$$
$$\frac{12 - 2l}{2} = w$$
$$6 - l = w$$
$$A = lw$$
$$A = l(6 - l)$$

(b)

$A = l(6 - l) = 6l - l^2$

(c) For $l = 4$, the graph indicates that A is approximately 8 square meters. The model in part (a) indicates that for $l = 4$, $A = 4(6 - 4) = 4(2) = 8$ square meters.

83. Graph the equations

$$y_1 = \tfrac{1}{2}(x - 4)$$

and

$$y_2 = \tfrac{1}{2}x - 2$$

on the same axes. If the graphs are identical, then $y_1 = y_2$.

$$\tfrac{1}{2}(x - 4) = \tfrac{1}{2}x - 2$$

This is an illustration of the Distributive Property.

85. Graph the equations

$$y_1 = x + (2x - 1)$$

and

$$y_2 = (x + 2x) - 1$$

on the same axes. If the graphs are identical, then $y_1 = y_2$.

$$x + (2x + 1) = (x + 2x) + 1$$

This is an illustration of the Associative Property of Addition.

87. The x-intercepts are $(3, 0)$ and $(-3, 0)$.

$$0 = x^2 - 9$$
$$0 = (x + 3)(x - 3)$$
$$x + 3 = 0 \Rightarrow x = -3$$
$$x - 3 = 0 \Rightarrow x = 3$$

The first coordinates of the x-intercepts are the solutions of the polynomial equation.

89. The x-intercepts are $(-1, 0)$ and $(3, 0)$.

$$0 = x^2 - 2x - 3$$
$$0 = (x - 3)(x + 1)$$
$$x - 3 = 0 \Rightarrow x = 3$$
$$x + 1 = 0 \Rightarrow x = -1$$

The first coordinates of the x-intercepts are the solutions of the polynomial equation.

91.

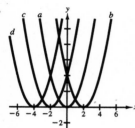

The graph of
$$y = (x + c)^2, c > 0,$$
is obtained by shifting the graph of $y = x^2$ to the *left* c units.

The graph of
$$y = (x - c)^2, \ c > 0,$$
is obtained by shifting the graph of $y = x^2$ to the *right* c units.

93.

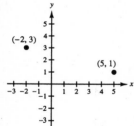

95.

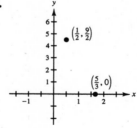

97. $\dfrac{3}{4}x + 8 = 0$

$$4\left(\dfrac{3}{4}x + 8\right) = 4(0)$$
$$3x + 32 = 0$$
$$3x = -32$$
$$x = -\dfrac{32}{3}$$

99. $y = \dfrac{9}{7}(14) - 1$

$$= 18 - 1$$
$$= 17$$

The missing coordinate is 17.

101. $-3 = 3.8 - 1.2x$

$$-6.8 = -1.2x$$
$$\dfrac{-6.8}{-1.2} = x$$
$$\dfrac{68}{12} = x$$
$$\dfrac{17}{3} = x$$

The missing coordinate is 17/3.

103.

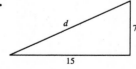

$$d = \sqrt{(15)^2 + 7^2}$$
$$= \sqrt{225 + 49}$$
$$= \sqrt{274}$$
$$\approx 16.55$$

Length of rafter $= 2 + d \approx 18.55$ feet.

Section 2.3   Slope: An Aid to Graphing Lines

1. $m = \frac{2}{3}$

(For example, from $(0, 2)$ to $(3, 4)$, the change in y is 2 and the change in x is 3.)

3. $m = -2$

(For example, from $(0, 8)$ to $(1, 6)$, the change in y is -2 and the change in x is 1.)

5. m is undefined. (For any two points on the line, the change in x is 0, so the slope is undefined.)

7. $m = \dfrac{0 - 12}{8 - 0} = \dfrac{-12}{8} = -\dfrac{3}{2}$

The line is falling.

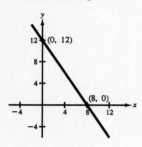

9. $m = \dfrac{1 - (-3)}{6 - (-2)} = \dfrac{4}{8} = \dfrac{1}{2}$

The line is rising.

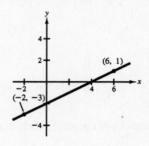

11. $m = \dfrac{-3 - 6}{5 - (-7)} = \dfrac{-9}{12} = -\dfrac{3}{4}$

The line is falling.

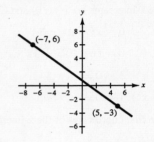

13. $m = \dfrac{-6 - 4}{-9 - (-5)} = \dfrac{-10}{-4} = \dfrac{5}{2}$

The line is rising.

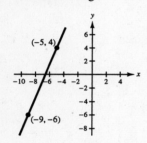

15. $m = \dfrac{-5 - (-5)}{7 - 2} = \dfrac{0}{5} = 0$

The line is horizontal.

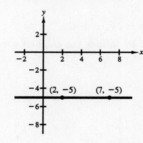

17. $m = \dfrac{5 - (-5)}{3 - (-3)} = \dfrac{10}{6} = \dfrac{5}{3}$

The line is rising.

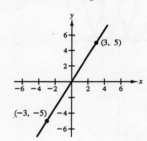

19. $m = \dfrac{\frac{1}{5} - \frac{2}{5}}{-\frac{1}{3} - \frac{5}{6}} = \dfrac{-\frac{1}{5}}{-\frac{7}{6}} = \dfrac{-1}{5} \cdot \dfrac{-6}{7} = \dfrac{6}{35}$

The line is rising.

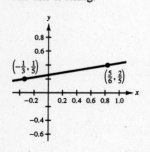

21. $m = \dfrac{-(5/2) - 2}{5 - (3/4)}$

$= \dfrac{-(5/2) - (4/2)}{(20/4) - (3/4)}$

$= -\dfrac{9}{2} \div \dfrac{17}{4}$

$= -\dfrac{9}{2} \cdot \dfrac{4}{17}$

$= -\dfrac{18}{17}$

The line is falling.

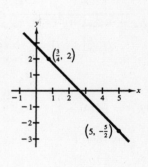

23. $m = \dfrac{6 - (-1)}{-4.2 - 4.2}$

$= \dfrac{7}{-8.4}$

$= \dfrac{70}{-84}$

$= -\dfrac{70}{84}$

$= -\dfrac{\cancel{(7)}(5)\cancel{(2)}}{\cancel{(7)}\cancel{(2)}(6)}$

$= -\dfrac{5}{6}$

The line is falling.

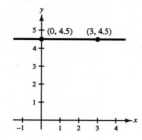

25. $m = \dfrac{4.5 - 4.5}{3 - 0} = \dfrac{0}{3} = 0$

The line is horizontal.

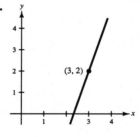

27. $-\dfrac{2}{3} = \dfrac{7 - 5}{x - 4}$

$\dfrac{-2}{3} = \dfrac{2}{x - 4}$

$-2(x - 4) = 3(2)$

$-2x + 8 = 6$

$-2x = -2$

$x = 1$

29. $\dfrac{3}{2} = \dfrac{3 - y}{9 - (-3)}$

$\dfrac{3}{2} = \dfrac{3 - y}{12}$

$3(12) = 2(3 - y)$

$36 = 6 - 2y$

$30 = -2y$

$-15 = y$

31.

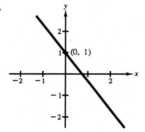

33.

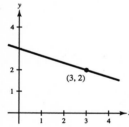

35.

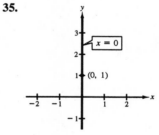

37.

39. $m = 0 \Longrightarrow$ the line is horizontal.

Every point on the horizontal line through $(5, 2)$ has a y-coordinate of 2.

$(6, 2), (10, 2), (-1, 2),$
$(8, 2), (-5, 2),$ etc.

41. $m = 3 = \dfrac{3}{1} \Longrightarrow \dfrac{\text{change in } y}{\text{change in } x} = \dfrac{3}{1}$

$(3 + 1, -4 + 3) = (4, -1)$

$(4 + 1, -1 + 3) = (5, 2)$

$(5 + 1, 2 + 3) = (6, 5)$

$(6 + 1, 5 + 3) = (7, 8),$ etc.

43. $m = -1 = \dfrac{-1}{1} \Longrightarrow \dfrac{\text{change in } y}{\text{change in } x} = \dfrac{-1}{1}$

$(0 + 1, 3 - 1) = (1, 2)$

$(1 + 1, 2 - 1) = (2, 1)$

$(2 + 1, 1 - 1) = (3, 0)$

$(3 + 1, 0 - 1) = (4, -1)$

$(4 + 1, -1 - 1) = (5, -2),$ etc.

45. $m = \frac{4}{3} \Rightarrow \dfrac{\text{change in } y}{\text{change in } x} = \frac{4}{3}$

$(-5 + 3, 0 + 4) = (-2, 4)$

$(-2 + 3, 4 + 4) = (1, 8)$

$(1 + 3, 8 + 4) = (4, 12)$

$(4 + 3, 12 + 4) = (7, 16)$

$(7 + 3, 16 + 4) = (10, 20)$, etc.

47. $6x - 3y = 9$

$-3y = -6x + 9$

$y = 2x - 3$

Slope: 2

y-intercept: $(0, -3)$

49. $4y - x = -4$

$4y = x - 4$

$y = \frac{1}{4}x - 1$

Slope: $\frac{1}{4}$

y-intercept: $(0, -1)$

51. $2x + 5y - 3 = 0$

$5y = -2x + 3$

$y = -\frac{2}{5}x + \frac{3}{5}$

Slope: $-\frac{2}{5}$

y-intercept: $\left(0, \frac{3}{5}\right)$

53. $x = 2y - 4$

$x - 2y = -4$

$-2y = -x - 4$

$y = \frac{1}{2}x + 2$

Slope: $\frac{1}{2}$

y-intercept: $(0, 2)$

55. $3x - y - 2 = 0$

$-y = -3x + 2$

$y = 3x - 2$

Slope: 3

y-intercept: $(0, -2)$

57. $x + y = 0$

$y = -x + 0$

Slope: -1

y-intercept: $(0, 0)$

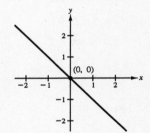

59. $3x + 2y - 2 = 0$

$2y = -3x + 2$

$y = -\frac{3}{2}x + 1$

Slope: $-\frac{3}{2}$

y-intercept: $(0, 1)$

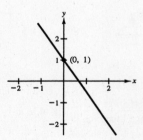

61. $x - 4y + 2 = 0$

$-4y = -x - 2$

$y = \frac{1}{4}x + \frac{1}{2}$

Slope: $\frac{1}{4}$

y-intercept: $\left(0, \frac{1}{2}\right)$

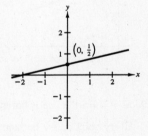

63. $y - 2 = 0$

$y = 2$

Slope: 0

y-intercept: $(0, 2)$

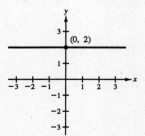

65. $x - 0.2y - 1 = 0$

$-0.2y = -x + 1$

$y = \dfrac{-x}{-0.2} + \dfrac{1}{-0.2}$

$y = 5x - 5$

Slope: 5

y-intercept: $(0, -5)$

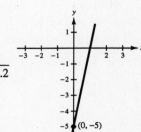

67. $y = \frac{1}{2}x - 2 \Rightarrow m_1 = \frac{1}{2}$

$y = \frac{1}{2}x + 3 \Rightarrow m_2 = \frac{1}{2}$

The slopes are equal, so the lines are parallel.

69. $y = \frac{3}{4}x - 3 \Rightarrow m_1 = \frac{3}{4}$

$y = -\frac{4}{3}x + 1 \Rightarrow m_2 = -\frac{4}{3}$

The slopes are negative reciprocals, so the lines are perpendicular.

71. $y = -\frac{2}{5}x + 1 \Rightarrow m_1 = -\frac{2}{5}$

$y = -\frac{5}{2}x + 1 \Rightarrow m_2 = -\frac{5}{2}$

The slopes are not equal and they are not negative reciprocals, so the lines are neither parallel nor perpendicular.

73. $x + 2y - 3 = 0 \Rightarrow y = -\frac{1}{2}x + \frac{3}{2} \Rightarrow m_1 = -\frac{1}{2}$

$-2x - 4y + 1 = 0 \Rightarrow y = -\frac{1}{2}x + \frac{1}{4} \Rightarrow m_2 = -\frac{1}{2}$

The slopes are equal, so the lines are parallel.

75. $m_1 = \frac{1-3}{2-1} = \frac{-2}{1} = -2$

$m_2 = \frac{2-0}{4-0} = \frac{2}{4} = \frac{1}{2}$

L_1 and L_2 are perpendicular because m_1 and m_2 are negative reciprocals.

77. $m_1 = \frac{4-0}{4-(-2)} = \frac{4}{6} = \frac{2}{3}$

$m_2 = \frac{0-(-2)}{4-1} = \frac{2}{3}$

L_1 and L_2 are parallel because $m_1 = m_2$.

79. $x + y = 9$

$y = -x + 9$

Slope: -1

(a) Slope of parallel line: -1

(b) Slope of perpendicular line: 1

81. $14x - 7y = 11$

$-7y = -14x + 11$

$y = 2x - \frac{11}{7}$

Slope: 2

(a) Slope of parallel line: 2

(b) Slope of perpendicular line: $-\frac{1}{2}$

83. $5x - 2y = -26$

$-2y = -5x - 26$

$y = \frac{5}{2}x + 13$

Slope: $\frac{5}{2}$

(a) Slope of parallel line: $\frac{5}{2}$

(b) Slope of perpendicular line: $-\frac{2}{5}$

85. $18x + 12y + 10 = 0$

$12y = -18x - 10$

$y = -\frac{3}{2}x - \frac{5}{6}$

Slope: $-\frac{3}{2}$

(a) Slope of parallel line: $-\frac{3}{2}$

(b) Slope of perpendicular line: $\frac{2}{3}$

87. Graph (b)

The slope is -10.

The slope represents the change in the unpaid loan per week. For every week that goes by, the amount owed on the loan decreases by $10.

89. Graph (a)

The slope is 0.25.

The slope represents the change in the amount the sales representative is paid per mile. For every mile traveled within a day, the daily pay increases by $0.25.

91. The Fahrenheit degrees increase by 180 as the Celsius degrees increase by 100. So the rate of change in Fahrenheit temperatures in relation to Celsius temperatures is 180/100. In simplified form, the rate of change can be simplified to 9/5 or 1.8.

93. (a) Perimeter $= 2(\text{Length}) + 2(\text{Width})$

$= 2(40 + 2x) + 2(30 + 2x)$

$= 80 + 4x + 60 + 4x$

$= 8x + 140$

(c) $P = 8x + 140$

The slope of the graph is 8. For each additional 1-foot increase in the width of the walkway, the outside perimeter increases by 8 feet.

(b)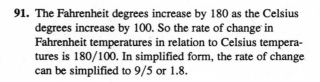

95. $-\dfrac{12}{100} = -\dfrac{2000}{x}$

$-12x = -200,000$

$x = \dfrac{-200,000}{-12}$

$\approx 16,667$ feet (or approximately 3.16 miles)

97. $m \approx \dfrac{180}{16} = \dfrac{45}{4}$ or 11.25

99. $\dfrac{1}{2}(30) = 15 \Longrightarrow \dfrac{3}{4} = \dfrac{h}{15}$

$3(15) = 4h$

$45 = 4h$

$\dfrac{45}{4} = h$

11.25 feet $= h$

101. A negative slope indicates that the values of y get smaller as x gets larger; the line falls from left to right. A zero slope indicates that the values of y remain unchanged as x changes; the line is horizontal. A positive slope indicates that the values of y get larger as x gets larger; the line rises from left to right.

103. No. Two lines with positive slopes could not be perpendicular to one another. Perpendicular lines have slopes that are negative reciprocals of one another. If the slope of a line is positive, every line perpendicular to it has a negative slope.

105. $\dfrac{a - b}{c - d} = \dfrac{5 - 2}{3 - (-1)} = \dfrac{3}{4}$

107. $y - 6 = 4[x - (-10)]$

$y - 6 = 4(x + 10)$

$y - 6 = 4x + 40$

$y = 4x + 46$

109. $P = 2L + 2W$

$P = 2(4x + 1) + 2(x + 3)$

$P = 8x + 2 + 2x + 6$

$P = 10x + 8$

Mid-Chapter Quiz for Chapter 2

1. (a)

(b) $d = \sqrt{(x_2 - x_1)^2 + (y_2 - y_1)^2}$

$= \sqrt{(3 + 1)^2 + (2 - 5)^2}$

$= \sqrt{4^2 + (-3)^2}$

$= \sqrt{16 + 9}$

$= \sqrt{25}$

$= 5$

(c) $M = \left(\dfrac{x_1 + x_2}{2}, \dfrac{y_1 + y_2}{2}\right)$

$= \left(\dfrac{-1 + 3}{2}, \dfrac{5 + 2}{2}\right)$

$= \left(1, \dfrac{7}{2}\right)$

(d) $m = \dfrac{y_2 - y_1}{x_2 - x_1}$

$= \dfrac{2 - 5}{3 - (-1)}$

$= \dfrac{-3}{4}$

$= -\dfrac{3}{4}$

2. (a)

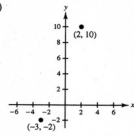

(b) $d = \sqrt{(x_2 - x_1)^2 + (y_2 - y_1)^2}$

$= \sqrt{(2 + 3)^2 + (10 + 2)^2}$

$= \sqrt{5^2 + 12^2}$

$= \sqrt{25 + 144}$

$= \sqrt{169}$

$= 13$

(c) $M = \left(\dfrac{x_1 + x_2}{2}, \dfrac{y_1 + y_2}{2}\right)$

$= \left(\dfrac{-3 + 2}{2}, \dfrac{-2 + 10}{2}\right)$

$= \left(-\dfrac{1}{2}, 4\right)$

(d) $m = \dfrac{y_2 - y_1}{x_2 - x_1}$

$= \dfrac{10 - (-2)}{2 - (-3)}$

$= \dfrac{12}{5}$

3. $(10, -3)$

4. (a) $4x - 3y = 10$

$4(2) - 3(1) \overset{?}{=} 10$

$8 - 3 \overset{?}{=} 10$

$5 \neq 10$

No

(b) $4x - 3y = 10$

$4(1) - 3(-2) \overset{?}{=} 10$

$4 + 6 \overset{?}{=} 10$

$10 = 10$

Yes

(c) $4x - 3y = 10$

$4(2.5) - 3(0) \overset{?}{=} 10$

$10 - 0 \overset{?}{=} 10$

$10 = 10$

Yes

(d) $4x - 3y = 10$

$4(2) - 3\left(-\dfrac{2}{3}\right) \overset{?}{=} 10$

$8 + \dfrac{6}{3} \overset{?}{=} 10$

$8 + 2 \overset{?}{=} 10$

$10 = 10$

Yes

5.

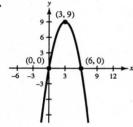

$y = 6x - x^2$

x	-1	0	2	3	5	6	7
y	-7	0	8	9	5	0	-7
(x, y)	$(-1, -7)$	$(0, 0)$	$(2, 8)$	$(3, 9)$	$(5, 5)$	$(6, 0)$	$(7, -7)$

6.

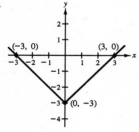

$y = |x| - 3$

x	-4	-3	-1	0	1	3	4
y	1	0	-2	-3	-2	0	1
(x, y)	$(-4, 1)$	$(-3, 0)$	$(-1, -2)$	$(0, -3)$	$(1, -2)$	$(3, 0)$	$(4, 1)$

7.

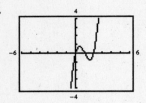

The x-intercepts appear to be located at $(0, 0)$, $(1, 0)$ and $(2, 0)$.
The y-intercept appears to be located at $(0, 0)$.

Verifying algebraically:

x-intercept y-intercept

Let $y = 0$.   Let $x = 0$.

$$0 = 2x^3 - 6x^2 + 4x$$ $$y = 2(0)^3 - 6(0)^2 + 4(0)$$

$$0 = 2x(x^2 - 3x + 2)$$ $$y = 0 - 0 + 0$$

$$0 = 2x(x - 2)(x - 1)$$ $$y = 0$$

$$2x = 0 \Rightarrow x = 0$$ $(0, 0)$

$$x - 2 = 0 \Rightarrow x = 2$$

$$x - 1 = 0 \Rightarrow x = 1$$

$(0, 0)$, $(2, 0)$ and $(1, 0)$

8.

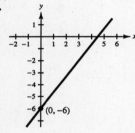

9. $3x + y - 6 = 0$

$$y = -3x + 6$$

Slope: -3

y-intercept: $(0, 6)$

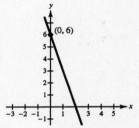

10. $8x - 6y = 30$

$$-6y = -8x + 30$$

$$y = \frac{-8}{-6}x + \frac{30}{-6}$$

$$y = \frac{4}{3}x - 5$$

Slope: $\frac{4}{3}$

y-intercept: $(0, -5)$

11. $y = 3x + 2 \Rightarrow m_1 = 3$

$$y = -\frac{1}{3}x - 4 \Rightarrow m_2 = -\frac{1}{3}$$

The slopes are negative reciprocals, so the lines are perpendicular.

12. $y = 2x - 3 \Rightarrow m_1 = 2$

$$y = -2x - 3 \Rightarrow m_2 = -2$$

The slopes are not equal and they are not negative reciprocals, so the lines are neither parallel nor perpendicular.

13. $y = 4x + 3 \Rightarrow m_1 = 4$

$$y = \frac{1}{2}(8x + 5) \Rightarrow y = 4x + \frac{5}{2} \Rightarrow m_2 = 4$$

The slopes are equal, so the lines are parallel.

14. $5x + 3y - 9 = 0$

$$3y = -5x + 9$$

$$y = -\frac{5}{3}x + 3$$

Slope: $-\frac{5}{3}$

(a) Slope of parallel line: $-\frac{5}{3}$

(b) Slope of perpendicular line: $\frac{3}{5}$

15. $\frac{4000 - 85,000}{10 - 0} = \frac{-81,000}{10} = -8100$

The average rate of change in the value of the printing press is $-\$8100$ per year.

16. The highest temperature was approximately 68°F.
The highest temperature occurred at 12:00 noon.
The lowest temperature was approximately 58°F.
The lowest temperature occurred at 8:00 A.M.

Section 2.4   Relations, Functions, and Function Notation

1. $\{(-2, 0), (0, 1), (1, 4), (0, -1)\}$

Domain $= \{-2, 0, 1\}$

Range $= \{-1, 0, 1, 4\}$

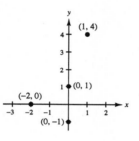

3. $\{(0, 0), (4, -3), (2, 8), (5, 5), (6, 5)\}$

Domain $= \{0, 2, 4, 5, 6\}$

Range $= \{-3, 0, 5, 8\}$

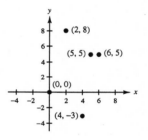

5. $d = rt$

$d = 50t$ with domain $= \left\{3, 2, 8, 6, \frac{1}{2}\right\}$

$\left\{(3, 150), (2, 100), (8, 400), (6, 300), \left(\frac{1}{2}, 25\right)\right\}$

7. $\{(1, 1), (2, 8), (3, 27), (4, 64), (5, 125), (6, 216), (7, 343)\}$

9. $\{(1997, \text{Florida Marlins}), (1998, \text{New York Yankees}), (1999, \text{New York Yankees}),$
$(2000, \text{New York Yankees}), (2001, \text{Arizona Diamondbacks})\}$

11. No, it is not a function. One element in the domain is matched with more than one element in the range.

13. Yes, it is a function. Every element in the domain is matched with exactly one element in the range.

15. No, it is not a function. Elements in the domain are matched with more than one element in the range.

17. No, it is not a function. Elements in the domain are matched with more than one element in the range.

19. (a) Yes, because every element in A is matched with exactly one element in B.

(b) No, because the element 1 in A is matched with two elements, -2 and 1, in B.

(c) Yes, because every element in A is matched with exactly one element in B.

(d) No, because not all elements in A are matched with an element in B.

21. $(0, 5)$: $0^2 + 5^2 \overset{?}{=} 25$

$\qquad\qquad 25 = 25$

$\quad (0, -5)$: $0^2 + (-5)^2 \overset{?}{=} 25$

$\qquad\qquad\quad 25 = 25$

In this equation, y is not a function of x because an x-value, 0, is matched with two different y-values.

23. $(1, 3)$: $|3| \overset{?}{=} 1 + 2$

$\qquad\qquad 3 = 3$

$\quad (1, -3)$: $|-3| \overset{?}{=} 1 + 2$

$\qquad\qquad\quad 3 = 3$

In this equation, y is not a function of x because an x-value, 1, is matched with two different y-values.

25. Yes, there is exactly one high school enrollment for each year. Yes, there is exactly one college enrollment for each year.

27. (a) $f(2) = 3(2) + 5 = 11$

(b) $f(-2) = 3(-2) + 5 = -1$

(c) $f(k) = 3(k) + 5 = 3k + 5$

(d) $f(k + 1) = 3(k + 1) + 5 = 3k + 8$

29. (a) $f(3) = 12(3) - 7$

$\qquad\quad = 36 - 7$

$\qquad\quad = 29$

(c) $f(a) + f(1) = [12(a) - 7] + [12(1) - 7]$

$\qquad\qquad\qquad = (12a - 7) + (12 - 7)$

$\qquad\qquad\qquad = 12a - 7 + 5$

$\qquad\qquad\qquad = 12a - 2$

(b) $f\left(\frac{3}{2}\right) = 12\left(\frac{3}{2}\right) - 7$

$\qquad\quad\ = 18 - 7$

$\qquad\quad\ = 11$

(d) $f(a + 1) = 12(a + 1) - 7$

$\qquad\qquad\ = 12a + 12 - 7$

$\qquad\qquad\ = 12a + 5$

31. (a) $f(-1) = \sqrt{-1 + 5} = \sqrt{4} = 2$

(c) $f\left(\frac{16}{3}\right) = \sqrt{\frac{16}{3} + 5} = \sqrt{\frac{31}{3}}$

(b) $f(4) = \sqrt{4 + 5} = \sqrt{9} = 3$

(d) $f(5z) = \sqrt{5z + 5}$

33. (a) $f(0) = \dfrac{3(0)}{0 - 5} = \dfrac{0}{-5} = 0$

(c) $f(2) - f(-1) = \dfrac{3(2)}{2 - 5} - \dfrac{3(-1)}{-1 - 5}$

$\qquad\qquad\qquad\ = \dfrac{6}{-3} - \dfrac{-3}{-6}$

$\qquad\qquad\qquad\ = -2 - \dfrac{1}{2}$

$\qquad\qquad\qquad\ = -\dfrac{5}{2}$

(b) $f\left(\dfrac{5}{3}\right) = \dfrac{3\left(\dfrac{5}{3}\right)}{\dfrac{5}{3} - 5}$

$\qquad\qquad = \dfrac{5}{-\dfrac{10}{3}}$

$\qquad\qquad = \dfrac{5}{1} \cdot -\dfrac{3}{10}$

$\qquad\qquad = -\dfrac{3}{2}$

(d) $f(x + 4) = \dfrac{3(x + 4)}{x + 4 - 5} = \dfrac{3x + 12}{x - 1}$

35. (a) $f(4) = 10 - 2(4) = 10 - 8 = 2$

(c) $f(0) = 10 - 2(0) = 10$

(b) $f(-10) = -10 + 8 = -2$

(d) $f(6) - f(-2) = [10 - 2(6)] - [-2 + 8]$

$\qquad\qquad\qquad\ = 10 - 12 - 6$

$\qquad\qquad\qquad\ = -8$

37. (a) $f(-2) = 5(-2) - 4 = -14$

(b) $f(2) = 10(2) = 20$

(c) $f(1) + f(3) = [5(1) - 4] + [10(3)]$
$$= 1 + 30$$
$$= 31$$

(d) $f(-5) + f(-4) = 12 + 12 = 24$

39. (a) $f(x + 2) - f(2) = [2(x + 2)^2 + 5] - [2(2)^2 + 5]$
$$= [2(x^2 + 4x + 4) + 5] - [2(4) + 5]$$
$$= 2x^2 + 8x + 8 + 5 - 8 - 5$$
$$= 2x^2 + 8x$$

(b) $\dfrac{f(x - 3) - f(3)}{x} = \dfrac{[2(x - 3)^2 + 5] - [2(3)^2 + 5]}{x}$
$$= \dfrac{[2(x^2 - 6x + 9) + 5] - [2(9) + 5]}{x}$$
$$= \dfrac{2x^2 - 12x + 18 + 5 - 18 - 5}{x}$$
$$= \dfrac{2x^2 - 12x}{x}$$
$$= 2x - 12$$

41. Domain = $\{0, 2, 4, 6\}$

Range = $\{0, 1, 8, 27\}$

43. Domain = $\{-5, -2, 0, 3, 6\}$

Range = $\{-2, 3, 9\}$

45. Domain = $\{r: r > 0\}$

Range = $\{C: C > 0\}$

47. Domain = $\{x: x \text{ is a real number}\}$

49. Domain = $\{x: x \neq 3\}$

51. Domain = $\{t: t \neq 0, -2\}$

53. Domain = $\{x: x \neq 2, 1\}$

55. $f(x) = \dfrac{9}{x^2 - 1}$
$$= \dfrac{9}{(x + 1)(x - 1)}$$
Domain = $\{x: x \neq -1, 1\}$

57. Domain = $\{x: x \geq 2\}$

59. Domain = $\{x: x \leq 3\}$

61. Domain = $\{t: t \text{ is a real number}\}$

63. Domain = $\{t: t \text{ is a real number}\}$

65. Perimeter = 4 (Side)
$$P = 4x$$

67. Volume = (Length)(Width)(Height)
$$V = (24 - 2x)(24 - 2x)(x)$$
$$= 2(12 - x)(2)(12 - x)(x)$$
$$= 4x(12 - x)^2$$

69. Area = (Length)(Width)
$$A = (32 - 2x)(32 - 2x)$$
$$= 2(16 - x)(2)(16 - x)$$
$$= 4(16 - x)^2$$

71. Distance = (Rate)(Time)
$$d = 230t$$

73. $P(x) = 50\sqrt{x} - 0.5x - 500$

(a) $P(1600) = 50\sqrt{1600} - 0.5(1600) - 500$
$$= 50(40) - 800 - 500$$
$$= 2000 - 800 - 500$$
$$= \$700$$

(b) $P(2500) = 50\sqrt{2500} - 0.5(2500) - 500$
$$= 50(50) - 1250 - 500$$
$$= 2500 - 1250 - 500$$
$$= \$750$$

75. A relation is any set of ordered pairs. A function is a relation in which no two ordered pairs have the same first component and different second components.

77. (a) Yes, this is correct.

(b) No, this is not correct.

79. $5(x + t) + 8 - (3x + 1) = 5x + 5t + 8 - 3x - 1$
$$= 2x + 5t + 7$$

81. $7(x + 4)^2 - 8 = 7(x^2 + 8x + 16) - 8$
$$= 7x^2 + 56x + 112 - 8$$
$$= 7x^2 + 56x + 104$$

83. (a) $8 - 5(1)^2 = 8 - 5(1)$
$$= 8 - 5$$
$$= 3$$

(b) $8 - 5(-2)^2 = 8 - 5(4)$
$$= 8 - 20$$
$$= -12$$

85. (a) $\dfrac{-5 + 5}{-5 - 12} = \dfrac{0}{-17} = 0$

(b) $\dfrac{12 + 5}{12 - 12} = \dfrac{17}{0}$ undefined

This expression is undefined for $x = 12$; $x = 12$ is not in the domain of this function because division by zero is undefined.

87.   (Length)(Width) = Area
$$(w + 8)(w) = 308$$
$$w^2 + 8w - 308 = 0$$
$$(w + 22)(w - 14) = 0$$
$$w + 22 = 0 \Rightarrow w = -22$$
$$w - 14 = 0 \Rightarrow w = 14$$

(Discard the negative answer.)
The width is 14 feet and the length is 14 + 8, or 22, feet.

Section 2.5   Graphs of Functions

1. Slope: 2

y-intercept: $(0, -6)$

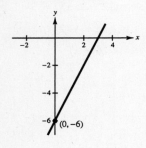

3. Slope: -1

y-intercept: $(0, -4)$

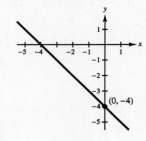

5. Slope: $-\frac{3}{4}$

y-intercept: $(0, 1)$

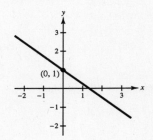

7. $3x - y + 10 = 0$
$$-y = -3x - 10$$
$$y = 3x + 10$$
$$f(x) = 3x + 10$$

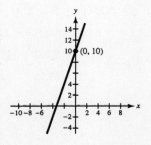

9. $0.2x + 0.8y - 4.5 = 0$

$10(0.2x + 0.8y - 4.5) = 10(0)$

$2x + 8y - 45 = 0$

$8y = -2x + 45$

$y = -\frac{2}{8}x + \frac{45}{8}$

$y = -\frac{1}{4}x + \frac{45}{8}$

$f(x) = -\frac{1}{4}x + \frac{45}{8}$

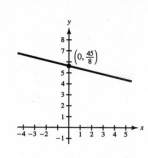

11. No vertical line intersects the graph more than once, so *y is a* function of *x*.

13. Some vertical lines intersect the graph more than once, so *y is not* a function of *x*.

15. Some vertical lines intersect the graph more than once, so *y is not* a function of *x*.

17. No vertical line intersects the graph more than once, so *y is a* function of *x*.

19. Domain = $\{x: x$ is a real number$\}$

Range = $\{y: y$ is a real number$\}$

21. Domain = $\{x: x \leq -2$ or $x \geq 2\}$

Range = $\{y: y \geq 0\}$

23. Domain = $\{x: -3 \leq x \leq 3\}$

Range = $\{y: 0 \leq y \leq 3\}$

25.

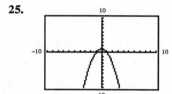

Domain = $\{x: x$ is a real number$\}$

Range = $\{y: y \leq 1\}$

27.

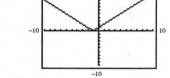

Domain = $\{x: x$ is a real number$\}$

Range = $\{y: y \geq 0\}$

29.

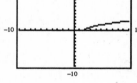

Domain = $\{x: x \geq 2\}$

Range = $\{y: y \geq 0\}$

31.

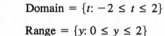

Domain = $\{t: -2 \leq t \leq 2\}$

Range = $\{y: 0 \leq y \leq 2\}$

33. Graph (b)

35. Graph (d)

37.

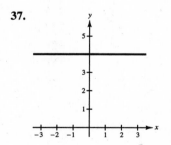

Domain = $\{x: x$ is a real number$\}$

Range = $\{4\}$

39.

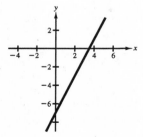

Domain = $\{x: x$ is a real number$\}$

Range = $\{y: y$ is a real number$\}$

41.

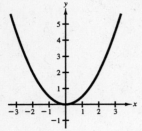

Domain = {x: x is a real number}
Range = {y: $y \geq 0$}

43.

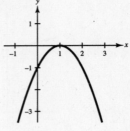

Domain = {x: x is a real number}
Range = {y: $y \leq 0$}

45.

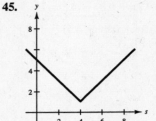

Domain = {x: x is a real number}
Range = {y: $y \leq 0$}

47.

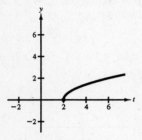

Domain = {t: $t \geq 2$}
Range = {y: $y \geq 0$}

49.

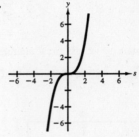

Domain = {s: s is a real number}
Range = {y: y is a real number}

51.

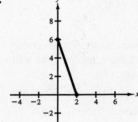

Domain = {x: $0 \leq x \leq 2$}
Range = {y: $0 \leq y \leq 6$}

53.

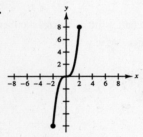

Domain = {x: $-2 \leq x \leq 2$}
Range = {y: $-8 \leq y \leq 8$}

55.

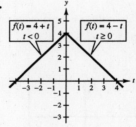

Domain = {t: t is a real number}
Range = {y: $y \leq 4$}

57.

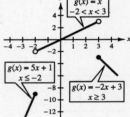

Domain = {x: x is a real number}
Range = {y: $y \leq -3$ or $-2 < y < 3$}

59. (a) Perimeter = 2(Length) + 2(Width)

$$200 = 2x + 2w$$

$$200 - 2x = 2w$$

$$100 - x = w$$

Area = (Length)(Width)

$$= x(100 - x)$$

(b)

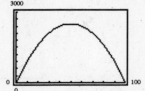

(c) The area is largest when $x = 50$. This indicates that the area is largest when the rectangle is a square.

61. (a)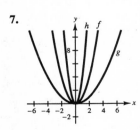

(b) $P(x) = 0.47x - 100$

$0 = 0.47x - 100$

$100 = 0.47x$

$\dfrac{100}{0.47} = x$

$212.8 \approx x$

To break even, approximately 213 units must be sold.

(c) $P(x) = 0.47x - 100$

$300 = 0.47x - 100$

$400 = 0.47x$

$\dfrac{400}{0.47} = x$

$851.1 \approx x$

To make a profit of $300, approximately 851 units must be sold.

63. Every value of y corresponds to exactly one value of x, so x *is* a function of y.

65. A graph shows y as a function of x if and only if no vertical line passes through the graph more than once.

67. $x^2 - 13x + 40 = 0$

$(x - 5)(x - 8) = 0$

$x - 5 = 0 \Rightarrow x = 5$

$x - 8 = 0 \Rightarrow x = 8$

69.

$3a^3 - 12a = 0$

$3a(a^2 - 4) = 0$

$3a(a + 2)(a - 2) = 0$

$3a = 0 \Rightarrow a = 0$

$a + 2 = 0 \Rightarrow a = -2$

$a - 2 = 0 \Rightarrow a = 2$

71. (a) $f(-3) = 4(-3)^2 = 4(9) = 36$

(b) $f(0) = 4(0)^2 = 4(0) = 0$

73. (a) $f(8) = \dfrac{8 + 2}{8 - 5} = \dfrac{10}{3}$

(b) $f(t + 1) = \dfrac{(t + 1) + 2}{(t + 1) - 5} = \dfrac{t + 3}{t - 4}$

75. $C = 2.05x + 9500$

Section 2.6   Transformations of Functions

1.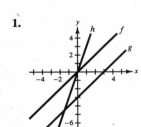

3.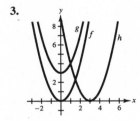

5.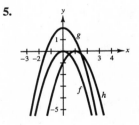

7.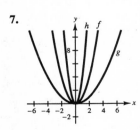

9.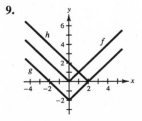

11.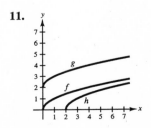

13. Basic function: $y = x^3$.

Transformation: Horizontal shift 2 units to the right

Equation: $y = (x - 2)^3$

15. Basic Function: $y = x^2$.

Transformation: Reflection in the x-axis

Equation: $y = -x^2$

17. Basic function: $y = \sqrt{x}$.

Transformation: Reflection in the x-axis and vertical shift upward of 1 unit

Equation: $y = -\sqrt{x} + 1$

19. Basic Function: $y = x^2$.

Transformation: Vertical shift downward of 1 unit

Equation: $y = x^2 - 1$

21. Basic function: $y = x^3$.

Transformation: Reflection in the x-axis (or y axis) and shift upward of 1 unit

Equation: $y = -x^3 + 1$ or $y = (-x)^3 + 1$

23. Basic function: $y = x$.

Transformation: Vertical shift upward of 3 units (or Horizontal shift of 3 units to the left)

Equation: $y = x + 3$

25. Basic function: $y = |x|$.

Transformation: Horizontal shift 2 units to the right and vertical shift downward of 2 units

Equation: $y = |x - 2| - 2$

27. Reflection in the x-axis

$$y = -\sqrt{x}$$

29. Horizontal shift to the left of 2 units

$$y = \sqrt{x + 2}$$

31. Reflection in the y-axis

$$y = \sqrt{-x}$$

33. Vertical shift upward of 3 units

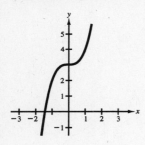

35. Horizontal shift to the right of 3 units

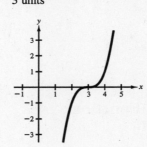

37. Reflection in the y-axis

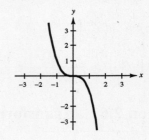

39. Reflection in the x-axis, horizontal shift to the right of 1 unit, and vertical shift upward of 2 units

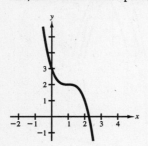

41. Horizontal shift 5 units to the right

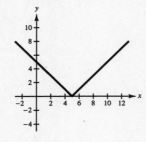

43. Vertical shift 5 units downward

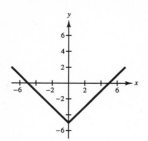

45. Reflection in the *x*-axis

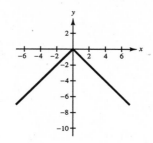

47. The graph of *g* is a horizontal shift of 2 units to the left.

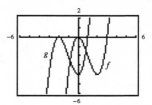

49. The graph of *g* is a reflection in the *y*-axis.

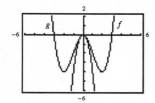

51. The graph is a reflection in the *x*-axis and a vertical shift upward of 1 unit.

$$g(x) = -(x^3 - 3x^2) + 1$$
$$= -x^3 + 3x^2 + 1$$

53. (a)

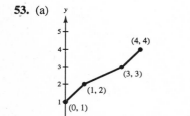

(b)

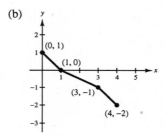

(c)

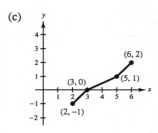

(d)

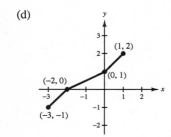

(e)

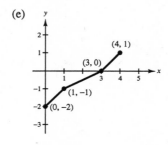

(f)

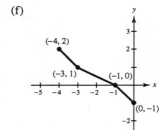

55. (a)

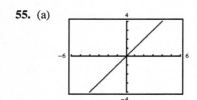

(b)

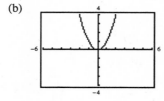

(c)

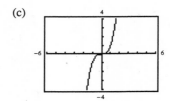

(d)

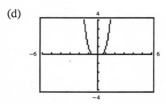

(e)

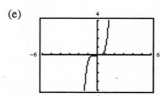

(f)

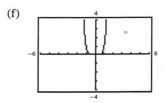

All graphs pass through the origin. The graphs of the even powers resemble the squaring function. The graphs of the odd powers resemble the cubing function. As the powers of *x* increase, the graphs become flatter for the *x*-values between −1 and 1.

57.

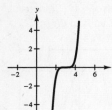

59. (a)

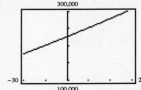

(c)

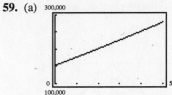

(b) In the transformation, $t = 0$ corresponds to the year 1980. The transformation is a horizontal shift 30 units to the left.

61. The graph has x-intercepts at $(0, 0)$ and $(6, 0)$. The graph touches but does not cross the x-axis at $(6, 0)$; the graph crosses the x-axis at $(0, 0)$.

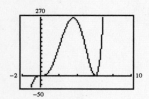

63. The graph has x-intercepts at $(0, 0)$ and $(6, 0)$. The graph crosses the x-axis at these intercepts.

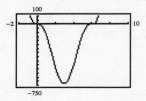

65. $4(x + 9) - 2 = 4x + 36 - 2$
$$= 4x + 34$$

67. $2(x + 6) - 5(x + 6) = 2x + 12 - 5x - 30$
$$= -3x - 18$$

69. $x(x + 7) - 4(x + 7) = x^2 + 7x - 4x - 28$
$$= x^2 + 3x - 28$$

71. $5(z - 2) + (z - 2)^2 = 5z - 10 + z^2 - 4z + 4$
$$= z^2 + z - 6$$

73. $8(x + 4) - (x + 4)^2 = 8x + 32 - (x^2 + 8x + 16)$
$$= 8x + 32 - x^2 - 8x - 16$$
$$= -x^2 + 16$$

75. Area $= 12(x + 6) - 7x$
$$= 12x + 72 - 7x$$
$$= 5x + 72$$

Review Exercises for Chapter 2

1.

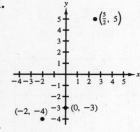

3.

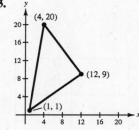

5. The point $(2, -6)$ is in Quadrant IV.

The point is located to the *right* of the vertical axis and *below* the horizontal axis in Quadrant IV.

7. The point $(4, y)$ may be located in Quadrants I or IV.

If $y > 0$, the point is located to the *right* of the vertical axis and *above* the horizontal axis in Quadrant I. If $y < 0$, the point is located to the *right* of the vertical axis and *below* the horizontal axis in Quadrant IV.

Note: If $y = 0$, the point would be located on the x-axis *between* the first and fourth quadrants.

9. (a) $(4, 2)$

$y = 4 - \frac{1}{2}x$

$2 \overset{?}{=} 4 - \frac{1}{2}(4)$

$2 \overset{?}{=} 4 - 2$

$2 = 2$

Yes, $(4, 2)$ is a solution.

(b) $(-1, 5)$

$y = 4 - \frac{1}{2}x$

$5 \overset{?}{=} 4 - \frac{1}{2}(-1)$

$5 \overset{?}{=} 4 + \frac{1}{2}$

$5 \neq 4\frac{1}{2}$

No, $(-1, 5)$ is not a solution.

(c) $(-4, 0)$

$y = 4 - \frac{1}{2}x$

$0 \overset{?}{=} 4 - \frac{1}{2}(-4)$

$0 \overset{?}{=} 4 + 2$

$0 \neq 6$

No, $(-4, 0)$ is not a solution.

(d) $(8, 0)$

$y = 4 - \frac{1}{2}x$

$0 \overset{?}{=} 4 - \frac{1}{2}(8)$

$0 \overset{?}{=} 4 - 4$

$0 = 0$

Yes, $(8, 0)$ is a solution.

11. (a)

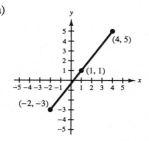

(b) $d = \sqrt{(4 - (-2))^2 + (5 - (-3))^2}$

$\quad = \sqrt{(6)^2 + (8)^2}$

$\quad = \sqrt{100}$

$\quad = 10$

(c) $M = \left(\dfrac{-2 + 4}{2}, \dfrac{-3 + 5}{2} \right) = (1, 1)$

13. (a)

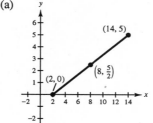

(b) $d = \sqrt{(14 - 2)^2 + (5 - 0)^2}$

$\quad = \sqrt{(12)^2 + (5)^2}$

$\quad = \sqrt{169}$

$\quad = 13$

(c) $M = \left(\dfrac{2 + 14}{2}, \dfrac{0 + 5}{2} \right) = \left(8, \dfrac{5}{2} \right)$

15.

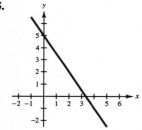

17.

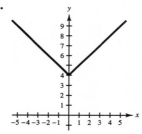

19. x-intercept

Let $y = 0$.

$0 = 6 - \dfrac{1}{3}x$

$\dfrac{1}{3}x = 6$

$x = 18$

$(18, 0)$

y-intercept

Let $x = 0$.

$y = 6 - \dfrac{1}{3}(0)$

$y = 6 - 0$

$y = 6$

$(0, 6)$

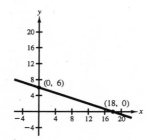

21. *x*-intercept

Let $y = 0$.

$3(0) - 2x - 3 = 0$

$-2x = 3$

$x = -\dfrac{3}{2}$

$\left(-\dfrac{3}{2}, 0\right)$

y-intercept

Let $x = 0$.

$3y - 2(0) - 3 = 0$

$3y = 3$

$y = 1$

$(0, 1)$

23. *x*-intercept

Let $y = 0$.

$x = |0 - 3|$

$x = |-3|$

$x = 3$

$(3, 0)$

y-intercept

Let $x = 0$.

$0 = |y - 3|$

$0 = y - 3$

$3 = y$

$(0, 3)$

25. *x*-intercept

Let $y = 0$.

$0 = 1 - x^2$

$x^2 - 1 = 0$

$(x + 1)(x - 1) = 0$

$x + 1 = 0 \Longrightarrow x = -1$

$x - 1 = 0 \Longrightarrow x = 1$

$(1, 0)$ and $(-1, 0)$

y-intercept

Let $x = 0$.

$y = 1 - 0^2$

$y = 1$

$(0, 1)$

27. $4x - 9 = 7$

$4x = 16$

$x = 4$

29. $x^2 - 3x = 18$

$x^2 - 3x - 18 = 0$

$(x - 6)(x + 3) = 0$

$x - 6 = 0 \Longrightarrow x = 6$

$x + 3 = 0 \Longrightarrow x = -3$

31. $(-1, 1), (6, 3)$

$m = \dfrac{3 - 1}{6 - (-1)} = \dfrac{2}{7}$

33. $(-1, 3), (4, 3)$

$m = \dfrac{3 - 3}{4 - (-1)} = \dfrac{0}{5} = 0$

35. $(0, 6), (8, 0)$

$m = \dfrac{0 - 6}{8 - 0}$

$= \dfrac{-6}{8}$

$= -\dfrac{3}{4}$

37. $m = -3 = \frac{-3}{1} \Rightarrow \frac{\text{change in } y}{\text{change in } x} = \frac{-3}{1}$

$(2, -4)$

$(2 + 1, -4 - 3) = (3, -7)$

$(3 + 1, -7 - 3) = (4, -10)$

$(4 + 1, -10 - 3) = (5, -13),$

etc.

39. $m = \frac{5}{4} \Rightarrow \frac{\text{change in } y}{\text{change in } x} = \frac{5}{4}$

$(3, 1)$

$(3 + 4, 1 + 5) = (7, 6)$

$(7 + 4, 6 + 5) = (11, 11)$

$(11 + 4, 11 + 5) = (15, 16),$

etc.

41. m undefined $\Rightarrow$ the line is vertical.

Every point on the vertical line through $(3, 7)$ has an x-coordinate of 3.

$(3, 0), (3, 5), (3, -4), (3, -9)$, etc.

43. $5x - 2y - 4 = 0$

$-2y = -5x + 4$

$y = \frac{5}{2}x - 2$

45. $x + 2y - 2 = 0$

$2y = -x + 2$

$y = -\frac{1}{2}x + 1$

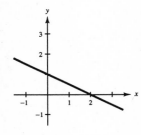

47. $y = \frac{3}{2}x + 1 \Rightarrow m_1 = \frac{3}{2}$

$y = \frac{2}{3}x - 1 \Rightarrow m_2 = \frac{2}{3}$

The slopes are not equal and they are not negative reciprocals, so the lines are neither parallel nor perpendicular.

49. $y = \frac{3}{2}x - 2 \Rightarrow m_1 = \frac{3}{2}$

$y = -\frac{2}{3}x + 1 \Rightarrow m_2 = -\frac{2}{3}$

The slopes are negative reciprocals, so the lines are perpendicular.

51. $5x + 20y = 30 \Rightarrow y = -\frac{1}{4}x + \frac{3}{2} \Rightarrow m_1 = -\frac{1}{4}$

$-12x + 3y = -10 \Rightarrow y = 4x - \frac{10}{3} \Rightarrow m_2 = 4$

The slopes are negative reciprocals, so the lines are perpendicular.

53. Domain = $\{1, 2, 3\}$

Range = $\{6, 7, 8, 10\}$

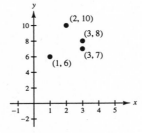

55. The relation *is not* a function because one element in the domain is matched with more than one element in the range.

57. The relation *is* a function because each element in the domain is matched with exactly one element in the range.

59. The relation *is* a function because each element in the domain is matched with exactly one element in the range.

61. The relation *is not* a function because one element in the domain is matched with more than one element in the range.

63. The relation *is* a function because each element in the domain is matched with exactly one element in the range.

65. $f(x) = 4 - \frac{5}{2}x$

(a) $f(-10) = 4 - \frac{5}{2}(-10)$

$= 4 + 25 = 29$

(b) $f\left(\frac{2}{5}\right) = 4 - \frac{5}{2}\left(\frac{2}{5}\right)$

$= 4 - 1 = 3$

(c) $f(t) + f(-4) = \left(4 - \frac{5}{2}t\right) + \left(4 - \frac{5}{2}(-4)\right)$

$= 4 - \frac{5}{2}t + 4 + 10 = 18 - \frac{5}{2}t$

(d) $f(x + h) = 4 - \frac{5}{2}(x + h)$

$= 4 - \frac{5}{2}x - \frac{5}{2}h$

67. $f(t) = \sqrt{5 - t}$

(a) $f(-4) = \sqrt{5 - (-4)} = \sqrt{9} = 3$

(b) $f(5) = \sqrt{5 - 5} = 0$

(c) $f(3) = \sqrt{5 - 3} = \sqrt{2}$

(d) $f(5z) = \sqrt{5 - 5z}$

69. $g(x) = \frac{|x + 4|}{4}$

(a) $g(0) = \frac{|0 + 4|}{4} = 1$

(b) $g(-8) = \frac{|-8 + 4|}{4} = 1$

(c) $g(2) - g(-5) = \frac{|2 + 4|}{4} - \frac{|-5 + 4|}{4}$

$= \frac{6}{4} - \frac{1}{4} = \frac{5}{4}$

(d) $g(x - 2) = \frac{|x - 2 + 4|}{4} = \frac{|x + 2|}{4}$

71. $f(x) = \begin{cases} -3x, & \text{if } x \le 0 \\ 1 - x^2, & \text{if } x > 0 \end{cases}$

(a) $f(2) = 1 - 2^2 = -3$

(b) $f\left(-\frac{2}{3}\right) = -3\left(-\frac{2}{3}\right) = 2$

(c) $f(1) = 1 - 1^2 = 0$

(d) $f(4) - f(3) = (1 - 4^2) - (1 - 3^2)$

$= 1 - 16 - 1 + 9 = -7$

73. Domain $= \{1, 3, 5, 7\}$

Range $= \{2, 7, 8\}$

75. Domain $= \{s: s > 0\}$

Range $= \{P: P > 0\}$

77. $h(x) = 4x^2 - 7$

Domain $= \{x: x \text{ is a real number}\}$

79. $f(x) = \sqrt{x - 2}$

Domain $= \{x: x \ge 2\}$

81. $g(x) = \frac{1}{8}x^2$

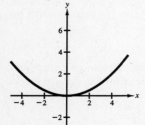

83. $y = 3(x - 2)^2$

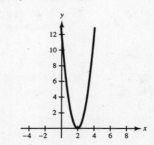

85. $y = 8 - 2|x|$

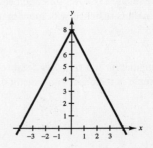

87. $g(x) = \frac{1}{4}x^3, -2 \le x \le 2$

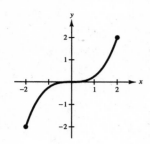

89. $f(x) = -\frac{3}{2}x + 5$

Slope: $-\frac{3}{2}$

y-intercept: $(0, 5)$

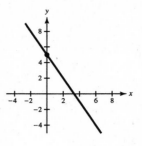

91. $f(x) = -0.4x - 3$

Slope: -0.4

y-intercept: $(0, -3)$

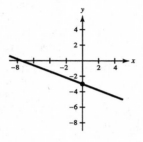

93. Some vertical lines intersect the graph more than once, so y *is not* a function of x.

95. No vertical line intersects the graph more than once, so y *is* a function of x.

97. No vertical line intersects the graph more than once, so y *is* a function of x.

99. Domain $= \{x: x$ is a real number$\}$

Range $= \{y: y \ge -3\}$

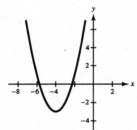

101. Domain $= \{x: x$ is a real number$\}$

Range $= \{y: y \ge 0\}$

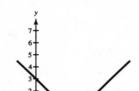

103.

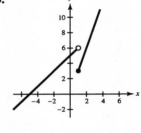

105. The basic function is $f(x) = x$.
Transformations: Reflection in x-axis; vertical shift of two units upward
Function: $h(x) = -x + 2$

107. The basic function is $f(x) = x^2$.
Transformations: Reflection in x-axis; horizontal shift of four units to the right
Function: $h(x) = -(x - 4)^2$

109. The function is a constant function $f(x) = c$.
Function: $h(x) = -6$

111. Reflection in the x-axis

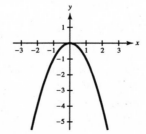

113. Horizontal shift one unit to the right

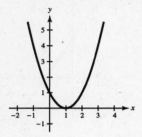

115. (a)

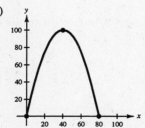

(b) The projectile strikes the ground when the height $y = 0$.

$$y = -\frac{1}{16}x^2 + 5x$$

$$0 = -\frac{1}{16}x^2 + 5x$$

$$0 = -x^2 + 80x$$

$$0 = -x(x - 80)$$

$$-x = 0 \Longrightarrow x = 0 \quad \text{Discard this answer.}$$

$$x - 80 = 0 \Longrightarrow x = 80$$

The projectile strikes the ground at a horizontal distance 80 feet from where it was launched.

117. (a)

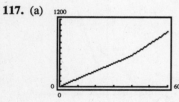

(b) $W(35) = 14(35) = \$490$

$W(40) = 14(40) = \$560$

$W(45) = 21(45 - 40) + 560$

$\qquad = 21(5) + 560$

$\qquad = \$665$

$W(50) = 21(50 - 40) + 560$

$\qquad = 21(10) + 560$

$\qquad = \$770$

(c) No. Values of $h < 0$ are not included in the domain of this function.

119. $\dfrac{33,900 - 28,500}{2001 - 1998} = \dfrac{5400}{3} = \$1800$

The average rate of change of salary is \$1800 per year.

Chapter Test for Chapter 2

1. The point (x, y), $x > 0, y < 0$, is in Quadrant IV.

The point is located to the *right* of the vertical axis and *below* the horizontal axis in Quadrant IV.

2. $x^2 - y = 3$

$(-2)^2 - 1 = 4 - 1$

$\qquad\qquad = 3$

Yes, $(-2, 1)$ is a solution.

3. (a) $d = \sqrt{(3-0)^2 + (1-9)^2}$

$= \sqrt{3^2 + (-8)^2}$

$= \sqrt{73}$

≈ 8.54

(b) $M = \left(\dfrac{0+3}{2}, \dfrac{9+1}{2}\right)$

$= \left(\dfrac{3}{2}, 5\right)$

4. $\qquad\qquad y = -3(x+1)$

x-intercept: $0 = -3(x+1)$

$0 = -3x - 3$

$3 = -3x$

$-1 = x \Longrightarrow (-1, 0)$ is the x-intercept

y-intercept: $y = -3(0+1)$

$y = -3(1)$

$y = -3 \Longrightarrow (0, -3)$ is the y-intercept

5.

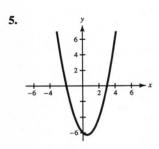

6. (a) $(-4, 7), (2, 3)$

$m = \dfrac{3-7}{2-(-4)} = \dfrac{-4}{6} = -\dfrac{2}{3}$

(b) $(3, -2), (3, 6)$

$m = \dfrac{6-(-2)}{3-3} = \dfrac{8}{0}$ (undefined)

7. $2x - 4y = 12$

$-4 = -2x + 12$

$y = \dfrac{-2}{-4}x + \dfrac{12}{-4}$

$y = \dfrac{1}{2}x - 3$

8. (a) The slope of a parallel line is 7/5.

(b) The slope of a perpendicular line is $-5/7$.

9. Some vertical lines intersect the graph at more than one point, so *y is not* a function of *x*.

10. $g(x) = \dfrac{x}{x-3}$

(a) $g(2) = \dfrac{2}{2-3}$

$= \dfrac{2}{-1}$

$= -2$

(b) $g\left(\dfrac{7}{2}\right) = \dfrac{7/2}{7/2 - 3}$

$= \dfrac{7}{2} \div \dfrac{1}{2}$

$= \dfrac{7}{2} \cdot \dfrac{2}{1}$

$= 7$

(c) $g(3) = \dfrac{3}{3-3}$

$= \dfrac{3}{0}$

$g(3)$ is undefined

(d) $g(x+2) = \dfrac{x+2}{(x+2)-3}$

$= \dfrac{x+2}{x+2-3}$

$= \dfrac{x+2}{x-1}$

11. $f(x) = \begin{cases} 3x - 1, & x < 5 \\ x^2 + 4, & x \geq 5 \end{cases}$

(a) $f(10) = 10^2 + 4$

$= 100 + 4$

$= 104$

(b) $f(-8) = 3(-8) - 1$

$= -24 - 1$

$= -25$

(c) $f(5) = 5^2 + 4$

$= 25 + 4$

$= 29$

(d) $f(0) = 3(0) - 1$

$= 0 - 1$

$= -1$

12. (a) $h(t) = \sqrt{t + 9}$

   Domain = $\{t: t \geq -9\}$

   (b) $f(x) = \dfrac{x + 1}{x - 4}$

   Domain = $\{x: x \neq 4\}$

   (c) $g(r) = 3r + 5$

   Domain = $\{r: r \text{ is a real number}\}$

13.

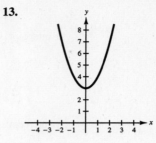

14. (a) horizontal shift to the right of 2 units; $y = |x - 2|$

   (b) vertical shift downward of 2 units; $y = |x| - 2$

   (c) reflection in the x-axis, and vertical shift upward of 2 units; $y = -|x| + 2$

Cumulative Test for Chapters 1–2

1. $\dfrac{a^2 - 2ab}{a + b}, a = -4, b = 7$

$\dfrac{(-4)^2 - 2(-4)(7)}{(-4) + 7} = \dfrac{16 + 56}{3}$

$= \dfrac{72}{3}$

$= 24$

2. (a) $(2a^2b)^3(-ab^2)^2 = 2^3 a^{2 \cdot 3} b^3 (-a)^2 b^{2 \cdot 2}$

$= 8a^6 b^3 a^2 b^4$

$= 8a^{6+2} b^{3+4}$

$= 8a^8 b^7$

(b) $3(x^2y)^2(-2xy^3)^3 = 3x^4y^2(-2^3x^3y^9)$

$= 3(-8)x^{4+3}y^{2+9}$

$= -24x^7y^{11}$

3. (a) $t(3t - 1) - 2t(t + 4) = 3t^2 - t - 2t^2 - 8t$

$= (3t^2 - 2t^2) + (-t - 8t)$

$= t^2 - 9t$

(b) $3x(x^2 - 2) - x(x^2 + 5) = 3x^3 - 6x - x^3 - 5x$

$= (3x^3 - x^3) + (-6x - 5x)$

$= 2x^3 - 11x$

4. (a) $(3x + 7)(x^2 - 2x + 5) = 3x^3 - 6x^2 + 15x + 7x^2 - 14x + 35$

$= 3x^3 + x^2 + x + 35$

(b) $[2 + (x - y)]^2 = 2^2 + 2(2)(x - y) + (x - y)^2$

$= 4 + 4(x - y) + x^2 - 2xy + y^2$

$= 4 + 4x - 4y + x^2 - 2xy + y^2$

$= x^2 - 2xy + y^2 + 4x - 4y + 4$

or $[2 + (x - y)]^2 = (2 + x - y)(2 + x - y)$

$= 4 + 2x - 2y + 2x + x^2 - xy - 2y - xy + y^2$

$= 4 + 4x - 4y + x^2 - 2xy + y^2$

$= x^2 - 2xy + y^2 + 4x - 4y + 4$

5. (a) $12 - 5(3 - x) = x + 3$

$12 - 15 + 5x = x + 3$

$-3 + 5x = x + 3$

$-3 + 4x = 3$

$4x = 6$

$x = \dfrac{6}{4}$

$x = \dfrac{3}{2}$

(b) $1 - \dfrac{x + 2}{4} = \dfrac{7}{8}$

$8\left(1 - \dfrac{x + 2}{4}\right) = 8\left(\dfrac{7}{8}\right)$

$8 - 2(x + 2) = 7$

$8 - 2x - 4 = 7$

$-2x + 4 = 7$

$-2x = 3$

$x = \dfrac{3}{-2}$

$x = -\dfrac{3}{2}$

6. (a) $y^2 - 64 = 0$

$(y + 8)(y - 8) = 0$

$y + 8 = 0 \Rightarrow y = -8$

$y - 8 = 0 \Rightarrow y = 8$

(b) $2t^2 - 5t - 3 = 0$

$(2t + 1)(t - 3) = 0$

$2t + 1 = 0 \Rightarrow 2t = -1 \Rightarrow t = -\dfrac{1}{2}$

$t - 3 = 0 \Rightarrow t = 3$

7. $2x - 3y + 9 = 0$

$-3y = -2x - 9$

$y = \dfrac{-2}{-3}x + \dfrac{-9}{-3}$

$y = \dfrac{2}{3}x + 3$

8. $y = -4x + 7$

(a) For the given line, the slope $m = -4$.

(b) For a parallel line, the slope $m = -4$ because the slopes of parallel lines are the same.

(c) For a perpendicular line, the slope $m = \frac{1}{4}$ because the slopes of perpendicular lines are negative reciprocals of each other.

9. (a) $8x \cdot \dfrac{1}{8x} = 1$

Multiplicative Inverse Property

(b) $5 + (-3 + x) = (5 - 3) + x$

Associative Property of Addition

10. Perimeter $= (x + 5) + x + x + (2x + 1) + (2x + 5) + (x + 1)$

$= 8x + 12$

11. $y^3 - 3y^2 - 9y + 27 = (y^3 - 3y^2) + (-9y + 27)$

$= y^2(y - 3) - 9(y - 3)$

$= (y - 3)(y^2 - 9)$

$= (y - 3)(y + 3)(y - 3)$

$= (y - 3)^2(y + 3)$

12. $3x^2 - 8x - 35 = (3x + 7)(x - 5)$

13. (a)

(b) $d = \sqrt{(1 - (-3))^2 + (5 - 8)^2}$

$= \sqrt{(4)^2 + (-3)^2}$

$= \sqrt{16 + 9}$

$= \sqrt{25}$

$= 5$

(c) midpoint $= \left(\dfrac{-3 + 1}{2}, \dfrac{8 + 5}{2}\right)$

$= \left(\dfrac{-2}{2}, \dfrac{13}{2}\right)$

$= \left(-1, \dfrac{13}{2}\right)$

(d) $m = \dfrac{8 - 5}{-3 - 1}$

$= \dfrac{3}{-4}$

$= -\dfrac{3}{4}$

14. No vertical line intersects the graph more than once, so *y is* a function of *x*.

15. $f(x) = \sqrt{x - 2}$

$x - 2 \geq 0$

$x \geq 2$

Domain: $= \{x: x \geq 2\}$

16. $f(x) = x^2 - 3x$

(a) $f(4) = 4^2 - 3(4)$

$= 16 - 12$

$= 4$

(b) $f(c + 3) = (c + 3)^2 - 3(c + 3)$

$= c^2 + 6c + 9 - 3c - 9$

$= c^2 + 3c$

17. $4x + 3y - 12 = 0$

$3y = -4x + 12$

$y = -\frac{4}{3}x + 4$

The graph is a line with slope of $-4/3$ and a *y*-intercept of (0, 4).

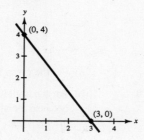

18. $y = -(x - 2)^2$

The graph is a transformation of the squaring function. The graph of the basic function $y = x^2$ is shifted horizontally 2 units to the right and reflected about the *x*-axis.

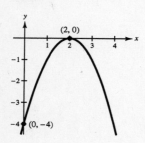

19. $y = |x + 3| + 2$

The graph is a transformation of the absolute value function. The graph of the basic function $y = |x|$ is shifted horizontally 3 units to the left and shifted vertically upward 2 units.

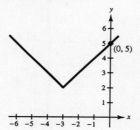

20. $y = x^3 - 1$

The graph is a transformation of the cubing function. The graph of the basic function $y = x^3$ is shifted vertically downward 1 unit.

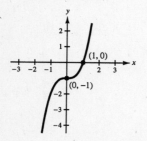

CHAPTER 3
Linear Functions, Equations, and Inequalities

CHAPTER 3
Linear Functions, Equations, and Inequalities

Section 3.1   Writing Equations of Lines

Solutions to Odd-Numbered Exercises

1. $y = \dfrac{2}{3}x - 2$

This equation is in slope-intercept form.

Slope: $m = \dfrac{2}{3}$

y-intercept: $(0, -2)$

3. $3x - 2y = 0$

$$-2y = -3x$$

$$y = \dfrac{-3}{-2}x$$

$$y = \dfrac{3}{2}x \quad \left(\text{or } y = \dfrac{3}{2}x + 0\right)$$

This equation is in slope-intercept form.

Slope: $m = \dfrac{3}{2}$

y-intercept $(0, 0)$

5. $5x - 2y + 24 = 0$

$$-2y = -5x - 24$$

$$y = \dfrac{-5}{-2}x + \dfrac{-24}{-2}$$

$$y = \dfrac{5}{2}x + 12$$

This equation is in slope—intercept form.

Slope: $m = \dfrac{5}{2}$

y-intercept: $(0, 12)$

7. $y - y_1 = m(x - x_1)$

Point: $(0, 0)$

Slope: $m = -\dfrac{1}{2}$

$$y - 0 = -\dfrac{1}{2}(x - 0)$$

$$y = -\dfrac{1}{2}x$$

9. $y - y_1 = m(x - x_1)$

Point: $(0, -4)$

Slope: $m = 3$

$$y - (-4) = 3(x - 0)$$

$$y + 4 = 3x$$

$$y = 3x - 4$$

11. Because the line is vertical and passes through the point $(10, -6)$, every point on the line has an x-coordinate of 10. So, the equation is $x = 10$.

13. Because the line is horizontal and passes through the point $(2, 3)$, every point on the line has a y-coordinate of 3. So, the equation is $y = 3$.

15. Graph (b)

17. Graph (a)

19. Graph (f)

21. $y - y_1 = m(x - x_1)$

Point: $(5, 6)$

Slope: $m = 2$

$$y - 6 = 2(x - 5)$$

$$y - 6 = 2x - 10$$

$$y = 2x - 4$$

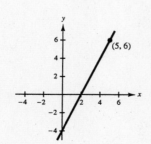

23. $y - y_1 = m(x - x_1)$

Point: $(-8, 1)$

Slope: $m = \dfrac{3}{4}$

$$y - 1 = \dfrac{3}{4}(x + 8)$$

$$y - 1 = \dfrac{3}{4}x + 6$$

$$y = \dfrac{3}{4}x + 7$$

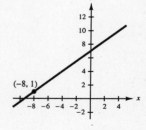

25. $y - y_1 = m(x - x_1)$

Point: $(5, -3)$

Slope: $m = \frac{2}{3}$

$y - (-3) = \frac{2}{3}(x - 5)$

$y + 3 = \frac{2}{3}x - \frac{10}{3}$

$y = \frac{2}{3}x - \frac{10}{3} - 3$

$y = \frac{2}{3}x - \frac{10}{3} - \frac{9}{3}$

$y = \frac{2}{3}x - \frac{19}{3}$

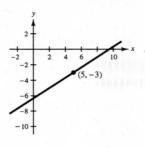

27. $y - y_1 = m(x - x_1)$

Point: $(-8, 5)$

Slope: $m = 0$

$y - 5 = 0(x + 8)$

$y - 5 = 0$

$y = 5$

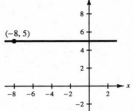

Note: The slope $m = 0 \Rightarrow$ the line is horizontal with an equation of the form $y = b$. The equation of the horizontal line through $(-8, 5)$ is $y = 5$.

29. $y - y_1 = m(x - x_1)$

Point: $(2, -1)$

Slope: m is undefined

$\Rightarrow$ vertical line, $x = a$

$x = 2$

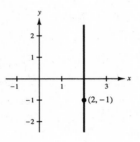

31. $y - y_1 = m(x - x_1)$

Point: $\left(\frac{3}{4}, \frac{5}{2}\right)$

Slope: $m = \frac{4}{3}$

$y - \frac{5}{2} = \frac{4}{3}\left(x - \frac{3}{4}\right)$

$y - \frac{5}{2} = \frac{4}{3}x - 1$

$y = \frac{4}{3}x - \frac{2}{2} + \frac{5}{2}$

$y = \frac{4}{3}x + \frac{3}{2}$

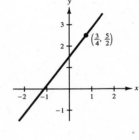

33. $(0, 0)$ and $(2, 3)$

Slope: $m = \dfrac{y_2 - y_1}{x_2 - x_1} = \dfrac{3 - 0}{2 - 0} = \dfrac{3}{2}$

Point: $(0, 0)$

$y - y_1 = m(x - x_1)$

$y - 0 = \frac{3}{2}(x - 0)$

$y = \frac{3}{2}x$

35. $(7, 3)$ and $(5, -5)$

Slope: $m = \dfrac{y_2 - y_1}{x_2 - x_1} = \dfrac{-5 - 3}{5 - 7} = \dfrac{-8}{-2} = 4$

Point: $(7, 3)$

$y - y_1 = m(x - x_1)$

$y - 3 = 4(x - 7)$

$y - 3 = 4x - 28$

$y = 4x - 25$

37. $(-5, 2)$ and $(5, -2)$

Slope: $m = \dfrac{y_2 - y_1}{x_2 - x_1} = \dfrac{-2 - 2}{5 - (-5)} = \dfrac{-4}{10} = -\dfrac{2}{5}$

Point: $(-5, 2)$

$y - y_1 = m(x - x_1)$

$y - 2 = -\frac{2}{5}(x + 5)$

$y - 2 = -\frac{2}{5}x - 2$

$y = -\frac{2}{5}x$

39. $(-2, 12)$ and $(6, 12)$

Slope: $m = \dfrac{y_2 - y_1}{x_2 - x_1} = \dfrac{12 - 12}{6 - (-2)} = \dfrac{0}{8} = 0$

$m = 0 \Rightarrow$ horizontal line, $y = b$

Point: $(-2, 12)$

$y = 12$

41. $(-2, 3)$ and $(5, 0)$

Slope: $m = \dfrac{y_2 - y_1}{x_2 - x_1} = \dfrac{0 - 3}{5 - (-2)} = \dfrac{-3}{7} = -\dfrac{3}{7}$

Point: $(-2, 3)$

$y - y_1 = m(x - x_1)$

$y - 3 = -\dfrac{3}{7}(x + 2)$

$y - 3 = -\dfrac{3}{7}x - \dfrac{6}{7}$

$y = -\dfrac{3}{7}x - \dfrac{6}{7} + 3$

$y = -\dfrac{3}{7}x - \dfrac{6}{7} + \dfrac{21}{7}$

$y = -\dfrac{3}{7}x + \dfrac{15}{7}$

43. $(1, -2)$ and $(1, 8)$

Slope: $m = \dfrac{y_2 - y_1}{x_2 - x_1} = \dfrac{8 - (-2)}{1 - 1} = \dfrac{10}{0}$ undefined

m undefined $\Rightarrow$ line is vertical, $x = a$

Point: $(1, -2)$

$x = 1$

45. $(-5, 0.6)$ and $(3, -3.4)$

Slope: $m = \dfrac{y_2 - y_1}{x_2 - x_1} = \dfrac{-3.4 - 0.6}{3 - (-5)} = \dfrac{-4}{8} = -\dfrac{1}{2}$

Point: $(-5, 0.6)$

$y - y_1 = m(x - x_1)$

$y - 0.6 = -\dfrac{1}{2}(x + 5)$

$y - 0.6 = -\dfrac{1}{2}x - \dfrac{5}{2}$

$y = -\dfrac{1}{2}x - \dfrac{5}{2} + 0.6$

$y = -\dfrac{1}{2}x - \dfrac{25}{10} + \dfrac{6}{10}$

$y = -\dfrac{1}{2}x - \dfrac{19}{10}$

47. $\left(\dfrac{3}{2}, 3\right)$ and $\left(\dfrac{9}{2}, -4\right)$

Slope: $m = \dfrac{y_2 - y_1}{x_2 - x_1} = \dfrac{-4 - 3}{(9/2 - 3/2)} = \dfrac{-7}{(6/2)} = -\dfrac{7}{3}$

Point: $\left(\dfrac{3}{2}, 3\right)$

$y - y_1 = m(x - x_1)$

$y - 3 = -\dfrac{7}{3}\left(x - \dfrac{3}{2}\right)$

$y - 3 = -\dfrac{7}{3}x + \dfrac{7}{2}$

$y = -\dfrac{7}{3}x + \dfrac{7}{2} + 3$

$y = -\dfrac{7}{3}x + \dfrac{7}{2} + \dfrac{6}{2}$

$y = -\dfrac{7}{3}x + \dfrac{13}{2}$

49. $\left(\dfrac{3}{5}, 9\right)$ and $\left(-4, \dfrac{1}{2}\right)$

Slope: $m = \dfrac{(1/2) - 9}{-4 - (3/5)} = \dfrac{-(17/2)}{-(23/5)} = -\dfrac{17}{2} \cdot -\dfrac{5}{23} = \dfrac{85}{46}$

Point: $\left(\dfrac{3}{5}, 9\right)$

$y - y_1 = m(x - x_1)$

$y - 9 = \dfrac{85}{46}\left(x - \dfrac{3}{5}\right)$

$y - 9 = \dfrac{85}{46}x - \dfrac{51}{46}$

$y = \dfrac{85}{46}x - \dfrac{51}{46} + 9$

$y = \dfrac{85}{46}x - \dfrac{51}{46} + \dfrac{414}{46}$

$y = \dfrac{85}{46}x + \dfrac{363}{46}$

51. $(-2, 2)$ and $(4, 5)$

Slope: $m = \dfrac{y_2 - y_1}{x_2 - x_1} = \dfrac{5 - 2}{4 - (-2)} = \dfrac{3}{6} = \dfrac{1}{2}$

Point: $(-2, 2)$

$y - y_1 = m(x - x_1)$

$y - 2 = \dfrac{1}{2}(x + 2)$

$y - 2 = \dfrac{1}{2}x + 1$

$y = \dfrac{1}{2}x + 3$

$f(x) = \dfrac{1}{2}x + 3$

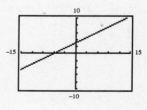

53. $(-2, 3)$ and $(4, 3)$

Slope: $m = \dfrac{y_2 - y_1}{x_2 - x_1} = \dfrac{3 - 3}{4 - (-2)} = \dfrac{0}{6} = 0$

$\qquad m = 0 \Rightarrow$ line is horizontal, $y = b$

Point: $(-2, 3)$

$\qquad y = 3$

$\qquad f(x) = 3$

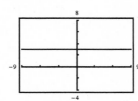

55. $-9x + 3y = 6$

$\qquad 3y = 9x + 6$

$\qquad y = 3x + 2$

(a) Point $(2, 1)$, $m = 3$

$\qquad y - y_1 = m(x - x_1)$

$\qquad y - 1 = 3(x - 2)$

$\qquad y - 1 = 3x - 6$

$\qquad y = 3x - 5$

(b) Point $(2, 1)$, $m = -\dfrac{1}{3}$

$\qquad y - y_1 = m(x - x_1)$

$\qquad y - 1 = -\dfrac{1}{3}(x - 2)$

$\qquad y - 1 = -\dfrac{1}{3}x + \dfrac{2}{3}$

$\qquad y = -\dfrac{1}{3}x + \dfrac{5}{3}$

57. $3x + 2y = 4$

$\qquad 2y = -3x + 4$

$\qquad y = -\dfrac{3}{2}x + 2$

(a) Point $(-6, 2)$, $m = -\dfrac{3}{2}$

$\qquad y - y_1 = m(x - x_1)$

$\qquad y - 2 = -\dfrac{3}{2}(x + 6)$

$\qquad y - 2 = -\dfrac{3}{2}x - 9$

$\qquad y = -\dfrac{3}{2}x - 7$

(b) Point $(-6, 2)$, $m = \dfrac{2}{3}$

$\qquad y - y_1 = m(x - x_1)$

$\qquad y - 2 = \dfrac{2}{3}(x + 6)$

$\qquad y - 2 = \dfrac{2}{3}x + 4$

$\qquad y = \dfrac{2}{3}x + 6$

59. $4x - 3y = 9$

$\qquad -3y = -4x + 9$

$\qquad y = \dfrac{4}{3}x - 3$

(a) Point $(1, -7)$, $m = \dfrac{4}{3}$

$\qquad y - y_1 = m(x - x_1)$

$\qquad y + 7 = \dfrac{4}{3}(x - 1)$

$\qquad y + 7 = \dfrac{4}{3}x - \dfrac{4}{3}$

$\qquad y = \dfrac{4}{3}x - \dfrac{4}{3} - 7$

$\qquad y = \dfrac{4}{3}x - \dfrac{25}{3}$

(b) Point $(1, -7)$, $m = -\dfrac{3}{4}$

$\qquad y - y_1 = m(x - x_1)$

$\qquad y + 7 = -\dfrac{3}{4}(x - 1)$

$\qquad y + 7 = -\dfrac{3}{4}x + \dfrac{3}{4}$

$\qquad y = -\dfrac{3}{4}x + \dfrac{3}{4} - 7$

$\qquad y = -\dfrac{3}{4}x - \dfrac{25}{4}$

61. $y = -5$

This is a horizontal line with slope $m = 0$.

(a) Point $(-1, 2)$, $m = 0$

Horizontal line: $y = 2$

(b) Point $(-1, 2)$, m undefined

Vertical line: $x = -1$

63. $y = mx + b$

Slope: $m = \dfrac{1}{5}$

y-intercept: $(0, 3)$

$y = \dfrac{1}{5}x + 3$

65. Slope: $m = -2$

Point: $(7, -4)$

$\qquad y - y_1 = m(x - x_1)$

$\qquad y - (-4) = -2(x - 7)$

$\qquad y + 4 = -2x + 14$

$\qquad y = -2x + 10$

67. $7x + 3y = 8$

$\qquad 3y = -7x + 8$

$\qquad y = -\dfrac{7}{3}x + \dfrac{8}{3} \Rightarrow m = -\dfrac{7}{3}$

Slope of perpendicular line: $m = \dfrac{3}{7}$

Point: $(-2, 3)$

$\qquad y - y_1 = m(x - x_1)$

$\qquad y - 3 = \dfrac{3}{7}(x - (-2))$

$\qquad y - 3 = \dfrac{3}{7}(x + 2)$

$\qquad y - 3 = \dfrac{3}{7}x + \dfrac{6}{7}$

$\qquad y = \dfrac{3}{7}x + \dfrac{6}{7} + 3$

$\qquad y = \dfrac{3}{7}x + \dfrac{6}{7} + \dfrac{21}{7}$

$\qquad y = \dfrac{3}{7}x + \dfrac{27}{7}$

69. $x - 4y = 10$

$\qquad -4y = -x + 10$

$\qquad y = \dfrac{-1}{-4}x + \dfrac{10}{-4}$

$\qquad y = \dfrac{1}{4}x - \dfrac{5}{2} \Rightarrow m = \dfrac{1}{4}$

Slope of parallel line: $m = \dfrac{1}{4}$

Point: $(5, -6)$

$\qquad y - y_1 = m(x - x_1)$

$\qquad y - (-6) = \dfrac{1}{4}(x - 5)$

$\qquad y + 6 = \dfrac{1}{4}(x - 5)$

$\qquad y + 6 = \dfrac{1}{4}x - \dfrac{5}{4}$

$\qquad y = \dfrac{1}{4}x - \dfrac{5}{4} - 6$

$\qquad y = \dfrac{1}{4}x - \dfrac{5}{4} - \dfrac{24}{4}$

$\qquad y = \dfrac{1}{4}x - \dfrac{29}{4}$

71. Because the line is horizontal and passes through the point $(-5, 1)$, every point on the line has a y-coordinate of 1. So, the equation is $y = 1$.

73. Because the line is vertical and passes through the point $(9, 2)$, every point on the line has an x-coordinate of 9. So, the equation is $x = 9$.

75. (a) $(0, 12{,}500)$, $(5, 1000)$

$$m = \frac{1000 - 12{,}500}{5 - 0} = \frac{11{,}500}{5} = -2300$$

$y = mx + b$ or $V = mt + b$

$V = -2300t + 12{,}500$

(b) $V = -2300(3) + 12{,}500$

$V = \$5600$

77. $(450, 50)$, $(525, 45)$

(a) $m = \dfrac{45 - 50}{525 - 450} = \dfrac{-5}{75} = -\dfrac{1}{15}$

$x - 50 = -\dfrac{1}{15}(p - 450)$

$x - 50 = -\dfrac{1}{15}p + 30$

$x = -\dfrac{1}{15}p + 80$

(b)

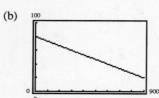

(c) $x = -\dfrac{1}{15}(570) + 80$

$x = -38 + 80$

$x = 42$

So, 42 apartments would be occupied if the rent were raised to \$570.

(d) $x = -\dfrac{1}{15}(480) + 80$

$x = -32 + 80$

$x = 48$

So, 48 apartments would be occupied if the rent were lowered to \$480.

79. (a) $N = 60t + 1500$

(b)

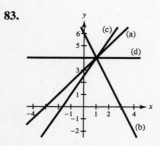

(c) $N = 1500 + 60t$, $t = 20$ for the year 2010

$N = 1500 + 60(20)$

$N = 1500 + 1200$

$N = 2700$

(d) $N = 1500 + 60t$, $t = 5$ for the year 1995

$N = 1500 + 60(5)$

$N = 1500 + 300$

$N = 1800$

81. Yes, any pair of points on a line can be used to determine an equation of the line. When different pairs of points are selected, the change in y and the change in x are lengths of the sides of similar triangles. Corresponding sides of similar triangles are proportional.

83.

85. $C = 5.75x + 12{,}000$

Section 3.2   Applications of Linear Equations

1. $8 + n$

3. $15 - 3n$

5. $\frac{1}{3}n$

7. $0.30L$

9. $\frac{x}{6}$

11. $\frac{3 + 4x}{8}$

13. $|n - 5|$

15. The sum of three times a number and two

17. Eight times the difference of a number and five

19. The ratio of a number to 8 *or* the quotient of a number and 8

21. The sum of a number and ten, all divided by three

23. *Verbal Model:* $\boxed{\text{Value of quarter}} \cdot \boxed{\text{Number of quarters}}$

   Labels: Value of quarter $= 0.25$ (dollars/quarter)

   Number of quarters $= n$ (quarters)

   Algebraic Expression: $0.25n$ (dollars)

25. *Verbal Model:* $\boxed{\text{Rate}} \cdot \boxed{\text{Time}}$

   Labels: Rate $= 55$ (miles per hour)

   Time $= t$ (hours)

   Algebraic Expression: $55t$ (miles)

27. *Verbal Model:* $\boxed{\text{Distance}} \div \boxed{\text{Rate}}$

   Labels: Distance $= 100$ (miles)

   Rate $= r$ (miles per hour)

   Algebraic Expression: $\dfrac{100}{r}$ (hours)

29. *Verbal Model:* $\boxed{\text{Percent of antifreeze}} \cdot \boxed{\text{Amount of coolant}}$

   Labels: Percent of antifreeze $= 0.45$ (decimal form)

   Amount of coolant $= y$ (gallons)

   Algebraic Expression: $0.45y$ (gallons)

31. *Verbal Model:* $\boxed{\text{Tax rate}} \cdot \boxed{\text{Taxable income}}$

   Labels: Tax rate $= 0.0125$ (decimal form)

   Taxable income $= I$ (dollars)

   Algebraic Expression: $0.0125I$ (dollars)

33. *Verbal Model:* $\boxed{\text{List price}} - \boxed{\text{Discount rate}} \cdot \boxed{\text{List price}}$

   Labels: List price $= L$ (dollars)

   Discount rate $= 0.20$ (decimal form)

   Algebraic Expression: $L - 0.20L = 0.80L$ (dollars)

35. *Verbal Model:* $\boxed{\text{Base pay}} + \boxed{\text{Pay per unit produced}} \cdot \boxed{\text{Number of units}}$

   Labels: Base pay $= 8.25$ (dollars)

   Pay per unit $= 0.60$ (dollars per unit)

   Number of units $= q$ (units)

   Algebraic Expression: $8.25 + 0.60q$ (dollars)

37. $\dfrac{x}{6} = \dfrac{2}{3}$

$3x = 12$

$x = 4$

39. $\dfrac{5}{4} = \dfrac{t}{6}$

$30 = 4t$

$\dfrac{30}{4} = t$

$\dfrac{15}{2} = t$

41. $\dfrac{y + 5}{6} = \dfrac{y - 2}{4}$

$4(y + 5) = 6(y - 2)$

$4y + 20 = 6y - 12$

$20 = 2y - 12$

$32 = 2y$

$16 = y$

43. $\dfrac{z - 3}{3} = \dfrac{z + 8}{12}$

$12(z - 3) = 3(z + 8)$

$12z - 36 = 3z + 24$

$9z - 36 = 24$

$9z = 60$

$z = \dfrac{60}{9}$

$z = \dfrac{20}{3}$

45. $\dfrac{2x - 3}{7} = \dfrac{x + 4}{2}$

$2(2x - 3) = 7(x + 4)$

$4x - 6 = 7x + 28$

$-3x - 6 = 28$

$-3x = 34$

$x = -\dfrac{34}{3}$

47. $\dfrac{4x - 2}{3} = \dfrac{2x + 1}{4}$

$4(4x - 2) = 3(2x + 1)$

$16x - 8 = 6x + 3$

$10x - 8 = 3$

$10x = 11$

$x = \dfrac{11}{10}$

49. *Verbal Model:* 2 ⬚ Length ⬚ $+ 2$ ⬚ Width ⬚

Labels: Width $= w$

Length $= 2w$

Expression: $2(2w) + 2w = 4w + 2w = 6w$

51. *Verbal Model:*   ⬚ Side 1 ⬚ $+$ ⬚ Side 2 ⬚ $+$ ⬚ Side 3 ⬚ $+$ ⬚ Side 4 ⬚ $+$ ⬚ Side 5 ⬚ $+$ ⬚ Side 6 ⬚

Labels: Side 1 $= x$

Side 2 $= 3$

Side 3 $= x$

Side 4 $= 3$

Side 5 $= 2x$

Side 5 $= 6$

Expression: $x + 3 + x + 3 + 2x + 6 = 4x + 12$

53. *Verbal Model:*   ⬚ Side ⬚^{2}

Labels: Side $= 2x - 3$

Expression: $(2x - 3)^2 = (2x)^2 - 2(2x)(3) + 3^2 = 4x^2 - 12x + 9$

55. *Verbal Model:*   ⬚ Length ⬚ $\cdot$ ⬚ Width ⬚

Labels: Length $= w + 12$

Width $= 2w$

Expression: $(w + 12)(2w) = 2w^2 + 24w$

57. *Verbal Model:*   ⬚ Width ⬚ $\cdot$ ⬚ Length ⬚

Labels: Width $= w$ (meters)

Length $= 6w$ (meters)

Expression: $w(6w) = 6w^2$

The units are square meters.

59. Perimeter:

Verbal Model: 2 | Length | + 2 | Width |

Labels: Width $= w$ (feet)

Length $= 3w + 2$ (feet)

Expression: $2(3w + 2) + 2(w) = 6w + 4 + 2w$

$= 8w + 4$ (feet)

Area:

Verbal Model: | Length | · | Width |

Labels: Width $= w$ (feet)

Length $= 3w + 2$ (feet)

Expression: $(3w + 2)(w) = 3w^2 + 2w$ (square feet)

61. *Verbal Model:* | First integer | + | Second consecutive integer | + | Third consecutive integer | $= 60$

Labels: Integer $= n$

Next consecutive integer $= n + 1$

Next consecutive integer $= n + 2$

Equation: $n + (n + 1) + (n + 2) = 60$

$3n + 3 = 60$

$3n = 57$

$n = 19$ and $n + 1 = 20, n + 2 = 21$

The three consecutive even integers are 19, 20, and 21.

63. *Verbal Model:* | First consecutive even integer | + | Second consecutive even integer | + | Third consecutive even integer | $= 138$

Labels: First consecutive even integer $= n$

Second consecutive even integer $= n + 2$

Third consecutive even integer $= n + 4$

Equation: $n + (n + 2) + (n + 4) = 138$

$3n + 6 = 138$

$3n = 132$

$n = 44$ and $n + 2 = 46, n + 4 = 48$

The three consecutive even integers are 44, 46, and 48.

65. *Verbal Model:* $8 \left(\boxed{\text{Number}} + 6 \right) = 128$

Label: Number $= x$

Equation: $8(x + 6) = 128$

$8x + 48 = 128$

$8x = 80$

$x = 10$

The number is 10.

67. *Verbal Model:* $\left(\boxed{\text{Number}} + 18 \right) \div 5 = 12$

Label: Number $= x$

Equation: $\dfrac{x + 18}{5} = 12$

$5\left(\dfrac{x + 18}{5} \right) = (12)5$

$x + 18 = 60$

$x = 42$

The number is 42.

69. *Verbal Model:* | Number | $+ 30 = 82$

Label: Number $= x$

Equation: $x + 30 = 82$

$x = 52$

The number is 52.

71. *Verbal Model:* $6\left(\boxed{\text{Number}} - 12\right) = 300$

Label: Number $= x$

Equation: $6(x - 12) = 300$

$$6x - 72 = 300$$

$$6x = 372$$

$$x = \frac{372}{6}$$

$$x = 62$$

The number is 62.

73. *Verbal Model:* $5\boxed{\begin{array}{c}\text{Odd}\\\text{integer}\end{array}} - 3\boxed{\begin{array}{c}\text{Next consecutive}\\\text{odd integer}\end{array}} = 24$

Labels: First consecutive odd integer $= n$

Next consecutive odd integer $= n + 2$

Equations: $5n - 3(n + 2) = 24$

$$5n - 3n - 6 = 24$$

$$2n - 6 = 24$$

$$2n = 30$$

$$n = 15 \text{ and } n + 2 = 17$$

The two consecutive odd integers are 15 and 17.

75. *Verbal Model:* $3\boxed{\text{Side}} = \boxed{\text{Perimeter}}$

Labels: Side $= s$ (cm)

Perimeter $= 129$ (cm)

Equation: $3s = 129$

$$s = 43$$

So, the length of each side is 43 centimeters.

77. *Verbal Model:* $\boxed{\begin{array}{c}\text{Costs for}\\\text{parts}\end{array}} + \boxed{\begin{array}{c}\text{Labor cost}\\\text{per hour}\end{array}} \cdot \boxed{\begin{array}{c}\text{Number of}\\\text{hours}\end{array}} = \boxed{\begin{array}{c}\text{Total}\\\text{cost}\end{array}}$

Labels: Costs for parts $= 275$ (dollars)

Labor cost per hour $= 35$ (dollars per hour)

Number of hours $= x$ (hours)

Total cost $= 380$ (dollars)

Equation: $275 + 35x = 380$

$$35x = 105$$

$$x = \frac{105}{35} = 3$$

So, 3 hours were spent in repairing the car.

79. *Verbal Model:* $\boxed{\begin{array}{c}\text{Number of people}\\\text{laid off}\end{array}} = \boxed{\begin{array}{c}\text{Percent of}\\\text{work force}\end{array}} \cdot \boxed{\begin{array}{c}\text{Number of}\\\text{employees}\end{array}}$

Labels: Number laid off $= 25$ (people)

Percentage of work force $= p$ (decimal form)

Number of employees $= 160$

Equation: $25 = p(160)$

$$\frac{25}{160} = p$$

$$0.15625 = p$$

So, 15.625% of the work force was laid off.

81. *Verbal Model:* | Number of defective parts in sample | = | Percentage of defective parts in sample | · | Size of sample |

      Labels: Number of defective parts in sample = 3 (parts)

                       Percentage of defective parts = 0.015 (decimal form)

                       Size of sample = x (parts)

      Equation: $3 = 0.015x$

$$\frac{3}{0.015} = x$$

$$200 = x$$

So, there were 200 parts in the sample.

83. *Verbal Model:* | Measure of first angle | + | Measure of supplement | = 180

      Labels: Measure of first angle = x (degrees)

                       Measure of supplement = $2x + 24$ (degrees)

      Equation: $x + 2x + 24 = 180$

$$3x + 24 = 180$$

$$3x = 156$$

$$x = 52 \text{ and } 2x + 24 = 128$$

The measures of the two angles are 52 degrees and 128 degrees.

85. *Verbal Model:* | Measure of first angle | + | Measure of supplement | = 180

      Labels: Measure of first angle = $7x + 2$ (degrees)

                       Measure of supplement = $11x - 2$ (degrees)

      Equation: $(7x + 2) + (11x - 2) = 180$

$$18x = 180$$

$$x = 10 \text{ so } 7x + 2 = 72, \ 11x - 2 = 108$$

The measures of the two angles are 72 degrees and 108 degrees.

87. *Verbal Model:* | Measure of first angle | + | Measure of second angle | + | Measure of third angle | = 180

      Labels: Measure of first angle = $4x - 1$ (degrees)

                       Measure of second angle = $x + 14$ (degrees)

                       Measure of third angle = $2x + 13$ (degrees)

      Equation: $(4x - 1) + (x + 14) + (2x + 13) = 180$

$$7x + 26 = 180$$

$$7x = 154$$

$$x = 22 \text{ so } 4x - 1 = 87, \ x + 14 = 36, \ 2x + 13 = 57$$

The measures of the three angles are 87 degrees, 36 degrees, and 57 degrees.

89. Proportion: $\dfrac{4}{5.6} = \dfrac{x}{7}$

$$5.6x = 4(7)$$

$$5.6x = 28$$

$$x = 5$$

91. (a) Population in 1980 = 365,000

Population in 1990 = 36,000(1.033) = 377,045

Population in 2000 = 377,045(1.05) = 395,897

(b) $\dfrac{395,897}{365,000} \approx 1.085$

The percent increase was approximately 8.5%.

Different bases are used for the two calculations.

93. Price-to-earnings ratio $= \dfrac{\text{Price of stock}}{\text{Earnings of stock}} = \dfrac{56.25}{6.25} = \dfrac{9}{1}$

95. $\dfrac{\text{Number of teeth in gear 1}}{\text{Number of teeth in gear 2}} = \dfrac{60 \text{ teeth}}{40 \text{ teeth}} = \dfrac{3}{2}$

97. *Verbal Model:* 12 $\boxed{\dfrac{\text{Votes for candidate in sample}}{\text{Voters in sample}}} = \boxed{\dfrac{\text{Votes for candidate in election}}{\text{Voters in election}}}$

Labels: Sample: Votes for candidate = 870 (votes); votes = 1500 (people)

Election: Votes for candidate = x (votes); votes = 80,000 (people)

Equations: $\dfrac{870}{1500} = \dfrac{x}{80,000}$

$870(80,000) = 1500x$

$69,600,000 = 1500x$

$\dfrac{69,600,000}{1500} = x$

$46,400 = x$

So, the candidate can expect to receive 46,400 votes.

99. *Verbal Model:* $\boxed{\dfrac{\text{First amount of flour}}{\text{First amount of cookies}}} = \boxed{\dfrac{\text{Second amount of flour}}{\text{Second amount of cookies}}}$

Labels: First: Amount of flour = 3 (cups); amount of cookies = 1 (batch)

Second: Amount of flour = x (cups); amount of cookies = $3\frac{1}{2}$ (batches)

Equations: $\dfrac{3}{1} = \dfrac{x}{7/2}$

$\dfrac{3}{1}\left(\dfrac{7}{2}\right) = 1(x)$

$\dfrac{21}{2} = x$

So, $10\frac{1}{2}$ cups of flour are required.

101. No, $\frac{1}{2}\% \neq 50\%$.

$\frac{1}{2}\% = 0.5\% = 0.005$ or $\frac{5}{100}$ or $\frac{1}{200}$

$50\% = 0.50$ or $\frac{50}{100}$ or $\frac{1}{2}$

103. If a and b have the same units, then a/b is the ratio of a to b. Examples of ratios include price earnings ratios and gear ratios.

105. $12 - 4(x - 1) = 12 - 4x + 4$

$= -4x + 16$

107. $9(3y + 2) - 4(2y - 1) = 27y + 18 - 8y + 4$

$= 19y + 22$

109. $3x + 7 = 13$

$3x = 6$

$x = 2$

111. $\dfrac{x}{4} - 1 = 2$

$$4\left(\dfrac{x}{4} - 2\right) = 4(2)$$

$$x - 4 = 8$$

$$x = 12$$

113. $5(x - 2) = 3(x + 2)$

$$5x - 10 = 3x + 6$$

$$2x - 10 = 6$$

$$2x = 16$$

$$x = 8$$

115. $2.5 - 1.8 = 0.7$

The last person runs 0.7 mile.

Section 3.3   Business and Scientific Problems

1. $d = rt$

$$d = (650)\left(\tfrac{7}{2}\right)$$

$$d = 2275 \text{ miles}$$

3. $d = rt$

$$250 = 32t$$

$$\tfrac{250}{32} = t$$

$$\tfrac{125}{16} = t \ \text{ or } \ t = 7\tfrac{13}{16} \text{ seconds}$$

5. *Verbal Model:* | Selling price | = | List price | − | Discount |

Labels: Selling price $= x$ (dollars)

Discount $= 0.20(279.95) = 55.99$ (dollars)

Equation: $x = 279.95 - 55.99 = \$223.96$ at department store

Verbal Model: | Selling price | = | List price | + | Shipping |

Label: Selling price $= x$ (dollars)

Equation: $x = 228.95 + 4.32 = \$233.27$ from catalog

The department store machine is the better buy.

7. *Verbal Model:* | Parts charges | + | Labor rate | · | Hours of labor | = | Total charges |

Label: Hours of labor $= x$ (hours)

Equation: $136.37 + 32x = 216.37$

$$32x = 80$$

$$x = \tfrac{80}{32} = 2.5 \text{ hours}$$

So, it took 2.5 hours to repair the auto.

9. *Verbal Model:* | Total cost | = | Cost of first minute | + | Cost of each additional minute | · | Number of additional minutes |

Label: Number of additional minutes $= x$ (minutes)

Equation: $5.15 = 0.75 + 0.55x$

$$4.40 = 0.55x$$

$$8 = x$$

So, there were 8 *additional* minutes plus the *first* minute, so the call was 9 minutes long.

Verbal Model: | Selling price | = | List price | − | Discount |

Label: Selling price $= x$ (dollars)

Discount $= 0.60(5.15) = 3.09$ (dollars)

Equation: $x = 5.15 - 3.09 = \$2.06$

So, the call would have cost $2.06 during the weekend.

11. *Verbal Model:* | Tip | = | Total paid | − | Meal cost |

  Label: Tip = x (dollars)

  Equation: $x = 10 - 8.45 = \$1.55$

  Verbal Model: | Tip | = | Tip rate | · | Meal cost |

  Label: Tip rate = p (decimal form)

  Equation: $1.55 = p \cdot 8.45$

$$\frac{1.55}{8.45} = p$$

$$0.183 \approx p$$

So, the tip rate was approximately 18.3%.

13. *Verbal Model:* | Commission | = | Commission rate | · | Sales |

  Label: commission rate = p (decimal form)

  Equation: $450 = p \cdot 5000$

$$\tfrac{450}{5000} = p$$

$$0.09 = p$$

The commission rate was 9%.

15. *Verbal Model:* | Weekly Salary | + | Commission Rate | · | Sales | = | Weekly pay |

  Label: Weekly pay = x (dollars)

  Equation: $250 + 0.06(5500) = x$

$$250 + 330 = x$$

$$580 = x$$

So, the weekly pay is $580.

17. *Verbal Model:* | Surcharge | = | Previous premium | · | Surcharge rate |

  Labels: Surcharge = x (dollars)

            Previous premium = 862 (dollars)

            Surcharge rate = 0.20 (percent in decimal form)

  Equation: $x = 862(0.20)$

            $x = 172.40$

The surcharge is $172.40.

  Verbal Model: | New premium | = | Previous premium | + | Surcharge |

  Labels: New premium = x (dollars)

            Previous premium = 862 (dollars)

            Surcharge = 172.40 (dollars)

  Equation: $x = 862 + 172.40$

            $x = 1034.40$

The new premium is $1034.40.

19.
$$F = \tfrac{9}{5}C + 32$$
$$60.8 = \tfrac{9}{5}C + 32$$
$$5(60.8) = 5\left(\tfrac{9}{5}C + 32\right)$$
$$304 = 9C + 160$$
$$144 = 9C$$
$$16 = C$$

The temperature is 16 degrees Celsius.

21.
$$V = \pi r^2 h$$
$$V = \pi(3.5)^2(12)$$
$$V \approx 461.8$$

The volume is approximately 461.8 cubic centimeters.

23. *Verbal Model:* | Distance | = | Rate | · | Time |

Label: Runner's speed $= x$ (meters per minute)

Equation:
$$500 = x \cdot 13\tfrac{20}{60}$$
$$5000 = x \cdot 13\tfrac{1}{3}$$
$$\frac{5000}{(40/3)} = x$$
$$5000 \cdot \frac{3}{40} = x$$
$$375 = x$$

So, the average speed of the runner is 375 meters per minute.

25.
$$d = rt$$
$$93,000,000 = 186,282.369t$$
$$\frac{93,000,000}{186,282.369} = t$$
$$499.24 \text{ seconds} \approx t$$
$$8.32 \text{ minutes} \approx t$$

27. *Verbal Model:* | Distance flown by first plane | + | Distance flown by second plane | = | Distance between planes |

Labels: First plane: Rate $= 480$ (mph); time $= 1\tfrac{1}{3}$ (hours)
Second plane: Rate $= 600$ (mph); time $= 1\tfrac{1}{3}$ (hours)
Distance between planes $= x$ (miles)

Equation:
$$480\left(\tfrac{4}{3}\right) + 600\left(\tfrac{4}{3}\right) = x$$
$$640 + 800 = x$$
$$1440 = x$$

So, the planes are 1440 miles apart after $1\tfrac{1}{3}$ hours.

29. *Verbal Model:* | Distance of first walker | = | Distance of second walker |

Labels: First walker: Rate $= 3$ (mph); time $= x + 1$ (hours)
Second walker: Rate $= 4$ (mph); time $= x$ (hours)

Equation:
$$3(x + 1) = 4x$$
$$3x + 3 = 4x$$
$$3 = x$$

The second person catches up to the first person in 3 hours.

31. *Verbal Model:* | Distance of bus | = | Distance of car |

Labels: Bus: Rate $= 50$ (mph); time $= x + \frac{1}{2}$ (hours)

Car: Rate $= 60$ (mph); time $= x$ (hours)

Equation:
$$50\left(x + \tfrac{1}{2}\right) = 60x$$
$$50x + 25 = 60x$$
$$25 = 10x$$
$$2.5 = x$$

The car catches up with the bus after 2.5 hours at 3:00 P.M.

33. *Verbal Model:* | Distance on first part of trip | + | Distance on second part of trip | = | Total distance |

Labels: First part of trip: Rate $= 58$ (mph); time $= x$ (hours)

Second part of trip: Rate $= 52$ (mph); time $= 5\frac{3}{4} - x$ (hours)

Total distance $= 317$ (miles)

Equation:
$$58x + 52(5.75 - x) = 317$$
$$58x + 299 - 52x = 317$$
$$6x + 299 = 317$$
$$6x = 18$$
$$x = 3 \text{ and } 5\tfrac{3}{4} - x = 2\tfrac{3}{4}$$

So, the sales representative drove for 3 hours at 58 miles per hour and for $2\frac{3}{4}$ hours at 52 miles per hour.

35. You can complete the project in 5 hours. $\implies$ You can complete $\frac{1}{5}$ of the project per hour.
Your friend can complete the product in 8 hours. $\implies$ Your friend can complete $\frac{1}{8}$ of the project per hour.

Verbal Model: | Portion done by you | + | Portion done by your friend | = | Work done |

Labels: You: Rate $= \frac{1}{5}$ (job per hour); time $= t$ (hours)

Friend: Rate $= \frac{1}{8}$ (job per hour); time $= t$ (hours)

Work done $= 1$ (job)

Equation:
$$\tfrac{1}{5}t + \tfrac{1}{8}t = 1$$
$$40\left(\tfrac{1}{5}t + \tfrac{1}{8}t\right) = 40(1)$$
$$8t + 5t = 40$$
$$13t = 40$$
$$t = \tfrac{40}{13}$$
$$t = 3\tfrac{1}{13}$$

So, if you and your friend work together, you can complete the project in $3\frac{1}{13}$ hours.

37. *Verbal Model:* | Portion done by you | + | Portion done by your friend | = | Work done |

    Labels: You: Rate $= \frac{1}{3}$ (job per hour); time $= t$ (hours)

               Your friend: Rate $= \frac{1}{4}$ (job per hour); time $= t$ (hours)

               Work done $= 1$ (job)

    Equation:

$$\frac{1}{3}t + \frac{1}{4}t = 1$$

$$12\left(\frac{1}{3}t + \frac{1}{4}t\right) = 12(1)$$

$$4t + 3t = 12$$

$$7t = 12$$

$$t = \frac{12}{7}$$

So, it will take $\frac{12}{7}$ hours or approximately 1 hour, 43 minutes to mow the lawn working together.

39. *Verbal Model:* | Portion done by you | + | Portion done by your sister | = | Work done |

    Labels: You: Rate $= \frac{1}{4}$ (job per hour); time $= 3$ (hours)

               Your sister: Rate $= \frac{1}{x}$ (job per hour); time $= 3$ (hours)

               Work done $= 1$ (job)

    Equation:

$$\frac{1}{4}(3) + \frac{1}{x}(3) = 1$$

$$\frac{3}{4} + \frac{3}{x} = 1$$

$$4x\left(\frac{3}{4} + \frac{3}{x}\right) = 4x(1)$$

$$3x + 12 = 4x$$

$$12 = x$$

So, it will take 12 hours for your sister to clean the house by herself.

41. *Verbal Model:* | Total value | = | Value of $0.20 stamps | + | Value of $0.32 stamps |

    Labels: $0.20 stamps: Number $= x$; value $= 0.20x$ (dollars)

             $0.32 stamps: Number $= 70 - x$; value $= 0.32(70 - x)$ (dollars)

    Equation: $20.96 = 0.20x + 0.32(70 - x) = 0.20x + 22.40 - 0.32x$

            $-1.44 = -0.12x$

            $12 = x$ and $70 - x = 58$

So, there are 12 $0.20 stamps and 58 $0.32 stamps.

43. *Verbal Model:* | Total sales | = | Adult sales | + | Children sales |

    Labels: Children's tickets: Number $= x$, value $= 4(x)$ (dollars)

             Adults' tickets: Number $= 3x$; value $= 6(3x)$ (dollars)

    Equation: $2200 = 6(3x) + 4(x)$

           $2200 = 22x$

           $100 = x$

So, 100 children's tickets were sold.

45. *Verbal Model:* | Interest on 4.5% investment | $+$ | Interest on 5.5% investment | $=$ | Total interest |

Labels: 4.5% investment: Amount invested $= x$ (dollars);
                                Interest $= 0.045x$ (dollars)

            5.5% investment: Amount invested $= 6000 - x$ (dollars);
                                Interest $= 0.055(6000 - x)$ (dollars)

Equation: $0.045x + 0.055(6000 - x) = 305$

            $0.045x + 330 - 0.055x = 305$

            $-0.01x + 330 = 305$

            $-0.01x = -25$

            $x = 2500$ and $6000 - x = 3500$

So, \$2500 is invested at 4.5% and \$3500 is invested at 5.5%.

47. (a) Interest accumulated on \$40,000 invested at 8% for 3 years $Prt = (40,000)(0.08)(3) = \$9600$

   Interest accumulated on \$40,000 invested at 10% for 3 years $Prt = (40,000)(0.10)(3) = \$12,000$

   (b) *Verbal Model:* | Interest from 8% investment | $+$ | Interest from 10% investment | $=$ | Total interest |

Labels: 8% investment: Amount invested $= x$ (dollars);
                                Interest $= 0.08x$ (dollars)

            10% investment: Amount invested $= 40,000 - x$ (dollars);
                                Interest $= 0.10(40,000 - x)$ (dollars)

Equation: $0.08x + 0.10(40,000 - x) = 3500$

            $0.08x + 4000 - 0.10x = 3500$

            $-0.02x + 4000 = 3500$

            $-0.02x = -500$

            $x = \dfrac{-500}{-0.02}$

            $x = 25,000$ and $40,000 - x = 15,000$

You need to invest \$15,000 (or more) at 10%.

49. *Verbal model:* | Solute in 25% solution | $+$ | Solute in 60% solution | $=$ | Solute in 30% solution |

Labels: 25% solution: Number of milliliters $= 300$
                            Amount of solute $= 0.25(300)$ (milliliters)

            60% solution: Number of milliliters $= x$
                            Amount of solute $= 0.60x$ (milliliters)

            30% solution: Number of milliliters $= 300 + x$
                            Amount of solute $= 0.30(300 + x)$ (milliliters)

Equation: $0.25(300) + 0.60x = 0.30(300 + x)$

            $75 + 0.60x = 90 + 0.30x$

            $75 + 0.30x = 90$

            $0.30x = 15$

            $x = 50$

So, 50 milliliters of the 60% solution must be used.

51. *Verbal Model:*

$$\boxed{\begin{array}{c}\text{Amount of alcohol}\\\text{in 20\% solution}\end{array}} + \boxed{\begin{array}{c}\text{Amount of alcohol}\\\text{in 60\% solution}\end{array}} = \boxed{\begin{array}{c}\text{Amount of alcohol}\\\text{in final 40\% solution}\end{array}}$$

Labels: 20% solution: Number of gallons $= x$; amount of alcohol $= 0.20x$ (gal)

60% solution: Number of gallons $= 100 - x$; amount of alcohol $= 0.60(100 - x)$ (gal)

40% solution: Number of gallons $= 100$; amount of alcohol $= 0.40(100)$ (gal)

Equation:
$$0.20x + 0.60(100 - x) = 0.40(100)$$
$$0.20x + 60 - 0.60x = 40$$
$$60 - 0.40x = 40$$
$$-0.40x = -20$$
$$x = \frac{-20}{-0.40}$$
$$x = 50 \text{ and } 100 - x = 50$$

So, 50 gallons of the 20% solution and 50 gallons of the 60% solution are needed.

53. *Verbal Model:*

$$\boxed{\begin{array}{c}\text{Amount of alcohol}\\\text{in 15\% solution}\end{array}} + \boxed{\begin{array}{c}\text{Amount of alcohol}\\\text{in 60\% solution}\end{array}} = \boxed{\begin{array}{c}\text{Amount of alcohol}\\\text{in final 45\% solution}\end{array}}$$

Labels: 15% solution: Number of quarts $= x$; amount of alcohol $= 0.15x$ (qt)

60% solution: Number of quarts $= 24 - x$; amount of alcohol $= 0.60(24 - x)$ (qt)

45% solution: Number of quarts $= 24$; amount of alcohol $= 0.45(24)$ (qt)

Equation:
$$0.15x + 0.60(24 - x) = 0.45(24)$$
$$0.15x + 14.4 - 0.60x = 10.8$$
$$14.4 - 0.45x = 10.8$$
$$-0.45x = -3.6$$
$$x = \frac{-3.6}{-0.45}$$
$$x = 8 \text{ and } 24 - x = 16$$

So, 8 quarts of the 15% solution and 16 quarts of the 60% solution are needed.

55. *Verbal Model:*

$$\boxed{\begin{array}{c}\text{Amount of antifreeze in}\\\text{original 40\% solution}\end{array}} - \boxed{\begin{array}{c}\text{Amount of antifreeze}\\\text{withdrawn}\end{array}} + \boxed{\begin{array}{c}\text{Amount of pure}\\\text{antifreeze added}\end{array}} = \boxed{\begin{array}{c}\text{Amount of antifreeze in}\\\text{final 50\% solution}\end{array}}$$

Labels: Original solution: Number of gallons $= 5$; amount of antifreeze $= 0.40(5)$

Withdrawn solution: Number of gallons $= x$; amount of antifreeze $- 0.40(x)$

Antifreeze added: Number of gallons $= x$; amount of antifreeze $= 1.00(x)$

Final solution: Number of gallons $= 5$; amount of antifreeze $= 0.50(5)$

Equation:
$$0.40(5) - 0.40(x) + 1.00(x) = 0.50(5)$$
$$2 + 0.60x = 2.50$$
$$0.60x = 0.50$$
$$x = \frac{0.50}{0.60} = \frac{5}{6}$$

So, 5/6 of a gallon of the original coolant must be withdrawn and replaced with pure antifreeze.

57. *Verbal Model:* | Number preferring A | + | Number preferring B | + | Number preferring C | = | Number polled |

  Labels: Number preferring A $= x$ (people)

  Number preferring B $= 2x$ (people)

  Number preferring C $= 3x$ (people)

  Number polled $= 1320$ (people)

  Equation: $x + 2x + 3x = 1320$

  $6x = 1320$

  $x = 220$

  There were 220 people in the sample that preferred candidate A.

59. Perimeter $= 2(\text{Height}) + 2(\text{Width})$

  $3 = 2h + 2(0.62h)$

  $3 = 2h + 1.24h$

  $3 = 3.24h$

  $\dfrac{3}{3.24} = h$

  $0.926 \approx h$

  The height of the picture frame is approximately 0.926 feet.

61. (a) $y = 13.46 + 0.371t$

  $14.38 = 13.46 + 0.371t$

  $0.92 = 0.371t$

  $\dfrac{0.92}{0.371} = t$

  $2.48 \approx t$

  The average hourly wage was $14.38 during the year 1992.

  (b) The average annual hourly raise was $0.371, the slope of the model $y = 13.46 + 0.371t$.

63. The average wages of the construction workers increased at a greater annual rate than the average wages of the retail workers.

65. (a) Area $= \frac{1}{2}(\text{Base})(\text{Height})$

  $A(x) = \frac{1}{2}x\left(\frac{2}{3}x + 1\right)$

  $= \frac{1}{3}x^2 + \frac{1}{2}x, \; x > 0$

  (b)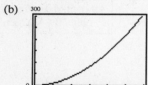

  (c) From the graph it appears that the area is 200 square units when the value of x is approximately 23.76 units. (You could use the zoom and trace features of your graphing utility to determine this answer.)

67. Side of square $= s \implies$ Perimeter $= 4s$

  Doubled side:
  Side of square $= 2s \implies$ Perimeter $= 4(2s) = 8s$

  Yes, if the sides of a square are doubled, the perimeter is also doubled.

69. $32 - 20x = 0$

  $-20x = -32$

  $x = \dfrac{-32}{-20}$

  $x = \dfrac{8}{5}$

71. $4(x + 7) - 1 = 0$

  $4x + 28 - 1 = 0$

  $4x + 27 = 0$

  $4x = -27$

  $x = -\dfrac{27}{4}$

73. $\dfrac{2x}{5} - \dfrac{x}{2} = 4$

  $10\left(\dfrac{2x}{5} - \dfrac{x}{2}\right) = 10(4)$

  $4x - 5x = 40$

  $-x = 40$

  $x = -40$

75. $\dfrac{5x}{9} + 3 = \dfrac{x}{2} - 1$

$18\left(\dfrac{5x}{9} + 3\right) = 18\left(\dfrac{x}{2} - 1\right)$

$10x + 54 = 9x - 18$

$x + 54 = -18$

$x = -72$

77. $\dfrac{5}{105} = \dfrac{x}{6(60)}$

$\dfrac{5}{105} = \dfrac{x}{360}$

$105x = 5(360)$

$105x = 1800$

$x = \dfrac{1800}{105}$

$x = \dfrac{\cancel{(3)}\cancel{(5)}(120)}{\cancel{(3)}\cancel{(5)}(7)}$

$x = \dfrac{120}{7}$

So, $\frac{120}{7}$ gallons of fuel, or approximately 17.1 gallons, would be used in 6 hours.

Mid-Chapter Quiz for Chapter 3

1. $y - y_1 = m(x - x_1)$

Point: $\left(0, -\dfrac{3}{2}\right)$

Slope: $m = 2$

$y - \left(-\dfrac{3}{2}\right) = 2(x - 0)$

$y + \dfrac{3}{2} = 2x$

$y = 2x - \dfrac{3}{2}$

2. $y - y_1 = m(x - x_1)$

Point: $(4, 7)$

Slope: $m = \dfrac{1}{2}$

$y - 7 = \dfrac{1}{2}(x - 4)$

$y - 7 = \dfrac{1}{2}x - 2$

$y = \dfrac{1}{2}x + 5$

3. $y - y_1 = m(x - x_1)$

Point: $\left(\dfrac{5}{2}, 6\right)$

Slope: $m = -\dfrac{3}{4}$

$y - 6 = -\dfrac{3}{4}\left(x - \dfrac{5}{2}\right)$

$y - 6 = -\dfrac{3}{4}x + \dfrac{15}{8}$

$y = -\dfrac{3}{4}x + \dfrac{15}{8} + \dfrac{48}{8}$

$y = -\dfrac{3}{4}x + \dfrac{63}{8}$

4. $y - y_1 = m(x - x_1)$

Point: $(-3.5, -1.8)$

Slope: $m = 3$

$y - (-1.8) = 3[x - (-3.5)]$

$y + 1.8 = 3(x + 3.5)$

$y + 1.8 = 3x + 10.5$

$y = 3x + 8.7$

5. $(2, 1)$ and $(4, 5)$

Slope: $m = \dfrac{y_2 - y_1}{x_2 - x_1} = \dfrac{5 - 1}{4 - 2} = \dfrac{4}{2} = 2$

Point: $(2, 1)$

$y - y_1 = m(x - x_1)$

$y - 1 = 2(x - 2)$

$y - 1 = 2x - 4$

$y = 2x - 3$

6. $(0, 0.8)$ and $(3, -2.3)$

Slope: $m = \dfrac{y_2 - y_1}{x_2 - x_1} = \dfrac{-2.3 - 0.8}{3 - 0} = \dfrac{-3.1}{3} = -\dfrac{31}{30}$

Point: $(0, 0.8)$

$y - y_1 = m(x - x_1)$

$y - 0.8 = -\dfrac{31}{30}(x - 0)$

$y - 0.8 = -\dfrac{31}{30}x$

$y = -\dfrac{31}{30}x + 0.8$ or $y = -\dfrac{31}{30}x + \dfrac{4}{5}$

7. $(3, -1)$ and $(10, -1)$

Slope: $m = \dfrac{y_2 - y_1}{x_2 - x_1} = \dfrac{-1 - (-1)}{10 - 3} = \dfrac{0}{7} = 0$

$m = 0 \Longrightarrow$ horizontal line, $y = b$

The horizontal line through $(3, -1)$ is $y = -1$.

8. $\left(4, \dfrac{5}{3}\right)$ and $(4, 8)$

Slope: $m = \dfrac{y_2 - y_1}{x_2 - x_1} = \dfrac{8 - 5/3}{4 - 4} = \dfrac{8 - 5/3}{0} \Longrightarrow m$ is undefined

m undefined $\Longrightarrow$ vertical line $x = a$

The vertical line through $(4, 8)$ is $x = 4$.

9. $2x - 3y = 1$

$\qquad -3y = -2x + 1$

$\qquad y = \dfrac{2}{3}x - \dfrac{1}{3}$

Point $(3, 5)$, $m = \dfrac{2}{3}$

$y - y_1 = m(x - x_1)$

$y - 5 = \dfrac{2}{3}(x - 3)$

$y - 5 = \dfrac{2}{3}x - 2$

$y = \dfrac{2}{3}x + 3$

10. $6x + 8y = -9$

$\qquad 8y = -6x - 9$

$\qquad y = \dfrac{-6}{8}x - \dfrac{9}{8}$

$\qquad y = -\dfrac{3}{4}x - \dfrac{9}{8}$

Point $(-1, 4)$, $m = \dfrac{4}{3}$

$y - y_1 = m(x - x_1)$

$y - 4 = \dfrac{4}{3}(x + 1)$

$y - 4 = \dfrac{4}{3}x + \dfrac{4}{3}$

$y = \dfrac{4}{3}x + \dfrac{16}{3}$

11. Profit = Revenue − Cost

$\qquad = 9.20x - (5.60x + 24{,}000)$

$\qquad = 9.20x - 5.60x - 24{,}000$

$\qquad = 3.60x - 24{,}000$

12. The product of a number n and 5 is decreased by 8.

$5n - 8$

13. Perimeter = 2(Length) + 2(Width)

$\qquad = 2(l) + 2(0.6l - 1)$

$\qquad = 2l + 1.2l - 2$

$\qquad = 3.2l - 2$

Area = (Length)(Width)

$\qquad = l(0.6l - 1)$

$\qquad = 0.6l^2 - l$

14. Three consecutive even integers n, $(n + 2)$, $(n + 4)$

$n + (n + 2) + (n + 4) = 3n + 6$

15. *Verbal Model:* $\boxed{\dfrac{\text{Defective Units in Sample}}{\text{Units in Sample}}} = \boxed{\dfrac{\text{Defective Units in Shipment}}{\text{Units in Shipment}}}$

Equation: $\dfrac{1}{300} = \dfrac{x}{600{,}000}$

$\qquad 1(600{,}000) = 300x$

$\qquad 600{,}000 = 300x$

$\qquad \dfrac{600{,}000}{300} = x$

$\qquad 2000 = x$

The expected number of defective units in the shipment is 2000.

16. *Verbal Model:* | Acid in 25% solution | + | Acid in 50% solution | = | Acid in 30% solution |

Labels: 25% solution: Amount of solution = x (gallons)
                    Amount of acid = $0.25x$ (gallons)

                50% solution: Amount of solution = $50 - x$ (gallons)
                    Amount of acid = $0.50(50 - x)$ (gallons)

                30% solution: Amount of solution = 50 (gallons)
                    Amount of acid = $0.30(50)$ (gallons)

Equation: $0.25x + 0.50(50 - x) = 0.30(50)$

$0.25x + 25 - 0.50x = 15$

$-0.25x + 25 = 15$

$-0.25x = -10$

$x = 40$ and $50 - x = 10$

So, 40 gallons of the 25% solution and 10 gallons of the 50% solution are required.

17. *Verbal Model:* | Portion done by you | + | Portion done by your friend | = | Work done |

Labels: You: Rate = $\frac{1}{3}$ (job per hour); time = t (hours)

                Your friend: Rate = $\frac{1}{4}$ (job per hour); time = t (hours)

                Work done = 1 (job)

Equation: $\frac{1}{3}t + \frac{1}{4}t = 1$

$12\left(\frac{1}{3}t + \frac{1}{4}t\right) = 12(1)$

$4t + 3t = 12$

$7t = 12$

$t = \frac{12}{7}$

So, it will take $\frac{12}{7}$ hours or approximately 1 hour, 43 minutes to paint the room working together.

18. *Verbal Model:* | Distance of faster car | − | Distance of slower car | = | Distance apart |

Labels: Faster car: Rate = 55 (miles per hour); time = t (hours)
                    Distance = $55t$ (miles)

                Slower car: Rate = 40 (miles per hour); time = t (hours)
                    Distance = $50t$ (miles)

                Distance apart = 5 (miles)

Equation: $55t - 40t = 5$

$15t = 5$

$t = \frac{5}{15}$

$t = \frac{1}{3}$

So, $\frac{1}{3}$ of an hour, or 20 minutes, must elapse before the cars are 5 miles apart.

Section 3.4   Linear Inequalities in One Variable

1. (a) $x = 3$

$7(3) - 10 \overset{?}{>} 0$

$21 - 10 \overset{?}{>} 0$

$11 > 0$

Yes, 3 *is* a solution.

(b) $x = -2$

$7(-2) - 10 \overset{?}{>} 0$

$-14 - 10 \overset{?}{>} 0$

$-24 \not> 0$

No, 2 is *not* a solution.

(c) $x = \frac{5}{2}$

$7\left(\frac{5}{2}\right) - 10 \overset{?}{>} 0$

$\frac{35}{2} - 10 \overset{?}{>} 0$

$\frac{35}{2} - \frac{20}{2} \overset{?}{>} 0$

$\frac{15}{2} > 0$

Yes, $\frac{5}{2}$ *is* a solution.

(d) $x = \frac{1}{2}$

$7\left(\frac{1}{2}\right) - 10 \overset{?}{>} 0$

$\frac{7}{2} - 10 \overset{?}{>} 0$

$\frac{7}{2} - \frac{20}{2} \overset{?}{>} 0$

$-\frac{13}{2} \not> 0$

No, $\frac{1}{2}$ is *not* a solution.

3. (a) $x = 10$

$$0 \overset{?}{\leq} \frac{10 + 5}{6} \overset{?}{<} 2$$

$$0 \overset{?}{\leq} \frac{15}{6} \overset{?}{<} 2$$

$$0 < 2\frac{1}{2} \not< 2$$

No, 10 is *not* a solution.

(b) $x = 4$

$$0 \overset{?}{\leq} \frac{4 + 5}{6} \overset{?}{<} 2$$

$$0 \overset{?}{\leq} \frac{9}{6} \overset{?}{<} 2$$

$$0 < 1\frac{1}{2} < 2$$

Yes, 4 *is* a solution.

(c) $x = 0$

$$0 \overset{?}{\leq} \frac{0 + 5}{6} \overset{?}{<} 2$$

$$0 < \frac{5}{6} < 2$$

Yes, 0 *is* a solution.

(d) $x = -6$

$$0 \overset{?}{\leq} \frac{-6 + 5}{6} \overset{?}{<} 2$$

$$0 \not< -\frac{1}{6} < 2$$

No, -6 is *not* a solution.

5. Graph (d)

7. Graph (c)

9. Graph (c)

11. Graph (d)

13. Graph (d)

15. Graph (c)

17. $(-\infty, 2]$

19. $(3.5, \infty)$

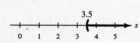

21. $(-5, 3]$

23. $\left(0, \frac{3}{2}\right]$

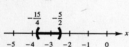

25. $\left(-\frac{15}{4}, -\frac{5}{2}\right)$

27. $(-\infty, -5) \cup (-1, \infty)$

29. $(-\infty, 3] \cup (7, \infty)$

31. $x + 7 \leq 9$

$x + 7 - 7 \leq 9 - 7$

$x \leq 2$

33. $4x < 22$

$$\frac{4x}{4} < \frac{22}{4}$$

$$x < \frac{11}{2}$$

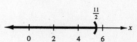

35. $-9x \geq 36$

$$\frac{-9x}{-9} \leq \frac{36}{-9}$$

$$x \leq -4$$

37. $-\frac{3}{4}x < -6$

$-\frac{4}{3}\left(-\frac{3}{4}x\right) > -\frac{4}{3}(-6)$

$x > 8$

39. $2x - 5 > 9$

$2x - 5 + 5 > 9 + 5$

$2x > 14$

$$\frac{2x}{2} > \frac{14}{2}$$

$x > 7$

41. $5 - x \leq -2$

$5 - x - 5 \leq -2 - 5$

$-x \leq -7$

$(-1)(-x) \geq (-7)(-1)$

$x \geq 7$

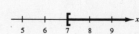

43.
$$5 - 3x < 7$$
$$5 - 3x - 5 < 7 - 5$$
$$-3x < 2$$
$$\frac{-3x}{-3} > \frac{2}{-3}$$
$$x > -\frac{2}{3}$$

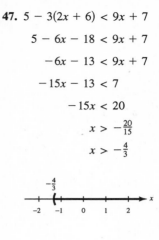

45.
$$3x - 11 > -x + 7$$
$$3x - 11 + x > -x + 7 + x$$
$$4x - 11 > 7$$
$$4x - 11 + 11 > 7 + 11$$
$$4x > 18$$
$$\frac{4x}{4} > \frac{18}{4}$$
$$x > \frac{9}{2}$$

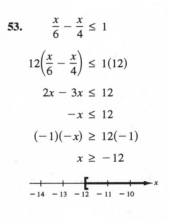

47.
$$5 - 3(2x + 6) < 9x + 7$$
$$5 - 6x - 18 < 9x + 7$$
$$-6x - 13 < 9x + 7$$
$$-15x - 13 < 7$$
$$-15x < 20$$
$$x > -\frac{20}{15}$$
$$x > -\frac{4}{3}$$

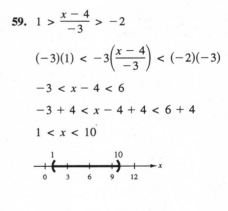

49.
$$16 < 4(y + 2) - 5(2 - y)$$
$$16 < 4y + 8 - 10 + 5y$$
$$16 < 9y - 2$$
$$16 + 2 < 9y - 2 + 2$$
$$18 < 9y$$
$$\frac{18}{9} < \frac{9y}{9}$$
$$2 < y \text{ or } y > 2$$

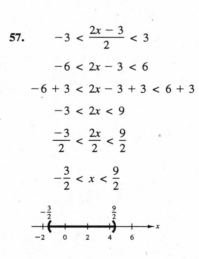

51.
$$-3(y + 10) \geq 4(y + 10)$$
$$-3y - 30 \geq 4y + 40$$
$$3y - 3y - 30 \geq 3y + 4y + 40$$
$$-30 \geq 7y + 40$$
$$-40 - 30 \geq 7y + 40 - 40$$
$$-70 \geq 7y$$
$$-10 \geq y \text{ or } y \leq -10$$

53.
$$\frac{x}{6} - \frac{x}{4} \leq 1$$
$$12\left(\frac{x}{6} - \frac{x}{4}\right) \leq 1(12)$$
$$2x - 3x \leq 12$$
$$-x \leq 12$$
$$(-1)(-x) \geq 12(-1)$$
$$x \geq -12$$

55.
$$0 < 2x - 5 < 9$$
$$0 + 5 < 2x - 5 + 5 < 9 + 5$$
$$5 < 2x < 14$$
$$\frac{5}{2} < \frac{2x}{2} < \frac{14}{2}$$
$$\frac{5}{2} < x < 7$$

57.
$$-3 < \frac{2x - 3}{2} < 3$$
$$-6 < 2x - 3 < 6$$
$$-6 + 3 < 2x - 3 + 3 < 6 + 3$$
$$-3 < 2x < 9$$
$$\frac{-3}{2} < \frac{2x}{2} < \frac{9}{2}$$
$$-\frac{3}{2} < x < \frac{9}{2}$$

59.
$$1 > \frac{x - 4}{-3} > -2$$
$$(-3)(1) < -3\left(\frac{x - 4}{-3}\right) < (-2)(-3)$$
$$-3 < x - 4 < 6$$
$$-3 + 4 < x - 4 + 4 < 6 + 4$$
$$1 < x < 10$$

61.
$$\frac{2}{5} < x + 1 < \frac{4}{5}$$
$$\frac{2}{5} - 1 < x + 1 - 1 < \frac{4}{5} - 1$$
$$\frac{2}{5} - \frac{5}{5} < x < \frac{4}{5} - \frac{5}{5}$$
$$-\frac{3}{5} < x < -\frac{1}{5}$$

63. $2x - 4 \leq 4$ and $2x + 8 > 6$
$$2x \leq 8 \qquad 2x > -2$$
$$x \leq 4 \qquad x > -1$$
$$-1 < x \leq 4$$

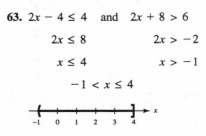

65. $7 + 4x < -5 + x$ and $2x + 10 \leq -2$

$7 + 3x < -5$ $2x \leq -12$

$3x < -12$ $x \leq -6$

$x < -4$

$x \leq -6$

$$-8 \quad -6 \quad -4 \quad -2 \quad 0$$

67. $6 - \dfrac{x}{2} \geq 1$ or $\dfrac{5}{4}x - 6 \geq 4$

$2\left(6 - \dfrac{x}{2}\right) \geq 2(1)$ $4\left(\dfrac{5}{4}x - 6\right) \geq 4(4)$

$12 - x \geq 2$ $5x - 24 \geq 16$

$-x \geq -10$ $5x \geq 40$

$x \leq 10$ $x \geq 8$

All real numbers

$$-2 \quad -1 \quad 0 \quad 1 \quad 2$$

69. $7x + 11 < 3 + 4x$ or $\dfrac{5}{2}x - 1 \geq 9 - \dfrac{3}{2}x$

$3x + 11 < 3$ $2\left(\dfrac{5}{2}x - 1\right) \geq 2\left(9 - \dfrac{3}{2}x\right)$

$3x < -8$ $5x - 2 \geq 18 - 3x$

$x < -\dfrac{8}{3}$ $8x - 2 \geq 18$

$8x \geq 20$

$x \geq \dfrac{20}{8}$

$x \geq \dfrac{5}{2}$

$x < -\dfrac{8}{3}$ or $x \geq \dfrac{5}{2}$

$$-4 \;\; -3 \;\; -2 \;\; -1 \;\; 0 \;\; 1 \;\; 2 \;\; 3 \;\; 4$$

71. $x \geq 0$

73. $z \geq 2$

75. $n \leq 16$

77. x is at least $\dfrac{5}{2}$.

79. z is greater than 0 and no more than π.

81. $12 \leq 4n \leq 30$

$\dfrac{12}{4} \leq \dfrac{4n}{4} \leq \dfrac{30}{4}$

$3 \leq n \leq \dfrac{15}{2}$

$\left[3, \dfrac{15}{2}\right]$

83. *Verbal Model:* | Transportation costs | $+$ | Other costs | $\leq$ | Total money for trip |

Inequality: $1900 + C \leq 4500$

$1900 + C - 1900 \leq 4500 - 1900$

$C \leq 2600$

So, all other costs must be no more than \$2600.

85. *Verbal Model:* | Temp in Miami | $>$ | Temp in Washington | and | Temp in Washington | $>$ | Temp in New York |

The average temperature in Miami, is greater than ($>$) the average temperature in New York.

87. *Verbal Model:* | Operating cost | $<$ | $10,000 |

Inequality: $0.28m + 2900 < 10,000$

$$0.28m + 2900 - 2900 < 10,000 - 2900$$

$$0.28m < 7100$$

$$\frac{0.28m}{0.28} < \frac{7100}{0.28}$$

$$m < 25,357.14286$$

Note: The answer could be written as $m \leq 25,357$.

So, the maximum number of miles is 25,357.

89. *Verbal Model:* | Revenue | $>$ | Cost |

Inequality: $89.95x > 61x + 875$

$$89.95x - 61x > 61x + 875 - 61x$$

$$28.95x > 875$$

$$\frac{28.95x}{28.95} > \frac{875}{28.95}$$

$$x > 30.224525$$

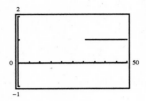

Note: The answer could be written as $x \geq 31$.

91. *Verbal Model:* | Cost of first minute | $+$ | Cost of additional minutes | $\leq$ | $5.00 |

Label: Number of *additional* minutes $= x$ (minutes)

Inequality: $0.96 + 0.75x \leq 5.00$

$$0.96 + 0.75x - 0.96 \leq 5.00 - 0.96$$

$$0.75x \leq 4.04$$

$$\frac{0.75x}{0.75} \leq \frac{4.04}{0.75}$$

$$x \leq 5.386667$$

Since x represents the additional minutes after the first minute, the call must be less than 6.38 minutes. If a portion of a minute is billed as a full minute, then the call must be less than or equal to 6 minutes.

93. Maximum distance: $5 + 3 = 8$ miles

Minimum distance: $5 - 3 = 2$ miles

$$2 \leq d \leq 8 \quad \text{or} \quad [2, 8]$$

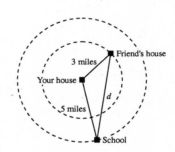

95. *Verbal Model:* $36 \leq$ | Perimeter | ≤ 64

Inequality: $36 \leq 2(16) + 2(x - 2) \leq 64$

$$36 \leq 32 + 2x - 4 \leq 64$$

$$36 \leq 2x + 28 \leq 64$$

$$8 \leq 2x \leq 36$$

$$4 \leq x \leq 18$$

$$[4, 18]$$

97. First plan-pay per hour: $12.50

Second plan-pay per hour: $8.00 + 0.75 per unit

$$8 + 0.75x > 12.50$$

$$0.75x > 4.50$$

$$x > \frac{4.50}{0.75}$$

$$x > 6$$

The second plan gives the greater hourly wage if 6 or more units are produced.

99. (a) The number of nurses was greater than 2 million during the years 1994, 1995, 1996, 1997, and 1998.

(b) The number of nurses was less than 1,900,000 during the years 1990 and 1991.

101. The multiplication and division properties differ. The inequality symbol is reversed if both sides of the inequality are multiplied or divided by a negative real number.

103. $t < 8 \implies -t > -8$

So, $-t$ is the interval $(-8, \infty)$.

105. $\frac{7}{8} < 4$

107. $-10 < \frac{1}{10}$

109. $-3x - 4 = 8$

$$-3x = 12$$
$$x = -4$$

111. $12 - 5(2 - x) = 6$

$$12 - 10 + 5x = 6$$
$$2 + 5x = 6$$
$$5x = 4$$
$$x = \frac{4}{5}$$

113. $A = \frac{1}{2}(7 + 4)(3.6)$

$$A = \frac{1}{2}(11)(3.6)$$
$$A = 19.8 \text{ square meters}$$

Section 3.5   Absolute Value Equations and Inequalities

1. $|4x + 5| = 10, x = -3$

$$|4(-3) + 5| \stackrel{?}{=} 10$$
$$|-12 + 5| \stackrel{?}{=} 10$$
$$|-7| \stackrel{?}{=} 10$$
$$7 \neq 10$$

The value $x = -3$ is not a solution of the equation.

3. $|6 - 2w| = 2, w = 4$

$$|6 - 2(4)| \stackrel{?}{=} 2$$
$$|6 - 8| \stackrel{?}{=} 2$$
$$|-2| \stackrel{?}{=} 2$$
$$2 = 2$$

The value $w = 4$ is a solution of the equation.

5. $|x - 10| = 17$

$x - 10 = -17$ or $x - 10 = 17$

7. $|4x + 1| = \frac{1}{2}$

$4x + 1 = -\frac{1}{2}$ or $4x + 1 = \frac{1}{2}$

9. $|t| = 45$

$t = -45$ or $t = 45$

11. $|h| = 0$

$h = 0$

13. $|x - 16| = 5$

$$x - 16 = -5 \text{ or } x - 16 = 5$$
$$x = 11 \qquad\quad x = 21$$

15. $|2s + 3| = 25$

$$2s + 3 = -25 \text{ or } 2s + 3 = 25$$
$$2s = -28 \qquad\quad 2s = 22$$
$$s = -14 \qquad\quad s = 11$$

17. $|32 - 3y| = 16$

$$32 - 3y = -16 \quad\text{or}\quad 32 - 3y = 16$$
$$-3y = -48 \qquad\qquad -3y = -16$$
$$y = 16 \qquad\qquad y = \frac{16}{3}$$

19. No solution; the absolute value of a real number cannot be negative.

21. $|2x + 13| = 4$

$$2x + 13 = -4 \quad\text{or}\quad 2x + 13 = 4$$
$$2x = -17 \qquad\qquad 2x = -9$$
$$x = -\frac{17}{2} \qquad\qquad x = -\frac{9}{2}$$

23. $|4 - 3x| = 0$

$$4 - 3x = 0$$
$$-3x = -4$$
$$x = \frac{-4}{-3}$$
$$x = \frac{4}{3}$$

25. $\left|\dfrac{2}{3}x + 4\right| = 9$

$\dfrac{2}{3}x + 4 = -9$ or $\dfrac{2}{3}x + 4 = 9$

$\dfrac{2}{3}x = -13$ $\qquad$ $\dfrac{2}{3}x = 5$

$2x = -39$ $\qquad$ $2x = 15$

$x = -\dfrac{39}{2}$ $\qquad$ $x = \dfrac{15}{2}$

27. $|0.32x - 2| = 4$

$0.32x - 2 = -4$ or $0.32x - 2 = 4$

$0.32x = -2$ $\qquad$ $0.32x = 6$

$x = \dfrac{-2}{0.32} = -6.25$ $\qquad$ $x = \dfrac{6}{0.32} = 18.75$

29. $|5x - 3| + 8 = 22$

$|5x - 3| = 14$

$5x - 3 = -14$ or $5x - 3 = 14$

$5x = -11$ $\qquad$ $5x = 17$

$x = -\dfrac{11}{5}$ $\qquad$ $x = \dfrac{17}{5}$

31. $|7x + 9| - 5 = 16$

$|7x + 9| = 21$

$7x + 9 = -21$ or $7x + 9 = 21$

$7x = -30$ $\qquad$ $7x = 12$

$x = -\dfrac{30}{7}$ $\qquad$ $x = \dfrac{12}{7}$

33. $|2x + 10| - 7 = -2$

$|2x + 10| = 5$

$2x + 10 = -5$ or $2x + 10 = 5$

$2x = -15$ $\qquad$ $2x = -5$

$x = -\dfrac{15}{2}$ $\qquad$ $x = -\dfrac{5}{2}$

35. $4|x + 5| = 9$

$|x + 5| = \dfrac{9}{4}$

$x + 5 = -\dfrac{9}{4}$ or $x + 5 = \dfrac{9}{4}$

$x = -\dfrac{9}{4} - 5$ $\qquad$ $x = \dfrac{9}{4} - 5$

$x = -\dfrac{9}{4} - \dfrac{20}{4}$ $\qquad$ $x = \dfrac{9}{4} - \dfrac{20}{4}$

$x = -\dfrac{29}{4}$ $\qquad$ $x = -\dfrac{11}{4}$

37. $8|x + 5| = 20$

$|x + 5| = \dfrac{20}{8}$

$|x + 5| = \dfrac{5}{2}$

$x + 5 = -\dfrac{5}{2}$ or $x + 5 = \dfrac{5}{2}$

$x = -\dfrac{15}{2}$ $\qquad$ $x = -\dfrac{5}{2}$

39. $5|8x + 11| = -10$

$|8x + 11| = -2$

No solution

41. $3|4x + 3| + 2 = 15$

$3|4x + 3| = 13$

$|4x + 3| = \dfrac{13}{3}$

$4x + 3 = -\dfrac{13}{3}$ or $4x + 3 = \dfrac{13}{3}$

$4x = -\dfrac{22}{3}$ $\qquad$ $4x = \dfrac{4}{3}$

$x = -\dfrac{22}{12}$ $\qquad$ $x = \dfrac{4}{12}$

$x = -\dfrac{11}{6}$ $\qquad$ $x = \dfrac{1}{3}$

43. $7|5x + 9| - 12 = -5$

$7|5x + 9| = 7$

$|5x + 9| = 1$

$5x + 9 = -1$ or $5x + 9 = 1$

$5x = -10$ $\qquad$ $5x = -8$

$x = -2$ $\qquad$ $x = -\dfrac{8}{5}$

45. $|x + 8| = |2x + 1|$

$x + 8 = -(2x + 1)$ or $x + 8 = 2x + 1$

$x + 8 = -2x - 1$ $\qquad\qquad 8 = x + 1$

$3x + 8 = -1$ $\qquad\qquad\qquad 7 = x$

$\qquad 3x = -9$

$\qquad\quad x = -3$

47. $|45 - 4x| = |32 - 3x|$

$45 - 4x = -(32 - 3x)$ or $45 - 4x = 32 - 3x$

$45 - 4x = -32 + 3x$ $\qquad\qquad 45 = 32 + x$

$\qquad 45 = -32 + 7x$ $\qquad\qquad 13 = x$

$\qquad 77 = 7x$

$\qquad 11 = x$

49. $|x + 2| = |3x - 1|$

$x + 2 = -(3x - 1)$ or $x + 2 = 3x - 1$

$x + 2 = -3x + 1$ $\qquad -2x + 2 = -1$

$4x + 2 = 1$ $\qquad\qquad -2x = -3$

$\quad 4x = -1$ $\qquad\qquad\quad x = \frac{3}{2}$

$\qquad x = -\frac{1}{4}$

51. $2x + 3 = 5,\, 2x + 3 = -5$

$|2x + 3| = 5$

53. (a) $x = 2$

$|x| < 3$

$|2| \overset{?}{<} 3$

$2 < 3$

Yes

(b) $x = -4$

$|x| < 3$

$|-4| \overset{?}{<} 3$

$4 \not< 3$

No

55. (a) $x = 2$

$|x| \geq 3$

$|2| \overset{?}{\geq} 3$

$2 \not\geq 3$

No

(b) $x = -4$

$|x| \geq 3$

$|-4| \overset{?}{\geq} 3$

$4 \geq 3$

Yes

57. (a) $x = 9$

$|x - 7| < 3$

$|9 - 7| \overset{?}{<} 3$

$|2| \overset{?}{<} 3$

$2 < 3$

Yes

(b) $x = -4$

$|x - 7| < 3$

$|-4 - 7| \overset{?}{\geq} 3$

$|-11| \overset{?}{<} 3$

$11 \not< 3$

No

59. (a) $x = 9$

$|x - 7| \geq 3$

$|9 - 7| \overset{?}{\geq} 3$

$|2| \overset{?}{\geq} 3$

$2 \not\geq 3$

No

(b) $x = -4$

$|x - 7| \geq 3$

$|-4 - 7| \overset{?}{\geq} 3$

$|-11| \overset{?}{\geq} 3$

$11 \geq 3$

Yes

61. $|y + 5| < 3$

$-3 < y + 5 < 3$

63. $|7 - 2h| \geq 9$

$7 - 2h \leq -9 \text{ or } 7 - 2h \geq 9$

65.

67.

69. $|y| < 4$

$-4 < y < 4$

71. $|2x| < 14$

$-14 < 2x < 14$

$-7 < x < 7$

73. $\left|\dfrac{y}{3}\right| \leq 3$

$-3 \leq \dfrac{y}{3} \leq 3$

$-9 \leq y \leq 9$

75. $|y - 2| \leq 4$

$-4 \leq y - 2 \leq 4$

$-2 \leq y \leq 6$

77. $|y + 5| < 8$

$-8 < y + 5 < 8$

$-13 < y < 3$

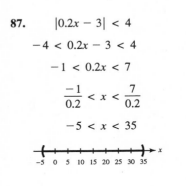

79. $|4 - y| > 6$

$4 - y < -6$ or $4 - y > 6$

$-y < -10$ $\qquad -y > 2$

$y > 10$ $\qquad y < -2$

81. $\left|y + \frac{1}{2}\right| \geq 4$

$y + \frac{1}{2} \leq -4$ or $y + \frac{1}{2} \geq 4$

$y \leq -\frac{9}{2}$ $\qquad y \geq \frac{7}{2}$

83. $|6t + 15| \geq 30$

$6t + 15 \leq -30$ or $6t + 15 \geq 30$

$6t \leq -45$ $\qquad 6t \geq 15$

$t \leq -\dfrac{45}{6}$ $\qquad t \geq \dfrac{15}{6}$

$t \leq -\dfrac{15}{2}$ $\qquad t \geq \dfrac{5}{2}$

85. $\left|\dfrac{z}{10} - 3\right| > 8$

$\dfrac{z}{10} - 3 < -8$ or $\dfrac{z}{10} - 3 > 8$

$\dfrac{z}{10} < -5$ $\qquad \dfrac{z}{10} > 11$

$z < -50$ $\qquad z > 110$

87. $|0.2x - 3| < 4$

$-4 < 0.2x - 3 < 4$

$-1 < 0.2x < 7$

$\dfrac{-1}{0.2} < x < \dfrac{7}{0.2}$

$-5 < x < 35$

89. $2|x + 4| > 8$

$|x + 4| > 4$

$x + 4 < -4$ or $x + 4 > 4$

$x < -8$ $\qquad x > 0$

91. $5|9 - 4x| < 25$

$|9 - 4x| < 5$

$-5 < 9 - 4x < 5$

$-14 < -4x < -4$

$\dfrac{-14}{-4} > x > \dfrac{-4}{-4}$

$\dfrac{7}{2} > x > 1$ or $1 < x < \dfrac{7}{2}$

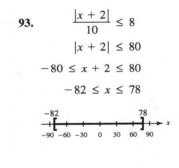

93. $\dfrac{|x + 2|}{10} \leq 8$

$|x + 2| \leq 80$

$-80 \leq x + 2 \leq 80$

$-82 \leq x \leq 78$

95. $\dfrac{|s - 3|}{5} > 4$

$|s - 3| > 20$

$s - 3 < -20$ or $s - 3 > 20$

$s < -17$ $\qquad s > 23$

97. $|12x + 5| - 2 \geq 7$

$|12x + 5| \geq 9$

$12x + 5 \leq -9$ or $12x + 5 \geq 9$

$12x \leq -14$ $\qquad 12x \geq 4$

$x \leq \dfrac{-14}{12}$ $\qquad x \geq \dfrac{4}{12}$

$x \leq -\dfrac{7}{6}$ $\qquad x \geq \dfrac{1}{3}$

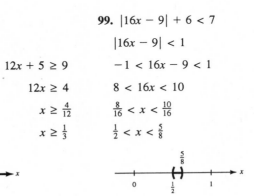

99. $|16x - 9| + 6 < 7$

$|16x - 9| < 1$

$-1 < 16x - 9 < 1$

$8 < 16x < 10$

$\dfrac{8}{16} < x < \dfrac{10}{16}$

$\dfrac{1}{2} < x < \dfrac{5}{8}$

101. $2|x + 9| - 1 > 15$

$2|x + 9| > 16$

$|x + 9| > 8$

$x + 9 < -8 \quad \text{or} \quad x + 9 > 8$

$x < -17 \qquad\qquad x > -1$

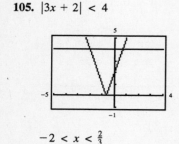

103. $5|4x + 18| - 7 \le -2$

$5|4x + 18| \le 5$

$|4x + 18| \le 1$

$-1 \le 4x + 18 \le 1$

$-19 \le 4x \le -17$

$-\frac{19}{4} \le x \le -\frac{17}{4}$

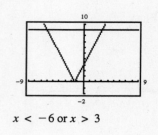

105. $|3x + 2| < 4$

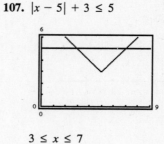

$-2 < x < \frac{2}{3}$

107. $|x - 5| + 3 \le 5$

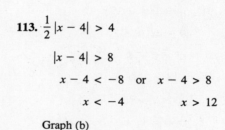

$3 \le x \le 7$

109. $|2x + 3| > 9$

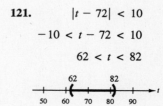

$x < -6 \text{ or } x > 3$

111. $\quad |x - 4| \le 4$

$-4 \le x - 4 \le 4$

$\quad 0 \le x \le 8$

Graph (d)

113. $\frac{1}{2}|x - 4| > 4$

$|x - 4| > 8$

$x - 4 < -8 \quad \text{or} \quad x - 4 > 8$

$x < -4 \qquad\qquad x > 12$

Graph (b)

115. $-2 \le x \le 2$

$|x| \le 2$

117. $\quad 7 < x < 13$

$-3 < x - 10 < 3$

$\quad |x - 10| < 3$

119. $\quad 16 < x < 22$

$-3 < x - 19 < 3$

$\quad |x - 19| < 3$

121. $\quad |t - 72| < 10$

$-10 < t - 72 < 10$

$\quad 62 < t < 82$

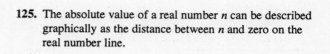

123. $\quad 97.6 < t < 99.6$

$97.6 - 98.6 < t - 98.6 < 99.6 - 98.6$

$-1 < t - 98.6 < 1$

$|t - 98.6| < 1$

125. The absolute value of a real number n can be described graphically as the distance between n and zero on the real number line.

127. Answers will vary.

129. $|-8| = 8$

131. $-|26| = -26$

133. $|35 - 43| = |-8| = 8$

135. $\quad -3 \le 5x + 9 \le 7$

$-12 \le 5x \le -2$

$-\frac{12}{5} \le x \le -\frac{2}{5}$

137. $2x + 1 > 5 \quad \text{or} \quad 5x - 2 < -2$

$2x > 4 \qquad\qquad 5x < 0$

$x > 2 \qquad\qquad x < 0$

139. $\left(\dfrac{1999 + 2001}{2}, \dfrac{696.5 + 1308.7}{2}\right) = (2000, \$1002.6 \text{ million})$

Review Exercises for Chapter 3

1. $2x + 6y = 18$

$\qquad 6y = -2x + 18$

$\qquad y = -\dfrac{2}{6}x + \dfrac{18}{6}$

$\qquad y = -\dfrac{1}{3}x + 3$

Slope: $-\dfrac{1}{3}$

y-intercept: $(0, 3)$

3. $7x - 9y + 27 = 0$

$\qquad -9y = -7x - 27$

$\qquad y = \dfrac{-7}{-9}x + \dfrac{-27}{-9}$

$\qquad y = \dfrac{7}{9}x + 3$

Slope: $\dfrac{7}{9}$

y-intercept: $(0, 3)$

5. $y = mx + b$

$\qquad y = \dfrac{3}{2}x - 7$

7. (a) Horizontal line though $(3, -4)$: $y = -4$

    (b) Vertical line through $(3, -4)$: $x = 3$

9. $(1, -4), m = 2$

$\qquad y - y_1 = m(x - x_1)$

$\qquad y + 4 = 2(x - 1)$

$\qquad y + 4 = 2x - 2$

$\qquad\qquad y = 2x - 6$

11. $(-1, 4), m = -4$

$\qquad y - y_1 = m(x - x_1)$

$\qquad y - 4 = -4(x + 1)$

$\qquad y - 4 = -4x - 4$

$\qquad\qquad y = -4x$

13. $\left(\dfrac{5}{2}, 4\right), m = -\dfrac{2}{3}$

$\qquad y - y_1 = m(x - x_1)$

$\qquad y - 4 = -\dfrac{2}{3}\left(x - \dfrac{5}{2}\right)$

$\qquad y - 4 = -\dfrac{2}{3}x + \dfrac{5}{3}$

$\qquad\qquad y = -\dfrac{2}{3}x + \dfrac{17}{3}$

15. $(7, 8), m$ is undefined

$\qquad m$ undefined $\Longrightarrow$ vertical line, $x = a$

$\qquad x = 7$

17. $(-6, 0), (0, -3)$

$\qquad m = \dfrac{-3 - 0}{0 - (-6)} = \dfrac{-3}{6} = -\dfrac{1}{2}$

$\qquad y = mx + b$

$\qquad y = -\dfrac{1}{2}x - 3$

or $y - y_1 = m(x - x_1)$

$\qquad y - 0 = -\dfrac{1}{2}(x + 6)$

$\qquad y = -\dfrac{1}{2}x - 3$

19. $(-10, 2), (4, -7)$

$$m = \frac{-7 - 2}{4 - (-10)} = \frac{-9}{14} = -\frac{9}{14}$$

$$y - y_1 = m(x - x_1)$$

$$y - 2 = -\frac{9}{14}(x + 10)$$

$$y - 2 = -\frac{9}{14}x - \frac{45}{7}$$

$$y = -\frac{9}{14}x - \frac{31}{7}$$

21. $\left(\frac{5}{2}, 0\right), \left(\frac{5}{2}, 5\right)$

$$m = \frac{5 - 0}{(5/2) - (5/2)} = \frac{5}{0} \text{ (undefined)}$$

m undefined $\Rightarrow$ vertical line, $x = a$

$$x = \frac{5}{2}$$

23. $2x + 4y = 1$

$$4y = -2x + 1$$

$$y = -\frac{1}{2}x + \frac{1}{4}$$

$$m = -\frac{1}{2}$$

parallel line

(a) $m = -\frac{1}{2}, (-1, 5)$

$$y - y_1 = m(x - x_1)$$

$$y - 5 = -\frac{1}{2}(x + 1)$$

$$y - 5 = -\frac{1}{2}x - \frac{1}{2}$$

$$y = -\frac{1}{2}x + \frac{9}{2}$$

perpendicular line

(b) $m = 2, (-1, 5)$

$$y - y_1 = m(x - x_1)$$

$$y - 5 = 2(x + 1)$$

$$y - 5 = 2x + 2$$

$$y = 2x + 7$$

25. $3x + y = 2$

$$y = -3x + 2$$

$$m = -3$$

parallel line

(a) $m = -3, \left(\frac{3}{5}, -\frac{4}{5}\right)$

$$y - y_1 = m(x - x_1)$$

$$y + \frac{4}{5} = -3\left(x - \frac{3}{5}\right)$$

$$y + \frac{4}{5} = -3x + \frac{9}{5}$$

$$y = -3x + 1$$

perpendicular line

(b) $m = \frac{1}{3}, \left(\frac{3}{5}, -\frac{4}{5}\right)$

$$y - y_1 = m(x - x_1)$$

$$y + \frac{4}{5} = \frac{1}{3}\left(x - \frac{3}{5}\right)$$

$$y + \frac{4}{5} = \frac{1}{3}x - \frac{1}{5}$$

$$y = \frac{1}{3}x - 1$$

27. Two hundred decreased by 3 times a number

$$200 - 3n$$

29. The sum of the square of a real number and 49

$$n^2 + 49$$

31. The absolute value of the sum of a real number and 10

$$|n + 10|$$

33. $2y + 7$

The sum of twice a number and seven

35. $\dfrac{x - 5}{4}$

The difference of a number and five, all divided by four

37. *Verbal Model:* $\boxed{\text{Tax Rate}} \cdot \boxed{\text{Taxable income}}$

Labels: Tax rate = 0.18 (decimal form)

Taxable income = I (dollars)

Algebraic expression: $0.18I$ (dollars)

39. *Verbal Model:* $\boxed{\text{Length}} \cdot \boxed{\text{Width}}$

Labels: Length = l (units)

Width = $l - 5$ (units)

Algebraic expression: $l(l - 5) = l^2 - 5l$ (square units)

41. *Verbal Model:* | Number of acres | · | Price per acre |

Labels: Number of acres = 30 (acre)

Price per acre = p (*dollars per acre*)

Algebraic expression: $30p$

43. $\dfrac{x + 9}{2} = 15$

45. $350 + 32x = 590$

47. z is no more than 10.

$z \leq 10$

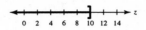

49. y is at least 7 but less than 14.

$7 \leq y < 14$

51. The volume V is less than 27 cubic feet.

$V < 27$

53. $x \leq 3$ or $x > 10$

55. $7x + 3 > 2x + 18$

$5x + 3 > 18$

$5x > 15$

$x > 3$

57. $\frac{1}{3} - \frac{1}{2}y < 12$

$2 - 3y < 72$

$-3y < 70$

$y > -\frac{70}{3}$

59. $-4 < \dfrac{x}{5} \leq 4$

$-20 < x \leq 20$

61. $5 > \dfrac{x + 1}{-3} > 0$

$-15 < x + 1 < 0$

$-16 < x < -1$

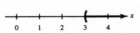

63. $8 - 2x < -3$ or $5x + 1 > 11$

$-2x < -11$ $5x > 10$

$x > \frac{11}{2}$ $x > 2$

$x > 2$

65. $3(x + 2) \leq 0$ or $5(x - 4) > 20$

$3x + 6 \leq 0$ $5x - 20 > 20$

$3x \leq -6$ $5x > 40$

$x \leq -2$ $x > 8$

67. $|x - 2| - 2 = 4$

$|x - 2| = 6$

$x - 2 = -6$ or $x - 2 = 6$

$x = -4$ $x = 8$

69. $4|2x + 3| = 20$

$|2x + 3| = 5$

$2x + 3 = -5$ or $2x + 3 = 5$

$2x = -8$ $2x = 2$

$x = -4$ $x = 1$

71. $7|x - 12| + 8 = 22$

$7|x - 12| = 14$

$|x - 12| = 2$

$x - 12 = -2$ or $x - 12 = 2$

$x = 10$ $x = 14$

73. $|3x - 4| = |x + 2|$

$3x - 4 = x + 2$ or $3x - 4 = -(x + 2)$

$2x - 4 = 2$ $\qquad$ $3x - 4 = -x - 2$

$\quad 2x = 6$ $\qquad$ $4x - 4 = -2$

$\qquad x = 3$ $\qquad\qquad$ $4x = 2$

$\qquad\qquad\qquad\qquad x = \frac{2}{4}$

$\qquad\qquad\qquad\qquad x = \frac{1}{2}$

75. $|2x - 7| < 15$

$-15 < 2x - 7 < 15$

$-8 < 2x < 22$

$-4 < x < 11$

77. $|x - 4| > 3$

$x - 4 < -3$ or $x - 4 > 3$

$\quad x < 1$ $\qquad\qquad$ $x > 7$

79. $|b + 2| - 6 \geq 1$

$|b + 2| \geq 7$

$b + 2 \leq -7$ or $b + 2 \geq 7$

$\quad b \leq -9$ $\qquad\qquad$ $b \geq 5$

81. (a) $(7, \$625)$ and $(10, \$800)$

$$m = \frac{800 - 625}{10 - 7} = \frac{175}{3}$$

$$R - 625 = \frac{175}{3}(t - 7)$$

$$R - 625 = \frac{175}{3}t - \frac{1225}{3}$$

$$R = \frac{175}{3}t + \frac{650}{3}$$

(b)

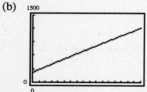

(c) For $t = 15$,

$$R = \frac{175}{3}(15) + \frac{650}{3}$$

$$R = \frac{3275}{3}$$

$$R \approx \$1091.67$$

(d) For $t = 9$,

$$R = \frac{175}{3}(9) + \frac{650}{3}$$

$$R = \frac{2225}{3}$$

$$R \approx \$741.67$$

83. *Verbal Model:* $\boxed{\dfrac{\text{Assessed value of property}}{\text{Taxes on first property}}} = \boxed{\dfrac{\text{Assessed value of second property}}{\text{Taxes on second property}}}$

Labels: $\qquad$ First property: Assessed value $= 80{,}000$ (dollars); taxes $= 1350$ (dollars)

$\qquad\qquad$ Second property: Assessed value $= 110{,}000$ (dollars); taxes $= x$ (dollars)

Equation: $\qquad \dfrac{80{,}000}{1350} = \dfrac{110{,}000}{x}$

$\qquad\qquad 80{,}000x = 1350(110{,}000)$

$\qquad\qquad\qquad x = \dfrac{1350(110{,}000)}{80{,}000}$

$\qquad\qquad\qquad x = 1856.25$

So, the tax is $\$1856.25$.

85. *Verbal Model:*

$$\boxed{\dfrac{\text{Length of map scale}}{\text{Distance represented on scale}}} = \boxed{\dfrac{\text{Distance between cities on map}}{\text{Actual distance between cities}}}$$

Labels: Scale: Length on map $= \dfrac{1}{3}$ (inch); distance represented $= 50$ (miles)

Cities: Map distance $= 3\dfrac{1}{4}$ (inches); actual distance $= x$ (miles)

Equation: $\dfrac{1/3}{50} = \dfrac{13/4}{x}$

$$\dfrac{1}{3}x = 50\left(\dfrac{13}{4}\right)$$

$$\dfrac{x}{3} = \dfrac{25(13)}{2}$$

$$3\left(\dfrac{x}{3}\right) = \left(\dfrac{325}{2}\right)3$$

$$x = \dfrac{975}{2}$$

$$x = 487.5$$

So, the cities are about 487.5 miles apart.

87. $\dfrac{2}{6} = \dfrac{x}{9}$

$6x = 18$

$x = 3$

89. *Verbal Model:*

$$\boxed{\dfrac{\text{Your height}}{\text{Your shadow's length}}} = \boxed{\dfrac{\text{Building's height}}{\text{Length of shadow of building}}}$$

Labels: You: Height $= 6$ (feet); Length of shadow $= 1.5$ (feet)

Building: Height $= x$ (feet); Length of shadow $= 20$ (feet)

Equation: $\dfrac{6}{1.5} = \dfrac{x}{20}$

$6(20) = 1.5x$

$120 = 1.5x$

$\dfrac{120}{1.5} = x$

$80 = x$

The building is 80 feet tall.

91. 1980: $40,000

2000: $110,000

Increase in expenses: $110{,}000 - 40{,}000 = 70{,}000$

Verbal Model: $\boxed{\text{Increase in expenses}} = \boxed{\text{Percent of increase}} \cdot \boxed{\text{1980 expenses}}$

Labels: Increse in expenses $= 70{,}000$ (dollars)

Percent of increase $= p$ (percent in decimal form)

1980 expenses $= 40{,}000$ (dollars)

Equation: $70{,}000 = p(40{,}000)$

$\dfrac{70{,}000}{40{,}000} = p$

$1.75 = p$

The increase in expenses was approximately 175%.

93. *Verbal Model:* | Current price | + | Price increase | = | Next year's price |

Labels: Current price = 25,750 (dollars)

Rate of increase = 0.055 (percent in decimal form)

Price increase = 25,750(0.055) (dollars)

Next year's price = P (dollars)

Equation: $25{,}750 + 25{,}750(0.055) = P$

$25{,}750 + 1416.25 = P$

$27{,}166.25 = P$

The price of the truck for the next model year is projected to be $27,166.25.

95. *Verbal Model:* | Measure of first angle | + | Measure of second angle | = 180

Labels: Measure of first angle = $3x - 8$ (degrees)

Measure of second angle = $x - 12$ (degrees)

Equation: $(3x - 8) + (x - 12) = 180$

$3x - 8 + x - 12 = 180$

$4x - 20 = 180$

$4x = 200$

$x = 50$ so $3x - 8 = 142$ and $x - 12 = 38$

The measure of the first angle is 142 degrees and the measure of the second angle is 38 degrees.

97. *Verbal Model:* | Amount of salt in 30% solution | + | Amount of salt in 60% solution | + | Amount of salt in final 50% solution |

Labels: 30% solution: Number of liters = x; amount of salt = $0.30x$ (liters)

60% solution: Number of liters = $10 - x$; amount of salt = $0.60(10 - x)$ (liters)

50% solution: Number of liters = 10; amount of salt = $0.50(10)$ (liters)

Equation: $0.30x + 0.60(10 - x) = 0.50(10)$

$0.30x + 6 - 0.60x = 5$

$-0.30x + 6 = 5$

$-0.30x = -1$

$x = \dfrac{-1}{-0.30}$

$x = \dfrac{10}{3} \Longrightarrow 10 - x = \dfrac{20}{3}$

So, $3\frac{1}{3}$ liters of the 30% solution and $6\frac{2}{3}$ liters of the 60% solution are needed.

99. *Verbal Model:* | Distance | = | Rate | · | Time |

Labels: Distance = 330 (miles)

Rate = 52 (miles per hour)

Time = x (hours)

Equation: $330 = 52x$

$$\frac{330}{52} = x$$

$$6.35 \approx x$$

So, approximately 6.35 hours are required.

101. *Verbal Model:* | Distance | = | Rate | · | Time |

Labels: Distance = x (miles)

Rate = 1200 (miles per hour)

Time = $2\frac{1}{3}$ (hours)

Equation: $x = 1200\left(2\frac{1}{3}\right)$

$$x = 1200\left(\frac{7}{3}\right)$$

$$x = 400(7)$$

$$x = 2800$$

So, the jet can travel 2800 miles.

103. *Verbal Model:* | Portion done by faster worker | + | Portion done by slower worker | = | Work done |

Labels: Faster worker: Rate = $\frac{1}{8}$ (task per hour); time = t (hours)

Slower worker: Rate = $\frac{1}{10}$ (task per hour); time = t (hours)

Work done = $\frac{1}{2}$ (task)

Equation: $\frac{1}{8}t + \frac{1}{10}t = \frac{1}{2}$

$$40\left(\tfrac{1}{8}t + \tfrac{1}{10}t\right) = 40\left(\tfrac{1}{2}\right)$$

$$5t + 4t = 20$$

$$9t = 20$$

$$t = \frac{20}{9}$$

It would take them $2\frac{2}{9}$ hours to complete half the task.

105. *Verbal Model:* | Simple interest | = | Principle | · | Rate | · | Time |

Labels: Simple Interest = I (dollars)

Principle = 1000 (dollars)

Rate = 0.085 (percent in decimal form)

Time = 4 (years)

Equation: $I = 1000(0.085)(4)$

$$I = 340$$

The total simple interest earned in 4 years is $340.

107. *Verbal Model:* | Simple interest | = | Principle | · | Rate | · | Time |

Labels: Simple Interest = 20,000 (dollars)

Principle = P (dollars)

Rate = 0.095 (percent in decimal form)

Time = 1 (years)

Equation: $20{,}000 = P(0.095)(1)$

$20{,}000 = 0.095P$

$$\frac{20{,}000}{0.095} = P$$

$210{,}526.32 \approx P$

The principle required is $210,526.32.

109. *Verbal Model:* | Interest on 8.5% investment | + | Interest on 10% investment | = | Total interest |

Labels: Amount invested at 8.5% = x (dollars)

Amount invested at 10% = $50{,}000 - x$ (dollars)

Total interest = 4700 (dollars)

Equation: $0.085x + 0.10(50{,}000 - x) = 4700$

$0.085x + 5000 - 0.10x = 4700$

$85x + 5{,}000{,}000 - 100x = 4{,}700{,}000$

$-15x = -300{,}000$

$$x = \frac{-300{,}000}{-15}$$

$x = 20{,}000 \text{ and } 50{,}000 - x = 30{,}000$

So, $30,000 is the minimum amount to invest at 10%.

111. *Verbal Model:* | Length | · | Width | = | Area |

Labels: Length = x (inches)

Width = $x + 2$ (inches)

Area = 48 (square inches)

Equation: $x(x + 2) = 48$

$x^2 + 2x = 48$

$x^2 + 2x - 48 = 0$

$(x + 8)(x - 6) = 0$

$x + 8 = 0 \implies x = -8$

$x - 6 = 0 \implies x = 6$

and $x + 2 = 8$

(Discard the negative solution.) The length is 6 inches and the width is 8 inches.

113. Maximum distance: $8 + 3 = 11$ miles

Minimum distance: $8 - 3 = 5$ miles

$5 \le d \le 11$

Chapter Test for Chapter 3

1. $(25, -15)$ and $(75, 10)$

Slope: $m = \dfrac{y_2 - y_1}{x_2 - x_1} = \dfrac{10 - (-15)}{75 - 25} = \dfrac{25}{50} = \dfrac{1}{2}$

Point: $(25, 15)$

$y - y_1 = m(x - x_1)$

$y - (-15) = \dfrac{1}{2}(x - 25)$

$y + 15 = \dfrac{1}{2}x - \dfrac{25}{2}$

$y = \dfrac{1}{2}x - \dfrac{25}{2} - \dfrac{30}{2}$

$y = \dfrac{1}{2}x - \dfrac{55}{2}$

2. $5x + 3y - 9 = 0$

$3y = -5x + 9$

$y = -\dfrac{5}{3}x + 3$

$m = -\dfrac{5}{3} \implies$ a perpendicular line has slope $m = \dfrac{3}{5}$

Point $(10, 2)$, $m = \dfrac{3}{5}$

$y - y_1 = m(x - x_1)$

$y - 2 = \dfrac{3}{5}(x - 10)$

$y - 2 = \dfrac{3}{5}x - 6$

$y = \dfrac{3}{5}x - 4$

3. $(0, 26{,}000)$ and $(4, 10{,}000)$

Slope: $m = \dfrac{V_2 - V_1}{t_2 - t_1} = \dfrac{10{,}000 - 26{,}000}{4 - 0} = \dfrac{-16{,}000}{4} = -4000$

Point: $(0, 26{,}000)$

$V - V_1 = m(t - t_1)$

$V - 26{,}000 = -4000(t - 0)$

$V - 26{,}000 = -4000t$

$V = -4000t + 26{,}000$

$16{,}000 = -4000t + 26{,}000$

$-10{,}000 = -4000t$

$2.5 = t$

The value of the car will be \$16,000 in 2.5 years.

4. $T = 3t + 72$

5. *Verbal Model:* $2\boxed{\text{Length}} + 2\boxed{\text{Width}} = \boxed{\text{Perimeter}}$

Labels: Length $= 3x - 1$ (inches)

Width $= x + 6$ (inches)

Perimeter $= 114$ (inches)

Equation: $2(3x - 1) + 2(x + 6) = 114$

$6x - 2 + 2x + 12 = 114$

$8x + 10 = 114$

$8x = 104$

$x = 13$

so $3x - 1 = 38$ and $x + 6 = 19$

The length of the rectangle is 38 inches and the width is 19 inches.

6. $5n - 8 = 27$

$5n = 35$

$n = 7$

The number is 7.

7. *Verbal Model:* | First even integer | $+$ | Second even integer | $=$ | 54 |

  Labels: First even integer $= n$

  Second even integer $= n + 2$

Equation: $n + (n + 2) = 54$

$$2n + 2 = 54$$

$$2n = 52$$

$$n = 26 \text{ and } n + 2 = 28$$

The two consecutive even integers are 26 and 28.

8. *Verbal Model:* $\dfrac{\text{Tax on first property}}{\text{Assessed value of first property}} = \dfrac{\text{Tax on second property}}{\text{Assessed value of second property}}$

Equation: $\dfrac{1200}{90{,}000} = \dfrac{x}{110{,}000}$

$$90{,}000x = 1200(110{,}000)$$

$$x = \dfrac{1200(110{,}000)}{90{,}000}$$

$$x \approx \$1466.67$$

9. *Verbal Model:* | Cost of parts | $+ 16$ | Number of half-hours of labor | $=$ | Total bill |

  Label: *Half*-hours of labor $= x$

  Equation: $85 + 16x = 165$

$$16x = 80$$

$$x = \dfrac{80}{16}$$

$$x = 5$$

So, 5 *half*-hours or $2\frac{1}{2}$ hours were spent on the repair of the appliance.

10. *Verbal Model:* | Amount of antifreeze in 10% solution | $+$ | Amount of antifreeze in 40% solution | $=$ | Amount of antifreeze in 30% solution |

  Labels: 10% solution: Number of liters $= x$; amount of antifreeze $= 0.10x$ (liters)

  40% solution: Number of liters $= 100 - x$; amount of antifreeze $= 0.40(100 - x)$(liters)

  30% solution: Number of liters $= 100$; amount of antifreeze $= 0.30(100)$(liters)

  Equation: $0.10x + 0.40(100 - x) = 0.30(100)$

$$0.10x + 40 - 0.40x = 30$$

$$40 - 0.30x = 30$$

$$-0.30x = -10$$

$$x = \dfrac{-10}{-0.30}$$

$$x = 33\tfrac{1}{3} \text{ and } 100 - x = 66\tfrac{2}{3}$$

So, $33\frac{1}{3}$ liters of the 10% solution and $66\frac{2}{3}$ liters of the 40% solution are required.

11. *Verbal Model:* | Distance of second car | − | Distance of first car | = 10

Labels: First car: Rate = 40 (miles per hour); time = t (hours)

             Second car: Rate = 55 (miles per hour); time = t (hours)

Equation: $55t - 40t = 10$

$$15t = 10$$

$$t = \tfrac{10}{15}$$

$$t = \tfrac{2}{3}$$

So, $\tfrac{2}{3}$ of an hour, or 40 minutes, must elapse before the two cars are 10 miles apart.

12. *Verbal Model:* | Simple Interest | = | Principle | · | Rate | · | Time |

Labels: Simple Interest = 300 (dollars)

             Principle = P (dollars)

             Rate = 0.075 (percent in decimal form)

             Time = 2 (years)

Equation: $300 = P(0.075)(2)$

$$300 = 0.15P$$

$$\tfrac{300}{0.15} = P$$

$$2000 = P$$

The principle required is $2000.

13. *Verbal Model:* | Measure of angle A | + | Measure of angle B | + | Measure of angle C | = 180

Labels: Measure of angle A = $x + 6$ (degrees)

             Measure of angle B = $2x + 4$ (degrees)

             Measure of angle C = $2x + 15$ (degrees)

Equation: $(x + 6) + (2x + 4) + (2x + 15) = 180$

$$5x + 25 = 180$$

$$5x = 155$$

$$x = 31, \text{ so } \quad x + 6 = 37$$

$$2x + 4 = 66$$

$$2x + 15 = 77$$

The measure of angle A is 37 degrees, the measure of angle B is 66 degrees, and the measure of angle C is 77 degrees.

14. $|x + 6| - 3 = 8$

$|x + 6| = 11$

$x + 6 = -11$ or $x + 6 = 11$

$\quad x = -17 \qquad x = 5$

15. $|4x - 1| = |2x + 7|$

$4x - 1 = -(2x + 7)$ or $4x - 1 = 2x + 7$

$4x - 1 = -2x - 7 \qquad 2x - 1 = 7$

$6x - 1 = -7 \qquad\qquad 2x = 8$

$6x = -6 \qquad\qquad\quad x = 4$

$x = -1$

16. $1 + 2x > 7 - x$

$\quad\ \ 1 + 3x > 7$

$\quad\ \ \ \ \ \ \ 3x > 6$

$\quad\ \ \ \ \ \ \ \ \ x > 2$

17. $0 \le \dfrac{1 - x}{4} < 2$

$\quad\ \ 0 \le 1 - x < 8$

$\quad\ \ -1 \le -x < 7$

$\quad\ \ 1 \ge x > -7 \ \ \text{or} \ \ -7 < x \le 1$

18. $5(x + 7) < -4 \quad \text{or} \quad 2(3x - 8) + 6 \ge 10$

$\quad\ \ \ \ 5x + 35 < -4 \qquad\quad 6x - 16 + 6 \ge 10$

$\quad\ \ \ \ \ \ \ \ \ 5x < -39 \qquad\qquad\ \ \ 6x - 10 \ge 10$

$\quad\ \ \ \ \ \ \ \ \ \ \ \ x < -\frac{39}{5} \qquad\qquad\qquad\ \ 6x \ge 20$

$\qquad\qquad\qquad\qquad\qquad\qquad\qquad\quad\ x \ge \frac{20}{6}$

$\qquad\qquad\qquad\qquad\qquad\qquad\qquad\quad\ x \ge \frac{10}{3}$

19. $|x - 3| \le 2$

$\quad\ \ -2 \le x - 3 \le 2$

$\quad\ \ \ \ \ 1 \le x \le 5$

20. $|3x + 13| > 4$

$\quad\ \ 3x + 13 < -4 \quad \text{or} \quad 3x + 13 > 4$

$\quad\ \ \ \ \ \ \ 3x < -17 \qquad\qquad\ \ 3x > -9$

$\quad\ \ \ \ \ \ \ \ \ x < -\frac{17}{3} \qquad\qquad\ \ \ \ x > -3$

Cumulative Test for Chapters 1–3

1. $5^2 x^3 \cdot 5x^4 = 5^{2+1} x^{3+4}$

$\qquad\qquad\quad = 5^3 x^7$

$\qquad\qquad\quad = 125 x^7$

2. $(-7z^3)^4 (2z^3) = (-7)^4 z^{12} (2z^3)$

$\qquad\qquad\qquad\ = 2401 z^{12} (2z^3)$

$\qquad\qquad\qquad\ = 4802 z^{12+3}$

$\qquad\qquad\qquad\ = 4802 z^{15}$

3. $\dfrac{(2a^2 b^7)^2}{-5(ab^4)^3} = \dfrac{4a^4 b^{14}}{-5a^3 b^{12}}$

$\qquad\qquad\ \ = -\dfrac{4}{5} a^{4-3} b^{14-12}$

$\qquad\qquad\ \ = -\dfrac{4}{5} ab^2$

4. $-\left(\dfrac{3x^6}{2y^2}\right)^3 = -\dfrac{3^3 x^{18}}{2^3 y^6}$

$\qquad\qquad\qquad = -\dfrac{27 x^{18}}{8y^6}$

5. $2(x + 5) - 3 - (2x - 3) = 2x + 10 - 3 - 2x + 3$

$\qquad\qquad\qquad\qquad\qquad = 10$

6. $4 - 2[3 + 4(x + 1)] = 4 - 2[3 + 4x + 4]$

$\qquad\qquad\qquad\qquad = 4 - 6 - 8x - 8$

$\qquad\qquad\qquad\qquad = -8x - 10$

7. $(2x - 1)(3x + 4) = 6x^2 + 8x - 3x - 4$
$= 6x^2 + 5x - 4$

8. $(5x + 1)(5x - 1) = (5x)^2 - 1^2$
$= 25x^2 - 1$

9. $x^3 + 5x^2 - 8x - 40 = x^2(x + 5) - 8(x + 5)$
$= (x + 5)(x^2 - 8)$

10. $4x^2 - 28xy + 49y^2 = (2x - 7y)(2x - 7y)$ or $(2x - 7y)^2$

11. $8x^3 + 27 = (2x)^3 + 3^3$
$= (2x + 3)(4x^2 - 6x + 9)$

12. $12x^2 - 75 = 3(4x^2 - 25)$
$= 3[(2x)^2 - 5^2]$
$= 3(2x + 5)(2x - 5)$

13. $5x - 7 = 2(x + 10)$
$5x - 7 = 2x + 20$
$3x - 7 = 20$
$3x = 27$
$x = 9$

14. $3(6x - 1) + 4 = 17 - (x - 6)$
$18x - 3 + 4 = 17 - x + 6$
$18x + 1 = -x + 23$
$19x + 1 = 23$
$19x = 22$
$x = \frac{22}{19}$

15. $3x + 6 = x + 4$
$2x + 6 = 4$
$2x = -2$
$x = -1$

16. $4x - 2 = 2(x - 8) + 1$
$4x - 2 = 2x - 16 + 1$
$4x - 2 = 2x - 15$
$2x - 2 = -15$
$2x = -13$
$x = -\frac{13}{2}$

17. $3x^2 + x - 24 = 0$
$(3x - 8)(x + 3) = 0$
$3x - 8 = 0 \Rightarrow 3x = 8 \Rightarrow x = \frac{8}{3}$
$x + 3 = 0 \Rightarrow x = -3$

18. $7x^3 - 63x = 0$
$7x(x^2 - 9) = 0$
$7x(x + 3)(x - 3) = 0$
$7x = 0 \Rightarrow x = 0$
$x + 3 = 0 \Rightarrow x = -3$
$x - 3 = 0 \Rightarrow x = 3$

19. $|2x + 6| = 16$
$2x + 6 = -16$ or $2x + 6 = 16$
$2x = -22 \qquad 2x = 10$
$x = -11 \qquad x = 5$

20. $|3x - 5| = |6x - 1|$
$3x - 5 = -(6x - 1)$ or $3x - 5 = 6x - 1$
$3x - 5 = -6x + 1 \qquad -3x - 5 = -1$
$9x - 5 = 1 \qquad\qquad -3x = 4$
$9x = 6 \qquad\qquad x = -\frac{4}{3}$
$x = \frac{6}{9}$
$x = \frac{2}{3}$

21. *x*-intercept
$5x + 4(0) - 20 = 0$
$5x - 20 = 0$
$5x = 20$
$x = 4$
$(4, 0)$

y-intercept
$5(0) + 4y - 20 = 0$
$4y - 20 = 0$
$4y = 20$
$y = 5$
$(0, 5)$

22. *x*-intercept
$0 = (x + 3)(x - 7)$
$x + 3 = 0 \Rightarrow x = -3$
$x - 7 = 0 \Rightarrow x = 7$
$(-3, 0), (7, 0)$

y-intercept
$y = (0 + 3)(0 - 7)$
$y = (3)(-7)$
$y = -21$
$(0, -21)$

23. $4x + 3y - 12 = 0$

$$3y = -4x + 12$$

$$y = -\frac{4}{3}x + 4$$

The graph is line with slope $m = -\frac{4}{3}$ and y-intercept $(0, 4)$.

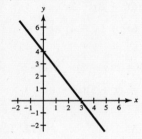

24. $y = 1 - (x - 2)^2$

$$y = -(x - 2)^2 + 1$$

The graph is a parabola with the same shape as the basic squaring function $y = x^2$ but reflected about the x-axis and shifted two units to the right and one unit upward.

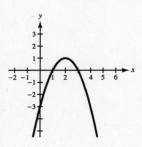

25. $f(x) = 3 - 2x$

(a) $f(5) = 3 - 2(5)$

$$= 3 - 10$$

$$= -7$$

(b) $f(x + 3) - f(3) = [3 - 2(x + 3)] - [3 - 2(3)]$

$$= (3 - 2x - 6) - (3 - 6)$$

$$= (-2x - 3) - (-3)$$

$$= -2x - 3 + 3$$

$$= -2x$$

26. A vertical line could cross the graph at more than one point, so the equation does not represent y as a function of x.

27. Domain $= \{x: x \geq 4\}$

Range $= \{y: y \geq 2\}$

28. The graph of $f(x) = \sqrt{x}$ is reflected about the x-axis shifted 2 units up. The equation for the graph is $f(x) = -\sqrt{x} + 2$.

29. $3x + 7y = 20$

$$7y = -3x + 20$$

$$y = -\frac{3}{7}x + \frac{20}{7}$$

$$m = -\frac{3}{7}$$

Perpendicular line has slope $m = \frac{7}{3}$.

30. $2x + 8y = 5$

$$8y = -2x + 5$$

$$y = -\frac{2}{8}x + \frac{5}{8}$$

$$y = \frac{1}{4}x + \frac{5}{8} \implies m = -\frac{1}{4} \implies parallel\ slope\ m = -\frac{1}{4}$$

Point: $(3, -6)$

Slope: $m = -\frac{1}{4}$

$$y - y_1 = m(x - x_1)$$

$$y - (-6) = -\frac{1}{4}(x - 3)$$

$$y + 6 = -\frac{1}{4}x + \frac{3}{4}$$

$$y = -\frac{1}{4}x - \frac{21}{4}$$

31. $(-2, 7)$ and $(5, 5)$

Slope: $m = \dfrac{y_2 - y_1}{x_2 - x_1} = \dfrac{5 - 7}{5 - (-2)} = \dfrac{-2}{7} = -\dfrac{2}{7}$

Point: $(-2, 7)$

$y - y_1 = m(x - x_1)$

$y - 7 = -\dfrac{2}{7}[x - (-2)]$

$y - 7 = -\dfrac{2}{7}(x + 2)$

$y - 7 = -\dfrac{2}{7}x - \dfrac{4}{7}$

$y = -\dfrac{2}{7}x - \dfrac{4}{7} + \dfrac{49}{7}$

$y = -\dfrac{2}{7}x + \dfrac{45}{7}$

32. $7 - 3x > 4 - x$

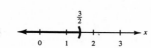

$7 - 2x > 4$

$-2x > -3$

$x < \dfrac{-3}{-2}$

$x < \dfrac{3}{2}$

33. $-1 < \dfrac{3x + 4}{5} < 5$

$5(-1) < 5\left(\dfrac{3x + 4}{5}\right) < 5(5)$

$-5 < 3x + 4 < 25$

$-9 < 3x < 21$

$-3 < x < 7$

34. $|x - 2| \geq 3$

$x - 2 \leq -3 \text{ or } x - 2 \geq 3$

$x \leq -1 \qquad x \geq 5$

35.

$R = C$

$x^2 - 36x = 232 + 18x$

$x^2 - 54x - 232 = 0$

$(x - 58)(x + 4) = 0$

$x - 58 = 0 \implies x = 58$

$x + 4 = 0 \implies x = -4$

(Discard the negative answer.) To break even, 58 units must be produced and sold.

36. $C = 5.75x + 12,000$

37. *Verbal Model:* | Portion done by first person | + | Portion done by second person | = | Work done |

Labels: First person: Rate $= \frac{1}{9}$ (job per hour); Time $= t$ (hours)

Second person: Rate $= \frac{1}{12}$ (job per hour); Time $= t$ (hours)

Work done $= \frac{1}{2}$ (job)

Equation: $\frac{1}{9}t + \frac{1}{12}t = \frac{1}{2}$

$36\left(\frac{1}{9}t + \frac{1}{12}t\right) = 36\left(\frac{1}{2}\right)$

$4t + 3t = 18$

$7t = 18$

$t = \dfrac{18}{7}$ or $t = 2\frac{4}{7}$

The two people working together will complete half of the task in $2\frac{4}{7}$ hours, or approximately 2 hours, 34 minutes.

38. $\dfrac{x}{13} = \dfrac{4.5}{9}$

$9x = 13(4.5)$

$9x = 58.5$

$x = \dfrac{58.5}{9}$

$x = 6.5$

The side of the triangle is 6.5 units.

39. *Verbal Model:* $\boxed{\begin{array}{c}\text{Interest from} \\ \text{7.5\% investment}\end{array}}$ $+$ $\boxed{\begin{array}{c}\text{Interest from} \\ \text{9\% investment}\end{array}}$ $=$ $\boxed{\begin{array}{c}\text{Total} \\ \text{interest}\end{array}}$

     Labels: 7.5% investment: Amount invested $= x$ (dollars)

                     Interest $= 0.075x$ (dollars)

           9% investment: Amount invested $= 24{,}000 - x$ (dollars)

                     Interest $= 0.09(24{,}000 - x)$ (dollars)

           Total interest $= 1935$

     Equation: $0.075x + 0.09(24{,}000 - x) = 1935$

$0.075x + 2160 - 0.09x = 1935$

$-0.015x + 2160 = 1935$

$-0.015x = -225$

$x = \dfrac{-225}{-0.015}$

$x = 15{,}000 \text{ and } 24{,}000 - x = 9000$

So, \$15,000 is invested in the 7.5% bond and \$9000 is invested in the 9% bond.

40. $|h - 50| \le 30$

$-30 \le h - 50 \le 30$

$20 \le h \le 80$

Minimum relative humidity: 20

Maximum relative humidity: 80

CHAPTER 4
Systems of Linear Equations and Inequalities

C H A P T E R   4
Systems of Linear Equations and Inequalities

Section 4.1   Systems of Linear Equations in Two Variables

Solutions to Odd-Numbered Exercises

1. $\begin{cases} x + 2y = 9 \\ -2x + 3y = 10 \end{cases}$

  (a) $(1, 4)$

$$1 + 2(4) \overset{?}{=} 9$$
$$1 + 8 = 9$$
$$-2(1) + 3(4) \overset{?}{=} 10$$
$$-2 + 12 = 10$$

  Yes, $(1, 4)$ *is* a solution.

  (b) $(3, -1)$

$$3 + 2(-1) \overset{?}{=} 9$$
$$3 - 2 \neq 9$$

  No, $(3, -1)$ is *not* a solution.

3. $\begin{cases} -2x + 7y = 46 \\ y = -3x \end{cases}$

  (a) $(-3, 2)$

$$-2(-3) + 7(2) = 46$$
$$6 + 14 \neq 46$$

  No, $(-3, 2)$ is *not* a solution.

  (b) $(-2, 6)$

$$-2(-2) + 7(6) \overset{?}{=} 46$$
$$4 + 42 = 46$$
$$6 \overset{?}{=} -3(-2)$$
$$6 = 6$$

  Yes, $(-2, 6)$ *is* a solution.

5. Because the lines are parallel, there is no point of intersection and, therefore, no solution.

7. The lines appear to intersect at $\left(1, \frac{1}{3}\right)$.

  Check:

$$5(1) - 3\left(\tfrac{1}{3}\right) \overset{?}{=} 4 \qquad 2(1) + 3\left(\tfrac{1}{3}\right) \overset{?}{=} 3$$
$$5 - 1 = 4 \qquad\qquad\quad 2 + 1 = 3$$

  The solution is $\left(1, \frac{1}{3}\right)$.

9. Because the lines coincide, there are infinitely many solutions. The solution set consists of all points (x, y) lying on the line $-x + 2y = 5$.

11. The solution is $(-1, -1)$.

13. The solution is $(10, 0)$.

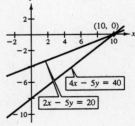

128

15. The solution is $(9, 12)$.

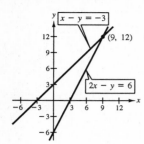

17. The solution is $(2, 1)$.

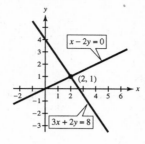

19. The solution is $(5, 3)$.

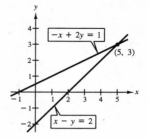

21. There is no solution.

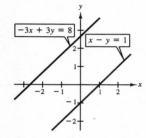

23. The solution is $\left(-\frac{3}{2}, \frac{5}{2}\right)$.

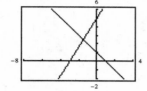

25. The solution is $(3, 5)$.

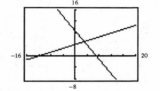

27. The solution is $\left(\frac{1}{2}, -1\right)$.

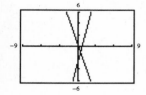

29.
$$4x - 5y = 3 \qquad\qquad -8x + 10y = -6$$
$$-5y = -4x + 3 \qquad\qquad 10y = 8x - 6$$
$$y = \tfrac{4}{5}x - \tfrac{3}{5} \qquad\qquad y = \tfrac{8}{10}x - \tfrac{6}{10}$$
$$\qquad\qquad\qquad\qquad = \tfrac{4}{5}x - \tfrac{3}{5}$$

The two equations are equivalent, so the two lines coincide and the system has infinitely many solutions. The system is consistent.

31.
$$-2x + 5y = 3 \qquad\qquad 5x + 2y = 8$$
$$5y = 2x + 3 \qquad\qquad 2y = -5x + 8$$
$$y = \tfrac{2}{5}x + \tfrac{3}{5} \qquad\qquad y = -\tfrac{5}{2}x + 4$$
$$m_1 = \tfrac{2}{5} \qquad\qquad m_2 = -\tfrac{5}{2}$$

$m_1 \neq m_2 \Longrightarrow$ The lines intersect. The system has one solution. The system is consistent.

33. The system is inconsistent.

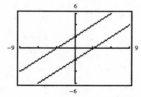

35. The system is consistent. The system has one solution.

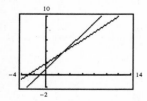

37.
$$\begin{cases} x - 100y = -200 \implies -3x + 300y = 600 \\ 3x - 275y = 198 \implies \underline{3x - 275y = 198} \end{cases}$$
$$25y = 798$$
$$y = \tfrac{798}{25}$$

$$x - 100\left(\tfrac{798}{25}\right) = -200$$
$$x - 4(798) = -200$$
$$x - 3192 = -200$$
$$x = 2992$$
$$\left(2992, \tfrac{798}{25}\right)$$

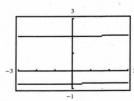

39. $\begin{cases} 3x - 25y = 50 \\ 9x - 100y = 50 \end{cases} \implies \begin{array}{r} -9x + 75y = -150 \\ 9x - 100y = 50 \\ \hline -25y = -100 \\ y = 4 \end{array}$

$3x - 25(4) = 50$

$3x - 100 = 50$

$3x = 150$

$x = 50$

$(50, 4)$

41. $\begin{cases} x = 4 \\ x - 2y = -2 \end{cases}$

$4 - 2y = -2$

$-2y = -6$

$y = 3 \quad \text{and} \quad x = 4$

$(4, 3)$

43. $\begin{cases} x + y = 3 \implies x = 3 - y \\ 2x - y = 0 \end{cases}$

$2x - y = 0$

$2(3 - y) - y = 0$

$6 - 2y - y = 0$

$6 - 3y = 0$

$-3y = -6$

$y = 2 \quad \text{and} \quad x = 3 - 2$

$x = 1$

$(1, 2)$

45. $\begin{cases} x + y = 2 \implies x = 2 - y \\ x - 4y = 12 \end{cases}$

$x - 4y = 12$

$(2 - y) - 4y = 12$

$2 - 5y = 12$

$-5y = 10$

$y = -2 \quad \text{and} \quad x = 2 - (-2) = 4$

$(4, -2)$

47. $\begin{cases} x + 6y = 19 \implies x = 19 - 6y \\ x - 7y = -7 \end{cases}$

$x - 7y = -7$

$(19 - 6y) - 7y = -7$

$19 - 13y = -7$

$-13y = -26$

$y = 2 \quad \text{and} \quad x = 19 - 6(2) = 7$

$(7, 2)$

49. $\begin{cases} 2x + 5y = 29 \implies 2x = -5y + 29 \implies x = -\dfrac{5}{2}y + \dfrac{29}{2} \\ 5x + 2y = 13 \end{cases}$

$5x + 2y = 13$

$5\left(-\dfrac{5}{2}y + \dfrac{29}{2}\right) + 2y = 13$

$-\dfrac{25}{2}y + \dfrac{145}{2} + 2y = 13$

$-25y + 145 + 4y = 26$

$-21y = -119$

$y = \dfrac{-119}{-21}$

$y = \dfrac{17}{3} \quad \text{and} \quad x = -\dfrac{5}{2}\left(\dfrac{17}{3}\right) + \dfrac{29}{2}$

$x = -\dfrac{85}{6} + \dfrac{87}{6}$

$x = \dfrac{1}{3}$

$\left(\dfrac{1}{3}, \dfrac{17}{3}\right)$

51. $\begin{cases} 4x - 14y = -15 \Rightarrow 4x = 14y - 15 \Rightarrow x = \frac{7}{2}y - \frac{15}{4} \\ 18x - 12y = \quad 9 \end{cases}$

$$18\left(\frac{7}{2}y - \frac{15}{4}\right) - 12y = 9$$

$$63y - \frac{135}{2} - 12y = 9$$

$$126y - 135 - 24y = 18$$

$$102y = 153$$

$$y = \frac{153}{102}$$

$$y = \frac{3}{2} \quad \text{and} \quad x = \frac{7}{2}\left(\frac{3}{2}\right) - \frac{15}{4}$$

$$x = \frac{21}{4} - \frac{15}{4}$$

$$x = \frac{6}{4}$$

$$x = \frac{3}{2}$$

$\left(\frac{3}{2}, \frac{3}{2}\right)$

53. $\begin{cases} 7x + 8y = 24 \\ x - 8y = \quad 8 \Rightarrow x = 8 + 8y \end{cases}$

$$7x + 8y = 24$$

$$7(8 + 8y) + 8y = 24$$

$$56 + 56y + 8y = 24$$

$$56 + 64y = 24$$

$$64y = -32$$

$$y = -\frac{32}{64}$$

$$y = -\frac{1}{2} \quad \text{and} \quad x = 8 + 8\left(-\frac{1}{2}\right)$$

$$x = 8 - 4 = 4$$

$\left(4, -\frac{1}{2}\right)$

55. $\begin{cases} 3x - 2y = \quad 5 \\ x + 2y = \quad 7 \end{cases}$

$$4x \qquad = 12$$

$$x \qquad = 3 \quad \text{and} \quad 3 + 2y = 7$$

$$2y = 4$$

$$y = 2$$

$(3, 2)$

57. $\begin{cases} 4x + \quad y = -3 \\ -4x + 3y = \quad 23 \end{cases}$

$$4y = \quad 20$$

$$y = \quad 5 \quad \text{and} \quad 4x + 5 = -3$$

$$4x = -8$$

$$x = -2$$

$(-2, 5)$

59. $\begin{cases} x - 3y = 2 \Rightarrow -3x + 9y = -6 \\ 3x - 7y = 4 \Rightarrow \quad 3x - 7y = \quad 4 \end{cases}$

$$2y = -2$$

$$y = -1 \quad \text{and} \quad x - 3(-1) = 2$$

$$x + 3 = 2$$

$$x = -1$$

$(-1, -1)$

61. $\begin{cases} 2u + 3v = \quad 8 \Rightarrow \quad 6u + 9v = \quad 24 \\ 3u + 4v = 13 \Rightarrow \quad -6u - 8v = -26 \end{cases}$

$$v = \quad -2 \quad \text{and} \quad 2u + 3(-2) = 8$$

$$2u - 6 = 8$$

$$2u = 14$$

$$u = 7$$

$(7, -2)$

63. $\begin{cases} 12x - 5y = 2 \implies \quad 24x - 10y = 4 \\ -24x + 10y = 6 \implies \underline{-24x + 10y = 6} \end{cases}$

$$0 = 10 \quad \text{False}$$

The system has *no* solution; it is inconsistent.

65. $\begin{cases} \frac{2}{3}r - s = 0 \implies \quad 10r - 15s = 0 \\ 10r + 4s = 19 \implies \underline{-10r - 4s = -19} \end{cases}$

$$-19s = -19$$

$$s = 1 \quad \text{and} \quad 10r - 15(1) = 0$$

$$10r = 15$$

$$r = \tfrac{15}{10} = \tfrac{3}{2}$$

$\left(\tfrac{3}{2}, 1\right)$

67. $\begin{cases} 0.7u - v = -0.4 \implies 7u - 10v = -4 \implies \quad 28u - 40v = -16 \\ 0.3u - 0.8v = 0.2 \implies 3u - 8v = 2 \implies \underline{-15u + 40v = -10} \end{cases}$

$$13u = -26$$

$$u = -2 \quad \text{and} \quad 7(-2) - 10v = -4$$

$$-14 - 10v = -4$$

$$-10v = 10$$

$$v = -1$$

$(-2, -1)$

69. $\begin{cases} 5x + 7y = 25 \implies -10x - 14y = -50 \\ x + 1.4y = 5 \implies \underline{10x + 14y = 50} \end{cases}$

$$0 = 0$$

The system has *infinitely* many solutions. The solution set consists of all ordered pairs (x, y) such that $5x + 7y = 25$.

71. $\begin{cases} 2x = 25 \implies \quad 4x = 50 \\ 4x - 10y = 52 \implies \underline{-4x + 10y = -0.52} \end{cases}$

$$10y = 49.48$$

$$y = 4.948 \quad \text{and} \quad 2x = 25$$

$$x = 12.5$$

$(12.5, 4.948)$

73. $\begin{cases} \frac{3}{2}x - y = 4 \implies \quad 3x - 2y = 8 \\ -x + \frac{2}{3}y = -1 \implies \underline{-3x + 2y = -3} \end{cases}$

$$0 = 5 \quad \text{False}$$

The system has *no* solution; it is inconsistent.

75. The solution is $(2, -3)$.

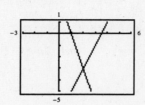

77. The solution is $(2, 7)$.

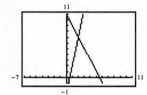

79. The solution is $(15, 10)$.

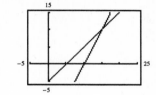

81. The solution is $(6, 11)$.

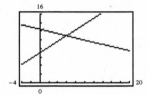

83. The solution is $(3{,}000, -2000)$.

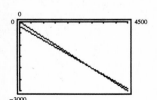

85. The solution is $\left(1, \frac{3}{2}\right)$.

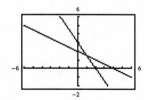

87. There is no solution.

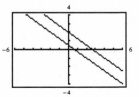

89. $\begin{cases} 5x - 10y = 40 \implies 10x - 20y = -80 \\ -2x + ky = 30 \implies \underline{-10x + 5ky = \ \ 150} \end{cases}$

When the two equations are added, the result is:

$$-20y + 5ky = 70$$

$$(-20 + 5k)y = 70$$

The system is inconsistent if:

$$-20 + 5k = 0$$

$$5k = 20$$

$$k = 4$$

If $k = 4$, the system is inconsistent.

91. $(5, 2)$

There are many correct answers. Here are some examples.

$\begin{aligned} 2(5) + 5(2) &= 20 \implies \\ 3(5) - 8(2) &= -1 \implies \end{aligned} \quad \begin{cases} 2x + 5y = 20 \\ 3x - 8y = -1 \end{cases}$

$\begin{aligned} -(5) + 3(2) &= 1 \implies \\ 4(5) - 9(2) &= 2 \implies \end{aligned} \quad \begin{cases} -x + 3y = 1 \\ 4x - 9y = 2 \end{cases}$

93. $\left(3, -\frac{3}{2}\right)$

There are many correct answers. Here are some examples.

$\begin{aligned} 1(3) + 2\left(-\tfrac{3}{2}\right) &= 0 \implies \\ 4(3) + 2\left(-\tfrac{3}{2}\right) &= 9 \implies \end{aligned} \quad \begin{cases} x + 2y = 0 \\ 4x + 2y = 9 \end{cases}$

$\begin{aligned} 3(3) - 2\left(-\tfrac{3}{2}\right) &= 12 \implies \\ 5(3) + 2\left(-\tfrac{3}{2}\right) &= 12 \implies \end{aligned} \quad \begin{cases} 3x - 2y = 12 \\ 5x + 2y = 12 \end{cases}$

95. $\begin{cases} 2l + 2w = 220 \\ l \ \ \ \ \ \ = 1.20w \end{cases}$

$$2(1.20w) + 2w = 220$$

$$2.40w + 2w = 220$$

$$4.40w = 220$$

$$w = 50 \quad \text{and} \quad l = 1.20(50)$$

$$l = 60$$

The length of the rectangle is 60 meters and the width is 50 meters.

97. *Verbal Model:* $\boxed{\text{Measure of first angle}} + \boxed{\text{Measure of second angle}} = \boxed{90}$

$\boxed{\text{Measure of first angle}} = 3\boxed{\text{Measure of second angle}} - \boxed{10}$

Labels: Measure of first angle $= x$ (degrees)

Measure of second angle $= y$ (degrees)

Equations: $\begin{cases} x + y = 90 \\ x = 3y - 10 \end{cases}$

$$(3y - 10) + y = 90$$

$$4y - 10 = 90$$

$$4y = 100$$

$$y = 25 \quad \text{and} \quad x = 3(25) - 10$$

$$x = 65$$

The measure of the first angle is 65 degrees and the measure of the second angle is 25 degrees.

99. *Verbal Model:* $12\boxed{\text{Price per gallon of regular gasoline}} + 8\boxed{\text{Price per gallon of premium gasoline}} = \boxed{\text{Total price}}$

$\boxed{\text{Price per gallon of premium gasoline}} = \boxed{\text{Price per gallon of regular gasoline}} + 0.11$

Labels: Price per gallon of regular gasoline $= x$ (dollars)

Price per gallon of premium gasoline $= y$ (dollars)

Total price $= 23.08$ (dollars)

Equations: $\begin{cases} 12x + 8y = 23.08 \\ y = x + 0.11 \end{cases}$

$$12x + 8y = 23.08$$

$$12x + 8(x + 0.11) = 23.08$$

$$12x + 8x + 0.88 = 23.08$$

$$20x + 0.88 = 23.08$$

$$20x = 22.20$$

$$x = 1.11 \quad \text{and} \quad y = 1.11 + 0.11$$

$$y = 1.22$$

$(1.11, 1.22)$

Regular unleaded gasoline is \$1.11 per gallon; premium unleaded gasoline is \$1.22 per gallon.

101. *Verbal Model:*

$$\boxed{\begin{array}{c}\text{Tons of}\\\$125\text{ hay}\end{array}} + \boxed{\begin{array}{c}\text{Tons of}\\\$75\text{ hay}\end{array}} = \boxed{\begin{array}{c}\text{Total}\\\text{tons}\end{array}}$$

$$\boxed{\begin{array}{c}\text{Value of}\\\$125\text{ hay}\end{array}} + \boxed{\begin{array}{c}\text{Value of}\\\$75\text{ hay}\end{array}} = \boxed{\begin{array}{c}\text{Total}\\\text{value}\end{array}}$$

Labels:

$125 hay: Number of tons $= x$, Value $= 125x$ (dollars)

$75 hay: Number of tons $= y$, Value $= 75y$ (dollars)

Totals: Total number of tons $= 100$, Total value $= 90(100)$ (dollars)

Equations:

$$\begin{cases} x + y = 100 \\ 125x + 75y = 90(100) \end{cases} \Longrightarrow \begin{array}{r} -125x - 125y = -12{,}500 \\ \underline{125x + 75y = 9000} \end{array}$$

$$\begin{aligned} -50y &= -3500 \\ y &= 70 \quad \text{and} \quad x + 70 = 100 \\ x &= 30 \end{aligned}$$

$(30, 70)$

Thus, 30 tons of the $125 per ton hay and 70 tons of the $75 per ton hay must be purchased.

103. *Verbal Model:*

$$\boxed{\begin{array}{c}\text{Number of}\\\text{adult tickets}\end{array}} + \boxed{\begin{array}{c}\text{Number of}\\\text{student tickets}\end{array}} = 800$$

$$\boxed{\begin{array}{c}\text{Receipts from}\\\text{adult tickets}\end{array}} + \boxed{\begin{array}{c}\text{Receipts from}\\\text{student tickets}\end{array}} = 8600$$

Labels:

Adult tickets: Number $= x$, Receipts $= 12.50x$ (dollars)

Student tickets: Number $= y$, Receipts $= 7.50y$ (dollars)

Equations:

$$\begin{cases} x + y = 800 \implies y = 800 - x \\ 12.50x + 7.50y = 8600 \end{cases}$$

$$\begin{aligned} 12.50x + 7.50y &= 8600 \\ 12.50x + 7.50(800 - x) &= 8600 \\ 12.50x + 6000 - 7.50x &= 8600 \\ 5x + 6000 &= 8600 \\ 5x &= 2600 \\ x &= 520 \quad \text{and} \quad y = 800 - 520 \\ y &= 280 \end{aligned}$$

$(520, 280)$

Thus, 520 adult tickets and 280 student tickets were sold.

105. *Verbal Model:*

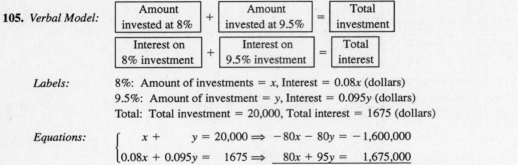

Labels: 8%: Amount of investments $= x$, Interest $= 0.08x$ (dollars)

9.5%: Amount of investment $= y$, Interest $= 0.095y$ (dollars)

Total: Total investment $= 20,000$, Total interest $= 1675$ (dollars)

Equations:

$$\begin{cases} x + y = 20,000 \Rightarrow -80x - 80y = -1,600,000 \\ 0.08x + 0.095y = 1675 \Rightarrow \underline{80x + 95y = 1,675,000} \end{cases}$$

$$15y = 75,000$$

$$y = 5000 \quad \text{and} \quad x + 5000 = 20,000$$

$$x = 15,000$$

(15,000, 5000)

Thus, \$15,000 is invested at 8% and \$5000 is invested at 9.5%.

107. $\begin{cases} C = 8000 + 1.20x \\ R = 2.00x \end{cases}$

$R = C$ (break-even point)

$2.00x = 8000 + 1.20x$

$0.80x = 8000$

$x = 10,000$

Thus, 10,000 items must be sold before the business breaks even.

109. *Verbal Model:*

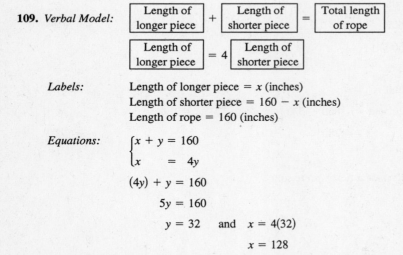

Labels: Length of longer piece $= x$ (inches)

Length of shorter piece $= 160 - x$ (inches)

Length of rope $= 160$ (inches)

Equations: $\begin{cases} x + y = 160 \\ x = 4y \end{cases}$

$(4y) + y = 160$

$5y = 160$

$y = 32 \quad \text{and} \quad x = 4(32)$

$x = 128$

One piece of rope is 128 inches long and the other piece is 32 inches long.

111. *Verbal Model:*

| Longer distance driven by a driver | + | Shorter distance driven by other driver | = | Total distance |

| Longer distance driven by a driver | = 4 · | Shorter distance driven by other driver |

Labels: Longer distance $= x$ (miles)

Shorter distance $= y$ (miles)

Total distance $= 300$ (miles)

Equations: $\begin{cases} x + y = 300 \\ x \quad\;\; = 3y \end{cases}$

$(3y) + y = 300$

$4y = 300$

$y = 75$ and $x = 3(75)$

$x = 225$

One driver drives a distance of 75 miles and the other drives a distance of 225 miles.

113. If the graphs of a system of equations do not intersect, the system has no solution.

115. No. The lines represented by a consistent system of linear equations either intersect at one point (one solution) or coincide (infinitely many solutions).

117.

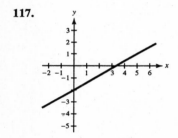

119.

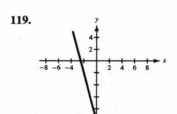

121. $5x + 2(x - 5) = 4$

$5x + 2x - 10 = 4$

$7x - 10 = 4$

$7x = 14$

$x = 2$

123. $(3, 7)$ and $(-1, -5)$

Point: $(3, 7)$

Slope: $m = \dfrac{-5 - 7}{-1 - 3} = \dfrac{-12}{-4} = 3$

$y - y_1 = m(x - x_1)$

$y - 7 = 3(x - 3)$

$y - 7 = 3x - 9$

$y = 3x - 2$

125. $C = 045m + 6200$
$C < 15{,}000$

$0.45m + 6200 < 15{,}000$

$0.45m < 8800$

$m < \dfrac{8800}{0.45}$

$m < 19{,}555$

The annual operating cost will be less than \$15,000 if the number of miles is 19,555 or less.

Section 4.2   Systems of Linear Equations in Three Variables

1. (a) $(0, 3, -2)$

$$0 + 3(3) + 2(-2) \overset{?}{=} 1$$
$$0 + 9 - 4 \neq 1$$

No, $(0, 3, -2)$ is not a solution.

(b) $(12, 5, -13)$

$$12 + 3(5) + 2(-13) \overset{?}{=} 1$$
$$12 + 15 - 26 = 1$$
$$5(12) - (5) + 3(-13) \overset{?}{=} 16$$
$$60 - 5 - 39 = 16$$
$$-3(12) + 7(5) + (-13) \overset{?}{=} -14$$
$$-36 + 35 - 13 = -14$$

Yes, $(12, 5, -13)$ is a solution.

(c) $(1, -2, 3)$

$$1 + 3(-2) + 2(3) \overset{?}{=} 1$$
$$1 - 6 + 6 = 1$$
$$5(1) - (-2) + 3(3) \overset{?}{=} 16$$
$$5 + 2 + 9 = 16$$
$$-3(1) + 7(-2) + (3) \overset{?}{=} -14$$
$$-3 - 14 + 3 = -14$$

Yes, $(1, -2, 3)$ is a solution.

(d) $(-2, 5, -3)$

$$-2 + 3(5) + 2(-3) \overset{?}{=} 1$$
$$-2 + 15 - 6 \neq 1$$

No, $(-2, 5, -3)$ is not a solution.

3. $\begin{cases} x - 2y + 4z = 4 \\ 3y - z = 2 \\ z = -5 \end{cases}$

$$3y - (-5) = 2 \qquad x - 2(-1) + 4(-5) = 4$$
$$3y + 5 = 2 \qquad x + 2 - 20 = 4$$
$$3y = -3 \qquad x - 18 = 4$$
$$y = -1 \qquad x = 22$$

$(22, -1, -5)$

5. $\begin{cases} x - 2y + 4z = 4 \\ y = 3 \\ y + z = 2 \end{cases}$

$$3 + z = 2 \qquad x - 2(3) + 4(-1) = 4$$
$$z = -1 \qquad x - 6 - 4 = 4$$
$$\qquad x - 10 = 4$$
$$\qquad x = 14$$

$(14, 3, -1)$

7. $\begin{cases} x + z = 4 \\ y = 2 \\ -3z = -9 \end{cases}$ (-4)Eqn. 1 + Eqn. 3

$\begin{cases} x + z = 4 \\ y = 2 \\ z = 3 \end{cases}$ $(-1/3)$Eqn. 3

$$x + 3 = 4$$
$$x = 1$$

$(1, 2, 3)$

9. $\begin{cases} x + y + z = 6 \\ -3y - z = -9 \\ -3y - 4z = -18 \end{cases}$ (-2)Eqn. 1 + Eqn. 2
(-3)Eqn. 1 + Eqn. 3

$\begin{cases} x + y + z = 6 \\ -3y - z = -9 \\ -3z = -9 \end{cases}$ (-1)Eqn. 2 + Eqn. 3

$\begin{cases} x + y + z = 6 \\ y + \frac{1}{3}z = 3 \\ z = 3 \end{cases}$ $(-1/3)$Eqn. 2
$(-1/3)$Eqn. 3

$$y + \frac{1}{3}(3) = 3 \qquad x + 2 + 3 = 6$$
$$y + 1 = 3 \qquad x + 5 = 6$$
$$y = 2 \qquad x = 1$$

$(1, 2, 3)$

11. $\begin{cases} x + y + z = -3 \\ \quad -3y - 7z = 23 \qquad (-4)\text{Eqn. 1} + \text{Eqn. 2} \\ \quad -5y \quad\;\; = 15 \qquad (-2)\text{Eqn. 1} + \text{Eqn. 3} \end{cases}$

$\begin{cases} x + y + z = -3 \\ \quad\; y \quad\;\; = -3 \qquad (-1/5)\text{Eqn. 3} \\ \quad -3y - 7z = 23 \qquad \text{Eqn. 2} \end{cases}$

$\begin{cases} x + y + z = -3 \\ \quad\; y \quad\;\; = -3 \\ \quad -7z = 14 \qquad (3)\text{Eqn. 2} + \text{Eqn. 3} \end{cases}$

$\begin{cases} x + y + z = -3 \\ \quad\; y \quad\;\; = -3 \\ \quad\quad\; z = -2 \qquad (-1/7)\text{Eqn. 3} \end{cases}$

$x + (-3) + (-2) = -3$

$\quad\quad\;\; x - 5 = -3$

$\quad\quad\quad\quad\; x = 2$

$(2, -3, -2)$

13. $\begin{cases} x + 2y + 6z = 5 \\ \quad\; 3y + 4z = 8 \qquad \text{Eqn. 1} + \text{Eqn. 2} \\ \quad -6y - 8z = -4 \qquad (-1)\text{Eqn. 1} + \text{Eqn. 3} \end{cases}$

$\begin{cases} x + 2y + 6z = 5 \\ \quad\; 3y + 4z = 8 \\ \quad\quad\quad 0 = 12 \qquad (2)\text{Eqn. 2} + \text{Eqn. 3} \end{cases}$

False

The system has *no* solution; it is inconsistent.

15. $\begin{cases} x \quad\;\; + z = 1 \qquad (1/2)\text{Eqn. 1} \\ 5x + 3y \quad\;\; = 4 \\ \quad\;\; 3y - 4z = 4 \end{cases}$

$\begin{cases} x \quad\; + z = 1 \\ \quad\; 3y - 5z = -1 \qquad (-5)\text{Eqn. 1} + \text{Eqn. 2} \\ \quad\; 3y - 4z = 4 \end{cases}$

$\begin{cases} x \quad\; + z = 1 \\ \quad\; y - \frac{5}{3}z = -\frac{1}{3} \qquad (1/3)\text{Eqn. 2} \\ \quad\; 3y - 4z = 4 \end{cases}$

$\begin{cases} x \quad\; + z = 1 \\ \quad\; y - \frac{5}{3}z = -\frac{1}{3} \\ \quad\quad\quad z = 5 \qquad (-3)\text{Eqn. 2} + \text{Eqn. 3} \end{cases}$

$y - \frac{5}{3}(5) = -\frac{1}{3} \qquad x + 5 = 1$

$\quad\; y - \frac{25}{3} = -\frac{1}{3} \qquad\quad x = -4$

$\quad\quad\quad y = \frac{24}{3} = 8$

$(-4, 8, 5)$

17. $\begin{cases} x + y + 8z = 3 \\ 2x + y + 11z = 4 \\ x \quad\;\; + 3z = 0 \end{cases}$

$\begin{cases} x + y + 8z = 3 \\ \quad -y - 5z = -2 \qquad (-2)\text{Eqn. 1} + \text{Eqn. 2} \\ \quad -y - 5z = -3 \qquad (-1)\text{Eqn. 1} + \text{Eqn. 2} \end{cases}$

$\begin{cases} x + y + 8z = 3 \\ \quad -y - 5z = -2 \\ \quad\quad\quad 0 = -1 \qquad (-1)\text{Eqn. 2} + \text{Eqn. 3} \end{cases}$

False

The system has no solution; it is inconsistent.

19. $\begin{cases} x \quad\;\; + 2z = \quad 2 \\ \quad\;\; y + z = \quad 5 \\ 2x - 3y \quad\;\; = -14 \end{cases}$ (1/2)Eqn. 2

             Eqn. 1

$\begin{cases} x \quad\;\; + 2z = \quad 2 \\ \quad\;\; y + z = \quad 5 \\ \quad -3y - 4z = -18 \end{cases}$ (−2)Eqn. 1 + Eqn. 3

$\begin{cases} x \quad\;\; + 2z = \quad 2 \\ \quad\;\; y + z = \quad 5 \\ \quad\quad\;\; -z = -3 \end{cases}$ (3)Eqn. 2 + Eqn. 3

$\begin{cases} x \quad\;\; + 2z = 2 \\ \quad\;\; y + z = 5 \\ \quad\quad\;\; z = 3 \end{cases}$ (−1)Eqn. 3

$\begin{aligned} y + 3 &= 5 \\ y &= 2 \end{aligned}$ $\begin{aligned} x + 2(3) &= 2 \\ x + 6 &= 2 \\ x &= -4 \end{aligned}$

$(-4, 2, 3)$

---

21. $\begin{cases} x - 2y + \frac{1}{2}z = \quad 0 \\ 3x \quad\;\; + 2z = -1 \\ -6x + 3y + 2z = -10 \end{cases}$ (1/2)Eqn. 1

$\begin{cases} x - 2y + \frac{1}{2}z = \quad 0 \\ \quad\;\; 6y + \frac{1}{2}z = -1 \\ \quad -9y + 5z = -10 \end{cases}$ (−3)Eqn. 1 + Eqn. 2

               (6)Eqn. 1 + Eqn. 3

$\begin{cases} x - 2y + \frac{1}{2}z = \quad 0 \\ \quad\;\; y + \frac{1}{12}z = -\frac{1}{6} \\ \quad -9y + 5z = -10 \end{cases}$ (1/6)Eqn. 2

$\begin{cases} x - 2y + \frac{1}{2}z = \quad 0 \\ \quad\;\; y + \frac{1}{12}z = -\frac{1}{6} \\ \quad\quad\;\; \frac{23}{4}z = -\frac{23}{2} \end{cases}$ (9)Eqn. 2 + Eqn. 3

$\begin{cases} x - 2y + \frac{1}{2}z = \quad 0 \\ \quad\;\; y + \frac{1}{12}z = -\frac{1}{6} \\ \quad\quad\;\; z = -2 \end{cases}$ (4/23)Eqn. 3

$\begin{aligned} y + \tfrac{1}{12}(-2) &= -\tfrac{1}{6} \\ y - \tfrac{1}{6} &= -\tfrac{1}{6} \\ y &= 0 \end{aligned}$ $\begin{aligned} x - 2(0) + \tfrac{1}{2}(-2) &= 0 \\ x - 1 &= 0 \\ x &= 1 \end{aligned}$

$(1, 0, -2)$

---

23. $\begin{cases} 2x + 6y - 4z = \quad 8 \\ 3x + 10y - 7z = \quad 12 \\ -2x - 6y + 5z = -3 \end{cases}$

$\begin{cases} \quad x + 3y - 2z = \quad 4 \\ 3x + 10y - 7z = \quad 12 \\ -2x - 6y + 5z = -3 \end{cases}$ (1/2)Eqn. 1

$\begin{cases} x + 3y - 2z = 4 \\ \quad\;\; y - z = 0 \\ \quad\quad\;\; z = 5 \end{cases}$ (−3)Eqn. 1 + Eqn. 2

               (2)Eqn. 1 + Eqn. 3

$\begin{aligned} y - 5 &= 0 \\ y &= 5 \end{aligned}$ $\begin{aligned} x + 3(5) - 2(5) &= 4 \\ x + 15 - 10 &= 4 \\ x + 5 &= 4 \\ x &= -1 \end{aligned}$

$(-1, 5, 5)$

---

25. $\begin{cases} x - 2y - z = \quad 3 \\ \quad\;\; 5y - z = -5 \\ \quad\;\; 10y - 2z = -10 \end{cases}$ (−2)Eqn. 1 + Eqn. 2

               (−1)Eqn. 1 + Eqn. 3

$\begin{cases} x - 2y - z = \quad 3 \\ \quad\;\; y - \frac{1}{5}z = -1 \\ \quad\;\; 10y - 2z = -10 \end{cases}$ (1/5)Eqn. 2

$\begin{cases} x - 2y - z = \quad 3 \\ \quad\;\; y - \frac{1}{5}z = -1 \\ \quad\quad\;\; 0 = \quad 0 \end{cases}$ (−10)Eqn. 2 + Eqn. 3

The system has *infinitely* many solutions. Let $z = a$.

$\begin{aligned} y - \tfrac{1}{5}a &= -1 \\ y &= \tfrac{1}{5}a - 1 \end{aligned}$ $\begin{aligned} x - 2\left(\tfrac{1}{5}a - 1\right) - a &= 3 \\ x - \tfrac{2}{5}a + 2 - \tfrac{5}{5}a &= 3 \\ x - \tfrac{7}{5}a &= 1 \\ x &= \tfrac{7}{5}a + 1 \end{aligned}$

$\left(\tfrac{7}{5}a + 1, \tfrac{1}{5}a - 1, a\right)$

Note: The answer could be written as $(7a + 8, a, 5a + 5)$ or $\left(a, \tfrac{1}{7}a - \tfrac{8}{7}, \tfrac{5}{7}a - \tfrac{5}{7}\right)$.

27. $\begin{cases} x - 2y + 4z = 12 \\ 2x - y + 5z = 18 \\ -x + 3y - 3z = -8 \end{cases}$

$\begin{cases} x - 2y + 4z = 12 \\ 3y - 3z = -6 \\ y + z = 4 \end{cases}$ $\quad$ (−2)Eqn. 1 + Eqn. 2

Eqn. 1 + Eqn. 3

$\begin{cases} x - 2y + 4z = 12 \\ y - z = -2 \\ y + z = 4 \end{cases}$ $\quad$ (1/3)Eqn. 2

$\begin{cases} x - 2y + 4z = 12 \\ y - z = -2 \\ 2z = 6 \end{cases}$ $\quad$ (−1)Eqn. 2 + Eqn. 3

$\begin{cases} x - 2y + 4z = 12 \\ y - z = -2 \\ z = 3 \end{cases}$ $\quad$ (1/2)Eqn. 3

$y - 3 = -2 \qquad x - 2(1) + 4(3) = 12$
$y = 1 \qquad\qquad x - 2 + 12 = 12$
$x + 10 = 12$
$x = 2$

$(2, 1, 3)$

29. $\begin{cases} 2x - y + 4z = 1 \\ x - y + 3z = 0 \\ -x + 2y - 4z = 2 \end{cases}$

$\begin{cases} x - y + 3z = 0 \\ 2x - y + 4z = 1 \\ -x + 2y - 4z = 2 \end{cases}$ $\quad$ Eqn. 2

Eqn. 1

$\begin{cases} x - y + 3z = 0 \\ y - 2z = 1 \\ y - z = 2 \end{cases}$ $\quad$ (−2)Eqn. 1 + Eqn. 2

Eqn. 1 + Eqn. 3

$\begin{cases} x - y + 3z = 0 \\ y - 2z = 1 \\ z = 1 \end{cases}$ $\quad$ (−1)Eqn. 2 + Eqn. 3

$y - 2(1) = 1 \qquad x - 3 + 3(1) = 0$
$y - 2 = 1 \qquad\qquad x = 0$
$y = 3$

$(0, 3, 1)$

31. $\begin{cases} 2x + z = 1 \\ 5y - 3z = 2 \\ 6x + 20y - 9z = 11 \end{cases}$

$\begin{cases} x + \frac{1}{2}z = \frac{1}{2} \\ 5y - 3z = 2 \\ 6x + 20y - 9z = 11 \end{cases}$ $\quad$ (1/2) Eqn. 1

$\begin{cases} x + \frac{1}{2}z = \frac{1}{2} \\ 5y - 3z = 2 \\ 20y - 12z = 8 \end{cases}$ $\quad$ (−6) Eqn. 1 + Eqn. 3

$\begin{cases} x + \frac{1}{2}z = \frac{1}{2} \\ 5y - 3z = 2 \\ 0 = 0 \end{cases}$ $\quad$ (−4) Eqn. 2 + Eqn. 3

$\begin{cases} x + \frac{1}{2}z = \frac{1}{2} \\ y - \frac{3}{5}z = \frac{2}{5} \\ 0 = 0 \end{cases}$ $\quad$ (1/5) Eqn. 2

The system has *infinitely* many solutions.

Let $z = a$.

$y - \frac{3}{5}a = \frac{2}{5} \qquad x + \frac{1}{2}a = \frac{1}{2}$
$y = \frac{3}{5}a + \frac{2}{5} \qquad x = -\frac{1}{2}a + \frac{1}{2}$

$\left(-\frac{1}{2}a + \frac{1}{2}, \frac{3}{5}a + \frac{2}{5}, a\right)$

33. $\begin{cases} 2x + 3z = 4 \\ 5x + y + z = 2 \\ 11x + 3y - 3z = 0 \end{cases}$

$\begin{cases} x + \frac{3}{2}z = 2 \\ 5x + y + z = 2 \\ 11x + 3y - 3z = 0 \end{cases}$ $\quad$ (1/2) Eqn. 1

$\begin{cases} x + \frac{3}{2}z = 2 \\ y - \frac{13}{2}z = -8 \\ 3y - \frac{39}{2}z = -22 \end{cases}$ $\quad$ (−5) Eqn. 1 + Eqn. 2

(−11) Eqn. 1 + Eqn. 3

$\begin{cases} x + \frac{3}{2}z = 2 \\ y - \frac{13}{2}z = -8 \\ 0 = 2 \end{cases}$ $\quad$ (−3) Eqn. 2 + Eqn. 3

False

The system has *no* solution; it is inconsistent.

35. $\begin{cases} \frac{1}{3}x + \frac{2}{3}y + 2z = -1 \\ x + 2y + \frac{3}{2}z = \frac{3}{2} \\ \frac{1}{2}x + 2y + \frac{12}{5}z = \frac{1}{10} \end{cases}$

$\begin{cases} x + 2y + 6z = -3 \\ 2x + 4y + 3z = 3 \\ 5x + 20y + 24z = 1 \end{cases}$ (3)Eqn. 1

      (2)Eqn. 2

      (10)Eqn. 3

$\begin{cases} x + 2y + 6z = -3 \\ \quad\quad\quad -9z = 9 \\ \quad 10y - 6z = 16 \end{cases}$ (−2)Eqn. 1 + Eqn. 2

      (−5)Eqn. 1 + Eqn. 3

$\begin{cases} x + 2y + 6z = -3 \\ \quad 10y - 6z = 16 \\ \quad\quad\quad\quad z = -1 \end{cases}$ Eqn. 3

      (−1/9)Eqn. 2

$10y - 6(-1) = 16$

$10y + 6 = 16$

$10y = 10$

$y = 1$

$(1, 1, -1)$

$x + 2(1) + 6(-1) = -3$

$x + 2 - 6 = -3$

$x - 4 = 1$

37. $\begin{cases} x \quad\quad + 3z = 7 \\ 2x + 13y + 6z = 1 \\ 2x + 10y + 8z = 8 \end{cases}$ (10)Eqn. 2

      (10)Eqn. 1

$\begin{cases} x \quad\quad + 3z = 7 \\ \quad 13y \quad\quad = -13 \\ \quad 10y + 2z = -6 \end{cases}$ (−2)Eqn. 1 + Eqn. 2

      (−2)Eqn. 1 + Eqn. 3

$\begin{cases} x \quad\quad + 3z = 7 \\ \quad y \quad\quad = -1 \\ \quad 10y + 2z = -6 \end{cases}$ (1/13)Eqn. 2

$\begin{cases} x \quad + 3z = 7 \\ \quad y \quad\quad = -1 \\ \quad\quad 2z = 4 \end{cases}$ (−10)Eqn. 2 + Eqn. 3

$\begin{cases} x \quad + 3z = 7 \\ \quad y \quad\quad = -1 \\ \quad\quad z = 2 \end{cases}$ (1/2)Eqn. 3

$x + 3(2) = 7$

$x + 6 = 7$

$x = 1$

$(1, -1, 2)$

39. There are many correct answers. Here are some examples.

$1(4) + 2(-3) - (2) = -4$

$\quad\quad 1(-3) + 2(2) = 1$

$3(4) + (-3) + 3(2) = 15$

$\begin{cases} x + 2y - z = -4 \\ \quad\quad y + 2z = 1 \\ 3x + y + 3z = 15 \end{cases}$

$1(4) - 1(-3) - 1(2) = 5$

$7(4) + 8(-3) + 5(2) = 14$

$3(4) + 1(-3) + 7(2) = 23$

$\begin{cases} x - y - z = 5 \\ 7x + 8y + 5z = 14 \\ 3x + y + 7z = 23 \end{cases}$

41. $(0, -4):\quad -4 = a(0)^2 + b(0) + c$

$(1, 1):\quad\quad 1 = a(1)^2 + b(1) + c$

$(2, 10):\quad\quad 10 = a(2)^2 + b(2) + c$

$\begin{cases} \quad\quad\quad\quad c = -4 \\ a + b + c = 1 \\ 4a + 2b + c = 10 \end{cases}$

$\begin{cases} a + b + c = 1 \\ 4a + 2b + c = 10 \\ \quad\quad\quad c = -4 \end{cases}$ Eqn. 2

      Eqn. 3

      Eqn. 1

$\begin{cases} a + b + c = 1 \\ \quad -2b - 3c = 6 \\ \quad\quad\quad c = -4 \end{cases}$ (−4)Eqn. 1 + Eqn. 2

$\begin{cases} a + b + c = 1 \\ \quad b + \frac{3}{2}c = -3 \\ \quad\quad\quad c = -4 \end{cases}$ (−1/2)Eqn. 2

$b + \frac{3}{2}(-4) = -3$ $a + 3 + (-4) = 1$

$b - 6 = -3$ $a - 1 = 1$

$b = 3$ $a = 2$

$(2, 3, -4)$

$y = 2x^2 + 3x - 4$

43. $(1, 0)$: $0 = a(1)^2 + b(1) + c$

$(2, -1)$: $-1 = a(2)^2 + b(2) + c$

$(3, 0)$: $0 = a(3)^2 + b(3) + c$

$$\begin{cases} a + b + c = 0 \\ 4a + 2b + c = -1 \\ 9a + 3b + c = 0 \end{cases}$$

$$\begin{cases} a + b + c = 0 \\ -2b - 3c = -1 \\ -6b - 8c = 0 \end{cases}$$ (-4)Eqn. 1 + Eqn. 2
  (-9)Eqn. 1 + Eqn. 3

$$\begin{cases} a + b + c = 0 \\ b + \frac{3}{2}c = \frac{1}{2} \\ -6b - 8c = 0 \end{cases}$$ $(-1/2)$Eqn. 2

$$\begin{cases} a + b + c = 0 \\ b + \frac{3}{2}c = \frac{1}{2} \\ c = 3 \end{cases}$$ (6)Eqn. 2 + Eqn. 3

$b + \frac{3}{2}(3) = \frac{1}{2}$ $a + (-4) + 3 = 0$

$b + \frac{9}{2} = \frac{1}{2}$ $a - 1 = 0$

$b = -\frac{8}{2}$ $a = 1$

$b = -4$

$(1, -4, 3)$

$y = x^2 - 4x + 3$

45. $(-1, -3)$: $-3 = a(-1)^2 + b(-1) + c$

$(1, 1)$: $1 = a(1)^2 + b(1) + c$

$(2, 0)$: $0 = a(2)^2 + b(2) + c$

$$\begin{cases} a - b + c = -3 \\ a + b + c = 1 \\ 4a + 2b + c = 0 \end{cases}$$

$$\begin{cases} a - b + c = -3 \\ 2b = 4 \\ 6b - 3c = 12 \end{cases}$$ (-1)Eqn. 1 + Eqn. 2
  (-4)Eqn. 1 + Eqn. 3

$$\begin{cases} a - b + c = -3 \\ b = 2 \\ 6b - 3c = 12 \end{cases}$$ $(1/2)$Eqn. 2

$$\begin{cases} a - b + c = -3 \\ b = 2 \\ -3c = 0 \end{cases}$$ (-6)Eqn. 2 + Eqn. 3

$$\begin{cases} a - b + c = -3 \\ b = 2 \\ c = 0 \end{cases}$$ $(-1/3)$Eqn. 3

$a - 2 + 0 = -3$

$a = -1$

$(-1, 2, 0)$

$y = -x^2 + 2x$

47. $(1, 128)$: $\quad 128 = \frac{1}{2}a(1)^2 + v_0(1) + s_0$

$\quad(2, 80)$: $\quad 80 = \frac{1}{2}a(2)^2 + v_0(2) + s_0$

$\quad(3, 0)$: $\quad\quad 0 = \frac{1}{2}a(3)^2 + v_0(3) + s_0$

$$\begin{cases} \frac{1}{2}a + v_0 + s_0 = 128 \\ 2a + 2v_0 + s_0 = 80 \\ \frac{9}{2}a + 3v_0 + s_0 = 0 \end{cases}$$

$$\begin{cases} a + 2v_0 + 2s_0 = 256 & \text{(2)Eqn. 1} \\ 2a + 2v_0 + s_0 = 80 \\ 9a + 6v_0 + 2s_0 = 0 & \text{(2)Eqn. 2} \end{cases}$$

$$\begin{cases} a + 2v_0 + 2s_0 = 256 \\ -2v_0 - 3s_0 = -432 & (-2)\text{Eqn. 1 + Eqn. 2} \\ -12v_0 - 16s_0 = -2304 & (-9)\text{Eqn. 1 + Eqn. 3} \end{cases}$$

$$\begin{cases} a + 2v_0 + 2s_0 = 256 \\ v_0 + \frac{3}{2}s_0 = 216 & (-1/2)\text{Eqn. 2} \\ -12v_0 - 16s_0 = -2304 \end{cases}$$

$$\begin{cases} a + 2v_0 + 2s_0 = 256 \\ v_0 + \frac{3}{2}s_0 = 216 \\ 2s_0 = 288 & (12)\text{Eqn. 2 + Eqn. 3} \end{cases}$$

$$\begin{cases} a + 2v_0 + 2s_0 = 256 \\ v_0 + \frac{3}{2}s_0 = 216 \\ s_0 = 144 & (1/2)\text{Eqn. 3} \end{cases}$$

$v_0 + \frac{3}{2}(144) = 216 \quad\quad a + 2(0) + 2(144) = 256$

$v_0 + 216 = 216 \quad\quad\quad a + 288 = 256$

$v_0 = 0 \quad\quad\quad\quad\quad\quad a = -32$

$(-32, 0, 144)$

$s = \frac{1}{2}(-32)t^2 + 0t + 144$

$s = -16t^2 + 144$

49. $(1, 32)$: $\quad 32 = \frac{1}{2}a(1)^2 + v_0(1) + s_0$

$\quad(2, 32)$: $\quad 32 = \frac{1}{2}a(2)^2 + v_0(2) + s_0$

$\quad(3, 0)$: $\quad\quad 0 = \frac{1}{2}a(3)^2 + v_0(3) + s_0$

$$\begin{cases} \frac{1}{2}a + v_0 + s_0 = 32 \\ 2a + 2v_0 + s_0 = 32 \\ \frac{9}{2}a + 3v_0 + s_0 = 0 \end{cases}$$

$$\begin{cases} a + 2v_0 + 2s_0 = 64 & \text{(2)Eqn. 1} \\ 2a + 2v_0 + s_0 = 32 \\ 9a + 6v_0 + 2s_0 = 0 & \text{(2)Eqn. 3} \end{cases}$$

$$\begin{cases} a + 2v_0 + 2s_0 = 64 \\ -2s_0 - 3s_0 = -96 & (-2)\text{Eqn. 1 + Eqn. 2} \\ -12s_0 - 16s_0 = -576 & (-9)\text{Eqn. 1 + Eqn. 3} \end{cases}$$

$$\begin{cases} a + 2v_0 + 2s_0 = 64 \\ v_0 + \frac{3}{2}s_0 = 48 & (-1/2)\text{Eqn. 2} \\ -12v_0 - 16s_0 = -576 \end{cases}$$

$$\begin{cases} a + 2v_0 + 2s_0 = 64 \\ v_0 + \frac{3}{2}s_0 = 48 \\ 2s_0 = 0 & (12)\text{Eqn. 2 + Eqn. 3} \end{cases}$$

$$\begin{cases} a + 2v_0 + 2s_0 = 64 \\ v_0 + \frac{3}{2}s_0 = 48 \\ s_0 = 0 & (1/2)\text{Eqn. 3} \end{cases}$$

$v_0 + \frac{3}{2}(0) = 48 \quad\quad a + 2(48) + 2(0) = 64$

$v_0 = 48 \quad\quad\quad\quad a + 96 = 64$

$a = -32$

$(-32, 48, 0)$

$s = \frac{1}{2}(-32)t^2 + 48t + 0 = -16t^2 + 48t$

51. Labels: Amount invested at 5% = x (dollars)

　　　　Amount invested at 6% = y (dollars)

　　　　Amount invested at 7% = z (dollars)

$$\begin{cases} x + y + z = 16{,}000 \implies x + y + z = 16{,}000 \\ 0.05x + 0.06y + 0.07z = 940 \implies 5x + 6y + 7z = 94{,}000 \\ y = x - 3000 \implies -x + y = -3000 \end{cases}$$

$$\begin{cases} x + y + z = 16{,}000 \\ y + 2z = 14{,}000 \quad (-5)\text{Eqn. 1} + \text{Eqn. 2} \\ 2y + z = 13{,}000 \quad \text{Eqn. 1} + \text{Eqn. 3} \end{cases}$$

$$\begin{cases} x + y + z = 16{,}000 \\ y + 2z = 14{,}000 \\ -3z = -15{,}000 \quad (-2)\text{Eqn. 2} + \text{Eqn. 3} \end{cases}$$

$$\begin{cases} x + y + z = 16{,}000 \\ y + 2z = 14{,}000 \\ z = 5000 \quad (-1/3)\text{Eqn. 3} \end{cases}$$

$$y + 2(5000) = 14{,}000 \qquad x + 4000 + 5000 = 16{,}000$$

$$y + 10{,}000 = 14{,}000 \qquad x + 9000 = 16{,}000$$

$$y = 4000 \qquad\qquad x = 7000$$

So, \$7000 is invested at 5%, \$4000 is invested at 6%, and \$5000 is invested at 7%.

53.

	X	Y	Z	Gal
A	20%		50%	12
B	40%		50%	16
C	40%	100%		26

$$\begin{cases} 0.20x + 0.50z = 12 \\ 0.40x + 0.50z = 16 \\ 0.40x + y = 26 \end{cases}$$

$$\begin{cases} 20x + 50z = 1200 \quad (100)\text{Eqn. 1} \\ 40x + 50z = 1600 \quad (100)\text{Eqn. 2} \\ 40x + 100y = 2600 \quad (100)\text{Eqn. 3} \end{cases}$$

$$\begin{cases} x + \tfrac{5}{2}z = 60 \quad (1/20)\text{Eqn. 1} \\ -50z = -800 \quad (-2)\text{Eqn. 1} + \text{Eqn. 2} \\ 100y - 100z = 200 \quad (-2)\text{Eqn. 1} + \text{Eqn. 3} \end{cases}$$

$$\begin{cases} x + \tfrac{5}{2}z = 60 \\ y - z = 2 \quad (1/100)\text{Eqn. 3} \\ z = 16 \quad (1/50)\text{Eqn. 2} \end{cases}$$

$$y - 16 = 2 \qquad x + \tfrac{5}{2}(16) = 60$$

$$y = 18 \qquad\qquad x + 40 = 60$$

$$ \qquad\qquad x = 20$$

$(20, 18, 16)$

Thus, 20 gallons of spray X, 18 gallons of spray Y, and 16 gallons of spray Z are needed.

55. Labels: Number of members of defense $= x$
Number of members of offense $= y$
Number of members of special teams $= z$

$$\begin{cases} \quad\ 0.10y + 0.20z = \ 5 \\ 0.10x + 0.20y + 0.50z = 13 \\ 0.10x \quad\quad\ + 0.20z = \ 4 \end{cases}$$

$$\begin{cases} 0.10x + 0.20y + 0.50z = 13 \quad\ \text{Eqn. 2} \\ \quad\quad\ + 0.10y + 0.20z = \ 5 \quad\ \text{Eqn. 1} \\ 0.10x \quad\quad\ + 0.20z = \ 4 \end{cases}$$

$$\begin{cases} 0.10x + 0.20y + 0.50z = \ 13 \\ \quad\quad\ 0.10y + 0.20z = \ \ 5 \\ \quad\quad\ -0.20y - 0.30z = -9 \quad (-1)\text{Eqn. 1 + Eqn. 3} \end{cases}$$

$$\begin{cases} 0.10x + 0.20y + 0.50z = 13 \\ \quad\quad\ 0.10y + 0.20z = \ 5 \\ \quad\quad\quad\quad\quad 0.10z = \ 1 \quad (2)\text{Eqn. 2 + Eqn. 3} \end{cases}$$

$$\begin{cases} 0.10x + 0.20y + 0.50z = 13 \\ \quad\quad\ 0.10y + 0.20z = \ 5 \\ \quad\quad\quad\quad\quad\quad\ z = 10 \quad (10)\text{Eqn. 3} \end{cases}$$

$0.10y + 0.20(10) = 5 \quad\quad\quad 0.10x + 0.20(30) + 0.50(10) = 13$

$0.10y + 2 = 5 \quad\quad\quad\quad\quad\quad 0.10x + 6 + 5 = 13$

$0.10y = 3 \quad\quad\quad\quad\quad\quad\quad\quad 0.10x = 2$

$y = 30 \quad\quad\quad\quad\quad\quad\quad\quad\quad x = 20$

So, there are 20 members of the defense, 30 members of the offense, and 10 members of the special teams.

57. Labels: Measure of first angle $= x$ (degrees)
Measure of second angle $= y$ (degrees)
Measure of third angle $= z$ (degrees)

$$\begin{cases} x + y + z = \quad\ 180 \Rightarrow x + y + \ z = 180 \\ x + y \quad\quad = \quad\quad 2z \Rightarrow x + y - 2z = \quad 0 \\ \quad\quad\quad\quad y = z - 28 \Rightarrow \quad\quad y - \ z = \ \ 28 \end{cases}$$

$$\begin{cases} x + y + \ z = \ 180 \\ \quad\quad\ - 3z = -180 \quad (-1)\text{Eqn. 1 + Eqn. 2} \\ \quad\quad y - \ z = \ -28 \end{cases}$$

$$\begin{cases} x + y + z = \ 180 \\ \quad\quad y - z = -28 \quad\ \text{Eqn. 3} \\ \quad\quad\quad\quad z = \ \ 60 \quad\ (-1/3)\text{Eqn. 2} \end{cases}$$

$y - 60 = -28 \quad\quad\quad y + 32 + 60 = 180$

$y = 32 \quad\quad\quad\quad\quad\quad y + 92 = 180$

$\quad\quad\quad\quad\quad\quad\quad\quad\quad\quad\quad\quad x = 88$

So, the measure of the first angle is 88 degrees, the measure of the second angle is 32 degrees, and the measure of the third angle is 60 degrees.

59.
$$\begin{cases} 4A - 2B + C = \ \ 0 \\ -4A + \ B \quad\quad = -9 \\ \quad A \quad\quad\quad = \ \ 2 \end{cases}$$

$$\begin{cases} 4A - 2B + C = \ \ 0 \\ \quad\quad - \ B + C = -9 \quad\ \text{Eqn. 1 + Eqn. 2} \\ \ A \quad\quad\quad\quad = \ \ 2 \end{cases}$$

$$\begin{cases} A \quad\quad\quad\quad = \ \ 2 \quad\ \text{Eqn. 3} \\ \quad - \ B + C = -9 \\ 4A - 2B + C = \ \ 0 \quad\ \text{Eqn. 1} \end{cases}$$

$$\begin{cases} A \quad\quad\quad\quad = \ \ 2 \\ \quad - \ B + C = -9 \\ \quad - 2B + C = -8 \quad (-4)\text{Eqn. 1 + Eqn. 3} \end{cases}$$

$$\begin{cases} A \quad\quad\quad\quad = \ \ 2 \\ \quad - \ B + C = -9 \\ \quad\quad\quad - \ C = \ 10 \quad (-2)\text{Eqn. 2 + Eqn. 3} \end{cases}$$

$$\begin{cases} A \quad\quad\quad\quad = \ \ 2 \\ \quad - \ B + C = \ -9 \\ \quad\quad\quad\quad C = -10 \quad (-1)\text{Eqn. 3} \end{cases}$$

$-B - 10 = -9 \quad\quad\quad A = 2$

$-B = 1$

$B = -1$

$$\frac{2x^2 - 9x}{(x-2)^3} = \frac{2}{x-2} - \frac{1}{(x-2)^2} - \frac{10}{(x-2)^3}$$

61. $y = ax^2 + bx + c$

$(3, 3) : 3 = a(3)^2 + b(3) + c$

$(4, 6) : 6 = a(4)^2 + b(4) + c$

$(5, 10) : 10 = a(5)^2 + b(5) + c$

$$\begin{cases} 9a + 3b + c = 3 \\ 16a + 4b + c = 6 \\ 25a + 5b + c = 10 \end{cases}$$

$$\begin{cases} a + \frac{1}{3}b + \frac{1}{9}c = \frac{1}{3} & \text{(1/9) Eqn. 1} \\ 16a + 4b + c = 6 \\ 25a + 5b + c = 10 \end{cases}$$

$$\begin{cases} a + \frac{1}{3}b + \frac{1}{9}c = \frac{1}{3} \\ -\frac{4}{3}b - \frac{7}{9}c = \frac{2}{3} & \text{(}-16\text{) Eqn. 1 + Eqn. 2} \\ -\frac{10}{3}b - \frac{16}{9}c = \frac{5}{3} & \text{(}-25\text{) Eqn. 1 + Eqn. 3} \end{cases}$$

$$\begin{cases} a + \frac{1}{3}b + \frac{1}{9}c = \frac{1}{3} \\ b + \frac{7}{12}c = -\frac{1}{2} & \text{(}-3/4\text{) Eqn. 2} \\ -\frac{10}{3}b - \frac{16}{9}c = \frac{5}{3} \end{cases}$$

$$\begin{cases} a + \frac{1}{3}b + \frac{1}{9}c = \frac{1}{3} \\ b + \frac{7}{12}c = -\frac{1}{2} \\ \frac{1}{6}c = 0 & \text{(10/3) Eqn. 2 + Eqn. 3} \end{cases}$$

$$\begin{cases} a + \frac{1}{3}b + \frac{1}{9}c = \frac{1}{3} \\ b + \frac{7}{12}c = -\frac{1}{2} \\ c = 0 & \text{(6) Eqn. 3} \end{cases}$$

$b + \frac{7}{12}(0) = -\frac{1}{2}$ $a + \left(\frac{1}{3}\right)\left(-\frac{1}{2}\right) + \frac{1}{9}(0) = \frac{1}{3}$

$\quad\quad\quad b = -\frac{1}{2}$ $a - \frac{1}{6} = \frac{1}{3}$

$\quad\quad\quad\quad\quad\quad\quad\quad\quad\quad\quad\quad\quad\quad a = \frac{1}{6} + \frac{2}{6}$

$\quad\quad\quad\quad\quad\quad\quad\quad\quad\quad\quad\quad\quad\quad a = \frac{1}{2}$

$\left(\frac{1}{2}, -\frac{1}{2}, 0\right)$

$y = \frac{1}{2}x^2 - \frac{1}{2}x$

Yes, the equation gives the correct answer for a polygon with six sides.

63. No, the two systems are not equivalent. The constant in the second equation in the second system is incorrect; it should be -11. When the first equation is multiplied by -2 and added to the second equation, the result should be $-7y + 4z = -11$.

65.
$$\begin{cases} x - 2y + 3z = 5 \\ -x + 3y - 5z = 4 \\ 2x \quad\quad - 3z = 0 \end{cases}$$

$$\begin{cases} x - 2y + 3z = 5 \\ y - 2z = 9 \\ 2x \quad\quad - 3z = 0 \end{cases}$$

This step eliminates the x-term in the second equation.

67. $4(6) - 2(5) + 7(-10) = 24 - 10 - 70$

$\quad\quad\quad\quad\quad\quad\quad\quad\quad\quad\quad = -56$

69. $\dfrac{5(-1) + 3(3) - 10(5)}{-6(7) + 5(-2) - 6(3)} = \dfrac{-5 + 9 - 50}{-42 - 10 - 18}$

$\qquad\qquad\qquad\qquad = \dfrac{-46}{-70}$

$\qquad\qquad\qquad\qquad = \dfrac{23}{35}$

71. $(5x - 3y + 10z) + 2(2x + y - 5z) = 5x - 3y + 10z + 4x + 2y - 10z$

$\qquad\qquad\qquad\qquad\qquad\qquad = 9x - y$

73. $2(3x - 12y + 9z) - 3(2x + 2y + 5z) = 6x - 24y + 18z - 6x - 6y - 15z$

$\qquad\qquad\qquad\qquad\qquad\qquad = -30y + 3z$

75. $40(2) + 55t = 50(t + 2)$

$\qquad 80 + 55t = 50t + 100$

$\qquad 80 + 5t = 100$

$\qquad\qquad 5t = 20$

$\qquad\qquad\ t = 4$ hours

Section 4.3   Matrices and Linear Systems

1. 3 rows, 2 columns, 3×2

3. 2 rows, 3 columns, 2×3

5. 3 rows, 4 columns, 3×4

7. $\begin{bmatrix} 4 & -5 & \vdots & -2 \\ -1 & 8 & \vdots & 10 \end{bmatrix}$

9. $\begin{bmatrix} 1 & 10 & -3 & \vdots & 2 \\ 5 & -3 & 4 & \vdots & 0 \\ 2 & 4 & 0 & \vdots & 6 \end{bmatrix}$

11. $\begin{bmatrix} 4 & 13 & 0 & \vdots & -2 \\ 7 & -6 & 1 & \vdots & 0 \end{bmatrix}$

13. $\begin{cases} 4x + 3y = 8 \\ x - 2y = 3 \end{cases}$

15. $\begin{cases} x \qquad\ + 2z = -10 \\ \quad 3y - z = 5 \\ 4x + 2y \qquad = 3 \end{cases}$

17. $\begin{cases} 15x - 4y + 12z + 3w = 6 \\ -x - 3y \qquad\quad + 10w = 14 \\ 9x + 2y - 7z \qquad = 5 \end{cases}$

19. $\begin{bmatrix} 1 & 4 & 3 \\ 2 & 10 & 5 \end{bmatrix}$

$-2R_1 + R_2 \begin{bmatrix} 1 & 4 & 3 \\ 0 & 2 & -1 \end{bmatrix}$

21. $\begin{bmatrix} 1 & 1 & 4 & -1 \\ 3 & 8 & 10 & 3 \\ -2 & 1 & 12 & 6 \end{bmatrix}$

$\begin{matrix} -3R_1 + R_2 \\ 2R_1 + R_3 \end{matrix} \begin{bmatrix} 1 & 1 & 4 & -1 \\ 0 & 5 & -2 & 6 \\ 0 & 3 & 20 & 4 \end{bmatrix}$

$\tfrac{1}{5}R_2 \begin{bmatrix} 1 & 1 & 4 & -1 \\ 0 & 1 & -\tfrac{2}{5} & \tfrac{6}{5} \\ 0 & 3 & 20 & 4 \end{bmatrix}$

23. $\begin{bmatrix} 1 & 2 & 3 \\ 2 & -1 & -4 \end{bmatrix}$

$-2R_1 + R_2 \begin{bmatrix} 1 & 2 & 3 \\ 0 & -5 & -10 \end{bmatrix}$

$-\tfrac{1}{5}R_2 \begin{bmatrix} 1 & 2 & 3 \\ 0 & 1 & 2 \end{bmatrix}$

25.
$$\begin{bmatrix} 4 & 6 & 1 \\ -2 & 2 & 5 \end{bmatrix}$$

$\frac{1}{4}R_1 \begin{bmatrix} 1 & \frac{3}{2} & \frac{1}{4} \\ -2 & 2 & 5 \end{bmatrix}$

$2R_1 + R_2 \begin{bmatrix} 1 & \frac{3}{2} & \frac{1}{4} \\ 0 & 5 & \frac{11}{2} \end{bmatrix}$

$\frac{1}{5}R_2 \begin{bmatrix} 1 & \frac{3}{2} & \frac{1}{4} \\ 0 & 1 & \frac{11}{10} \end{bmatrix}$

27.
$$\begin{bmatrix} 1 & 1 & 0 & 5 \\ -2 & -1 & 2 & -10 \\ 3 & 6 & 7 & 14 \end{bmatrix}$$

$\begin{matrix} \\ 2R_1 + R_2 \\ -3R_1 + R_3 \end{matrix} \begin{bmatrix} 1 & 1 & 0 & 5 \\ 0 & 1 & 2 & 0 \\ 0 & 3 & 7 & -1 \end{bmatrix}$

$\begin{matrix} \\ \\ -3R_2 + R_3 \end{matrix} \begin{bmatrix} 1 & 1 & 0 & 5 \\ 0 & 1 & 2 & 0 \\ 0 & 0 & 1 & -1 \end{bmatrix}$

29.
$$\begin{bmatrix} 1 & -1 & -1 & 1 \\ 4 & -4 & 1 & 8 \\ -6 & 8 & 18 & 0 \end{bmatrix}$$

$\begin{matrix} \\ -4R_1 + R_2 \\ 6R_1 + R_3 \end{matrix} \begin{bmatrix} 1 & -1 & -1 & 1 \\ 0 & 0 & 5 & 4 \\ 0 & 2 & 12 & 6 \end{bmatrix}$

$\begin{matrix} \\ R_3 \\ R_2 \end{matrix} \begin{bmatrix} 1 & -1 & -1 & 1 \\ 0 & 2 & 12 & 6 \\ 0 & 0 & 5 & 4 \end{bmatrix}$

$\begin{matrix} \\ \frac{1}{2}R_2 \\ \frac{1}{5}R_3 \end{matrix} \begin{bmatrix} 1 & -1 & -1 & 1 \\ 0 & 1 & 6 & 3 \\ 0 & 0 & 1 & \frac{4}{5} \end{bmatrix}$

31.
$$\begin{bmatrix} 1 & 1 & -1 & 3 \\ 2 & 1 & 2 & 5 \\ 3 & 2 & 1 & 8 \end{bmatrix}$$

$\begin{matrix} \\ -2R_1 + R_2 \\ -3R_1 + R_3 \end{matrix} \begin{bmatrix} 1 & 1 & -1 & 3 \\ 0 & -1 & 4 & -1 \\ 0 & -1 & 4 & -1 \end{bmatrix}$

$\begin{matrix} \\ -1R_2 \\ \\ \end{matrix} \begin{bmatrix} 1 & 1 & -1 & 3 \\ 0 & 1 & -4 & 1 \\ 0 & -1 & 4 & -1 \end{bmatrix}$

$\begin{matrix} \\ \\ R_2 + R_3 \end{matrix} \begin{bmatrix} 1 & 1 & -1 & 3 \\ 0 & 1 & -4 & 1 \\ 0 & 0 & 0 & 0 \end{bmatrix}$

33. $\begin{cases} x - 2y = 4 \\ y = -3 \end{cases}$

$x - 2(-3) = 4$

$x + 6 = 4$

$x = -2$

$(-2, -3)$

35. $\begin{cases} x - y + 2z = 4 \\ \phantom{x - {}}y - z = 2 \\ z = -2 \end{cases}$

$\begin{aligned} y - (-2) &= 2 \\ y + 2 &= 2 \\ y &= 0 \end{aligned}$ $\qquad \begin{aligned} x - 0 + 2(-2) &= 4 \\ x - 4 &= 4 \\ x &= 8 \end{aligned}$

$(8, 0, -2)$

37.
$$\begin{bmatrix} 1 & 2 & \vdots & 7 \\ 3 & 1 & \vdots & 11 \end{bmatrix}$$

$-3R_1 + R_2 \begin{bmatrix} 1 & 2 & \vdots & 7 \\ 0 & -5 & \vdots & -10 \end{bmatrix}$

$-\frac{1}{5}R_2 \begin{bmatrix} 1 & 2 & \vdots & 7 \\ 0 & 1 & \vdots & 2 \end{bmatrix}$

$\begin{cases} x + 2y = 7 \\ y = 2 \end{cases}$

$x + 2(2) = 7$

$x = 3$

$(3, 2)$

39.
$$\begin{bmatrix} 6 & -4 & \vdots & 2 \\ 5 & 2 & \vdots & 7 \end{bmatrix}$$

$$\tfrac{1}{6}R_1 \begin{bmatrix} 1 & -\tfrac{2}{3} & \vdots & \tfrac{1}{3} \\ 5 & 2 & \vdots & 7 \end{bmatrix}$$

$$-5R_1 + R_2 \begin{bmatrix} 1 & -\tfrac{2}{3} & \vdots & \tfrac{1}{3} \\ 0 & \tfrac{16}{3} & \vdots & \tfrac{16}{3} \end{bmatrix}$$

$$\tfrac{3}{16}R_2 \begin{bmatrix} 1 & -\tfrac{2}{3} & \vdots & \tfrac{1}{3} \\ 0 & 1 & \vdots & 1 \end{bmatrix}$$

$$\begin{cases} x - \tfrac{2}{3}y = \tfrac{1}{3} \\ \qquad y = 1 \end{cases}$$

$$x - \tfrac{2}{3}(1) = \tfrac{1}{3}$$
$$x = 1$$

$$(1, 1)$$

41.
$$\begin{bmatrix} 1 & -2 & \vdots & 5 \\ 3 & 1 & \vdots & 1 \end{bmatrix}$$

$$-3R_1 + R_2 \begin{bmatrix} 1 & -2 & \vdots & 5 \\ 0 & 7 & \vdots & -14 \end{bmatrix}$$

$$-\tfrac{1}{7}R_2 \begin{bmatrix} 1 & -2 & \vdots & 5 \\ 0 & 1 & \vdots & -2 \end{bmatrix}$$

$$\begin{cases} x - 2y = 5 \\ \qquad y = -2 \end{cases}$$

$$x - 2(-2) = 5$$
$$x + 4 = 5$$
$$x = 1$$

$$(1, -2)$$

43.
$$\begin{bmatrix} -1 & 2 & \vdots & 1.5 \\ 2 & -4 & \vdots & 3 \end{bmatrix}$$

$$-R_1 \begin{bmatrix} 1 & -2 & \vdots & -1.5 \\ 2 & -4 & \vdots & 3 \end{bmatrix}$$

$$-2R_1 + R_2 \begin{bmatrix} 1 & -2 & \vdots & -1.5 \\ 0 & 0 & \vdots & 6 \end{bmatrix}$$

$$\begin{cases} x - 2y = -1.5 \\ \qquad 0 = 6 \quad \text{False} \end{cases}$$

The system has *no* solution; it is inconsistent.

45.
$$\begin{bmatrix} 1 & -2 & -1 & \vdots & 6 \\ 0 & 1 & 4 & \vdots & 5 \\ 4 & 2 & 3 & \vdots & 8 \end{bmatrix}$$

$$-4R_1 + R_3 \begin{bmatrix} 1 & -2 & -1 & \vdots & 6 \\ 0 & 1 & 4 & \vdots & 5 \\ 0 & 10 & 7 & \vdots & -16 \end{bmatrix}$$

$$-10R_2 + R_3 \begin{bmatrix} 1 & -2 & -1 & \vdots & 6 \\ 0 & 1 & 4 & \vdots & 5 \\ 0 & 0 & -33 & \vdots & -66 \end{bmatrix}$$

$$-\tfrac{1}{33}R_3 \begin{bmatrix} 1 & -2 & -1 & \vdots & 6 \\ 0 & 1 & 4 & \vdots & 5 \\ 0 & 0 & 1 & \vdots & 2 \end{bmatrix}$$

$$\begin{cases} x - 2y - z = 6 \\ \qquad y + 4z = 5 \\ \qquad\qquad z = 2 \end{cases}$$

$$y + 4(2) = 5 \qquad\qquad x - 2(-3) - 2 = 6$$
$$y + 8 = 5 \qquad\qquad\quad x + 6 - 2 = 6$$
$$y = -3 \qquad\qquad\qquad x + 4 = 6$$
$$\qquad\qquad\qquad\qquad\qquad\quad x = 2$$

$$(2, -3, 2)$$

47.
$$\begin{bmatrix} 1 & 2 & -1 & : & 1 \\ 2 & -1 & 1 & : & 3 \\ -1 & 2 & 3 & : & 7 \end{bmatrix}$$

$$\begin{matrix} \\ -2R_1 + R_2 \\ R_1 + R_2 \end{matrix} \begin{bmatrix} 1 & 2 & -1 & : & 1 \\ 0 & -5 & 3 & : & 1 \\ 0 & 4 & 2 & : & 8 \end{bmatrix}$$

$$\begin{matrix} \\ -\frac{1}{5}R_2 \\ \\ \end{matrix} \begin{bmatrix} 1 & 2 & -1 & : & 1 \\ 0 & 1 & -0.6 & : & -0.2 \\ 0 & 4 & 2 & : & 8 \end{bmatrix}$$

$$\begin{matrix} \\ \\ -4R_2 \to R_3 \end{matrix} \begin{bmatrix} 1 & 2 & -1 & : & 1 \\ 0 & 1 & -0.6 & : & -0.2 \\ 0 & 0 & 4.4 & : & 8.8 \end{bmatrix}$$

$$\begin{matrix} \\ \\ \frac{1}{4.4}R_2 \end{matrix} \begin{bmatrix} 1 & 2 & -1 & : & 1 \\ 0 & 1 & -0.6 & : & -0.2 \\ 0 & 0 & 1 & : & 2 \end{bmatrix}$$

$$\begin{cases} x + 2y - z = 1 \\ y - 0.6z = -0.2 \\ z = 2 \end{cases}$$

$y - 0.6(2) = -0.2$

$ y - 1.2 = -0.2$ $x + 2(1) - 2 = 1$

$ y = 1$ $x + 2 - 2 = 1$

$ x = 1$

$(1, 1, 2)$

49.
$$\begin{bmatrix} 3 & 6 & -3 & : & 6 \\ -2 & -4 & -3 & : & -1 \\ 3 & 6 & -2 & : & 10 \end{bmatrix}$$

$$\frac{1}{3}R_1 \begin{bmatrix} 1 & 2 & -1 & : & 2 \\ -2 & -4 & -3 & : & -1 \\ 3 & 6 & -2 & : & 10 \end{bmatrix}$$

$$\begin{matrix} \\ 2R_1 + R_2 \\ -3R_1 + R_3 \end{matrix} \begin{bmatrix} 1 & 2 & -1 & : & 2 \\ 0 & 0 & -5 & : & 3 \\ 0 & 0 & 1 & : & 4 \end{bmatrix}$$

$$\begin{matrix} \\ -\frac{1}{5}R_2 \\ \\ \end{matrix} \begin{bmatrix} 1 & 2 & -1 & : & 2 \\ 0 & 0 & 1 & : & -\frac{3}{5} \\ 0 & 0 & 1 & : & 4 \end{bmatrix}$$

$$\begin{matrix} \\ \\ -R_2 + R_3 \end{matrix} \begin{bmatrix} 1 & 2 & -1 & : & 2 \\ 0 & 1 & 1 & : & -\frac{3}{5} \\ 0 & 0 & 0 & : & \frac{23}{5} \end{bmatrix}$$

$$\begin{cases} x + 2y - z = 2 \\ z = -\frac{3}{5} \\ 0 = \frac{23}{5} \quad \text{False} \end{cases}$$

The system has no solution; it is inconsistent.

51.
$$\begin{bmatrix} 1 & 1 & -5 & : & 3 \\ 1 & 0 & -2 & : & 1 \\ 2 & -1 & -1 & : & 0 \end{bmatrix}$$

$$\begin{matrix} \\ -R_1 + R_2 \\ -2R_1 + R_3 \end{matrix} \begin{bmatrix} 1 & 1 & -5 & : & 3 \\ 0 & -1 & 3 & : & -2 \\ 0 & -3 & 9 & : & -6 \end{bmatrix}$$

$$\begin{matrix} \\ -R_2 \\ \\ \end{matrix} \begin{bmatrix} 1 & 1 & -5 & : & 3 \\ 0 & 1 & -3 & : & 2 \\ 0 & -3 & 9 & : & -6 \end{bmatrix}$$

$$\begin{matrix} \\ \\ 3R_2 + R_3 \end{matrix} \begin{bmatrix} 1 & 1 & -5 & : & 3 \\ 0 & 1 & -3 & : & 2 \\ 0 & 0 & 0 & : & 0 \end{bmatrix}$$

$$\begin{cases} x + y - 5z = 3 \\ y - 3z = 2 \\ 0 = 0 \end{cases}$$

The system has *infinitely* many solutions. Let $z = a$.

$y - 3a = 2$ $x + (3a + 2) - 5a = 3$

$ y = 3a + 2$ $x - 2a + 2 = 3$

$ x = 2a + 1$

$(2a + 1, 3a + 2, a)$

53.
$$\begin{bmatrix} 2 & 4 & 0 & : & 10 \\ 2 & 2 & 3 & : & 3 \\ -3 & 1 & 2 & : & -3 \end{bmatrix}$$

$$\frac{1}{2}R_1 \begin{bmatrix} 1 & 2 & 0 & : & 5 \\ 2 & 2 & 3 & : & 3 \\ -3 & 1 & 2 & : & -3 \end{bmatrix}$$

$$\begin{matrix} \\ -2R_1 + R_2 \\ 3R_1 + R_3 \end{matrix} \begin{bmatrix} 1 & 2 & 0 & : & 5 \\ 0 & -2 & 3 & : & -7 \\ 0 & 7 & 2 & : & 12 \end{bmatrix}$$

$$\begin{matrix} \\ -\frac{1}{2}R_2 \\ \\ \end{matrix} \begin{bmatrix} 1 & 2 & 0 & : & 5 \\ 0 & 1 & -\frac{3}{2} & : & \frac{7}{2} \\ 0 & 7 & 2 & : & 12 \end{bmatrix}$$

$$\begin{matrix} \\ \\ -7R_2 + R_3 \end{matrix} \begin{bmatrix} 1 & 2 & 0 & : & 5 \\ 0 & 1 & -\frac{3}{2} & : & \frac{7}{2} \\ 0 & 0 & \frac{25}{2} & : & -\frac{25}{2} \end{bmatrix}$$

$$\begin{matrix} \\ \\ \frac{2}{25}R_3 \end{matrix} \begin{bmatrix} 1 & 2 & 0 & : & 5 \\ 0 & 1 & -\frac{3}{2} & : & \frac{7}{2} \\ 0 & 0 & 1 & : & -1 \end{bmatrix}$$

$$\begin{cases} x + 2y \phantom{ - \frac{3}{2}z} = 5 \\ y - \frac{3}{2}z = \frac{7}{2} \\ \phantom{x + 2y - \frac{3}{2}}z = -1 \end{cases}$$

$y - \frac{3}{2}(-1) = \frac{7}{2}$ $x + 2(2) = 5$

$ y + \frac{3}{2} = \frac{7}{2}$ $x + 4 = 5$

$\phantom{y + \frac{3}{2}} y = 2$ $x = 1$

$(1, 2, -1)$

55.
$$\begin{bmatrix} 1 & -3 & 2 & \vdots & 8 \\ 0 & 2 & -1 & \vdots & -4 \\ 1 & 0 & 1 & \vdots & 3 \end{bmatrix}$$

$$-R_1 + R_3 \begin{bmatrix} 1 & -3 & 2 & \vdots & 8 \\ 0 & 2 & -1 & \vdots & -4 \\ 0 & 3 & -1 & \vdots & -5 \end{bmatrix}$$

$$\tfrac{1}{2}R_2 \begin{bmatrix} 1 & -3 & 2 & \vdots & 8 \\ 0 & 1 & -\tfrac{1}{2} & \vdots & -2 \\ 0 & 3 & -1 & \vdots & -5 \end{bmatrix}$$

$$-3R_2 + R_3 \begin{bmatrix} 1 & -3 & 2 & \vdots & 8 \\ 0 & 1 & -\tfrac{1}{2} & \vdots & -2 \\ 0 & 0 & \tfrac{1}{2} & \vdots & 1 \end{bmatrix}$$

$$2R_3 \begin{bmatrix} 1 & -3 & 2 & \vdots & 8 \\ 0 & 1 & -\tfrac{1}{2} & \vdots & -2 \\ 0 & 0 & 1 & \vdots & 2 \end{bmatrix}$$

$$\begin{cases} x - 3y + 2z = 8 \\ y - \tfrac{1}{2}z = -2 \\ z = 2 \end{cases}$$

$y - \tfrac{1}{2}(2) = -2 \qquad x - 3(-1) + 2(2) = 8$

$y - 1 = -2 \qquad x + 3 + 4 = 8$

$y = -1 \qquad x + 7 = 8$

$x = 1$

$(1, -1, 2)$

59.
$$\begin{bmatrix} 2 & 0 & 4 & \vdots & 1 \\ 1 & 1 & 3 & \vdots & 0 \\ 1 & 3 & 5 & \vdots & 0 \end{bmatrix}$$

$$\tfrac{1}{2}R_1 \begin{bmatrix} 1 & 0 & 2 & \vdots & \tfrac{1}{2} \\ 1 & 1 & 3 & \vdots & 0 \\ 1 & 3 & 5 & \vdots & 0 \end{bmatrix}$$

$$\begin{matrix} -R_1 + R_2 \\ -R_1 + R_3 \end{matrix} \begin{bmatrix} 1 & 0 & 2 & \vdots & \tfrac{1}{2} \\ 0 & 1 & 1 & \vdots & -\tfrac{1}{2} \\ 0 & 3 & 3 & \vdots & -\tfrac{1}{2} \end{bmatrix}$$

$$-3R_2 + R_3 \begin{bmatrix} 1 & 0 & 2 & \vdots & \tfrac{1}{2} \\ 0 & 1 & 1 & \vdots & -\tfrac{1}{2} \\ 0 & 0 & 0 & \vdots & 1 \end{bmatrix}$$

$$\begin{cases} x + 2z = \tfrac{1}{2} \\ y + z = -\tfrac{1}{2} \\ 0 = 1 \quad \text{False} \end{cases}$$

The system has *no* solution, it is inconsistent.

57.
$$\begin{bmatrix} -2 & -2 & -15 & \vdots & 0 \\ 1 & 2 & 2 & \vdots & 18 \\ 3 & 3 & 22 & \vdots & 2 \end{bmatrix}$$

$$-\tfrac{1}{2}R_1 \begin{bmatrix} 1 & 1 & \tfrac{15}{2} & \vdots & 0 \\ 1 & 2 & 2 & \vdots & 18 \\ 3 & 3 & 22 & \vdots & 2 \end{bmatrix}$$

$$\begin{matrix} -R_1 + R_2 \\ -3R_1 + R_3 \end{matrix} \begin{bmatrix} 1 & 1 & \tfrac{15}{2} & \vdots & 0 \\ 0 & 1 & -\tfrac{11}{2} & \vdots & 18 \\ 0 & 0 & -\tfrac{1}{2} & \vdots & 2 \end{bmatrix}$$

$$-2R_3 \begin{bmatrix} 1 & 1 & \tfrac{15}{2} & \vdots & 0 \\ 0 & 1 & -\tfrac{11}{2} & \vdots & 18 \\ 0 & 0 & 1 & \vdots & -4 \end{bmatrix}$$

$$\begin{cases} x + y + \tfrac{15}{2}z = 0 \\ y - \tfrac{11}{2}z = 18 \\ z = -4 \end{cases}$$

$y - \tfrac{11}{2}(-4) = 18 \qquad x + (-4) + \tfrac{15}{2}(-4) = 0$

$y + 22 = 18 \qquad x - 4 - 30 = 0$

$y = -4 \qquad x = 34$

$(34, -4, -4)$

61.
$$\begin{bmatrix} 1 & 3 & 0 & \vdots & 2 \\ 2 & 6 & 0 & \vdots & 4 \\ 2 & 5 & 4 & \vdots & 3 \end{bmatrix}$$

$$\begin{matrix} -2R_1 + R_2 \\ -2R_1 + R_3 \end{matrix} \begin{bmatrix} 1 & 1 & 0 & \vdots & 2 \\ 0 & 0 & 0 & \vdots & 0 \\ 0 & -1 & 4 & \vdots & -1 \end{bmatrix}$$

$$\begin{matrix} R_3 \\ R_2 \end{matrix} \begin{bmatrix} 1 & 3 & 0 & \vdots & 2 \\ 0 & -1 & 4 & \vdots & -1 \\ 0 & 0 & 0 & \vdots & 0 \end{bmatrix}$$

$$-1R_2 \begin{bmatrix} 1 & 3 & 0 & \vdots & 2 \\ 0 & 1 & -4 & \vdots & 1 \\ 0 & 0 & 0 & \vdots & 0 \end{bmatrix}$$

$$\begin{cases} x + 3y = 2 \\ y - 4z = 1 \\ 0 = 0 \end{cases}$$

The system has *infinitely* many solutions.

Let $z = a$

$y - 4a = 1 \qquad x + 3(4a + 1) = 2$

$y = 4a + 1 \qquad x + 12a + 3 = 2$

$x = -12a - 1$

$(-12a - 1, 4a + 1, a)$

63. *Verbal Model:*

$$\boxed{\begin{array}{c}\text{Amount}\\\text{borrowed at 8\%}\end{array}} + \boxed{\begin{array}{c}\text{Amount}\\\text{borrowed at 9\%}\end{array}} + \boxed{\begin{array}{c}\text{Amount}\\\text{borrowed at 12\%}\end{array}} = \boxed{\begin{array}{c}\text{Total}\\\text{borrowed}\end{array}}$$

$$\boxed{\begin{array}{c}\text{Interest on}\\\text{8\% loan}\end{array}} + \boxed{\begin{array}{c}\text{Interest on}\\\text{9\% loan}\end{array}} + \boxed{\begin{array}{c}\text{Interest on}\\\text{12\% loan}\end{array}} = \boxed{\begin{array}{c}\text{Total}\\\text{interest}\end{array}}$$

$$\boxed{\begin{array}{c}\text{Amount borrowed}\\\text{at 8\%}\end{array}} = 4 \cdot \boxed{\begin{array}{c}\text{Amount borrowed}\\\text{at 12\%}\end{array}}$$

Labels:

8%: Amount borrowed $= x$, Interest $= 0.08x$ (dollars)

9%: Amount borrowed $= y$, Interest $= 0.09y$ (dollars)

12%: Amount borrowed $= z$, Interest $= 0.12z$ (dollars)

Totals: Total borrowed $= 1,500,000$, Total interest $= 133,000$ (dollars)

Equations:

$$\begin{cases} x + \quad y + \quad z = 1,500,000 \\ 0.08x + 0.09y + 0.12z = 133,000 \\ \qquad\qquad\qquad x = 4z \end{cases}$$

$$\begin{cases} x + \ y + \ z = \ 1,500,000 \\ 8x + 9y + 12z = 13,300,000 \\ x \qquad - \ 4z = 0 \end{cases}$$

$$\begin{bmatrix} 1 & 1 & 1 & \vdots & 1,500,000 \\ 8 & 9 & 12 & \vdots & 13,300,000 \\ 1 & 0 & -4 & \vdots & 0 \end{bmatrix}$$

$$\begin{matrix} \\ -8R_1 + R_2 \\ -R_1 + R_2 \end{matrix} \begin{bmatrix} 1 & 1 & 1 & \vdots & 1,500,000 \\ 0 & 1 & 4 & \vdots & 1,300,000 \\ 0 & -1 & -5 & \vdots & -1,500,000 \end{bmatrix}$$

$$\begin{matrix} \\ \\ R_2 + R_3 \end{matrix} \begin{bmatrix} 1 & 1 & 1 & \vdots & 1,500,000 \\ 0 & 1 & 4 & \vdots & 1,300,000 \\ 0 & 0 & -1 & \vdots & -200,000 \end{bmatrix}$$

$$\begin{matrix} \\ \\ -R_3 \end{matrix} \begin{bmatrix} 1 & 1 & 1 & \vdots & 1,500,000 \\ 0 & 1 & 4 & \vdots & 1,300,000 \\ 0 & 0 & 1 & \vdots & 200,000 \end{bmatrix}$$

$$\begin{cases} x + y + \ z = 1,500,000 \\ \quad y + 4z = 1,300,000 \\ \qquad\quad z = \ 200,000 \end{cases}$$

$$y + 4(200,000) = 1,300,000 \qquad\qquad x + 500,000 + 200,000 = 1,500,000$$

$$y + 800,000 = 1,300,000 \qquad\qquad\qquad x + 700,000 = 1,500,000$$

$$y = 500,000 \qquad\qquad\qquad\qquad x = 800,000$$

$$(800,000, \ 500,000, \ 200,000)$$

Thus, \$800,000 was borrowed at 8%, \$500,000 was borrowed at 9%, and \$200,000 was borrowed at 12%.

65. $(1, 7)$: $7 = a(1)^2 + b(1) + c$

$(2, 12)$: $12 = a(2)^2 + b(2) + c$

$(3, 19)$: $19 = a(3)^2 + b(3) + c$

$$\begin{cases} a + b + c = 7 \\ 4a + 2b + c = 12 \\ 9a + 3b + c = 19 \end{cases}$$

$$\begin{bmatrix} 1 & 1 & 1 & \vdots & 7 \\ 4 & 2 & 1 & \vdots & 12 \\ 9 & 3 & 1 & \vdots & 19 \end{bmatrix}$$

$$\begin{matrix} \\ -4R_1 + R_2 \\ -9R_1 + R_3 \end{matrix} \begin{bmatrix} 1 & 1 & 1 & \vdots & 7 \\ 0 & -2 & -3 & \vdots & -16 \\ 0 & -6 & -8 & \vdots & -44 \end{bmatrix}$$

$$\begin{matrix} \\ -\frac{1}{2}R_2 \\ \\ \end{matrix} \begin{bmatrix} 1 & 1 & 1 & \vdots & 7 \\ 0 & 1 & \frac{3}{2} & \vdots & 8 \\ 0 & -6 & -8 & \vdots & -44 \end{bmatrix}$$

$$\begin{matrix} \\ \\ 6R_2 + R_3 \end{matrix} \begin{bmatrix} 1 & 1 & 1 & \vdots & 7 \\ 0 & 1 & \frac{3}{2} & \vdots & 8 \\ 0 & 0 & 1 & \vdots & 4 \end{bmatrix}$$

$$\begin{cases} a + b + c = 7 \\ b + \frac{3}{2}c = 8 \\ c = 4 \end{cases}$$

$b + \frac{3}{2}(4) = 8$ $a + 2 + 4 = 7$

$b + 6 = 8$ $a + 6 = 7$

$b = 2$ $a = 1$

$(1, 2, 4)$

$y = x^2 + 2x + 4$

67. (a) $y = at^2 + bt + c$

1996: $57.3 = a(6)^2 + b(6) + c \Longrightarrow 36a + 6b + c = 57.3$

1997: $59.4 = a(7)^2 + b(7) + c \Longrightarrow 49a + 7b + c = 59.4$

1998: $62.2 = a(8)^2 + b(8) + c \Longrightarrow 64a + 8b + c = 62.2$

$$\begin{bmatrix} 36 & 6 & 1 & \vdots & 57.3 \\ 49 & 7 & 1 & \vdots & 59.4 \\ 64 & 8 & 1 & \vdots & 62.2 \end{bmatrix}$$

$$\begin{matrix} \frac{1}{36}R_1 \\ \\ \end{matrix} \begin{bmatrix} 1 & \frac{1}{6} & \frac{1}{36} & \vdots & \frac{57.3}{36} \\ 49 & 7 & 1 & \vdots & 59.4 \\ 64 & 8 & 1 & \vdots & 62.2 \end{bmatrix}$$

$$\begin{matrix} \\ -49R_1 + R_2 \\ -64R_1 + R_3 \end{matrix} \begin{bmatrix} 1 & \frac{1}{6} & \frac{1}{36} & \vdots & \frac{57.3}{36} \\ 0 & -\frac{7}{6} & -\frac{13}{36} & \vdots & -\frac{669.3}{36} \\ 0 & -\frac{8}{3} & -\frac{7}{9} & \vdots & -\frac{119}{3} \end{bmatrix}$$

$$\begin{matrix} \\ -\frac{6}{7}R_2 \\ \\ \end{matrix} \begin{bmatrix} 1 & \frac{1}{6} & \frac{1}{36} & \vdots & \frac{57.3}{36} \\ 0 & 1 & \frac{13}{42} & \vdots & \frac{669.3}{42} \\ 0 & -\frac{8}{3} & -\frac{7}{9} & \vdots & -\frac{119}{3} \end{bmatrix}$$

$$\begin{matrix} \\ \\ \frac{8}{3}R_2 + R_3 \end{matrix} \begin{bmatrix} 1 & \frac{1}{6} & \frac{1}{36} & \vdots & \frac{57.3}{36} \\ 0 & 1 & \frac{13}{42} & \vdots & \frac{669.3}{42} \\ 0 & 0 & \frac{1}{21} & \vdots & -\frac{59.4}{21} \end{bmatrix}$$

$$\begin{matrix} \\ \\ 21R_3 \end{matrix} \begin{bmatrix} 1 & \frac{1}{6} & \frac{1}{36} & \vdots & \frac{57.3}{36} \\ 0 & 1 & \frac{13}{42} & \vdots & \frac{669.3}{42} \\ 0 & 0 & 1 & \vdots & 59.4 \end{bmatrix}$$

$$\begin{bmatrix} 1 & \frac{1}{6} & \frac{1}{36} & \vdots & \frac{57.3}{36} \\ 0 & 1 & \frac{13}{42} & \vdots & \frac{669.3}{42} \\ 0 & 0 & 1 & \vdots & \frac{297}{5} \end{bmatrix}$$

$$\begin{cases} x + \frac{1}{6}y + \frac{1}{36}z = \frac{57.3}{36} \\ y + \frac{13}{42}z = \frac{669.3}{42} \\ z = \frac{297}{5} \end{cases}$$

$$y + \left(\frac{13}{42}\right)\left(\frac{297}{5}\right) = \frac{669.3}{42}$$

$$210y + 3861 = 3346.5$$

$$210y = -514.5$$

$$y = -\frac{49}{20}$$

$$y = \frac{7}{20}t^2 - \frac{49}{20}t + \frac{297}{5}$$

$$x + \left(\frac{1}{6}\right)\left(-\frac{49}{20}\right) + \left(\frac{1}{36}\right)\left(\frac{297}{5}\right) = \left(\frac{57.3}{36}\right)$$

$$360x - 147 + 594 = 573$$

$$360x + 447 = 573$$

$$360x = 126$$

$$x = \frac{7}{20}$$

—CONTINUED—

67. —CONTINUED—

(b)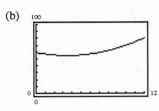

(c) $y = \dfrac{7}{20}(16)^2 - \dfrac{49}{20}(16) + \dfrac{297}{5}$

$y = 89.6 - 39.2 + 59.4$

$y = 109.8$ million short tons

69. (a) $y = ax^2 + bx + c$

$(50, 26)$: $26 = a(50)^2 + b(50) + c \Rightarrow 2500a + 50b + c = 26$

$(25, 18.5)$: $18.5 = a(25)^2 + b(25) + c \Rightarrow 625a + 25b + c = 18.5$

$(0, 6)$: $6 = a(0)^2 + b(0) + c \Rightarrow$ $c = 6$

$$\begin{bmatrix} 2500 & 50 & 1 & \vdots & 26 \\ 625 & 25 & 1 & \vdots & 18.5 \\ 0 & 0 & 1 & \vdots & 6 \end{bmatrix}$$

$$\begin{bmatrix} 625 & 25 & 1 & \vdots & 18.5 \\ 2500 & 50 & 1 & \vdots & 26 \\ 0 & 0 & 1 & \vdots & 6 \end{bmatrix}$$

$$\begin{bmatrix} 1 & 0.04 & 0.0016 & \vdots & 0.0296 \\ 2500 & 50 & 1 & \vdots & 26 \\ 0 & 0 & 1 & \vdots & 0 \end{bmatrix}$$

$$\begin{bmatrix} 1 & 0.04 & 0.0016 & \vdots & 0.0296 \\ 0 & -50 & -3 & \vdots & -48 \\ 0 & 0 & 1 & \vdots & 6 \end{bmatrix}$$

$$\begin{bmatrix} 1 & 0.04 & 0.0016 & \vdots & 0.0296 \\ 0 & 1 & .06 & \vdots & 0.96 \\ 0 & 0 & 1 & \vdots & 6 \end{bmatrix}$$

$$\begin{cases} a + 0.04b + 0.0016c = 0.0296 \\ \qquad b + \quad 0.06c = \quad 0.96 \\ \qquad\qquad\quad c = \quad 6 \end{cases}$$

$b + 0.06(6) = 0.96$ $a + 0.04(0.6) + 0.0016(6) = 0.0296$

$b + 0.36 = 0.96$ $a + 0.024 + 0.0096 = 0.0296$

$b = 0.6$ $a + 0.0336 = 0.0296$

$a = -0.004$

$(-0.004, 0.6, 6)$

$y = -0.004x^2 + 0.6x + 6$

(b)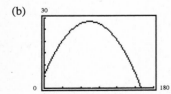

(c) Using the graphing utility, the maximum height is 28.5 feet and the point at which the ball struck the ground is approximately 159.4 feet from where it was thrown.

71. There will be a row in the matrix with a nonzero entry in the last column and zero entries in all other columns.

73. Two matrices are row-equivalent if one can be obtained from the other by a series of elementary row operations.

75. The three elementary row operations that can be performed on an augmented matrix are interchanging two rows, multiplying a row by a nonzero constant, and adding a multiple of a row to another row.

77.

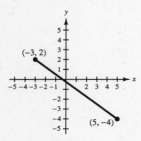

$$m = \frac{-4 - 2}{5 - (-3)} = \frac{-6}{8} = -\frac{3}{4}$$

79.

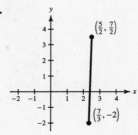

$$m = \frac{-2 - \frac{7}{2}}{\frac{7}{3} - \frac{5}{2}} = \frac{-\frac{4}{2} - \frac{7}{2}}{\frac{14}{6} - \frac{15}{6}} = -\frac{-\frac{11}{2}}{-\frac{1}{6}} = -\frac{11}{2}\left(-\frac{6}{1}\right) = 33$$

81.

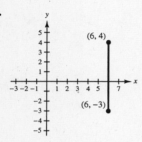

$$m = \frac{-3 - 4}{6 - 6} = \frac{-7}{0} \text{ undefined}$$

83. $5 - \left(\frac{1}{3}\right)(-6) = 5 + 2$

$\qquad\qquad = 7$

85. $4(-2) - \frac{4}{5}(-3) = -8 + \frac{12}{5}$

$$= -\frac{40}{5} + \frac{12}{5}$$

$$= -\frac{28}{5}$$

87. *Verbal Model:* $\boxed{\begin{array}{c}\text{Original}\\\text{membership}\end{array}} + 0.10 \boxed{\begin{array}{c}\text{Original}\\\text{membership}\end{array}} = \boxed{\begin{array}{c}\text{Current}\\\text{membership}\end{array}}$

Labels: Original membership $= x$ (people)

Current membership $= 8415$ (people)

Equation: $x + 0.10x = 8415$

$1.10x = 8415$

$x = 7650$

The station had 7650 members before the membership drive.

Mid-Chapter Quiz for Chapter 4

1. The solution is $(2, 1)$.

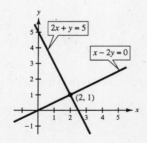

2. The solution is $(3, -2)$.

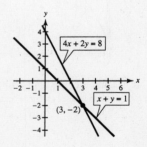

3. $\begin{cases} 5x - y = 32 \Rightarrow -y = -5x + 32 \Rightarrow y = 5x - 32 \\ 6x - 9y = 15 \end{cases}$

$$6x - 9(5x - 32) = 15$$

$$6x - 45x + 288 = 15$$

$$-39x + 288 = 15$$

$$-39x = -273$$

$$x = 7 \quad \text{and} \quad y = 5(7) - 32$$

$$y = 3$$

$(7, 3)$

4. $\begin{cases} 0.2x + 0.7y = 8 \\ -x + 2y = 15 \Rightarrow -x = -2y + 15 \Rightarrow x = 2y - 15 \end{cases}$

$$0.2(2y - 15) + 0.7y = 8$$

$$0.4y - 3 + 0.7y = 8$$

$$1.1y - 3 = 8$$

$$1.1y = 11$$

$$y = \frac{11}{1.1}$$

$$y = 10 \quad \text{and} \quad x = 2(10) - 15$$

$$x = 20 - 15$$

$$x = 5$$

$(5, 10)$

5. $\begin{cases} x + 10y = 18 \\ 5x + 2y = 42 \end{cases}$

$\begin{cases} x + 10y = 18 \\ -48y = -48 \end{cases}$ -5 Eqn. 1 + Eqn. 2

$\begin{cases} x + 10y = 18 \\ y = 1 \end{cases}$ $(-1/48)$Eqn. 2

$$x + 10(1) = 18$$

$$x = 8$$

$(8, 1)$

6. $\begin{cases} 3x + 11y = 38 \\ 7x - 5y = -34 \end{cases}$

$\begin{cases} x + \frac{11}{3}y = \frac{38}{3} \\ 7x - 5y = -34 \end{cases}$ $1/3$ Eqn. 1

$\begin{cases} x + \frac{11}{3}y = \frac{38}{3} \\ -\frac{92}{3}y = -\frac{368}{3} \end{cases}$ -7 Eqn. 1 + Eqn. 2

$\begin{cases} x + \frac{11}{3}y = \frac{38}{3} \\ \phantom{x + \frac{11}{3}} y = 4 \end{cases}$ $-3/97$ Eqn. 2

$$x + \frac{11}{3}(4) = \frac{38}{3}$$

$$x + \frac{44}{3} = \frac{38}{3}$$

$$x = -\frac{6}{3}$$

$$x = -2$$

$(-2, 4)$

7. $\begin{cases} x \qquad\; + 4z = \;\;\; 17 \\ -3x + 2y - \;\; z = -20 \\ x - 5y + 3z = \;\;\; 19 \end{cases}$

$\begin{cases} x \qquad\; + \;\; 4z = 17 \\ 2y + 11z = 31 \\ -5y - \;\;\; z = 2 \end{cases}$ 3 Eqn. 1 + Eqn. 2

               (−1)Eqn. 1 + Eqn. 3

$\begin{cases} x \qquad\; + \;\; 4z = 17 \\ y + \frac{11}{2}z = \frac{31}{2} \\ -5y - \;\;\; z = 2 \end{cases}$ 1/2 Eqn. 2

$\begin{cases} x \;\; + \;\; 4z = 17 \\ y + \frac{11}{2}z = \frac{31}{2} \\ \frac{53}{2}z = \frac{159}{2} \end{cases}$ 5 Eqn. 2 + Eqn. 3

$\begin{cases} x \;\; + \;\; 4z = 17 \\ y + \frac{11}{2}z = \frac{31}{2} \\ z = \;\; 3 \end{cases}$ (2/53)Eqn. 3

$y + \frac{11}{2}(3) = \frac{31}{2}$ and $x + 4(3) = 17$

    $y + \frac{33}{2} = \frac{31}{2}$ $x + 12 = 17$

        $y = -1$ $x = 5$

$(5, -1, 3)$

8. $\begin{bmatrix} 1 & -3 & 1 & \vdots & -3 \\ 3 & 2 & -5 & \vdots & 18 \\ 0 & 1 & 1 & \vdots & -1 \end{bmatrix}$

$-3R_1 + R_2 \begin{bmatrix} 1 & -3 & 1 & \vdots & -3 \\ 0 & 11 & -8 & \vdots & 27 \\ 0 & 1 & 1 & \vdots & -1 \end{bmatrix}$

$\frac{1}{11}R_2 \begin{bmatrix} 1 & -3 & 1 & \vdots & -3 \\ 0 & 1 & -\frac{8}{11} & \vdots & \frac{27}{11} \\ 0 & 1 & 1 & \vdots & -1 \end{bmatrix}$

$(-1)R_2 + R_3 \begin{bmatrix} 1 & -3 & 1 & \vdots & -3 \\ 0 & 1 & -\frac{8}{11} & \vdots & \frac{27}{11} \\ 0 & 0 & \frac{19}{11} & \vdots & -\frac{38}{11} \end{bmatrix}$

$\left(\frac{11}{19}\right)R_3 \begin{bmatrix} 1 & -3 & 1 & \vdots & -3 \\ 0 & 1 & -\frac{8}{11} & \vdots & \frac{27}{11} \\ 0 & 0 & 1 & \vdots & -2 \end{bmatrix}$

$\begin{cases} x - 3y + \;\; z = -3 \\ y - \frac{8}{11}z = \frac{27}{11} \\ z = -2 \end{cases}$

$y - \frac{8}{11}(-2) = \frac{27}{11}$ and $x - 3(1) + (-2) = -3$

    $y + \frac{16}{11} = \frac{27}{11}$ $x - 5 = -3$

          $y = \frac{11}{11}$ $x = 2$

          $y = 1$

$(2, 1, -2)$

9. $(10, -12)$; There are many correct answers. Here are some examples.

$1(10) + 1(-12) = -2 \Longrightarrow \begin{cases} x + y = -2 \\ 2x + y = \;\;\; 8 \end{cases}$
$2(10) + 1(-12) = \;\;\; 8 \Longrightarrow$

$2(10) + 1(-12) = \;\;\; 8 \Longrightarrow \begin{cases} 2x + y = \;\;\; 8 \\ 3x - y = 42 \end{cases}$
$3(10) - 1(-12) = 42 \Longrightarrow$

10. $(2, -3, 1)$; There are many correct answers. Here are some examples.

$1(2) + 1(-3) + 1(1) = 0 \Longrightarrow \begin{cases} x + y + \;\; z = 0 \\ x - y + \;\; z = 6 \\ x - y - 3z = 2 \end{cases}$
$1(2) - 1(-3) + 1(1) = 6 \Longrightarrow$
$1(2) - 1(-3) - 3(1) = 2 \Longrightarrow$

$1(2) + 1(-3) - 1(1) = -2 \Longrightarrow \begin{cases} x + \;\; y - \;\; z = -2 \\ 5x + 4y + 2z = \;\;\; 0 \\ 4x + 2y + 5z = \;\;\; 7 \end{cases}$
$5(2) + 4(-3) + 2(1) = \;\;\; 0 \Longrightarrow$
$4(2) + 2(-3) + 5(1) = \;\;\; 7 \Longrightarrow$

11. *Verbal Model:*

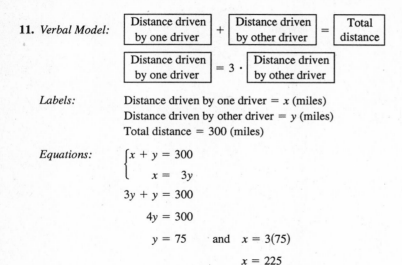

Labels: Distance driven by one driver $= x$ (miles)
Distance driven by other driver $= y$ (miles)
Total distance $= 300$ (miles)

Equations:
$$\begin{cases} x + y = 300 \\ \quad x = 3y \end{cases}$$

$$3y + y = 300$$

$$4y = 300$$

$$y = 75 \quad \text{and} \quad x = 3(75)$$

$$x = 225$$

$(225, 75)$

So, one driver drove 225 miles and the other driver drove 75 miles.

12. *Verbal Model:*

$$\boxed{\begin{array}{c}\text{Gallons of}\\\text{20\% solution}\end{array}} + \boxed{\begin{array}{c}\text{Gallons of}\\\text{50\% solution}\end{array}} = \boxed{\begin{array}{c}\text{Gallons of}\\\text{30\% solution}\end{array}}$$

$$\boxed{\begin{array}{c}\text{Amount of salt}\\\text{in 20\% solution}\end{array}} + \boxed{\begin{array}{c}\text{Amount of salt}\\\text{in 50\% solution}\end{array}} = \boxed{\begin{array}{c}\text{Amount of salt}\\\text{in 30\% solution}\end{array}}$$

Labels: 20% solution: Number of gallons $= x$; Amount of salt $= 0.20x$
50% solution: Number of gallons $= y$; Amount of salt $= 0.50y$
30% solution: Number of gallons $= 20$; Amount of salt $= 0.30(20)$

Equations:
$$\begin{cases} x + y = 20 \\ 0.20x + 0.50y = 0.30(20) \end{cases}$$

$$\begin{cases} -0.20x - 0.20y = -4 \\ 0.20x + 0.50y = 6 \end{cases}$$

$$0.30y = 2$$

$$3y = 20$$

$$y = \tfrac{20}{3} \quad or \quad 6\tfrac{2}{3} \quad \text{and} \quad x + \tfrac{20}{3} = 20$$

$$3x + 20 = 60$$

$$3x = 40$$

$$x = \tfrac{40}{3} \quad or \quad 13\tfrac{1}{3}$$

$\left(13\tfrac{1}{3}, 6\tfrac{2}{3}\right)$

So, $13\tfrac{1}{3}$ gallons of the 20% solution and $6\tfrac{2}{3}$ gallons of the 50% solution are required for the mixture.

13. *Verbal Model:* $\boxed{\begin{array}{c}\text{Measure of}\\\text{first angle}\end{array}} + \boxed{\begin{array}{c}\text{Measure of}\\\text{second angle}\end{array}} = 180$

$\boxed{\begin{array}{c}\text{Measure of}\\\text{second angle}\end{array}} = 2 \cdot \boxed{\begin{array}{c}\text{Measure of}\\\text{first angle}\end{array}} - 45$

Labels: Measure of first angle $= x$ (degrees)

Measure of second angle $= y$ (degrees)

Equations: $\begin{cases} x + y = 180 \\ y = 2x - 45 \end{cases}$

$$x + (2x - 45) = 180$$

$$3x - 45 = 180$$

$$3x = 225$$

$$x = 75 \quad \text{and} \quad x = 2(75) - 45$$

$$y = 105$$

$(75, 105)$

The measure of the first angle is 75 degrees and the measure of the second angle is 105 degrees.

14. $y = ax^2 + bx + c$

$(1, 2):$ $\quad 2 = a(1)^2 + b(1) + c$

$(-1, -4):$ $\quad -4 = a(-1)^2 + b(-1) + c$

$(2, 8):$ $\quad 8 = a(2)^2 + b(2) + c$

$\begin{cases} a + b + c = 2 \\ a - b + c = -4 \\ 4a + 2b + c = 8 \end{cases}$

$\begin{cases} a + b + c = 2 \\ \quad -2b = -6 \quad (-1)\text{Eqn. 1} + \text{Eqn. 2} \\ \quad -2b - 3c = 0 \quad (-4)\text{Eqn. 1} + \text{Eqn. 3} \end{cases}$

$\begin{cases} a + b + c = 2 \\ \quad b = 3 \quad (-1/2)\text{Eqn. 2} \\ \quad -2b - 3c = 0 \end{cases}$

$\begin{cases} a + b + c = 2 \\ \quad b = 3 \\ \quad -3c = 6 \quad (2)\text{Eqn. 2} + \text{Eqn. 3} \end{cases}$

$\begin{cases} a + b + c = 2 \\ \quad b = 3 \\ \quad c = -2 \quad (-1/3)\text{Eqn. 3} \end{cases}$

$a + 3 - 2 = 2$

$a + 1 = 2$

$a = 1$

$(1, 3, -2)$

The equation of the parabola is $y = x^2 + 3x - 2$.

Section 4.4   Determinants and Linear Systems

1. $\begin{vmatrix} 2 & 1 \\ 3 & 4 \end{vmatrix} = 2(4) - 3(1) = 8 - 3 = 5$

3. $\begin{vmatrix} 5 & 2 \\ -6 & 3 \end{vmatrix} = 5(3) - (-6)(2)$

$= 15 + 12 = 27$

5. $\begin{vmatrix} 5 & -4 \\ -10 & 8 \end{vmatrix} = 5(8) - (-10)(-4)$

$= 40 - 40 = 0$

7. $\begin{vmatrix} 2 & 6 \\ 0 & 3 \end{vmatrix} = 2(3) - 0(6) = 6 - 0 = 6$

9. $\begin{vmatrix} -7 & 6 \\ \frac{1}{2} & 3 \end{vmatrix} = -7(3) - \frac{1}{2}(6)$

$= -21 - 3 = -24$

11. $\begin{vmatrix} 0.3 & 0.5 \\ 0.5 & 0.3 \end{vmatrix} = 0.3(0.3) - 0.5(0.5)$

$= 0.09 - 0.25 = -0.16$

13. $\begin{vmatrix} 2 & 3 & -1 \\ 6 & 0 & 0 \\ 4 & 1 & 1 \end{vmatrix}$

First row: $2\begin{vmatrix} 0 & 0 \\ 1 & 1 \end{vmatrix} - 3\begin{vmatrix} 6 & 0 \\ 4 & 1 \end{vmatrix} + (-1)\begin{vmatrix} 6 & 0 \\ 4 & 1 \end{vmatrix} = 2(0) - 3(6) - 1(6) = 0 - 18 - 6 = -24$

Second row: $-6\begin{vmatrix} 3 & -1 \\ 1 & 1 \end{vmatrix} + 0\begin{vmatrix} 2 & -1 \\ 4 & 1 \end{vmatrix} - 0\begin{vmatrix} 2 & 3 \\ 4 & 1 \end{vmatrix} = -6(4) + 0 - 0 = -24$

Third row: $4\begin{vmatrix} 3 & -1 \\ 0 & 0 \end{vmatrix} - 1\begin{vmatrix} 2 & -1 \\ 6 & 0 \end{vmatrix} + 1\begin{vmatrix} 2 & 3 \\ 6 & 0 \end{vmatrix} = 4(0) - 1(6) + 1(-18) = 0 - 6 - 18 = -24$

First column: $2\begin{vmatrix} 0 & 0 \\ 1 & 1 \end{vmatrix} - 6\begin{vmatrix} 3 & -1 \\ 1 & 1 \end{vmatrix} + 4\begin{vmatrix} 3 & -1 \\ 0 & 0 \end{vmatrix} = 2(0) - 6(4) + 4(0) = 0 - 24 + 0 = -24$

Second column: $-3\begin{vmatrix} 6 & 0 \\ 4 & 1 \end{vmatrix} + 0\begin{vmatrix} 2 & -1 \\ 4 & 1 \end{vmatrix} - 1\begin{vmatrix} 2 & -1 \\ 6 & 0 \end{vmatrix} = -3(6) + 0 - 1(6) = -18 + 0 - 6 = -24$

Third column: $-1\begin{vmatrix} 6 & 0 \\ 4 & 1 \end{vmatrix} - 0\begin{vmatrix} 2 & 3 \\ 4 & 1 \end{vmatrix} + 1\begin{vmatrix} 2 & 3 \\ 6 & 0 \end{vmatrix} = -1(6) - 0 + 1(-18) = -6 - 0 - 18 = -24$

15. $\begin{vmatrix} 1 & 1 & 2 \\ 3 & 1 & 0 \\ -2 & 0 & 3 \end{vmatrix}$

First column: $1\begin{vmatrix} 1 & 0 \\ 0 & 3 \end{vmatrix} - 3\begin{vmatrix} 1 & 2 \\ 0 & 3 \end{vmatrix} + (-2)\begin{vmatrix} 1 & 2 \\ 1 & 0 \end{vmatrix} = 1(3) - 3(3) - 2(-2) = 3 - 9 + 4 = -2$

Second column: $-1\begin{vmatrix} 3 & 0 \\ -2 & 3 \end{vmatrix} + 1\begin{vmatrix} 1 & 2 \\ -2 & 3 \end{vmatrix} - 0\begin{vmatrix} 1 & 2 \\ 3 & 0 \end{vmatrix} = -1(9) + 1(7) - 0 = -9 + 7 - 0 = -2$

Third column: $2\begin{vmatrix} 3 & 1 \\ -2 & 0 \end{vmatrix} - 0\begin{vmatrix} 1 & 1 \\ -2 & 0 \end{vmatrix} + 3\begin{vmatrix} 1 & 1 \\ 3 & 1 \end{vmatrix} = 2(2) - 0 + 3(-2) = 4 - 0 - 6 = -2$

First row: $1\begin{vmatrix} 1 & 0 \\ 0 & 3 \end{vmatrix} - 1\begin{vmatrix} 3 & 0 \\ -2 & 3 \end{vmatrix} + 2\begin{vmatrix} 3 & 1 \\ -2 & 0 \end{vmatrix} = 1(3) - 1(9) + 2(2) = 3 - 9 + 4 = -2$

Second row: $-3\begin{vmatrix} 1 & 2 \\ 0 & 3 \end{vmatrix} + 1\begin{vmatrix} 1 & 2 \\ -2 & 3 \end{vmatrix} - 0\begin{vmatrix} 1 & 1 \\ -2 & 0 \end{vmatrix} = -3(3) + 1(7) - 0 = -9 + 7 - 0 = -2$

Third row: $-2\begin{vmatrix} 1 & 2 \\ 1 & 0 \end{vmatrix} - 0\begin{vmatrix} 1 & 2 \\ 3 & 0 \end{vmatrix} + 3\begin{vmatrix} 1 & 1 \\ 3 & 1 \end{vmatrix} = -2(-2) - 0 + 3(-2) = 4 - 0 - 6 = -2$

17. $\begin{vmatrix} 2 & 4 & 6 \\ 0 & 3 & 1 \\ 0 & 0 & -5 \end{vmatrix} = 2\begin{vmatrix} 3 & 1 \\ 0 & -5 \end{vmatrix} - 0\begin{vmatrix} 4 & 6 \\ 0 & -5 \end{vmatrix} + 0\begin{vmatrix} 4 & 6 \\ 3 & 1 \end{vmatrix}$

$\qquad = 2(-15) - 0 + 0 = -30$

19. $\begin{vmatrix} -2 & 2 & 3 \\ 1 & -1 & 0 \\ 0 & 1 & 4 \end{vmatrix} = -2\begin{vmatrix} -1 & 0 \\ 1 & 4 \end{vmatrix} - 1\begin{vmatrix} 2 & 3 \\ 1 & 4 \end{vmatrix} + 0\begin{vmatrix} 2 & 3 \\ -1 & 0 \end{vmatrix}$

$\qquad = -2(-4) - 1(5) + 0 = 8 - 5 + 0 = 3$

21. $\begin{vmatrix} 1 & 4 & -2 \\ 3 & 6 & -6 \\ -2 & 1 & 4 \end{vmatrix} = 1\begin{vmatrix} 6 & -6 \\ 1 & 4 \end{vmatrix} - 3\begin{vmatrix} 4 & -2 \\ 1 & 4 \end{vmatrix} + (-2)\begin{vmatrix} 4 & -2 \\ 6 & -6 \end{vmatrix}$

$\qquad = 1(30) - 3(18) - 2(-12) = 30 - 54 + 24 = 0$

23. $\begin{vmatrix} -3 & 2 & 1 \\ 4 & 5 & 6 \\ 2 & -3 & 1 \end{vmatrix} = -3\begin{vmatrix} 5 & 6 \\ -3 & 1 \end{vmatrix} - 4\begin{vmatrix} 2 & 1 \\ -3 & 1 \end{vmatrix} + 2\begin{vmatrix} 2 & 1 \\ 5 & 6 \end{vmatrix}$

$\qquad = -3(23) - 4(5) + 2(7) = -69 - 20 + 14 = -75$

25. $\begin{vmatrix} 1 & 4 & -2 \\ 3 & 2 & 0 \\ -1 & 4 & 3 \end{vmatrix} = -2\begin{vmatrix} 3 & 2 \\ -1 & 4 \end{vmatrix} - 0\begin{vmatrix} 1 & 4 \\ -1 & 4 \end{vmatrix} + 3\begin{vmatrix} 1 & 4 \\ 3 & 2 \end{vmatrix}$

$\qquad = -2(14) - 0 + 3(-10) = -28 - 30 = -58$

27. $\begin{vmatrix} 0.1 & 0.2 & 0.3 \\ -0.3 & 0.2 & 0.2 \\ 5 & 4 & 4 \end{vmatrix} = 5\begin{vmatrix} 0.2 & 0.3 \\ 0.2 & 0.2 \end{vmatrix} - 4\begin{vmatrix} 0.1 & 0.3 \\ -0.3 & 0.2 \end{vmatrix} + 4\begin{vmatrix} 0.1 & 0.2 \\ -0.3 & 0.2 \end{vmatrix}$

$\qquad = 5(-0.02) - 4(0.11) + 4(0.08) = -0.10 - 0.44 + 0.32 = -0.22$

29. $\begin{vmatrix} x & y & 1 \\ 3 & 1 & 1 \\ -2 & 0 & 1 \end{vmatrix} = -y\begin{vmatrix} 3 & 1 \\ -2 & 1 \end{vmatrix} + 1\begin{vmatrix} x & 1 \\ -2 & 1 \end{vmatrix} - 0\begin{vmatrix} x & 1 \\ 3 & 1 \end{vmatrix}$

$\qquad = -y(5) + 1(x + 2) - 0 = -5y + x + 2 = x - 5y + 2$

31. $\begin{vmatrix} 5 & -3 & 2 \\ 7 & 5 & -7 \\ 0 & 6 & -1 \end{vmatrix} = 248$

33. $\begin{vmatrix} 35 & 15 & 70 \\ -8 & 20 & 3 \\ -5 & 6 & 20 \end{vmatrix} = 19{,}185$

35. $\begin{vmatrix} 0.4 & 0.3 & 0.3 \\ -0.2 & 0.6 & 0.6 \\ 3 & 1 & 1 \end{vmatrix} = 0$

37. $\begin{vmatrix} 2 & -3 & 3 \\ \frac{3}{4} & 1 & -\frac{1}{4} \\ 12 & 3 & -\frac{1}{2} \end{vmatrix} = -20.875 \text{ or } -\frac{167}{8}$

39. $\begin{vmatrix} \frac{3}{2} & -\frac{3}{4} & 1 \\ 10 & 8 & 7 \\ 12 & -4 & 12 \end{vmatrix} = 77$

41. $\begin{vmatrix} 0.2 & 0.8 & -0.3 \\ 0.1 & 0.8 & 0.6 \\ -10 & -5 & 1 \end{vmatrix} = -6.37$

43. $D = \begin{vmatrix} 1 & 2 \\ -1 & 1 \end{vmatrix} = 1 - (-2) = 1 + 2 = 3$

$x = \dfrac{D_x}{D} = \dfrac{\begin{vmatrix} 5 & 2 \\ 1 & 1 \end{vmatrix}}{3} = \dfrac{5 - 2}{3} = \dfrac{3}{3} = 1$

$y = \dfrac{D_y}{D} = \dfrac{\begin{vmatrix} 1 & 5 \\ -1 & 1 \end{vmatrix}}{3} = \dfrac{1 - (-5)}{3} = \dfrac{6}{3} = 2$

$(1, 2)$

45. $D = \begin{vmatrix} 3 & 4 \\ 5 & 3 \end{vmatrix} = 9 - 20 = -11$

$x = \dfrac{D_x}{D} = \dfrac{\begin{vmatrix} -2 & 4 \\ 4 & 3 \end{vmatrix}}{-11} = \dfrac{-6 - 16}{-11} = \dfrac{-22}{-11} = 2$

$y = \dfrac{D_y}{D} = \dfrac{\begin{vmatrix} 3 & -2 \\ 5 & 4 \end{vmatrix}}{-11} = \dfrac{12 - (-10)}{-11} = \dfrac{22}{-11} = -2$

$(2, -2)$

47. $D = \begin{vmatrix} 4 & 8 \\ 2 & -4 \end{vmatrix} = -16 - 16 = -32$

$x = \dfrac{D_x}{D} = \dfrac{\begin{vmatrix} 16 & 8 \\ 5 & -4 \end{vmatrix}}{-32} = \dfrac{-64 - 40}{-32} = \dfrac{-104}{-32} = \dfrac{13}{4}$

$y = \dfrac{D_y}{D} = \dfrac{\begin{vmatrix} 4 & 16 \\ 2 & 5 \end{vmatrix}}{-32} = \dfrac{20 - 32}{-32} = \dfrac{-12}{-32} = \dfrac{3}{8}$

$\left(\dfrac{13}{4}, \dfrac{3}{8} \right)$

49. $D = \begin{vmatrix} 20 & 8 \\ 12 & -24 \end{vmatrix} = -480 - 96 = -576$

$x = \dfrac{D_x}{D} = \dfrac{\begin{vmatrix} 11 & 8 \\ 21 & -24 \end{vmatrix}}{-576} = \dfrac{-432}{-576} = \dfrac{\cancel{9}\cancel{(16)}(3)}{\cancel{9}\cancel{(16)}(4)} = \dfrac{3}{4}$

$y = \dfrac{D_y}{D} = \dfrac{\begin{vmatrix} 20 & 11 \\ 12 & 21 \end{vmatrix}}{-576} = \dfrac{288}{-576} = -\dfrac{1}{2}$

$\left(\dfrac{3}{4}, -\dfrac{1}{2} \right)$

51. $D = \begin{vmatrix} 3 & 6 \\ 6 & 14 \end{vmatrix} = 42 - 36 = 6$

$x = \dfrac{D_x}{D} = \dfrac{\begin{vmatrix} 5 & 6 \\ 11 & 14 \end{vmatrix}}{6} = \dfrac{70 - 66}{6} = \dfrac{4}{6} = \dfrac{2}{3}$

$y = \dfrac{D_y}{D} = \dfrac{\begin{vmatrix} 3 & 5 \\ 6 & 11 \end{vmatrix}}{6} = \dfrac{33 - 30}{6} = \dfrac{3}{6} = \dfrac{1}{2}$

$\left(\dfrac{2}{3}, \dfrac{1}{2} \right)$

53. $D = \begin{vmatrix} 4 & -1 & 1 \\ 2 & 2 & 3 \\ 5 & -2 & 6 \end{vmatrix} = 4\begin{vmatrix} 2 & 3 \\ -2 & 6 \end{vmatrix} - 2\begin{vmatrix} -1 & 1 \\ -2 & 6 \end{vmatrix} + 5\begin{vmatrix} -1 & 1 \\ 2 & 3 \end{vmatrix}$

$$= 4(18) - 2(-4) + 5(-5) = 72 + 8 - 25 = 55$$

$x = \dfrac{D_x}{D} = \dfrac{\begin{vmatrix} -5 & -1 & 1 \\ 10 & 2 & 3 \\ 1 & -2 & 6 \end{vmatrix}}{55} = \dfrac{-5\begin{vmatrix} 2 & 3 \\ -2 & 6 \end{vmatrix} - 10\begin{vmatrix} -1 & 1 \\ -2 & 6 \end{vmatrix} + 1\begin{vmatrix} -1 & 1 \\ 2 & 3 \end{vmatrix}}{55}$

$$= \frac{-5(18) - 10(-4) + 1(-5)}{55} = \frac{-55}{55} = -1$$

$y = \dfrac{D_y}{D} = \dfrac{\begin{vmatrix} 4 & -5 & 1 \\ 2 & 10 & 3 \\ 5 & 1 & 6 \end{vmatrix}}{55} = \dfrac{4\begin{vmatrix} 10 & 3 \\ 1 & 6 \end{vmatrix} - 2\begin{vmatrix} -5 & 1 \\ 1 & 6 \end{vmatrix} + 5\begin{vmatrix} -5 & 1 \\ 10 & 3 \end{vmatrix}}{55}$

$$= \frac{4(57) - 2(-31) + 5(-25)}{55} = \frac{165}{55} = 3$$

$z = \dfrac{D_z}{D} = \dfrac{\begin{vmatrix} 4 & -1 & -5 \\ 2 & 2 & 10 \\ 5 & -2 & 1 \end{vmatrix}}{55} = \dfrac{4\begin{vmatrix} 2 & 10 \\ -2 & 1 \end{vmatrix} - 2\begin{vmatrix} -1 & -5 \\ -2 & 1 \end{vmatrix} + 5\begin{vmatrix} -1 & -5 \\ 2 & 10 \end{vmatrix}}{55}$

$$= \frac{4(22) - 2(-11) + 5(0)}{55} = \frac{110}{55} = 2$$

$(-1, 3, 2)$

55. $D = \begin{vmatrix} 3 & 4 & 4 \\ 4 & -4 & 6 \\ 6 & -6 & 0 \end{vmatrix} = 4\begin{vmatrix} 4 & -4 \\ 6 & -6 \end{vmatrix} - 6\begin{vmatrix} 3 & 4 \\ 6 & -6 \end{vmatrix} + 0\begin{vmatrix} 3 & 4 \\ 4 & -4 \end{vmatrix}$

$$= 4(0) - 6(-42) + 0 = 252$$

$x = \dfrac{D_x}{D} = \dfrac{\begin{vmatrix} 11 & 4 & 4 \\ 11 & -4 & 6 \\ 3 & -6 & 0 \end{vmatrix}}{252} = \dfrac{4\begin{vmatrix} 11 & -4 \\ 3 & -6 \end{vmatrix} - 6\begin{vmatrix} 11 & 4 \\ 3 & -6 \end{vmatrix} + 0}{252}$

$$= \frac{4(-54) - 6(-78) + 0}{252} = \frac{252}{252} = 1$$

$y = \dfrac{D_y}{D} = \dfrac{\begin{vmatrix} 3 & 11 & 4 \\ 4 & 11 & 6 \\ 6 & 3 & 0 \end{vmatrix}}{252} = \dfrac{4\begin{vmatrix} 4 & 11 \\ 6 & 3 \end{vmatrix} - 6\begin{vmatrix} 3 & 11 \\ 6 & 3 \end{vmatrix} + 0}{252}$

$$= \frac{4(-54) - 6(-57) + 0}{252} = \frac{126}{252} = \frac{1}{2}$$

$z = \dfrac{D_z}{D} = \dfrac{\begin{vmatrix} 3 & 4 & 11 \\ 4 & -4 & 11 \\ 6 & -6 & 3 \end{vmatrix}}{252} = \dfrac{3\begin{vmatrix} -4 & 11 \\ -6 & 3 \end{vmatrix} - 4\begin{vmatrix} 4 & 11 \\ -6 & 3 \end{vmatrix} + 6\begin{vmatrix} 4 & 11 \\ -4 & 11 \end{vmatrix}}{252}$

$$= \frac{3(54) - 4(78) + 6(88)}{252} = \frac{378}{252} = \frac{3}{2}$$

$\left(1, \dfrac{1}{2}, \dfrac{3}{2}\right)$

57. $D = \begin{vmatrix} 3 & 3 & 4 \\ 3 & 5 & 9 \\ 5 & 9 & 17 \end{vmatrix} = 3\begin{vmatrix} 5 & 9 \\ 9 & 17 \end{vmatrix} - 3\begin{vmatrix} 3 & 4 \\ 9 & 17 \end{vmatrix} + 5\begin{vmatrix} 3 & 4 \\ 5 & 9 \end{vmatrix}$

$$= 3(4) - 3(15) + 5(7) = 2$$

$$a = \frac{D_a}{D} = \frac{\begin{vmatrix} 1 & 3 & 4 \\ 2 & 5 & 9 \\ 4 & 9 & 17 \end{vmatrix}}{2} = \frac{1\begin{vmatrix} 5 & 9 \\ 9 & 17 \end{vmatrix} - 2\begin{vmatrix} 3 & 4 \\ 9 & 17 \end{vmatrix} + 4\begin{vmatrix} 3 & 4 \\ 5 & 9 \end{vmatrix}}{2}$$

$$= \frac{1(4) - 2(15) + 4(7)}{2} = \frac{2}{2} = 1$$

$$b = \frac{D_b}{D} = \frac{\begin{vmatrix} 3 & 1 & 4 \\ 3 & 2 & 9 \\ 5 & 4 & 17 \end{vmatrix}}{2} = \frac{3\begin{vmatrix} 2 & 9 \\ 4 & 17 \end{vmatrix} - 3\begin{vmatrix} 1 & 4 \\ 4 & 17 \end{vmatrix} + 5\begin{vmatrix} 1 & 4 \\ 2 & 9 \end{vmatrix}}{2}$$

$$= \frac{3(-2) - 3(1) + 5(1)}{2} = \frac{-4}{2} = -2$$

$$c = \frac{D_c}{D} = \frac{\begin{vmatrix} 3 & 3 & 1 \\ 3 & 5 & 2 \\ 5 & 9 & 4 \end{vmatrix}}{2} = \frac{3\begin{vmatrix} 5 & 2 \\ 9 & 4 \end{vmatrix} - 3\begin{vmatrix} 3 & 1 \\ 9 & 4 \end{vmatrix} + 5\begin{vmatrix} 3 & 1 \\ 5 & 2 \end{vmatrix}}{2}$$

$$= \frac{3(2) - 3(3) + 5(1)}{2} = \frac{2}{2} = 1$$

$(1, -2, 1)$

59. $D = \begin{vmatrix} -3 & 10 \\ 9 & -3 \end{vmatrix} = -81$

$$x = \frac{\begin{vmatrix} 22 & 10 \\ 0 & -3 \end{vmatrix}}{-81} = \frac{-66}{-81} = \frac{22}{27}$$

$$y = \frac{\begin{vmatrix} -3 & 22 \\ 9 & 0 \end{vmatrix}}{-81} = \frac{-198}{-81} = \frac{22}{9}$$

$\left(\dfrac{22}{27}, \dfrac{22}{9}\right)$

61. $D = \begin{vmatrix} 4 & -1 \\ -2 & 1 \end{vmatrix} = 2$

$$x = \frac{\begin{vmatrix} -2 & -1 \\ 3 & 1 \end{vmatrix}}{2} = \frac{1}{2}$$

$$y = \frac{\begin{vmatrix} 4 & -2 \\ -2 & 3 \end{vmatrix}}{2} = \frac{8}{2} = 4$$

$\left(\dfrac{1}{2}, 4\right)$

63. $D = \begin{vmatrix} 1 & 1 & -1 \\ 6 & 4 & 3 \\ 3 & 0 & 6 \end{vmatrix} = 9$

$x = \dfrac{\begin{vmatrix} 2 & 1 & -1 \\ 4 & 4 & 3 \\ -3 & 0 & 6 \end{vmatrix}}{9} = \dfrac{3}{9} = \dfrac{1}{3}$

$y = \dfrac{\begin{vmatrix} 1 & 2 & -1 \\ 6 & 4 & 3 \\ 3 & -3 & 6 \end{vmatrix}}{9} = \dfrac{9}{9} = 1$

$z = \dfrac{\begin{vmatrix} 1 & 1 & 2 \\ 6 & 4 & 4 \\ 3 & 0 & -3 \end{vmatrix}}{9} = \dfrac{-6}{9} = -\dfrac{2}{3}$

$\left(\dfrac{1}{3}, 1, -\dfrac{2}{3}\right)$

65. $D = \begin{vmatrix} 3 & 1 & 1 \\ 1 & -4 & 2 \\ 1 & -3 & 1 \end{vmatrix} = 8$

$x = \dfrac{\begin{vmatrix} 6 & 1 & 1 \\ -1 & -4 & 2 \\ 0 & -3 & 1 \end{vmatrix}}{8} = \dfrac{16}{8} = 2$

$y = \dfrac{\begin{vmatrix} 3 & 6 & 1 \\ 1 & -1 & 2 \\ 1 & 0 & 1 \end{vmatrix}}{8} = \dfrac{4}{8} = \dfrac{1}{2}$

$z = \dfrac{\begin{vmatrix} 3 & 1 & 6 \\ 1 & -4 & -1 \\ 1 & -3 & 0 \end{vmatrix}}{8} = \dfrac{-4}{8} = -\dfrac{1}{2}$

$\left(2, \dfrac{1}{2}, -\dfrac{1}{2}\right)$

67. $D = \begin{vmatrix} 3 & -2 & 3 \\ 1 & 3 & 6 \\ 1 & 2 & 9 \end{vmatrix} = 48$

$x = \dfrac{\begin{vmatrix} 8 & -2 & 3 \\ -3 & 3 & 6 \\ -5 & 2 & 9 \end{vmatrix}}{48} = \dfrac{153}{48} = \dfrac{51}{16}$

$y = \dfrac{\begin{vmatrix} 3 & 8 & 3 \\ 1 & -3 & 6 \\ 1 & -5 & 9 \end{vmatrix}}{48} = \dfrac{-21}{48} = -\dfrac{7}{16}$

$z = \dfrac{\begin{vmatrix} 3 & -2 & 8 \\ 1 & 3 & -3 \\ 1 & 2 & -5 \end{vmatrix}}{48} = \dfrac{-39}{48} = -\dfrac{13}{16}$

$\left(\dfrac{51}{16}, -\dfrac{7}{16}, -\dfrac{13}{16}\right)$

69. $\begin{vmatrix} 5 - x & 4 \\ 1 & 2 - x \end{vmatrix} = 0$

$(5 - x)(2 - x) - 1(4) = 0$

$10 - 7x + x^2 - 4 = 0$

$x^2 - 7x + 6 = 0$

$(x - 6)(x - 1) = 0$

$x - 6 = 0 \Longrightarrow x = 6$

$x - 1 = 0 \Longrightarrow x = 1$

71. $\begin{vmatrix} x_1 & y_1 & 1 \\ x_2 & y_2 & 1 \\ x_3 & y_3 & 1 \end{vmatrix} = \begin{vmatrix} 0 & 3 & 1 \\ 4 & 0 & 1 \\ 8 & 5 & 1 \end{vmatrix} = 0 - 4\begin{vmatrix} 3 & 1 \\ 5 & 1 \end{vmatrix} + 8\begin{vmatrix} 3 & 1 \\ 0 & 1 \end{vmatrix}$

$= 0 - 4(-2) + 8(3)$

$= 8 + 24$

$= 32$

$A = +\dfrac{1}{2}(32) = 16$

73. $\begin{vmatrix} x_1 & y_1 & 1 \\ x_2 & y_2 & 1 \\ x_3 & y_3 & 1 \end{vmatrix} = \begin{vmatrix} -2 & 1 & 1 \\ 3 & -1 & 1 \\ 1 & 6 & 1 \end{vmatrix} = -2\begin{vmatrix} -1 & 1 \\ 6 & 1 \end{vmatrix} - 3\begin{vmatrix} 1 & 1 \\ 6 & 1 \end{vmatrix} + 1\begin{vmatrix} 1 & 1 \\ -1 & 1 \end{vmatrix}$

$$= -2(-7) - 3(-5) + 1(2)$$

$$= 14 + 15 + 2$$

$$= 31$$

$A = +\frac{1}{2}(31) = \frac{31}{2}$

75. $\begin{vmatrix} x_1 & y_1 & 1 \\ x_2 & y_2 & 1 \\ x_3 & y_3 & 1 \end{vmatrix} = \begin{vmatrix} 10 & 1 & 1 \\ 0 & -2 & 1 \\ 3 & 3 & 1 \end{vmatrix} = 10\begin{vmatrix} -2 & 1 \\ 3 & 1 \end{vmatrix} - 0\begin{vmatrix} 1 & 1 \\ 3 & 1 \end{vmatrix} + 3\begin{vmatrix} 1 & 1 \\ -2 & 1 \end{vmatrix}$

$$= 10(-5) - 0 + 3(3)$$

$$= -50 - 0 + 9$$

$$= -41$$

$A = -\frac{1}{2}(-41) = \frac{41}{2}$

77. $\begin{vmatrix} x_1 & y_1 & 1 \\ x_2 & y_2 & 1 \\ x_3 & y_3 & 1 \end{vmatrix} = \begin{vmatrix} -3 & 4 & 1 \\ 0 & \frac{3}{2} & 1 \\ 0 & \frac{7}{2} & 1 \end{vmatrix} = -3\begin{vmatrix} \frac{3}{2} & 1 \\ \frac{7}{2} & 1 \end{vmatrix} - 0\begin{vmatrix} 4 & 1 \\ \frac{7}{2} & 1 \end{vmatrix} + 0\begin{vmatrix} 4 & 1 \\ \frac{7}{2} & 1 \end{vmatrix}$

$$= -3(-2) - 0 + 0$$

$$= 6$$

$A = +\frac{1}{2}(6) = 3$

79. Area of shaded region = Area of rectangle − Area of triangle

The remaining vertices of the rectangle are $(6, -2)$, $(-3, -2)$, and $(-3, 2)$.

Rectangle: The length of the rectangle is $|6 - (-3)| = 9$.

               The width of the rectangle is $|2 - (-2)| = 4$.

               The area of the rectangle is $9(4) = 36$.

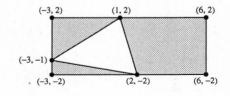

Triangle: $\begin{vmatrix} x_1 & y_1 & 1 \\ x_2 & y_2 & 1 \\ x_3 & y_3 & 1 \end{vmatrix} = \begin{vmatrix} 1 & 2 & 1 \\ 2 & -2 & 1 \\ -3 & -1 & 1 \end{vmatrix} = 1\begin{vmatrix} -2 & 1 \\ -1 & 1 \end{vmatrix} - 2\begin{vmatrix} 2 & 1 \\ -1 & 1 \end{vmatrix} + (-3)\begin{vmatrix} 2 & 1 \\ -2 & 1 \end{vmatrix}$

$$= 1(-1) - 2(3) - 3(4)$$

$$= -1 - 6 - 12$$

$$= -19$$

Area of triangle $= -\frac{1}{2}(-19) = \frac{19}{2}$

Area of shaded region $= 36 - \frac{19}{2} = \frac{72}{2} - \frac{19}{2} = \frac{53}{2}$

81. Area I:

$$\begin{vmatrix} x_1 & y_1 & 1 \\ x_2 & y_2 & 1 \\ x_3 & y_3 & 1 \end{vmatrix} = \begin{vmatrix} -1 & 2 & 1 \\ 5 & 2 & 1 \\ 0 & 0 & 1 \end{vmatrix} = 0 - 0 + 1\begin{vmatrix} -1 & 2 \\ 5 & 2 \end{vmatrix} = 1(-12) = -12$$

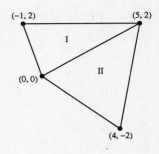

Area I $= -\frac{1}{2}(-12) = 6$

Area II:

$$\begin{vmatrix} x_1 & y_1 & 1 \\ x_2 & y_2 & 1 \\ x_3 & y_3 & 1 \end{vmatrix} = \begin{vmatrix} 0 & 0 & 1 \\ 5 & 2 & 1 \\ 4 & -2 & 1 \end{vmatrix} = 0 - 0 + 1\begin{vmatrix} 5 & 2 \\ 4 & -2 \end{vmatrix} = 1(-18) = -18$$

Area II $= -\frac{1}{2}(-18) = 9$

Area of shaded region $= 6 + 9 = 15$

83.
$$\begin{vmatrix} x_1 & y_1 & 1 \\ x_2 & y_2 & 1 \\ x_3 & y_3 & 1 \end{vmatrix} = \begin{vmatrix} -1 & 11 & 1 \\ 0 & 8 & 1 \\ 2 & 2 & 1 \end{vmatrix} = -1\begin{vmatrix} 8 & 1 \\ 2 & 1 \end{vmatrix} - 0 + 2\begin{vmatrix} 11 & 1 \\ 8 & 1 \end{vmatrix}$$

$$= -1(6) + 2(3)$$

$$= -6 + 6$$

$$= 0$$

Yes, the three points *are* collinear.

85.
$$\begin{vmatrix} x_1 & y_1 & 1 \\ x_2 & y_2 & 1 \\ x_3 & y_3 & 1 \end{vmatrix} = \begin{vmatrix} 1 & 2 & 1 \\ 5 & 0 & 1 \\ 10 & -2 & 1 \end{vmatrix} = 1\begin{vmatrix} 0 & 1 \\ -2 & 1 \end{vmatrix} - 5\begin{vmatrix} 0 & 1 \\ -2 & 1 \end{vmatrix} + 10\begin{vmatrix} 2 & 1 \\ 0 & 1 \end{vmatrix}$$

$$= 1(2) - 5(2) + 10(2)$$

$$= 2 - 10 + 20$$

$$= 12$$

No, the three points *are not* collinear.

87.
$$\begin{vmatrix} x_1 & y_1 & 1 \\ x_2 & y_2 & 1 \\ x_3 & y_3 & 1 \end{vmatrix} = \begin{vmatrix} -2 & \frac{1}{3} & 1 \\ 2 & 1 & 1 \\ 3 & \frac{1}{5} & 1 \end{vmatrix} = -2\begin{vmatrix} 1 & 1 \\ \frac{1}{5} & 1 \end{vmatrix} - 2\begin{vmatrix} \frac{1}{3} & 1 \\ \frac{1}{5} & 1 \end{vmatrix} + 3\begin{vmatrix} \frac{1}{3} & 1 \\ 1 & 1 \end{vmatrix}$$

$$= -2\left(\frac{4}{5}\right) - 2\left(\frac{2}{15}\right) + 3\left(-\frac{2}{3}\right)$$

$$= -\frac{8}{5} - \frac{4}{15} - 2$$

$$= -\frac{24}{15} - \frac{4}{15} - \frac{30}{15} = -\frac{58}{15}$$

No, the three points *are not* collinear.

89.
$$\begin{vmatrix} x & y & 1 \\ 0 & 0 & 1 \\ 5 & 3 & 1 \end{vmatrix} = 0$$

$$\begin{vmatrix} x & y & 1 \\ 0 & 0 & 1 \\ 5 & 3 & 1 \end{vmatrix} = -0 + 0 - 1\begin{vmatrix} x & y \\ 5 & 3 \end{vmatrix} = -1(3x - 5y) = -3x + 5y = 0 \quad \text{or } 3x - 5y = 0$$

So, an equation of the line through $(0, 0)$ and $(5, 3)$ is $3x - 5y = 0$.

91. $\begin{vmatrix} x & y & 1 \\ 10 & 7 & 1 \\ -2 & -7 & 1 \end{vmatrix} = 0$

$\begin{vmatrix} x & y & 1 \\ 10 & 7 & 1 \\ -2 & -7 & 1 \end{vmatrix} = x\begin{vmatrix} 7 & 1 \\ -7 & 1 \end{vmatrix} - 10\begin{vmatrix} y & 1 \\ -7 & 1 \end{vmatrix} - 2\begin{vmatrix} y & 1 \\ 7 & 1 \end{vmatrix} = 0$

$x(14) - 10(y + 7) - 2(y - 7) = 0$

$14x - 10y - 70 - 2y + 14 = 0$

$14x - 12y - 56 = 0$ or $7x - 6y - 28 = 0$

So, an equation of the line through $(10, 7)$ and $(-2, -7)$ is $14x - 12y - 56 = 0$ (or $7x - 6y - 28 = 0$).

93. $\begin{vmatrix} x & y & 1 \\ 2 & 7 & 1 \\ 10 & -1 & 1 \end{vmatrix} = 0$

$\begin{vmatrix} x & y & 1 \\ 2 & 7 & 1 \\ 10 & -1 & 1 \end{vmatrix} = x\begin{vmatrix} 7 & 1 \\ -1 & 1 \end{vmatrix} - 2\begin{vmatrix} y & 1 \\ -1 & 1 \end{vmatrix} + 10\begin{vmatrix} y & 1 \\ 7 & 1 \end{vmatrix} = 0$

$x(8) + (-2)(y + 1) + 10(y - 7) = 0$

$8x - 2y - 2 + 10y - 70 = 0$

$8x + 8y - 72 = 0$ or $x + y - 9 = 0$

95. $\begin{vmatrix} x & y & 1 \\ 6 & \frac{3}{2} & 1 \\ -9 & -\frac{7}{2} & 1 \end{vmatrix} = 0$

$\begin{vmatrix} x & y & 1 \\ 6 & \frac{3}{2} & 1 \\ -9 & -\frac{7}{2} & 1 \end{vmatrix} = x\begin{vmatrix} \frac{3}{2} & 1 \\ -\frac{7}{2} & 1 \end{vmatrix} - 6\begin{vmatrix} y & 1 \\ -\frac{7}{2} & 1 \end{vmatrix} + (-9)\begin{vmatrix} y & 1 \\ \frac{3}{2} & 1 \end{vmatrix} = 0$

$x(5) - 6\left(y + \frac{7}{2}\right) + (-9)\left(y - \frac{3}{2}\right) = 0$

$5x - 6y - 21 - 9y + \frac{27}{2} = 0$

$5x - 15y - \frac{15}{2} = 0$ or $x - 3y - \frac{3}{2} = 0$ or $2x - 6y - 3 = 0$

97. $(0, 1)$: $\quad 1 = a(0)^2 + b(0) + c \Rightarrow \qquad\qquad c = 1$

$(1, -3)$: $\quad -3 = a(1)^2 + b(1) + c \Rightarrow \quad a + b + c = -3$

$(-2, 21)$: $\quad 21 = a(-2)^2 + b(-2) + c \Rightarrow 4a - 2b + c = 21$

$$D = \begin{vmatrix} 0 & 0 & 1 \\ 1 & 1 & 1 \\ 4 & -2 & 1 \end{vmatrix} = 0 - 0 + 1\begin{vmatrix} 1 & 1 \\ 4 & -2 \end{vmatrix} = 1(-6) = -6$$

$$a = \frac{D_a}{D} = \frac{\begin{vmatrix} 1 & 0 & 1 \\ -3 & 1 & 1 \\ 21 & -2 & 1 \end{vmatrix}}{-6} = \frac{1\begin{vmatrix} 1 & 1 \\ -2 & 1 \end{vmatrix} - 0 + 1\begin{vmatrix} -3 & 1 \\ 21 & -2 \end{vmatrix}}{-6} = \frac{1(3) - 0 + 1(-15)}{-6} = \frac{-12}{-6} = 2$$

$$b = \frac{D_b}{D} = \frac{\begin{vmatrix} 0 & 1 & 1 \\ 1 & -3 & 1 \\ 4 & 21 & 1 \end{vmatrix}}{-6} = \frac{0 - 1\begin{vmatrix} 1 & 1 \\ 4 & 1 \end{vmatrix} + 1\begin{vmatrix} 1 & -3 \\ 4 & 21 \end{vmatrix}}{-6} = \frac{0 - 1(-3) + 1(33)}{-6} = \frac{36}{-6} = -6$$

$$c = \frac{D_c}{D} = \frac{\begin{vmatrix} 0 & 0 & 1 \\ 1 & 1 & -3 \\ 4 & -2 & 21 \end{vmatrix}}{-6} = \frac{0 - 0 + 1\begin{vmatrix} 1 & 1 \\ 4 & -2 \end{vmatrix}}{-6} = \frac{0 - 0 + 1(-6)}{-6} = \frac{-6}{-6} = 1$$

$(2, -6, 1)$

$y = 2x^2 - 6x + 1$

99. $(1, -1)$: $\quad -1 = a(1)^2 + b(1) + c \Rightarrow a + b + c = -1 \Rightarrow$

$(-1, -5)$: $\quad -5 = a(-1)^2 + b(-1) + c \Rightarrow a - b + c = -5 \Rightarrow$

$\left(\frac{1}{2}, \frac{1}{4}\right)$: $\quad \frac{1}{4} = a\left(\frac{1}{2}\right)^2 + b\left(\frac{1}{2}\right) + c \Rightarrow \frac{1}{4}a + \frac{1}{2}b + c = \frac{1}{4} \Rightarrow$

$$\begin{cases} a + b + c = -1 \\ a - b + c = -5 \\ a + 2b + 4c = 1 \end{cases}$$

$$D = \begin{vmatrix} 1 & 1 & 1 \\ 1 & -1 & 1 \\ 1 & 2 & 4 \end{vmatrix} = 1\begin{vmatrix} -1 & 1 \\ 2 & 4 \end{vmatrix} - 1\begin{vmatrix} 1 & 1 \\ 2 & 4 \end{vmatrix} + 1\begin{vmatrix} 1 & 1 \\ -1 & 1 \end{vmatrix} = 1(-6) - 1(2) + 1(2) = -6$$

$$a = \frac{D_a}{D} = \frac{\begin{vmatrix} -1 & 1 & 1 \\ -5 & -1 & 1 \\ 1 & 2 & 4 \end{vmatrix}}{-6} = \frac{-1\begin{vmatrix} -1 & 1 \\ 2 & 4 \end{vmatrix} - (-5)\begin{vmatrix} 1 & 1 \\ 2 & 4 \end{vmatrix} + 1\begin{vmatrix} 1 & 1 \\ -1 & 1 \end{vmatrix}}{-6} = \frac{-1(-6) + 5(2) + 1(2)}{-6} = \frac{18}{-6} = -3$$

$$b = \frac{D_b}{D} = \frac{\begin{vmatrix} 1 & -1 & 1 \\ 1 & -5 & 1 \\ 1 & 1 & 4 \end{vmatrix}}{-6} = \frac{1\begin{vmatrix} -5 & 1 \\ 1 & 4 \end{vmatrix} - 1\begin{vmatrix} -1 & 1 \\ 1 & 4 \end{vmatrix} + 1\begin{vmatrix} -1 & 1 \\ -5 & 1 \end{vmatrix}}{-6} = \frac{1(-21) - 1(-5) + 1(4)}{-6} = \frac{-12}{-6} = 2$$

$$c = \frac{D_c}{D} = \frac{\begin{vmatrix} 1 & 1 & -1 \\ 1 & -1 & -5 \\ 1 & 2 & 1 \end{vmatrix}}{-6} = \frac{1\begin{vmatrix} -1 & -5 \\ 2 & 1 \end{vmatrix} - 1\begin{vmatrix} 1 & -1 \\ 2 & 1 \end{vmatrix} + 1\begin{vmatrix} 1 & -1 \\ -1 & -5 \end{vmatrix}}{-6} = \frac{1(9) - 1(3) + 1(-6)}{-6} = \frac{0}{-6} = 0$$

$(-3, 2, 0)$

$y = -3x^2 + 2x + 0 = -3x^2 + 2x$

101. $\begin{cases} I_1 - I_2 + I_3 = 0 \\ 3I_1 + 2I_2 \qquad = 7 \\ \qquad 2I_2 + 4I_3 = 8 \end{cases}$

$D = \begin{vmatrix} 1 & -1 & 1 \\ 3 & 2 & 0 \\ 0 & 2 & 4 \end{vmatrix} = 26$

$I_1 = \dfrac{\begin{vmatrix} 0 & -1 & 1 \\ 7 & 2 & 0 \\ 8 & 2 & 4 \end{vmatrix}}{26} = \dfrac{26}{26} = 1$

$I_2 = \dfrac{\begin{vmatrix} 1 & 0 & 1 \\ 3 & 7 & 0 \\ 0 & 8 & 4 \end{vmatrix}}{26} = \dfrac{52}{26} = 2$

$I_3 = \dfrac{\begin{vmatrix} 1 & -1 & 0 \\ 3 & 2 & 7 \\ 0 & 2 & 8 \end{vmatrix}}{26} = \dfrac{26}{26} = 1$

$(1, 2, 1)$

$I_1 = 1, \ I_2 = 2, \ I_3 = 1$

103. (a) $D = \begin{vmatrix} k & 1 - k \\ 1 - k & k \end{vmatrix} = k^2 - (1 - k)^2$

$\qquad\qquad\qquad = k^2 - (1 - 2k + k^2)$

$\qquad\qquad\qquad = k^2 - 1 + 2k - k^2$

$\qquad\qquad\qquad = 2k - 1$

$x = \dfrac{D_x}{D} = \dfrac{\begin{vmatrix} 1 & 1 - k \\ 3 & k \end{vmatrix}}{2k - 1} = \dfrac{k - 3(1 - k)}{2k - 1}$

$\qquad\qquad = \dfrac{k - 3 + 3k}{2k - 1} = \dfrac{4k - 3}{2k - 1}$

$y = \dfrac{D_y}{D} = \dfrac{\begin{vmatrix} k & 1 \\ 1 - k & 3 \end{vmatrix}}{2k - 1} = \dfrac{3k - (1 - k)}{2k - 1}$

$\qquad\qquad = \dfrac{3k - 1 + k}{2k - 1} = \dfrac{4k - 1}{2k - 1}$

$\left(\dfrac{4k - 3}{2k - 1}, \dfrac{4k - 1}{2k - 1} \right)$

(b) Cramer's Rule cannot be used when $D = 0$ and $D = 2k - 1$.

$2k - 1 = 0$

$2k = 1$

$k = \dfrac{1}{2}$

If $k = \frac{1}{2}$, Cramer's Rule cannot be used.

105. No, the matrix must be square.

107. The minor of an entry of a square matrix is the determinant of the matrix that remains after the deletion of the row and column in which the entry occurs.

109. $f(x) = \dfrac{1}{3}x^2$

(a) $f(6) = \dfrac{1}{3}(6)^2 = \dfrac{1}{3}(36) = 12$

(b) $f\left(\dfrac{3}{4}\right) = \dfrac{1}{3}\left(\dfrac{3}{4}\right)^2 = \dfrac{1}{3}\left(\dfrac{9}{16}\right) = \dfrac{3}{16}$

111. Vertical shift 2 units downward

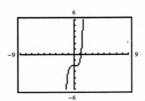

113. Reflection in the x-axis

115. $(9, -2)$ and $(5, 8)$

Point: $(9, -2)$

Slope: $m = \dfrac{8 - (-2)}{5 - 9} = \dfrac{10}{-4} = -\dfrac{5}{2}$

$y - y_1 = m(x - x_1)$

$y - (-2) = -\dfrac{5}{2}(x - 9)$

$y + 2 = -\dfrac{5}{2}x + \dfrac{45}{2}$

$y = -\dfrac{5}{2}x + \dfrac{41}{2}$

117. *Verbal Model:* $\boxed{\text{First number}} + \boxed{\text{Second number}} + \boxed{\text{Third number}} = 33$

  Labels: First number $= x$

     Second number $= x + 3$

     Third number $= 4x$

  Equation: $x + (x + 3) + 4x = 33$

       $6x + 3 = 33$

       $6x = 30$

       $x = 5$ and $x + 3 = 8, 4x = 20$

The three numbers are 5, 8, and 20.

Section 4.5   Linear Inequalities in Two Variables

1. $y \geq -2$

  Graph (b)

3. $3x - 2y < 0$

   $-2y < -3x$

   $y > \frac{3}{2}x$

  Graph (d)

5. $x + y < 4$

   $y < -x + 4$

  Graph (f)

7. (a) $(0, 0)$

    $0 - 2(0) \overset{?}{<} 4$

       $0 < 4$

  Yes, $(0, 0)$ *is* a solution.

  (b) $(2, -1)$

     $2 - 2(-1) \overset{?}{<} 4$

      $2 + 2 \overset{?}{<} 4$

        $4 \not< 4$

   No, $(2, -1)$ is *not* a solution.

  (c) $(3, 4)$

     $3 - 2(4) \overset{?}{<} 4$

      $3 - 8 \overset{?}{<} 4$

       $-5 < 4$

  Yes, $(3, 4)$ *is* a solution.

  (d) $(5, 1)$

     $5 - 2(1) \overset{?}{<} 4$

      $5 - 2 \overset{?}{<} 4$

        $3 < 4$

   Yes, $(5, 1)$ *is* a solution.

9. (a) $(1, 3)$

     $3(1) + 3 \overset{?}{\geq} 10$

        $6 \not\geq 10$

  No, $(1, 3)$ is *not* a solution.

  (b) $(-3, 1)$

     $3(-3) + 1 \overset{?}{\geq} 10$

      $-9 + 1 \overset{?}{\geq} 10$

        $-8 \not\geq 10$

   No, $(-3, 1)$ is *not* a solution.

  (c) $(3, 1)$

     $3(3) + 1 \overset{?}{\geq} 10$

       $10 \geq 10$

  Yes, $(3, 1)$ *is* a solution.

  (d) $(2, 15)$

     $3(2) + 15 \overset{?}{\geq} 10$

       $21 \geq 10$

   Yes, $(2, 15)$ *is* a solution.

11. (a) $(0, 2)$

$$2 \overset{?}{>} 0.2(0) - 1$$

$$2 > -1$$

Yes, $(0, 2)$ *is* a solution.

(b) $(6, 0)$

$$0 \overset{?}{>} 0.2(6) - 1$$

$$0 \overset{?}{>} 1.2 - 1$$

$$0 \not> 0.2$$

No, $(6, 0)$ is *not* a solution.

(c) $(4, -1)$

$$-1 \overset{?}{>} 0.2(4) - 1$$

$$-1 \overset{?}{>} 0.8 - 1$$

$$-1 \not> -0.2$$

No, $(4, -1)$ is *not* a solution.

(d) $(-2, 7)$

$$7 \overset{?}{>} 0.2(-2) - 1$$

$$7 \overset{?}{>} -0.4 - 1$$

$$7 > -1.4$$

Yes, $(-2, 7)$ *is* a solution.

13. (a) $(-1, 4)$

$$4 \overset{?}{\leq} 3 - |-1|$$

$$4 \overset{?}{\leq} 3 - 1$$

$$4 \not\leq 2$$

No, $(-1, 4)$ is *not* a solution.

(b) $(2, -2)$

$$-2 \overset{?}{\leq} 3 - |-2|$$

$$-2 \overset{?}{\leq} 3 - 2$$

$$-2 \leq 1$$

Yes, $(2, -2)$ *is* a solution.

(c) $(6, 0)$

$$0 \overset{?}{\leq} 3 - |6|$$

$$0 \overset{?}{\leq} 3 - 6$$

$$0 \not\leq -3$$

No, $(6, 0)$ is *not* a solution.

(d) $(5, -2)$

$$-2 \overset{?}{\leq} 3 - |5|$$

$$-2 \overset{?}{\leq} 3 - 5$$

$$-2 \leq -2$$

Yes, $(5, -2)$ *is* a solution.

15.

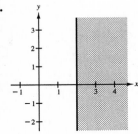

17.

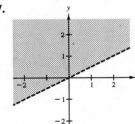

19.

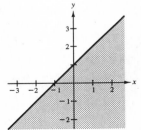

21. Note: The equation $y - 1 = -\frac{1}{2}(x - 2)$ in point-slope form is an equation of the line through the point $(2, 1)$ with slope $m = -\frac{1}{2}$.

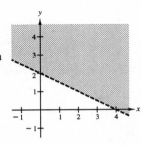

23. $\dfrac{x}{3} + \dfrac{y}{4} \leq 1$

$$\frac{y}{4} \leq -\frac{x}{3} + 1$$

$$y \leq -\frac{4}{3}x + 4$$

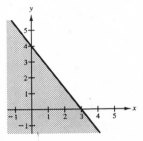

25. $x - 2y \geq 6$

$\qquad -2y \geq -x + 6$

$\qquad\quad y \leq \frac{1}{2}x - 3$

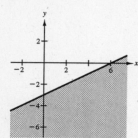

27. $3x - 2y \geq 4$

$\qquad -2y \geq -3x + 4$

$\qquad\quad y \leq \frac{3}{2}x - 2$

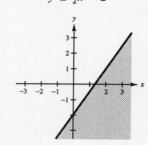

29. $0.2x + 0.3y < 2$

$\qquad 2x + 3y < 20$

$\qquad\quad 3y < -2x + 20$

$\qquad\qquad y < -\frac{2}{3}x + \frac{20}{3}$

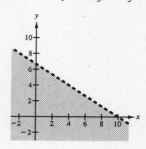

31.

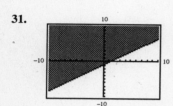

33.

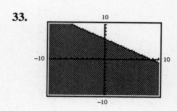

35. $x - 2y - 4 \geq 0$

$\qquad -2y \geq -x + 4$

$\qquad\quad y \leq \frac{1}{2}x - 2$

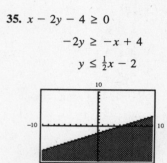

37. $2x + 3y - 12 \leq 0$

$\qquad 3y \leq -2x + 12$

$\qquad\quad y \leq -\frac{2}{3}x + 4$

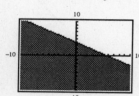

39. Line through $(-1, 5)$ and $(3, 2)$:

$$m = \frac{y_2 - y_1}{x_2 - x_1}$$

$$m = \frac{2 - 5}{3 - (-1)} = \frac{-3}{4} = -\frac{3}{4}$$

$$y - y_1 = m(x - x_1)$$

$$y - 5 = -\frac{3}{4}(x + 1) = -\frac{3}{4}x - \frac{3}{4}$$

$$y = -\frac{3}{4}x + \frac{17}{4}$$

The shaded region is *above* the line, and the boundary line is dashed. So, the inequality is $y > -\frac{3}{4}x + \frac{17}{4}$ (or $3x + 4y > 17$).

41. Horizontal line through $(0, 2)$:

$\qquad y = 2$

The shaded region is *below* the line, and the boundary line is dashed. So, the inequality is $y < 2$.

43. Line through $(0, 0)$ and $(2, 1)$:

$$m = \frac{y_2 - y_1}{x_2 - x_1}$$

$$m = \frac{1 - 0}{2 - 0} = \frac{1}{2}$$

$$y = mx + b = \frac{1}{2}x + 0 = \frac{1}{2}x$$

The shaded region is *above* the line, and the boundary line is dashed. So, the inequality is $y > \frac{1}{2}x$ (or $-x + 2y > 0$ or $x - 2y < 0$).

45. Graph (c)

47. Graph (f)

49. Graph (a)

51.

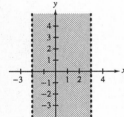

53.

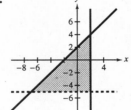

55.

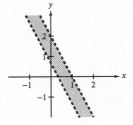

57.

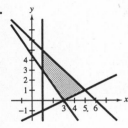

59.

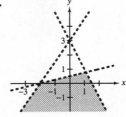

61.

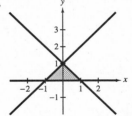

63.

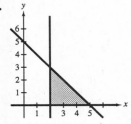

65.

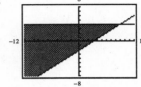

67.

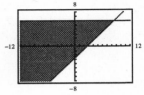

69.

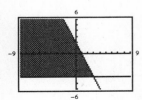

71. Lines: $x = 1$

        $x = 8$

        $y = 3$

        $y = -5$

System of inequalities:

$$\begin{cases} x & \geq & 1 \\ x & \leq & 8 \\ y & \leq & 3 \\ y & \geq & -5 \end{cases}$$

73. Line through $(-6, 3)$ and $(4, 12)$:

$$m = \frac{y_2 - y_1}{x_2 - x_1}$$

$$m = \frac{12 - 3}{4 - (-6)} = \frac{9}{10}$$

$$y - y_1 = m(x - x_1)$$

$$y - 3 = \frac{9}{10}(x + 6) = \frac{9}{10}x + \frac{27}{5}$$

$$y = \frac{9}{10}x + \frac{42}{5}$$

Line through $(4, 12)$ and $(3, 9)$:

$$m = \frac{y_2 - y_1}{x_2 - x_1}$$

$$m = \frac{9 - 12}{3 - 4} = \frac{-3}{-1} = 3$$

$$y - y_1 = m(x - x_1)$$

$$y - 12 = 3(x - 4) = 3x - 12$$

$$y = 3x$$

Line through $(3, 9)$ and $(-6, 3)$:

$$m = \frac{y_2 - y_1}{x_2 - x_1}$$

$$m = \frac{3 - 9}{-6 - 3} = \frac{-6}{-9} = \frac{2}{3}$$

$$y - y_1 = m(x - x_1)$$

$$y - 9 = \frac{2}{3}(x - 3) = \frac{2}{3}x - 2$$

$$y = \frac{2}{3}x + 7$$

—CONTINUED—

73. —CONTINUED—

Lines: System of inequalities:

$$y = \frac{9}{10}x + \frac{42}{5}$$

$$y = 3x$$

$$y = \frac{2}{3}x + 7$$

$$\begin{cases} y \le \dfrac{9}{10}x + \dfrac{42}{5} \\[2mm] y \ge 3x \\[2mm] y \ge \dfrac{2}{3}x + 7 \end{cases}$$

75. *Verbal Model:* $2\;\boxed{\text{Length}}\; + 2\;\boxed{\text{Width}}\; \le 500$

Inequality: $2x + 2y \le 500$

$$2y \le -2x + 500$$

$$y \le -x + 250$$

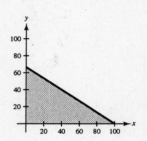

Note: $x \ge 0$
$\phantom{\text{Note:}}\;\; y \ge 0$

77. *Verbal Model:* $10\;\boxed{\text{Number of chairs}}\; + 15\;\boxed{\text{Number of tables}}\; \le 1{,}000$

Inequality: $10x + 15y \le 1{,}000$

$$15y \le -10x + 1{,}000$$

$$y \le -\tfrac{2}{3}x + \tfrac{200}{3}$$

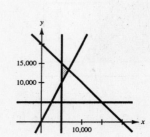

Note: $x \ge 0$
$\phantom{\text{Note:}}\;\; y \ge 0$

79. *Verbal Model:*

$$\boxed{\begin{array}{c}\text{Amount in}\\\text{first account}\end{array}} + \boxed{\begin{array}{c}\text{Amount in}\\\text{second account}\end{array}} \le 20{,}000$$

$$\boxed{\text{Amount in first account}} \ge 5000$$

$$\boxed{\text{Amount in second account}} \ge 5000$$

$$\boxed{\begin{array}{c}\text{Amount in}\\\text{second account}\end{array}} \ge 2 \boxed{\begin{array}{c}\text{Amount in}\\\text{first account}\end{array}}$$

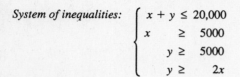

Labels: Amount in first account $= x$ (dollars)

Amount in second account $= y$ (dollars)

System of inequalities:
$$\begin{cases} x + y \le 20{,}000 \\ x \ge 5000 \\ \phantom{x+{}} y \ge 5000 \\ \phantom{x+{}} y \ge 2x \end{cases}$$

81. *Verbal Model:*

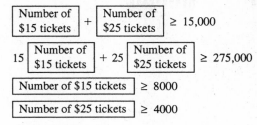

$$\boxed{\text{Number of } \$15 \text{ tickets}} + \boxed{\text{Number of } \$25 \text{ tickets}} \geq 15{,}000$$

$$15\,\boxed{\text{Number of } \$15 \text{ tickets}} + 25\,\boxed{\text{Number of } \$25 \text{ tickets}} \geq 275{,}000$$

$$\boxed{\text{Number of } \$15 \text{ tickets}} \geq 8000$$

$$\boxed{\text{Number of } \$25 \text{ tickets}} \geq 4000$$

Labels: Number of \$15 tickets = x

Number of \$25 tickets = y

System of inequalities:
$$\begin{cases} x + y \geq 15{,}000 \\ 15x + 25y \geq 275{,}000 \\ x \geq 8000 \\ y \geq 4000 \end{cases}$$

83. Line through $(0, 0)$ and $(70, -10)$:   Lines: System of inequalities:

$$m = \frac{y_2 - y_1}{x_2 - x_1}$$

$$m = \frac{-10 - 0}{70 - 0} = \frac{-10}{70} = -\frac{1}{7}$$

$$y = mx + b$$

$$y = -\frac{1}{7}x + 0 = -\frac{1}{7}x$$

Lines:

$$y = -\frac{1}{7}x$$

$$y = 0$$

$$y = -10$$

$$x = 90$$

System of inequalities:
$$\begin{cases} y \geq -\frac{1}{7}x \\ y \leq 0 \\ y \geq -10 \\ x < 90 \end{cases}$$

85. A line on a plane divides the plane into two regions—the set of points on one side of the line and the set of points on the other side of the line. Each of these regions is a half-plane. Here is an example of an inequality whose graph is a half-plane:

$$x - 2y > 1$$

87. You can test a point in one of the half-planes to determine whether the coordinates of that point satisfy the inequality. If they do, the half-plane containing the test point is the solution set. If not, the half-plane on the opposite side of the line is the solution set.

89. $2x + 4y = 8$

$$4y = -2x + 8$$

$$y = -\frac{1}{2}x + 2$$

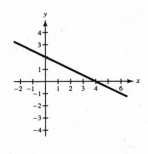

91. $y = 4(x + 1) - 3$

$$y = 4x + 4 - 3$$

$$y = 4x + 1$$

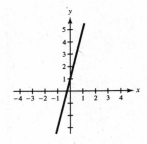

93. $x + 4y = 5$

$$4y = -x + 5$$

$$y = -\frac{1}{4}x + \frac{5}{4}$$

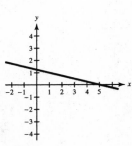

95. $y = -5$ Horizontal line

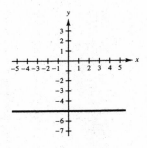

97. $42(4) + 60(x) = 55(4 + x)$

$$168 + 60x = 220 + 55x$$

$$168 + 5x = 220$$

$$5x = 52$$

$$x = 10.4$$

The truck must travel at 60 miles per hour for 10.4 hours.

Review Exercises for Chapter 4

1. (a) $(5, -2)$

$$-3(5) + 2(-2) \stackrel{?}{=} -19$$

$$-15 - 4 = -19$$

$$5(5) - (-2) \stackrel{?}{=} 27$$

$$25 + 2 = 27$$

Yes, $(5, -2)$ *is* a solution.

(b) $(3, 4)$

$$-3(3) + 2(4) \stackrel{?}{=} -19$$

$$-9 + 8 \neq -19$$

No, $(3, 4)$ is *not* a solution.

3. The solution is $(1, 1)$.

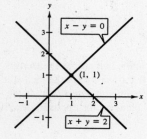

5. The system has *no* solution; it is inconsistent.

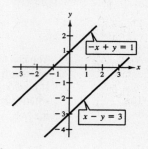

7. The solution is $(4, 8)$.

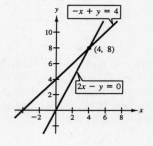

9. $(3, 4)$

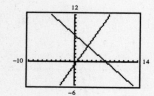

11. $(2, 1)$

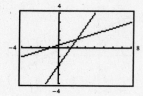

13. $\begin{cases} 2x + 3y = 1 \\ x + 4y = -2 \Longrightarrow x = -4y - 2 \end{cases}$

$$2x + 3y = 1$$
$$2(-4y - 2) + 3y = 1$$
$$-8y - 4 + 3y = 1$$
$$-5y - 4 = 1$$
$$-5y = 5$$
$$y = -1 \quad \text{and} \quad x = -4(-1) - 2$$
$$x = 4 - 2$$
$$x = 2$$

$(2, -1)$

15. $\begin{cases} -5x + 2y = 4 \Longrightarrow 2y = 5x + 4 \Longrightarrow y = \frac{5}{2}x + 2 \\ 10x - 4y = 7 \end{cases}$

$$10x - 4y = 7$$
$$10x - 4\left(\frac{5}{2}x + 2\right) = 7$$
$$10x - 10x - 8 = 7$$
$$-8 = 7 \quad \text{False}$$

The system has no solution; it is inconsistent.

17. $\begin{cases} 3x - 7y = 5 \Longrightarrow 3x = 7y + 5 \Longrightarrow x = \frac{7}{3}y + \frac{5}{3} \\ 5x - 9y = -5 \end{cases}$

$$5x - 9y = -5$$
$$5\left(\frac{7}{3}y + \frac{5}{3}\right) - 9y = -5$$
$$\frac{35}{3}y + \frac{25}{3} - 9y = -5$$
$$35y + 25 - 27y = -15$$
$$8y + 25 = -15$$
$$8y = -40$$
$$y = -5 \quad \text{and} \quad x = \frac{7}{3}(-5) + \frac{5}{3}$$
$$x = -\frac{35}{3} + \frac{5}{3}$$
$$x = -\frac{30}{3}$$
$$x = -10$$

$(-10, -5)$

19. $\begin{cases} x + y = 0 \Longrightarrow -x - y = 0 \\ 2x + y = 0 \Longrightarrow \underline{\quad 2x + y = 0 \quad} \end{cases}$

$$x \quad\quad = 0 \quad \text{and} \quad 0 + y = 0$$
$$y = 0$$

$(0, 0)$

21. $\begin{cases} 0.2x - 0.1y = 0.2 \Longrightarrow 2x - y = 2 \Longrightarrow 16x - 8y = 16 \\ 0.6x + 0.8y = 3.9 \Longrightarrow 6x + 8y = 39 \Longrightarrow \underline{\quad 6x + 8y = 39 \quad} \end{cases}$

$$22x = 55$$
$$x = \frac{55}{22}$$
$$x = \frac{5}{2} \quad \text{and} \quad 2\left(\frac{5}{2}\right) - y = 2$$
$$5 - y = 2$$
$$-y = -3$$
$$y = 3$$

$\left(\frac{5}{2}, 3\right)$

23. (a) $(8, 1, -3)$

$$3(8) - (1) + 5(-3) \overset{?}{=} 6$$
$$24 - 1 - 15 \overset{?}{=} 6$$
$$8 \neq 6$$

No, $(8, 1, -3)$ is *not* a solution.

(b) $(0, 4, 2)$

$$3(0) - (4) + 5(2) \overset{?}{=} 6$$
$$0 - 4 + 10 \overset{?}{=} 6$$
$$6 = 6$$
$$-0 + 8(2) \overset{?}{=} 16$$
$$16 = 16$$
$$10(0) + 3(4) - 5(2) \overset{?}{=} 2$$
$$0 + 12 - 10 \overset{?}{=} 2$$
$$2 = 2$$

Yes, $(0, 4, 2)$ *is* a solution.

25. $\begin{cases} x - 6y + 2z = 13 \\ \quad\quad y - z = -4 \\ \quad\quad\quad\quad z = 3 \end{cases}$

$$y - 3 = -4 \qquad\qquad x - 6(-1) + 2(3) = 13$$
$$y = -1 \qquad\qquad\quad x + 6 + 6 = 13$$
$$x + 12 = 13$$
$$x = 1$$

$(1, -1, 3)$

27. $\begin{cases} 2x + 3y + z = 10 \\ 2x - 3y - 3z = 22 \\ 4x - 2y + 3z = -2 \end{cases}$

$\begin{cases} x + \frac{3}{2}y + \frac{1}{2}z = 5 \qquad (1/2)\text{Eqn. 1} \\ 2x - 3y - 3z = 22 \\ 4x - 2y + 3z = -2 \end{cases}$

$\begin{cases} x + \frac{3}{2}y + \frac{1}{2}z = 5 \\ \quad -6y - 4z = 12 \qquad (-2)\text{Eqn. 1} + \text{Eqn. 2} \\ \quad -8y + z = -22 \qquad (-4)\text{Eqn. 1} + \text{Eqn. 3} \end{cases}$

$\begin{cases} x + \frac{3}{2}y + \frac{1}{2}z = 5 \\ \quad y + \frac{2}{3}z = -2 \qquad (-1/6)\text{Eqn. 2} \\ \quad -8y + z = -22 \end{cases}$

$\begin{cases} x + \frac{3}{2}y + \frac{1}{2}z = 5 \\ \quad y + \frac{2}{3}z = -2 \\ \quad\quad \frac{19}{3}z = -38 \qquad (8)\text{Eqn. 2} + \text{Eqn. 3} \end{cases}$

$\begin{cases} x + \frac{3}{2}y + \frac{1}{2}z = 5 \\ \quad y + \frac{2}{3}z = -2 \\ \quad\quad z = -6 \qquad (3/19)\text{Eqn. 3} \end{cases}$

$$y + \tfrac{2}{3}(-6) = -2 \qquad\qquad x + \tfrac{3}{2}(2) + \tfrac{1}{2}(-6) = 5$$
$$y - 4 = -2 \qquad\qquad\qquad x + 3 - 3 = 5$$
$$y = 2 \qquad\qquad\qquad\qquad x = 5$$

$(5, 2, -6)$

29. 1 row, 2 columns
This is a 1×2 matrix.

31. 3 rows, 1 column
This is a 3×1 matrix.

33. (a) $\begin{bmatrix} 12 & -2 \\ 3 & 8 \end{bmatrix}$

(b) $\begin{bmatrix} 12 & -2 & \vdots & 6 \\ 3 & 8 & \vdots & -5 \end{bmatrix}$

35. (a) $\begin{bmatrix} 3 & -3 & 1 \\ 10 & 7 & 0 \\ 6 & 14 & -9 \end{bmatrix}$

(b) $\begin{bmatrix} 3 & -3 & 1 & \vdots & 1 \\ 10 & 7 & 0 & \vdots & 4 \\ 6 & 14 & -9 & \vdots & 10 \end{bmatrix}$

37. $\begin{cases} 6x + 4y + z = 0 \\ 12x + 9y - 2z = 6 \\ x + 8y - 4z = -5 \end{cases}$

39. $\begin{cases} x + y = 4 \\ x - y = 2 \end{cases}$

$$\begin{bmatrix} 1 & 1 & 4 \\ 1 & -1 & 2 \end{bmatrix}$$

$-R_1 + R_2 \begin{bmatrix} 1 & 1 & 4 \\ 0 & -2 & -2 \end{bmatrix}$

$-1/2\, R_2 \begin{bmatrix} 1 & 1 & 4 \\ 0 & 1 & 1 \end{bmatrix}$

$x + y = 4$ and $x + 1 = 4$

$\qquad y = 1 \qquad\qquad x = 3$

$(3, -1)$

41. $\begin{cases} -x + y + 2z = 4 \\ 4x + 2y + z = -2 \\ x + 4y + 2z = 3 \end{cases}$

$$\begin{bmatrix} -1 & 1 & 2 & \vdots & 4 \\ 4 & 2 & 1 & \vdots & -2 \\ 1 & 4 & 2 & \vdots & 3 \end{bmatrix}$$

$-R_1 \begin{bmatrix} 1 & -1 & -2 & \vdots & -4 \\ 4 & 2 & 1 & \vdots & -2 \\ 1 & 4 & 2 & \vdots & 3 \end{bmatrix}$

$\begin{matrix} \\ -4R_1 + R_2 \\ -R_1 + R_3 \end{matrix} \begin{bmatrix} 1 & -1 & -2 & \vdots & -4 \\ 0 & 6 & 9 & \vdots & 14 \\ 0 & 5 & 4 & \vdots & 7 \end{bmatrix}$

$1/6\, R_2 \begin{bmatrix} 1 & -1 & -2 & \vdots & -4 \\ 0 & 1 & \frac{3}{2} & \vdots & \frac{7}{3} \\ 0 & 5 & 4 & \vdots & 7 \end{bmatrix}$

$-5R_2 + R_3 \begin{bmatrix} 1 & -1 & -2 & \vdots & -4 \\ 0 & 1 & \frac{3}{2} & \vdots & \frac{7}{3} \\ 0 & 0 & -\frac{7}{2} & \vdots & -\frac{14}{3} \end{bmatrix}$

$-2/7R_3 \begin{bmatrix} 1 & -1 & -2 & \vdots & -4 \\ 0 & 1 & \frac{3}{2} & \vdots & \frac{7}{3} \\ 0 & 0 & 1 & \vdots & \frac{4}{3} \end{bmatrix}$

$\begin{cases} x - y - 2z = -4 \\ y + \frac{3}{2}z = \frac{7}{3} \\ z = \frac{4}{3} \end{cases}$

$y + \frac{3}{2}\left(\frac{4}{3}\right) = \frac{7}{3} \qquad x - \frac{1}{3} - 2\left(\frac{4}{3}\right) = -4$

$\qquad y + 2 = \frac{7}{3} \qquad\qquad x - 3 = -4$

$\qquad\quad y = \frac{1}{3} \qquad\qquad\qquad x = -1$

$\left(-1, \frac{1}{3}, \frac{4}{3}\right)$

43. $\begin{cases} 5x + 4y = 2 \\ -x + y = -22 \end{cases}$

$$\begin{bmatrix} 5 & 4 & \vdots & 2 \\ -1 & 1 & \vdots & -22 \end{bmatrix}$$

$\begin{matrix} R_2 \\ R_1 \end{matrix} \begin{bmatrix} -1 & 1 & \vdots & -22 \\ 5 & 4 & \vdots & 2 \end{bmatrix}$

$-1R_1 \begin{bmatrix} 1 & -1 & \vdots & 22 \\ 5 & 4 & \vdots & 2 \end{bmatrix}$

$-5R_1 + R_2 \begin{bmatrix} 1 & -1 & \vdots & 22 \\ 0 & 9 & \vdots & -108 \end{bmatrix}$

$\frac{1}{9}R_2 \begin{bmatrix} 1 & -1 & \vdots & 22 \\ 0 & 1 & \vdots & -12 \end{bmatrix}$

$\begin{cases} x - y = 22 \\ y = -12 \end{cases}$

$x - (-12) = 22$

$x + 12 = 22$

$\qquad x = 10$

$(10, -12)$

45.
$$\begin{cases} x + 2y + 6z = 4 \\ -3x + 2y - z = -4 \\ 4x + 2z = 16 \end{cases}$$

$$\begin{cases} x + 2y + 6z = 4 \\ y + \frac{17}{8}z = 1 \\ z = -\frac{8}{5} \end{cases}$$

$$\begin{bmatrix} 1 & 2 & 6 & \vdots & 4 \\ -3 & 2 & -1 & \vdots & -4 \\ 4 & 0 & 2 & \vdots & 16 \end{bmatrix}$$

$$\begin{matrix} \\ 3R_1 + R_2 \\ -4R_1 + R_3 \end{matrix}\begin{bmatrix} 1 & 2 & 6 & \vdots & 4 \\ 0 & 8 & 17 & \vdots & 8 \\ 0 & -8 & -22 & \vdots & 0 \end{bmatrix}$$

$$\begin{matrix} \\ \frac{1}{8}R_2 \\ \\ \end{matrix}\begin{bmatrix} 1 & 2 & 6 & \vdots & 4 \\ 0 & 1 & \frac{17}{8} & \vdots & 1 \\ 0 & -8 & -22 & \vdots & 0 \end{bmatrix}$$

$$\begin{matrix} \\ \\ 8R_2 + R_3 \end{matrix}\begin{bmatrix} 1 & 2 & 6 & \vdots & 4 \\ 0 & 1 & \frac{17}{8} & \vdots & 1 \\ 0 & 0 & -5 & \vdots & 8 \end{bmatrix}$$

$$\begin{matrix} \\ \\ -\frac{1}{5}R_3 \end{matrix}\begin{bmatrix} 1 & 2 & 6 & \vdots & 4 \\ 0 & 1 & \frac{17}{8} & \vdots & 1 \\ 0 & 0 & 1 & \vdots & -\frac{8}{5} \end{bmatrix}$$

$$y + \frac{17}{8}\left(-\frac{8}{5}\right) = 1 \qquad x + 2\left(\frac{22}{5}\right) + 6\left(-\frac{8}{5}\right) = 4$$
$$y - \frac{17}{5} = 1 \qquad\qquad x + \frac{44}{5} - \frac{48}{5} = 4$$
$$y = \frac{22}{5} \qquad\qquad\qquad x - \frac{4}{5} = 4$$
$$\qquad\qquad\qquad\qquad x = \frac{24}{5}$$

$$\left(\frac{24}{5}, \frac{22}{5}, -\frac{8}{5}\right)$$

47.
$$\begin{cases} 2x_1 + 3x_2 + 3x_3 = 3 \\ 6x_1 + 6x_2 + 12x_3 = 13 \\ 12x_1 + 9x_2 - x_3 = 2 \end{cases}$$

$$\begin{cases} x_1 + x_2 + 2x_3 = \frac{13}{6} \\ x_2 - x_3 = -\frac{4}{3} \\ x_3 = 1 \end{cases}$$

$$\begin{bmatrix} 2 & 3 & 3 & \vdots & 3 \\ 6 & 6 & 12 & \vdots & 13 \\ 12 & 9 & -1 & \vdots & 2 \end{bmatrix}$$

$$\begin{matrix} R_2 \\ R_1 \\ \\ \end{matrix}\begin{bmatrix} 6 & 6 & 12 & \vdots & 13 \\ 2 & 3 & 3 & \vdots & 3 \\ 12 & 9 & -1 & \vdots & 2 \end{bmatrix}$$

$$\begin{matrix} \frac{1}{6}R_1 \\ \\ \\ \end{matrix}\begin{bmatrix} 1 & 1 & 2 & \vdots & \frac{13}{6} \\ 2 & 3 & 3 & \vdots & 3 \\ 12 & 9 & -1 & \vdots & 2 \end{bmatrix}$$

$$\begin{matrix} \\ -2R_1 + R_2 \\ -12R_1 + R_3 \end{matrix}\begin{bmatrix} 1 & 1 & 2 & \vdots & \frac{13}{6} \\ 0 & 1 & -1 & \vdots & -\frac{4}{3} \\ 0 & -3 & -25 & \vdots & -24 \end{bmatrix}$$

$$\begin{matrix} \\ \\ 3R_2 + R_3 \end{matrix}\begin{bmatrix} 1 & 1 & 2 & \vdots & \frac{13}{6} \\ 0 & 1 & -1 & \vdots & -\frac{4}{3} \\ 0 & 0 & -28 & \vdots & -28 \end{bmatrix}$$

$$\begin{matrix} \\ \\ -\frac{1}{28}R_3 \end{matrix}\begin{bmatrix} 1 & 1 & 2 & \vdots & \frac{13}{6} \\ 0 & 1 & -1 & \vdots & -\frac{4}{3} \\ 0 & 0 & 1 & \vdots & 1 \end{bmatrix}$$

$$x_2 - 1 = -\frac{4}{3} \qquad x_1 + \left(-\frac{1}{3}\right) + 2(1) = \frac{13}{6}$$
$$x_2 = -\frac{1}{3} \qquad\qquad x_1 + \frac{5}{3} = \frac{13}{6}$$
$$\qquad\qquad\qquad x_1 = \frac{13}{6} - \frac{10}{6}$$
$$\qquad\qquad\qquad x_1 = \frac{1}{2}$$

$$\left(\frac{1}{2}, -\frac{1}{3}, 1\right)$$

49. $\begin{cases} 0.2x - 0.1y = 0.07 \\ 0.4x - 0.5y = -0.01 \end{cases}$

$$\begin{bmatrix} 0.2 & -0.1 & \vdots & 0.07 \\ 0.4 & -0.5 & \vdots & -0.01 \end{bmatrix}$$

$\begin{matrix} 100R_1 \\ 100R_2 \end{matrix} \begin{bmatrix} 20 & -10 & \vdots & 7 \\ 40 & -50 & \vdots & -1 \end{bmatrix}$

$\frac{1}{20}R_1 \begin{bmatrix} 1 & -\frac{1}{2} & \vdots & \frac{7}{20} \\ 40 & -50 & \vdots & -1 \end{bmatrix}$

$-40R_1 + R_2 \begin{bmatrix} 1 & -\frac{1}{2} & \vdots & \frac{7}{20} \\ 0 & -30 & \vdots & -15 \end{bmatrix}$

$-\frac{1}{30}R_2 \begin{bmatrix} 1 & -\frac{1}{2} & \vdots & \frac{7}{20} \\ 0 & 1 & \vdots & \frac{1}{2} \end{bmatrix}$

$\begin{cases} x - \frac{1}{2}y = \frac{7}{20} \\ \quad\quad y = \frac{1}{2} \end{cases}$

$x - \frac{1}{2}\left(\frac{1}{2}\right) = \frac{7}{20}$

$x - \frac{1}{4} = \frac{7}{20}$

$x = \frac{12}{20}$

$x = \frac{3}{5}$

$\left(\frac{3}{5}, \frac{1}{2}\right)$ or $(0.6, 0.5)$

51.
$$\begin{bmatrix} 5 & -3 & 2 & \vdots & 2 \\ 2 & 2 & -3 & \vdots & 3 \\ 1 & -7 & 8 & \vdots & -4 \end{bmatrix}$$

$\frac{1}{5}R_1 \begin{bmatrix} 1 & -0.6 & 0.4 & \vdots & 0.4 \\ 2 & 2 & -3 & \vdots & 3 \\ 1 & -7 & 8 & \vdots & -4 \end{bmatrix}$

$\begin{matrix} \\ -2R_1 + R_2 \\ -R_1 + R_3 \end{matrix} \begin{bmatrix} 1 & -0.6 & 0.4 & \vdots & 0.4 \\ 0 & 3.2 & -3.8 & \vdots & 2.2 \\ 0 & -6.4 & 7.6 & \vdots & -4.4 \end{bmatrix}$

$\frac{1}{3.2}R_2 \begin{bmatrix} 1 & -0.6 & 0.4 & \vdots & 0.4 \\ 0 & 1 & -1.1875 & \vdots & 0.6875 \\ 0 & -6.4 & 7.6 & \vdots & -4.4 \end{bmatrix}$

$6.4R_2 + R_3 \begin{bmatrix} 1 & -0.6 & 0.4 & \vdots & 0.4 \\ 0 & 1 & -1.1875 & \vdots & 0.6875 \\ 0 & 0 & 0 & \vdots & 0 \end{bmatrix}$

$\begin{cases} x - 0.6y + \quad 0.4z = 0.4 \\ \quad\quad y - 1.1875z = 0.6875 \\ \quad\quad\quad\quad 0 = 0 \end{cases}$

The system has infinitely many solutions. Let $z = a$.

$y - 1.1875a = 0.6875$

$y = 1.1875a + 0.6875$

or $y = \frac{19}{16}a + \frac{11}{16}$

$x - 0.6(1.1875a + 0.6875) + 0.4a = 0.4$

$x - 0.7125a - 0.4125 + 0.4a = 0.4$

$x = 0.3125a + 0.8125$

or $x = \frac{5}{16}a + \frac{13}{16}$

$\left(\frac{5}{16}a + \frac{13}{16}, \frac{19}{16}a + \frac{11}{16}, a\right)$

53. $\begin{vmatrix} 7 & 10 \\ 10 & 15 \end{vmatrix} = 105 - 100 = 5$

55. $\begin{vmatrix} 8 & 3 & 2 \\ 1 & -2 & 4 \\ 6 & 0 & 5 \end{vmatrix} = 6\begin{vmatrix} 3 & 2 \\ -2 & 4 \end{vmatrix} - 0 + 5\begin{vmatrix} 8 & 3 \\ 1 & -2 \end{vmatrix}$

$= 6(16) - 0 + 5(-19)$

$= 96 - 95$

$= 1$

57. $\begin{vmatrix} 2 & -5 & 0 \\ 4 & 7 & 0 \\ -7 & 25 & 3 \end{vmatrix} = 102$

59. $\begin{cases} 7x + 12y = 63 \\ 2x + 3y = 15 \end{cases}$

$$D = \begin{vmatrix} 7 & 12 \\ 2 & 3 \end{vmatrix} = 21 - 24 = -3$$

$$x = \frac{D_x}{D} = \frac{\begin{vmatrix} 63 & 12 \\ 15 & 3 \end{vmatrix}}{-3} = \frac{189 - 180}{-3} = \frac{9}{-3} = -3$$

$$y = \frac{D_y}{D} = \frac{\begin{vmatrix} 7 & 63 \\ 2 & 15 \end{vmatrix}}{-3} = \frac{105 - 126}{-3} = \frac{-21}{-3} = 7$$

$(-3, 7)$

61. $\begin{cases} 3x - 2y = 16 \\ 12x - 8y = -5 \end{cases}$

$$D = \begin{vmatrix} 3 & -2 \\ 12 & -8 \end{vmatrix} = -24 + 24 = 0$$

This system cannot be solved by Cramer's Rule because $D = 0$.

63. $\begin{cases} -x + y + 2z = 1 \\ 2x + 3y + z = -2 \\ 5x + 4y + 2z = 4 \end{cases}$

$$D = \begin{vmatrix} -1 & 1 & 2 \\ 2 & 3 & 1 \\ 5 & 4 & 2 \end{vmatrix} = -1\begin{vmatrix} 3 & 1 \\ 4 & 2 \end{vmatrix} - 2\begin{vmatrix} 1 & 2 \\ 4 & 2 \end{vmatrix} + 5\begin{vmatrix} 1 & 2 \\ 3 & 1 \end{vmatrix} = -1(2) - 2(-6) + 5(-5) = -2 + 12 - 25 = -15$$

$$x = \frac{D_x}{D} = \frac{\begin{vmatrix} 1 & 1 & 2 \\ -2 & 3 & 1 \\ 4 & 4 & 2 \end{vmatrix}}{-15} = \frac{1\begin{vmatrix} 3 & 1 \\ 4 & 2 \end{vmatrix} - (-2)\begin{vmatrix} 1 & 2 \\ 4 & 2 \end{vmatrix} + 4\begin{vmatrix} 1 & 2 \\ 3 & 1 \end{vmatrix}}{-15} = \frac{1(2) + 2(-6) + 4(-5)}{-15} = \frac{-30}{-15} = 2$$

$$y = \frac{D_y}{D} = \frac{\begin{vmatrix} -1 & 1 & 2 \\ 2 & -2 & 1 \\ 5 & 4 & 2 \end{vmatrix}}{-15} = \frac{-1\begin{vmatrix} -2 & 1 \\ 4 & 2 \end{vmatrix} - 2\begin{vmatrix} 1 & 2 \\ 4 & 2 \end{vmatrix} + 5\begin{vmatrix} 1 & 2 \\ -2 & 1 \end{vmatrix}}{-15} = \frac{-1(-8) - 2(-6) + 5(5)}{-15} = \frac{45}{-15} = -3$$

$$z = \frac{D_z}{D} = \frac{\begin{vmatrix} -1 & 1 & 1 \\ 2 & 3 & -2 \\ 5 & 4 & 4 \end{vmatrix}}{-15} = \frac{-1\begin{vmatrix} 3 & -2 \\ 4 & 4 \end{vmatrix} - 2\begin{vmatrix} 1 & 1 \\ 4 & 4 \end{vmatrix} + 5\begin{vmatrix} 1 & 1 \\ 3 & -2 \end{vmatrix}}{-15} = \frac{-1(20) - 2(0) + 5(-5)}{-15} = \frac{-45}{-15} = 3$$

$(2, -3, 3)$

65. $(1, 0), (5, 0), (5, 8)$

$$\begin{vmatrix} x_1 & y_1 & 1 \\ x_2 & y_2 & 1 \\ x_3 & y_3 & 1 \end{vmatrix} = \begin{vmatrix} 1 & 0 & 1 \\ 5 & 0 & 1 \\ 5 & 8 & 1 \end{vmatrix} = -0 + 0 - 8\begin{vmatrix} 1 & 1 \\ 5 & 1 \end{vmatrix} = -8(-4) = 32$$

$A = +\frac{1}{2}(32) = 16$

67. $(1, 2), (4, -5), (3, 2)$

$$\begin{vmatrix} x_1 & y_1 & 1 \\ x_2 & y_2 & 1 \\ x_3 & y_3 & 1 \end{vmatrix} = \begin{vmatrix} 1 & 2 & 1 \\ 4 & -5 & 1 \\ 3 & 2 & 1 \end{vmatrix} = 1\begin{vmatrix} -5 & 1 \\ 2 & 1 \end{vmatrix} - 4\begin{vmatrix} 2 & 1 \\ 2 & 1 \end{vmatrix} + 3\begin{vmatrix} 2 & 1 \\ -5 & 1 \end{vmatrix}$$

$$= 1(-7) - 4(0) + 3(7) = -7 - 0 + 21 = 14$$

$A = +\frac{1}{2}(14) = 7$

69. $\begin{vmatrix} x_1 & y_1 & 1 \\ x_2 & y_2 & 1 \\ x_3 & y_3 & 1 \end{vmatrix} = \begin{vmatrix} -1 & -5 & 1 \\ 1 & -1 & 1 \\ 4 & 5 & 1 \end{vmatrix} = -1\begin{vmatrix} -1 & 1 \\ 5 & 1 \end{vmatrix} - 1\begin{vmatrix} -5 & 1 \\ 5 & 1 \end{vmatrix} + 4\begin{vmatrix} -5 & 1 \\ -1 & 1 \end{vmatrix}$

$$= -1(-6) - 1(-10) + 4(-4)$$

$$= 6 + 10 - 16$$

$$= 0$$

Yes, the three points *are* collinear.

71. $\begin{vmatrix} x_1 & y_1 & 1 \\ x_2 & y_2 & 1 \\ x_3 & y_3 & 1 \end{vmatrix} = \begin{vmatrix} -8 & 2 & 1 \\ -3 & 1 & 1 \\ 5 & -3 & 1 \end{vmatrix} = -8\begin{vmatrix} 1 & 1 \\ -3 & 1 \end{vmatrix} - (-3)\begin{vmatrix} 2 & 1 \\ -3 & 1 \end{vmatrix} + 5\begin{vmatrix} 2 & 1 \\ 1 & 1 \end{vmatrix}$

$$= -8(4) + 3(5) + 5(1)$$

$$= -32 + 15 + 5$$

$$= -12$$

No, the three points *are not* collinear.

73. $(-4, 0), (4, 4)$

$$\begin{vmatrix} x & y & 1 \\ -4 & 0 & 1 \\ 4 & 4 & 1 \end{vmatrix} = 0$$

$$\begin{vmatrix} x & y & 1 \\ -4 & 0 & 1 \\ 4 & 4 & 1 \end{vmatrix} = x\begin{vmatrix} 0 & 1 \\ 4 & 1 \end{vmatrix} - y\begin{vmatrix} -4 & 1 \\ 4 & 1 \end{vmatrix} + 1\begin{vmatrix} -4 & 0 \\ 4 & 4 \end{vmatrix} = 0$$

$$x(-4) - y(-8) + 1(-16) = 0$$

$$-4x + 8y - 16 = 0 \quad \text{or} \quad x - 2y + 4 = 0$$

So, an equation of the line through $(-4, 0)$ and $(4, 4)$ is $x - 2y + 4 = 0$.

75. $\left(-\frac{5}{2}, 3\right), \left(\frac{7}{2}, 1\right)$

$$\begin{vmatrix} x & y & 1 \\ -\frac{5}{2} & 3 & 1 \\ \frac{7}{2} & 1 & 1 \end{vmatrix} = 0$$

$$\begin{vmatrix} x & y & 1 \\ -\frac{5}{2} & 3 & 1 \\ \frac{7}{2} & 1 & 1 \end{vmatrix} = x\begin{vmatrix} 3 & 1 \\ 1 & 1 \end{vmatrix} - y\begin{vmatrix} -\frac{5}{2} & 1 \\ \frac{7}{2} & 1 \end{vmatrix} + 1\begin{vmatrix} -\frac{5}{2} & 3 \\ \frac{7}{2} & 1 \end{vmatrix} = 0$$

$$x(2) - y(-6) + 1(-13) = 0$$

$$2x + 6y - 13 = 0$$

So, an equation of the line through $\left(-\frac{5}{2}, 3\right)$ and $\left(\frac{7}{2}, 1\right)$ is $2x + 6y - 13 = 0$.

77. $-6x + 2y \le 14$

(a) $(4, 1)$

$$-6(4) + 2(1) \overset{?}{\le} 14$$

$$-24 + 2 \overset{?}{\le} 14$$

$$-22 \le 14$$

Yes, $(4, 1)$ *is* a solution.

(b) $(-3, 2)$

$$-6(-3) + 2(2) \overset{?}{\le} 14$$

$$18 + 4 \overset{?}{\le} 14$$

$$22 \not\le 14$$

No, $(-3, 2)$ is *not* a solution.

(c) $(-2, 1)$

$$-6(-2) + 2(1) \overset{?}{\le} 14$$

$$12 + 2 \overset{?}{\le} 14$$

$$14 \le 14$$

Yes, $(-2, 1)$ *is* a solution.

(d) $(0, 9)$

$$-6(0) + 2(9) \overset{?}{\le} 14$$

$$0 + 18 \overset{?}{\le} 14$$

$$18 \not\le 14$$

No, $(0, 9)$ is *not* a solution.

79. $x - 2 \ge 0$

$x \ge 2$

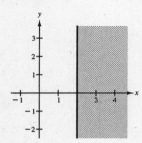

81. $2x + y < 1$

$y < -2x + 1$

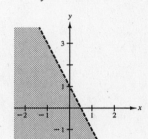

83. $x \le 4y - 2$

$x + 2 \le 4y$

$\frac{1}{4}x + \frac{1}{2} \le y$

$y \ge \frac{1}{4}x + \frac{1}{2}$

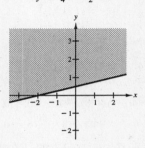

85.

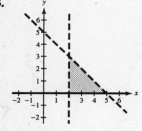

87.

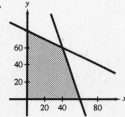

89.

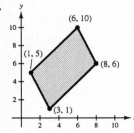

Line through $(1, 5)$ and $(6, 10)$:

$$m = \frac{y_2 - y_1}{x_2 - x_1}$$

$$m = \frac{10 - 5}{6 - 1} = \frac{5}{5} = 1$$

$$y - y_1 = m(x - x_1)$$

$$y - 5 = 1(x - 1)$$

$$y - 5 = x - 1$$

$$y = x + 4$$

Line through $(6, 10)$ and $(8, 6)$:

$$m = \frac{y_2 - y_1}{x_2 - x_1}$$

$$m = \frac{6 - 10}{8 - 6} = \frac{-4}{2} = -2$$

$$y - y_1 = m(x - x_1)$$

$$y - 10 = -2(x - 6)$$

$$y - 10 = -2x + 12$$

$$y = -2x + 22$$

Line through $(8, 6)$ and $(3, 1)$:

$$m = \frac{y_2 - y_1}{x_2 - x_1}$$

$$m = \frac{1 - 6}{3 - 8} = \frac{-5}{-5} = 1$$

$$y - y_1 = m(x - x_1)$$

$$y - 6 = 1(x - 8)$$

$$y - 6 = x - 8$$

$$y = x - 2$$

Line through $(3, 1)$ and $(1, 5)$:

$$m = \frac{y_2 - y_1}{x_2 - x_1}$$

$$m = \frac{5 - 1}{1 - 3} = \frac{4}{-2} = -2$$

$$y - y_1 = m(x - x_1)$$

$$y - 1 = -2(x - 3)$$

$$y - 1 = -2x + 6$$

$$y = -2x + 7$$

Lines:

$$y = x + 4$$
$$y = -2x + 22$$
$$y = x - 2$$
$$y = -2x + 7$$

System of inequalities:

$$\begin{cases} y \le x + 4 \\ y \le -2x + 22 \\ y \ge x - 2 \\ y \ge -2x + 7 \end{cases} \quad \text{or} \quad \begin{cases} x - y \ge -4 \\ 2x + y \le 22 \\ x - y \le 2 \\ 2x + y \ge 7 \end{cases}$$

91. *Verbal Model:* $\boxed{\text{Revenue}} = \boxed{\text{Cost}}$

Labels: Number of items $= x$

Revenue $= 5.25x$ (dollars)

Cost $= 25{,}000 + 3.75x$ (dollars)

Equation: $5.25x = 25{,}000 + 3.75x$

$$1.5x = 25{,}000$$

$$x = \frac{25{,}000}{1.5}$$

$$x = 16{,}666\tfrac{2}{3}$$

$$x \approx 16{,}667$$

So, 16,667 items must be sold before the business breaks even.

93. *Verbal Model:* $2\,\boxed{\text{Length}} + 2\,\boxed{\text{Width}} = \boxed{\text{Perimeter}}$

$\boxed{\text{Length}} = 1.5\,\boxed{\text{Width}}$

Labels: Length $= x$ (meters)

Width $= y$ (meters)

Perimeter $= 480$ (meters)

Equations: $\begin{cases} 2x + 2y = 480 \\ x = 1.5y \end{cases}$

$$2x + 2y = 480$$

$$2(1.5y) + 2y = 480$$

$$3y + 2y = 480$$

$$5y = 480$$

$$y = 96 \quad \text{and} \quad x = 1.5(96)$$

$$x = 144$$

$$(144, 96)$$

So, the length of the rectangle is 144 meters and the width is 96 meters.

95. *Verbal Model:* $\boxed{\text{Number of \$9.95 tapes}}$ + $\boxed{\text{Number of \$14.95 tapes}}$ = 650

$9.95 \boxed{\text{Number of \$9.95 tapes}}$ + $14.95 \boxed{\text{Number of \$14.95 tapes}}$ = 7717.50

Labels: Number \$9.95 tapes = x

Number of \$14.95 tapes = y

Equations:

$$\begin{cases} x + y = 650 \\ 9.95x + 14.95y = 7717.50 \end{cases} \begin{aligned} \Rightarrow \quad & 995x + 995y = 646{,}750 \\ \Rightarrow \quad & \underline{-995x - 1495y = -771{,}750} \\ & -500y = -125{,}000 \end{aligned}$$

$$y = \frac{-125{,}000}{-500}$$

$$y = 250 \quad \text{and} \quad x + 250 = 650$$

$$x = 400$$

(400, 250)

So, 400 of the \$9.95 tapes and 250 of the \$14.95 tapes were sold.

97. *Verbal Model:* $\boxed{\text{Measure of first angle}}$ + $\boxed{\text{Measure of second angle}}$ + $\boxed{\text{Measure of third angle}}$ = 180

$\boxed{\text{Measure of first angle}}$ + $\boxed{\text{Measure of second angle}}$ = $2\boxed{\text{Measure of third angle}}$

$\boxed{\text{Measure of second angle}}$ = $\boxed{\text{Measure of third angle}}$ − 10

Labels: Measure of first angle = x (degrees)

Measure of second angle = y (degrees)

Measure of third angle = z (degrees)

Equations:

$$\begin{cases} x + y + z = 180 \\ x + y = 2z \\ y = z - 10 \end{cases}$$

$$\begin{cases} x + y + z = 180 \\ x + y - 2z = 0 \\ y - z = -10 \end{cases}$$

$$\begin{bmatrix} 1 & 1 & 1 & 180 \\ 1 & 1 & -2 & 0 \\ 0 & 1 & -1 & -10 \end{bmatrix}$$

$$-R_1 + R_2 \begin{bmatrix} 1 & 1 & 1 & 180 \\ 0 & 0 & -3 & -180 \\ 0 & 1 & -1 & -10 \end{bmatrix}$$

$$\begin{matrix} \\ R_3 \\ R_2 \end{matrix} \begin{bmatrix} 1 & 1 & 1 & 180 \\ 0 & 1 & -1 & -10 \\ 0 & 0 & -3 & -180 \end{bmatrix}$$

$$-\tfrac{1}{3}R_3 \begin{bmatrix} 1 & 1 & 1 & 180 \\ 0 & 1 & -1 & -10 \\ 0 & 0 & 1 & 60 \end{bmatrix}$$

$$\begin{cases} x + y + z = 180 \\ y - z = -10 \\ z = 60 \end{cases}$$

$$\begin{array}{ll} y - 60 = -10 & x + 50 + 60 = 180 \\ y = 50 & x + 110 = 180 \\ & x = 70 \end{array}$$

(70, 50, 60)

The measure of the first angle is 70 degrees, the measure of the second angle is 50 degrees, and the measure of the third angle is 60 degrees.

99. $y = ax^2 + bx + c$

$(0, -6)$: $\quad -6 = a(0)^2 + b(0) + c \Longrightarrow \qquad c = -6$

$(1, -3)$: $\quad -3 = a(1)^2 + b(1) + c \Longrightarrow a + b + c = -3$

$(2, 4)$: $\quad\;\; 4 = a(2)^2 + b(2) + c \Longrightarrow 4a + 2b + c = 4$

$$\begin{cases} a + b + c = -3 \\ 4a + 2b + c = 4 \\ \qquad\qquad c = -6 \end{cases}$$

$$\begin{cases} a + b + c = -3 \\ \quad -2b - 3c = 16 \qquad (-4)\text{Eqn. 1} + \text{Eqn. 2} \\ \qquad\qquad c = -6 \end{cases}$$

$$\begin{cases} a + b + c = -3 \\ \quad b + \frac{3}{2}c = -8 \qquad (-1/2)\text{Eqn. 2} \\ \qquad\quad c = -6 \end{cases}$$

$b + \frac{3}{2}(-6) = -8 \qquad\quad a + 1 - 6 = -3$

$\qquad b - 9 = -8 \qquad\qquad a - 5 = -3$

$\qquad\qquad b = 1 \qquad\qquad\qquad a = 2$

$(2, 1, -6)$

$y = 2x^2 + x - 6$

101. (a) Points $(0, 10)$, $(15, 15)$, $(30, 10)$ shifted upward one unit $\Longrightarrow$ $(0, 11)$, $(15, 16)$, $(30, 11)$

$\qquad y = ax^2 + bx + c$

$\qquad (0, 11)$: $11 = a(0)^2 + b(0) + c$

$\qquad (15, 16)$: $16 = a(15)^2 + b(15) + c$

$\qquad (30, 11)$: $11 = a(30)^2 + b(30) + c$

$$\begin{cases} \qquad\qquad\quad c = 11 \\ 225a + 15b + c = 16 \\ 900a + 30b + c = 11 \end{cases}$$

$D = \begin{vmatrix} 0 & 0 & 1 \\ 225 & 15 & 1 \\ 900 & 30 & 1 \end{vmatrix} = -6750$

$a = \dfrac{\begin{vmatrix} 11 & 0 & 1 \\ 16 & 15 & 1 \\ 11 & 30 & 1 \end{vmatrix}}{-6750} = \dfrac{150}{-6750} = -\dfrac{1}{45}$

$b = \dfrac{\begin{vmatrix} 0 & 11 & 1 \\ 225 & 16 & 1 \\ 900 & 11 & 1 \end{vmatrix}}{-6750} = \dfrac{-4500}{-6750} = \dfrac{2}{3}$

$\left(-\dfrac{1}{45}, \dfrac{2}{3}, 11 \right)$

$y = -\dfrac{1}{45}x^2 + \dfrac{2}{3}x + 11$

(b)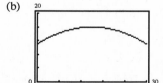

(c) From the graphing utility you can see that when $y = 5$, x is approximately equal to 7.25. So, the child is approximately 7.25 feet from the edge of the garage.

103. *Verbal Model:* | Bushels marketed at Harrisburg | + | Bushels marketed at Philadelphia | ≤ 1500

| Bushels marketed at Harrisburg | ≥ 400

| Bushels marketed at Philadelphia | ≥ 600

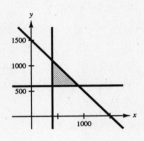

Labels: Bushels marketed at Harrisburg $= x$

Bushels marketed at Philadelphia $= y$

Inequalities: $\begin{cases} x + y \le 1500 \\ x \qquad\ \ge 400 \\ \qquad y \ge 600 \end{cases}$

Chapter Test for Chapter 4

1. $\begin{cases} 5x - y = 6 \implies -y = -5x + 6 \implies y = 5x - 6 \\ 4x - 3y = -4 \end{cases}$

$$4x - 3y = -4$$

$$4x - 3(5x - 6) = -4$$

$$4x - 15x + 18 = -4$$

$$-11x + 18 = -4$$

$$-11x = -22$$

$$x = 2 \quad \text{and} \quad y = 5(2) - 6$$

$$y = 10 - 6$$

$$y = 4$$

$(2, 4)$

2. $\begin{cases} x + 2y - 3z = 0 \\ 3x + y + z = 5 \\ 3x - y + 5z = 7 \end{cases}$

$\begin{cases} x + 2y - 3z = 0 \\ \quad -5y + 10z = 5 \qquad (-3)\text{Eqn. 1 + Eqn. 2} \\ \quad -7y + 14z = 7 \qquad (-3)\text{Eqn. 1 + Eqn. 3} \end{cases}$

$\begin{cases} x + 2y - 3z = 0 \\ \quad y - 2z = -1 \qquad (-\frac{1}{5})\text{Eqn. 2} \\ \quad -7y + 14z = 7 \end{cases}$

$\begin{cases} x + 2y - 3z = 0 \\ \quad y - 2z = -1 \\ \qquad\qquad 0 = 0 \qquad (7)\text{Eqn. 2 + Eqn. 3} \end{cases}$

There are infinitely many solutions.

Let $z = a$

$y - 2a = -1 \qquad\qquad x + 2(2a - 1) - 3a = 0$

$\quad y = 2a - 1 \qquad\qquad x + 4a - 2 - 3a = 0$

$\qquad\qquad\qquad\qquad\qquad x + a - 2 = 0$

$\qquad\qquad\qquad\qquad\qquad\qquad x = -a + 2$

$(-a + 2, 2a - 1, a)$

3.

$$\begin{bmatrix} 1 & 4 & 3 & \vdots & 1 \\ 2 & 8 & 11 & \vdots & 7 \\ 1 & 6 & 7 & \vdots & 3 \end{bmatrix}$$

$\begin{matrix} \\ -2R_1 + R_2 \\ -R_1 + R_3 \end{matrix} \begin{bmatrix} 1 & 4 & 3 & \vdots & 1 \\ 0 & 0 & 5 & \vdots & 5 \\ 0 & 2 & 4 & \vdots & 2 \end{bmatrix}$

$\begin{matrix} \\ R_3 \\ R_2 \end{matrix} \begin{bmatrix} 1 & 4 & 3 & \vdots & 1 \\ 0 & 2 & 4 & \vdots & 2 \\ 0 & 0 & 5 & \vdots & 5 \end{bmatrix}$

$\begin{matrix} \\ \frac{1}{2}R_2 \\ \frac{1}{5}R_3 \end{matrix} \begin{bmatrix} 1 & 4 & 3 & \vdots & 1 \\ 0 & 1 & 2 & \vdots & 1 \\ 0 & 0 & 1 & \vdots & 1 \end{bmatrix}$

$\begin{cases} x + 4y + 3z = 1 \\ \quad y + 2z = 1 \\ \qquad z = 1 \end{cases}$

$y + 2(1) = 1 \qquad\qquad x + 4(-1) + 3(1) = 1$

$\quad y + 2 = 1 \qquad\qquad\qquad x - 4 + 3 = 1$

$\qquad y = -1 \qquad\qquad\qquad\qquad x - 1 = 1$

$\qquad\qquad\qquad\qquad\qquad\qquad\qquad x = 2$

$(2, -1, 1)$

4. $D = \begin{vmatrix} 1 & 3 & 1 \\ 2 & 5 & 1 \\ 1 & 2 & 3 \end{vmatrix} = 1 \begin{vmatrix} 5 & 1 \\ 2 & 3 \end{vmatrix} - 2 \begin{vmatrix} 3 & 1 \\ 2 & 3 \end{vmatrix} + 1 \begin{vmatrix} 3 & 1 \\ 5 & 1 \end{vmatrix}$

$$= 1(13) - 2(7) + 1(-2) = 13 - 14 - 2 = -3$$

$x = \dfrac{D_x}{D} = \dfrac{\begin{vmatrix} -2 & 3 & 1 \\ -5 & 5 & 1 \\ 6 & 2 & 3 \end{vmatrix}}{-3} = \dfrac{-2\begin{vmatrix} 5 & 1 \\ 2 & 3 \end{vmatrix} - (-5)\begin{vmatrix} 3 & 1 \\ 2 & 3 \end{vmatrix} + 6\begin{vmatrix} 3 & 1 \\ 5 & 1 \end{vmatrix}}{-3}$

$$= \frac{-2(13) + 5(7) + 6(-2)}{-3} = \frac{-26 + 35 - 12}{-3} = \frac{-3}{-3} = 1$$

$y = \dfrac{D_y}{D} = \dfrac{\begin{vmatrix} 1 & -2 & 1 \\ 2 & -5 & 1 \\ 1 & 6 & 3 \end{vmatrix}}{-3} = \dfrac{1\begin{vmatrix} -5 & 1 \\ 6 & 3 \end{vmatrix} - 2\begin{vmatrix} -2 & 1 \\ 6 & 3 \end{vmatrix} + 1\begin{vmatrix} -2 & 1 \\ -5 & 1 \end{vmatrix}}{-3}$

$$= \frac{1(-21) - 2(-12) + 1(3)}{-3} = \frac{-21 + 24 + 3}{-3} = \frac{6}{-3} = -2$$

$z = \dfrac{D_z}{D} = \dfrac{\begin{vmatrix} 1 & 3 & -2 \\ 2 & 5 & -5 \\ 1 & 2 & 6 \end{vmatrix}}{-3} = \dfrac{1\begin{vmatrix} 5 & -5 \\ 2 & 6 \end{vmatrix} - 2\begin{vmatrix} 3 & -2 \\ 2 & 6 \end{vmatrix} + 1\begin{vmatrix} 3 & -2 \\ 5 & -5 \end{vmatrix}}{-3}$

$$= \frac{1(40) - 2(22) + 1(-5)}{-3} = \frac{40 - 44 - 5}{-3} = \frac{-9}{-3} = 3$$

$(1, -2, 3)$

5. $\begin{bmatrix} 2 & 2 & -2 & \vdots & 1 \\ 2 & 5 & 1 & \vdots & 9 \\ 1 & 3 & 4 & \vdots & 9 \end{bmatrix}$

$\begin{bmatrix} 1 & 2 & -2 & \vdots & 1 \\ 0 & 1 & 5 & \vdots & 7 \\ 0 & 1 & 6 & \vdots & 8 \end{bmatrix}$

$\begin{bmatrix} 1 & 2 & -2 & \vdots & 1 \\ 0 & 1 & 5 & \vdots & 7 \\ 0 & 0 & 1 & \vdots & 1 \end{bmatrix}$

$\begin{cases} x + 2y - 2z = 1 \\ \quad\;\; y + 5z = 7 \\ \quad\qquad z = 1 \end{cases}$

$y + 5(1) = 7 \qquad x + 2(2) - 2(1) = 1$

$\quad y + 5 = 7 \qquad\quad x + 4 - 2 = 1$

$\qquad\quad y = 2 \qquad\qquad x + 2 = 1$

$\qquad\qquad\qquad\qquad\qquad x = -1$

$(-1, 2, 1)$

6. $\begin{vmatrix} 3 & -2 & 0 \\ -1 & 5 & 3 \\ 2 & 7 & 1 \end{vmatrix} = 3\begin{vmatrix} 5 & 3 \\ 7 & 1 \end{vmatrix} - (-2)\begin{vmatrix} -1 & 3 \\ 2 & 1 \end{vmatrix} + 0$

$$= 3(-16) + 2(-7) + 0$$

$$= -48 - 14$$

$$= -62$$

7. $y = ax^2 + bx + c$

$(0, 4):$ $4 = a(0)^2 + b(0) + c \Rightarrow$ $c = 4$

$(1, 3):$ $3 = a(1)^2 + b(1) + c \Rightarrow a + b + c = 3$

$(2, 6):$ $6 = a(2)^2 + b(2) + c \Rightarrow 4a + 2b + c = 6$

$\begin{cases} a + b + c = 3 \\ 4a + 2b + c = 6 \\ \phantom{4a + 2b + {}} c = 4 \end{cases}$

$\begin{cases} a + b + c = 3 \\ \phantom{a + {}} -2b - 3c = -6 \\ \phantom{a + b + {}} c = 4 \end{cases}$ (-4)Eqn. 1 + Eqn. 2

$\begin{cases} a + b + c = 3 \\ \phantom{a + {}} b + \frac{3}{2}c = 3 \\ \phantom{a + b + {}} c = 4 \end{cases}$ $(-1/2)$Eqn. 2

$b + \frac{3}{2}(4) = 3$ $a + (-3) + 4 = 3$

$\phantom{b + {}} b + 6 = 3$ $ a + 1 = 3$

$ b = -3$ $ a = 2$

$(2, -3, 4)$

$y = 2x^2 - 3x + 4$

8. $(0, 0), (5, 4), (6, 0)$

$\begin{vmatrix} x_1 & y_1 & 1 \\ x_2 & y_2 & 1 \\ x_3 & y_3 & 1 \end{vmatrix} = \begin{vmatrix} 0 & 0 & 1 \\ 5 & 4 & 1 \\ 6 & 0 & 1 \end{vmatrix} = 0 - 0 + 1 \begin{vmatrix} 5 & 4 \\ 6 & 0 \end{vmatrix}$

$ = 0 - 0 + 1(-24)$

$ = -24$

$A = -\frac{1}{2}(-24) = 12$

9. $x + 2y \le 4$

$\phantom{x + {}} 2y \le -x + 4$

$ y \le -\frac{1}{2}x + 2$

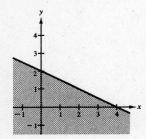

10. $\begin{cases} 3x - y < 4 \\ x > 0 \\ \phantom{3x -{}} y > 0 \end{cases}$

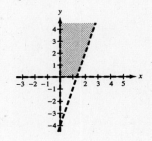

11. $\begin{cases} x + y < 6 \\ 2x + 3y > 9 \\ x \ge 0 \\ \phantom{2x +{}} y \ge 0 \end{cases}$

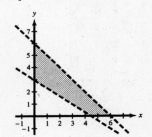

12. *Verbal Model:* $\boxed{\begin{array}{c}\text{Distance driven} \\ \text{by first person}\end{array}} + \boxed{\begin{array}{c}\text{Distance driven} \\ \text{by second person}\end{array}} = 200$

$\boxed{\begin{array}{c}\text{Distance driven} \\ \text{by first person}\end{array}} = 4\,\boxed{\begin{array}{c}\text{Distance driven} \\ \text{by second person}\end{array}}$

Labels: Distance driven by first person $= x$ (miles)

Distance driven by second person $= y$ (miles)

Equations: $\begin{cases} x + y = 200 \\ \quad x = 4y \end{cases}$

$x + y = 200$

$(4y) + y = 200$

$5y = 200$

$y = 40 \quad$ and $\quad x = 4(40)$

$x = 160$

$(160, 40)$

So, one person drives 160 miles and the other person drives 40 miles.

13. *Verbal Model:* $8\,\boxed{\begin{array}{c}\text{Price per gallon} \\ \text{of regular}\end{array}} + 12\,\boxed{\begin{array}{c}\text{Price per gallon} \\ \text{of premium}\end{array}} = 27.84$

$\boxed{\begin{array}{c}\text{Price per gallon} \\ \text{of premium}\end{array}} = \boxed{\begin{array}{c}\text{Price per gallon} \\ \text{of regular}\end{array}} + 0.17$

Labels: Price per gallon of regular $= r$ (dollars)

Price per gallon of premium $= p$ (dollars)

Equations: $8r + 12p = 27.84$

$p = r + 0.17$

$8r + 12(r + 0.17) = 27.84$

$8r + 12r + 2.04 = 27.84$

$20r + 2.04 = 27.84$

$20r = 25.80$

$r = 1.29 \quad$ and $\quad p = 1.29 + 0.17$

$p = 1.46$

So, the price per gallon of regular gasoline is \$1.29 and the price per gallon of premium gasoline is \$1.46.

14. *Labels:* Amount in Program A $= x$ (dollars)

Amount in Program B $= y$ (dollars)

Inequalities: $\begin{cases} x + y \leq 1000 \\ \quad\; x \geq \;\;300 \\ \quad\; y \geq \;\;400 \end{cases}$

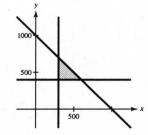

Cumulative Test for Chapters 1–4

1. $(-3a^5)^2 \cdot (-6a^8) = (-3)^2 a^{5 \cdot 2}(-6)a^8$

$\qquad\qquad\qquad\quad = 9(-6)a^{10+8}$

$\qquad\qquad\qquad\quad = -54a^{18}$

2. $\dfrac{(2x^3y)^4}{6x^2y^3} = \dfrac{2^4 x^{3 \cdot 4} y^4}{6x^2y^3}$

$\qquad\qquad = \dfrac{16x^{12}y^4}{6x^2y^3}$

$\qquad\qquad = \dfrac{8x^{12-2}y^{4-3}}{3}$

$\qquad\qquad = \dfrac{8x^{10}y}{3} \ \text{ or } \ \dfrac{8}{3}x^{10}y$

3. $-\left(\dfrac{6x^2y}{5z^3}\right)^2 = -\dfrac{6^2 x^4 y^2}{5^2 z^6}$

$\qquad\qquad\quad\ = -\dfrac{36x^4y^2}{25z^6}$

4. $\dfrac{2a^{2n}b^{m+5}}{a^n b^5} = 2a^{2n-n}b^{m+5-5}$

$\qquad\qquad\quad = 2a^n b^m$

5. *Labels:*   Base $= 8b$

$\qquad\qquad$ Height $= b^2 + 1$

$\qquad\qquad$ Hypotenuse $= 9b - 1$

Perimeter $=$ Base $+$ Height $+$ Hypotenuse

$\qquad\quad = (8b) + (b^2 + 1) + (9b - 1)$

$\qquad\quad = 8b + b^2 + 1 + 9b - 1$

$\qquad\quad = b^2 + 17b$

When $b = 3$:

Perimeter $= b^2 + 17b$

$\qquad\qquad = 3^2 + 17(3)$

$\qquad\qquad = 9 + 51$

$\qquad\qquad = 60$

Area $= \left(\frac{1}{2}\right)(\text{Base})(\text{Height})$

$\qquad = \left(\frac{1}{2}\right)(8b)(b^2 + 1)$

$\qquad = 4b(b^2 + 1)$

$\qquad = 4b^3 + 4b$

Area $= 4b^3 + 4b$

$\qquad = 4(3)^3 + 4(3)$

$\qquad = 4(27) + 12$

$\qquad = 120$

6. *Labels:*   Larger rectangle: Length $= x + 8$; Width $= 2x - 1$

$\qquad\qquad\qquad\qquad\qquad\ \ $ Area $= (x + 8)(2x - 1)$

$\qquad\qquad\ $ Smaller rectangle: Length $= x + 5$; Width $= x$

$\qquad\qquad\qquad\qquad\qquad\ \ $ Area $= (x + 5)x$

Area of shaded region $=$ Area of larger rectangle $-$ Area of smaller rectangle

$\qquad\qquad\qquad = (x + 8)(2x - 1) - (x + 5)x$

$\qquad\qquad\qquad = 2x^2 + 15x - 8 - (x^2 + 5x)$

$\qquad\qquad\qquad = 2x^2 + 15x - 8 - x^2 - 5x$

$\qquad\qquad\qquad = x^2 + 10x - 8$

7. $10a - 15a^4 = 5a(2 - 3a^3)$

8. $2a^2 + ab - 6b^2 = (2a - 3b)(a + 2b)$

9. $x^3 - 3x^2 - x + 3 = (x^3 - 3x^2) + (-x + 3)$

$\qquad\qquad\qquad\quad = x^2(x - 3) - 1(x - 3)$

$\qquad\qquad\qquad\quad = (x - 3)(x^2 - 1)$

$\qquad\qquad\qquad\quad = (x - 3)(x + 1)(x - 1)$

10. $y^3 - 64 = y^3 - 4^3$

$\qquad\qquad\ = (y - 4)(y^2 + 4y + 16)$

11. $x + \dfrac{x}{2} = 4$

$2\left(x + \dfrac{x}{2}\right) = 2(4)$

$2x + x = 8$

$3x = 8$

$x = \dfrac{8}{3}$

12. $5(x - 2) = 3 - 2(x + 3)$

$5x - 10 = 3 - 2x - 6$

$5x - 10 = -2x - 3$

$7x - 10 = -3$

$7x = 7$

$x = 1$

13. $8(x - 1) + 14 = 3(x + 7)$

$8x - 8 + 14 = 3x + 21$

$8x + 6 = 3x + 21$

$5x + 6 = 21$

$5x = 15$

$x = 3$

14. $x^2 + x - 42 = 0$

$(x + 7)(x - 6) = 0$

$x + 7 = 0 \Longrightarrow x = -7$

$x - 6 = 0 \Longrightarrow x = 6$

15. $\dfrac{x - 2}{2} = \dfrac{4x + 1}{14}$

$14(x - 2) = 2(4x + 1)$

$14x - 28 = 8x + 2$

$6x - 28 = 2$

$6x = 30$

$x = 5$

16. $\left|\dfrac{2}{x}x + 2\right| = 10$

$\dfrac{2}{3}x + 2 = -10$ *or* $\dfrac{2}{3}x + 2 = 10$

$2x + 6 = -30$ $2x + 6 = 30$

$2x = -36$ $2x = 24$

$x = -18$ $x = 12$

17. $f(x) = 3x + 5$

Domain $= \{x\colon x$ is a real number$\}$

18. $g(x) = x^2 + 4x + 4$

Domain $= \{x\colon x$ is a real number$\}$

19. $h(x) = \dfrac{4}{x - 8}$

Domain $= \{x\colon x \neq 8\}$

20. $f(t) = \sqrt{t - 6}$

Domain $= \{t\colon t \geq 6\}$

21. $g(x) = x^4 - 2$

Vertical shift 2 units down

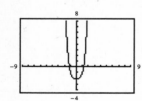

22. $g(x) = (x - 2)^4$

Horizontal shift 2 units to the right

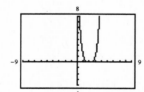

23. $g(x) = -x^4$

Reflection in the x-axis

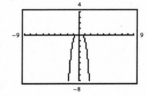

24. (a) $2(\text{Length}) + 2(\text{Width}) = \text{Perimeter}$

$2x + 2y = 500$

$2y = 500 - 2x$

$y = 250 - x$

(b) Area $= (\text{Length})(\text{Width})$

$A = xy$

$A = x(250 - x)$

25. $5x - y = 8$

$-y = -5x + 8$

$y = 5x - 8$

$m = 5$ and slope of parallel line is 5.

Point: $(7, -2)$

Slope: $m = 5$

$y - y_1 = m(x - x_1)$

$y - (-2) = 5(x - 7)$

$y + 2 = 5x - 35$

$y = 5x - 37$

26. A vertical line has an equation of the form $x = a$, so the equation of the vertical line through $(-2, 4)$ is $x = -2$.

27. $(0, 117,000)$ and $(5, 52,000)$

Point: $(0, 117,000)$

Slope: $m = \dfrac{V_2 - V_1}{t_2 - t_1} = \dfrac{52,000 - 117,000}{5 - 0} = \dfrac{-65,000}{5} = -13,000$

$V - 117,000 = -13,000(t - 0)$

$V - 117,000 = -13,000t$

$V = -13,000t + 117,000$

For $t = 3$, $V = -13,000(3) + 117,000$

$ = -39,000 + 117,000$

$ = 78,000$

After 3 years, the value of the system will be \$78,000.

28. *Verbal Model:* $\boxed{\text{Car payment}} = 0.40 \boxed{\text{Monthly income}}$

Labels: Car payment $= 260$ (dollars)

 Monthly income $= x$ (dollars)

Equation: $260 = 0.40x$

$\dfrac{260}{0.40} = x$

$\$650 = x$

So, the monthly income is \$650.

29. *Verbal Model:* $14.50 \boxed{\text{Number of adult tickets}} + 6.50 \boxed{\text{Number of student tickets}} = 1806$

$\boxed{\text{Number of adult tickets}} = 4 \boxed{\text{Number of student tickets}}$

Labels: Number of adult tickets $= x$

 Number of student tickets $= y$

Equations: $\begin{cases} 14.50x + 6.50y = 1806 \\ x = 4y \end{cases}$

$14.50(4y) + 6.50y = 1806$

$58y + 6.50y = 1806$

$64.50y = 1806$

$y = \dfrac{1806}{64.50}$

$y = 28$ and $x = 4(28)$

$x = 112$

So, 112 adult tickets and 28 student tickets were sold.

30. $-16 < 6x + 2 \le 5$

$\quad\;\; -18 < \quad 6x \quad \le 3$

$\quad\;\;\;\; -3 < \quad\;\; x \quad\;\; \le \frac{1}{2}$

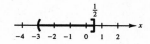

31. $|9 - 2x| - 2 < -1$

$\quad\;\;\; |9 - 2x| < 1$

$\quad\; -1 < 9 - 2x < 1$

$\quad\; -10 < -2x < -8$

$\quad\; \dfrac{-10}{-2} > \dfrac{-2x}{-2} > \dfrac{-8}{-2}$

$\quad\quad 5 > x > 4 \quad \text{or} \quad 4 < x < 5$

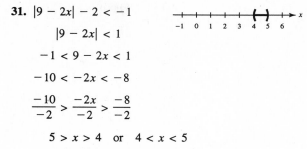

32.

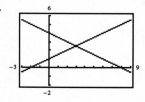

From the graph, the solution appears to be $\left(3, \frac{5}{2}\right)$.

$x + 2y = \quad 8 \implies -2x - 4y = -16$

$2x - 4y = -4 \implies \quad 2x - 4y = \quad -4$

$\quad\quad\quad\quad\quad\quad\quad\quad\quad -8y = -20$

$\quad\quad\quad\quad\quad\quad\quad\quad\quad\;\; y = \dfrac{-20}{-8}$

$\quad\quad\quad\quad\quad\quad\quad\quad\quad\;\; y = \dfrac{5}{2} \quad \text{and} \quad x + 2\left(\dfrac{5}{2}\right) = 8$

$\quad\quad\quad\quad\quad\quad\quad\quad\quad\quad\quad\quad\quad\quad\;\; x + 5 = 8$

$\quad\quad\quad\quad\quad\quad\quad\quad\quad\quad\quad\quad\quad\quad\quad\quad\; x = 3$

$\left(3, \frac{5}{2}\right)$

33.

From the graph, the solution appears to be $(-5, 7)$.

$x + y = 2 \implies \quad\quad x + y = \quad 2$

$0.2x + y = 6 \implies -0.2x - y = -6$

$\quad\quad\quad\quad\quad\quad\quad\quad 0.8x \quad\quad = -4$

$\quad\quad\quad\quad\quad\quad\quad\quad\quad\;\; x = \dfrac{-4}{0.8}$

$\quad\quad\quad\quad\quad\quad\quad\quad\quad\;\; x = -5 \quad \text{and} \quad (-5) + y = 2$

$\quad\quad\quad\quad\quad\quad\quad\quad\quad\quad\quad\quad\quad\quad\quad\quad\quad\quad\; y = 7$

$(-5, 7)$

34. $\begin{cases} 2x + y - 2z = 1 \\ x \quad\;\; - z = 1 \\ 3x + 3y + z = 12 \end{cases}$

$\begin{cases} x \quad\quad - z = 1 \\ 2x + y - 2z = 1 \\ 3x + 3y + z = 12 \end{cases}$ Interchange Eqn. 1 and Eqn. 2

$\begin{cases} x \quad\; - z = \quad 1 \\ \quad\; y \quad\quad = -1 \\ \quad\; 3y + 4z = \quad 9 \end{cases}$ (-2)Eqn. 1 + Eqn. 2

$\quad\quad\quad\quad\quad\quad\quad\quad\;$ (-3)Eqn. 1 + Eqn. 3

$\begin{cases} x \quad\; - z = \quad 1 \\ \quad\; y \quad\quad = -1 \\ \quad\quad\; 4z = 12 \end{cases}$ (-3)Eqn. 2 + Eqn. 3

$\begin{cases} x \quad\; - z = \quad 1 \\ \quad\; y \quad\quad = -1 \\ \quad\quad\;\; z = \quad 3 \end{cases}$ $\left(\frac{1}{4}\right)$Eqn. 3

$x - 3 = 1$

$\quad\;\; x = 4$

$(4, -1, 3)$

35. *Verbal Model:* $2\boxed{\text{Length}} + 2\boxed{\text{Width}} = \boxed{\text{Perimeter}}$

$\quad\quad\quad\quad\quad\quad\;\; \boxed{\text{Width}} = \frac{8}{9}\boxed{\text{Length}}$

Labels: Length $= x$ (feet)

$\quad\quad\quad$ Width $= y$ (feet)

$\quad\quad\quad$ Perimeter $= 68$ (feet)

Equations: $\begin{cases} 2x + 2y = 68 \\ \quad\quad\; y = \frac{8}{9}x \end{cases}$

$\quad\quad\quad 2x + 2\left(\frac{8}{9}x\right) = 68$

$\quad\quad\quad\quad\; 2x + \frac{16}{9}x = 68$

$\quad\quad\quad\quad 18x + 16x = 612$

$\quad\quad\quad\quad\quad\quad\;\; 34x = 612$

$\quad\quad\quad\quad\quad\quad\quad\;\; x = 18 \quad \text{and} \quad y = \frac{8}{9}(18) = 16$

The length of the rectangle is 18 feet and the width is 16 feet.

36. *Verbal Model:* | Amount invested at 8% | + | Amount invested at 9% | + | Amount invested at 10% | = 800,000

0.08 | Amount invested at 8% | + 0.09 | Amount invested at 9% | + 0.10 | Amount invested at 10% | = 67,000

| Amount invested at 8% | = 5 | Amount invested at 10% |

Labels: Amount invested at 8% = x (dollars)

Amount invested at 9% = y (dollars)

Amount invested at 10% = z (dollars)

Equations:
$$\begin{cases} x + y + z = 800,000 \\ 0.08x + 0.09y + 0.10z = 67,000 \\ x = 5z \end{cases} \Rightarrow \begin{cases} x + y + z = 800,000 \\ 0.08x + 0.09y + 0.10z = 67,000 \\ x - 5z = 0 \end{cases}$$

$$\begin{bmatrix} 1 & 1 & 1 & 800,000 \\ 0.08 & 0.09 & 0.10 & 67,000 \\ 1 & 0 & -5 & 0 \end{bmatrix}$$

$$\begin{matrix} \\ -0.08R_1 + R_2 \\ -R_1 + R_2 \end{matrix} \begin{bmatrix} 1 & 1 & 1 & 800,000 \\ 0 & 0.01 & 0.02 & 3000 \\ 0 & -1 & -6 & -800,000 \end{bmatrix}$$

$$100R_2 \begin{bmatrix} 1 & 1 & 1 & 800,000 \\ 0 & 1 & 2 & 300,000 \\ 0 & -1 & -6 & -800,000 \end{bmatrix}$$

$$R_2 + R_3 \begin{bmatrix} 1 & 1 & 1 & 800,000 \\ 0 & 1 & 2 & 300,000 \\ 0 & 0 & -4 & -500,000 \end{bmatrix}$$

$$-\tfrac{1}{4}R_3 \begin{bmatrix} 1 & 1 & 1 & 800,000 \\ 0 & 1 & 2 & 300,000 \\ 0 & 0 & 1 & 125,000 \end{bmatrix}$$

$$\begin{cases} x + y + z = 800,000 \\ y + 2z = 300,000 \\ z = 125,000 \end{cases}$$

$$y + 2(125,000) = 300,000$$

$$y + 250,000 = 300,000$$

$$y = 50,000$$

(625,000, 50,000, 125,000)   So, \$625,000 is invested at 8%, \$50,000 is invested at 9%, and \$125,000 is invested at 10%.

37.
$$\begin{vmatrix} x & y & 1 \\ -2 & 5 & 1 \\ 0 & 6 & 1 \end{vmatrix} = 0$$

$$\begin{vmatrix} x & y & 1 \\ -2 & 5 & 1 \\ 0 & 6 & 1 \end{vmatrix} = x\begin{vmatrix} 5 & 1 \\ 6 & 1 \end{vmatrix} - (-2)\begin{vmatrix} y & 1 \\ 6 & 1 \end{vmatrix} + 0\begin{vmatrix} y & 1 \\ 5 & 1 \end{vmatrix} = 0$$

$$x(-1) + 2(y - 6) + 0(y - 5) = 0$$

$$-x + 2y - 12 = 0$$

$$\text{or} \quad x - 2y + 12 = 0$$

38. $\begin{vmatrix} -4 & -2 & 1 \\ -1 & 1 & 1 \\ 3 & 4 & 1 \end{vmatrix} = -4\begin{vmatrix} 1 & 1 \\ 4 & 1 \end{vmatrix} - (-1)\begin{vmatrix} -2 & 1 \\ 4 & 1 \end{vmatrix} + 3\begin{vmatrix} -2 & 1 \\ 1 & 1 \end{vmatrix}$

$$= -4(-3) + 1(-6) + 3(-3) = 12 - 6 - 9 = -3$$

No, the points are *not* collinear.

39.

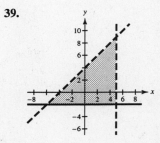

CHAPTER 5
Radicals and Complex Numbers

CHAPTER 5
Radicals and Complex Numbers

Section 5.1 Integer Exponents and Scientific Notation
Solutions to Odd-Numbered Exercises

1. $5^{-2} = \dfrac{1}{5^2} = \dfrac{1}{25}$

3. $\dfrac{1}{4^{-3}} = 1 \cdot 4^3 = 64$

5. $\left(\dfrac{3}{16}\right)^0 = 1$

7. $27 \cdot 3^{-3} = 27\left(\dfrac{1}{3^3}\right) = 27\left(\dfrac{1}{27}\right) = 1$

9. $\dfrac{10^3}{10^{-2}} = 10^3 \cdot 10^2 = 10^5 = 100{,}000$

11. $(4^2 \cdot 4^{-1})^{-2} = (4^1)^{-2}$

$$= 4^{-2} = \dfrac{1}{4^2} = \dfrac{1}{16}$$

or

$$(4^2 \cdot 4^{-1})^{-2} = 4^{-4} \cdot 4^2$$

$$= 4^{-2} = \dfrac{1}{16}$$

13. $7x^{-3} = 7\left(\dfrac{1}{x^3}\right) = \dfrac{7}{x^3}$

15. $y^4 \cdot y^{-2} = y^{4+(-2)} = y^2$ or $y^4 \cdot y^{-2} = \dfrac{y^4}{y^2} = y^2$

17. $x^{-2} \cdot x^{-5} = x^{-7} = \dfrac{1}{x^7}$ or $x^{-2} \cdot x^{-5} = \dfrac{1}{x^2} \cdot \dfrac{1}{x^5} = \dfrac{1}{x^7}$

19. $\dfrac{1}{x^{-6}} = x^6$

21. $\dfrac{x^{-3}}{y^{-1}} = \dfrac{y}{x^3}$

23. $\left(\dfrac{x}{10}\right)^{-1} = \left(\dfrac{10}{x}\right)^1 = \dfrac{10}{x}$

25. $\dfrac{8t^{-3}}{6t^{-4}} = \dfrac{8t^4}{6t^3} = \dfrac{4t}{3}$

27. $\dfrac{(4t)^0}{t^{-2}} = \dfrac{1}{t^{-2}} = t^2$

29. $(2x^2)^{-2} = 2^{-2}x^{-4} = \dfrac{1}{2^2 x^4} = \dfrac{1}{4x^4}$

31. $(-3x^{-3}y^2)(4x^2 y^{-5}) = -12x^{-1}y^{-3} = -\dfrac{12}{xy^3}$

33. $-\left(\dfrac{4x^2}{y^3}\right)^{-3} = -\dfrac{4^{-3}x^{2(-3)}}{y^{3(-3)}} = -\dfrac{4^{-3}x^{-6}}{y^{-9}} = -\dfrac{y^9}{4^3 x^6} = -\dfrac{y^9}{64x^6}$

35. $(3x^2y^{-2})^{-2} = 3^{-2}x^{-4}y^4 = \dfrac{y^4}{3^2 x^4} = \dfrac{y^4}{9x^4}$

37. $(4xy^{-3})^{-3}(9x^2y^{-1}) = 4^{-3}x^{-3}y^{-3(-3)}9x^2y^{-1} = \dfrac{9x^2y^9}{4^3 x^3 y^1} = \dfrac{9y^8}{64x}$

39. $\dfrac{6^2 x^3 y^{-3}}{12x^{-2}y} = \dfrac{36x^3 x^2}{12y^3 y} = \dfrac{3x^5}{y^4}$

41. $\left(\dfrac{3u^2 v^{-1}}{3^3 u^{-1} v^3}\right)^{-2} = \left(\dfrac{u^3}{3^2 v^4}\right)^{-2} = \left(\dfrac{9v^4}{u^3}\right)^2 = \dfrac{81v^8}{u^6}$

or

$$\left(\dfrac{3u^2 v^{-1}}{3^3 u^{-1} v^3}\right)^{-2} = \dfrac{3^{-2}u^{-4}v^2}{3^{-6}u^2 v^{-6}} = \dfrac{3^6 v^2 v^6}{3^2 u^2 u^4} = \dfrac{3^4 v^8}{u^6} = \dfrac{81v^8}{u^6}$$

43. $\left[\left(\dfrac{2y^2}{x^3}\right)^2\right]^{-2} = \left(\dfrac{2y^2}{x^3}\right)^{-4} = \dfrac{2^{-4}y^{-8}}{x^{-12}} = \dfrac{x^{12}}{2^4 y^8} = \dfrac{x^{12}}{16y^8}$

45. $[(x^{-4}y^{-6})^{-1}]^2 = [x^{-4}y^{-6}]^{-2} = x^8 y^{12}$

or

$$[(x^{-4}y^{-6})^{-1}]^2 = (x^4 y^6)^2 = x^8 y^{12}$$

47. $(4m)^3 \left(\dfrac{4}{3m}\right)^{-2} = 4^3 m^3 \cdot \dfrac{4^{-2}}{3^{-2}m^{-2}}$

$\qquad = 4^3 m^3 \cdot \dfrac{3^2 m^2}{4^2}$

$\qquad = \dfrac{64(9)m^3 m^2}{16}$

$\qquad = \dfrac{\cancel{16}(4)(9)m^{3+2}}{\cancel{16}}$

$\qquad = 36m^5$

49. $\left(\dfrac{6x^4}{7y^{-2}}\right)(14x^{-1}y^5) = \left(\dfrac{6x^4 y^2}{7}\right)\left(\dfrac{14y^5}{x}\right) = \dfrac{6(7)(2)x^4 y^7}{7x} = 12x^3 y^7$

51. $(5x^2 y^4)^3 (xy^{-5})^{-3} = 5^3 x^{2\cdot 3} y^{4\cdot 3} \cdot x^{-3} y^{-5(-3)}$

$\qquad = 125x^6 y^{12} \cdot x^{-3} y^{15}$

$\qquad = 125x^{6+(-3)} y^{12+15}$

$\qquad = 125x^3 y^{27}$

53. $\dfrac{(2a^{-2}b^4)^3}{(10a^3 b)^2} = \dfrac{2^3 a^{-2(3)} b^{4(3)}}{10^2 a^{3(2)} b^2}$

$\qquad = \dfrac{8a^{-6} b^{12}}{100a^6 b^2}$

$\qquad = \dfrac{\cancel{4}(2)b^{12}}{\cancel{4}(25)a^6 a^6 b^2}$

$\qquad = \dfrac{2b^{12-2}}{25a^{6+6}}$

$\qquad = \dfrac{2b^{10}}{25a^{12}}$

55. $(2x^0 y^{-1})(4xy^{-6}) = \dfrac{2(1)4x}{y(y^6)} = \dfrac{8x}{y^7}$

57. $(18x)^0 (4xy)^2 (3x^{-1}) = 1(4^2 x^2 y^2)\dfrac{3}{x} = \dfrac{16(3)x^2 y^2}{x} = 48xy^2$

59. $3{,}600{,}000 = 3.6 \times 10^6$

61. $0.00381 = 3.81 \times 10^{-3}$

63. $139{,}400{,}000 = 1.394 \times 10^8$

65. $6 \times 10^7 = 60{,}000{,}000$

67. $1.359 \times 10^{-7} = 0.0000001359$

69. $3.5 \times 10^8 = 350{,}000{,}000$

71. $(2 \times 10^9)(3.4 \times 10^{-4}) = 6.8 \times 10^5$

73. $\dfrac{3.6 \times 10^9}{9 \times 10^5} = 0.4 \times 10^4$

$\qquad = 4 \times 10^3$

75. $(4{,}500{,}000)(2{,}000{,}000{,}000) = (4.5 \times 10^6)(2 \times 10^9)$

$\qquad = 9 \times 10^{15}$

77. $\dfrac{1.357 \times 10^{12}}{(4.2 \times 10^2)(6.87 \times 10^{-3})} \approx 4.70 \times 10^{11}$

79. $\dfrac{(0.0000565)(2{,}850{,}000{,}000{,}000)}{0.00465} = \dfrac{(5.65 \times 10^{-5})(2.85 \times 10^{12})}{4.65 \times 10^{-3}} \approx 3.46 \times 10^{10}$

81. $\dfrac{1.49 \times 10^{11}}{9.46 \times 10^{15}} \approx 0.158 \times 10^{-4} = 1.58 \times 10^{-5}$

The time for light to travel from the sun to Earth is approximately 1.58×10^{-5} year. *Note:* 1.58×10^{-5} year is approximately 8.3 minutes.

83. $\dfrac{1.99 \times 10^{30}}{5.975 \times 10^{24}} \approx 0.333 \times 10^6 \approx 3.33 \times 10^5 \approx 333{,}000$

The mass of the sun is approximately 333,000 times the mass of the earth.

85. $(-2x)^{-4} = (-2)^{-4}x^{-4} = \frac{1}{16}x^{-4}$

$(-2x)^{-4} \neq -2x^{-4}$

In the first expression, the base of the exponent is $-2x$, but in the second expression, the base of the exponent is x.

87. The denominator, $x + 1$, cannot be equal to zero.

Domain $= \{x: x \neq -1\}$

89. The radicand, $x - 4$, must be greater than or equal to zero.

Domain $= \{x: x \geq 4\}$

91. *Labels:*   Amount invested at $7.5\% = x$ (dollars)

Amount invested at $9\% = y$ (dollars)

Equations:
$$x + y = 24{,}000 \implies -0.075x - 0.075y = -1800$$
$$0.075x + 0.09y = 1935 \implies \underline{0.075x + 0.09\ y = \ \ \ 1935}$$
$$0.015y = \ \ \ 135$$
$$y = \ \ 9000$$

$$x + 9000 = 24{,}000$$
$$x = 15{,}000$$
$$(15{,}000, 9000)$$

So, \$15,000 is invested at 7.5% and \$9000 is invested at 9%.

Section 5.2   Rational Exponents and Radicals

1. Because $7^2 = 49$, $\boxed{7}$ is a square root of 49.

3. Because $4.2^3 = 74.088$, $\boxed{4.2}$ is a cube root of 74.088.

5. Because $45^2 = 2025$, 45 is a $\boxed{\text{square root}}$ of 2025.

7. $\sqrt{64} = 8$

9. $\sqrt{-100}$ is not a real number.

11. $-\sqrt{\frac{4}{9}} = -\frac{2}{3}$

13. $\sqrt{0.09} = 0.3$

15. $\sqrt[3]{125} = 5$

17. $\sqrt[3]{1000} = 10$

19. $\sqrt[3]{-\frac{1}{64}} = -\frac{1}{4}$

21. $\sqrt[4]{81} = 3$

23. $\sqrt[5]{-0.00243} = -0.3$

25. Irrational number

$\sqrt{6}$

27. Irrational number

$\sqrt{\frac{24}{25}}$

29. Rational number

$\sqrt{900} = 30$

31. $16^{1/2} = 4$

33. $\sqrt[3]{125} = 5$

35. $25^{1/2} = \sqrt{25} = 5$

37. $32^{-2/5} = \frac{1}{32^{2/5}} = \frac{1}{(\sqrt[5]{32})^2} = \frac{1}{2^2} = \frac{1}{4}$

39. $\left(\frac{8}{27}\right)^{2/3} = \left(\sqrt[3]{\frac{8}{27}}\right)^2 = \left(\frac{2}{3}\right)^2 = \frac{4}{9}$

41. $\left(\frac{121}{9}\right)^{-1/2} = \left(\frac{9}{121}\right)^{1/2}$

$= \sqrt{\frac{9}{121}} = \frac{3}{11}$

43. $\sqrt{73} \approx 8.5440$

45. $1698^{-3/4} \approx 0.0038$

47. $\sqrt[4]{342} = 342^{1/4} \approx 4.3004$

49. $\sqrt[3]{545^2} \approx 66.7213$

51. $\sqrt{t^2} = |t^{2/2}| = |t^1| = |t|$

53. $\sqrt[3]{y^9} = y^{9/3} = y^3$

55. $t\sqrt[3]{t^6} = t(t^{6/3}) = t(t^2) = t^3$

57. $x^2\sqrt{x^8} = x^2(x^{8/2}) = x^2(x^4) = x^6$

59. $\frac{\sqrt{x}}{\sqrt{x^3}} = \frac{x^{1/2}}{x^{3/2}} = x^{1/2 - 3/2} = x^{-2/2} = x^{-1} = \frac{1}{x}$

61. $\sqrt[3]{x^2} \cdot \sqrt[3]{x^7} = x^{2/3} \cdot x^{7/3} = x^{(2/3)+(7/3)} = x^{9/3} = x^3$

63. $\sqrt[4]{x^3y} = (x^3y)^{1/4} = x^{3/4}y^{1/4}$

65. $z^2\sqrt{y^5z^4} = z^2(y^5z^4)^{1/2} = z^2(y^{5/2}z^2) = y^{5/2}z^4$

67. $(3^{1/4} \cdot 3^{3/4}) = 3^{1/4+3/4} = 3^1 = 3$

69. $\dfrac{2^{1/5}}{2^{6/5}} = 2^{1/5-6/5} = 2^{-1} = \dfrac{1}{2}$

71. $\left(\dfrac{2}{3}\right)^{5/3} \cdot \left(\dfrac{2}{3}\right)^{1/3} = \left(\dfrac{2}{3}\right)^{5/3+1/3}$

$= \left(\dfrac{2}{3}\right)^{6/3}$

$= \left(\dfrac{2}{3}\right)^2$

$= \dfrac{4}{9}$

73. $(6^{2/5})^{10/3} = 6^{(2/5)(10/3)} = 6^{4/3}$

75. $x^{2/3} \cdot x^{7/3} = x^{(2/3)+(7/3)}$

$= x^{9/3}$

$= x^3$

77. $(3x^{-1/3}y^{3/4})^2 = 3^2x^{(-1/3)2}y^{(3/4)2}$

$= 9x^{-2/3}y^{3/2}$

$= \dfrac{9y^{3/2}}{x^{2/3}}$

79. $\dfrac{a^{3/4} \cdot a^{1/2}}{a^{5/2}} = \dfrac{a^{(3/4)+(1/2)}}{a^{5/2}} = \dfrac{a^{(3/4)+(2/4)}}{a^{5/2}} = \dfrac{a^{5/4}}{a^{10/4}} = a^{(5/4)-(10/4)} = a^{-5/4} = \dfrac{1}{a^{5/4}}$

81. $\dfrac{3a^{-1/2}b^{2/5}}{6a^{5/2}b^{-1/5}} = \dfrac{3b^{2/5}b^{1/5}}{6a^{5/2}a^{1/2}} = \dfrac{3b^{(2/5)+(1/5)}}{6a^{(5/2)+(1/2)}} = \dfrac{b^{3/5}}{2a^3}$

83. $\left(\dfrac{x^{1/4}}{x^{1/6}}\right)^3 = (x^{(1/4)-(1/6)})^3 = (x^{(3/12)-(2/12)})^3$

$= (x^{1/12})^3 = x^{(1/12)3} = x^{3/12} = x^{1/4}$

85. $\left(\dfrac{7x^{2/3}y^{3/4}}{14xy^{-1/4}}\right)^3 = \left(\dfrac{x^{2/3}y^{3/4}y^{1/4}}{2x}\right)^3 = \left(\dfrac{y^{(3/4)+(1/4)}}{2x^{1/3}}\right)^3 = \left(\dfrac{y}{2x^{1/3}}\right)^3 = \dfrac{y^3}{2^3x^{3(1/3)}} = \dfrac{y^3}{8x}$

87. $(c^{3/2})^{1/3} = c^{(3/2)(1/3)}$

$= c^{3/6} = c^{1/2}$

89. $(x^{3/5}y^{-1/2})^{2/3} = x^{(3/5)(2/3)}y^{(-1/2)(2/3)} = x^{2/5}y^{-1/3} = \dfrac{x^{2/5}}{y^{1/3}}$

91. $\dfrac{x^{4/3}y^{2/3}}{(xy)^{1/3}} = \dfrac{x^{4/3}y^{2/3}}{x^{1/3}y^{1/3}}$

$= x^{(4/3)-(1/3)}y^{(2/3)-(1/3)}$

$= x^1y^{1/3}$

$= xy^{1/3}$

93. $\sqrt{\sqrt[4]{y}} = \sqrt{y^{1/4}} = (y^{1/4})^{1/2}$

$= y^{(1/4)(1/2)} = y^{1/8} \text{ or } \sqrt[8]{y}$

95. $\sqrt{\sqrt[3]{x}} = \sqrt{x^{1/3}} = (x^{1/3})^{1/2}$

$= x^{(1/3)(1/2)} = x^{1/6} \text{ or } \sqrt[6]{x}$

97. $\dfrac{(x+y)^{3/4}}{\sqrt[4]{x+y}} = \dfrac{(x+y)^{3/4}}{(x+y)^{1/4}} = (x+y)^{(3/4)-(1/4)} = (x+y)^{1/2}$

99. $\dfrac{(3u-2v)^{2/3}}{\sqrt{(3u-2v)^3}} = \dfrac{(3u-2v)^{2/3}}{(3u-2v)^{3/2}} = (3u-2v)^{(2/3)-(3/2)} = (3u-2v)^{-5/6} = \dfrac{1}{(3u-2v)^{5/6}}$

101. $g(x) = 3\sqrt{2x}$

   (a) $g(0) = 3\sqrt{2(0)} = 3\sqrt{0} = 3(0) = 0$

   (b) $g(2) = 3\sqrt{2(2)} = 3\sqrt{4} = 3(2) = 6$

   (c) $g(8) = 3\sqrt{2(8)} = 3\sqrt{16} = 3(4) = 12$

   (d) $g(50) = 3\sqrt{2(50)} = 3\sqrt{100} = 3(10) = 30$

103. $f(x) = \sqrt[3]{x+1}$

   (a) $f(7) = \sqrt[3]{7+1} = \sqrt[3]{8} = 2$

   (b) $f(0) = \sqrt[3]{0+1} = \sqrt[3]{1} = 1$

   (c) $f(63) = \sqrt[3]{63+1} = \sqrt[3]{64} = 4$

   (d) $f(-28) = \sqrt[3]{-28+1} = \sqrt[3]{-27} = -3$

105. The radicand must be greater than or equal to zero.

Domain $= \{x: x \geq 0\}$

107. The radicand must be greater than or equal to zero, but the denominator cannot be zero, so the domain includes only positive values of the variable.

Domain $= \{x: x > 0\}$

109. The radicand must be greater than or equal to zero.

Domain $= \{x: x \geq -10\}$

111. The radicand must be greater than or equal to zero.

Domain $= \left\{x: x \geq -\frac{8}{3}\right\}$

113. The radicand must be greater than or equal to zero, but the denominator cannot be zero, so the domain includes only positive values of the variable.

Domain $= \{x: x > 0\}$

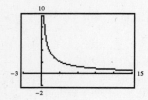

115. The base of the exponent can be any real number.

Domain $= \{x: x$ is a real number$\}$

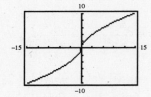

117. $x^{1/2}(2x - 3) = 2x^{(1/2)+1} - 3x^{1/2} = 2x^{3/2} - 3x^{1/2}$

119. $y^{-1/3}(y^{1/3} + 5y^{4/3}) = y^{(-1/3)+(1/3)} + 5y^{(-1/3)+(4/3)}$

$$= y^0 + 5y^{3/3} = 1 + 5y$$

121. $r = 1 - \left(\dfrac{S}{C}\right)^{1/n}$

$r = 1 - \left(\dfrac{25{,}000}{75{,}000}\right)^{1/8} = 1 - \left(\dfrac{1}{3}\right)^{1/8}$

$r \approx 0.128 \approx 12.8\%$

The annual depreciation rate is approximately 12.8%.

123. $\sqrt[3]{2197} = 13$

The inside dimensions of the oven are 13 in. $\times$ 13 in. $\times$ 13 in.

125. $0.03\sqrt{v} = 0.03\sqrt{\frac{3}{4}} = 0.03\sqrt{0.75} \approx 0.026$ in.

127. If a and b are real numbers, n is an integer greater than or equal to two, and $a = b^n$, then b is the nth root of a.

129. If $x < 0$, $\sqrt{x^2} \neq x$.

   Note: $\sqrt{x^2} = \begin{cases} x, & x \geq 0 \\ -x, & x < 0 \end{cases}$

131. $(2x + 1)(x - 4) = 2x^2 - 8x + x - 4$

$$= 2x^2 - 7x - 4$$

133. $(x + 2y)(x - 2y) = x^2 - (2y)^2$

$$= x^2 - 4y^2$$

135. $3x^2 - 7x + 4 = (3x - 4)(x - 1)$

137. $2x^3 - 16x^2 + 2x - 16 = 2(x^3 - 8x^2 + x - 8)$

$$= 2[x^2(x - 8) + 1(x - 8)]$$

$$= 2(x - 8)(x^2 + 1)$$

139. Average speed = (Distance)/(Time)

Distance = 200 miles

Time = Time for first trip + Time for return trip $= \dfrac{100}{54} + \dfrac{100}{45} = \dfrac{4500 + 5400}{54(45)} = \dfrac{9900}{2430} = \dfrac{110}{27}$

Average speed $= 200 \div \dfrac{110}{27} = 200\left(\dfrac{27}{110}\right) \approx 49.09$ miles per hour

The average speed for the round trip was approximately 49.09 miles per hour.

Section 5.3   Simplifying and Combining Radicals

1. $\sqrt{20} = \sqrt{4 \cdot 5}$

$= \sqrt{2^2 \cdot 5}$

$= 2\sqrt{5}$

3. $\sqrt{27} = \sqrt{9 \cdot 3}$

$= \sqrt{3^2 \cdot 3}$

$= 3\sqrt{3}$

5. $\sqrt{0.04} = 0.2$

7. $\sqrt[3]{24} = \sqrt[3]{8 \cdot 3}$

$= \sqrt[3]{2^3 \cdot 3}$

$= 2\sqrt[3]{3}$

9. $\sqrt[4]{30,000} = \sqrt[4]{10,000 \cdot 3}$

$= \sqrt[4]{10^4 \cdot 3}$

$= 10\sqrt[4]{3}$

11. $\sqrt{\dfrac{15}{4}} = \dfrac{\sqrt{15}}{2}$

13. $\sqrt[3]{\dfrac{35}{64}} = \dfrac{\sqrt[3]{35}}{4}$

15. $\sqrt[5]{\dfrac{15}{243}} = \dfrac{\sqrt[5]{15}}{3}$

17. $\sqrt{\dfrac{13}{25}} = \dfrac{\sqrt{13}}{5}$

19. $\sqrt{a^7} = \sqrt{a^6(a)} = a^3\sqrt{a}$

21. $\sqrt{9x^5} = \sqrt{3^2x^4(x)}$

$= 3x^2\sqrt{x}$

23. $\sqrt{48y^4} = \sqrt{4^2 \cdot 3y^4}$

$= 4y^2\sqrt{3}$

25. $\sqrt[3]{x^{10}} = \sqrt[3]{x^9(x)} = x^3\sqrt[3]{x}$

27. $\sqrt[3]{8a^7} = \sqrt[3]{2^3a^6(a)} = 2a^2\sqrt[3]{a}$

29. $\sqrt[3]{54x^5} = \sqrt[3]{27x^3(2x^2)} = \sqrt[3]{3^3x^3(2x^2)} = 3x\sqrt[3]{2x^2}$

31. $\sqrt{a^9b^5} = \sqrt{a^8b^4(ab)} = a^4b^2\sqrt{ab}$

33. $\sqrt[3]{x^4y^3} = \sqrt[3]{x^3y^3(x)} = xy\sqrt[3]{x}$

35. $\sqrt[4]{128u^4v^7} = \sqrt[4]{16u^4v^4(8v^3)}$

$= \sqrt[4]{2^4u^4v^4(8v^3)}$

$= 2|uv|\sqrt[4]{8v^3}$

37. $\sqrt[5]{32x^5y^6} = \sqrt[5]{2^5x^5y^5(y)} = 2xy\sqrt[5]{y}$

39. $\sqrt[5]{\dfrac{32x^2}{y^5}} = \dfrac{\sqrt[5]{2^5(x^2)}}{\sqrt[5]{y^5}} = \dfrac{2\sqrt[5]{x^2}}{y}$

41. $\sqrt[3]{\dfrac{54a^4}{b^9}} = \dfrac{\sqrt[3]{27a^3(2a)}}{\sqrt[3]{b^9}}$

$= \dfrac{\sqrt[3]{3^3a^3(2a)}}{\sqrt[3]{b^9}} = \dfrac{3a\sqrt[3]{2a}}{b^3}$

43. $\sqrt{\dfrac{32a^4}{b^2}} = \dfrac{\sqrt{16a^4(2)}}{\sqrt{b^2}}$

$= \dfrac{\sqrt{4^2a^4(2)}}{\sqrt{b^2}} = \dfrac{4a^2\sqrt{2}}{|b|}$

45. $\sqrt[4]{(3x^2)^4} = 3x^2$

47. $y_1 = \sqrt{12x^2}$

$y_2 = 2x\sqrt{3}$

$y_1 \neq y_2$ because $\sqrt{12x^2} = 2|x|\sqrt{3}$

From the graphs, you can see that the range of y_1 is $[0, \infty)$, and the range of y_2 is all real numbers.

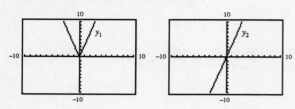

49. $y_1 = \sqrt[3]{16x^3} = \sqrt[3]{2^3 \cdot 2 \cdot x^3} = 2x\sqrt[3]{2}$

$y_2 = 2x\sqrt[3]{2}$

$y_1 = y_2$

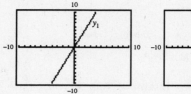 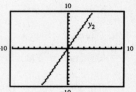

51. $\sqrt{\dfrac{1}{3}} = \dfrac{\sqrt{1}}{\sqrt{3}} = \dfrac{1 \cdot \sqrt{3}}{\sqrt{3} \cdot \sqrt{3}}$

$= \dfrac{\sqrt{3}}{\sqrt{3^2}} = \dfrac{\sqrt{3}}{3}$

53. $\dfrac{12}{\sqrt{3}} = \dfrac{12\sqrt{3}}{\sqrt{3} \cdot \sqrt{3}} = \dfrac{12\sqrt{3}}{\sqrt{3^2}}$

$= \dfrac{12\sqrt{3}}{3} = \dfrac{\cancel{3}(4)\sqrt{3}}{\cancel{3}} = 4\sqrt{3}$

55. $\sqrt[4]{\dfrac{5}{4}} = \dfrac{\sqrt[4]{5}}{\sqrt[4]{2^2}} = \dfrac{\sqrt[4]{5} \cdot \sqrt[4]{2^2}}{\sqrt[4]{2^2} \cdot \sqrt[4]{2^2}}$

$= \dfrac{\sqrt[4]{20}}{\sqrt[4]{2^4}} = \dfrac{\sqrt[4]{20}}{2}$

57. $\dfrac{6}{\sqrt[3]{32}} = \dfrac{6\sqrt[3]{2}}{\sqrt[3]{2^5} \cdot \sqrt[3]{2}} = \dfrac{6\sqrt[3]{2}}{\sqrt[3]{2^6}}$

$= \dfrac{6\sqrt[3]{2}}{2^2} = \dfrac{3\cancel{(2)}\sqrt[3]{2}}{2\cancel{(2)}} = \dfrac{3\sqrt[3]{2}}{2}$

59. $\dfrac{1}{\sqrt{y}} = \dfrac{1\sqrt{y}}{\sqrt{y} \cdot \sqrt{y}} = \dfrac{\sqrt{y}}{\sqrt{y^2}} = \dfrac{\sqrt{y}}{y}$

61. $\sqrt{\dfrac{4}{x}} = \dfrac{\sqrt{4}}{\sqrt{x}} = \dfrac{2 \cdot \sqrt{x}}{\sqrt{x} \cdot \sqrt{x}}$

$= \dfrac{2\sqrt{x}}{\sqrt{x^2}} = \dfrac{2\sqrt{x}}{x}$

63. $\sqrt{\dfrac{4}{x^3}} = \dfrac{\sqrt{4}\sqrt{x}}{\sqrt{x^3}\sqrt{x}} = \dfrac{2\sqrt{x}}{\sqrt{x^4}} = \dfrac{2\sqrt{x}}{x^2}$

65. $\sqrt[3]{\dfrac{2x}{3y}} = \dfrac{\sqrt[3]{2x} \cdot \sqrt[3]{9y^2}}{\sqrt[3]{3y} \cdot \sqrt[3]{9y^2}}$

$= \dfrac{\sqrt[3]{18xy^2}}{\sqrt[3]{3^3y^3}} = \dfrac{\sqrt[3]{18xy^2}}{3y}$

67. $\dfrac{a^3}{\sqrt[3]{ab^2}} = \dfrac{a^3\sqrt[3]{a^2b}}{\sqrt[3]{ab^2}\sqrt[3]{a^2b}}$

$= \dfrac{a^3\sqrt[3]{a^2b}}{\sqrt[3]{a^3b^3}} = \dfrac{a^3\sqrt[3]{a^2b}}{ab}$

$= \dfrac{a^2\cancel{(a)}\sqrt[3]{a^2b}}{\cancel{a}(b)} = \dfrac{a^2\sqrt[3]{a^2b}}{b}$

69. $\dfrac{6}{\sqrt{3b^3}} = \dfrac{6\sqrt{3b}}{\sqrt{3b^3}\sqrt{3b}} = \dfrac{6\sqrt{3b}}{\sqrt{9b^4}} = \dfrac{6\sqrt{3b}}{\sqrt{3^2b^4}} = \dfrac{6\sqrt{3b}}{3b^2} = \dfrac{\cancel{3}(2)\sqrt{3b}}{\cancel{3}b^2} = \dfrac{2\sqrt{3b}}{b^2}$

71. $\sqrt{\dfrac{6a^3}{15b^7}} = \sqrt{\dfrac{2a^3}{5b^7}} = \dfrac{\sqrt{2a^3}}{\sqrt{5b^7}} \cdot \dfrac{\sqrt{5b}}{\sqrt{5b}} = \dfrac{\sqrt{a^2(10ab)}}{\sqrt{5^2b^8}} = \dfrac{|a|\sqrt{10ab}}{5b^4}$

73. $\sqrt[3]{\dfrac{8u^6}{3v}} = \dfrac{\sqrt[3]{2^3u^6}}{\sqrt[3]{3v}} \cdot \dfrac{\sqrt[3]{9v^2}}{\sqrt[3]{9v^2}} = \dfrac{\sqrt[3]{2^3u^6(9v^2)}}{\sqrt[3]{3^3v^3}} = \dfrac{2u^2\sqrt[3]{9v^2}}{3v}$

75. $\sqrt{\dfrac{12x^3y^4}{4z^7}} = \sqrt{\dfrac{3x^3y^4}{z^7}} = \dfrac{\sqrt{x^2y^4(3x)}}{\sqrt{z^7}} \cdot \dfrac{\sqrt{z}}{\sqrt{z}} = \dfrac{\sqrt{x^2y^4(3xz)}}{\sqrt{z^8}} = \dfrac{|x|y^2\sqrt{3xz}}{z^4}$

77. $\sqrt[3]{\dfrac{7x^{12}y^5}{4z^{10}}} = \dfrac{\sqrt[3]{x^{12}y^3(7y^2)}}{\sqrt[3]{2^2z^{10}}} \cdot \dfrac{\sqrt[3]{2z^2}}{\sqrt[3]{2z^2}} = \dfrac{\sqrt[3]{x^{12}y^3(14y^2z^2)}}{\sqrt[3]{2^3z^{12}}} = \dfrac{x^4y\sqrt[3]{14y^2z^2}}{2z^4}$

79. $y_1 = \sqrt{\dfrac{3}{x}} = \dfrac{\sqrt{3}}{\sqrt{x}} = \dfrac{\sqrt{3}}{\sqrt{x}} \cdot \dfrac{\sqrt{x}}{\sqrt{x}}$

$= \dfrac{\sqrt{3x}}{x} = y_2$

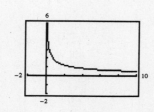

81. $3\sqrt{2} - \sqrt{2} = 2\sqrt{2}$

83. $12\sqrt{8} - 3\sqrt{8} = 9\sqrt{8}$
$$= 9\sqrt{4(2)}$$
$$= 9(2)\sqrt{2}$$
$$= 18\sqrt{2}$$

85. $\sqrt{12} + \sqrt{75} = \sqrt{4(3)} + \sqrt{25(3)}$
$$= 2\sqrt{3} + 5\sqrt{3}$$
$$= 7\sqrt{3}$$

87. $5\sqrt{54} + 3\sqrt{24} = 5\sqrt{9(6)} + 3\sqrt{4(6)}$
$$= 5(3)\sqrt{6} + 3(2)\sqrt{6}$$
$$= 15\sqrt{6} + 6\sqrt{6} = 21\sqrt{6}$$

89. $2\sqrt[3]{54} + 12\sqrt[3]{16} = 2\sqrt[3]{27 \cdot 2} + 12\sqrt[3]{8 \cdot 2}$
$$= 2 \cdot 3\sqrt[3]{2} + 12 \cdot 2\sqrt[3]{2}$$
$$= 6\sqrt[3]{2} + 24\sqrt[3]{2} = 30\sqrt[3]{2}$$

91. $5\sqrt{9x} - 3\sqrt{x} = 5\sqrt{3^2 x} - 3\sqrt{x}$
$$= 5(3)\sqrt{x} - 3\sqrt{x}$$
$$= 15\sqrt{x} - 3\sqrt{x} = 12\sqrt{x}$$

93. $\sqrt{25y} + \sqrt{64y} = \sqrt{5^2 y} + \sqrt{8^2 y}$
$$= 5\sqrt{y} + 8\sqrt{y}$$
$$= 13\sqrt{y}$$

95. $\sqrt{9a^3} + \sqrt{36a^7} = \sqrt{9a^2(a)} + \sqrt{36a^6(a)}$
$$= 3a\sqrt{a} + 6a^3\sqrt{a}$$
$$= (3a + 6a^3)\sqrt{a}$$

97. $10\sqrt[3]{z} - \sqrt[3]{z^4} = 10\sqrt[3]{z} - \sqrt[3]{z^3(z)}$
$$= 10\sqrt[3]{z} - z\sqrt[3]{z}$$
$$= (10 - z)\sqrt[3]{z}$$

99. $2y^2\sqrt{y} + 5\sqrt{y^5} = 2y^2\sqrt{y} + 5\sqrt{y^4(y)}$
$$= 2y^2\sqrt{y} + 5y^2\sqrt{y}$$
$$= 7y^2\sqrt{y}$$

101. $5\sqrt[4]{2x^7} + x\sqrt[4]{2x^3} = 5\sqrt[4]{x^4(2x^3)} + x\sqrt[4]{2x^3}$
$$= 5x\sqrt[4]{2x^3} + x\sqrt[4]{2x^3}$$
$$= 6x\sqrt[4]{2x^3}$$

103. $7\sqrt{8x^3y^5} + xy\sqrt{2xy^3} = 7\sqrt{4x^2y^4(2xy)} + xy\sqrt{y^2(2xy)}$
$$= 7(2xy^2)\sqrt{2xy} + xy(y)\sqrt{2xy}$$
$$= 14xy^2\sqrt{2xy} + xy^2\sqrt{2xy}$$
$$= 15xy^2\sqrt{2xy}$$

105. $7\sqrt{80x^3y} - 2x\sqrt{125xy} + 3\sqrt{45x^3y} = 7\sqrt{16x^2(5xy)} - 2x\sqrt{25(5xy)} + 3\sqrt{9x^2(5xy)}$
$$= 7(4x)\sqrt{5xy} - 2x(5)\sqrt{5xy} + 3(3x)\sqrt{5xy}$$
$$= 28x\sqrt{5xy} - 10x\sqrt{5xy} + 9x\sqrt{5xy}$$
$$= 27x\sqrt{5xy}$$

107. $y_1 = 7\sqrt{x^3} - 2x\sqrt{4x}$
$$= 7x\sqrt{x} - 4x\sqrt{x}$$
$$= 3x\sqrt{x}$$
$$= y_2$$

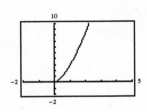

109. $\sqrt{5} - \dfrac{3}{\sqrt{5}} = \dfrac{\sqrt{5}}{1} - \dfrac{3\sqrt{5}}{\sqrt{5} \cdot \sqrt{5}} = \dfrac{\sqrt{5}}{1} - \dfrac{3\sqrt{5}}{5}$
$$= \frac{5\sqrt{5}}{5} - \frac{3\sqrt{5}}{5} = \frac{5\sqrt{5} - 3\sqrt{5}}{5} = \frac{2\sqrt{5}}{5}$$

111. $\dfrac{x}{\sqrt{3x}} + \sqrt{27x} = \dfrac{x\sqrt{3x}}{\sqrt{3x} \cdot \sqrt{3x}} + \sqrt{9(3x)} = \dfrac{x\sqrt{3x}}{3x} + 3\sqrt{3x} = \dfrac{x\sqrt{3x}}{3(x)} + \dfrac{3\sqrt{3x}}{1} = \dfrac{\sqrt{3x}}{3} + \dfrac{3 \cdot 3\sqrt{3x}}{3 \cdot 1} = \dfrac{\sqrt{3x} + 9\sqrt{3x}}{3} = \dfrac{10\sqrt{3x}}{3}$

113. $\sqrt{7} + \sqrt{18} = \sqrt{7} + 3\sqrt{2} \approx 6.888$

$\sqrt{7 + 18} = \sqrt{25} = 5$

So, $\sqrt{7} + \sqrt{18} > \sqrt{7 + 18}$

115. $\sqrt{3^2 + 2^2} = \sqrt{9 + 4} = \sqrt{13} \approx 3.606$

So, $5 > \sqrt{3^2 + 2^2}$

117. $c = \sqrt{a^2 + b^2}$

$c = \sqrt{3^2 + 6^2}$

$c = \sqrt{9 + 36}$

$c = \sqrt{45}$

$c = 3\sqrt{5}$

The length of the hypotenuse is $3\sqrt{5}$.

119. $S = \pi r \sqrt{r^2 + h^2}$

$= \pi(4)\sqrt{4^2 + 8^2}$

$= 4\pi\sqrt{16 + 64}$

$= 4\pi\sqrt{80}$

$= 4\pi\sqrt{16(5)}$

$= 4\pi(4)\sqrt{5}$

$= 16\pi\sqrt{5}$

≈ 112.40 square feet

121. Let $x =$ the length of the hypotenuse of each triangular cut-out.

$c = \sqrt{a^2 + b^2}$

$x = \sqrt{2^2 + 2^2}$

$x = \sqrt{4 + 4}$

$x = \sqrt{8}$

$x = 2\sqrt{2}$

Perimeter $= 2$ (Length of longest sides) $+ 4$ (Length of each hypotenuse)

Perimeter $= 2(8 - 2 \cdot 2) + 4(2\sqrt{2})$

$= 2(4) + 4(2\sqrt{2})$

$= 8 + 8\sqrt{2}$

The perimeter of the remaining piece of plywood is $8 + 8\sqrt{2}$ feet.

123. $f = \dfrac{1}{100}\sqrt{\dfrac{400 \times 10^6}{5}} \approx 89.44$ cycles per second

125. A simplified radical expression has all possible nth powered factors removed from the radical, it does not contain a fraction, and no denominator of a fraction contains a radical.

127. Two or more radical expressions are alike if they have the same radicand and the same index.

129. $(-4, 2), (1, 12)$

Distance: $d = \sqrt{[1 - (-4)]^2 + (12 - 2)^2}$

$= \sqrt{5^2 + 10^2}$

$= \sqrt{125}$

$= \sqrt{25(5)}$

$= 5\sqrt{5} \approx 11.18$

Midpoint: $\left(\dfrac{-4 + 1}{2}, \dfrac{2 + 12}{2}\right) = \left(\dfrac{-3}{2}, \dfrac{14}{2}\right) = \left(-\dfrac{3}{2}, 7\right)$

131. $(3, 6), (-5, -8)$

Distance: $d = \sqrt{(-5 - 3)^2 + (-8 - 6)^2}$

$= \sqrt{(-8)^2 + (-14)^2}$

$= \sqrt{64 + 196}$

$= \sqrt{260}$

$= \sqrt{4(65)}$

$= 2\sqrt{65} \approx 16.12$

Midpoint: $\left(\dfrac{3 + (-5)}{2}, \dfrac{6 + (-8)}{2}\right) = \left(\dfrac{-2}{2}, \dfrac{-2}{2}\right) = (-1, -1)$

133. $10 - 3x \le 0$

$-3x \le -10$

$\dfrac{-3x}{-3} \ge \dfrac{-10}{-3}$

$x \ge \dfrac{10}{3}$

135. $|4 - (x - 2)| < 20$

$|4 - x + 2| < 20$

$|-x + 6| < 20$

$-20 < -x + 6 < 20$

$-26 < -x < 14$

$\dfrac{-26}{-1} > \dfrac{-x}{-1} > \dfrac{14}{-1}$

$26 > x > -14$ or $-14 < x < 26$

137. (Length)(Width) = Area

Labels: Length $= x$ (inches)

Width $= x$ (inches)

Equation: $\qquad\qquad x(x) = 1024$

$x^2 = 1024$

$x^2 - 1024 = 0$

$(x + 32)(x - 32) = 0$

$x + 32 = 0 \Longrightarrow x = -32$

$x - 32 = 0 \Longrightarrow x = 32$

Discard the negative solution. The dimensions of the square are 32 inches by 32 inches.

Mid-Chapter Quiz for Chapter 5

1. $-12^{-2} = -\dfrac{1}{12^2}$

$= -\dfrac{1}{144}$

2. $\left(\dfrac{3}{4}\right)^{-3} = \dfrac{3^{-3}}{4^{-3}}$

$= \dfrac{4^3}{3^3}$

$= \dfrac{64}{27}$

3. $\sqrt{\dfrac{25}{9}} = \dfrac{\sqrt{25}}{\sqrt{9}}$

$= \dfrac{5}{3}$

4. $(-64)^{2/3} = \left(\sqrt[3]{-64}\right)^2$

$= (-4)^2$

$= 16$

5. $(t^3)^{-1/2}(3t^3) = t^{3(-1/2)}(3)(t^3)$

$= 3t^{-3/2}t^3$

$= 3t^{(-3/2)+3}$

$= 3t^{3/2}$

6. $\dfrac{(10x)^0}{(4x^{-2})^{3/2}} = \dfrac{1}{4^{3/2}x^{(-2)(3/2)}}$

$= \dfrac{1}{\left(\sqrt{4}\right)^3 x^{-3}}$

$= \dfrac{x^3}{8}$

7. $\dfrac{10u^{-2}}{15u} = \dfrac{\cancel{5}(2)}{\cancel{5}(3)u \cdot u^2}$

$= \dfrac{2}{3u^3}$

8. $(3x^2y^{-1})(4x^{-2}y)^{-2} = \dfrac{3x^2}{y} \cdot 4^{-2}x^{(-2)(-2)}y^{-2}$

$= \dfrac{3x^2}{y} \cdot \dfrac{x^4}{4^2y^2}$

$= \dfrac{3x^6}{16y^3}$

9. (a) $13{,}400{,}000 = 1.34 \times 10^7$

(b) $0.00075 = 7.5 \times 10^{-4}$

10. (a) $\sqrt{150} = \sqrt{25(6)}$

$= 5\sqrt{6}$

(b) $\sqrt[3]{54} = \sqrt[3]{27(2)}$

$= 3\sqrt[3]{2}$

11. (a) $\sqrt[3]{27x^7} = \sqrt[3]{27x^6(x)} = 3x^2\sqrt[3]{x}$

(b) $\sqrt[4]{81x^{11}y^5} = \sqrt[4]{81x^8y^4(x^3y)} = 3x^2y\sqrt[4]{x^3y}$

12. (a) $\sqrt[4]{\dfrac{5}{16}} = \dfrac{\sqrt[4]{5}}{\sqrt[4]{16}}$ (b) $\sqrt{\dfrac{24}{49}} = \dfrac{\sqrt{24}}{\sqrt{49}}$

$= \dfrac{\sqrt[4]{5}}{2}$ $= \dfrac{\sqrt{4 \cdot 6}}{7}$

$= \dfrac{2\sqrt{6}}{7}$

13. (a) $\sqrt{\dfrac{40u^3}{9z^{10}}} = \sqrt{\dfrac{4u^2(10u)}{\sqrt{9z^{10}}}}$

$= \dfrac{2u\sqrt{10u}}{3z^5}$

(b) $\sqrt[3]{\dfrac{16a^5}{b^{12}}} = \dfrac{\sqrt[3]{8a^3(2a^2)}}{\sqrt[3]{b^{12}}}$

$= \dfrac{2a\sqrt[3]{2a^2}}{b^4}$

14. $f(x) = \sqrt{x - 12}$

(a) $f(12) = \sqrt{12 - 12} = \sqrt{0} = 0$

(b) $f(40) = \sqrt{40 - 12} = \sqrt{28} = \sqrt{4(7)} = 2\sqrt{7}$

15. $g(x) = \sqrt{x + 10}$

(a) $g(26) = \sqrt{26 + 10} = \sqrt{36} = 6$

(b) $g(8) = \sqrt{8 + 10} = \sqrt{18} = \sqrt{9(2)} = 3\sqrt{2}$

16. (a) The radicand $12 - 4x$ must be greater than or equal to zero.

$12 - 4x \geq 0$

$-4x \geq -12$

$x \leq 3$

Domain $= \{x: x \leq 3\}$

(b) Domain $= \{x: x \text{ is a real number}\}$

17. (a) $\sqrt{\dfrac{2}{3}} = \dfrac{\sqrt{2}}{\sqrt{3}} \cdot \dfrac{\sqrt{3}}{\sqrt{3}}$

$= \dfrac{\sqrt{6}}{(\sqrt{3})^2}$

$= \dfrac{\sqrt{6}}{3}$

(b) $\dfrac{24}{\sqrt{12}} = \dfrac{24}{\sqrt{12}} \cdot \dfrac{\sqrt{3}}{\sqrt{3}}$

$= \dfrac{24\sqrt{3}}{\sqrt{36}}$

$= \dfrac{4(6)\sqrt{3}}{6}$

$= 4\sqrt{3}$

18. (a) $\dfrac{10}{\sqrt{5x}} = \dfrac{10}{\sqrt{5x}} \cdot \dfrac{\sqrt{5x}}{\sqrt{5x}}$

$= \dfrac{10\sqrt{5x}}{(\sqrt{5x})^2}$

$= \dfrac{10\sqrt{5x}}{5x}$

$= \dfrac{2(5)\sqrt{5x}}{5x}$

$= \dfrac{2\sqrt{5x}}{x}$

(b) $\sqrt[3]{\dfrac{3}{2a^2}} = \dfrac{\sqrt[3]{3}}{\sqrt[3]{2a^2}} \cdot \dfrac{\sqrt[3]{2^2a}}{\sqrt[3]{2^2a}}$

$= \dfrac{\sqrt[3]{12a}}{\sqrt[3]{2^3a^3}}$

$= \dfrac{\sqrt[3]{12a}}{2a}$

19. $\sqrt{200y} - 3\sqrt{8y} = \sqrt{100 \cdot 2y} - 3\sqrt{4 \cdot 2y}$

$= 10\sqrt{2y} - 3(2)\sqrt{2y}$

$= 10\sqrt{2y} - 6\sqrt{2y}$

$= 4\sqrt{2y}$

20. $6x\sqrt[3]{5x} + 2\sqrt[3]{40x^4} = 6x\sqrt[3]{5x} + 2\sqrt[3]{8x^3 \cdot 5x}$

$= 6x\sqrt[3]{5x} + 2(2x)\sqrt[3]{5x}$

$= 6x\sqrt[3]{5x} + 4x\sqrt[3]{5x}$

$= 10x\sqrt[3]{5x}$

21. $\sqrt{\dfrac{5x}{2}} + \sqrt{10x} = \dfrac{\sqrt{5x}}{\sqrt{2}} \cdot \dfrac{\sqrt{2}}{\sqrt{2}} + \sqrt{10x}$

$= \dfrac{\sqrt{10x}}{\sqrt{4}} + \dfrac{\sqrt{10x}}{1}$

$= \dfrac{\sqrt{10x}}{2} + \dfrac{2\sqrt{10x}}{2(1)}$

$= \dfrac{3\sqrt{10x}}{2}$

22. $\sqrt{5^2 + 12^2} = \sqrt{25 + 144}$

$= \sqrt{169}$

$= 13$

23. Let $x =$ the length of the hypotenuse of each triangular cut-out.

$c = \sqrt{a^2 + b^2}$ (Pythagorean Theorem)

$x = \sqrt{2^2 + 2^2}$

$x = \sqrt{4 + 4}$

$x = \sqrt{8}$

$x = \sqrt{4(2)}$

$x = 2\sqrt{2}$

Let $x =$ length of horizontal side,

$\quad y =$ length of vertical side, and

$\quad h =$ length of hypotenuse of triangular cut–out

Perimeter $= 2x + 2y + 4h$

Perimeter $= 2(11 - 2 \cdot 2) + 2(8.5 - 2 \cdot 2) + 4\left(2\sqrt{2}\right)$

$\quad\quad = 2(7) + 2(4.5) + 4\left(2\sqrt{2}\right)$

$\quad\quad = 14 + 9 + 8\sqrt{2}$

$\quad\quad = 23 + 8\sqrt{2}$

The perimeter of the remaining piece of paper is $23 + 8\sqrt{2}$ inches.

Section 5.4   **Multiplying and Dividing Radicals**

1. $\sqrt{2} \cdot \sqrt{8} = \sqrt{16} = 4$

3. $\sqrt{3} \cdot \sqrt{6} = \sqrt{18}$
$\quad\quad\quad\quad = \sqrt{9 \cdot 2}$
$\quad\quad\quad\quad = 3\sqrt{2}$

5. $\sqrt{5}\left(2 - \sqrt{3}\right) = 2\sqrt{5} - \sqrt{5}\sqrt{3}$
$\quad\quad\quad\quad\quad\quad = 2\sqrt{5} - \sqrt{15}$

7. $\sqrt{2}\left(\sqrt{20} + 8\right) = \sqrt{2}\sqrt{20} + 8\sqrt{2}$
$\quad\quad\quad\quad\quad = \sqrt{40} + 8\sqrt{2}$
$\quad\quad\quad\quad\quad = \sqrt{4 \cdot 10} + 8\sqrt{2}$
$\quad\quad\quad\quad\quad = 2\sqrt{10} + 8\sqrt{2}$

9. $\left(\sqrt{3} + 2\right)\left(\sqrt{3} - 2\right) = \left(\sqrt{3}\right)^2 - 2^2$
$\quad\quad\quad\quad\quad\quad\quad = 3 - 4 = -1$

11. $\left(3 - \sqrt{5}\right)\left(3 + \sqrt{5}\right) = 3^2 - \left(\sqrt{5}\right)^2$
$\quad\quad\quad\quad\quad\quad\quad = 9 - 5$
$\quad\quad\quad\quad\quad\quad\quad = 4$

13. $\left(2\sqrt{2} + \sqrt{4}\right)\left(2\sqrt{2} - \sqrt{4}\right) = \left(2\sqrt{2}\right)^2 - \left(\sqrt{4}\right)^2$
$\quad\quad\quad\quad\quad\quad\quad\quad = 4(2) - 4 = 8 - 4 = 4$

15. $\left(\sqrt{20} + 2\right)^2 = \left(\sqrt{20}\right)^2 + 2\left(\sqrt{20}\right)(2) + 2^2$
$\quad\quad\quad = 20 + 4\sqrt{4 \cdot 5} + 4 = 20 + 4 \cdot 2\sqrt{5} + 4$
$\quad\quad\quad = 24 + 8\sqrt{5}$

17. $\left(10 + \sqrt{2x}\right)^2 = 10^2 + 2(10)\sqrt{2x} + \left(\sqrt{2x}\right)^2$
$\quad\quad\quad = 100 + 20\sqrt{2x} + 2x$

19. $\sqrt{y}\left(\sqrt{y} + 4\right) = \left(\sqrt{y}\right)^2 + 4\sqrt{y}$
$\quad\quad\quad\quad = y + 4\sqrt{y}$

21. $\left(\sqrt{5} + 3\right)\left(\sqrt{3} - 5\right) = \sqrt{5}\sqrt{3} - 5\sqrt{5} + 3\sqrt{3} - 15$
$\quad\quad\quad\quad = \sqrt{15} - 5\sqrt{5} + 3\sqrt{3} - 15$

23. $\left(9\sqrt{x} + 2\right)\left(5\sqrt{x} - 3\right) = 45\left(\sqrt{x}\right)^2 - 27\sqrt{x} + 10\sqrt{x} - 6$
$\quad\quad\quad\quad\quad\quad = 45x - 17\sqrt{x} - 6$

25. $\left(\sqrt{x} + \sqrt{y}\right)\left(\sqrt{x} - \sqrt{y}\right) = \left(\sqrt{x}\right)^2 - \left(\sqrt{y}\right)^2 = x - y$

27. $\sqrt[3]{4}\left(\sqrt[3]{2} - 7\right) = \sqrt[3]{4}\,\sqrt[3]{2} - 7\sqrt[3]{4} = \sqrt[3]{8} - 7\sqrt[3]{4} = 2 - 7\sqrt[3]{4}$

29. $\left(\sqrt[3]{2x} + 5\right)^2 = \left(\sqrt[3]{2x}\right)^2 + 2\left(\sqrt[3]{2x}\right)(5) + 5^2 = \sqrt[3]{4x^2} + 10\sqrt[3]{2x} + 25$

31. $\left(\sqrt[3]{2y} + 10\right)\left(\sqrt[3]{4y^2} - 10\right) = \sqrt[3]{2y}\,\sqrt[3]{4y^2} - 10\sqrt[3]{2y} + 10\sqrt[3]{4y^2} - 100$

$$= \sqrt[3]{8y^3} - 10\sqrt[3]{2y} + 10\sqrt[3]{4y^2} - 100$$

$$= 2y - 10\sqrt[3]{2y} + 10\sqrt[3]{4y^2} - 100$$

33. $5x\sqrt{3} + 15\sqrt{3} = 5\sqrt{3}(x + 3)$

The missing factor is $(x + 3)$.

35. $4\sqrt{12} - 2x\sqrt{27} = 4\sqrt{4 \cdot 3} - 2x\sqrt{9 \cdot 3}$

$$= 4 \cdot 2\sqrt{3} - 2x \cdot 3\sqrt{3}$$

$$= 8\sqrt{3} - 6x\sqrt{3}$$

$$= 2\sqrt{3}(4 - 3x)$$

The missing factor is $(4 - 3x)$.

37. $6u^2 + \sqrt{18u^3} = 6u^2 + \sqrt{9u^2(2u)}$

$$= 6u^2 + 3u\sqrt{2u}$$

$$= 3u\left(2u + \sqrt{2u}\right)$$

The missing factor is $\left(2u + \sqrt{2u}\right)$.

39. $\dfrac{4 - 8\sqrt{x}}{12} = \dfrac{\cancel{4}\left(1 - 2\sqrt{x}\right)}{\cancel{4}(3)}$

$$= \dfrac{1 - 2\sqrt{x}}{3}$$

41. $\dfrac{-2y + \sqrt{12y^3}}{8y} = \dfrac{-2y + \sqrt{4y^2(3y)}}{8y}$

$$= \dfrac{-2y + 2y\sqrt{3y}}{8y}$$

$$= \dfrac{2\cancel{y}\left(-1 + \sqrt{3y}\right)}{2\cancel{y}(4)}$$

$$= \dfrac{-1 + \sqrt{3y}}{4}$$

43. $\dfrac{-4x^2 + \sqrt{28x^3}}{2x} = \dfrac{-4x^2 + \sqrt{4x^2(7x)}}{2x}$

$$= \dfrac{-4x^2 + 2x\sqrt{7x}}{2x}$$

$$= \dfrac{2\cancel{x}\left(-2x + \sqrt{7x}\right)}{2\cancel{x}(1)}$$

$$= -2x + \sqrt{7x}$$

45. Conjugate: $2 - \sqrt{5}$

Product: $2^2 - \left(\sqrt{5}\right)^2 = 4 - 5 = -1$

47. Conjugate: $\sqrt{6} - 10$

Product: $\left(\sqrt{6}\right)^2 - 10^2 = 6 - 100 = -94$

49. Conjugate: $\sqrt{11} + \sqrt{3}$

Product: $\left(\sqrt{11}\right)^2 - \left(\sqrt{3}\right)^2 = 11 - 3 = 8$

51. Conjugate: $\sqrt{x} + 3$

Product: $\left(\sqrt{x}\right)^2 - 3^2 = x - 9$

53. Conjugate: $\sqrt{2u} + \sqrt{3}$

Product: $\left(\sqrt{2u}\right)^2 - \left(\sqrt{3}\right)^2 = 2u - 3$

55. Conjugate: $\sqrt{6x} - \sqrt{y}$

Product: $\left(\sqrt{6x}\right)^2 - \left(\sqrt{y}\right)^2 = 6x - y$

57. $\dfrac{6}{\sqrt{11} - 2} = \dfrac{6}{\sqrt{11} - 2} \cdot \dfrac{\sqrt{11} + 2}{\sqrt{11} + 2}$

$$= \dfrac{6\left(\sqrt{11} + 2\right)}{\left(\sqrt{11}\right)^2 - 2^2}$$

$$= \dfrac{6\sqrt{11} + 12}{11 - 4}$$

$$= \dfrac{6\sqrt{11} + 12}{7}$$

59. $\dfrac{7}{5 + \sqrt{3}} = \dfrac{7}{5 + \sqrt{3}} \cdot \dfrac{5 - \sqrt{3}}{5 - \sqrt{3}}$

$$= \dfrac{7\left(5 - \sqrt{3}\right)}{5^2 - \left(\sqrt{3}\right)^2}$$

$$= \dfrac{35 - 7\sqrt{3}}{25 - 3}$$

$$= \dfrac{35 - 7\sqrt{3}}{22}$$

61. $\dfrac{8}{\sqrt{7}+3} = \dfrac{8(\sqrt{7}-3)}{(\sqrt{7}+3)(\sqrt{7}-3)}$

$= \dfrac{8(\sqrt{7}-3)}{(\sqrt{7})^2 - 3^2}$

$= \dfrac{8(\sqrt{7}-3)}{7-9}$

$= \dfrac{8(\sqrt{7}-3)}{-2}$

$= -\dfrac{4(2)(\sqrt{7}-3)}{2}$

$= -4(\sqrt{7}-3)$ or $-4\sqrt{7}+12$

63. $\dfrac{2}{\sqrt{6}+\sqrt{2}} = \dfrac{2}{\sqrt{6}+\sqrt{2}} \cdot \dfrac{\sqrt{6}-\sqrt{2}}{\sqrt{6}-\sqrt{2}}$

$= \dfrac{2(\sqrt{6}-\sqrt{2})}{(\sqrt{6})^2 - (\sqrt{2})^2}$

$= \dfrac{2(\sqrt{6}-\sqrt{2})}{6-2}$

$= \dfrac{2(\sqrt{6}-\sqrt{2})}{4}$

$= \dfrac{2(\sqrt{6}-\sqrt{2})}{2(2)}$

$= \dfrac{\sqrt{6}-\sqrt{2}}{2}$

65. $\dfrac{2}{6+\sqrt{2}} = \dfrac{2}{6+\sqrt{2}} \cdot \dfrac{6-\sqrt{2}}{6-\sqrt{2}}$

$= \dfrac{12-2\sqrt{2}}{6^2 - (\sqrt{2})^2}$

$= \dfrac{12-2\sqrt{2}}{36-2}$

$= \dfrac{12-2\sqrt{2}}{34}$

$= \dfrac{2(6-\sqrt{2})}{2(17)}$

$= \dfrac{6-\sqrt{2}}{17}$

67. $(\sqrt{7}+2) \div (\sqrt{7}-2) = \dfrac{\sqrt{7}+2}{\sqrt{7}-2}$

$= \dfrac{(\sqrt{7}+2)(\sqrt{7}+2)}{(\sqrt{7}-2)(\sqrt{7}+2)}$

$= \dfrac{(\sqrt{7})^2 + 2(\sqrt{7})(2) + 2^2}{(\sqrt{7})^2 - 2^2}$

$= \dfrac{7 + 4\sqrt{7} + 4}{7-4}$

$= \dfrac{11 + 4\sqrt{7}}{3}$

69. $\dfrac{3x}{\sqrt{15}-\sqrt{3}} = \dfrac{3x(\sqrt{15}+\sqrt{3})}{(\sqrt{15}-\sqrt{3})(\sqrt{15}+\sqrt{3})} = \dfrac{3x(\sqrt{15}+\sqrt{3})}{(\sqrt{15})^2 - (\sqrt{3})^2} = \dfrac{3x(\sqrt{15}+\sqrt{3})}{15-3}$

$= \dfrac{3x(\sqrt{15}+\sqrt{3})}{12} = \dfrac{3x(\sqrt{15}+\sqrt{3})}{3(4)} = \dfrac{x(\sqrt{15}+\sqrt{3})}{4}$ or $\dfrac{x\sqrt{15}+x\sqrt{3}}{4}$

71. $\dfrac{2t^2}{\sqrt{5t}-\sqrt{t}} = \dfrac{2t^2(\sqrt{5t}+\sqrt{t})}{(\sqrt{5t}-\sqrt{t})(\sqrt{5t}+\sqrt{t})} = \dfrac{2t^2(\sqrt{5t}+\sqrt{t})}{(\sqrt{5t})^2 - (\sqrt{t})^2} = \dfrac{2t^2(\sqrt{5t}+\sqrt{t})}{5t-t}$

$= \dfrac{2t^2(\sqrt{5t}+\sqrt{t})}{4t} = \dfrac{2t(t)(\sqrt{5t}+\sqrt{t})}{2(2t)} = \dfrac{t(\sqrt{5t}+\sqrt{t})}{2}$ or $\dfrac{t\sqrt{5t}+t\sqrt{t}}{2}$

73. $\dfrac{7z}{\sqrt{5z}-\sqrt{z}} = \dfrac{7z}{\sqrt{5z}-\sqrt{z}} \cdot \dfrac{\sqrt{5z}+\sqrt{z}}{\sqrt{5z}+\sqrt{z}}$

$= \dfrac{7z(\sqrt{5z}+\sqrt{z})}{(\sqrt{5z})^2 - (\sqrt{z})^2}$

$= \dfrac{7z(\sqrt{5z}+\sqrt{z})}{5z-z}$

$= \dfrac{7z(\sqrt{5z}+\sqrt{z})}{4z}$

$= \dfrac{7\sqrt{5z}+7\sqrt{z}}{4}$

75. $\dfrac{2\sqrt{x}-1}{\sqrt{x}+2} = \dfrac{2\sqrt{x}-1}{\sqrt{x}+2} \cdot \dfrac{\sqrt{x}-2}{\sqrt{x}-2}$

$= \dfrac{2x - 4\sqrt{x} - \sqrt{x} + 2}{(\sqrt{x})^2 - 2^2}$

$= \dfrac{2x - 5\sqrt{x} + 2}{x-4}$

77. $\left(\sqrt{x} - 5\right) \div \left(2\sqrt{x} - 1\right) = \dfrac{\sqrt{x} - 5}{2\sqrt{x} - 1}$

$$= \frac{\left(\sqrt{x} - 5\right)\left(2\sqrt{x} + 1\right)}{\left(2\sqrt{x} - 1\right)\left(2\sqrt{x} + 1\right)}$$

$$= \frac{2\left(\sqrt{x}\right)^2 + \sqrt{x} - 10\sqrt{x} - 5}{\left(2\sqrt{x}\right)^2 - 1^2}$$

$$= \frac{2x - 9\sqrt{x} - 5}{4x - 1}$$

79. $\dfrac{\sqrt{u + v}}{\sqrt{u - v} - \sqrt{u}} = \dfrac{\sqrt{u + v}}{\sqrt{u - v} - \sqrt{u}} \cdot \dfrac{\sqrt{u - v} + \sqrt{u}}{\sqrt{u - v} + \sqrt{u}}$

$$= \frac{\sqrt{u + v}\left(\sqrt{u - v} + \sqrt{u}\right)}{\left(\sqrt{u - v}\right)^2 - \left(\sqrt{u}\right)^2}$$

$$= \frac{\sqrt{u + v}\left(\sqrt{u - v} + \sqrt{u}\right)}{u - v - u}$$

$$= \frac{\sqrt{u + v}\left(\sqrt{u - v} + \sqrt{u}\right)}{-v}$$

$$= -\frac{\sqrt{u + v}\left(\sqrt{u - v} + \sqrt{u}\right)}{v}$$

81. $\qquad f(x) = x^2 - 6x + 1$

(a) $f\left(2 - \sqrt{3}\right) = \left(2 - \sqrt{3}\right)^2 - 6\left(2 - \sqrt{3}\right) + 1$

$\qquad = 2^2 - 2(2)\sqrt{3} + \left(\sqrt{3}\right)^2 - 12 + 6\sqrt{3} + 1$

$\qquad = 4 - 4\sqrt{3} + 3 - 12 + 6\sqrt{3} + 1$

$\qquad = 2\sqrt{3} - 4$

(b) $f\left(3 - 2\sqrt{2}\right) = \left(3 - 2\sqrt{2}\right)^2 - 6\left(3 - 2\sqrt{2}\right) + 1$

$\qquad = 3^2 - 2(3)\left(2\sqrt{2}\right) + \left(2\sqrt{2}\right)^2 - 18 + 12\sqrt{2} + 1$

$\qquad = 9 - 12\sqrt{2} + 4(2) - 18 + 12\sqrt{2} + 1$

$\qquad = 0$

83. $\qquad f(x) = x^2 - 2x - 1$

(a) $f\left(1 + \sqrt{2}\right) = \left(1 + \sqrt{2}\right)^2 - 2\left(1 + \sqrt{2}\right) - 1$

$\qquad = 1^2 + 2(1)\sqrt{2} + \left(\sqrt{2}\right)^2 - 2 - 2\sqrt{2} - 1$

$\qquad = 1 + 2\sqrt{2} + 2 - 2 - 2\sqrt{2} - 1$

$\qquad = 0$

(b) $f\left(\sqrt{4}\right) = f(2)$

$\qquad = (2)^2 - 2(2) - 1$

$\qquad = 4 - 4 - 1$

$\qquad = -1$

85. $y_1 = \dfrac{10}{\sqrt{x} + 1}$

$$= \frac{10}{\sqrt{x} + 1} \cdot \frac{\sqrt{x} - 1}{\sqrt{x} - 1}$$

$$= \frac{10\left(\sqrt{x} - 1\right)}{\left(\sqrt{x}\right)^2 - 1^2}$$

$$= \frac{10\left(\sqrt{x} - 1\right)}{x - 1}$$

$$= y_2$$

87. $y_1 = \dfrac{2\sqrt{x}}{2 - \sqrt{x}}$

$$= \frac{2\sqrt{x}}{2 - \sqrt{x}} \cdot \frac{2 + \sqrt{x}}{2 + \sqrt{x}}$$

$$= \frac{2\sqrt{x}\left(2 + \sqrt{x}\right)}{2^2 - \left(\sqrt{x}\right)^2}$$

$$= \frac{4\sqrt{x} + 2x}{4 - x}$$

$$= \frac{2\left(2\sqrt{x} + x\right)}{4 - x}$$

$$= y_2$$

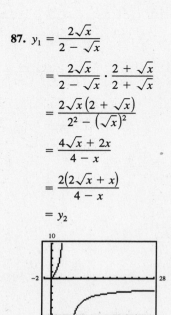

89. $b = \sqrt{c^2 - a^2}$ (Pythagorean Theorem)

Length $= \sqrt{14^2 - 8^2}$

$= \sqrt{196 - 64}$

$= \sqrt{132}$

$= \sqrt{4(33)}$

$= 2\sqrt{33}$

Area $=$ (Length)(Width)

$= \left(2\sqrt{33}\right)(8)$

$= 16\sqrt{33}$

The area is $16\sqrt{33}$ square inches.

91. Area $=$ (Width)(Height)

$= \left(8\sqrt{3}\right)\left(\sqrt{24^2 - \left(8\sqrt{3}\right)^2}\right)$

$= \left(8\sqrt{3}\right)\left(\sqrt{24^2 - 64(3)}\right)$

$= \left(8\sqrt{3}\right)\left(\sqrt{576 - 192}\right)$

$= \left(8\sqrt{3}\right)\left(\sqrt{384}\right)$

$= \left(8\sqrt{3}\right)\left(\sqrt{64(6)}\right)$

$= \left(8\sqrt{3}\right)\left(8\sqrt{6}\right)$

$= 64\sqrt{18}$

$= 64\sqrt{9(2)}$

$= 64 \cdot 3\sqrt{2} = 192\sqrt{2}$

The area is $192\sqrt{2}$ in².

93. Let r = radius of smaller circle and R = radius of larger circle.

$\dfrac{\text{Area of smaller circle}}{\text{Area of larger circle}} = \dfrac{\pi r^2}{\pi R^2}$

$\dfrac{15}{20} = \dfrac{\pi r^2}{\pi R^2}$

$\dfrac{3(5)}{4(5)} = \dfrac{\pi(r^2)}{\pi(R^2)}$

$\dfrac{3}{4} = \dfrac{r^2}{R^2}$

$\dfrac{3}{4} = \left(\dfrac{r}{R}\right)^2$

$\sqrt{\dfrac{3}{4}} = \dfrac{r}{R}$

$\dfrac{\sqrt{3}}{\sqrt{4}} = \dfrac{r}{R}$

$\dfrac{\sqrt{3}}{2} = \dfrac{r}{R}$

95. $\dfrac{500}{\sqrt{k^2 + 1}} = \dfrac{500k}{\sqrt{k^2 + 1}} \cdot \dfrac{\sqrt{k^2 + 1}}{\sqrt{k^2 + 1}}$

$= \dfrac{500k\sqrt{k^2 + 1}}{\left(\sqrt{k^2 + 1}\right)^2}$

$= \dfrac{500k\sqrt{k^2 + 1}}{k^2 + 1}$

97. Answers will vary.

The FOIL method is the same for polynomial expressions and radical expressions.

99. $12 + 4(x - 7) = 12 + 4x - 28$

$= 4x - 16$

101. $\left(x + \frac{1}{2}\right)^2 = (x)^2 + 2(x)\left(\frac{1}{2}\right) + \left(\frac{1}{2}\right)^2$

$= x^2 + x + \frac{1}{4}$

103. $(3x + 8)(3x - 8) = (3x)^2 - 8^2$

$= 9x^2 - 64$

105. Vertical shift 8 units upward

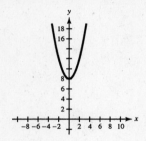

107. Vertical shift 1 unit upward and horizontal shift 1 unit to the right

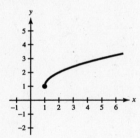

109. $\sqrt{128x^4y^7} = \sqrt{64x^4y^6(2y)} = 8x^2y^3\sqrt{2y}$

111. $\sqrt{32x^3} + 5x\sqrt{2x} = \sqrt{16x^2(2x)} + 5x\sqrt{2x} = 4x\sqrt{2x} + 5x\sqrt{2x} = 9x\sqrt{2x}$

113. $-16t^2 - 56t + 120 = 0$

$\qquad -8(2t^2 + 7t - 15) = 0$

$\qquad -8(2t - 3)(t + 5) = 0$

$\qquad\qquad 2t - 3 = 0 \Longrightarrow 2t = 3 \Longrightarrow t = \frac{3}{2}$

$\qquad\qquad t + 5 = 0 \Longrightarrow t = -5$

Discard the negative answer. The object reaches the ground in 1.5 seconds.

115. *Labels:*   Amount of 30% solution $= x$ (gallons)

$\qquad\qquad\qquad$ Amount of 60% solution $= y$ (gallons)

Equations: $\qquad x + \quad y = 20 \qquad \Longrightarrow \quad -0.30x - 0.30y = -6$

$\qquad\qquad\quad 0.30x + 0.60y = 0.40(20) \quad \Longrightarrow \quad \underline{0.30x + 0.60y = 8}$

$\qquad\qquad\qquad\qquad\qquad\qquad\qquad\qquad\qquad 0.30y = 2$

$\qquad\qquad\qquad\qquad\qquad\qquad\qquad\qquad y = 6\frac{2}{3} \quad$ and $\quad x + 6\frac{2}{3} = 20$

$\qquad\qquad\qquad\qquad\qquad\qquad\qquad\qquad\qquad\qquad\qquad x = 13\frac{1}{3}$

So, $13\frac{1}{3}$ gallons of the 30% solution and $6\frac{2}{3}$ gallons of the 60% solution must be used.

Section 5.5   Solving Radical Equations

1. $\sqrt{x} - 10 = 0$

(a) $x = -4$
$\quad \sqrt{-4} - 10 \neq 0$
$\quad \sqrt{-4}$ is not a real number.
$\quad$ No, -4 is not a real solution.

(b) $x = -100$
$\quad \sqrt{-100} - 10 \neq 0$
$\quad \sqrt{-100}$ is not a real number.
$\quad$ No, -100 is not a solution.

(c) $x = \sqrt{10}$
$\quad \sqrt{\sqrt{10}} - \sqrt{10} \neq 0$
$\quad$ No, $\sqrt{10}$ is not a solution.

(d) $x = 100$
$\quad \sqrt{100} - 10 = 0$
$\quad$ Yes, 100 is a solution.

3. $\sqrt[3]{x - 4} = 4$

(a) $x = -60$
$\quad \sqrt[3]{-60 - 4} \overset{?}{=} 4$
$\quad \sqrt[3]{-64} \overset{?}{=} 4$
$\quad -4 \neq 4$
$\quad$ No, -60 is not a solution.

(b) $x = 68$
$\quad \sqrt[3]{68 - 4} \overset{?}{=} 4$
$\quad \sqrt[3]{64} = 4$
$\quad$ Yes, 68 is a solution.

(c) $x = 20$
$\quad \sqrt[3]{20 - 4} \overset{?}{=} 4$
$\quad \sqrt[3]{16} \neq 4$
$\quad$ No, 20 is not a solution.

(d) $x = 0$
$\quad \sqrt[3]{0 - 4} \overset{?}{=} 4$
$\quad \sqrt[3]{-4} \neq 4$
$\quad$ No, 0 is not a solution.

5. $\sqrt{x} = 20$

$(\sqrt{x})^2 = 20^2$

$x = 400$

Check: $\sqrt{400} \overset{?}{=} 20$

$20 = 20$

7. $\sqrt{y} - 7 = 0$

$\sqrt{y} = 7$

$(\sqrt{y})^2 = 7^2$

$y = 49$

Check: $\sqrt{49} - 7 \overset{?}{=} 0$

$7 - 7 \overset{?}{=} 0$

$0 = 0$

9. $\sqrt{u} + 13 = 0$

$\sqrt{u} = -13$

$(\sqrt{u})^2 = (-13)^2$

$u = 169$ (Extraneous)

Check: $\sqrt{169} + 13 \overset{?}{=} 0$

$13 + 13 \overset{?}{=} 0$

$26 \neq 0$

The equation has *no* solution.

11. $\sqrt{a + 100} = 25$

$(\sqrt{a + 100})^2 = 25^2$

$a + 100 = 625$

$a = 525$

Check: $\sqrt{525 + 100} \overset{?}{=} 25$

$\sqrt{625} \overset{?}{=} 25$

$25 = 25$

13. $\sqrt{10x} = 30$

$(\sqrt{10x})^2 = 30^2$

$10x = 900$

$x = \frac{900}{10}$

$x = 90$

Check: $\sqrt{10(90)} \overset{?}{=} 30$

$\sqrt{900} \overset{?}{=} 30$

$30 = 30$

15. $\sqrt{3y + 5} - 3 = 4$

$\sqrt{3y + 5} = 7$

$(\sqrt{3y + 5})^2 = 7^2$

$3y + 5 = 49$

$3y = 44$

$y = \frac{44}{3}$

Check: $\sqrt{3\left(\frac{44}{3}\right) + 5} - 3 \overset{?}{=} 4$

$\sqrt{49} - 3 = 4$

$7 - 3 \overset{?}{=} 4$

$4 = 4$

17. $5\sqrt{x + 2} = 8$

$(5\sqrt{x + 2})^2 = 8^2$

$25(x + 2) = 64$

$25x + 50 = 64$

$25x = 14$

$x = \frac{14}{25}$

Check: $\sqrt{\frac{14}{25} + 2} \overset{?}{=} 8$

$5\sqrt{\frac{64}{25}} \overset{?}{=} 8$

$5\left(\frac{8}{5}\right) \overset{?}{=} 8$

$8 = 8$

19. $\sqrt{x^2 + 5} = x + 3$

$(\sqrt{x^2 + 5})^2 = (x + 3)^2$

$x^2 + 5 = x^2 + 6x + 9$

$5 = 6x + 9$

$-4 = 6x$

$\frac{-4}{6} = x$

$-\frac{2}{3} = x$

Check: $\sqrt{\left(-\frac{2}{3}\right)^2 + 5} \overset{?}{=} -\frac{2}{3} + 3$

$\sqrt{\frac{4}{9} + 5} \overset{?}{=} -\frac{2}{3} + 3$

$\sqrt{\frac{49}{9}} \overset{?}{=} \frac{7}{3}$

$\frac{7}{3} = \frac{7}{3}$

21. $\sqrt{2x} = x - 4$

$(\sqrt{2x})^2 = (x - 4)^2$

$2x = x^2 - 8x + 16$

$0 = x^2 - 10x + 16$

$0 = (x - 8)(x - 2)$

$x - 8 = 0 \Rightarrow x = 8$

$x - 2 = 0 \Rightarrow x = 2$ (Extraneous)

Check: $\sqrt{2(8)} \overset{?}{=} 8 - 4$

$\sqrt{16} \overset{?}{=} 4$

$4 = 4$

Check: $\sqrt{2(2)} \overset{?}{=} 2 - 4$

$\sqrt{4} \overset{?}{=} -2$

$2 \neq -2$

23. $\sqrt{4y} = 3 - y$

$\left(\sqrt{4y}\right)^2 = (3 - y)^2$

$4y = 9 - 6y + y^2$

$0 = y^2 - 10y + 9$

$0 = (y - 9)(y - 1)$

$y - 9 = 0 \Longrightarrow y = 9$ (Extraneous)

$y - 1 = 0 \Longrightarrow y = 1$

Check: $\sqrt{4(9)} \stackrel{?}{=} 3 - 9$

$\sqrt{36} \stackrel{?}{=} -6$

$6 \neq -6$

$\sqrt{4(1)} \stackrel{?}{=} 3 - 1$

$\sqrt{4} \stackrel{?}{=} 2$

$2 = 2$

25. $\sqrt{3y + 1} - 4 = 0$

$\sqrt{3y + 1} = 4$

$\left(\sqrt{3y + 1}\right)^2 = 4^2$

$3y + 1 = 16$

$3y = 15$

$y = \frac{15}{3}$

$y = 5$

Check: $\sqrt{3(5) + 1} - 4 \stackrel{?}{=} 0$

$\sqrt{16} - 4 \stackrel{?}{=} 0$

$4 - 4 \stackrel{?}{=} 0$

$0 = 0$

27. $\sqrt{3x + 2} + 5 = 0$

$\sqrt{3x + 2} = -5$

$\left(\sqrt{3x + 2}\right)^2 = (-5)^2$

$3x + 2 = 25$

$3x = 23$

$x = \frac{23}{3}$ (Extraneous)

Check: $\sqrt{3\left(\frac{23}{3}\right) + 2} + 5 \stackrel{?}{=} 0$

$\sqrt{23 + 2} + 5 \stackrel{?}{=} 0$

$\sqrt{25} + 5 \stackrel{?}{=} 0$

$5 + 5 \neq 0$

The equation has *no* solution.

29. $\sqrt{x + 3} = \sqrt{2x - 1}$

$\left(\sqrt{x + 3}\right)^2 = \left(\sqrt{2x - 1}\right)^2$

$x + 3 = 2x - 1$

$3 = x - 1$

$4 = x$

Check: $\sqrt{4 + 3} \stackrel{?}{=} \sqrt{2(4) - 1}$

$\sqrt{7} \stackrel{?}{=} \sqrt{8 - 1}$

$\sqrt{7} = \sqrt{7}$

31. $\sqrt{2x + 5} = \sqrt{x + 7}$

$\left(\sqrt{2x + 5}\right)^2 = \left(\sqrt{x + 7}\right)^2$

$2x + 5 = x + 7$

$x + 5 = 7$

$x = 2$

Check: $\sqrt{2(2) + 5} \stackrel{?}{=} \sqrt{2 + 7}$

$\sqrt{9} \stackrel{?}{=} \sqrt{9}$

$3 = 3$

33. $\sqrt{3y - 5} = 3\sqrt{y}$

$\left(\sqrt{3y - 5}\right)^2 = \left(3\sqrt{y}\right)^2$

$3y - 5 = 9y$

$-5 = 6y$

$-\frac{5}{6} = y$ (Extraneous)

Check: $\sqrt{3\left(-\frac{5}{6}\right) - 5} - 3\sqrt{-\frac{5}{6}} \stackrel{?}{=} 0$

The solution $y = -\frac{5}{6}$ is extraneous because it yields square roots of negative radicands. Thus, the equation has *no* solution.

35. $\sqrt{10x - 4} - 3\sqrt{x} = 0$

$\sqrt{10x - 4} = 3\sqrt{x}$

$\left(\sqrt{10x - 4}\right)^2 = \left(3\sqrt{x}\right)^2$

$10x - 4 = 9x$

$-4 = -x$

$4 = x$

Check: $\sqrt{10(4) - 4} - 3\sqrt{4} \stackrel{?}{=} 0$

$\sqrt{36} - 3(2) \stackrel{?}{=} 0$

$6 - 6 = 0$

37. $\sqrt[3]{3x - 4} = \sqrt[3]{x + 10}$

$\left(\sqrt[3]{3x - 4}\right)^3 = \left(\sqrt[3]{x + 10}\right)^3$

$3x - 4 = x + 10$

$2x - 4 = 10$

$2x = 14$

$x = \frac{14}{2}$

$x = 7$

Check: $\sqrt[3]{3(7) - 4} \stackrel{?}{=} \sqrt[3]{7 + 10}$

$\sqrt{17} = \sqrt{17}$

39. $\sqrt[3]{2x + 15} - \sqrt[3]{x} = 0$

$\qquad \sqrt[3]{2x + 15} = \sqrt[3]{x}$

$\qquad (\sqrt[3]{2x + 15})^3 = (\sqrt[3]{x})^3$

$\qquad 2x + 15 = x$

$\qquad 15 = -x$

$\qquad -15 = x$

Check: $\sqrt[3]{2(-15) + 15} - \sqrt[3]{-15} \stackrel{?}{=} 0$

$\qquad \sqrt[3]{-15} - \sqrt[3]{-15} \stackrel{?}{=} 0$

$\qquad 0 = 0$

41. $\sqrt{8x + 1} = x + 2$

$\qquad (\sqrt{8x + 1})^2 = (x + 2)^2$

$\qquad 8x + 1 = x^2 + 4x + 4$

$\qquad 0 = x^2 - 4x + 3$

$\qquad 0 = (x - 3)(x - 1)$

$\qquad x - 3 = 0 \Rightarrow x = 3$

$\qquad x - 1 = 0 \Rightarrow x = 1$

Check: $\sqrt{8(3) + 1} \stackrel{?}{=} 3 + 2$

$\qquad \sqrt{25} \stackrel{?}{=} 5$

$\qquad 5 = 5$

Check: $\sqrt{8(1) + 1} \stackrel{?}{=} 1 + 2$

$\qquad \sqrt{9} \stackrel{?}{=} 3$

$\qquad 3 = 3$

43. $\sqrt{5x - 4} = 2 - \sqrt{5x}$

$\qquad (\sqrt{5x - 4})^2 = (2 - \sqrt{5x})^2$

$\qquad 5x - 4 = 2^2 - 2(2)\sqrt{5x} + (\sqrt{5x})^2$

$\qquad 5x - 4 = 4 - 4\sqrt{5x} + 5x$

$\qquad -4 = 4 - 4\sqrt{5x}$

$\qquad -8 = -4\sqrt{5x}$

$\qquad 2 = \sqrt{5x}$

$\qquad 2^2 = (\sqrt{5x})^2$

$\qquad 4 = 5x$

$\qquad \frac{4}{5} = x$

Check: $\sqrt{5\left(\frac{4}{5}\right) - 4} \stackrel{?}{=} 2 - \sqrt{5\left(\frac{4}{5}\right)}$

$\qquad \sqrt{4 - 4} \stackrel{?}{=} 2 - \sqrt{4}$

$\qquad \sqrt{0} \stackrel{?}{=} 2 - 2$

$\qquad 0 = 0$

45. $\sqrt{z + 2} = 1 + \sqrt{z}$

$\qquad (\sqrt{z + 2})^2 = (1 + \sqrt{z})^2$

$\qquad z + 2 = 1 + 2\sqrt{z} + z$

$\qquad 1 = 2\sqrt{z}$

$\qquad 1^2 = (2\sqrt{z})^2$

$\qquad 1 = 4z$

$\qquad \frac{1}{4} = z$

Check: $\sqrt{\frac{1}{4} + 2} \stackrel{?}{=} 1 + \sqrt{\frac{1}{4}}$

$\qquad \sqrt{\frac{9}{4}} \stackrel{?}{=} 1 + \sqrt{\frac{1}{4}}$

$\qquad \frac{3}{2} \stackrel{?}{=} 1 + \frac{1}{2}$

$\qquad \frac{3}{2} = \frac{3}{2}$

47. $\sqrt{x} + \sqrt{x+2} = 2$

$$\sqrt{x} = 2 - \sqrt{x+2}$$
$$\left(\sqrt{x}\right)^2 = \left(2 - \sqrt{x-2}\right)^2$$
$$x = 4 - 4\sqrt{x+2} + x + 2$$
$$0 = 6 - 4\sqrt{x+2}$$
$$-6 = -4\sqrt{x+2}$$
$$3 = 2\sqrt{x+2}$$
$$3^2 = \left(2\sqrt{x+2}\right)^2$$
$$9 = 4(x+2)$$
$$9 = 4x + 8$$
$$1 = 4x$$
$$\tfrac{1}{4} = x$$

Check: $\sqrt{\tfrac{1}{4}} + \sqrt{\tfrac{1}{4} + 2} \overset{?}{=} 2$

$$\tfrac{1}{2} + \sqrt{\tfrac{9}{4}} \overset{?}{=} 2$$
$$\tfrac{1}{2} + \tfrac{3}{2} \overset{?}{=} 2$$
$$2 = 2$$

49. $\sqrt{2t+3} = 3 - \sqrt{2t}$

$$\left(\sqrt{2t+3}\right)^2 = \left(3 - \sqrt{2t}\right)^2$$
$$2t + 3 = 3^2 - 2(3)\left(\sqrt{2t}\right) + \left(\sqrt{2t}\right)^2$$
$$2t + 3 = 9 - 6\sqrt{2t} + 2t$$
$$3 = 9 - 6\sqrt{2t}$$
$$-6 = -6\sqrt{2t}$$
$$1 = \sqrt{2t}$$
$$1^2 = \left(\sqrt{2t}\right)^2$$
$$1 = 2t$$
$$\tfrac{1}{2} = t$$

Check: $\sqrt{2\left(\tfrac{1}{2}\right) + 3} \overset{?}{=} 3 - \sqrt{2\left(\tfrac{1}{2}\right)}$

$$\sqrt{4} \overset{?}{=} 3 - \sqrt{1}$$
$$2 \overset{?}{=} 3 - 1$$
$$2 = 2$$

51. $\sqrt{x+3} + \sqrt{x-2} = 5$

$$\left(\sqrt{x+3}\right)^2 = \left(5 - \sqrt{x-2}\right)^2$$
$$x + 3 = 25 - 10\sqrt{x-2} + x - 2$$
$$3 = 23 - 10\sqrt{x-2}$$
$$-20 = -10\sqrt{x-2}$$
$$2 = \sqrt{x-2}$$
$$(2)^2 = \left(\sqrt{x-2}\right)^2$$
$$4 = x - 2$$
$$6 = x$$

Check: $\sqrt{6+3} + \sqrt{6-2} \overset{?}{=} 5$

$$\sqrt{9} + \sqrt{4} \overset{?}{=} 5$$
$$3 + 2 \overset{?}{=} 5$$
$$5 = 5$$

53. The solution is approximately 1.347.

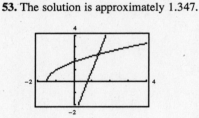

55. The solution is approximately 1.569.

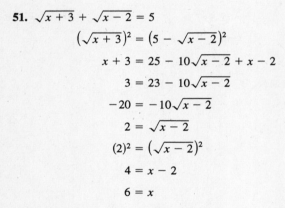

57. The solution is approximately 4.840.

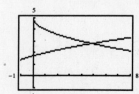

59. The solution is approximately 2.513.

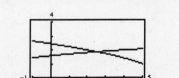

61.
$$t = 2\pi\sqrt{\frac{L}{32}}$$

$$1.5 = 2\pi\sqrt{\frac{L}{32}}$$

$$\frac{1.5}{2\pi} = \sqrt{\frac{L}{32}}$$

$$\left(\frac{1.5}{2\pi}\right)^2 = \left(\sqrt{\frac{L}{32}}\right)^2$$

$$\frac{2.25}{4\pi^2} = \frac{L}{32}$$

$$32\left(\frac{2.25}{4\pi^2}\right) = L$$

$$\frac{8(2.25)}{\pi^2} = L$$

$$\frac{18}{\pi^2} = L \text{ or } L \approx 1.82$$

The pendulum is approximately
1.82 feet long.

63.
$$S = \pi r\sqrt{r^2 + h^2}$$

$$\frac{S}{\pi r} = \sqrt{r^2 + h^2}$$

$$\left(\frac{S}{\pi r}\right)^2 = \left(\sqrt{r^2 + h^2}\right)^2$$

$$\frac{S^2}{\pi^2 r^2} = r^2 + h^2$$

$$\frac{S^2}{\pi^2 r^2} - r^2 = h^2 \text{ or } h^2 = \frac{S^2 - \pi^2 r^4}{\pi^2 r^2}$$

65. $v = \sqrt{2gh} = \sqrt{2(32)(50)} = \sqrt{3200}$

$\quad = \sqrt{1600(2)} = 40\sqrt{2} \text{ or } v \approx 56.57$

The velocity of the object is approximately 56.57 ft/sec when it hits the ground.

67.
$$v = \sqrt{2gh}$$

$$60 = \sqrt{2(32)h} = \sqrt{64h}$$

$$60^2 = \left(\sqrt{64h}\right)^2$$

$$3600 = 64h$$

$$\frac{3600}{64} = h$$

$$\frac{16(225)}{16(4)} = h$$

$$\frac{225}{4} = h \text{ or } h = 56.25$$

The object was dropped from a height of 56.25 feet.

69.
$$p = 50 - \sqrt{0.8(x - 1)}$$

$$30.02 = 50 - \sqrt{0.8(x - 1)}$$

$$-19.98 = -\sqrt{0.8(x - 1)}$$

$$19.98 = \sqrt{0.8(x - 1)}$$

$$(19.98)^2 = \left(\sqrt{0.8(x - 1)}\right)^2$$

$$399.2004 = 0.8(x - 1) = 0.8x - 0.8$$

$$400.0004 = 0.8x$$

$$\frac{400.0004}{0.8} = x$$

$$500 \approx x$$

Thus, the demand is approximately 500 units per day.

71. No. Raising each side of an equation to the *n*th power sometimes introduces extraneous solutions, which indicates that the new equation is not equivalent to the original equation. That is why it is always necessary to check any solutions in the original equation.

73. $9x - 4 = 41$

$\quad 9x = 45$

$\quad x = 5$

75. $14 + 2(x - 1) = 0$

$\quad 14 + 2x - 2 = 0$

$\quad 2x + 12 = 0$

$\quad 2x = -12$

$\quad x = -6$

77. $4(x + 7)(x - 11) = 0$

$\quad x + 7 = 0 \Rightarrow x = -7$

$\quad x - 11 = 0 \Rightarrow x = 11$

79. $\$12{,}000 + \$12{,}000(0.12)\left(\frac{1}{2}\right) = 12{,}000 + 720 = \$12{,}720$

The amount of the payment will be $12,720.

Section 5.6   Complex Numbers

1. $\sqrt{-4} = \sqrt{4(-1)} = \sqrt{4}\sqrt{-1} = 2i$

3. $\sqrt{-\frac{4}{25}} = \sqrt{\frac{4}{25}(-1)} = \sqrt{\frac{4}{25}}\sqrt{-1} = \frac{2}{5}i$

5. $\sqrt{-8} = \sqrt{8(-1)} = \sqrt{8}\sqrt{-1}$

$\qquad = \sqrt{4(2)}i = 2\sqrt{2}i$

7. $\sqrt{-7} = \sqrt{7(-1)}$

$\qquad = \sqrt{7}\sqrt{-1} = \sqrt{7}i$

9. $\sqrt{-16} + \sqrt{-36} = \sqrt{16(-1)} + \sqrt{36(-1)}$

$\qquad = \sqrt{16}\sqrt{-1} + \sqrt{36}\sqrt{-1}$

$\qquad = 4i + 6i$

$\qquad = 10i$

11. $\sqrt{-8}\sqrt{-2} = \sqrt{8}i\sqrt{2}i = \sqrt{16}i^2$

$\qquad = 4(-1) = -4$

13. $\sqrt{-18}\sqrt{-3} = \left(\sqrt{18}i\right)\left(\sqrt{3}i\right)$

$\qquad = \sqrt{54}i^2 = \sqrt{9(6)}i^2$

$\qquad = 3\sqrt{6}(-1) = -3\sqrt{6}$

15. $\sqrt{-3}\left(\sqrt{-3} + \sqrt{-4}\right) = \sqrt{3}i\left(\sqrt{3}i + 2i\right)$

$\qquad = \left(\sqrt{3}\right)^2 i^2 + 2\sqrt{3}i^2$

$\qquad = 3(-1) + 2\sqrt{3}(-1)$

$\qquad = -3 - 2\sqrt{3}$

17. $\sqrt{-5}\left(\sqrt{-16} - \sqrt{-10}\right) = \sqrt{5}i(4i - \sqrt{10}i)$

$\qquad = 4\sqrt{5}i^2 - \sqrt{50}i^2$

$\qquad = 4\sqrt{5}(-1) - \sqrt{25(2)}(-1)$

$\qquad = -4\sqrt{5} + 5\sqrt{2}$

19. $\left(\sqrt{-16}\right)^2 = \left(\sqrt{16}i\right)^2 = \left(\sqrt{16}\right)^2 i^2$

$\qquad = 16(-1) = -16$

21. $a = 3$ and $b = -4$

23. $5 - 4i = (a + 3) + (b - 1)i$

$\qquad 5 = a + 3 \quad$ and $\quad -4 = b - 1$

$\qquad 2 = a \qquad\qquad -3 = b$

25. $-4 - \sqrt{-8} = a + bi$

$\qquad -4 - \sqrt{4(2)}i = a + bi$

$\qquad -4 - 2\sqrt{2}i = a + bi$

$\qquad a = -4$ and $b = -2\sqrt{2}$

27. $a + 5 = 7$ and $b - 1 = -3$

$\qquad a = 2 \qquad\qquad b = -2$

29. $(4 - 3i) + (6 + 7i) = (4 + 6) + (-3 + 7)i$

$\qquad = 10 + 4i$

31. $(-4 - 7i) + (-10 - 33i) = (-4 - 10) + (-7 - 33)i$

$\qquad = -14 - 40i$

33. $13i - (14 - 7i) = -14 + [13 - (-7)]i$

$\qquad = -14 + 20i$

35. $(30 - i) - (18 + 6i) + 3i^2 = (30 - 18) + (-1 - 6)i + 3(-1)$

$\qquad = 12 - 7i - 3$

$\qquad = 9 - 7i$

37. $15i - (3 - 25i) + \sqrt{-81} = 15i - (3 - 25i) + 9i$

$\qquad = -3 + [15 - (-25) + 9]i$

$\qquad = -3 + 49i$

39. $(3i)(12i) = 36i^2 = 36(-1) = -36$

41. $(-6i)(-i)(6i) = 36i^3 = 36(-i) = -36i$

43. $(-3i)^3 = (-3)^3 i^3 = -27i^3$
$$= -27(-i) = 27i$$

45. $-5(13 + 2i) = -65 - 10i$

47. $4i(-3 - 5i) = -12i - 20i^2 = -12i - 20(-1)$
$$= -12i + 20 = 20 - 12i$$

49. $(4 + 3i)(-7 + 4i) = -28 + 16i - 21i + 12i^2$
$$= -28 - 5i + 12(-1)$$
$$= -40 - 5i$$

51. $(-7 + 7i)(4 - 2i) = -28 + 14i + 28i - 14i^2$
$$= -28 + 42i - 14(-1)$$
$$= -14 + 42i$$

53. $(3 - 4i)^2 = 3^2 - 2(3)(4i) + (4i)^2$
$$= 9 - 24i + 16i^2$$
$$= 9 - 24i + 16(-1)$$
$$= -7 - 24i$$

55. $(2 + 5i)^2 = 2^2 + 2(2)(5i) + (5i)^2$
$$= 4 + 20i + 25i^2$$
$$= 4 + 20i + 25(-1)$$
$$= -21 + 20i$$

57. $\left(-2 + \sqrt{-5}\right)\left(-2 - \sqrt{-5}\right) = \left(-2 + \sqrt{5(-1)}\right)\left(-2 - \sqrt{5(-1)}\right)$
$$= \left(-2 + \sqrt{5}i\right)\left(-2 - \sqrt{5}i\right)$$
$$= (-2)^2 - \left(\sqrt{5}i\right)^2$$
$$= 4 - 5i^2$$
$$= 4 - 5(-1)$$
$$= 4 + 5$$
$$= 9$$

59. $(2 + i)^3 = (2 + i)^2(2 + i)$
$$= [2^2 + 2(2)i + i^2](2 + i)$$
$$= [4 + 4i + (-1)](2 + i)$$
$$= (3 + 4i)(2 + i)$$
$$= 6 + 3i + 8i + 4i^2$$
$$= 6 + 11i + 4(-1)$$
$$= 2 + 11i$$

61. Number: $2 + i$

Conjugate: $2 - i$

Product: $2^2 + 1^2 = 5$

Note: $(2 + i)(2 - i) = 4 - 2i + 2i - i^2 = 5$

63. Number: $-2 - 8i$

Conjugate: $-2 + 8i$

Product: $(-2)^2 + 8^2 = 68$

Note: $(-2 - 8i)(-2 + 8i) = 4 - 16i + 16i - 64i^2 = 68$

65. Number: $5 - \sqrt{6}i$

Conjugate: $5 + \sqrt{6}i$

Product: $5^2 + \left(\sqrt{6}\right)^2 = 31$

Note: $\left(5 + \sqrt{6}i\right)\left(5 - \sqrt{6}i\right) = 25 + 5\sqrt{6}i - 5\sqrt{6}i - 6i^2 = 31$

67. Number: $10i$

Conjugate: $-10i$

Product: $0^2 + 10^2 = 100$

Note: $(10i)(-10i) = -100i^2 = 100$

69. Number: $1 + \sqrt{-3} = 1 + \sqrt{3}i$

Conjugate: $1 - \sqrt{-3} = 1 - \sqrt{3}i$

Product: $1^2 + \left(\sqrt{3}\right)^2 = 4$

Note: $\left(1 + \sqrt{3}i\right)\left(1 - \sqrt{3}i\right) = 1 - \sqrt{3}i + \sqrt{3}i - 3i^2 = 4$

71. $\dfrac{4}{1 - i} = \dfrac{4(1 + i)}{(1 - i)(1 + i)}$

$= \dfrac{4 + 4i}{1^2 + 1^2} = \dfrac{4 + 4i}{2} = 2 + 2i$

73. $\dfrac{-12}{2 + 7i} = \dfrac{-12(2 - 7i)}{(2 + 7i)(2 - 7i)}$

$= \dfrac{-24 + 84i}{2^2 + 7^2} = \dfrac{-24 + 84i}{53}$

$= -\dfrac{24}{53} + \dfrac{84}{53}i$

75. $\dfrac{12i}{7 - 6i} = \dfrac{12i(7 + 6i)}{(7 - 6i)(7 + 6i)}$

$= \dfrac{84i + 72i^2}{7^2 + 6^2}$

$= \dfrac{84i + 72(-1)}{49 + 36}$

$= \dfrac{-72 + 84i}{85}$

$= -\dfrac{72}{85} + \dfrac{84}{85}i$

77. $\dfrac{20}{2i} = \dfrac{20(-2i)}{2i(-2i)} = \dfrac{-40i}{-4i^2} = \dfrac{-40i}{-4(-1)}$

$= \dfrac{-40i}{4} = -10i$ or $0 - 10i$

79. $\dfrac{9 - 1}{5i} = \dfrac{(9 - i)(-5i)}{(5i)(-5i)}$

$= \dfrac{-45i + 5i^2}{-25i^2}$

$= \dfrac{-45i + 5i^2}{-25(-1)}$

$= \dfrac{-45i + 5(-1)}{25}$

$= \dfrac{-5 - 45i}{25}$

$= -\dfrac{5}{25} - \dfrac{45}{25}i$

$= -\dfrac{1}{5} - \dfrac{9}{5}i$

81. $\dfrac{4i}{1 - 3i} = \dfrac{4i(1 + 3i)}{(1 - 3i)(1 + 3i)} = \dfrac{4i + 12i^2}{1^2 + 3^2}$

$= \dfrac{4i + 12(-1)}{10} = \dfrac{-12}{10} + \dfrac{4i}{10}$

$= -\dfrac{6}{5} + \dfrac{2}{5}i$

83. $\dfrac{-8}{4(i - 5)} = \dfrac{-8(i + 5)}{4(i - 5)(i + 5)}$

$= \dfrac{-8i - 40}{4(i^2 - 5^2)}$

$= \dfrac{-40 - 8i}{4(-26)}$

$= \dfrac{40}{104} + \dfrac{8i}{104}$

$= \dfrac{5}{13} + \dfrac{1}{13}i$

85. $\dfrac{2 + 3i}{1 + 2i} = \dfrac{(2 + 3i)(1 - 2i)}{(1 + 2i)(1 - 2i)}$

$= \dfrac{2 - 4i + 3i - 6i^2}{1^2 + 2^2} = \dfrac{2 - i - 6(-1)}{5} = \dfrac{8 - i}{5} = \dfrac{8}{5} - \dfrac{1}{5}i$

87. $\dfrac{1}{1 - 2i} + \dfrac{4}{1 + 2i} = \dfrac{1(1 + 2i)}{(1 - 2i)(1 + 2i)} + \dfrac{4(1 - 2i)}{(1 + 2i)(1 - 2i)} = \dfrac{1 + 2i}{1^2 + 2^2} + \dfrac{4 - 8i}{1^2 + 2^2}$

$= \dfrac{(1 + 2i)}{5} + \dfrac{(4 - 8i)}{5} = \dfrac{5 - 6i}{5} = \dfrac{5}{5} - \dfrac{6}{5}i = 1 - \dfrac{6}{5}i$

89. $\dfrac{i}{4-3i} - \dfrac{5}{2+i} = \dfrac{i(4+3i)}{(4-3i)(4+3i)} - \dfrac{5(2-i)}{(2+i)(2-i)}$

$\qquad = \dfrac{4i+3i^2}{4^2+3^2} - \dfrac{10-5i}{2^2+1^2} = \dfrac{4i+3(-1)}{25} - \dfrac{10-5i}{5}$

$\qquad = \dfrac{-3+4i}{25} - \dfrac{10-5i}{5} = \dfrac{(-3+4i)-5(10-5i)}{25}$

$\qquad = \dfrac{-3+4i-50+25i}{25} = \dfrac{-53+29i}{25} = -\dfrac{53}{25} + \dfrac{29}{25}i$

91. $x^2 + 2x + 5 = 0$

(a) $x = -1 + 2i$

$\qquad (-1+2i)^2 + 2(-1+2i) + 5 \stackrel{?}{=} 0$

$\qquad 1 - 4i + 4i^2 - 2 + 4i + 5 \stackrel{?}{=} 0$

$\qquad 1 + 4(-1) - 2 + 5 - 4i \stackrel{?}{=} 0$

$\qquad (1 - 4 - 2 + 5) + (-4 + 4)i \stackrel{?}{=} 0$

$\qquad\qquad\qquad 0 = 0$

Yes, $-1 + 2i$ is a solution.

(b) $x = -1 - 2i$

$\qquad (-1-2i)^2 + 2(-1-2i) + 5 \stackrel{?}{=} 0$

$\qquad 1 + 4i + 4i^2 - 2 - 4i + 5 \stackrel{?}{=} 0$

$\qquad 1 + 4(-1) - 2 + 5 + 4i - 4i \stackrel{?}{=} 0$

$\qquad (1 - 4 - 2 + 5) + (4 - 4)i \stackrel{?}{=} 0$

$\qquad\qquad\qquad 0 = 0$

Yes, $-1 - 2i$ is a solution.

93. $x^3 + 4x^2 + 9x + 36 = 0$

(a) $x = -4$

$\qquad (-4)^3 + 4(-4)^2 + 9(-4) + 36 \stackrel{?}{=} 0$

$\qquad\qquad -64 + 64 - 36 + 36 \stackrel{?}{=} 0$

$\qquad\qquad\qquad\qquad\qquad 0 = 0$

Yes, -4 is a solution.

(b) $x = -3i$

$\qquad (-3i)^3 + 4(-3i)^2 + 9(-3i) + 36 \stackrel{?}{=} 0$

$\qquad\qquad -27i^3 + 4(9i^2) - 27i + 36 \stackrel{?}{=} 0$

$\qquad -27(-i) + 36(-1) - 27i + 36 \stackrel{?}{=} 0$

$\qquad\qquad 27i - 36 - 27i + 36 \stackrel{?}{=} 0$

$\qquad\qquad\qquad\qquad\qquad 0 = 0$

Yes, $-3i$ is a solution.

95. $(a + bi) + (a - bi) = (a + a) + (b - b)i$

$\qquad\qquad\qquad\qquad = 2a + 0i = 2a$

97. $(a + bi) - (a - bi) = (a - a) + [b - (-b)]i$

$\qquad\qquad\qquad\qquad = 0 + 2bi = 2bi$

99. The imaginary unit i is the square root of -1.

$\qquad i = \sqrt{-1}$

101. $\sqrt{-3}\,\sqrt{-3} = \left(\sqrt{3}i\right)\left(\sqrt{3}i\right) = 3i^2 = 3(-1) = -3$

The two imaginary numbers need to be written in i-form before they are multiplied.

103. $x + 21 = 0$

$\qquad x = -21$

105. $3x^2 - 8x - 16 = 0$

$\qquad (3x + 4)(x - 4) = 0$

$\qquad\qquad 3x + 4 = 0 \Longrightarrow 3x = -4 \Longrightarrow x = -\dfrac{4}{3}$

$\qquad\qquad x - 4 = 0 \Longrightarrow x = 4$

107. Time $= \dfrac{\text{Distance}}{\text{Rate}}$

$\qquad t = \dfrac{360}{r}$

Review Exercises for Chapter 5

1. $(2^3 \cdot 3^2)^{-1} = (8 \cdot 9)^{-1} = 72^{-1} = \dfrac{1}{72}$

3. $\left(\dfrac{2}{5}\right)^{-3} = \left(\dfrac{5}{2}\right)^3 = \dfrac{125}{8}$

5. $(x^3 y^{-4})^2 (x^{-2}y)^{-3} = x^6 y^{-8} x^6 y^{-3} = x^{12} y^{-11} = \dfrac{x^{12}}{y^{11}}$

7. $-3(a^4 b^{-6})^0 (-2a^{-3}b^2)^2 = -3(1)(-2)^2 a^{-6}b^4 = -3(4)a^{-6}b^4 = -\dfrac{12b^4}{a^6}$

9. $\dfrac{7x^5 y^{-2}}{14x^3 y^8} = \dfrac{x^{5-3}}{2y^2 y^8} = \dfrac{x^2}{2y^{10}}$

11. $\dfrac{15b^6 c}{3ab^2 c^{-4}} = \dfrac{5b^{6-2}c^1 c^4}{a} = \dfrac{5b^4 c^5}{a}$

13. $1{,}460{,}000{,}000 = 1.46 \times 10^9$

15. $0.000000641 = 6.41 \times 10^{-7}$

17. $4.09 \times 10^{-12} = 0.00000000000409$

19. $9.58 \times 10^{10} = 95{,}800{,}000{,}000$

21. $(6 \times 10^3)^2 = 36 \times 10^6$

$\qquad = 36{,}000{,}000 \text{ or } 3.6 \times 10^7$

23. $\dfrac{3.5 \times 10^7}{7 \times 10^4} = 0.5 \times 10^3 = 500$

25. $-\sqrt{16} = -4$

27. $\sqrt[3]{-8} = -2$

29. $\sqrt{\dfrac{4}{25}} = \dfrac{2}{5}$

31. $\sqrt[3]{0.001} = 0.1$

33. $\sqrt[4]{-625}$

Not a real number

35. $27^{4/3} = \left(\sqrt[3]{27}\right)^4$

$\qquad = 3^4 = 81$

37. $25^{3/2} = \left(\sqrt{25}\right)^3$

$\qquad = 5^3$

$\qquad = 125$

39. $16^{-1/4} = \dfrac{1}{16^{1/4}} = \dfrac{1}{2}$

41. $x^{3/4} \cdot x^{-1/6} = x^{(3/4) + (-1/6)}$

$\qquad = x^{(18/24) - (4/24)}$

$\qquad = x^{14/24} = x^{7/12}$

43. $(4a^0)^{2/3}(4a^{-2})^{1/3} = 4^{2/3}a^0(4^{1/3}a^{-2/3}) = 4^{3/3}(1)a^{-2/3} = \dfrac{4}{a^{2/3}}$

45. $\dfrac{15x^{1/4}y^{3/5}}{5x^{1/2}y} = 3x^{(1/4)-(1/2)}y^{(3/5)-1}$

$\qquad = 3x^{(1/4)-(2/4)}y^{(3/5)-(5/5)}$

$\qquad = 3x^{-1/4}y^{-2/5}$

$\qquad = \dfrac{3}{x^{1/4}y^{2/5}}$

47. $\dfrac{(3x+2)^{2/3}}{\sqrt[3]{3x+2}} = \dfrac{(3x+2)^{2/3}}{(3x+2)^{1/3}} = (3x+2)^{1/3} \text{ or } \sqrt[3]{3x+2}$

49. $75^{-3/4} \approx 0.04$

51. $\qquad g(x) = \sqrt{3x-2}$

(a) $g(1) = \sqrt{3(1)-2} = \sqrt{1} = 1$

(b) $g(2) = \sqrt{3(2)-2} = \sqrt{4} = 2$

(c) $g(9) = \sqrt{3(9)-2} = \sqrt{25} = 5$

(d) $g(6) = \sqrt{3(6)-2} = \sqrt{16} = 4$

53. $f(x) = \dfrac{1}{\sqrt{2x}}$

The domain of $\sqrt{2x}$ is the set of all real numbers that are greater than or equal to zero, but the denominator of this function cannot be equal to zero because division by zero is undefined. So, the domain of $f(x)$ includes only positive values of x.

Domain $= \{x : x > 0\}$

55. $h(x) = \sqrt{2x + 6}$

The radicand $2x + 6$ must be greater than or equal to zero.

Domain $= \{x: x \geq -3\}$

57. $\sqrt{360} = \sqrt{36 \cdot 10} = 6\sqrt{10}$

59. $\sqrt{50x^4} = \sqrt{25(2)x^4} = 5x^2\sqrt{2}$

61. $\sqrt[3]{48a^3b^4} = \sqrt[3]{8a^3b^3(6b)}$
$= 2ab\sqrt[3]{6b}$

63. $\sqrt[4]{81x^{12}y^{15}} = \sqrt[4]{81x^{12}y^{12}(y^3)} = 3x^3y^3\sqrt[4]{y^3}$

65. $\sqrt[5]{32a^{11}b^{16}} = \sqrt[5]{32a^{10}b^{15}(ab)} = 2a^2b^3\sqrt[5]{ab}$

67. $\sqrt{\dfrac{5}{6}} = \dfrac{\sqrt{5} \cdot \sqrt{6}}{\sqrt{6} \cdot \sqrt{6}} = \dfrac{\sqrt{30}}{\sqrt{6^2}}$
$= \dfrac{\sqrt{30}}{6}$

69. $\dfrac{3}{\sqrt{12x}} = \dfrac{3\sqrt{3x}}{\sqrt{12x} \cdot \sqrt{3x}} = \dfrac{3\sqrt{3x}}{\sqrt{36x^2}}$
$= \dfrac{3\sqrt{3x}}{6x} = \dfrac{\cancel{3}\sqrt{3x}}{\cancel{3}(2)(x)} = \dfrac{\sqrt{3x}}{2x}$

71. $\dfrac{2}{\sqrt[3]{2x}} = \dfrac{2\sqrt[3]{4x^2}}{\sqrt[3]{2x} \cdot \sqrt[3]{4x^2}} = \dfrac{2\sqrt[3]{4x^2}}{\sqrt[3]{2^3x^3}}$
$= \dfrac{2\sqrt[3]{4x^2}}{2x} = \dfrac{\sqrt[3]{4x^2}}{x}$

73. $\sqrt{\dfrac{12x^2y^3}{5z}} = \dfrac{\sqrt{12x^2y^3}}{\sqrt{5z}}$
$= \dfrac{\sqrt{4x^2y^2(3y)}}{\sqrt{5z}}$
$= \dfrac{2xy\sqrt{3y}}{\sqrt{5z}} \cdot \dfrac{\sqrt{5z}}{\sqrt{5z}}$
$= \dfrac{2xy\sqrt{15yz}}{\sqrt{25z^2}}$
$= \dfrac{2xy\sqrt{15yz}}{5z}$

75. $\dfrac{x^3}{\sqrt[3]{4xy^2}} = \dfrac{x^3}{\sqrt[3]{2^2xy^2}} \cdot \dfrac{\sqrt[3]{2x^2y}}{\sqrt[3]{2x^2y}}$
$= \dfrac{x^3\sqrt[3]{2x^2y}}{\sqrt[3]{8x^3y^3}}$
$= \dfrac{x^3\sqrt[3]{2x^2y}}{2xy}$
$= \dfrac{x^2\sqrt[3]{2x^2y}}{2y}$

77. $\dfrac{-4x}{\sqrt[4]{8x^3y}} = \dfrac{-4x}{\sqrt[4]{2^3x^3y}} \cdot \dfrac{\sqrt[4]{2xy^3}}{\sqrt[4]{2xy^3}}$
$= \dfrac{-4x\sqrt[4]{2xy^3}}{\sqrt[4]{16x^4y^4}}$
$= \dfrac{-4x\sqrt[4]{2xy^3}}{2xy}$
$= \dfrac{-2\sqrt[4]{2xy^3}}{y}$

79. $3\sqrt{40} - 10\sqrt{90} = 3\sqrt{4 \cdot 10} - 10\sqrt{9 \cdot 10}$
$= 3 \cdot 2\sqrt{10} - 10 \cdot 3\sqrt{10}$
$= 6\sqrt{10} - 30\sqrt{10} = -24\sqrt{10}$

81. $10\sqrt[4]{y + 3} - 3\sqrt[4]{y + 3} = 7\sqrt[4]{y + 3}$

83. $\sqrt{5x} - \dfrac{3}{\sqrt{5x}} = \sqrt{5x} - \dfrac{3}{\sqrt{5x}} \cdot \dfrac{\sqrt{5x}}{\sqrt{5x}}$
$= \sqrt{5x} - \dfrac{3\sqrt{5x}}{\sqrt{25x^2}}$
$= \sqrt{5x} - \dfrac{3\sqrt{5x}}{5x}$
$= \sqrt{5x}\left(1 - \dfrac{3}{5x}\right)$

85. $\left(\sqrt{5} + 6x\right)^2 = \left(\sqrt{5}\right)^2 + 2\left(\sqrt{5}\right)(6x) + (6x)^2$
$= 5 + 12x\sqrt{5} + 36x^2$

87. $\left(2\sqrt{x} + 7\right)\left(\sqrt{x} - 5\right) = 2\sqrt{x^2} - 10\sqrt{x} + 7\sqrt{x} - 35$
$= 2x - 3\sqrt{x} - 35$

89. Conjugate: $\sqrt{3} + \sqrt{x}$
Product: $\left(\sqrt{3}\right)^2 - \left(\sqrt{x}\right)^2 = 3 - x$

91. Conjugate: $\sqrt{5u} - \sqrt{v}$

Product: $\left(\sqrt{5u}\right)^2 - \left(\sqrt{v}\right)^2 = 5u - v$

93. $\dfrac{15}{\sqrt{x} + 3} = \dfrac{15(\sqrt{x} - 3)}{(\sqrt{x} + 3)(\sqrt{x} - 3)}$

$\qquad = \dfrac{15\sqrt{x} - 45}{\left(\sqrt{x}\right)^2 - 3^2}$

$\qquad = \dfrac{15\sqrt{x} - 45}{x - 9}$

95. $\dfrac{\sqrt{x}}{5 + \sqrt{6x}} = \dfrac{\sqrt{x}}{5 + \sqrt{6x}} \cdot \dfrac{5 - \sqrt{6x}}{5 - \sqrt{6x}}$

$\qquad = \dfrac{\sqrt{x}(5 - \sqrt{6x})}{5^2 - \left(\sqrt{6x}\right)^2}$

$\qquad = \dfrac{5\sqrt{x} - \sqrt{6x^2}}{25 - 6x}$

$\qquad = \dfrac{5\sqrt{x} - x\sqrt{6}}{25 - 6x}$

97. $\left(\sqrt{x} + 10\right) \div \left(\sqrt{x} - 10\right) = \dfrac{\sqrt{x} + 10}{\sqrt{x} - 10}$

$\qquad = \dfrac{\left(\sqrt{x} + 10\right)\left(\sqrt{x} + 10\right)}{\left(\sqrt{x} - 10\right)\left(\sqrt{x} + 10\right)}$

$\qquad = \dfrac{\left(\sqrt{x}\right)^2 + 2\left(\sqrt{x}\right)(10) + 10^2}{\left(\sqrt{x}\right)^2 - 10^2}$

$\qquad = \dfrac{x + 20\sqrt{x} + 100}{x - 100}$

99. $\sqrt{y} = 15$

$\left(\sqrt{y}\right)^2 = 15^2$

$y = 225$

Check: $\sqrt{225} \stackrel{?}{=} 15$

$15 = 15$

101. $\sqrt{2(a - 7)} = 14$

$\left(\sqrt{2(a - 7)}\right)^2 = 14^2$

$2(a - 7) = 196$

$2a - 14 = 196$

$2a = 210$

$a = \dfrac{210}{2}$

$a = 105$

Check: $\sqrt{2(105 - 7)} \stackrel{?}{=} 14$

$\sqrt{2(98)} \stackrel{?}{=} 14$

$\sqrt{196} \stackrel{?}{=} 14$

$14 = 14$

103. $\sqrt{2(x + 5)} = x + 5$

$\left(\sqrt{2(x + 5)}\right)^2 = (x + 5)^2$

$2(x + 5) = x^2 + 10x + 25$

$2x + 10 = x^2 + 10x + 25$

$0 = x^2 + 8x + 15$

$0 = (x + 3)(x + 5)$

$x + 3 = 0 \Longrightarrow x = -3$

$x + 5 = 0 \Longrightarrow x = -5$

Check: $\sqrt{2(-3 + 5)} \stackrel{?}{=} -3 + 5$

$\sqrt{2(2)} \stackrel{?}{=} 2$

$\sqrt{4} \stackrel{?}{=} 2$

$2 = 2$

Check: $\sqrt{2(-5 + 5)} \stackrel{?}{=} -5 + 5$

$\sqrt{2(0)} \stackrel{?}{=} 0$

$\sqrt{0} \stackrel{?}{=} 0$

$0 = 0$

105. $\sqrt[3]{5x + 2} - \sqrt[3]{7x - 8} = 0$

$\sqrt[3]{5x + 2} = \sqrt[3]{7x = 8}$

$\left(\sqrt[3]{5x + 2}\right)^3 = \left(\sqrt[3]{7x - 8}\right)^3$

$5x + 2 = 7x - 8$

$2 = 2x - 8$

$10 = 2x$

$\dfrac{10}{2} = x$

$5 = x$

Check: $\sqrt[3]{5(5) + 2} - \sqrt[3]{7(5) - 8} \overset{?}{=} 0$

$\sqrt[3]{27} - \sqrt[3]{27} \overset{?}{=} 0$

$0 = 0$

107. $\sqrt{1 + 6x} = 2 - \sqrt{6x}$

$\left(\sqrt{1 + 6x}\right)^2 = \left(2 - \sqrt{6x}\right)^2$

$1 + 6x = 4 - 2(2)\left(\sqrt{6x}\right) + \left(\sqrt{6x}\right)^2$

$1 + 6x = 4 - 4\sqrt{6x} + 6x$

$1 = 4 - 4\sqrt{6x}$

$-3 = -4\sqrt{6x}$

$(-3)^2 = \left(-4\sqrt{6x}\right)^2$

$9 = 16(6x)$

$9 = 96x$

$\dfrac{9}{96} = x$

$\dfrac{3}{32} = x$

Check: $\sqrt{1 + 6\left(\dfrac{3}{32}\right)} \overset{?}{=} 2 - \sqrt{6\left(\dfrac{3}{32}\right)}$

$\sqrt{\dfrac{32}{32} + \dfrac{18}{32}} \overset{?}{=} 2 - \sqrt{\dfrac{18}{32}}$

$\sqrt{\dfrac{50}{32}} \overset{?}{=} 2 - \sqrt{\dfrac{9}{16}}$

$\sqrt{\dfrac{25}{16}} \overset{?}{=} 2 - \dfrac{3}{4}$

$\dfrac{5}{4} \overset{?}{=} \dfrac{8}{4} - \dfrac{3}{4}$

$\dfrac{5}{4} = \dfrac{5}{4}$

109. The solution appears to be $x = -16/3$.

$\sqrt{5(4 - 3x)} = 10$

$\left(\sqrt{5(4 - 3x)}\right)^2 = 10^2$

$5(4 - 3x) = 100$

$20 - 15x = 100$

$-15x = 80$

$x = \dfrac{80}{-15}$

$x = -\dfrac{16}{3}$

111. $\sqrt{-16} = \sqrt{16(-1)} = \sqrt{16}\sqrt{-1} = 4i$

113. $\sqrt{-48} = \sqrt{48}\sqrt{-1}$

$= \sqrt{4^2(3)}\,i$

$= 4\sqrt{3}\,i$

115. $\sqrt{-0.16} = \sqrt{(0.16)(-1)}$

$= 0.4i$

117. $\sqrt{-25} + \sqrt{-4} = \sqrt{25(-1)} + \sqrt{4(-1)}$

$= 5i + 2i$

$= 7i$

119. $\sqrt{-10}\sqrt{-5} = \sqrt{10(-1)}\sqrt{5(-1)}$

$\qquad = \sqrt{10}\,i\left(\sqrt{5}\,i\right)$

$\qquad = \sqrt{50}\,i^2$

$\qquad = \sqrt{25}\sqrt{2}\,(-1)$

$\qquad = -5\sqrt{2}$

121. $\sqrt{-3}\sqrt{-10} + \sqrt{-15} = \sqrt{3}\,i\left(\sqrt{10}\,i + \sqrt{15}\,i\right)$

$\qquad = \sqrt{30}\,i^2 + \sqrt{45}\,i^2$

$\qquad = \sqrt{30}(-1) + \sqrt{9(5)}\,(-1)$

$\qquad = -\sqrt{30} - 3\sqrt{5}$

123. $\left(\sqrt{-7}\right)^2 = \left(\sqrt{7(-1)}\right)^2$

$\qquad = \left(\sqrt{7}\,i\right)^2$

$\qquad = \left(\sqrt{7}\right)^2 i^2$

$\qquad = 7(-1)$

$\qquad = -7$

125. $12 - 5i = (a + 2) + (b - 1)i$

$\quad 12 = a + 2 \quad \text{and} \quad -5 = b - 1$

$\quad 10 = a \qquad\qquad\qquad -4 = b$

127. $\sqrt{-49} + 4 = a + bi$

$\qquad 7i + 4 = a + bi$

$\qquad 4 + 7i = a + bi$

$\qquad a = 4 \quad \text{and} \quad b = 7$

129. $(-4 + 5i) - (-12 + 8i) = (-4 + 12) + (5 - 8)i = 8 - 3i$

131. $(-2)(15i)(-3i) = 90i^2 = 90(-1) = -90$

133. $(4 - 3i)(4 + 3i) = 4^2 + 3^2 \quad$ or $\quad (4 - 3i)(4 + 3i) = 16 + 12i - 12i - 9i^2$

$\qquad\qquad\qquad = 16 + 9 \qquad\qquad\qquad\qquad = 16 - 9(-1)$

$\qquad\qquad\qquad = 25 \qquad\qquad\qquad\qquad\qquad = 16 + 9$

$\qquad\qquad\qquad\qquad\qquad\qquad\qquad\qquad\qquad = 25$

135. $(12 - 5i)(2 + 7i) = 24 + 84i - 10i - 35i^2$

$\qquad\qquad\qquad = 24 + 74i - 35(-1)$

$\qquad\qquad\qquad = 59 + 74i$

137. $\dfrac{4}{5i} = \dfrac{4(-5i)}{(5i)(-5i)} = \dfrac{-20i}{-25i^2}$

$\qquad = \dfrac{-20i}{-25(-1)} = \dfrac{-20i}{25} = -\dfrac{4}{5}i$

139. $\dfrac{5i}{2 + 9i} = \dfrac{5i(2 - 9i)}{(2 + 9i)(2 - 9i)} = \dfrac{10i - 45i^2}{2^2 + 9^2}$

$\qquad = \dfrac{10i - 45(-1)}{85} = \dfrac{45 + 10i}{85}$

$\qquad = \dfrac{45}{85} + \dfrac{10}{85}i = \dfrac{9}{17} + \dfrac{2}{17}i$

141. $75(200)(10 \times 10^{-6}) = 150,000 \times 10^{-6} = 0.15$

The increase in length is 0.15 foot.

143. Hypotenuse $= \sqrt{10^2 + 12^2}$

$\qquad\qquad = \sqrt{100 + 144}$

$\qquad\qquad = \sqrt{244}$

$\qquad\qquad = \sqrt{4(61)}$

$\qquad\qquad = 2\sqrt{61}$

$\qquad\qquad \approx 15.62$

The length of the hypotenuse is approximately 15.62 inches.

145. $t = 2\pi\sqrt{\dfrac{L}{32}}$

$\qquad 1.3 = 2\pi\sqrt{\dfrac{L}{32}}$

$\qquad \dfrac{1.3}{2\pi} = \sqrt{\dfrac{L}{32}}$

$\qquad \left(\dfrac{1.3}{2\pi}\right)^2 = \left(\sqrt{\dfrac{L}{32}}\right)^2$

$\qquad \dfrac{1.69}{4\pi^2} = \dfrac{L}{32}$

$\qquad 32\left(\dfrac{1.69}{4\pi^2}\right) = L$

$\qquad \dfrac{8(1.69)}{\pi^2} = L \text{ or } L \approx 1.37$

The length of the pendulum is approximately 1.37 feet.

Chapter Test for Chapter 5

1. (a) $2^{-2} + 2^{-3} = \dfrac{1}{2^2} + \dfrac{1}{2^3} = \dfrac{1}{4} + \dfrac{1}{8} = \dfrac{2}{8} + \dfrac{1}{8} = \dfrac{3}{8}$

    (b) $\dfrac{6.3 \times 10^{-3}}{2.1 \times 10^2} = \dfrac{6.3}{2.1} \times 10^{-3-2} = 3 \times 10^{-5}$

2. (a) $27^{-2/3} = \dfrac{1}{27^{2/3}} = \dfrac{1}{(\sqrt[3]{27})^2} = \dfrac{1}{3^2} = \dfrac{1}{9}$

    (b) $\left(\dfrac{81}{16}\right)^{1/4} = \dfrac{81^{1/4}}{16^{1/4}} = \dfrac{3}{2}$

3. (a) $0.000032 = 3.2 \times 10^{-5}$

    (b) $3.04 \times 10^7 = 30{,}400{,}000$

4. (a) $\dfrac{12s^5t^{-2}}{20s^{-2}t^{-1}} = \dfrac{3s^5s^2t}{5t^2} = \dfrac{3s^7}{5t}$

    (b) $(-2x^5y^{-2}z^0)^{-1} = -2^{-1}x^{-5}y^2(1) = -\dfrac{y^2}{2x^5}$

5. (a) $\left(\dfrac{4x^{1/2}y^{-1/3}}{5x^{1/3}y^{5/6}z}\right)^2 = \dfrac{4^2xy^{-2/3}}{5^2x^{2/3}y^{5/3}z^2} = \dfrac{16x^{1/3}}{25y^{7/3}z^2}$

    (b) $(5xy^2)^{1/4}(5xy^2)^{7/4} = (5xy^2)^{(1/4)+(7/4)} = (5xy^2)^2 = 25x^2y^4$

6. (a) $\sqrt{\dfrac{32x^5y}{9x^2y^3}} = \sqrt{\dfrac{32x^3}{9y^2}} = \dfrac{\sqrt{16x^2(2x)}}{\sqrt{9y^2}} = \dfrac{4x\sqrt{2x}}{3y}$

    (b) $\sqrt[3]{24u^8v^{14}} = \sqrt[3]{8u^6v^{12}(3u^2v^2)} = 2u^2v^4\sqrt[3]{3u^2v^2}$

7. (a) $\dfrac{10}{\sqrt{6} - \sqrt{2}} = \dfrac{10}{\sqrt{6} - \sqrt{2}} \cdot \dfrac{\sqrt{6} + \sqrt{2}}{\sqrt{6} + \sqrt{2}}$

        $= \dfrac{10(\sqrt{6} + \sqrt{2})}{(\sqrt{6})^2 - (\sqrt{2})^2}$

        $= \dfrac{10(\sqrt{6} + \sqrt{2})}{6 - 2}$

        $= \dfrac{10(\sqrt{6} + \sqrt{2})}{4}$

        $= \dfrac{5(2)(\sqrt{6} + \sqrt{2})}{2(2)}$

        $= \dfrac{5(\sqrt{6} + \sqrt{2})}{2}$

    (b) $\dfrac{2}{\sqrt[3]{9y}} = \dfrac{2}{\sqrt[3]{9y}} \cdot \dfrac{\sqrt[3]{3y^2}}{\sqrt[3]{3y^2}}$

        $= \dfrac{2\sqrt[3]{3y^2}}{\sqrt[3]{27y^3}}$

        $= \dfrac{2\sqrt[3]{3y^2}}{3y}$

8. $5\sqrt{3x} - 3\sqrt{75x} = 5\sqrt{3x} - 3\sqrt{25(3x)}$

                $= 5\sqrt{3x} - 3(5)\sqrt{3x}$

                $= 5\sqrt{3x} - 15\sqrt{3x}$

                $= -10\sqrt{3x}$

9. $\sqrt{5}(\sqrt{15x} + 3) = \sqrt{75x} + 3\sqrt{5}$

                $= \sqrt{25(3x)} + 3\sqrt{5}$

                $= 5\sqrt{3x} + 3\sqrt{5}$

10. $(4 - \sqrt{2x})^2 = 16 - 2(4)\sqrt{2x} + (\sqrt{2x})^2$

           $= 16 - 8\sqrt{2x} + 2x$

11. $\sqrt{x^2 - 1} = x - 2$

$\left(\sqrt{x^2 - 1}\right)^2 = (x - 2)^2$

$x^2 - 1 = x^2 - 4x + 4$

$-1 = -4x + 4$

$-5 = -4x$

$\dfrac{-5}{-4} = x$

$\dfrac{5}{4} = x$ (Extraneous)

Check: $\sqrt{\left(\dfrac{5}{4}\right)^2 - 1} \overset{?}{=} \dfrac{5}{4} - 2$

$\sqrt{\dfrac{25}{16} - \dfrac{16}{16}} \overset{?}{=} \dfrac{5}{4} - \dfrac{8}{4}$

$\sqrt{\dfrac{9}{16}} \overset{?}{=} -\dfrac{3}{4}$

$\dfrac{3}{4} \neq -\dfrac{3}{4}$

The equation has *no* solution.

13. $\sqrt{x - 4} = \sqrt{x + 7} - 1$

$\left(\sqrt{x - 4}\right)^2 = \left(\sqrt{x + 7} - 1\right)^2$

$x - 4 = x + 7 - 2\sqrt{x + 7} + 1$

$x - 4 = x + 8 - 2\sqrt{x + 7}$

$-4 = 8 - 2\sqrt{x + 7}$

$-12 = -2\sqrt{x + 7}$

$6 = \sqrt{x + 7}$

$(6)^2 = \left(\sqrt{x + 7}\right)^2$

$36 = x + 7$

$29 = x$

Check: $\sqrt{29 - 4} \overset{?}{=} \sqrt{29 + 7} - 1$

$\sqrt{25} \overset{?}{=} \sqrt{36} - 1$

$5 = 6 - 1$

15. $\sqrt{-16}\left(1 + \sqrt{-4}\right) = \sqrt{16}\sqrt{-1}\left(1 + \sqrt{4}\sqrt{-1}\right)$

$= 4i(1 + 2i) = 4i + 8i^2$

$= 4i + 8(-1) = -8 + 4i$

17. $(2 - 3i)^2 = 2^2 - 2(2)(3i) + (3i)^2$

$= 4 - 12i + 9i^2$

$= 4 - 12i + 9(-1) = -5 - 12i$

12. $\sqrt{x} - x + 6 = 0$

$\sqrt{x} = x - 6$

$\left(\sqrt{x}\right)^2 = (x - 6)^2$

$x = x^2 - 12x + 36$

$0 = x^2 - 13x + 36$

$0 = (x - 9)(x - 4)$

$x - 9 = 0 \Rightarrow x = 9$

$x - 4 = 0 \Rightarrow x = 4$ (Extraneous)

Check: $\sqrt{9} - 9 + 6 \overset{?}{=} 0$

$3 - 9 + 6 \overset{?}{=} 0$

$0 = 0$

Check: $\sqrt{4} - 4 + 6 \overset{?}{=} 0$

$2 - 4 + 6 \overset{?}{=} 0$

$4 \neq 0$

14. $\left(2 + \sqrt{-9}\right) - \sqrt{-25} = 2 + \sqrt{9}\sqrt{-1} - \sqrt{25}\sqrt{-1}$

$= 2 + 3i - 5i$

$= 2 - 2i$

16. $(15 - 3i) + (-8 + 12i) = (15 - 8) + (-3i + 12i)$

$= 7 + 9i$

18. $(3 - 2i)(1 + 5i) = 3 + 15i - 2i - 10i^2$

$= 3 + 13i - 10(-1) = 13 + 13i$

19. $\dfrac{5 - 2i}{3 + i} = \dfrac{5 - 2i}{3 + i} \cdot \dfrac{3 - i}{3 - i}$

$\qquad = \dfrac{15 - 5i - 6i + 2i^2}{3^2 - i^2}$

$\qquad = \dfrac{15 - 11i + 2(-1)}{9 - (-1)}$

$\qquad = \dfrac{13 - 11i}{10}$

$\qquad = \dfrac{13}{10} - \dfrac{11}{10}i$

20. $\qquad v = \sqrt{2gh}$

$\qquad 80 = \sqrt{2(32)h}$

$\qquad 80 = \sqrt{64h}$

$\qquad (80)^2 = \left(\sqrt{64h}\right)^2$

$\qquad 6400 = 64h$

$\qquad \dfrac{6400}{64} = h$

$\qquad 100 = h$

The rock was dropped from a height of 100 feet.

Cumulative Test for Chapters 1–5

1. $y \le 45$

2. $x \ge 15$

3. Additive Identity Property

4. Distributive Property

5. $9.35 + 0.75q$

6. $D = 48t$

7. $16x^2 - 121 = (4x)^2 - 11^2$
$\qquad\qquad\qquad = (4x + 11)(4x - 11)$

8. $9t^2 - 24t + 16 = (3t)^2 - 2(3t)(4) + 4^2$
$\qquad\qquad\qquad\quad = (3t - 4)^2$

9. $x(x - 10) - 4(x - 10) = (x - 10)(x - 4)$

10. $4x^3 - 12x^2 + 16x = 4x(x^2 - 3x + 4)$

11. $\qquad \dfrac{x}{4} - \dfrac{2}{3} = 0$

$\qquad 12\left(\dfrac{x}{4} - \dfrac{2}{3}\right) = 12(0)$

$\qquad 12\left(\dfrac{x}{4}\right) - 12\left(\dfrac{2}{3}\right) = 0$

$\qquad\qquad 3x - 8 = 0$

$\qquad\qquad 3x = 8$

$\qquad\qquad x = \dfrac{8}{3}$

12. $2x - 3[1 + (4 - x)] = 0$

$\qquad 2x - 3[5 - x] = 0$

$\qquad 2x - 15 + 3x = 0$

$\qquad\qquad 5x - 15 = 0$

$\qquad\qquad 5x = 15$

$\qquad\qquad x = 3$

13. $3x^2 - 13x - 10 = 0$

$\qquad (3x + 2)(x - 5) = 0$

$\qquad 3x + 2 = 0 \Rightarrow 3x = -2 \Rightarrow x = -\tfrac{2}{3}$

$\qquad x - 5 = 0 \Rightarrow x = 5$

14.
$$x(x - 3) = 40$$
$$x^2 - 3x = 40$$
$$x^2 - 3x - 40 = 0$$
$$(x - 8)(x + 5) = 0$$
$$x - 8 = 0 \Rightarrow x = 8$$
$$x + 5 = 0 \Rightarrow x = -5$$

15. $|2x + 5| = 11$

$2x + 5 = -11$ or $2x + 5 = 11$
$$2x = -16 \qquad 2x = 6$$
$$x = -8 \qquad x = 3$$

16. $\sqrt{x - 5} - 6 = 0$
$$\sqrt{x - 5} = 6$$
$$\left(\sqrt{x - 5}\right)^2 = 6^2$$
$$x - 5 = 36$$
$$x = 41$$

Check: $\sqrt{41 - 5} - 6 \overset{?}{=} 0$
$$\sqrt{36} - 6 \overset{?}{=} 0$$
$$6 - 6 = 0$$

17. $y = -x + 6$

x-intercept	y-intercept
$0 = -x + 6$	$y = 0 + 6$
$-6 = -x$	$y = 6$
$6 = x$	$(0, 6)$
$(6, 0)$	

18. $y = x^2 + 2x - 8$

x-intercept	y-intercept
$0 = x^2 + 2x - 8$	$y = 0^2 + 2(0) - 8$
$0 = (x + 4)(x - 2)$	$y = 0 + 0 - 8$
$x + 4 = 0 \Rightarrow x = -4$	$y = -8$
$x - 2 = 0 \Rightarrow x = 2$	$(0, -8)$
$(-4, 0), (2, 0)$	

19. $y = 2|x|$

x-intercept	y-intercept
$0 = 2|x|$	$y = 2|0|$
$0 = |x|$	$y = 2(0)$
$0 = x$	$y = 0$
$(0, 0)$	$(0, 0)$

20.
$$f(x) = 3x - x^2$$
$$f(2 + t) - f(2) = 3(2 + t) - (2 + t)^2 - [3(2) - 2^2]$$
$$= 6 + 3t - (4 + 4t + t^2) - 6 + 4$$
$$= 6 + 3t - 4 - 4t - t^2 - 6 + 4$$
$$= -t^2 - t$$

21. (a) $g(x) = -\sqrt{x}$

(b) $g(x) = \sqrt{x} + 2$

(c) $g(x) = \sqrt{x - 2}$

22. Slope $m = \dfrac{-4 - 2}{5 - (-3)} = \dfrac{-6}{8} = -\dfrac{3}{4}$

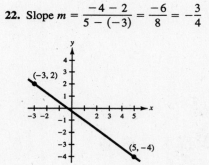

23. Slope $m = \dfrac{-3 - 8}{7 - 2} = \dfrac{-11}{5} = -\dfrac{11}{5}$

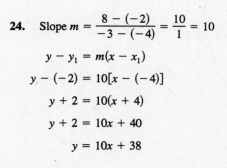

24. Slope $m = \dfrac{8 - (-2)}{-3 - (-4)} = \dfrac{10}{1} = 10$

$$y - y_1 = m(x - x_1)$$
$$y - (-2) = 10[x - (-4)]$$
$$y + 2 = 10(x + 4)$$
$$y + 2 = 10x + 40$$
$$y = 10x + 38$$

25. Slope $m = \dfrac{2-5}{-8-1} = \dfrac{-3}{-9} = \dfrac{1}{3}$

$y - y_1 = m(x - x_1)$

$y - 5 = \dfrac{1}{3}(x - 1)$

$y - 5 = \dfrac{1}{3}x - \dfrac{1}{3}$

$y = \dfrac{1}{3}x - \dfrac{1}{3} + 5$

$y = \dfrac{1}{3}x + \dfrac{14}{3}$

26. Slope $m = \dfrac{4-(-1)}{-\frac{1}{3}-\frac{3}{2}} = \dfrac{5}{-\frac{2}{6}-\frac{9}{6}} = \dfrac{5}{-\frac{11}{6}} = -\dfrac{30}{11}$

$y - y_1 = m(x - x_1)$

$y - (-1) = -\dfrac{30}{11}\left(x - \dfrac{3}{2}\right)$

$y + 1 = -\dfrac{30}{11}x + \dfrac{45}{11}$

$y = -\dfrac{30}{11}x + \dfrac{45}{11} - 1$

$y = -\dfrac{30}{11}x + \dfrac{34}{11}$

27. $\dfrac{x-3}{32} = \dfrac{18}{24}$

$24(x - 3) = 32(18)$

$24x - 72 = 576$

$24x = 648$

$x = 27$

28. $\dfrac{x+6}{7} = \dfrac{16}{8}$

$8(x + 6) = 7(16)$

$8x + 48 = 112$

$8x = 64$

$x = 8$

29. $3(x - 5) < 4x - 7$

$3x - 15 < 4x - 7$

$-x - 15 < -7$

$-x < 8$

$x > -8$

30. $|x + 8| - 1 \geq 15$

$|x + 8| \geq 16$

$x + 8 \leq -16 \quad or \quad x + 8 \geq 16$

$\quad x \leq -24 \qquad\qquad x \geq 8$

31. $\begin{cases} 4x - y = 0 \implies 8x - 2y = 0 \\ -3x + 2y = 2 \implies -3x + 2y = 2 \end{cases}$

$\qquad\qquad\qquad 5x \qquad = 2$

$x = \dfrac{2}{5} \quad \text{and} \quad 4\left(\dfrac{2}{5}\right) - y = 0$

$\qquad\qquad\qquad \dfrac{8}{5} - y = 0$

$\qquad\qquad\qquad\qquad -y = -\dfrac{8}{5}$

$\qquad\qquad\qquad\qquad\quad y = \dfrac{8}{5}$

$\left(\dfrac{2}{5}, \dfrac{8}{5}\right)$

32. $\begin{cases} x - y & = -1 \\ x + 2y - 2z = 3 \\ 3x - y + 2z = 3 \end{cases}$

$$\begin{bmatrix} 1 & -1 & 0 & : & -1 \\ 1 & 2 & -2 & : & 3 \\ 3 & -1 & 2 & : & 3 \end{bmatrix}$$

$\begin{matrix} \\ -R_1 + R_2 \\ -3R_1 + R_3 \end{matrix}\begin{bmatrix} 1 & -1 & 0 & : & -1 \\ 0 & 3 & -2 & : & 4 \\ 0 & 2 & 2 & : & 6 \end{bmatrix}$

$-\frac{1}{3}R_1\begin{bmatrix} 1 & -1 & 0 & : & -1 \\ 0 & 1 & -\frac{2}{3} & : & \frac{4}{3} \\ 0 & 2 & 2 & : & 6 \end{bmatrix}$

$-2R_2 + R_3\begin{bmatrix} 1 & -1 & 0 & : & -1 \\ 0 & 1 & -\frac{2}{3} & : & \frac{4}{3} \\ 0 & 0 & \frac{10}{3} & : & \frac{10}{3} \end{bmatrix}$

$\frac{3}{10}R_3\begin{bmatrix} 1 & -1 & 0 & : & -1 \\ 0 & 1 & -\frac{2}{3} & : & \frac{4}{3} \\ 0 & 0 & 1 & : & 1 \end{bmatrix}$

$\begin{cases} x - y & = -1 \\ y - \frac{2}{3}z = \frac{4}{3} \\ \qquad z = 1 \text{ and } y - \frac{2}{3}(1) = \frac{4}{3} \quad x - (2) = -1 \\ \qquad\qquad\qquad\qquad y = 2 \qquad\quad x = 1 \end{cases}$

$(1, 2, 1)$

33. *Verbal Model:*   Gallons of 75% solution + Gallons of 50% solution = Gallons of mixture

Acid in 75% solution + Acid in 50% solution = Acid in mixture

Labels: Gallons of 75% solution = x (gallons)

Gallons of 50% solution = y (gallons)

Equations: $x + \quad y = 100 \qquad \Rightarrow -0.50x - 0.50y = -50$

$0.75x + 0.50y = 0.60(100) \Rightarrow \underline{\quad 0.75x + 0.50y = 60 \quad}$

$0.25x \qquad\quad = 10$

$x \qquad\qquad = 40$

$40 + y = 100$

$y = 60$

So, 40 gallons of the 75% solution and 60 gallons of the 50% solution are used for the mixture.

34. $\begin{vmatrix} 4 & 0 & 5 \\ 0 & -7 & 2 \\ 9 & 1 & -1 \end{vmatrix} = 4\begin{vmatrix} -7 & 2 \\ 1 & -1 \end{vmatrix} - 0 + 9\begin{vmatrix} 0 & 5 \\ -7 & 2 \end{vmatrix}$

$$= 4(5) - 0 + 9(35)$$

$$= 20 + 315$$

$$= 335$$

35. $(-5, 8), (10, 0),$ and $(3, -4)$

$\begin{vmatrix} -5 & 8 & 1 \\ 10 & 0 & 1 \\ 3 & -4 & 1 \end{vmatrix} = -8\begin{vmatrix} 10 & 1 \\ 3 & 1 \end{vmatrix} + 0 - (-4)\begin{vmatrix} -5 & 1 \\ 10 & 1 \end{vmatrix}$

$$= -8(7) + 0 + 4(-15)$$

$$= -56 - 60$$

$$= -116$$

$$\text{Area} = -\tfrac{1}{2}(-116) = 58$$

36. $y \le 5 - \tfrac{1}{2}x$

$\quad y \le -\tfrac{1}{2}x + 5$

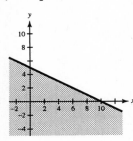

37. $3y - x \ge 7$

$\quad 3y \ge x + 7$

$\quad y \ge \tfrac{1}{3}x + \tfrac{7}{3}$

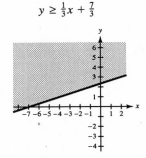

38. $\sqrt{24x^2y^3} = \sqrt{4 \cdot x^2 \cdot y^2 \cdot 6y}$

$\qquad\qquad = 2|x|y\sqrt{6y}$

39. $\sqrt[3]{80a^{15}b^8} = \sqrt[3]{8a^{15}b^6(10b^2)}$

$\qquad\qquad = 2a^5b^2\sqrt[3]{10b^2}$

40. $(12a^{-4}b^6) = \sqrt{\dfrac{12b^6}{a^4}}$

$\qquad\qquad = \dfrac{\sqrt{4 \cdot b^6 \cdot 3}}{\sqrt{a^4}}$

$\qquad\qquad = \dfrac{2b^3\sqrt{3}}{a^2}$

41. $(16^{1/3})^{3/4} = 16^{(1/3)(3/4)}$

$\qquad\qquad = 16^{1/4}$

$\qquad\qquad = \sqrt[4]{16}$

$\qquad\qquad = 2$

42. $(-4 + 11i) - (3 - 5i) = -4 + 11i - 3 + 5i$

$\qquad\qquad\qquad\qquad\quad = -7 + 16i$

43. $\dfrac{6 + 4i}{1 - 3i} = \dfrac{6 + 4i}{1 - 3i} \cdot \dfrac{1 + 3i}{1 + 3i}$

$\qquad = \dfrac{6 + 18i + 4i + 12i^2}{1^2 + 3^2}$

$\qquad = \dfrac{6 + 22i - 12}{1 + 9}$

$\qquad = \dfrac{-6 + 22i}{10}$

$\qquad = \dfrac{-6}{10} + \dfrac{22i}{10}$

$\qquad = -\dfrac{3}{5} + \dfrac{11}{5}i$

CHAPTER 6
Quadratic Functions, Equations, and Inequalities

CHAPTER 6
Quadratic Functions, Equations, and Inequalities

Section 6.1   The Factoring and Square Root Methods

Solutions to Odd-Numbered Exercises

1. $x^2 - 12x + 35 = 0$

$(x - 7)(x - 5) = 0$

$x - 7 = 0 \Rightarrow x = 7$

$x - 5 = 0 \Rightarrow x = 5$

3. $x^2 + x - 72 = 0$

$(x + 9)(x - 8) = 0$

$x + 9 = 0 \Rightarrow x = -9$

$x - 8 = 0 \Rightarrow x = 8$

5. $x^2 + 4x = 45$

$x^2 + 4x - 45 = 0$

$(x + 9)(x - 5) = 0$

$x + 9 = 0 \Rightarrow x = -9$

$x - 5 = 0 \Rightarrow x = 5$

7. $4x^2 - 12x = 0$

$4x(x - 3) = 0$

$4x = 0 \Rightarrow x = 0$

$x - 3 = 0 \Rightarrow x = 3$

9. $u(u - 9) - 12(u - 9) = 0$

$(u - 9)(u - 12) = 0$

$u - 9 = 0 \Rightarrow u = 9$

$u - 12 = 0 \Rightarrow u = 12$

11. $4x^2 - 25 = 0$

$(2x + 5)(2x - 5) = 0$

$2x + 5 = 0 \Rightarrow x = -\frac{5}{2}$

$2x - 5 = 0 \Rightarrow x = \frac{5}{2}$

13. $8x^2 - 10 + 3 = 0$

$(4x - 3)(2x - 1) = 0$

$4x - 3 = 0 \Rightarrow x = \frac{3}{4}$

$2x - 1 = 0 \Rightarrow x = \frac{1}{2}$

15. $x^2 + 60x + 900 = 0$

$(x + 30)(x + 30) = 0$

$x + 30 = 0 \Rightarrow x = -30$

17. $(y - 4)(y - 3) = 6$

$y^2 - 7y + 12 = 6$

$y^2 - 7y + 6 = 0$

$(y - 6)(y - 1) = 0$

$y - 6 = 0 \Rightarrow y = 6$

$y - 1 = 0 \Rightarrow y = 1$

19. $2x(3x + 2) = 5 - 6x^2$

$6x^2 + 4x = 5 - 6x^2$

$12x^2 + 4x - 5 = 0$

$(6x + 5)(2x - 1) = 0$

$6x + 5 = 0 \Rightarrow x = -\frac{5}{6}$

$2x - 1 = 0 \Rightarrow x = \frac{1}{2}$

21. $(3w + 2)(3w - 2) = -w^2 + 26w - 16$

$9w^2 - 4 = -w^2 + 26w - 16$

$10w^2 - 26w + 12 = 0$

$2(5w^2 - 13w + 6) = 0$

$2(5w - 3)(w - 2) = 0$

$2 \neq 0$

$5w - 3 = 0 \Rightarrow 5w = 3 \Rightarrow w = \frac{3}{5}$

$w - 2 = 0 \Rightarrow w = 2$

23. $3x(x - 6) - 5(x - 6) = 0$

$(x - 6)(3x - 5) = 0$

$x - 6 = 0 \Rightarrow x = 6$

$3x - 5 = 0 \Rightarrow x = \frac{5}{3}$

25. $x^2 = 64$

$x = \pm\sqrt{64}$

$x = \pm 8$

27. $6x^2 = 54$

$6x^2 - 54 = 0$

$6(x^2 - 9) = 0$

$6(x + 3)(x - 3) = 0$

$6 \neq 0$

$x + 3 = 0 \Rightarrow x = -3$

$x - 3 = 0 \Rightarrow x = 3$

29. $\dfrac{y^2}{2} = 32$

$2\left(\dfrac{y^2}{2}\right) = 2(32)$

$y^2 = 64$

$y^2 - 64 = 0$

$(y + 8)(y - 8) = 0$

$y + 8 = 0 \Rightarrow y = -8$

$y - 8 = 0 \Rightarrow y = 8$

31. $25x^2 = 16$

$x^2 = \frac{16}{25}$

$x = \pm\sqrt{\frac{16}{25}}$

$x = \pm\frac{4}{5}$

33. $2x^2 = 48$

$x^2 = 24$

$x = \pm\sqrt{24}$

$x = \pm\sqrt{4(6)}$

$x = \pm 2\sqrt{6}$

35. $4u^2 - 225 = 0$

$4u^2 = 225$

$u^2 = \frac{225}{4}$

$u = \pm\sqrt{\frac{225}{4}}$

$u = \pm\frac{15}{2}$

37. $(x + 4)^2 = 169$

$x + 4 = \pm\sqrt{169}$

$x + 4 = \pm 13$

$x = -4 \pm 13$

$x = -4 + 13 = 9$

$x = -4 - 13 = -17$

39. $(x - 3)^2 = 0.25$

$x - 3 = \pm\sqrt{0.25}$

$x - 3 = \pm 0.5$

$x = 3 \pm 0.5$

$x = 3 + 0.5 = 3.5$

$x = 3 - 0.5 = 2.5$

41. $(x - 2)^2 = 7$

$x - 2 = \pm\sqrt{7}$

$x = 2 \pm \sqrt{7}$

43. $(2x + 1)^2 = 50$

$2x + 1 = \pm\sqrt{50}$

$2x + 1 = \pm 5\sqrt{2}$

$2x = -1 \pm 5\sqrt{2}$

$x = \dfrac{-1 \pm 5\sqrt{2}}{2}$

45. $(x - 5)^2 - 36 = 0$

$(x - 5)^2 = 36$

$x - 5 = \pm\sqrt{36}$

$x - 5 = \pm 6$

$x = 5 \pm 6$

$x = 5 + 6 \Rightarrow x = 11$

$x = 5 - 6 \Rightarrow x = -1$

47. $(4x - 3)^2 - 98 = 0$

$(4x - 3)^2 = 98$

$4x - 3 = \pm\sqrt{98} = \pm 7\sqrt{2}$

$4x = 3 \pm 7\sqrt{2}$

$x = \dfrac{3 \pm 7\sqrt{2}}{4}$

49. $z^2 = -36$

$z = \pm\sqrt{-36} = \pm 6i$

51. $x^2 + 4 = 0$

$\qquad x^2 = -4$

$\qquad x = \pm\sqrt{-4}$

$\qquad x = \pm 2i$

53. $(t - 3)^2 = -25$

$\qquad t - 3 = \pm\sqrt{-25}$

$\qquad t - 3 = \pm 5i$

$\qquad t = 3 \pm 5i$

55. $(t - 1)^2 + 169 = 0$

$\qquad (t - 1)^2 = -169$

$\qquad t - 1 = \pm\sqrt{-169}$

$\qquad t - 1 = \pm 13i$

$\qquad t = 1 \pm 13i$

57. $(x + 4)^2 + 121 = 0$

$\qquad (x + 4)^2 = -121$

$\qquad x + 4 = \pm\sqrt{-121}$

$\qquad x + 4 = \pm 11i$

$\qquad x = -4 \pm 11i$

59. $(2x - 1)^2 + 4 = 0$

$\qquad (2x - 1)^2 = -4$

$\qquad 2x - 1 = \pm\sqrt{-4}$

$\qquad 2x - 1 = \pm 2i$

$\qquad 2x = 1 \pm 2i$

$\qquad x = \dfrac{1 \pm 2i}{2}$

$\qquad x = \dfrac{1}{2} \pm i$

61. $(3z + 4)^2 + 144 = 0$

$\qquad (3z + 4)^2 = -144$

$\qquad 3x + 4 = \pm\sqrt{-144} = \pm 12i$

$\qquad 3z = -4 \pm 12i$

$\qquad z = \dfrac{-4 \pm 12i}{3} \text{ or } z = -\dfrac{4}{3} \pm 4i$

63. $9(x + 6)^2 = -121$

$\qquad (x + 6)^2 = -\dfrac{121}{9}$

$\qquad x + 6 = \pm\sqrt{-\dfrac{121}{9}} = \pm\dfrac{11}{3}i$

$\qquad x = -6 \pm \dfrac{11}{3}i \text{ or } x = \dfrac{-18 \pm 11i}{3}$

65. $(x - 1)^2 = -27$

$\qquad x - 1 = \pm\sqrt{-27}$

$\qquad = \pm\sqrt{27}\,i = \pm 3\sqrt{3}\,i$

$\qquad x = 1 \pm 3\sqrt{3}\,i$

67. $(x + 1)^2 + 0.04 = 0$

$\qquad (x + 1)^2 = -0.04$

$\qquad x + 1 = \pm\sqrt{-0.04} = \pm 0.2i$

$\qquad x = -1 \pm 0.2i$

69. $\left(c - \dfrac{2}{3}\right)^2 + \dfrac{1}{9} = 0$

$\qquad \left(c - \dfrac{2}{3}\right)^2 = -\dfrac{1}{9}$

$\qquad c - \dfrac{2}{3} = \pm\sqrt{-\dfrac{1}{9}} = \pm\dfrac{1}{3}i$

$\qquad c = \dfrac{2}{3} \pm \dfrac{1}{3}i \text{ or } c = \dfrac{2 \pm i}{3}$

71. $\left(x + \dfrac{7}{3}\right)^2 = -\dfrac{38}{9}$

$\qquad x + \dfrac{7}{3} = \pm\sqrt{-\dfrac{38}{9}} = \pm\dfrac{\sqrt{38}}{3}i$

$\qquad x = -\dfrac{7}{3} \pm \dfrac{\sqrt{38}}{3}i \text{ or } x = \dfrac{-7 \pm \sqrt{38}\,i}{3}$

73. The x-intercepts appear to be $(3, 0)$ and $(-3, 0)$.

$\qquad x^2 - 9 = 9$

$\qquad x^2 = 9$

$\qquad x = \pm\sqrt{9}$

$\qquad x = \pm 3$

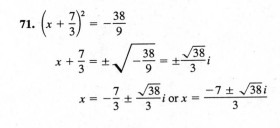

The solutions of the equation are the x-coordinates of the x-intercepts.

75. The x-intercepts appear to be $(5, 0)$ and $(-3, 0)$.

$$x^2 - 2x - 15 = 0$$

$$(x - 5)(x + 3) = 0$$

$$x - 5 = 0 \implies x = 5$$

$$x + 3 = 0 \implies x = -3$$

The solutions of the equation are the x-coordinates of the x-intercepts.

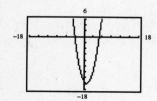

77. The x-intercepts appear to be $(1, 0)$ and $(5, 0)$.

$$4 - (x - 3)^2 = 0$$

$$4 = (x - 3)^2$$

$$\pm\sqrt{4} = (x - 3)$$

$$\pm 2 = x - 3$$

$$3 \pm 2 = x$$

$$x = 3 + 2 \implies x = 5$$

$$x = 3 - 2 \implies x = 1$$

The solutions of the equation are the x-coordinates of the x-intercepts.

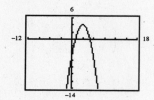

79. The x-intercepts appear to be $(2, 0)$ and $\left(-\frac{3}{2}, 0\right)$.

$$2x^2 - x - 6 = 0$$

$$(2x + 3)(x - 2) = 0$$

$$2x + 3 = 0 \implies x = -\frac{3}{2}$$

$$x - 2 = 0 \implies x = 2$$

The solutions of the equation are the x-coordinates of the x-intercepts.

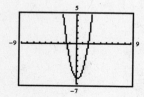

81. The x-intercepts appear to be $(4, 0)$ and $\left(-\frac{4}{3}, 0\right)$.

$$3x^2 - 8x - 16 = 0$$

$$(3x + 4)(x - 4) = 0$$

$$3x + 4 = 0 \implies x = -\frac{4}{3}$$

$$x - 4 = 0 \implies x = 4$$

The solutions of the equation are the x-coordinates of the x-intercepts.

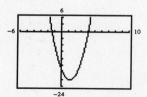

83. There are no x-intercepts. The roots of the equation are imaginary.

$$(x - 1)^2 + 1 = 0$$

$$(x - 1)^2 = -1$$

$$x - 1 = \pm\sqrt{-1}$$

$$x - 1 = \pm i$$

$$x = 1 \pm i$$

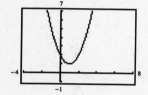

85. There are no x-intercepts. The roots of the equation are imaginary.

$$(x + 3)^2 + 5 = 0$$

$$(x + 3)^2 = -5$$

$$x + 3 = \pm\sqrt{-5}$$

$$x + 3 = \pm\sqrt{5}i$$

$$x = -3 \pm \sqrt{5}i$$

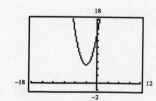

87. There are no x-intercepts. The roots of the equation are imaginary.

$$x^2 + 7 = 0$$
$$x^2 = -7$$
$$x = \pm\sqrt{-7}$$
$$x = \pm\sqrt{7}\,i$$

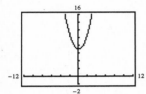

89. $3x^2 + 12 = 0$
$$3x^2 = -12$$
$$x^2 = -4$$
$$x = \pm\sqrt{-4}$$
$$x = \pm 2i$$

91. $3x^2 - 7x = 0$
$$x(3x - 7) = 0$$
$$x = 0$$
$$3x - 7 = 0 \implies 3x = 7 \implies x = \frac{7}{3}$$

93. $3x^2 + 8x - 16 = 0$
$$(3x - 4)(x + 4) = 0$$
$$3x - 4 = 0 \implies x = \tfrac{4}{3}$$
$$x + 4 = 0 \implies x = -4$$

95. $t^2 - 225 = 0$
$$t^2 = 225$$
$$t = \pm\sqrt{225}$$
$$t = \pm 15$$

97. $(x + 1)^2 = 121$
$$x + 1 = \pm\sqrt{121}$$
$$x + 1 = \pm 11$$
$$x = -1 \pm 11$$
$$x = -1 + 11 \implies x = 10$$
$$x = -1 - 11 \implies x = -12$$

99. $(y + 12)^2 + 400 = 0$
$$(y + 12)^2 = -400$$
$$y + 12 = \pm\sqrt{-400}$$
$$y + 12 = \pm 20i$$
$$y = -12 \pm 20i$$

101. $14x = x^2 + 49$
$$0 = x^2 - 14x + 49$$
$$0 = (x - 7)^2$$
$$x - 7 = \pm\sqrt{0}$$
$$x - 7 = 0$$
$$x = 7$$

103. $(x - 9)^2 - 324 = 0$
$$(x - 9)^2 = 324$$
$$x - 9 = \pm\sqrt{324}$$
$$x - 9 = \pm 18$$
$$x = 9 \pm 18$$
$$x = 9 + 18 \implies x = 27$$
$$x = 9 - 18 \implies x = -9$$

105. $x^2 + y^2 = 4$
$$y^2 = 4 - x^2$$
$$y = \pm\sqrt{4 - x^2}$$
$$f(x) = \sqrt{4 - x^2}$$
$$g(x) = -\sqrt{4 - x^2}$$

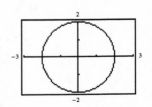

107. $x^2 + 4y^2 = 4$

$4y^2 = 4 - x^2$

$y^2 = \dfrac{4 - x^2}{4}$

$y = \pm\sqrt{\dfrac{4 - x^2}{4}}$

$y = \pm\sqrt{\dfrac{4 - x^2}{2}}$

$f(x) = \dfrac{\sqrt{4 - x^2}}{2}$ or $\dfrac{1}{2}\sqrt{4 - x^2}$

$g(x) = -\dfrac{\sqrt{4 - x^2}}{2}$ or $-\dfrac{1}{2}\sqrt{4 - x^2}$

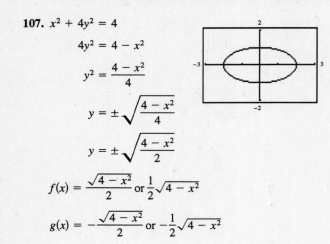

109. $x^4 - 5x^2 + 4 = 0$ Let $u = x^2$.

$u^2 - 5u + 4 = 0$

$(u - 4)(u - 1) = 0$

$u - 4 = 0 \Rightarrow u = 4$

$u - 1 = 0 \Rightarrow u = 1$

$u = 4 \Rightarrow x^2 = 4$

$x = \pm 2$

$u = 1 \Rightarrow x^2 = 1$

$x = \pm 1$

111. $x^4 - 5x^2 + 6 = 0$ Let $u = x^2$.

$u^2 - 5u + 6 = 0$

$(u - 2)(u - 3) = 0$

$u - 2 = 0 \Rightarrow u = 2$

$u - 3 = 0 \Rightarrow u = 3$

$u = 2 \Rightarrow x^2 = 2$

$x = \pm\sqrt{2}$

$u = 3 \Rightarrow x^2 = 3$

$x = \pm\sqrt{3}$

113. $x^4 - 3x^2 - 4 = 0$ Let $u = x^2$.

$u^2 - 3u - 4 = 0$

$(u - 4)(u + 1) = 0$

$u - 4 = 0 \Rightarrow u = 4$

$u + 1 = 0 \Rightarrow u = -1$

$u = 4 \Rightarrow x^2 = 4$

$x = \pm 2$

$u = -1 \Rightarrow x^2 = -1$

$x = \pm i$

115. $x^4 - 9x^2 + 8 = 0$ Let $u = x^2$..

$u^2 - 9u + 8 = 0$

$(u - 1)(u - 8) = 0$

$u - 1 = 0 \Rightarrow u = 1$

$u - 8 = 0 \Rightarrow u = 8$

$u = 1 \Rightarrow x^2 = 1$

$x = \pm\sqrt{1}$

$x = \pm 1$

$u = 8 \Rightarrow x^2 = 8$

$x = \pm\sqrt{8}$

$x = \pm 2\sqrt{2}$

117. $(x^2 - 4)^2 + 2(x^2 - 4) - 3 = 0$ Let $u = x^2 - 4$.

$u^2 + 2u - 3 = 0$

$(u - 1)(u + 3) = 0$

$u - 1 = 0 \Rightarrow u = 1$

$u + 3 = 0 \Rightarrow u = -3$

$u = 1 \Rightarrow x^2 - 4 = 1$

$x^2 = 5$

$x = \pm\sqrt{5}$

$u = -3 \Rightarrow x^2 - 4 = -3$

$x^2 = 1$

$x = \pm 1$

119. $x^{2/3} - x^{1/3} - 6 = 0$ Let $u = x^{1/3}$.

$$u^2 - u - 6 = 0$$
$$(u - 3)(u + 2) = 0$$
$$u - 3 = 0 \implies u = 3$$
$$u + 2 = 0 \implies u = -2$$
$$u = 3 \implies x^{1/3} = 3$$
$$x = 3^3$$
$$x = 27$$
$$u = -2 \implies x^{1/3} = -2$$
$$x = (-2)^3$$
$$x = -8$$

121. $x^{2/3} + 2x^{1/3} - 8 = 0$ Let $u = x^{1/3}$.

$$u^2 + 2u - 8 = 0$$
$$(u + 4)(u - 2) = 0$$
$$u + 4 = 0 \implies u = -4$$
$$u - 2 = 0 \implies u = 2$$
$$u = -4 \implies x^{1/3} = -4$$
$$x = (-4)^3$$
$$x = -64$$
$$u = 2 \implies x^{1/3} = 2$$
$$x = 2^3$$
$$x = 8$$

123. $2x^{2/3} - 7x^{1/3} + 5 = 0$ Let $u = x^{1/3}$.

$$2u^2 - 7u + 5 = 0$$
$$(2u - 5)(u - 1) = 0$$
$$2u - 5 = 0 \implies u = \tfrac{5}{2}$$
$$u - 1 = 0 \implies u = 1$$
$$u = \tfrac{5}{2} \implies x^{1/3} = \tfrac{5}{2}$$
$$x = \left(\tfrac{5}{2}\right)^3$$
$$x = \tfrac{125}{8}$$
$$u = 1 \implies x^{1/3} = 1$$
$$x = 1^3$$
$$x = 1$$

125. $x^{2/5} + 7x^{1/5} + 12 = 0$ Let $u = x^{1/5}$.

$$u^2 + 7u + 12 = 0$$
$$(u + 4)(u + 3) = 0$$
$$u + 4 = 0 \implies u = -4$$
$$u + 3 = 0 \implies u = -3$$
$$u = -4 \implies x^{1/5} = -4$$
$$x = (-4)^5$$
$$x = -1024$$
$$u = -3 \implies x^{1/5} = -3$$
$$x = (-3)^5$$
$$x = -243$$

127. $x^{2/5} - 3x^{1/5} + 2 = 0$ Let $u = x^{1/5}$.

$$u^2 - 3u + 2 = 0$$
$$(u - 1)(u - 2) = 0$$
$$u - 1 = 0 \implies u = 1$$
$$u - 2 = 0 \implies u = 2$$
$$u = 1 \implies x^{1/5} = 1$$
$$x = 1^5$$
$$x = 1$$
$$u = 2 \implies x^{1/5} = 2$$
$$x = 2^5$$
$$x = 32$$

129. $2x^{2/5} - 7x^{1/5} + 3 = 0$ Let $u = x^{1/5}$.

$$2u^2 - 7u + 3 = 0$$
$$(2u - 1)(u - 3) = 0$$
$$2u - 1 = 0 \implies u = \tfrac{1}{2}$$
$$u - 3 = 0 \implies u = 3$$
$$u = \tfrac{1}{2} \implies x^{1/5} = \tfrac{1}{2}$$
$$x = \left(\tfrac{1}{2}\right)^5$$
$$x = \tfrac{1}{32}$$
$$u = 3 \implies x^{1/5} = 3$$
$$x = 3^5$$
$$x = 243$$

131. $6^2 + 8^2 = x^2$

$36 + 64 = x^2$

$100 = x^2$

$\pm\sqrt{100} = x$

$\pm 10 = x$ (Choose the positive solution.)

The side of the triangle is 10 units long.

133. $7^2 + x^2 = 9^2$

$49 + x^2 = 81$

$x^2 = 32$

$x = \pm\sqrt{32} = \pm 4\sqrt{2}$ (Choose the positive solution.)

$x \approx 5.66$

The side of the triangle is $4\sqrt{2}$ or approximately 5.66 units long.

135. Let d = length of diagonal

$d^2 = 50^2 + 94^2$

$d^2 = 2500 + 8836$

$d^2 = 11{,}336$

$d = \pm\sqrt{11{,}336}$ (Choose the positive solution.)

$= \sqrt{4 \cdot 2834}$

$= 2\sqrt{2834}$

$d \approx 106.47$

The length of the diagonal is approximately 106.47 feet.

137. Let x = length of a side of the square.

$x^2 + x^2 = 25^2$

$2x^2 = 625$

$x^2 = \dfrac{625}{2}$

$x = \pm\sqrt{\dfrac{625}{2}} = \dfrac{\sqrt{625}}{\sqrt{2}} = \dfrac{\sqrt{625}}{\sqrt{2}} \cdot \dfrac{\sqrt{2}}{\sqrt{2}} = \dfrac{25\sqrt{2}}{2}$

(Choose the positive solution.)

Each side of the square has a length of $(25\sqrt{2})/2$ inches or approximately 17.68 inches.

139. $4\pi r^2 = S$

$4\pi r^2 = \dfrac{81}{\pi}$

$r^2 = \dfrac{81}{4\pi^2}$

$r = \pm\sqrt{\dfrac{81}{4\pi^2}}$

$r = \pm\dfrac{9}{2\pi}$

Discarding the negative solution and choosing the positive root, you obtain the radius of the baseball $r = \dfrac{9}{2\pi}$ inches.

141. $h = -16t^2 + s_0,\ s_0 = 128$

$h = -16t^2 + 128$

$16t^2 = 128$

$t^2 = 8$

$t = \sqrt{8}$

$t = \pm 2\sqrt{2}$ (Choose the positive solution.)

$t \approx 2.83$

The object reaches the ground in approximately 2.83 seconds.

143. When the object reaches the ground, $h = 0$.

$0 = 144 + 128t - 16t^2$

$0 = 16(9 + 8t - t^2)$

$0 = 16(9 - t)(1 + t)$

$16 \neq 0$

$9 - t = 0 \Rightarrow t = 9$

$1 + t = 0 \Rightarrow t = -1$

(Choose the positive solution.) The object reaches the ground in 9 seconds.

145.
$$y = (0.77t + 27.7)^2 \quad \text{Note: Expenditures} = \$900 \text{ billion} \Rightarrow y = 900$$
$$(0.77t + 27.7)^2 = 900$$
$$0.77t + 27.7 = \pm\sqrt{900}$$
$$0.77t + 27.7 = 30 \quad \text{(Discard negative solution.)}$$
$$0.77t = 2.3$$
$$t = \frac{2.3}{0.77}$$
$$t \approx 3$$

So, the expenditures were approximately $900 billion in 1993.
(The graph confirms the result.)

147. $\frac{1}{2}$ (Base)(Height) = Area

$$\frac{1}{2}(2x)(x + 1) = 6$$
$$x(x + 1) = 6$$
$$x^2 + x - 6 = 0$$
$$(x + 3)(x - 2) = 0$$
$$x + 3 = 0 \Rightarrow x = -3$$
$$x - 2 = 0 \Rightarrow x = 2$$

Discard the negative solution. So, $x = 2$.

149.
$$5000 = 3500(1 + r)^2$$
$$\frac{5000}{3500} = (1 + r)^2$$
$$\frac{10}{7} = (1 + r)^2$$
$$\pm\sqrt{\frac{10}{7}} = 1 + r$$
$$1.2 \approx 1 + r \qquad \text{(Discard negative solution.)}$$
$$0.2 \approx r$$

The interest rate r is approximately 20%.

151. According to the Zero-Factor Property, if a product of factors is equal to zero, then one or more of the factors must be equal to zero. If a quadratic polynomial is equal to zero and that polynomial can be factored, then one or more of those factors must be equal to zero. By factoring the quadratic polynomial, setting each linear factor equal to zero and solving the linear equations, you find the solution(s) of the quadratic equation.

153. Yes, it is possible for a quadratic equation to have only one solution. For example, $x^2 - 10x + 25 = 0$ has only one solution, $x = 5$.

155.
$$16r^4 - t^4 = (4r^2)^2 - (t^2)^2$$
$$= (4r^2 + t^2)(4r^2 - t^2)$$
$$= (4r^2 + t^2)(2r + t)(2r - t)$$

157. $57y^2 + y - 6 = (19y - 6)(3y + 1)$

159.
$$5x(2x + 7) - (2x + 7)^2 = (2x + 7)[5x - (2x + 7)]$$
$$= (2x + 7)(5x - 2x - 7)$$
$$= (2x + 7)(3x - 7)$$

161.
$$-3y + 20 = 2$$
$$-3y = -18$$
$$y = 6$$

163.
$$(x - 3)(x + 4) = 0$$
$$x - 3 = 0 \Rightarrow x = 3$$
$$x + 4 = 0 \Rightarrow x = -4$$

165. *Verbal Model:* | First runner's distance | = | Second runner's distance |

Labels: First runner: Rate = 6 (miles per hour)

Time = $x + \frac{1}{12}$ (hours)

Distance = $6\left(x + \frac{1}{12}\right)$ (miles)

Second runner: Rate = 8 (miles per hour)

Time = x (hours)

Distance = $8x$ (miles)

Equation:

$$6\left(x + \frac{1}{12}\right) = 8x$$

$$6x + \frac{1}{2} = 8x$$

$$2\left(6x + \frac{1}{2}\right) = 2(8x)$$

$$12x + 1 = 16x$$

$$1 = 4x$$

$$\frac{1}{4} = x$$

The second runner overtakes the first in $\frac{1}{4}$ hour, or 15 minutes. Each runner will have run 2 miles at that time.

Section 6.2   Completing the Square

1. $x^2 + 8x + \boxed{16}$

$\left(\frac{8}{2}\right)^2 = 4^2 = 16$

3. $y^2 - 20y + \boxed{100}$

$\left(-\frac{20}{2}\right)^2 = (-10)^2 = 100$

5. $x^2 + 4x + \boxed{4}$

$\left(\frac{4}{2}\right)^2 = 2^2 = 4$

7. $t^2 + 5t + \boxed{\frac{25}{4}}$

$\left(\frac{5}{2}\right)^2 = \frac{25}{4}$

9. $z^2 - 9z + \boxed{\frac{81}{4}}$

$\left(-\frac{9}{2}\right)^2 = \frac{81}{4}$

11. $x^2 - \frac{6}{5}x + \boxed{\frac{9}{25}}$

$\left(\frac{-6/5}{2}\right)^2 = \left(\frac{-3}{5}\right)^2$

$= \frac{9}{25}$

13. $y^2 - \frac{3}{5}y + \boxed{\frac{9}{100}}$

$\left(\frac{-3/5}{2}\right)^2 = \left(-\frac{3}{10}\right)^2$

$= \frac{9}{100}$

15. $r^2 - 0.4r + \boxed{0.04}$

$\left(-\frac{0.4}{2}\right)^2 = (-0.2)^2$

$= 0.04$

17. (a) $x^2 + 6x = 0$

$x^2 + 6x + (3)^2 = 0 + (3)^2$

$(x + 3)^2 = 9$

$x + 3 = \pm\sqrt{9}$

$x + 3 = \pm 3$

$x = -3 \pm 3$

$x = -3 + 3 \Rightarrow x = 0$

$x = -3 - 3 \Rightarrow x = -6$

(b) $x^2 + 6x = 0$

$x(x + 6) = 0$

$x = 0$

$x + 6 = 0 \Rightarrow x = -6$

19. (a) $t^2 - 8t + 7 = 0$

$t^2 - 8t = -7$

$t^2 - 8t + (-4)^2 = -7 + 16$

$(t - 4)^2 = 9$

$t - 4 = \pm\sqrt{9} = \pm 3$

$t = 4 \pm 3$

$t = 4 + 3 = 7$

$t = 4 - 3 = 1$

(b) $t^2 - 8t + 7 = 0$

$(t - 7)(t - 1) = 0$

$t - 7 = 0 \Rightarrow t = 7$

$t - 1 = 0 \Rightarrow t = 1$

21. (a) $x^2 + 2x - 24 = 0$

$$x^2 + 2x = 24$$

$$x^2 + 2x + (1)^2 = 24 + 1$$

$$(x + 1)^2 = 25$$

$$x + 1 = \pm\sqrt{25} = \pm 5$$

$$x = -1 \pm 5$$

$$x = -1 + 5 \implies x = 4$$

$$x = -1 - 5 \implies x = -6$$

(b) $x^2 + 2x - 24 = 0$

$$(x + 6)(x - 4) = 0$$

$$x + 6 = 0 \implies x = -6$$

$$x - 4 = 0 \implies x = 4$$

23. (a) $x^2 + 7x + 12 = 0$

$$x^2 + 7x = -12$$

$$x^2 + 7x + \left(\tfrac{7}{2}\right)^2 = -12 + \left(\tfrac{49}{4}\right)$$

$$\left(x + \tfrac{7}{2}\right)^2 = \tfrac{1}{4}$$

$$x + \tfrac{7}{2} = \pm\sqrt{\tfrac{1}{4}} = \pm\tfrac{1}{2}$$

$$x = -\tfrac{7}{2} \pm \tfrac{1}{2}$$

$$x = -\tfrac{7}{2} + \tfrac{1}{2} \implies x = -3$$

$$x = -\tfrac{7}{2} - \tfrac{1}{2} \implies x = -4$$

(b) $x^2 + 7x + 12 = 0$

$$(x + 3)(x + 4) = 0$$

$$x + 3 = 0 \implies x = -3$$

$$x + 4 = 0 \implies x = -4$$

25. (a) $5x^2 - 3x - 8 = 0$

$$5x^2 - 3x = 8$$

$$x^2 - \tfrac{3}{5}x = \tfrac{8}{5}$$

$$x^2 - \tfrac{3}{5}x + \left(-\tfrac{3}{10}\right)^2 = \tfrac{8}{5} + \tfrac{9}{100}$$

$$\left(x - \tfrac{3}{10}\right)^2 = \tfrac{169}{100}$$

$$x - \tfrac{3}{10} = \pm\sqrt{\tfrac{169}{100}} = \pm\tfrac{13}{10}$$

$$x = \tfrac{3}{10} \pm \tfrac{13}{10}$$

$$x = \tfrac{3}{10} + \tfrac{13}{10} \implies x = \tfrac{8}{5}$$

$$x = \tfrac{3}{10} - \tfrac{13}{10} \implies x = -1$$

(b) $5x^2 - 3x - 8 = 0$

$$(5x - 8)(x + 1) = 0$$

$$5x - 8 = 0 \implies x = \tfrac{8}{5}$$

$$x + 1 = 0 \implies x = -1$$

27. $x^2 - 4x - 3 = 0$

$$x^2 - 4x = 3$$

$$x^2 - 4x + (-2)^2 = 3 + 4$$

$$(x - 2)^2 = 7$$

$$x - 2 = \pm\sqrt{7}$$

$$x = 2 \pm \sqrt{7}$$

$$x = 2 + \sqrt{7} \implies x \approx 4.65$$

$$x = 2 - \sqrt{7} \implies x \approx -0.65$$

29. $x^2 + 6x + 7 = 0$

$$x^2 + 6x = -7$$

$$x^2 + 6x + (3)^2 = -7 + 9$$

$$(x + 3)^2 = 2$$

$$x + 3 = \pm\sqrt{2}$$

$$x = -3 \pm \sqrt{2}$$

$$x = -3 + \sqrt{2} \implies x \approx -1.59$$

$$x = -3 - \sqrt{2} \implies x \approx -4.41$$

31. $u^2 - 4u + 1 = 0$

$$u^2 - 4u = -1$$

$$u^2 - 4u + (-2)^2 = -1 + 4$$

$$(u - 2)^2 = 3$$

$$u - 2 = \pm\sqrt{3}$$

$$u = 2 \pm \sqrt{3}$$

$$u = 2 + \sqrt{3} \implies u \approx 3.73$$

$$u = 2 - \sqrt{3} \implies u \approx 0.27$$

33. $x^2 + 2x + 3 = 0$

$x^2 + 2x = -3$

$x^2 + 2x + 1^2 = -3 + 1$

$(x + 1)^2 = -2$

$x + 1 = \pm\sqrt{-2}$

$x = -1 \pm \sqrt{2}i$

$x = -1 + \sqrt{2}i \Rightarrow x \approx -1 + 1.41i$

$x = -1 - \sqrt{2}i \Rightarrow x \approx -1 - 1.41i$

35. $x^2 - 10x - 2 = 0$

$x^2 - 10x = 2$

$x^2 - 10x + (-5)^2 = 2 + 25$

$(x - 5)^2 = 27$

$x - 5 = \pm\sqrt{27}$

$x = 5 \pm 3\sqrt{3}$

$x = 5 + 3\sqrt{3} \Rightarrow x \approx 10.20$

$x = 5 - 3\sqrt{3} \Rightarrow x \approx -0.20$

37. $y^2 + 20y + 10 = 0$

$y^2 + 20y = -10$

$y^2 + 20y + 10^2 = -10 + 100$

$(y + 10)^2 = 90$

$y + 10 = \pm\sqrt{90}$

$y = -10 \pm 3\sqrt{10}$

$y = -10 + 3\sqrt{10} \Rightarrow y \approx -0.51$

$y = -10 - 3\sqrt{10} \Rightarrow y \approx -19.49$

39. $t^2 + 5t + 3 = 0$

$t^2 + 5t = -3$

$t^2 + 5t + \left(\dfrac{5}{2}\right)^2 = -3 + \dfrac{25}{4}$

$\left(t + \dfrac{5}{2}\right)^2 = \dfrac{13}{4}$

$t + \dfrac{5}{2} = \pm\dfrac{\sqrt{13}}{2}$

$t = -\dfrac{5}{2} \pm \dfrac{\sqrt{13}}{2}$

$t = \dfrac{-5 + \sqrt{13}}{2} \Rightarrow t \approx -0.70$

$t = \dfrac{-5 - \sqrt{13}}{2} \Rightarrow t \approx -4.30$

41. $v^2 + 3v - 2 = 0$

$v^2 + 3v = 2$

$v^2 + 3v + \left(\dfrac{3}{2}\right)^2 = 2 + \dfrac{9}{4}$

$\left(v + \dfrac{3}{2}\right)^2 = \dfrac{17}{4}$

$v + \dfrac{3}{2} = \pm\dfrac{\sqrt{17}}{2}$

$v = -\dfrac{3}{2} \pm \dfrac{\sqrt{17}}{2}$

$v = \dfrac{-3 + \sqrt{17}}{2} \Rightarrow v \approx 0.56$

$v = \dfrac{-3 - \sqrt{17}}{2} \Rightarrow v \approx -3.56$

43. $x^2 - 11x + 3 = 0$

$x^2 - 11x = -3$

$x^2 - 11x + \left(-\dfrac{11}{2}\right)^2 = -3 + \dfrac{121}{4}$

$\left(x - \dfrac{11}{2}\right)^2 = \dfrac{109}{4}$

$x - \dfrac{11}{2} = \pm\dfrac{\sqrt{109}}{2}$

$x = \dfrac{11}{2} \pm \dfrac{\sqrt{109}}{2}$

$x = \dfrac{11 + \sqrt{109}}{2} \Rightarrow x \approx 10.72$

$x = \dfrac{11 - \sqrt{109}}{2} \Rightarrow x \approx 0.28$

45. $-x^2 + x - 1 = 0$

$$-x^2 + x = 1$$

$$x^2 - x = -1$$

$$x^2 - x + \left(-\frac{1}{2}\right)^2 = -1 + \frac{1}{4}$$

$$\left(x - \frac{1}{2}\right)^2 = -\frac{3}{4}$$

$$x - \frac{1}{2} = \pm\frac{\sqrt{3}i}{2}$$

$$x = \frac{1}{2} \pm \frac{\sqrt{3}i}{2} \text{ or } x = \frac{1 \pm \sqrt{3}i}{2}$$

$$x = \frac{1}{2} + \frac{\sqrt{3}}{2}i \implies x \approx 0.5 + 0.87i$$

$$x = \frac{1}{2} - \frac{\sqrt{3}}{2}i \implies x \approx 0.5 - 0.87i$$

47. $x^2 - 5x + 5 = 0$

$$x^2 - 5x = -5$$

$$x^2 - 5x + \left(-\frac{5}{2}\right)^2 = -5 + \frac{25}{4}$$

$$\left(x - \frac{5}{2}\right)^2 = \frac{5}{4}$$

$$x - \frac{5}{2} = \pm\frac{\sqrt{5}}{2}$$

$$x = \frac{5}{2} \pm \frac{\sqrt{5}}{2}$$

$$x = \frac{5 + \sqrt{5}}{2} \implies x \approx 3.62$$

$$x = \frac{5 - \sqrt{5}}{2} \implies x \approx 1.38$$

49. $x^2 - \frac{2}{3}x - 3 = 0$

$$x^2 - \frac{2}{3}x = 3$$

$$x^2 - \frac{2}{3}x + \left(-\frac{1}{3}\right)^2 = 3 + \frac{1}{9}$$

$$\left(x - \frac{1}{3}\right)^2 = \frac{28}{9}$$

$$x - \frac{1}{3} = \pm\frac{2\sqrt{7}}{3}$$

$$x = \frac{1}{3} \pm \frac{2\sqrt{7}}{3}$$

$$x = \frac{1 + 2\sqrt{7}}{3} \implies x \approx 2.10$$

$$x = \frac{1 - 2\sqrt{7}}{3} \implies x \approx -1.43$$

51. $v^2 + \frac{3}{4}v - 2 = 0$

$$v^2 + \frac{3}{4}v = 2$$

$$v^2 + \frac{3}{4}v + \left(\frac{3}{8}\right)^2 = 2 + \frac{9}{64}$$

$$\left(v + \frac{3}{8}\right)^2 = \frac{137}{64}$$

$$v + \frac{3}{8} = \pm\frac{\sqrt{137}}{8}$$

$$x = -\frac{3}{2} \pm \frac{\sqrt{137}}{8}$$

$$x = \frac{-3 + \sqrt{137}}{8} \implies x \approx 1.09$$

$$x = \frac{-3 - \sqrt{137}}{8} \implies x \approx -1.84$$

53. $2x^2 + 8x + 3 = 0$

$$2x^2 + 8x = -3$$

$$x^2 + 4x = -\frac{3}{2}$$

$$x^2 + 4x + 2^2 = -\frac{3}{2} + 4$$

$$(x + 2)^2 = \frac{5}{2}$$

$$x + 2 = \pm\sqrt{\frac{5}{2}} = \pm\frac{\sqrt{10}}{2}$$

$$x = -2 \pm \frac{\sqrt{10}}{2}$$

$$x = \frac{-4 + \sqrt{10}}{2} \implies x \approx -0.42$$

$$x = \frac{-4 - \sqrt{10}}{2} \implies x \approx -3.58$$

55. $3x^2 + 9x + 5 = 0$

$$3x^2 + 9x = -5$$

$$x^2 + 3x = -\frac{5}{3}$$

$$x^2 + 3x + \left(\frac{3}{2}\right)^2 = -\frac{5}{3} + \frac{9}{4}$$

$$\left(x + \frac{3}{2}\right)^2 = \frac{7}{12}$$

$$x + \frac{3}{2} = \pm\sqrt{\frac{7}{12}} = \pm\frac{\sqrt{21}}{\sqrt{36}}$$

$$x = -\frac{3}{2} \pm \frac{\sqrt{21}}{6}$$

$$x = \frac{-9 + \sqrt{21}}{6} \implies x \approx -0.74$$

$$x = \frac{-9 - \sqrt{21}}{6} \implies x \approx -2.26$$

57. $4y^2 + 4y - 9 = 0$

$$4y^2 + 4y = 9$$

$$y^2 + y = \frac{9}{4}$$

$$y^2 + y + \left(\frac{1}{2}\right)^2 = \frac{9}{4} + \frac{1}{4}$$

$$\left(y + \frac{1}{2}\right)^2 = \frac{10}{4}$$

$$y + \frac{1}{2} = \pm\frac{\sqrt{10}}{2}$$

$$y = -\frac{1}{2} \pm \frac{\sqrt{10}}{2}$$

$$y = \frac{-1 + \sqrt{10}}{2} \Rightarrow y \approx 1.08$$

$$y = \frac{-1 - \sqrt{10}}{2} \Rightarrow y \approx -2.08$$

59. $2x^2 + 5x - 8 = 0$

$$2x^2 + 5x = 8$$

$$x^2 + \frac{5}{2}x = 4$$

$$x^2 + \frac{5}{2}x + \left(\frac{5}{4}\right)^2 = 4 + \frac{25}{16}$$

$$\left(x + \frac{5}{4}\right)^2 = \frac{80}{16}$$

$$x + \frac{5}{4} = \pm\frac{\sqrt{89}}{4}$$

$$x = -\frac{5}{4} \pm \frac{\sqrt{89}}{4}$$

$$x = \frac{-5 + \sqrt{89}}{4} \Rightarrow x \approx 1.11$$

$$x = \frac{-5 - \sqrt{89}}{4} \Rightarrow x \approx -3.61$$

61. $x^2 - \frac{3}{5}x + 2 = 0$

$$x^2 - \frac{3}{5}x = -2$$

$$x^2 - \frac{3}{5}x + \left(-\frac{3}{10}\right)^2 = -2 + \frac{9}{100}$$

$$\left(x - \frac{3}{10}\right)^2 = \frac{-191}{100}$$

$$x - \frac{3}{10} = \pm\frac{\sqrt{191}i}{10}$$

$$x = \frac{3}{10} \pm \frac{\sqrt{191}}{10}i$$

$$x = \frac{3}{10} \pm \frac{\sqrt{191}}{10}i \Rightarrow x \approx 0.3 \pm 1.38i$$

63. $x(x - 7) = 2$

$$x^2 - 7x = 2$$

$$x^2 - 7x + \left(-\frac{7}{2}\right)^2 = 2 + \frac{49}{4}$$

$$\left(x - \frac{7}{2}\right)^2 = \frac{57}{4}$$

$$x - \frac{7}{2} = \pm\frac{\sqrt{57}}{2}$$

$$x = \frac{7}{2} \pm \frac{\sqrt{57}}{2}$$

$$x = \frac{7 + \sqrt{57}}{2} \Rightarrow x \approx 7.27$$

$$x = \frac{7 - \sqrt{57}}{2} \Rightarrow x \approx -0.27$$

65. $\frac{1}{2}t^2 + t + 2 = 0$

$$t^2 + 2t + 4 = 0$$

$$t^2 + 2t = -4$$

$$t^2 + 2t + (1)^2 = -4 + (1)^2$$

$$(t + 1)^2 = -3$$

$$t + 1 = \pm\sqrt{-3}$$

$$t + 1 = \pm\sqrt{3}i$$

$$t = -1 \pm \sqrt{3}i$$

$$t = -1 + \sqrt{3}i \Rightarrow t = -1 + 1.73i$$

$$t = -1 - \sqrt{3}i \Rightarrow t = -1 - 1.73i$$

67. $0.1x^2 + 0.2x + 0.5 = 0$

$$x^2 + 2x + 5 = 0$$

$$x^2 + 2x = -5$$

$$x^2 + 2x + (1)^2 = -5 + (1)^2$$

$$(x + 1)^2 = -4$$

$$x + 1 = \pm\sqrt{-4}$$

$$x + 1 = \pm 2i$$

$$x = -1 \pm 2i$$

69.
$$\frac{x^2}{4} = \frac{x+1}{2}$$
$$2x^2 = 4(x+1)$$
$$2x^2 = 4x + 4$$
$$2x^2 - 4x = 4$$
$$x^2 - 2x = 2$$
$$x^2 - 2x + (1)^2 = 2 + (1)^2$$
$$(x-1)^2 = 3$$
$$x - 1 = \pm\sqrt{3}$$
$$x = 1 \pm \sqrt{3}$$

71.
$$\sqrt{2x+1} = x - 3$$
$$\left(\sqrt{2x+1}\right)^2 = (x-3)^2$$
$$2x + 1 = x^2 - 6x + 9$$
$$0 = x^2 - 8x + 8$$
$$-8 = x^2 - 8x$$
$$-8 + 16 = x^2 - 8x + (-4)^2$$
$$8 = (x-4)^2$$
$$\pm\sqrt{8} = x - 4$$
$$\pm 2\sqrt{2} = x - 4$$
$$4 \pm 2\sqrt{2} = x$$
$$x = 4 + 2\sqrt{2} \implies x \approx 6.83$$
$$x = 4 - 2\sqrt{2} \implies x \approx 1.17 \text{ (Extraneous)}$$

Check: $\sqrt{2(6.83)+1} \overset{?}{\approx} 6.83 - 3 \approx ?$
$$\sqrt{14.66} \overset{?}{\approx} 3.83$$
$$3.83 \approx 3.83$$

Check: $\sqrt{2(1.17)+1} \overset{?}{\approx} 1.17 - 3$
$$\sqrt{3.34} \overset{?}{\approx} -1.83$$
$$1.83 \neq -1.83$$

73.
$$x^2 + 6x - 4 = 0$$
$$x^2 + 6x = 4$$
$$x^2 + 6x + (3)^2 = 4 + (3)^2$$
$$(x+3)^2 = 13$$
$$x + 3 = \pm\sqrt{13}$$
$$x = -3 \pm \sqrt{13}$$

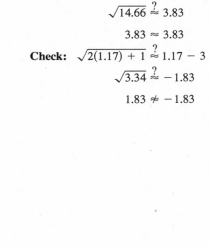

Note: $-3 + \sqrt{13} \approx 0.61$ and $-3 - \sqrt{13} \approx -6.61$

The solutions of the equation are the x-coordinates of the x-intercepts.

75. $\frac{1}{3}x^2 + 2x - 6 = 0$
$$x^2 + 6x - 18 = 0$$
$$x^2 + 6x = 18$$
$$x^2 + 6x + (3)^2 = 18 + (3)^2$$
$$(x+3)^2 = 27$$
$$x + 3 = \pm\sqrt{27}$$
$$x + 3 = \pm 3\sqrt{3}$$
$$x = -3 \pm 3\sqrt{3}$$

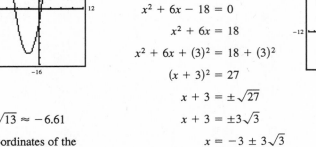

Note: $-3 + 3\sqrt{3} \approx 2.20$ and $-3 - 3\sqrt{3} \approx -8.20$

The solutions of the equation are the x-coordinates of the x-intercepts, $\left(-3 \pm 3\sqrt{3}, 0\right)$.

77. (a) Area of square: x^2

Area of rectangles: $2(4x) = 8x$

Total area: $x^2 + 8x$

(b) Area of small square: $4(4) = 16$

Entire area: $x^2 + 8x + 16$

(c) Dimensions: $(x+4)(x+4)$

Area: $(x+4)^2$ or $x^2 + 8x + 16$

79. Area $= \frac{1}{2}$(Base)(Height)

$$12 = \frac{1}{2}x(x + 2)$$

$$24 = x(x + 2)$$

$$24 = x^2 + 2x$$

$$0 = x^2 + 2x - 24 = (x + 6)(x - 4)$$

$x + 6 = 0 \Rightarrow x = -6$ (Discard this negative solution.)

$x - 4 = 0 \Rightarrow x = 4$ and $x + 2 = 6$

The base of the triangle is 4 centimeters and the height is 6 centimeters.

81. (Length)(Width)(Height) = Volume

$$(x + 4)(x)(6) = 840$$

$$6x^2 + 24x = 840$$

$$x^2 + 4x = 140$$

$$x^2 + 4x - 140 = 0$$

$$(x + 14)(x - 10) = 0$$

$x + 14 = 0 \Rightarrow x = -14$
(Discard negative answer.)

$x - 10 = 0 \Rightarrow x = 10$

and $x + 4 = 14$

The dimensions of the box are 10 inches by 14 inches by 6 inches.

83. *Verbal Model:* $\boxed{\text{First consecutive even integer}} \cdot \boxed{\text{Second consecutive even integer}} = 288$

Labels: First consecutive odd integer $= n$

Second consecutive odd integer $= n + 2$

Equation:

$$n(n + 2) = 288$$

$$n^2 + 2n = 288$$

$$n^2 + 2n + 1^2 = 288 + 1$$

$$(n + 1)^2 = 289$$

$$n + 1 = \pm\sqrt{289} \quad \text{(Discard negative solution.)}$$

$$n + 1 = 17$$

$$n = 16 \Rightarrow n + 2 = 18$$

The two consecutive positive even integers are 16 and 18.

85.

$$R = x\left(50 - \tfrac{1}{2}x\right)$$

$$1218 = 50x - \tfrac{1}{2}x^2$$

$$2436 = 100x - x^2$$

$$x^2 - 100x + 2436 = 0$$

$$x^2 - 100x = -2436$$

$$x^2 - 100x + (-50)^2 = -2436 + 2500$$

$$(x - 50)^2 = 64$$

$$x - 50 = \pm\sqrt{64} = \pm 8$$

$$x = 50 \pm 8$$

$$x = 50 + 8 = 58$$

$$x = 50 - 8 = 42$$

Thus, either 58 units or 42 units must be sold to produce a revenue of $1218.

Note: The equation $x^2 - 100x + 2436 = 0$ could also be solved by factoring.

87.

$$1.904t^2 - 1.36t + 81.0 = 200$$

$$1.904t^2 - 1.36t = 119$$

$$t^2 - \frac{1.36}{1.904}t = \frac{119}{1.904}$$

$$t^2 - \frac{1.36}{1.904}t + \left(-\frac{0.68}{1.904}\right)^2 = \frac{119}{1.904} + \frac{(0.68)^2}{(1.904)^2}$$

$$\left(t - \frac{0.68}{1.904}\right)^2 = \frac{227.0384}{(1.904)^2}$$

$$t - \frac{0.68}{1.904} = \pm\frac{\sqrt{227.0384}}{1.904}$$

$$t = \frac{0.68}{1.904} \pm \frac{\sqrt{227.0384}}{1.904}$$

$$t = \frac{0.68 \pm \sqrt{227.0384}}{1.904}$$

(Discard negative solution.)

$$t \approx 8$$

According to the model, the number of terminals reached 200,000 in 1998.

89. The first step is to divide both sides of the equation by the leading coefficient. Yes, dividing both sides of an equation by a nonzero constant produces an equivalent equation.

91. Yes, because a quadratic equation can be solved by factoring if the solutions are rational numbers.

93. $(x + 6)^2 - 10 = x^2 + 12x + 36 - 10$
$$= x^2 + 12x + 26$$

95. $(2x + 1)^2 - 4 = 4x^2 + 4x + 1 - 4$
$$= 4x^2 + 4x - 3$$

97. $y^2 = \dfrac{4}{225}$

$y = \pm\sqrt{\dfrac{4}{225}}$

$\quad = \pm\dfrac{2}{15}$

99. $(t - 3)^2 = -64$

$t - 3 = \pm\sqrt{-64}$

$t - 3 = \pm 8i$

$t = 3 \pm 8i$

101. $5(3 - t) + 2t(3 - t) = 0$

$(3 - t)(5 + 2t) = 0$

$3 - t = 0 \implies -t = -3 \implies t = 3$

$5 + 2t = 0 \implies 2t = -5 \implies t = -\dfrac{5}{2}$

103. $75 - \sqrt{1.2(x - 10)} = 59.90$

$-\sqrt{1.2(x - 10)} = -15.1$

$\sqrt{1.2(x - 10)} = 15.1$

$1.2(x - 10) = 15.1^2$

$1.2x - 12 = 228.01$

$1.2x = 240.01$

$x = \dfrac{240.01}{1.2}$

$x \approx 200$

The demand is approximately 200 units.

Section 6.3   The Quadratic Formula

1. $2x^2 = 7 - 2x$
$$2x^2 + 2x - 7 = 0$$

3. $x(10 - x) = 5$
$$10x - x^2 = 5$$
$$-x^2 + 10x - 5 = 0 \text{ or } x^2 - 10x + 5 = 0$$

5. (a) $x^2 - 11x + 28 = 0$

$a = 1,\ b = -11,\ c = 28$

$x = \dfrac{-(-11) \pm \sqrt{(-11)^2 - 4(1)(28)}}{2(1)}$

$= \dfrac{11 \pm \sqrt{121 - 112}}{2} = \dfrac{11 \pm \sqrt{9}}{2} = \dfrac{11 \pm 3}{2}$

$x = \dfrac{11 + 3}{2} \implies x = 7$

$x = \dfrac{11 - 3}{2} \implies x = 4$

(b) $x^2 - 11x + 28 = 0$

$(x - 7)(x - 4) = 0$

$x - 7 = 0 \implies x = 7$

$x - 4 = 0 \implies x = 4$

7. (a) $4x^2 + 4x + 1 = 0$

$a = 4, \; b = 4, \; c = 1$

$$x = \frac{-4 \pm \sqrt{4^2 - 4(4)(1)}}{2(4)}$$

$$= \frac{-4 \pm \sqrt{0}}{8} = -\frac{4}{8} = -\frac{1}{2}$$

(b) $4x^2 + 4x + 1 = 0$

$(2x + 1)(2x + 1) = 0$

$2x + 1 = 0 \implies 2x = -1 \implies x = -\frac{1}{2}$

9. (a) $6x^2 - x - 2 = 0$

$a = 6, \; b = -1, \; c = -2$

$$x = \frac{-(-1) \pm \sqrt{(-1)^2 - 4(6)(-2)}}{2(6)}$$

$$= \frac{1 \pm \sqrt{1 + 48}}{12} = \frac{1 \pm \sqrt{49}}{12} = \frac{1 \pm 7}{12}$$

$x = \frac{1 + 7}{12} \implies x = \frac{2}{3}$

$x = \frac{1 - 7}{12} \implies x = -\frac{1}{2}$

(b) $6x^2 - x - 2 = 0$

$(3x - 2)(2x + 1) = 0$

$3x - 2 = 0 \implies 3x = 2 \implies x = \frac{2}{3}$

$2x + 1 = 0 \implies 2x = -1 \implies x = -\frac{1}{2}$

11. (a) $x^2 - 5x - 300 = 0$

$a = 1, \; b = -5, \; c = -300$

$$x = \frac{-(-5) \pm \sqrt{(-5)^2 - 4(1)(-300)}}{2(1)}$$

$$= \frac{5 \pm \sqrt{25 + 1200}}{2} = \frac{5 \pm \sqrt{1225}}{2} = \frac{5 \pm 35}{2}$$

$x = \frac{5 + 35}{2} \implies x = 20$

$x = \frac{5 - 35}{2} \implies x = -15$

(b) $x^2 - 5x - 300 = 0$

$(x - 20)(x + 15) = 0$

$x - 20 = 0 \implies x = 20$

$x + 15 = 0 \implies x = -15$

13. $a = 1, \; b = -2, \; c = -4$

$$x = \frac{-(-2) \pm \sqrt{(-2)^2 - 4(1)(-4)}}{2(1)}$$

$$= \frac{2 \pm \sqrt{4 + 16}}{2} = \frac{2 \pm \sqrt{20}}{2} = \frac{2 \pm 2\sqrt{5}}{2} = \frac{2\left(1 \pm \sqrt{5}\right)}{2} = 1 \pm \sqrt{5}$$

15. $a = 1, \; b = 4, \; c = 1$

$$t = \frac{-4 \pm \sqrt{4^2 - 4(1)(1)}}{2(1)}$$

$$= \frac{-4 \pm \sqrt{16 - 4}}{2} = \frac{-4 \pm \sqrt{12}}{2} = \frac{-4 \pm 2\sqrt{3}}{2} = \frac{2\left(-2 \pm \sqrt{3}\right)}{2} = -2 \pm \sqrt{3}$$

17. $a = 1, \; b = 6, \; c = -3$

$$x = \frac{-6 \pm \sqrt{6^2 - 4(1)(-3)}}{2(1)}$$

$$= \frac{-6 \pm \sqrt{36 + 12}}{2} = \frac{-6 \pm \sqrt{48}}{2} = \frac{-6 \pm 4\sqrt{3}}{2} = \frac{2\left(-3 \pm 2\sqrt{3}\right)}{2} = -3 \pm 2\sqrt{3}$$

19. $a = 1,\ b = -10,\ c = 23$

$$x = \frac{-(-10) \pm \sqrt{(-10)^2 - 4(1)(23)}}{2(1)}$$

$$= \frac{10 \pm \sqrt{100 - 92}}{2} = \frac{10 \pm \sqrt{8}}{2} = \frac{10 \pm 2\sqrt{2}}{2} = \frac{\cancel{2}(5 \pm \sqrt{2})}{\cancel{2}} = 5 \pm \sqrt{2}$$

21. $a = 1,\ b = 3,\ c = 3$

$$x = \frac{-3 \pm \sqrt{3^2 - 4(1)(3)}}{2(1)}$$

$$= \frac{-3 \pm \sqrt{9 - 12}}{2} = \frac{-3 \pm \sqrt{-3}}{2} = \frac{-3 \pm \sqrt{3}i}{2} \text{ or } x = -\frac{3}{2} \pm \frac{\sqrt{3}}{2}i$$

23. $a = 2,\ b = -2,\ c = -1$

$$v = \frac{-(-2) \pm \sqrt{(-2)^2 - 4(2)(-1)}}{2(2)}$$

$$= \frac{2 \pm \sqrt{4 + 8}}{4} = \frac{2 \pm \sqrt{12}}{4} = \frac{2 \pm 2\sqrt{3}}{4} = \frac{\cancel{2}(1 \pm \sqrt{3})}{\cancel{2}(2)} = \frac{1 \pm \sqrt{3}}{2}$$

25. $a = 2,\ b = 4,\ c = -3$

$$x = \frac{-4 \pm \sqrt{4^2 - 4(2)(-3)}}{2(2)}$$

$$= \frac{-4 \pm \sqrt{16 + 24}}{4} = \frac{-4 \pm \sqrt{40}}{4} = \frac{-4 \pm 2\sqrt{10}}{4} = \frac{\cancel{2}(-2 \pm \sqrt{10})}{\cancel{2}(2)} = \frac{-2 \pm \sqrt{10}}{2}$$

27. $a = 2,\ b = 4,\ c = 1$

$$x = \frac{-4 \pm \sqrt{4^2 - 4(2)(1)}}{2(2)}$$

$$x = \frac{-4 \pm \sqrt{16 - 8}}{4} = \frac{-4 \pm \sqrt{8}}{4} = \frac{-4 \pm 2\sqrt{2}}{4} = \frac{\cancel{2}(-2 \pm \sqrt{2})}{\cancel{2}(2)} = \frac{-2 \pm \sqrt{2}}{2}$$

29. $a = 3,\ b = -5,\ c = 3$

$$x = \frac{-(-5) \pm \sqrt{(-5)^2 - 4(3)(3)}}{2(3)}$$

$$x = \frac{5 \pm \sqrt{25 - 36}}{6} = \frac{5 \pm \sqrt{-11}}{6} = \frac{5 \pm \sqrt{11}i}{6} = \frac{5}{6} \pm \frac{\sqrt{11}}{6}i$$

31. $a = 4,\ b = -4,\ c = 5$

$$x = \frac{-(-4) \pm \sqrt{(-4)^2 - 4(4)(5)}}{2(4)}$$

$$x = \frac{4 \pm \sqrt{16 - 80}}{8} = \frac{4 \pm \sqrt{-64}}{8} = \frac{4 \pm 8i}{8} = \frac{\cancel{4}(1 \pm 2i)}{\cancel{4}(2)} = \frac{1}{2} \pm i$$

33. $a = 9,\ b = 6,\ c = -4$

$$z = \frac{-6 \pm \sqrt{6^2 - 4(9)(-4)}}{2(9)}$$

$$= \frac{-6 \pm \sqrt{36 + 144}}{18} = \frac{-6 \pm \sqrt{180}}{18} = \frac{-6 \pm 6\sqrt{5}}{18} = \frac{\cancel{6}(-1 \pm \sqrt{5})}{\cancel{6}(3)} = \frac{-1 \pm \sqrt{5}}{3}$$

35. $a = 1$, $b = -0.4$, $c = -0.16$

$$x = \frac{-(-0.4) \pm \sqrt{(-0.4)^2 - 4(1)(-0.16)}}{2(1)}$$

$$= \frac{0.4 \pm \sqrt{0.16 + 0.64}}{2} = \frac{0.4 \pm \sqrt{0.80}}{2} = \frac{0.4 \pm 0.4\sqrt{5}}{2} = \frac{\cancel{2}(0.2 \pm 0.2\sqrt{5})}{\cancel{2}} = 0.2 \pm 0.2\sqrt{5}$$

$$\left(\text{Note: } 0.2 \pm 0.2\sqrt{5} = \frac{2 \pm 2\sqrt{5}}{10} = \frac{1 \pm \sqrt{5}}{5}, \text{ so this solution could also be written as } x = \frac{1 \pm \sqrt{5}}{5}.\right)$$

37. $a = 2.5$, $b = 1$, $c = -0.9$

$$x = \frac{-1 \pm \sqrt{1^2 - 4(2.5)(-0.9)}}{2(2.5)} = \frac{-1 \pm \sqrt{1 + 9}}{5} = \frac{-1 \pm \sqrt{10}}{5}$$

39. $4x^2 - 6x + 3 = 0$

$a = 4$, $b = -6$, $c = 3$

$$x = \frac{-(-6) \pm \sqrt{(-6)^2 - 4(4)(3)}}{2(4)}$$

$$= \frac{6 \pm \sqrt{36 - 48}}{8} = \frac{6 \pm \sqrt{-12}}{8} = \frac{6 \pm 2\sqrt{3}i}{8} = \frac{\cancel{2}(3 \pm \sqrt{3}i)}{\cancel{2}(4)} = \frac{3 \pm \sqrt{3}i}{4} \text{ or } x = \frac{3}{4} \pm \frac{\sqrt{3}}{4}i$$

41. $9x^2 = 1 + 9x$

$9x^2 - 9x - 1 = 0$

$a = 9$, $b = -9$, $c = -1$

$$x = \frac{-(-9) \pm \sqrt{(-9)^2 - 4(9)(-1)}}{2(9)}$$

$$= \frac{9 \pm \sqrt{81 + 36}}{18} = \frac{9 \pm \sqrt{117}}{18} = \frac{9 \pm 3\sqrt{13}}{18} = \frac{\cancel{3}(3 \pm \sqrt{13})}{\cancel{3}(6)} = \frac{3 \pm \sqrt{13}}{6}$$

43. $a = 1$, $b = 1$, $c = 1$

$$b^2 - 4ac = 1^2 - 4(1)(1)$$

$$= 1 - 4 = -3$$

Because the discriminant is negative, the equation has two distinct imaginary solutions.

45. $a = 2$, $b = -5$, $c = -4$

$$b^2 - 4ac = (-5)^2 - 4(2)(-4)$$

$$= 25 + 32 = 57$$

Because the discriminant is positive but is not a perfect square, the equation has two distinct irrational solutions.

47. $a = 1$, $b = 7$, $c = 15$

$$b^2 - 4ac = 7^2 - 4(1)(15)$$

$$= 49 - 60 = -11$$

Because the discriminant is negative, the equation has two distinct imaginary solutions.

49. $a = 4$, $b = -12$, $c = 9$

$$b^2 - 4ac = (-12)^2 - 4(4)(9)$$

$$= 144 - 144 = 0$$

Because the discriminant is zero, the equation has one (repeated) rational solution.

51. $3x^2 - x + 2 = 0$

$a = 3$, $b = -1$, $c = 2$

$$b^2 - 4ac = (-1)^2 - 4(3)(2)$$

$$= 1 - 24$$

$$= -23$$

Because the discriminant is negative, the equation has two distinct imaginary solutions.

53. $z^2 - 169 = 0$

$z^2 = 169$

$z = \pm\sqrt{169} = \pm 13$

55. $y^2 + 15y = 0$

$y(y + 15) = 0$

$y = 0$

$y + 15 = 0 \implies y = -15$

57. $9t^2 + 25 = 0$

$9t^2 = -25$

$t^2 = -\dfrac{25}{9}$

$t = \pm\sqrt{-\dfrac{25}{9}}$

$t = \pm\dfrac{5}{3}i$

59.

$25(x - 3)^2 - 36 = 0$

$[5(x - 3) + 6][5(x - 3) - 6] = 0$

$(5x - 15 + 6)(5x - 15 - 6) = 0$

$(5x - 9)(5x - 21) = 0$

$5x - 9 = 0 \implies x = \dfrac{9}{5}$

$5x - 21 = 0 \implies x = \dfrac{21}{5}$

or $25(x - 3)^2 - 36 = 0$

$25(x - 3)^2 = 36$

$(x - 3)^2 = \dfrac{36}{25}$

$x - 3 = \pm\sqrt{\dfrac{36}{25}}$

$x = 3 \pm \dfrac{6}{5}$

$x = \dfrac{15}{5} + \dfrac{6}{5} = \dfrac{21}{5}$

$x = \dfrac{15}{5} - \dfrac{6}{5} = \dfrac{9}{5}$

61. $3x(x + 10) - 4(x + 10) = 0$

$(x + 10)(3x - 4) = 0$

$x + 10 = 0 \implies x = -10$

$3x - 4 = 0 \implies 3x = 4 \implies x = \dfrac{4}{3}$

63. $(x + 4)^2 + 16 = 0$

$(x + 4)^2 = -16$

$x + 4 = \pm\sqrt{-16} = \pm 4i$

$x = -4 \pm 4i$

65. $t^2 + 7t + 12 = 0$

$(t + 3)(t + 4) = 0$

$t + 3 = 0 \implies t = -3$

$t + 4 = 0 \implies t = -4$

67. $a = 18, \; b = 15, \; c = -50$

$x = \dfrac{-15 \pm \sqrt{15^2 - 4(18)(-50)}}{2(18)}$

$= \dfrac{-15 \pm \sqrt{225 + 3600}}{36} = \dfrac{-15 \pm \sqrt{3825}}{36} = \dfrac{-15 \pm 15\sqrt{17}}{36} = \dfrac{\cancel{3}(-5 \pm 5\sqrt{17})}{\cancel{3}(12)} = \dfrac{-5 \pm 5\sqrt{17}}{12}$

69. $x^2 - 24x + 128 = 0$

$(x - 16)(x - 8) = 0$

$x - 16 = 0 \implies x = 16$

$x - 8 = 0 \implies x = 8$

71. $1.2x^2 - 0.8x - 5.5 = 0$

$a = 1.2, b = -0.8, c = -5.5$

$$x = \frac{-(-0.8) \pm \sqrt{(-0.8)^2 - 4(1.2)(-5.5)}}{2(1.2)}$$

$$x = \frac{0.8 \pm \sqrt{0.64 + 26.4}}{2.4}$$

$$x = \frac{0.8 \pm \sqrt{27.04}}{2.4}$$

$$x = \frac{0.8 \pm 5.2}{2.4}$$

$$x = \frac{0.8 + 5.2}{2.4} = \frac{6}{2.4} = \frac{60}{24} = \frac{5}{2} \text{ or } x = 2.5$$

$$x = \frac{0.8 - 5.2}{2.4} = \frac{-4.4}{2.4} = -\frac{44}{24} = -\frac{11}{6} \text{ or } x = -1.8\overline{3}$$

or
$$1.2x^2 - 0.8x - 5.5 = 0$$
$$12x^2 - 8x - 55 = 0$$
$$(6x + 11)(2x - 5) = 0$$
$$6x + 11 = 0 \implies x = -\frac{11}{6}$$
$$2x - 5 = 0 \implies x = \frac{5}{2}$$

73. $y = 3x^2 - 6x + 1$

$3x^2 - 6x + 1 = 0$

$a = 3, b = -6, c = 1$

$$x = \frac{-(-6) \pm \sqrt{(-6)^2 - 4(3)(1)}}{2(3)}$$

$$= \frac{6 \pm \sqrt{36 - 12}}{6}$$

$$= \frac{6 \pm \sqrt{24}}{6}$$

$$= \frac{6 \pm 2\sqrt{6}}{6}$$

$$= \frac{2(3 \pm \sqrt{6})}{2(3)}$$

$$= \frac{3 \pm \sqrt{6}}{3}$$

$$x = \frac{3 + \sqrt{6}}{3} \approx 1.82$$

$$x = \frac{3 - \sqrt{6}}{3} \approx 0.18$$

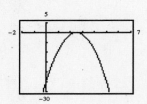

$(1.82, 0)$ and $(0.18, 0)$

The solutions of the equation are the x-coordinates of the x-intercepts.

75. $y = -(4x^2 - 20x + 25)$

$-(4x^2 - 20x + 25) = 0$

$4x^2 - 20x + 25 = 0$

$(2x - 5)(2x - 5) = 0$

$$2x - 5 = 0 \implies x = \frac{5}{2}$$

$\left(\frac{5}{2}, 0\right)$

The solution of the equation is the x-coordinate of the x-intercept.

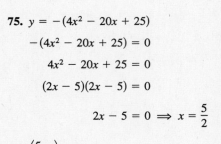

77. $y = 5x^2 - 18x + 6$

$5x^2 - 18x + 6 = 0$

$a = 5,\ b = -18,\ c = 6$

$$x = \frac{-(-18) \pm \sqrt{(-18)^2 - 4(5)(6)}}{2(5)}$$

$$= \frac{18 \pm \sqrt{324 - 120}}{10}$$

$$= \frac{18 \pm \sqrt{204}}{10}$$

$$= \frac{18 \pm 2\sqrt{51}}{10}$$

$$= \frac{2(9 \pm \sqrt{51})}{2(5)}$$

$$= \frac{9 \pm \sqrt{51}}{5}$$

$$x = \frac{9 + \sqrt{51}}{5} \approx 3.23$$

$$x = \frac{9 - \sqrt{51}}{5} \approx 0.37$$

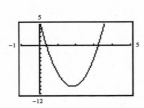

$(3.23, 0)$ and $(0.37, 0)$

The solutions of the equation are the x-coordinates of the x-intercepts.

79. $y = -0.04x^2 + 4x - 0.8$

$-0.04x^2 + 4x - 0.8 = 0$

$0.04x^2 - 4x + 0.8 = 0$

$a = 0.04,\ b = -4,\ c = 0.8$

$$x = \frac{-(-4) \pm \sqrt{(-4)^2 - 4(0.04)(0.8)}}{2(0.04)}$$

$$= \frac{4 \pm \sqrt{16 - 0.128}}{0.08}$$

$$= \frac{4 \pm \sqrt{15.872}}{0.08}$$

$$x = \frac{4 + \sqrt{15.872}}{0.08} \approx 99.80$$

$$x = \frac{4 - \sqrt{15.872}}{0.08} \approx 0.20$$

$(99.80, 0)$ and $(0.20, 0)$

The solutions of the equation are the x-coordinates of the x-intercepts.

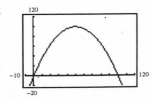

81. There are no real solutions; there are two imaginary solutions.

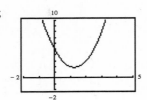

83. There are two real solutions.

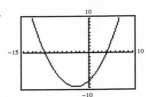

85. $$\frac{2x^2}{5} - \frac{x}{2} = 1$$

$$10\left(\frac{2x^2}{5} - \frac{x}{2}\right) = 10(1)$$

$$4x^2 - 5x = 10$$

$4x^2 - 5x - 10 = 0$

$a = 4,\ b = -5,\ c = -10$

$$x = \frac{-(-5) \pm \sqrt{(-5)^2 - 4(4)(-10)}}{2(4)}$$

$$= \frac{5 \pm \sqrt{25 + 160}}{8} = \frac{5 \pm \sqrt{185}}{8}$$

$$x = \frac{5 + \sqrt{185}}{8} \implies x \approx 2.33$$

$$x = \frac{5 - \sqrt{185}}{8} \implies x \approx -1.08$$

87. $\sqrt{x + 3} = x - 1$

$\left(\sqrt{x + 3}\right)^2 = (x - 1)^2$

$x + 3 = x^2 - 2x + 1$

$0 = x^2 - 3x - 2$

$a = 1,\ b = -3,\ c = -2$

$$x = \frac{-(-3) \pm \sqrt{(-3)^2 - 4(1)(-2)}}{2(1)} = \frac{3 \pm \sqrt{17}}{2}$$

$$x = \frac{3 + \sqrt{17}}{2} \implies x \approx 3.56$$

$$x = \frac{3 - \sqrt{17}}{2} \implies x \approx -0.56 \ \text{(Extraneous)}$$

—CONTINUED—

87. **—CONTINUED—**

Check: $\sqrt{3.56 + 3} \overset{?}{\approx} 3.56 - 1$

$\qquad \sqrt{6.56} \overset{?}{\approx} 2.56$

$\qquad 2.56 \approx 2.56$

Check: $\sqrt{-0.56 + 3} \overset{?}{\approx} -0.56 - 1$

$\qquad \sqrt{2.44} \overset{?}{\approx} -1.56$

$\qquad 1.56 \neq -1.56$

89.   (Length)(Width) = Area

$$(x + 6.3)(x) = 58.14$$

$$x^2 + 6.3x = 58.14$$

$$x^2 + 6.3x - 58.14 = 0$$

$$a = 1, \ b = 6.3, \ c = -58.14$$

$$x = \frac{-6.3 \pm \sqrt{(6.3)^2 - 4(1)(-58.14)}}{2(1)}$$

$$= \frac{-6.3 \pm \sqrt{39.69 + 232.56}}{2} = \frac{-6.3 \pm \sqrt{272.25}}{2}$$

$$x = \frac{-6.3 + \sqrt{272.25}}{2} = 5.1 \ \text{and} \ x + 6.3 = 11.4$$

$$x = \frac{-6.3 - \sqrt{272.25}}{2} = -11.4$$

(Discard negative solution.)

The dimensions of the rectangle are 5.1 centimeters by 11.4 centimeters.

91. (a)

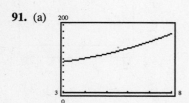

(c) The graph verifies the year of 1994.

(b) $y = 81.8 - 1.64t + 1.648t^2$

$$1.648t^2 - 1.64t + 81.8 = 100$$

$$1.648t^2 - 1.64t - 18.2 = 0$$

$$a = 1.648, \ b = -1.64, \ c = -18.2$$

$$t = \frac{-(-1.64) \pm \sqrt{(-1.64)^2 - 4(1.648)(-18.2)}}{2(1.648)}$$

$$t = \frac{1.64 \pm \sqrt{122.664}}{3.296} \quad \text{(Discard negative solution.)}$$

$$t = \frac{1.64 + \sqrt{122.664}}{3.296} \Rightarrow t \approx 4$$

The number of scientists employed in this industry was approximately 100,000 in 1994.

93. The solutions of the quadratic equation $ax^2 + bx + c = 0$ are the opposite of b, plus or minus the square root of b squared minus $4ac$, all divided by $2a$.

95. Discriminant: $b^2 - 4ac = (-6)^2 - 4(1)(c) = 36 - 4c$

(a) Two real-number solutions $\Rightarrow$ Discriminant > 0

$$36 - 4c > 0$$

$$-4c > -36$$

$$c < 9$$

(b) One real-number solution $\Rightarrow$ Discriminant $= 0$

$$36 - 4c = 0$$

$$-4c = -36$$

$$c = 9$$

(c) Two imaginary-number solutions $\Rightarrow$ Discriminant > 0

$$36 - 4c < 0$$

$$-4c < -36$$

$$c > 9$$

The equation has two real-number solutions when c is less than 9, one real-number solution when c equals 9, and two imaginary-number solutions when c is greater then 9.

97. Discriminant: $b^2 - 4ac = 8^2 - 4(1)(c) = 64 - 4c$

(a) Two real-number solutions $\Rightarrow$ Discriminant > 0

$$64 - 4c > 0$$
$$-4c > -64$$
$$c < 16$$

(b) One real-number solution $\Rightarrow$ Discriminant $= 0$

$$64 - 4c = 0$$
$$-4c = -64$$
$$c = 16$$

(c) Two imaginary-number solutions $\Rightarrow$ Discriminant < 0

$$64 - 4c < 0$$
$$-4c < -64$$
$$c > 16$$

The equation has two real-number solutions when c is less than 16, one real-number solution when c equals 16, and two imaginary-number solutions when c is greater then 16.

99. $x^2 + 6x + 8 = 0$

$(x + 2)(x + 4) = 0$

$$x + 2 = 0 \Rightarrow x = -2$$
$$x + 4 = 0 \Rightarrow x = -4$$

101. $4x^2 + 12x + 9 = 0$

$(2x + 3)(2x + 3) = 0$

$$2x + 3 = 0 \Rightarrow 2x = -3 \Rightarrow x = -\frac{3}{2}$$

103. $x^2 + 7x - 2 = 0$

$$x^2 + 7x = 2$$
$$x^2 + 7x + \left(\frac{7}{2}\right)^2 = 2 + \frac{49}{4}$$
$$\left(x + \frac{7}{2}\right)^2 = \frac{57}{4}$$
$$x + \frac{7}{2} = \pm\sqrt{\frac{57}{4}}$$
$$x = \frac{-7 \pm \sqrt{57}}{2}$$

105. $\sqrt{16 - 4(3)(1)} = \sqrt{4} = 2$

107. $\sqrt{9 - 4(-1)(3)} = \sqrt{21}$

109. $\begin{array}{rcl} x + y = 42 & \Rightarrow & -5x - 5y = -210 \\ 8x + 5y = 246 & \Rightarrow & \underline{8x + 5y = 246} \\ & & 3x = 36 \\ & & x = 12 \end{array}$

$$12 + y = 42$$
$$y = 30$$

There are 12 gold coins and 30 silver coins.

Mid-Chapter Quiz for Chapter 6

1.
$$2x^2 - 72 = 0$$
$$2(x^2 - 36) = 0$$
$$2(x + 6)(x - 6) = 0$$
$$2 \neq 0$$
$$x + 6 = 0 \implies x = -6$$
$$x - 6 = 0 \implies x = 6$$

2. $2x^2 + 3x - 20 = 0$
$$(2x - 5)(x + 4) = 0$$
$$2x - 5 = 0 \implies x = \tfrac{5}{2}$$
$$x + 4 = 0 \implies x = -4$$

3. $t^2 = 12$
$$t = \pm\sqrt{12}$$
$$t = \pm\sqrt{4 \cdot 3}$$
$$t = \pm 2\sqrt{3}$$

4. $(u - 3)^2 - 16 = 0$
$$(u - 3)^2 = 16$$
$$u - 3 = \pm\sqrt{16}$$
$$u - 3 = \pm 4$$
$$u = 3 \pm 4$$
$$u = 3 + 4 \implies u = 7$$
$$u = 3 - 4 \implies u = -1$$

5. $s^2 + 10s + 1 = 0$
$$s^2 + 10s = -1$$
$$s^2 + 10s + 5^2 = -1 + 5^2$$
$$(s + 5)^2 = 24$$
$$x + 5 = \pm\sqrt{24}$$
$$x + 5 = \pm\sqrt{4 \cdot 6}$$
$$x + 5 = \pm 2\sqrt{6}$$
$$x = -5 \pm 2\sqrt{6}$$

6. $2y^2 + 6y - 5 = 0$
$$y^2 + 3y - \frac{5}{2} = 0$$
$$y^2 + 3y = \frac{5}{2}$$
$$y^2 + 3y + \left(\frac{3}{2}\right)^2 = \frac{5}{2} + \left(\frac{3}{2}\right)^2$$
$$\left(y + \frac{3}{2}\right)^2 = \frac{10}{4} + \frac{9}{4}$$
$$y + \frac{3}{2} = \pm\sqrt{\frac{19}{4}}$$
$$y = -\frac{3}{2} \pm \frac{\sqrt{19}}{2} \text{ or } y = \frac{-3 \pm \sqrt{19}}{2}$$

7. $x^2 + 4x - 6 = 0$
$$a = 1,\ b = 4,\ c = -6$$
$$x = \frac{-4 \pm \sqrt{4^2 - 4(1)(-6)}}{2(1)}$$
$$= \frac{-4 \pm \sqrt{40}}{2}$$
$$= \frac{-4 \pm 2\sqrt{10}}{2}$$
$$= \frac{\cancel{2}\left(-2 \pm \sqrt{10}\right)}{\cancel{2}(1)}$$
$$= -2 \pm \sqrt{10}$$

8. $6v^2 - 3v - 4 = 0$
$$a = 6,\ b = -3,\ c = -4$$
$$v = \frac{-(-3) \pm \sqrt{(-3)^2 - 4(6)(-4)}}{2(6)}$$
$$= \frac{3 \pm \sqrt{105}}{12}$$

9. $x^2 + 5x + 7 = 0$

$a = 1, \ b = 5, \ c = 7$

$x = \dfrac{-5 \pm \sqrt{5^2 - 4(1)(7)}}{2(1)}$

$= \dfrac{-5 \pm \sqrt{-3}}{2}$

$= -\dfrac{5}{2} \pm \dfrac{\sqrt{3}i}{2}$

10. $36 - (t - 4)^2 = 0$

$-(t - 4)^2 = -36$

$(t - 4)^2 = 36$

$t - 4 = \pm\sqrt{36}$

$t - 4 = \pm 6$

$t = 4 \pm 6$

$t = 4 + 6 \implies t = 10$

$t = 4 - 6 \implies t = -2$

11. $x(x - 10) + 3(x - 10) = 0$

$(x - 10)(x + 3) = 0$

$x - 10 = 0 \implies x = 10$

$x + 3 = 0 \implies x = -3$

12. $x(x - 3) = 10$

$x^2 - 3x = 10$

$x^2 - 3x - 10 = 0$

$(x - 5)(x + 2) = 0$

$x - 5 = 0 \implies x = 5$

$x + 2 = 0 \implies x = -2$

13. $4b^2 - 12b + 9 = 0$

$(2b - 3)(2b - 3) = 0$

$2b - 3 = 0 \implies b = \dfrac{3}{2}$

14. $3m^2 + 10m + 5 = 0$

$a = 3, \ b = 10, \ c = 5$

$m = \dfrac{-10 \pm \sqrt{10^2 - 4(3)(5)}}{2(3)}$

$= \dfrac{-10 \pm \sqrt{40}}{6}$

$= \dfrac{-10 \pm 2\sqrt{10}}{6}$

$= \dfrac{2\left(-5 \pm \sqrt{10}\right)}{2(3)}$

$= \dfrac{-5 \pm \sqrt{10}}{3}$

15. $x^4 + 5x^2 - 14 = 0$ Let $u = x^2$.

$u^2 + 5u - 14 = 0$

$(u + 7)(u - 2) = 0$

$u + 7 = 0 \implies u = -7$

$u - 2 = 0 \implies u = 2$

$u = -7 \implies x^2 = -7$

$x = \pm\sqrt{-7}$

$x = \pm\sqrt{7}i$

$u = 2 \implies x^2 = 2$

$x = \pm\sqrt{2}$

16. $x^{2/3} - 8x^{13} + 15 = 0$ Let $u = x^{1/3}$.

$u^2 - 8u + 15 = 0$

$(u - 3)(u - 5) = 0$

$u - 3 = 0 \implies u = 3$

$u - 5 = 0 \implies u = 5$

$u = 3 \implies x^{1/3} = 3$

$x = 3^3$

$x = 27$

$u = 5 \implies x^{1/3} = 5$

$x = 5^3$

$x = 125$

17. $y = \frac{1}{2}x^2 - 3x - 1$

$\frac{1}{2}x^2 - 3x - 1 = 0$

$x^2 - 6x - 2 = 0$

$a = 1,\ b = -6,\ c = -2$

$x = \dfrac{-(-6) \pm \sqrt{(-6)^2 - 4(1)(-2)}}{2(1)}$

$= \dfrac{6 \pm \sqrt{44}}{2}$

$= \dfrac{6 \pm 2\sqrt{11}}{2}$

$= \dfrac{2(3 \pm \sqrt{11})}{2(1)}$

$= 3 \pm \sqrt{11}$

$x = 3 + \sqrt{11} \Rightarrow x \approx 6.32$

$x = 3 - \sqrt{11} \Rightarrow x \approx -0.32$

$(6.32, 0)$ and $(-0.32, 0)$

The solutions are the x-coordinates of the x-intercepts.

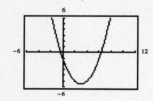

18. $y = x^2 + 0.45x - 4$

$x^2 + 0.45x - 4 = 0$

$a = 1,\ b = 0.45,\ c = -4$

$x = \dfrac{-0.45 \pm \sqrt{(0.45)^2 - 4(1)(-4)}}{2(1)}$

$= \dfrac{-0.45 \pm \sqrt{16.2025}}{2}$

$x = \dfrac{-0.45 + \sqrt{16.2025}}{2} \Rightarrow x \approx 1.79$

$x = \dfrac{-0.45 - \sqrt{16.2025}}{2} \Rightarrow x \approx -2.24$

$(1.79, 0)$ and $(-2.24, 0)$

The solutions are the x-coordinates of the x-intercepts.

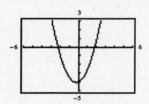

19. $\qquad R = x(20 - 0.2x)$ and $R = \$500$

$500 = x(20 - 0.2x)$

$500 = 20x - 0.2x^2$

$0.2x^2 - 20x + 500 = 0$

$x^2 - 100x + 2500 = 0$

$(x - 50)(x - 50) = 0$

$x - 50 = 0 \Rightarrow x = 50$

To produce a revenue of \$500, 50 units must be sold.

20. $x^2 + 28^2 = 34^2$

$x^2 = 34^2 - 28^2$

$x^2 = 1156 - 784$

$x^2 = 372$

$x = \pm\sqrt{372}$ (Discard negative solution.)

$x = 2\sqrt{93} \Rightarrow x \approx 19.29$

21. $x^2 + 15^2 = 26^2$

$x^2 = 26^2 - 15^2$

$x^2 = 676 - 225$

$x^2 = 451$

$x = \pm\sqrt{451}$ (Discard negative solution.)

$x = \sqrt{451} \Rightarrow x \approx 21.24$

Section 6.4   Applications of Quadratic Equations

1. *Verbal Model:* $\boxed{\begin{array}{c}\text{Smaller positive}\\\text{consecutive integer}\end{array}} \cdot \boxed{\begin{array}{c}\text{Larger positive}\\\text{consecutive integer}\end{array}} = 10 \boxed{\begin{array}{c}\text{Smaller positive}\\\text{consecutive integer}\end{array}} - 8$

    Labels: Smaller positive consecutive integer $= n$

                Larger positive consecutive integer $= n + 1$

    Equation:

$$n(n + 1) = 10n - 8$$
$$n^2 + n = 10n - 8$$
$$n^2 - 9n + 8 = 0$$
$$(n - 8)(n - 1) = 0$$
$$n - 8 = 0 \implies n = 8 \text{ and } n + 1 = 9$$
$$n - 1 = 0 \implies n = 1 \text{ and } n + 1 = 2$$

The consecutive integers are 8 and 9 or 1 and 2.

3. *Verbal Model:* $\boxed{\begin{array}{c}\text{Smaller positive}\\\text{consecutive even integer}\end{array}} \cdot \boxed{\begin{array}{c}\text{Larger positive}\\\text{consecutive even integer}\end{array}} = 3 \boxed{\begin{array}{c}\text{Larger positive}\\\text{consecutive even integer}\end{array}} + 50$

    Labels: Smaller positive consecutive integer $= n$

                Larger positive consecutive integer $= n + 2$

    Equation:

$$n(n + 2) = 3(n + 2) + 50$$
$$n^2 + 2n = 3n + 6 + 50$$
$$n^2 - n = 56$$
$$n^2 - n - 56 = 0$$
$$(n - 8)(n + 7) = 0$$
$$n - 8 = 0 \implies n = 8 \text{ and } n + 2 = 10$$
$$n + 7 = 0 \implies n = -7 \quad \text{(Discard negative solution.)}$$

The consecutive even integers are 8 and 10.

5. *Verbal Model:* $\boxed{\text{Length}} \cdot \boxed{\text{Width}} = \boxed{\text{Area}}$

    Labels: Width $= w$ (centimeters)

                Length $= 2.5w$ (centimeters)

    Equation:

$$(2.5w)(w) = 250$$
$$2.5w^2 = 250$$
$$w^2 = 100$$
$$w = \pm\sqrt{100} \quad \text{(Choose the positive solution.)}$$
$$w = 10 \quad \text{and} \quad 2.5w = 25$$

    Verbal Model: $\boxed{\text{Perimeter}} = 2\boxed{\text{Length}} + 2\boxed{\text{Width}}$

    Equation: $P = 2(25) + 2(10) = 50 + 20 = 70$

The width is 10 centimeters, the length is 25 centimeters, and the perimeter is 70 centimeters.

7. *Verbal Model:* | Length | · | Width | = | Area |

Labels: Length $= l$ (miles)

Width $= l - 6$ (miles)

Equation: $(1)(1 - 6) = 720$

$$l^2 - 6l = 720$$

$$l^2 - 6l - 720 = 0$$

$$l = \frac{6 \pm \sqrt{6^2 - 4(1)(-720)}}{2(1)}$$

$$l = \frac{6 \pm \sqrt{2916}}{2}$$

$$l = \frac{6 \pm 54}{2}$$ (Choose the positive solution.)

$$l = \frac{60}{2}$$

$$l = 30 \quad \text{and} \quad l - 6 = 24$$

Verbal Model: | Perimeter | = 2 | Length | + 2 | Width |

Equation: $P = 2(30) + 2(24) = 60 + 48 = 108$

The length is 30 miles, the width is 24 miles, and the perimeter is 108 miles.

9. *Verbal Model:* | Length | · | Width | = | Area |

Labels: Width $= x$ (feet)

Length $= 50 - 2x$ (feet)

Equation: $(50 - 2x)(x) = 312$

$$50x - 2x^2 = 312$$

$$-2x^2 + 50x - 312 = 0$$

$$-2(x^2 - 25x + 156) = 0$$

$$-2(x - 12)(x - 13) = 0$$

$$x - 12 = 0 \implies x = 12 \quad \text{and} \quad 50 - 2x = 50 - 24 = 26$$

$$x - 13 = 0 \implies x = 13 \quad \text{and} \quad 50 - 2x = 50 - 26 = 24$$

The dimensions of the pen could be 12 feet by 26 feet or it could be 13 feet by 24 feet.

11. $2l + 2w = 100$

$\quad\quad 2w = 100 - 2l$

$\quad\quad\quad w = 50 - l$

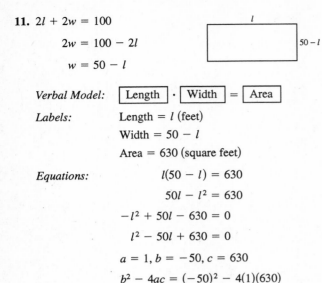

Verbal Model: $\boxed{\text{Length}} \cdot \boxed{\text{Width}} = \boxed{\text{Area}}$

Labels: Length $= l$ (feet)

$\quad\quad\quad$ Width $= 50 - l$

$\quad\quad\quad$ Area $= 630$ (square feet)

Equations:

$$l(50 - l) = 630$$

$$50l - l^2 = 630$$

$$-l^2 + 50l - 630 = 0$$

$$l^2 - 50l + 630 = 0$$

$$a = 1, b = -50, c = 630$$

$$b^2 - 4ac = (-50)^2 - 4(1)(630)$$

$$= 2500 - 2520$$

$$= -20$$

Because the discriminant is negative, the equation has *no real* solution. Thus, there is *not* enough fencing to enclose a rectangular region of 630 square feet.

Verbal Model: $\pi(\boxed{\text{Radius}})^2 = \boxed{\text{Area}}$

Labels: Radius $= r$ (feet)

$\quad\quad\quad$ Area $= 630$ (square feet)

Equations: $\pi r^2 = 630$

$$r^2 = \frac{630}{\pi}$$

$$r = \pm\sqrt{\frac{630}{\pi}} \quad \text{(Choose the positive solution.)}$$

$$r \approx 14.2$$

$2\pi r = \text{Circumference}$

$2\pi(14.2) \approx 89.2$

A circle with an area of 630 square feet has a circumference of approximately 89.2 feet. Thus, the 100 feet of fencing is enough to enclose a circular region of 630 square feet.

13. *Verbal Model:* $\boxed{\text{Area of first circle}} + \boxed{\text{Area of second circle}} = \boxed{\text{Total area}}$

Equation:

$$\pi r^2 + \pi(7 - r)^2 = 29\pi$$

$$\pi r^2 + \pi(7 - r)^2 = 29\pi$$

$$\pi r^2 + \pi(49 - 14r + r^2) - 29\pi = 0$$

$$2\pi r^2 - 14\pi r + 20\pi = 0$$

$$2\pi(r^2 - 7r + 10) = 0$$

$$2\pi(r - 2)(r - 5) = 0$$

$$r - 2 = 0 \Rightarrow r = 2 \quad \text{and} \quad 7 - r = 5$$

$$r - 5 = 0 \Rightarrow r = 5 \quad \text{and} \quad 7 - r = 2$$

The radii of the two circles are 5 yards and 2 yards.

15. (a) $d = \sqrt{(x + 3)^2 + (x + 4)^2}$

(b)

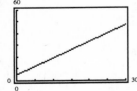

(c) From the graph it appears that when $d = 10$, x is approximately 3.55.

—CONTINUED—

15. —CONTINUED—

(d) $\quad d = \sqrt{(x+4)^2 + (x+3)^2}$

$\qquad 10 = \sqrt{x^2 + 8x + 16 + x^2 + 6x + 9}$

$\qquad 10 = \sqrt{2x^2 + 14x + 25}$

$\qquad (10)^2 = \left(\sqrt{2x^2 + 14x + 25}\right)^2$

$\qquad 100 = 2x^2 + 14x + 25$

$\qquad 0 = 2x^2 + 14x - 75$

$\qquad a = 2, b = 14, c = -75$

$$x = \frac{-14 \pm \sqrt{14^2 - 4(2)(-75)}}{2(2)}$$

$$x = \frac{-14 \pm \sqrt{196 + 600}}{4}$$

$$x = \frac{-14 \pm \sqrt{796}}{4}$$

$$x = \frac{-14 \pm 2\sqrt{199}}{4}$$

$$x = \frac{\cancel{2}\left(-7 \pm \sqrt{199}\right)}{\cancel{2}(2)}$$

$$x = \frac{-7 \pm \sqrt{199}}{2}$$

$$x = \frac{-7 + \sqrt{199}}{2} \approx 3.55$$

$$x = \frac{-7 - \sqrt{199}}{2} \approx -10.55 \quad \text{(Discard negative solution.)}$$

When $d = 10$, x is approximately 3.55 feet.

17. *Verbal Model:* $\left(\boxed{\begin{array}{c}\text{Distance from}\\\text{A to B}\end{array}}\right)^2 + \left(\boxed{\begin{array}{c}\text{Distance from}\\\text{B to C}\end{array}}\right)^2 = \left(\boxed{\begin{array}{c}\text{Distance from}\\\text{C to A}\end{array}}\right)^2$

Labels: $\qquad$ Distance from A to B $= x$ (miles)

$\qquad\qquad$ Distance from B to C $= 18 - x$ (miles) *(Note: $x > 18 - x$)*

$\qquad\qquad$ Distance from C to A $= 16$ (miles)

Equation: $\qquad x^2 + (18 - x)^2 = 16^2$

$\qquad\qquad x^2 + 324 - 36x + x^2 = 256$

$\qquad\qquad\quad 2x^2 - 36x + 68 = 0$

$\qquad\qquad\qquad x^2 - 18x + 34 = 0$

$\qquad a = 1, \ b = -18, \ c = 34$

$$x = \frac{-(-18) \pm \sqrt{(-18)^2 - 4(1)(34)}}{2(1)} = \frac{18 \pm \sqrt{188}}{2} = \frac{18 \pm 2\sqrt{47}}{2} = 9 \pm \sqrt{47}$$

$\qquad x = 9 + \sqrt{47} \approx 15.86$

$\qquad x = 9 - \sqrt{47} \approx 2.14$

Thus, the distance from A to B is approximately 15.86 miles or 2.14 miles.

19. *Verbal Model:* $\frac{1}{2}$ ⬚Base⬚ · ⬚Height⬚ = ⬚Area⬚

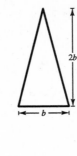

    Labels: Base $= b$ (inches)

                 Height $= 2b$ (inches)

                 Area $= 625$ (square inches)

    Equations: $\frac{1}{2}b(2b) = 625$

                  $b^2 = 625$

                  $b = \pm\sqrt{625}$ (Choose the positive solution.)

                  $b = 25$ and $2b = 50$

The base of the triangle is 25 inches and the height is 50 inches.

21. *Verbal Model:*   ⬚Length⬚ · ⬚Width⬚ = ⬚Area⬚

    Equation: $(2x)\left(\dfrac{200-4x}{3}\right) = 1400$

               $3(2x)\left(\dfrac{200-4x}{3}\right) = 3(1400)$

                    $2x(200-4x) = 4200$

                      $400x - 8x^2 = 4200$

              $-8x^2 + 400x - 4200 = 0$

              $-8(x^2 - 50x + 525) = 0$

              $-(x - 35)(x - 15) = 0$

                          $x - 35 = 0 \Longrightarrow 35$ and $\dfrac{200 - 4(35)}{3} = 20$

                          $x - 15 = 0 \Longrightarrow 15$ and $\dfrac{200 - 4(15)}{3} \approx 46.67$

The dimensions of each corral are 35 meters by 20 meters or 15 meters by approximately 46.67 meters.

23. *Verbal Model:* $a^2 + b^2 = c^2$ (Pythagorean Theorem)

    Equation: $x^2 + 15^2 = 75^2$

             $x^2 + 225 = 5625$

                   $x^2 = 5400$

                    $x = \pm\sqrt{5400}$ (Choose positive answer.)

                    $x = \sqrt{36(150)}$

                    $x = 6\sqrt{150}$

                    $x \approx 73.48$

The distance from the boat to the dock is approximately 73.48 feet.

25.
$$A = P(1 + r)^2$$
$$3499.20 = 3000(1 + r)^2$$
$$\frac{3499.20}{3000} = (1 + r)^2$$
$$1.1664 = (1 + r)^2$$
$$\sqrt{1.1664} = 1 + r$$
$$1.08 = 1 + r$$
$$0.08 = r$$

The interest rate is 8%.

27.
$$A = P(1 + r)^2$$
$$280.90 = 250(1 + r)^2$$
$$\frac{280.90}{250} = (1 + r)^2$$
$$1.1236 = (1 + r)^2$$
$$\sqrt{1.1236} = 1 + r$$
$$1.06 = 1 + r$$
$$0.06 = r$$

The interest rate is 6%.

29.
$$A = P(1 + r)^2$$
$$8420.20 = 8000(1 + r)^2$$
$$\frac{8420.20}{8000} = (1 + r)^2$$
$$1.0525 \approx (1 + r)^2$$
$$\sqrt{1.0525} \approx 1 + r$$
$$1.0259 \approx 1 + r$$
$$0.0259 \approx r$$

The interest rate is approximately 2.59%.

31.
$$h = h_0 - 16t^2$$
$$0 = 144 - 16t^2$$
$$= 16(9 - t^2) = 16(3 + t)(3 - t)$$
$$16 \neq 0$$
$$3 + t = 0 \implies t = -3 \quad \text{(Discard negative solution.)}$$
$$3 - t = 0 \implies t = 3$$

Thus, it takes 3 seconds for the object to fall to ground level.

33.
$$h = h_0 - 16t^2$$
$$0 = 1454 - 16t^2$$
$$16t^2 = 1454$$
$$t^2 = \frac{1454}{16}$$
$$t = \pm\frac{\sqrt{1454}}{4} \quad \text{(Choose the positive solution.)}$$
$$t = \frac{\sqrt{1454}}{4} \approx 9.5$$

Thus, it takes approximately 9.5 seconds for the object to fall to ground level.

35. $h = -16t^2 + 40t + 50$

(a)
$$-16t^2 + 40t + 50 = 50$$
$$-16t^2 + 40t = 0$$
$$-8t(2t - 5) = 0$$
$$-8t = 0 \implies t = 0$$
$$2t - 5 = 0 \implies 2t = 5 \implies t = \frac{5}{2}$$

The ball is again 50 feet above the water in 2.5 seconds.

(b)
$$-16t^2 + 40t + 50 = 0$$
$$t = \frac{-40 \pm \sqrt{40^2 - 4(-16)(50)}}{2(-16)}$$
$$t = \frac{-40 \pm \sqrt{4800}}{-32}$$
$$t = \frac{-40 \pm 40\sqrt{3}}{-32}$$
$$t = \frac{-8(5 \pm 5\sqrt{3})}{-8(4)}$$
$$t = \frac{5 \pm 5\sqrt{3}}{4} \quad \text{(Choose positive answer.)}$$
$$t \approx 3.42$$

The ball strikes the water in approximately 3.42 seconds.

37. Revenue − Cost = Profit

$R = x(90 - x)$

$C = 100 + 30x$

$P = \$800$

$x(90 - x) - (100 + 30x) = 800$

$90x - x^2 - 100 - 30x = 800$

$-x^2 + 60x - 900 = 0$

$x^2 - 60x + 900 = 0$

$(x - 30)(x - 30) = 0$

$x - 30 = 0 \Rightarrow x = 30$

The profit will be $800 when the number of units $x = 30$.

41. Answers will vary.
One strategy for solving word problems involves writing a verbal model, assigning labels to each part of the verbal model, using the labels to write an algebraic model, solving the resulting equation, answering the original equation, and checking that your answer satisfies the original problem as stated.

45. $2(x - 3) - 5 = 4(x - 5)$

$2x - 6 - 5 = 4x - 20$

$2x - 11 = 4x - 20$

$-2x - 11 = -20$

$-2x = -9$

$x = \dfrac{-9}{-2}$

$x = \dfrac{9}{2}$

49. $2(x + 4)^2 = 242$

$(x + 4)^2 = 121$

$x + 4 = \pm\sqrt{121}$

$x + 4 = \pm 11$

$x = -4 \pm 11$

$x = -4 + 11 \Rightarrow x = 7$

$x = -4 - 11 \Rightarrow x = -15$

53. $2w^2 = 5$

$w^2 = \dfrac{5}{2}$

$w = \pm\sqrt{\dfrac{5}{2}}$

$w = \pm\dfrac{\sqrt{5}}{\sqrt{2}} \cdot \dfrac{\sqrt{2}}{\sqrt{2}}$

$w = \pm\dfrac{\sqrt{10}}{2}$

39. Distance from station to farthest listener = r, the radius of the circle

Area of circle = πr^2

$25,000 = \pi r^2$

$\dfrac{25,000}{\pi} = r^2$

$7957.7 \approx r^2$

$\sqrt{7957.7} \approx r$

$89.2 \approx r$

The station is approximately 89.2 miles from its farthest listener.

43. $\dfrac{9 \text{ dollars}}{\text{hour}} \cdot (20 \text{ hours}) = 180 \text{ dollars}$

The units of the product are dollars.

47. $3t(t - 7) + 5(t - 7) = 0$

$(t - 7)(3t + 5) = 0$

$t - 7 = 0 \Rightarrow t = 7$

$3t + 5 = 0 \Rightarrow 3t = -5 \Rightarrow t = -\dfrac{5}{3}$

51. $v^2 - 5v - 24 = 0$

$(v - 8)(v + 3) = 0$

$v - 8 = 0 \Rightarrow v = 8$

$v + 3 = 0 \Rightarrow v = -3$

55. $6a + 2c = 2528 \Rightarrow \quad 6a + 2c = 2528$

$a + c = 454 \Rightarrow \underline{-2a - 2c = -908}$

$ 4a = 1620$

$ a = 405$

$405 + c = 454 \Rightarrow c = 49$

So, 405 adult tickets and 49 children's tickets were sold.

Section 6.5   Graphs of Quadratic Functions

1. Graph (b)

3. $y = (x + 1)^2 - 3$

Graph (e)

5. Graph (d)

7. $y = 2(x - 0)^2 + 2$

$a = 2$

Because $a > 0$, the parabola opens upward.
Vertex (h, k): $(0, 2)$

9. $f(x) = -4(x + 2)^2 - 3$

$a = -4$

Because $a < 0$, the parabola opens downward.
Vertex (h, k): $(-2, -3)$

11. $y = 4 - (x - 10)^2$ or $y = -(x - 10)^2 + 4$

$a = -1$

Because $a < 0$, the parabola opens downward.
Vertex (h, k): $(10, 4)$

13. $y = x^2 - 6$ or $y = 1(x - 0)^2 - 6$

$a = 1$

Because $a > 0$, the parabola opens upward.
Vertex (h, k): $(0, -6)$

15. $y = -(x - 3)^2$

$\quad = -1(x - 3)^2 + 0$

$a = -1$

Because $a < 0$, the parabola opens downward.
Vertex (h, k): $(3, 0)$

17. $f(x) = x^2 + 2 = (x - 0)^2 + 2$

Vertex (h, k): $(0, 2)$

19. $y = x^2 - 4x + 7$

$\quad = (x^2 - 4x + (-2)^2 - (-2)^2) + 7 = (x^2 - 4x + 4) - 4 + 7 = (x - 2)^2 + 3$

Vertex (h, k): $(2, 3)$

21. $y = x^2 + 6x + 5$

$\quad = x^2 + 6x + 3^2 - 3^2 + 5 = (x^2 + 6x + 9) - 9 + 5 = (x + 3)^2 - 4$

Vertex (h, k): $(-3, -4)$

23. $y = -x^2 + 6x - 10$

$y = -1(x^2 - 6x) - 10$

$\quad = -1(x^2 - 6x + (-3)^2 - (-3)^2) - 10$

$\quad = -1(x^2 - 6x + 9) + 9 - 10$

$\quad = -1(x - 3)^2 - 1$

Vertex (h, k): $(3, -1)$

25. $y = -x^2 + 2x - 7$

$\quad = -1(x^2 - 2x) - 7$

$\quad = -1(x^2 - 2x + (-1)^2 - (-1)^2) - 7$

$\quad = -1(x^2 - 2x + 1) + 1 - 7$

$\quad = -1(x - 1)^2 - 6$ or $= -(x - 1)^2 - 6$

Vertex (h, k): $(1, -6)$

27. $y = 2x^2 + 6x + 2$

$= 2(x^2 + 3x) + 2$

$= 2\left[x^2 + 3x + \left(\dfrac{3}{2}\right)^2 - \left(\dfrac{3}{2}\right)^2\right] + 2$

$= 2\left(x^2 + 3x + \dfrac{9}{4}\right) - 2\left(\dfrac{9}{4}\right) + 2$

$= 2\left(x + \dfrac{3}{2}\right)^2 - \dfrac{9}{2} + \dfrac{4}{2}$

$= 2\left(x + \dfrac{3}{2}\right)^2 - \dfrac{5}{2}$

Vertex (h, k): $\left(-\dfrac{3}{2}, -\dfrac{5}{2}\right)$

29. $y = x^2 - 10x$

$x = \dfrac{-b}{2a} = \dfrac{-(-10)}{2(1)} = \dfrac{10}{2} = 5$

$y = 5^2 - 10(5) = 25 - 50 = -25$

Vertex: $(5, -25)$

31. $g(x) = x^2 - 10$

$x = \dfrac{-b}{2a} = \dfrac{-0}{2(1)} = \dfrac{0}{2} = 0$

$g(0) = 0^2 - 10 = 0 - 10 = -10$

Vertex: $(0, -10)$

33. $y = -x^2 + 6x + 1$

$x = \dfrac{-b}{2a} = \dfrac{-6}{2(-1)} = \dfrac{-6}{-2} = 3$

$y = -3^2 + 6(3) + 1 = -9 + 18 + 1 = 10$

Vertex: $(3, 10)$

35. $f(x) = 2x^2 - 3x + 7$

$x = \dfrac{-b}{2a} = \dfrac{-(-3)}{2(2)} = \dfrac{3}{4}$

$f\left(\dfrac{3}{4}\right) = 2\left(\dfrac{3}{4}\right)^2 - 3\left(\dfrac{3}{4}\right) + 7 = \dfrac{9}{8} - \dfrac{9}{4} + 7 = \dfrac{47}{8}$

Vertex: $\left(\dfrac{3}{4}, \dfrac{47}{8}\right)$

37. $y = -4x^2 + x + 1$

$x = \dfrac{-b}{2a} = \dfrac{-1}{2(-4)} = \dfrac{-1}{-8} = \dfrac{1}{8}$

$y = -4\left(\dfrac{1}{8}\right)^2 + \left(\dfrac{1}{8}\right) + 1 = -\dfrac{1}{16} + \dfrac{1}{8} + 1 = \dfrac{17}{16}$

Vertex: $\left(\dfrac{1}{8}, \dfrac{17}{16}\right)$

39. $f(x) = -3x^2 + 3x + 7$

$x = \dfrac{-b}{2a} = \dfrac{-3}{2(-3)} = \dfrac{-3}{-6} = \dfrac{1}{2}$

$f\left(\dfrac{1}{2}\right) = -3\left(\dfrac{1}{2}\right)^2 + 3\left(\dfrac{1}{2}\right) + 7 = -\dfrac{3}{4} + \dfrac{3}{2} + 7 = \dfrac{31}{4}$

Vertex: $\left(\dfrac{1}{2}, \dfrac{31}{4}\right)$

41. $y = x^2 - 4 = (x - 0)^2 - 4$

Vertex (h, k): $(0, -4)$

y-intercept: $(0, -4)$

x-intercepts:

$0 = x^2 - 4 = (x + 2)(x - 2)$

$x + 2 = 0 \implies x = -2$

$x - 2 = 0 \implies x = 2$

$(-2, 0)$ and $(2, 0)$

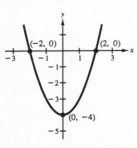

43. $y = -x^2 + 4 = -(x - 0)^2 + 4$

Vertex (h, k): $(0, 4)$

y-intercept: $(0, 4)$

x-intercepts:

$0 = -x^2 + 4 = 4 - x^2 = (2 + x)(2 - x)$

$2 + x = 0 \implies x = -2$

$2 - x = 0 \implies x = 2$

$(-2, 0)$ and $(2, 0)$

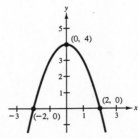

45. $f(x) = x^2 - 3x = \left(x^2 - 3x + \left(-\frac{3}{2}\right)^2 - \left(-\frac{3}{2}\right)^2\right)$

$\quad = \left(x^2 - 3x + \frac{9}{4}\right) - \frac{9}{4} = \left(x - \frac{3}{2}\right)^2 - \frac{9}{4}$

Vertex (h, k): $\left(\frac{3}{2}, -\frac{9}{4}\right)$

y-intercept: $y = 0^2 - 3(0) = 0$

$\qquad (0, 0)$

x-intercepts:

$\quad 0 = x^2 - 3x = x(x - 3)$

$\quad x = 0$

$\quad x - 3 = 0 \implies x = 3$

$(0, 0)$ and $(3, 0)$

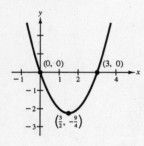

47. $g(x) = -x^2 + 3x = -(x^2 - 3x)$

$\quad = -\left[x^2 - 3x + \left(-\frac{3}{2}\right)^2 - \left(-\frac{3}{2}\right)^2\right]$

$\quad = -\left(x^2 - 3x + \frac{9}{4}\right) + \frac{9}{4}$

$\quad = -\left(x - \frac{3}{2}\right)^2 + \frac{9}{4}$

Vertex (h, k): $\left(\frac{3}{2}, \frac{9}{4}\right)$

y-intercept: $y = -0^2 - 3(0) = 0$

$\qquad (0, 0)$

x-intercepts:

$\quad 0 = -x^2 + 3x = x(-x + 3)$

$\quad x = 0$

$\quad -x + 3 = 0 \implies x = 3$

$(0, 0)$ and $(3, 0)$

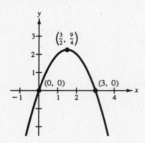

49. $y = (x - 4)^2 = (x - 4)^2 + 0$

Vertex (h, k): $(4, 0)$

y-intercept: $y = (0 - 4)^2 = 16$

$\qquad (0, 16)$

x-intercept:

$\quad 0 = (x - 4)^2$

$\quad x - 4 = 0 \implies x = 4$

$(4, 0)$

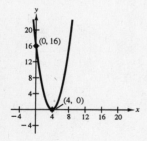

51. $y = x^2 - 8x + 15 = x^2 - 8x + (-4)^2 - (-4)^2 + 15$

$\quad = (x^2 - 8x + 16) - 16 + 15 = (x - 4)^2 - 1$

Vertex (h, k): $(4, -1)$

y-intercept: $y = 0^2 - 8(0) + 15 = 15$

$\qquad (0, 15)$

x-intercepts:

$\quad 0 = x^2 - 8x + 15 = (x - 5)(x - 3)$

$\quad x - 5 = 0 \implies x = 5$

$\quad x - 3 = 0 \implies x = 3$

$(5, 0)$ and $(3, 0)$

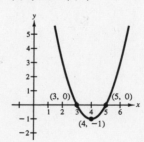

53. $y = -(x^2 + 6x + 5) = -(x^2 + 6x) - 5$

$\qquad = -(x^2 + 6x + 3^2 - 3^2) - 5$

$\qquad = -(x^2 + 6x + 9) + 9 - 5$

$\qquad = -(x + 3)^2 + 4$

Vertex (h, k): $(-3, 4)$

y-intercept: $y = -[0^2 + 6(0) + 5] = -5$

$\qquad\qquad (0, -5)$

x-intercepts:

$\qquad 0 = -(x^2 + 6x + 5) = -(x + 5)(x + 1)$

$\qquad x + 5 = 0 \implies x = -5$

$\qquad x + 1 = 0 \implies x = -1$

$(-5, 0)$ and $(-1, 0)$

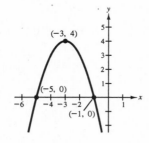

55. $y = -x^2 + 6x - 7 = -(x^2 - 6x) - 7$

$\qquad = -(x^2 - 6x + 3^2 - 3^2) - 7$

$\qquad = -(x^2 - 6x + 9) + 9 - 7 = -(x - 3)^2 + 2$

Vertex (h, k): $(3, 2)$

y-intercept: $y = -0^2 + 6(0) - 7 = -7$

$\qquad\qquad (0, -7)$

x-intercepts:

$\qquad 0 = -x^2 + 6x - 7$

$\qquad (-1)(0) = -1(-x^2 + 6x - 7)$

$\qquad 0 = x^2 - 6x + 7$

$\qquad a = 1, \ b = -6, \ c = 7$

$\qquad x = \dfrac{-(-6) \pm \sqrt{(-6)^2 - 4(1)(7)}}{2(1)}$

$\qquad = \dfrac{6 \pm \sqrt{8}}{2} = \dfrac{6 \pm 2\sqrt{2}}{2} = \dfrac{2(3 \pm \sqrt{2})}{2} = 3 \pm \sqrt{2}$

$\left(3 + \sqrt{2}, 0\right)$ and $\left(3 - \sqrt{2}, 0\right)$

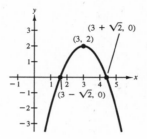

57. $y = 2(x^2 + 6x + 8)$

$y = 2(x^2 + 6x) + 16$

$y = 2(x^2 + 6x + 3^2 - 3^2) + 16$

$y = 2(x^2 + 6x + 9) - 18 + 16$

$y = 2(x + 3)^2 - 2$

Vertex (h, k): $(-3, -2)$

y-intercept: $y = 2[0^2 + 6(0) + 8] = 16$

$\qquad\qquad (0, 16)$

x-intercepts:

$\qquad 0 = 2(x^2 + 6x + 8)$

$\qquad 0 = 2(x + 4)(x + 2)$

$\qquad 2 \neq 0$

$\qquad x + 4 = 0 \implies x = -4$

$\qquad x + 2 = 0 \implies x = -2$

$(-4, 0)$ and $(-2, 0)$

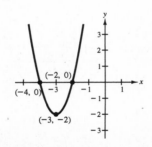

59. $y = \frac{1}{2}(x^2 - 2x - 3) = \frac{1}{2}(x^2 - 2x) - \frac{3}{2}$

$\qquad = \frac{1}{2}[x^2 - 2x + (-1)^2 - (-1)^2] - \frac{3}{2}$

$\qquad = \frac{1}{2}(x^2 - 2x + 1) - \frac{1}{2} - \frac{3}{2} = \frac{1}{2}(x - 1)^2 - 2$

Vertex (h, k): $(1, -2)$

y-intercept: $y = \frac{1}{2}[0^2 - 2(0) - 3] = -\frac{3}{2}$

$\qquad\qquad \left(0, -\frac{3}{2}\right)$

x-intercepts:

$\qquad 0 = \frac{1}{2}(x^2 - 2x - 3) = \frac{1}{2}(x - 3)(x + 1)$

$\qquad \frac{1}{2} \neq 0$

$\qquad x - 3 = 0 \implies x = 3$

$\qquad x + 1 = 0 \implies x = -1$

$(3, 0)$ and $(-1, 0)$

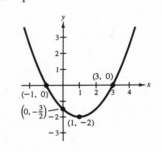

61. $y = 5 - \dfrac{x^2}{3} = -\dfrac{1}{3}x^2 + 5 = -\dfrac{1}{3}(x - 0)^2 + 5$

Vertex (h, k): $(0, 5)$

y-intercept: $y = 5 - \dfrac{0^2}{3} = 5$

$\qquad\qquad (0, 5)$

x-intercepts:

$\qquad 0 = 5 - \dfrac{x^2}{3}$

$\qquad 0 = 15 - x^3$

$\qquad x^2 = 15$

$\qquad x = \pm\sqrt{15}$

$\left(\sqrt{15}, 0\right)$ and $\left(-\sqrt{15}, 0\right)$

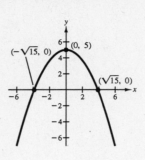

63. $y = \dfrac{1}{5}(3x^2 - 24x + 38) = \dfrac{3}{5}(x^2 - 8x) + \dfrac{38}{5}$

$\qquad = \dfrac{3}{5}[x^2 - 8x + (-4)^2 - (-4)^2] + \dfrac{38}{5}$

$\qquad = \dfrac{3}{5}(x^2 - 8x + 16) - \dfrac{3}{5}(16) + \dfrac{38}{5} = \dfrac{3}{5}(x - 4)^2 - 2$

Vertex (h, k): $(4, -2)$

y-intercept: $y = \dfrac{1}{5}[3(0)^2 - 24(0) + 38] = \dfrac{38}{5}$

$\qquad\qquad \left(0, \dfrac{38}{5}\right)$

x-intercepts:

$\qquad 0 = \dfrac{1}{5}(3x^2 - 24x + 38)$

$\qquad 5(0) = 5 \cdot \dfrac{1}{5}(3x^2 - 24x + 38)$

$\qquad 0 = 3x^2 - 24x + 38$

$\qquad a = 3,\ b = -24,\ c = 38$

$\qquad x = \dfrac{-(-24) \pm \sqrt{(-24)^2 - 4(3)(38)}}{2(3)}$

$\qquad = \dfrac{24 \pm \sqrt{120}}{6} = \dfrac{24 \pm 2\sqrt{30}}{6} = \dfrac{12 \pm \sqrt{30}}{3}$

$\left(\dfrac{12 + \sqrt{30}}{3}, 0\right)$ and $\left(\dfrac{12 - \sqrt{30}}{3}, 0\right)$

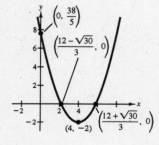

65. $y = \dfrac{1}{6}(2x^2 - 8x + 11)$

$\qquad = \dfrac{1}{3}x^2 - \dfrac{4}{3}x + \dfrac{11}{6}$

Vertex $(h, k) = \left(\dfrac{-b}{2a}, f\!\left(\dfrac{-b}{2a}\right)\right)$

$\dfrac{-b}{2a} = \dfrac{-(-4/3)}{2(1/3)}$

$\qquad = \dfrac{4/3}{2/3}$

$\qquad = 2$

$f(2) = \dfrac{1}{3}(2)^2 - \dfrac{4}{3}(2) + \dfrac{11}{6}$

$\qquad = \dfrac{4}{3} - \dfrac{8}{3} + \dfrac{11}{6}$

$\qquad = \dfrac{8}{6} - \dfrac{16}{6} + \dfrac{11}{6}$

$\qquad = \dfrac{3}{6}$

$\qquad = \dfrac{1}{2}$

Vertex: $\left(2, \dfrac{1}{2}\right)$

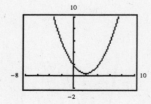

67. $y = -0.7x^2 - 2.7x + 2.3$

Vertex $(h, k) = \left(\dfrac{-b}{2a}, f\!\left(\dfrac{-b}{2a} \right) \right)$

$\dfrac{-b}{2a} = \dfrac{-(-2.7)}{2(-0.7)}$

$\quad = \dfrac{2.7}{-1.4} = -\dfrac{27}{14} \approx -1.9$

$f\!\left(-\dfrac{27}{14} \right) = -0.7\!\left(-\dfrac{27}{14} \right)^2 - 2.7\!\left(-\dfrac{27}{14} \right) + 2.3$

$\quad \approx 4.9$

Vertex: $(-1.9, 4.9)$

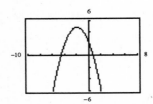

69.

$y_1 = -x^2 + 6$

$y_2 = 2$

$y_1 = y_2$

$-x^2 + 6 = 2$

$-x^2 = -4$

$x^2 = 4$

$x = \pm\sqrt{4}$

$x = \pm 2$

$(2, 2)$ and $(-2, 2)$

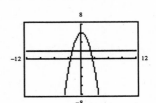

71.

$y_1 = \dfrac{1}{2}x^2 - 3x + \dfrac{13}{2}$

$y_2 = 3$

$y_1 = y_2$

$\dfrac{1}{2}x^2 - 3x + \dfrac{13}{2} = 3$

$x^2 - 6x + 13 = 6$

$x^2 - 6x = -7$

$x^2 - 6x + 9 = -7 + 9$

$(x - 3)^2 = 2$

$x - 3 = \pm\sqrt{2}$

$x = 3 \pm \sqrt{2}$

$\left(3 + \sqrt{2}, 3 \right)$ and $\left(3 - \sqrt{2}, 3 \right)$

Approximately $(4.4, 3)$ and $(1.6, 3)$

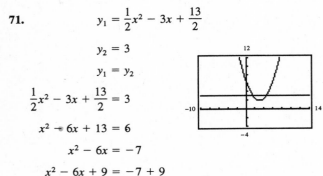

73. $y = a(x - h)^2 + k$ with $(h, k) = (2, 0)$

$y = a(x - 2)^2 + 0$ and $(x, y) = (0, 4)$

$4 = a(0 - 2)^2 + 0 = a(4)$

$1 = a \implies y = 1(x - 2)^2 + 0$

$\qquad = 1(x^2 - 4x + 4) = x^2 - 4x + 4$

75. $y = a(x - h)^2 + k$ and $(h, k) = (-2, 4)$

$y = a(x + 2)^2 + 4$ and $(x, y) = (0, 0)$

$0 = a(0 + 2)^2 + 4 = 4a + 4$

$-4 = 4a$

$-1 = a \implies y = -1(x + 2)^2 + 4$

$\qquad = -1(x^2 + 4x + 4) + 4$

$\qquad = -x^2 - 4x - 4 + 4 = -x^2 - 4x$

77. $y = a(x - h)^2 + k$ and $(h, k) = (-3, 3)$

$y = a(x + 3)^2 + 3$ and $(x, y) = (-2, 1)$

$1 = a(-2 + 3)^2 + 3 = a(1) + 3$

$-2 = a \implies y = -2(x + 3)^2 + 3$

$\qquad = -2(x^2 + 6x + 9) + 3$

$\qquad = -2x^2 - 12x - 18 + 3$

$\qquad = -2x^2 - 12x - 15$

79. $y = a(x - h)^2 + k$ with $(h, k) = (2, 1)$ and $a = 1$

$y = 1(x - 2)^2 + 1$

$\quad = x^2 - 4x + 4 + 1$

$\quad = x^2 - 4x + 5$

81. $y = a(x - h)^2 + k$ with $(h, k) = (-3, 4)$ and $a = -1$

$y = -1(x + 3)^2 + 4$

$\quad = -(x^2 + 6x + 9) + 4$

$\quad = -x^2 - 6x - 5$

83. $y = a(x - h)^2 + k$ with $(h, k) = (2, -4)$

$y = a(x - 2)^2 - 4$ and $(x, y) = (0, 0)$

$\quad 0 = a(0 - 2)^2 - 4 = a(4) - 4$

$\quad 4 = 4a$

$\quad 1 = a \implies y = 1(x - 2)^2 - 4$

$\qquad\qquad\qquad = 1(x^2 - 4x + 4) - 4$

$\qquad\qquad\qquad = x^2 - 4x + 4 - 4$

$\qquad\qquad\qquad = x^2 - 4x$

85. $y = a(x - h)^2 + k$ with $(h, k) = (3, 2)$

$y = a(x - 3)^2 + 2$ and $(x, y) = (1, 4)$

$\quad 4 = a(1 - 3)^2 + 2 = a(4) + 2$

$\quad 2 = 4a$

$\quad \frac{1}{2} = a \implies y = \frac{1}{2}(x - 3)^2 + 2$

$\qquad\qquad\qquad = \frac{1}{2}(x^2 - 6x + 9) + 2$

$\qquad\qquad\qquad = \frac{1}{2}x^2 - 3x + \frac{9}{2} + 2$

$\qquad\qquad\qquad = \frac{1}{2}x^2 - 3x + \frac{13}{2}$

87. $y = a(x - h)^2 + k$ with $(h, k) = (-1, 5)$

$y = a(x + 1)^2 + 5$ and $(x, y) = (0, 1)$

$\quad 1 = a(0 + 1)^2 + 5 = a(1) + 5$

$\quad -4 = a \implies y = -4(x + 1)^2 + 5$

$\qquad\qquad\qquad = -4(x^2 + 2x + 1) + 5$

$\qquad\qquad\qquad = -4x^2 - 8x - 4 + 5$

$\qquad\qquad\qquad = -4x^2 - 8x + 1$

89. $y = a(x - h)^2 + k$ with $(h, k) = (5, 2)$

$y = a(x - 5)^2 + 2$ and $(x, y) = (10, 3)$

$\quad 3 = a(10 - 5)^2 + 2 = a(25) + 2$

$\quad 1 = 25a$

$\quad \frac{1}{25} = a \implies y = \frac{1}{25}(x - 5)^2 + 2$

$\qquad\qquad\qquad = \frac{1}{25}(x^2 - 10x + 25) + 2$

$\qquad\qquad\qquad = \frac{1}{25}x^2 - \frac{2}{5}x + 1 + 2$

$\qquad\qquad\qquad = \frac{1}{25}x^2 - \frac{2}{5}x + 3$

91. (a)

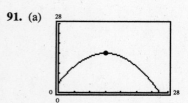

(c) The ball reached its maximum height at the vertex of the parabola.

$y = -\dfrac{1}{12}x^2 + 2x + 4$

$\quad = -\dfrac{1}{12}(x^2 - 24x) + 4$

$\quad = -\dfrac{1}{12}[x^2 - 24x + (-12)^2 - (-12)^2] + 4$

$\quad = -\dfrac{1}{12}(x^2 - 24x + 144) + 12 + 4$

$\quad = -\dfrac{1}{12}(x - 12)^2 + 16$

Vertex: $(12, 16)$

Thus, the maximum height of the ball was 16 feet.

(b) When the ball left the child's hand, the horizontal distance $x = 0$.

$$y = -\frac{1}{12}(0)^2 + 2(0) + 4 = 4$$

Thus, the ball was 4 feet high when it left the child's hand.

(d) When the ball struck the ground, the height $y = 0$.

$0 = -\dfrac{1}{12}x^2 + 2x + 4$

$0 = x^2 - 24x - 48$

$a = 1, \ b = -24, \ c = -48$

$x = \dfrac{-(-24) \pm \sqrt{(-24)^2 - 4(1)(-48)}}{2(1)}$

$\quad = \dfrac{24 \pm \sqrt{768}}{2}$

$\quad = \dfrac{24 \pm 16\sqrt{3}}{2}$

$\quad = 12 \pm 8\sqrt{3}$ (Choose the positive solution.)

Thus, the ball struck the ground at a distance of $12 + 8\sqrt{3}$ or approximately 25.9 feet from the child.

93. $P = 230 + 20s - \frac{1}{2}s^2$

$$P = -\frac{1}{2}s^2 + 20s + 230$$

$$P = -\frac{1}{2}(s^2 - 40s) + 230$$

$$P = -\frac{1}{2}(s^2 - 40s + 400 - 400) + 230$$

$$P = -\frac{1}{2}(s^2 - 40s + 400) + 200 + 230$$

$$P = -\frac{1}{2}(s - 20)^2 + 430$$

Vertex (h, k): $(20, 430)$

Thus, $20(100)$ or $2000 of advertising yields a maximum profit.

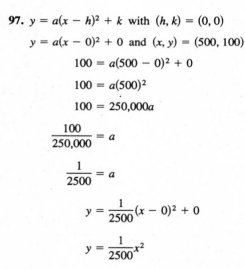

95. (a)

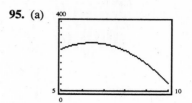

(b) $S = -471.6 + 232.62t - 18.100t^2$

Maximum at vertex

Vertex $(h, k) = \left(\dfrac{-b}{2a}, S\left(\dfrac{-b}{2a} \right) \right)$

$$\frac{-b}{2a} = \frac{-232.62}{2(-18.1)}$$

$$= \frac{-232.62}{-36.2}$$

$$\approx 6.43 \implies \text{the year was 1996}$$

$S = -471.6 + 232.62(6.43) - 18.1(6.43)^2$

$S \approx 275.804$ billion

$S \approx \$275,804,000,000$

97. $y = a(x - h)^2 + k$ with $(h, k) = (0, 0)$

$y = a(x - 0)^2 + 0$ and $(x, y) = (500, 100)$

$$100 = a(500 - 0)^2 + 0$$

$$100 = a(500)^2$$

$$100 = 250,000a$$

$$\frac{100}{250,000} = a$$

$$\frac{1}{2500} = a$$

$$y = \frac{1}{2500}(x - 0)^2 + 0$$

$$y = \frac{1}{2500}x^2$$

99. The graph of the quadratic function $y = ax^2 + bx + c$ opens upward if $a > 0$, and it opens downward if $a < 0$.

101. $f(x) = 2x - 3$

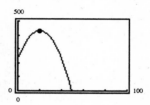

103. $y = -x + 4$

105. $6(x + 3) = 10$

$$6x + 18 = 10$$

$$6x = -8$$

$$x = \frac{-8}{6}$$

$$x = -\frac{4}{3}$$

107. $5x(x - 4) = 0$

$5x = 0 \Rightarrow x = 0$

$x - 4 = 0 \Rightarrow x = 4$

109. $-6(x + 5)^2 + 12 = -6(x^2 + 10x + 25) + 12$

$= -6x^2 - 60x - 150 + 12$

$= -6x^2 - 60x - 138$

111.   (Length)(Width) = Area

$(2w + 1)(w) = 36$

$2w^2 + w = 36$

$2w^2 + w - 36 = 0$

$(2w + 9)(w - 4) = 0$

$2w + 9 = 0 \Rightarrow 2w = -9 \Rightarrow w = -\dfrac{9}{2}$ (Discard negative answer.)

$x - 4 = 0 \Rightarrow w = 4$ and $2w + 1 = 9$

The dimensions of the rectangle are 4 centimeters by 9 centimeters.

Section 6.6   Quadratic Inequalities in One Variable

1. $4x^2 - 81 = 0$

$(2x + 9)(2x - 9) = 0$

$2x + 9 = 0 \Rightarrow x = -\dfrac{9}{2}$

$2x - 9 = 0 \Rightarrow x = \dfrac{9}{2}$

Critical numbers: $-\dfrac{9}{2}, \dfrac{9}{2}$

3. $x(2x - 5) = 0$

$x = 0$

$2x - 5 = 0 \Rightarrow x = \dfrac{5}{2}$

Critical numbers: $0, \dfrac{5}{2}$

5. $x^2 - 4x + 3 = 0$

$(x - 3)(x - 1) = 0$

$x - 3 = 0 \Rightarrow x = 3$

$x - 1 = 0 \Rightarrow x = 1$

Critical numbers: 1, 3

7. $4x^2 - 20x + 25 = 0$

$(2x - 5)(2x - 5) = 0$

$2x - 5 = 0 \Rightarrow x = \dfrac{5}{2}$

Critical number: $\dfrac{5}{2}$

9. $x - 4 = 0$

$x = 4$

Critical number: 4

Test Interval	Representative x-value	Value of Polynomial	Conclusion
$(-\infty, 4)$	$x = 0$	$0 - 4 = -4$	Polynomial is negative.
$(4, \infty)$	$x = 5$	$5 - 4 = 1$	Polynomial is positive.

11. $2x(x - 4) = 0$

$2x = 0 \Rightarrow x = 0$

$x - 4 = 0 \Rightarrow x = 4$

Critical numbers: 0, 4

Test Interval	Representative x-value	Value of Polynomial	Conclusion
$(-\infty, 0)$	$x = -1$	$2(-1)(-1 - 4) = 10$	Polynomial is positive.
$(0, 4)$	$x = 1$	$2(1)(1 - 4) = -6$	Polynomial is negative.
$(4, \infty)$	$x = 5$	$2(5)(5 - 4) = 10$	Polynomial is positive.

13. $x^2 - 9 = 0$

$(x + 3)(x - 3) = 0$

$x + 3 = 0 \implies x = -3$

$x - 3 = 0 \implies x = 3$

Critical numbers: $-3, 3$

Test Interval	Representative x-value	Value of Polynomial	Conclusion
$(-\infty, -3)$	$x = -5$	$(-5)^2 - 9 = 16$	Polynomial is positive.
$(-3, 3)$	$x = 0$	$(0)^2 - 9 = -9$	Polynomial is negative.
$(3, \infty)$	$x = 5$	$(5)^2 - 9 = 16$	Polynomial is positive.

15. $x^2 - 4x - 5 = 0$

$(x - 5)(x + 1) = 0$

$x - 5 = 0 \implies x = 5$

$x + 1 = 0 \implies x = -1$

Critical numbers: $-1, 5$

Test Interval	Representative x-value	Value of Polynomial	Conclusion
$(-\infty, -1)$	$x = -2$	$(-2)^2 - 4(-2) - 5 = 7$	Polynomial is positive.
$(-1, 5)$	$x = 0$	$0^2 - 4(0) - 5 = -5$	Polynomial is negative.
$(5, \infty)$	$x = 6$	$6^2 - 4(6) - 5 = 7$	Polynomial is positive.

17. $2x + 6 \geq 0$

$2x + 6 = 0$

$2x = -6$

$x = -3$

Critical number: -3

The solution is $[-3, \infty)$.

Test Interval	Representative x-value	Is inequality satisfied?
$(-\infty, -3)$	$x = -5$	$2(-5) + 6 \overset{?}{\geq} 0$ $-4 \not\geq 0$
$(-3, \infty)$	$x = 0$	$2(0) + 6 \overset{?}{\geq} 0$ $6 \geq 0$

19. $-\frac{3}{4}x + 6 < 0$

$-\frac{3}{4}x + 6 = 0$

$-3x + 24 = 0$

$-3x = -24$

$x = 8$

Critical number: 8

The solution is $(8, \infty)$.

Test Interval	Representative x-value	Is inequality satisfied?
$(-\infty, 8)$	$x = 0$	$\left(-\frac{3}{4}\right)(0) + 6 \overset{?}{<} 0$ $6 \not< 0$
$(8, \infty)$	$x = 12$	$\left(-\frac{3}{4}\right)(12) + 6 \overset{?}{<} 0$ $-3 < 0$

21. $3x(x - 2) < 0$

$3x(x - 2) = 0$

$3x = 0 \implies x = 0$

$x - 2 = 0 \implies x = 2$

Critical numbers: 0, 2

The solution is $(0, 2)$.

Test Interval	Representative x-value	Is inequality satisfied?
$(-\infty, 0)$	$x = -2$	$3(-2)(-2 - 2) \overset{?}{<} 0$ $24 \not< 0$
$(0, 2)$	$x = 1$	$3(1)(1 - 2) \overset{?}{<} 0$ $-3 < 0$
$(2, \infty)$	$x = 4$	$3(4)(4 - 2) \overset{?}{<} 0$ $24 \not< 0$

23. $3x(2-x) < 0$

$3x(2-x) = 0$

$3x = 0 \implies x = 0$

$2 - x = 0 \implies x = 2$

Critical numbers: 0, 2

The solution is $(-\infty, 0) \cup (2, \infty)$.

Test Interval	Representative x-value	Is inequality satisfied?
$(-\infty, 0)$	$x = -2$	$3(-2)[2 - (-2)] \overset{?}{<} 0$ $-24 < 0$
$(0, 2)$	$x = 1$	$3(1)(2-1) \overset{?}{<} 0$ $3 \not< 0$
$(2, \infty)$	$x = 4$	$3(4)(2-4) \overset{?}{<} 0$ $-24 < 0$

25. $x^2 - 4 \geq 0$

$x^2 - 4 = 0$

$(x+2)(x-2) = 0$

$x + 2 = 0 \implies x = -2$

$x - 2 = 0 \implies x = 2$

Critical numbers: $-2, 2$

The solution is $(-\infty, -2] \cup [2, \infty)$.

Test Interval	Representative x-value	Is inequality satisfied?
$(-\infty, -2)$	$x = -5$	$(-5)^2 - 4 \overset{?}{\geq} 0$ $21 \geq 0$
$(-2, 2)$	$x = 0$	$0^2 - 4 \overset{?}{\geq} 0$ $-4 \not\geq 0$
$(2, \infty)$	$x = 5$	$5^2 - 4 \overset{?}{\geq} 0$ $21 \geq 0$

27. $x^2 + 3x \leq 10$

$x^2 + 3x - 10 \leq 0$

$x^2 + 3x - 10 = 0$

$(x+5)(x-2) = 0$

$x + 5 = 0 \implies x = -5$

$x - 2 = 0 \implies x = 2$

Critical numbers: $-5, 2$

The solution is $[-5, 2]$.

Test Interval	Representative x-value	Is inequality satisfied?
$(-\infty, -5)$	$x = -8$	$(-8)^2 + 3(-8) \overset{?}{\leq} 10$ $40 \not\leq 10$
$(-5, 2)$	$x = 0$	$0^2 + 3(0) \overset{?}{\leq} 10$ $0 \leq 10$
$(2, \infty)$	$x = 5$	$5^2 + 3(5) \overset{?}{\leq} 10$ $40 \not\leq 10$

29. $-2u^2 + 7u + 4 < 0$

$2u^2 - 7u - 4 > 0$

$2u^2 - 7u - 4 = 0$

$(2u+1)(u-4) = 0$

$2u + 1 = 0 \implies u = -\frac{1}{2}$

$u - 4 = 0 \implies u = 4$

Critical numbers: $-\frac{1}{2}, 4$

The solution is $\left(-\infty, -\frac{1}{2}\right) \cup (4, \infty)$.

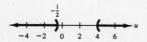

Test Interval	Representative u-value	Is inequality satisfied?
$\left(-\infty, -\frac{1}{2}\right)$	$u = -1$	$2(-1)^2 - 7(-1) - 4 \overset{?}{<} 0$ $5 \not< 0$
$\left(-\frac{1}{2}, 4\right)$	$u = 0$	$2(0)^2 - 7(0) - 4 \overset{?}{<} 0$ $-4 < 0$
$(4, \infty)$	$u = 5$	$2(5)^2 - 7(5) - 4 \overset{?}{<} 0$ $11 \not< 0$

31. $x^2 + 4x + 5 < 0$

$x^2 + 4x + 5 = 0$

$$x = \frac{-4 \pm \sqrt{4^2 - 4(1)(5)}}{2(1)}$$

$$= \frac{-4 \pm \sqrt{-4}}{2}$$

$$= \frac{-4 \pm 2i}{2} = -2 \pm i$$

No critical numbers

There is no solution.

Note: $x^2 + 4x + 5 > 0$ for all real numbers x.

Test Interval	Representative x-value	Is inequality satisfied?
$(-\infty, \infty)$	$x = 0$	$0^2 + 4(0) + 5 \overset{?}{<} 0$ $5 \not< 0$

33. $x^2 + 2x + 1 \geq 0$

$x^2 + 2x + 1 = 0$

$(x + 1)^2 = 0$

$x + 1 = 0 \implies x = -1$

Critical number: -1

The solution is $(-\infty, \infty)$.

Note: $(-\infty, -1] \cup [-1, \infty) = (-\infty, \infty)$

Test Interval	Representative x-value	Is inequality satisfied?
$(-\infty, -1)$	$x = -2$	$(-2)^2 + 2(-2) + 1 \overset{?}{\geq} 0$ $1 \geq 0$
$(-1, \infty)$	$x = 0$	$0^2 + 2(0) + 1 \overset{?}{\geq} 0$ $1 \geq 0$

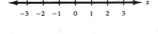

35. $x^2 - 4x + 2 > 0$

$x^2 - 4x + 2 = 0$

$$x = \frac{-(-4) \pm \sqrt{(-4)^2 - 4(1)(2)}}{2(1)}$$

$$= \frac{4 \pm \sqrt{8}}{2}$$

$$= \frac{4 \pm 2\sqrt{2}}{2}$$

$$= 2 \pm \sqrt{2}$$

Critical numbers: $2 - \sqrt{2}, 2 + \sqrt{2}$

The solution is $\left(-\infty, 2 - \sqrt{2}\right) \cup \left(2 + \sqrt{2}, \infty\right)$.

Test Interval	Representative x-value	Is inequality satisfied?
$\left(-\infty, 2 - \sqrt{2}\right)$	$x = 0$	$0^2 - 4(0) + 2 \overset{?}{>} 0$ $2 > 0$
$\left(2 - \sqrt{2}, 2 + \sqrt{2}\right)$	$x = 2$	$2^2 - 4(2) + 2 \overset{?}{>} 0$ $-2 \not> 0$
$\left(2 + \sqrt{2}, \infty\right)$	$x = 4$	$4^2 - 4(4) + 2 \overset{?}{>} 0$ $2 > 0$

37. $(x - 5)^2 < 0$

$(x - 5)^2 = 0$

$x - 5 = 0 \implies x = 5$

Critical number: 5

There is no solution.

Note: $(x - 5)^2 \geq 0$ for all real numbers x.

Test Interval	Representative x-value	Is inequality satisfied?
$(-\infty, 5)$	$x = 0$	$(0 - 5)^2 \overset{?}{<} 0$ $25 \not< 0$
$(5, \infty)$	$x = 10$	$(10 - 5)^2 \overset{?}{<} 0$ $25 \not< 0$

39.
$$6 - (x - 5)^2 < 0$$
$$6 - (x^2 - 10x + 25) < 0$$
$$-x^2 + 10x - 19 < 0$$
$$x^2 - 10x + 19 > 0$$
$$x^2 - 10x + 19 = 0$$
$$x = \frac{-(-10) \pm \sqrt{(-10)^2 - 4(1)(19)}}{2(1)}$$
$$= \frac{10 \pm \sqrt{24}}{2} = \frac{10 \pm 2\sqrt{6}}{2} = 5 \pm \sqrt{6}$$

Critical numbers: $5 - \sqrt{6}, 5 + \sqrt{6}$

The solution is $(-\infty, 5 - \sqrt{6}) \cup (5 + \sqrt{6}, \infty)$.

Test Interval	Representative x-value	Is inequality satisfied?
$(-\infty, 5 - \sqrt{6})$	$x = 0$	$0^2 - 10(0) + 19 \overset{?}{>} 0$ $19 > 0$
$(5 - \sqrt{6}, 5 + \sqrt{6})$	$x = 5$	$5^2 - 10(5) + 19 \overset{?}{>} 0$ $-6 \not> 0$
$(5 + \sqrt{6}, \infty)$	$x = 8$	$8^2 - 10(8) + 19 \overset{?}{>} 0$ $3 > 0$

41.
$$x^2 - 6x + 9 \ge 0$$
$$x^2 - 6x + 9 = 0$$
$$(x - 3)(x - 3) = 0$$
$$x - 3 = 0 \implies x = 3$$

Critical number: 3

The solution is $(-\infty, \infty)$.

Test Interval	Representative x-value	Is inequality satisfied?
$(-\infty, 3)$	$x = 0$	$0^2 - 6(0) + 9 \overset{?}{\ge} 0$ $9 \ge 0$
$(3, \infty)$	$x = 10$	$(10)^2 - 6(10) + 9 \overset{?}{\ge} 0$ $49 \ge 0$

43. $y = 0.5x^2 + 1.25x - 3, \ y > 0$

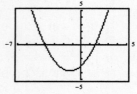

The solution is $(-\infty, -4) \cup \left(\frac{3}{2}, \infty\right)$.

45. $y = x^2 + 4x + 4, \ y \ge 9$

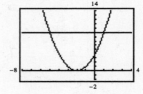

The solution is $(-\infty, -5] \cup [1, \infty)$.

47. $y = 9 - 0.2(x - 2)^2, \ y < 4$

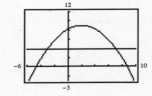

The solution is $(-\infty, -3) \cup (7, \infty)$.

49.
$$-16t^2 + 128t > 240$$
$$-16t^2 + 128t - 240 > 0$$
$$t^2 - 8t + 15 < 0$$
$$t^2 - 8t + 15 = 0$$
$$(t - 5)(t - 3) = 0$$
$$t - 5 = 0 \implies t = 5$$
$$t - 3 = 0 \implies t = 3$$

Critical numbers: 3, 5

Solution: $(3, 5)$

Test Interval	Representative t-value	Is inequality satisfied?
$(-\infty, 3)$	$t = 0$	$0^2 - 8(0) + 15 \overset{?}{<} 0$ $15 \not< 0$
$(3, 5)$	$t = 4$	$4^2 - 8(4) + 15 \overset{?}{<} 0$ $-1 < 0$
$(5, \infty)$	$t = 6$	$6^2 - 8(6) + 15 \overset{?}{<} 0$ $3 \not< 0$

The height will exceed 240 feet for values of t in the interval $(3, 5)$ or for 3 seconds $< t < 5$ seconds.

51. $(1000)(1 + r)^2 > 1150$

$$(1 + r)^2 > 1.15$$

$$r^2 + 2r + 1 > 1.15$$

$$r^2 + 2r - 0.15 > 0$$

$$r^2 + 2r - 0.15 = 0$$

$$r = \frac{-2 \pm \sqrt{2^2 - 4(1)(-0.15)}}{2(1)}$$

$$r = \frac{-2 \pm \sqrt{4.6}}{2}$$

Test Interval	Representative r-value	Is inequality satisfied?
$\left(-\infty, \dfrac{-2 - \sqrt{4.6}}{2}\right)$	$r = -3$	$[1 + (-3)]^2 \overset{?}{>} 1.15$ $4 > 1.15$
$\left(\dfrac{-2 - \sqrt{4.6}}{2}, \dfrac{-2 + \sqrt{4.6}}{2}\right)$	$r = 0$	$(1 + 0)^2 \overset{?}{>} 1.15$ $1 \not> 1.15$
$\left(\dfrac{-2 + \sqrt{4.6}}{2}, \infty\right)$	$r = 1$	$(1 + 1)^2 \overset{?}{>} 1.15$ $4 > 1.15$

Critical numbers: $\dfrac{-2 - \sqrt{4.6}}{2}, \dfrac{-2 + \sqrt{4.6}}{2}$

Solution: $\left(-\infty, \dfrac{-2 - \sqrt{4.6}}{2}\right) \cup \left(\dfrac{-2 + \sqrt{4.6}}{2}, \infty\right)$ (Discard negative values.)

The interest rate must exceed $\left(-2 + \sqrt{4.6}\right)/2$ or $r > 0.0724$.

Thus, the interest will exceed $150 if $r > 7.24\%$.

53. $A = l(50 - l)$

$$l(50 - l) \geq 500$$

$$50l - l^2 - 500 \geq 0$$

$$-l^2 + 50l - 500 \geq 0$$

$$l^2 - 50l + 500 \leq 0$$

$$l = \frac{-(-50) \pm \sqrt{(-50)^2 - 4(1)(500)}}{2(1)}$$

$$l = \frac{50 \pm \sqrt{500}}{2}$$

$$l = \frac{50 \pm 10\sqrt{5}}{2}$$

$$l = 25 \pm 5\sqrt{5}$$

Test Interval	Representative l-value	Is inequality satisfied?
$\left(-\infty, 25 - 5\sqrt{5}\right)$	$l = 0$	$0(50 - 0) \overset{?}{\geq} 500$ $0 \not\geq 500$
$\left(25 - 5\sqrt{5}, 25 + 5\sqrt{5}\right)$	$l = 20$	$20(50 - 20) \overset{?}{\geq} 500$ $600 \geq 500$
$\left(25 + 5\sqrt{5}, \infty\right)$	$l = 40$	$40(50 - 40) \overset{?}{\geq} 500$ $400 \not\geq 500$

Critical numbers: $25 - 5\sqrt{5}, 25 + 5\sqrt{5}$

Solution: $\left[25 - 5\sqrt{5}, 25 + 5\sqrt{5}\right]$

The length of the field must be in the interval $\left[25 - 5\sqrt{5}, 25 + 5\sqrt{5}\right]$ or $25 - 5\sqrt{5}$ meters $\leq l \leq 25 + 5\sqrt{5}$ meters.

55. Answers will vary. One example is $x^2 + 2x + 6 < 0$.

57. $4x + 20 \leq 0$

$\qquad 4x \leq -20$

$\qquad x \leq -5$

$\qquad (-\infty, -5]$

59. $8 - 4x < 9 - x$

$\qquad 8 - 3x < 9$

$\qquad -3x < 1$

$\qquad x > -\dfrac{1}{3}$

$\qquad \left(-\dfrac{1}{3}, \infty\right)$

61. $|x + 8| > 9$

$\qquad x + 8 < -9 \quad or \quad x + 8 > 9$

$\qquad x < -17 \qquad x > 1$

$\qquad (-\infty, -17) \cup (1, \infty)$

63. $(2x - 1)(2x + 1) = (2x)^2 - 1^2$

$\qquad\qquad\qquad\qquad = 4x^2 - 1$

65. $(x + 5)(x^2 + 10x + 15) = x^3 + 10x^2 + 15x + 5x^2 + 50x + 75$

$\qquad\qquad\qquad\qquad\qquad = x^3 + 15x^2 + 65x + 75$

67. (a) $|s - 4.25| \leq 0.01$

(b) $\qquad |s - 4.25| \leq 0.01$

$\qquad -0.01 \leq s - 4.25 \leq 0.01$

$\qquad\quad 4.24 \leq s \leq 4.26$

Review Exercises for Chapter 6

1. $x^2 + 12x = 0$

$\quad x(x + 12) = 0$

$\qquad x = 0$

$\quad x + 12 = 0 \implies x = -12$

3. $3z(z + 10) - 8(z + 10) = 0$

$\qquad (z + 10)(3z - 8) = 0$

$\qquad z + 10 = 0 \implies z = -10$

$\qquad 3z - 8 = 0 \implies z = \frac{8}{3}$

5. $\qquad 4y^2 - 1 = 0$

$\quad (2y + 1)(2y - 1) = 0$

$\qquad 2y + 1 = 0 \implies y = -\frac{1}{2}$

$\qquad 2y - 1 = 0 \implies y = \frac{1}{2}$

7. $4y^2 + 20y + 25 = 0$

$\quad (2y + 5)(2y + 5) = 0$

$\qquad 2y + 5 = 0 \implies y = -\frac{5}{2}$

9. $t^2 - t - 20 = 0$

$\quad (t - 5)(t + 4) = 0$

$\qquad t - 5 = 0 \implies t = 5$

$\qquad t + 4 = 0 \implies t = -4$

11. $2x^2 - 2x - 180 = 0$

$\quad 2(x^2 - x - 90) = 0$

$\quad 2(x - 10)(x + 9) = 0$

$\qquad\qquad 2 \neq 0$

$\quad x - 10 = 0 \implies x = 10$

$\quad x + 9 = 0 \implies x = -9$

13. $x^2 = 10{,}000$

$\quad x = \pm\sqrt{10{,}000}$

$\quad x = \pm 100$

15. $y^2 - 8 = 0$

$\qquad y^2 = 8$

$\qquad y = \pm\sqrt{8}$

$\qquad y = \pm 2\sqrt{2}$

17. $(x - 16)^2 = 400$

$x - 16 = \pm\sqrt{400}$

$x - 16 = \pm 20$

$x = 16 \pm 20$

$x = 16 + 20 \implies x = 36$

$x = 16 - 20 \implies x = -4$

19. $t^2 + 28 = 0$

$t^2 = -28$

$t = \pm\sqrt{-28}$

$t = \pm\sqrt{4(7)(-1)}$

$t = \pm 2\sqrt{7}i$

21. $(2x - 3)^2 = -75$

$2x - 3 = \pm\sqrt{-75}$

$2x - 3 = \pm\sqrt{25(3)(-1)}$

$2x - 3 = \pm 5\sqrt{3}i$

$2x = 3 \pm 5\sqrt{3}i$

$x = \dfrac{3}{2} \pm \dfrac{5\sqrt{3}}{2}i$

23. $(x + 6)^2 + 20 = 0$

$(x + 6)^2 = -20$

$x + 6 = \pm\sqrt{-20}$

$x + 6 = \pm\sqrt{4(5)(-1)}$

$x = -6 \pm 2\sqrt{5}i$

25. $x^4 - 4x^2 - 5 = 0$ Let $u = x^2$.

$u^2 - 4u - 5 = 0$

$(u - 5)(u + 1) = 0$

$u - 5 = 0 \implies u = 5$

$u + 1 = 0 \implies u = -1$

$u = 5 \implies x^2 = 5$

$x = \pm\sqrt{5}$

$u = -1 \implies x^2 = -1$

$x = \pm\sqrt{-1}$

$x = \pm i$

27. $x + 2\sqrt{x} - 3 = 0$ Let $u = \sqrt{x}$.

$u^2 + 2u - 3 = 0$

$(u - 1)(u + 3) = 0$

$u - 1 = 0 \implies u = 1$

$u + 3 = 0 \implies u = -3$

$u = 1 \implies \sqrt{x} = 1$

$x = 1^2$

$x = 1$

$u = -3 \implies \sqrt{x} = -3$

$x = (-3)^2$

$x = 9$ (Extraneous)

Check: $1 + 2\sqrt{1} - 3 = 0$

$1 + 2 - 3 = 0$

Check: $9 + 2\sqrt{9} - 3 = 0$

$9 + 6 - 3 \neq 0$

29. $x^{2/5} + 4x^{1/5} + 3 = 0$ Let $u = x^{1/5}$.

$u^2 + 4u + 3 = 0$

$(u + 3)(u + 1) = 0$

$u + 3 = 0 \implies u = -3$

$u + 1 = 0 \implies u = -1$

$u = -3 \implies x^{1/5} = -3$

$x = (-3)^5$

$x = -243$

$u = -1 \implies x^{1/5} = -1$

$x = (-1)^5$

$x = -1$

31. $6\left(\dfrac{1}{x}\right)^2 + 7\left(\dfrac{1}{x}\right) - 3 = 0$ Let $u = \dfrac{1}{x}$.

$6u^2 + 7u - 3 = 0$

$(3u - 1)(2u + 3) = 0$

$3u - 1 = 0 \implies u = \dfrac{1}{3}$

$2u + 3 = 0 \implies u = -\dfrac{3}{2}$

$u = \dfrac{1}{3} \implies \dfrac{1}{x} = \dfrac{1}{3}$

$x = 3$

$u = -\dfrac{3}{2} \implies \dfrac{1}{x} = -\dfrac{3}{2}$

$x = -\dfrac{2}{3}$

33. $b = 6 \implies \left(\dfrac{b}{2}\right)^2 = \left(\dfrac{6}{2}\right)^2 = (3)^2 = 9$

$x^2 + 6x + \boxed{9}$

35. $b = 7 \implies \left(\dfrac{b}{2}\right)^2 = \left(\dfrac{7}{2}\right)^2 = \dfrac{49}{4}$

$x^2 + 7x + \boxed{\dfrac{49}{4}}$

37. $b = \dfrac{3}{4} \implies \left(\dfrac{b}{2}\right)^2 = \left(\dfrac{1}{2} \cdot \dfrac{3}{4}\right)^2 = \dfrac{9}{64}$

$y^2 + \dfrac{3}{4} + \boxed{\dfrac{9}{64}}$

39.
$$x^2 - 6x - 3 = 0$$
$$x^2 - 6x = 3$$
$$x^2 - 6x + (-3)^2 = 3 + 9$$
$$(x - 3)^2 = 12$$
$$x - 3 = \pm\sqrt{12}$$
$$x - 3 = \pm 2\sqrt{3}$$
$$x = 3 \pm 2\sqrt{3}$$

41.
$$x^2 - 3x + 3 = 0$$
$$x^2 - 3x = -3$$
$$x^2 - 3x + \left(-\frac{3}{2}\right)^2 = -3 + \frac{9}{4}$$
$$\left(x - \frac{3}{2}\right)^2 = -\frac{3}{4}$$
$$x - \frac{3}{2} = \pm\frac{\sqrt{3}i}{2}$$
$$x = \frac{3 \pm \sqrt{3}i}{2} \text{ or } x = \frac{3}{2} \pm \frac{\sqrt{3}}{2}i$$

43.
$$2y^2 + 10y + 3 = 0$$
$$2y^2 + 10y = -3$$
$$y^2 + 5y = -\frac{3}{2}$$
$$y^2 + 5y + \left(\frac{5}{2}\right)^2 = -\frac{3}{2} + \frac{25}{4}$$
$$\left(y + \frac{5}{2}\right)^2 = \frac{19}{4}$$
$$y + \frac{5}{2} = \pm\frac{\sqrt{19}}{2}$$
$$y = \frac{-5 \pm \sqrt{19}}{2}$$

45.
$$2y^2 + 3y + 1 = 0$$
$$y^2 + \frac{3}{2}y + \frac{1}{2} = 0$$
$$y^2 + \frac{3}{2}y = -\frac{1}{2}$$
$$y^2 + \frac{3}{2}y + \left(\frac{3}{4}\right)^2 = -\frac{1}{2} + \left(\frac{3}{4}\right)^2$$
$$\left(y + \frac{3}{4}\right)^2 = \frac{1}{16}$$
$$y + \frac{3}{4} = \pm\sqrt{\frac{1}{16}}$$
$$y + \frac{3}{4} = \pm\frac{1}{4}$$
$$y = -\frac{3}{4} \pm \frac{1}{4}$$
$$y = -\frac{3}{4} + \frac{1}{4} \implies y = -\frac{1}{2}$$
$$y = -\frac{3}{4} - \frac{1}{4} \implies y = -1$$

47.
$$y^2 + y - 30 = 0$$
$$a = 1, \ b = 1, \ c = -30$$
$$y = \frac{-1 \pm \sqrt{1^2 - 4(1)(-30)}}{2(1)}$$
$$y = \frac{-1 \pm \sqrt{121}}{2}$$
$$y = \frac{-1 \pm 11}{2}$$
$$y = \frac{-1 + 11}{2} \implies y = 5$$
$$y = \frac{-1 - 11}{2} \implies y = -6$$

49.
$$2y^2 + y - 21 = 0$$
$$a = 2, \ b = 1, \ c = -21$$
$$y = \frac{-1 \pm \sqrt{1^2 - 4(2)(-21)}}{2(2)}$$
$$y = \frac{-1 \pm \sqrt{169}}{4}$$
$$y = \frac{-1 \pm 13}{4}$$
$$y = \frac{-1 + 13}{4} \implies y = 3$$
$$y = \frac{-1 - 13}{4} \implies y = -\frac{7}{2}$$

51. $-x^2 + 5x + 84 = 0$

$$a = -1, \ b = 5, \ c = 84$$

$$x = \frac{-5 \pm \sqrt{5^2 - 4(-1)(84)}}{2(-1)}$$

$$x = \frac{-5 \pm \sqrt{361}}{-2}$$

$$x = \frac{-5 \pm 19}{-2}$$

$$x = \frac{-5 + 19}{-2} = \frac{14}{-2} = -7$$

$$x = \frac{-5 - 19}{-2} = \frac{-24}{-2} = 12$$

53. $0.3t^2 - 2t + 5 = 0$

$$a = 0.3, \ b = -2, \ c = 5$$

$$t = \frac{-(-2) \pm \sqrt{(-2)^2 - 4(0.3)(5)}}{2(0.3)}$$

$$t = \frac{2 \pm \sqrt{-2}}{0.6}$$

$$t = \frac{2 \pm \sqrt{2}i}{0.6} \quad \text{or} \quad t = \frac{10}{3} \pm \frac{5\sqrt{2}}{3}i$$

Note: $\dfrac{2 \pm \sqrt{2}i}{0.6} = \dfrac{(2 \pm \sqrt{2}i)(10)}{(0.6)(10)}$

$$= \frac{20 \pm 10\sqrt{2}i}{6} = \frac{10}{3} \pm \frac{5\sqrt{2}}{3}i$$

55. $2x^2 - 5x - 7 = 0$

$$b^2 - 4ac = (-5)^2 - 4(2)(-7)$$

$$= 25 + 56$$

$$= 81$$

Because 81 is a positive perfect square, there are two distinct rational solutions.

57. $3x^2 - 2x + 10 = 0$

$$b^2 - 4ac = (-2)^2 - 4(3)(10)$$

$$= 4 - 120$$

$$= -116$$

Because -116 is a negative number, there are two distinct imaginary solutions.

59. $-x^2 - 20x + 100 = 0$

$$b^2 - 4ac = (-20)^2 - 4(-1)(100)$$

$$= 400 + 400$$

$$= 800$$

Because 800 is a positive number but not a perfect square, there are two distinct irrational solutions.

61. $y = 4(x - 3)^2 + 6$

$a = 4$

Because $a > 0$, the parabola opens upward.

Vertex: $(h, k) = (3, 6)$

63. $y = -3(x + 4)^2$

$a = -3$

Because $a < 0$, the parabola opens downward.

Vertex: $(h, k) = (-4, 0)$

65. $y = 3x^2 + 6x$

$$x = \frac{-b}{2a} = \frac{-6}{2(3)} = \frac{-6}{6} = -1$$

$$y = 3(-1)^2 + 6(-1) = 3 - 6 = -3$$

Vertex: $(-1, -3)$

67. $y = -x^2 + 7x + 4$

$$x = \frac{-b}{2a} = \frac{-7}{2(-1)} = \frac{-7}{-2} = \frac{7}{2}$$

$$y = -\left(\frac{7}{2}\right)^2 + 7\left(\frac{7}{2}\right) + 4 = -\frac{49}{4} + \frac{49}{2} + 4 = \frac{65}{4}$$

Vertex: $\left(\dfrac{7}{2}, \dfrac{65}{4}\right)$

69. $y = x^2 + 4$

$$y = (x - 0)^2 + 4$$

$a = 1$

Because $a > 0$, the parabola opens upward.

Vertex: (h, k): $(0, 4)$

x-intercepts: None

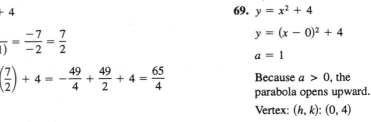
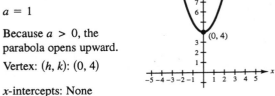

71. $y = x^2 + 8x$

$y = x^2 + 8x + 4^2 - 4^2$

$y = (x^2 + 8x + 16) - 16$

$y = (x + 4)^2 - 16$

$a = 1$

Because $a > 0$, the parabola opens upward.

Vertex: $(h, k) = (-4, -16)$

x-intercepts: Let $y = 0$

$x^2 + 8x = 0$

$x(x + 8) = 0$

$x = 0$

$x + 8 = 0 \implies x = -8$

$(0, 0)$ and $(-8, 0)$

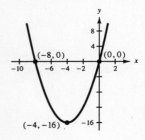

73. $y = x^2 - 6x + 5$

$y = x^2 - 6x + (-3)^2 - (-3)^2 + 5$

$y = (x^2 - 6x + 9) - 9 + 5$

$y = (x - 3)^2 - 4$

$a = 1$

Because $a > 0$, the parabola opens upward.

Vertex: $(h, k) = (3, -4)$

x-intercepts: Let $y = 0$

$x^2 - 6x + 5 = 0$

$(x - 5)(x - 1) = 0$

$x - 5 = 0 \implies x = 5$

$x - 1 = 0 \implies x = 1$

$(5, 0)$ and $(1, 0)$

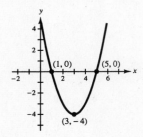

75. $y = a(x - h)^2 + k$, with $(h, k) = (3, 5)$ and $a = -2$.

$y = -2(x - 3)^2 + 5$

$y = -2(x^2 - 6x + 9) + 5$

$y = -2x^2 + 12x - 18 + 5$

$y = -2x^2 + 12x - 13$

77. $y = a(x - h)^2 + k$, with $(h, k) = (5, 0)$.

$y = a(x - 5)^2 + 0$, and $(x, y) = (1, 1)$

$1 = a(1 - 5)^2 + 0$

$1 = a(16) + 0$

$0 = 16a$

$\frac{1}{16} = a \implies y = \frac{1}{16}(x - 5)^2 + 0$

$y = \frac{1}{16}(x^2 - 10x + 25)$

$y = \frac{1}{16}x^2 - \frac{5}{8}x + \frac{25}{16}$

79. $12 - x$

Critical number: $12 - x = 0$

$-x = -12$

$x = 12$

Test Interval	Representative x-value	Value of expression	Conclusion
$(-\infty, 12)$	$x = 4$	$12 - 4 = 8$	Polynomial is positive.
$(12, \infty)$	$x = 15$	$12 - 15 = -3$	Polynomial is negative.

81. $4x(x + 1)$

Critical numbers: $4x(x + 1) = 0$

$$4x = 0 \implies x = 0$$
$$x + 1 = 0 \implies x = -1$$

Test Interval	Representative x-value	Value of expression	Conclusion
$(-\infty, -1)$	$x = -3$	$4(-3)(-3 + 1) = 24$	Polynomial is positive.
$(-1, 0)$	$x = -\frac{1}{2}$	$4\left(-\frac{1}{2}\right)\left(-\frac{1}{2} + 1\right) = -1$	Polynomial is negative.
$(0, \infty)$	$x = 5$	$4(5)(5 + 1) = 120$	Polynomial is positive.

83. $x^2 - 10x + 21$

Critical numbers: $x^2 - 10x + 21 = 0$

$$(x - 7)(x - 3) = 0$$
$$x - 7 = 0 \implies x = 7$$
$$x - 3 = 0 \implies x = 3$$

Test Interval	Representative x-value	Value of expression	Conclusion
$(-\infty, 3)$	$x = 0$	$0^2 - 10(0) + 21 = 21$	Polynomial is positive.
$(3, 7)$	$x = 5$	$5^2 - 10(5) + 21 = -4$	Polynomial is negative.
$(7, \infty)$	$x = 8$	$8^2 - 10(8) + 21 = 5$	Polynomial is positive.

85. $4x - 12 < 0$

$4(x - 3) < 0$

$4(x - 3) = 0$

$4 \neq 0$

$x - 3 = 0 \implies x = 3$

Critical number: 3

Solution: $(-\infty, 3)$

Test Interval	Representative x-value	Is inequality satisfied?
$(-\infty, 3)$	$x = 0$	$4(0) - 12 \overset{?}{<} 0$ $-12 < 0$
$(3, \infty)$	$x = 5$	$4(5) - 12 \overset{?}{<} 0$ $8 \not< 0$

87. $5x(7 - x) > 0$

$5x(7 - x) = 0$

$5x = 0 \implies x = 0$

$7 - x = 0 \implies x = 7$

Critical numbers: 0, 7

Solution: $(0, 7)$

Test Interval	Representative x-value	Is inequality satisfied?
$(-\infty, 0)$	$x = -1$	$5(-1)[7 - (-1)] \overset{?}{>} 0$ $-40 \not> 0$
$(0, 7)$	$x = 1$	$5(1)[7 - 1] \overset{?}{>} 0$ $30 > 0$
$(7, \infty)$	$x = 8$	$5(8)[7 - 8] \overset{?}{>} 0$ $-40 \not> 0$

89. $(x - 5)^2 - 36 > 0$

$(x - 5)^2 - 36 = 0$

$(x - 5)^2 = 36$

$x - 5 = \pm\sqrt{36}$

$x = 5 \pm 6$

$x = 5 + 6 = 11$

$x = 5 - 6 = -1$

Critical numbers: $-1, 11$

Solution: $(-\infty, -1) \cup (11, \infty)$

Test Interval	Representative x-value	Is inequality satisfied?
$(-\infty, -1)$	$x = -2$	$(-2 - 5)^2 - 36 \overset{?}{>} 0$ $13 \not> 0$
$(-1, 11)$	$x = 0$	$(0 - 5)^2 - 36 \overset{?}{>} 0$ $-11 > 0$
$(11, \infty)$	$x = 12$	$(12 - 5)^2 - 36 \overset{?}{>} 0$ $13 \not> 0$

91. $2x^2 + 3x - 20 < 0$

$2x^2 + 3x - 20 = 0$

$(2x - 5)(x + 4) = 0$

$2x - 5 = 0 \implies x = \frac{5}{2}$

$x + 4 = 0 \implies x = -4$

Critical numbers: $-4, \frac{5}{2}$

Solution: $\left(-4, \frac{5}{2}\right)$

Test Interval	Representative x-value	Is inequality satisfied?
$(-\infty, -4)$	$x = -5$	$2(-5)^2 + 3(-5) - 20 \overset{?}{<} 0$ $15 \not< 0$
$\left(-4, \frac{5}{2}\right)$	$x = 0$	$2(0)^2 + 3(0) - 20 \overset{?}{<} 0$ $-20 < 0$
$\left(\frac{5}{2}, \infty,\right)$	$x = 3$	$2(3)^2 + 3(3) - 20 \overset{?}{<} 0$ $7 \not< 0$

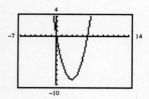

93. $y = x^2 - 6x, \quad y < 0$

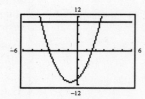

Use the graph of $y = x^2 - 6x$ to determine what portions of the graph lie below the x-axis.

$(0, 6)$

95. $f(x) = 2x^2 + 3x - 10, \quad y \le 10$

Determine what portions of the graph of $y = 2x^2 + 3x - 10$ lie below or on the line $y = 10$.

$[-4, 5/2]$

97. (a) $h = 200 - 16t^2$, $t = 0$

$h = 200 - 16(0)^2$

$= 200 - 0$

$= 200$

When $t = 0$, the height of the object is 200 feet.

(b) The object was dropped because the coefficient of the first-degree term is 0.

(c) $h = 200 - 16t^2$, $t > 0$

When the object strikes the ground, the height $h = 0$.

$0 = 200 - 16t^2$

$16t^2 = 200$

$t^2 = \dfrac{200}{16}$

$t = \pm\dfrac{\sqrt{200}}{\sqrt{16}}$

$t = \pm\dfrac{10\sqrt{2}}{4}$

$t = \pm\dfrac{5\sqrt{2}}{2}$ (Choose the positive solution.)

The object strikes the ground in $\dfrac{5\sqrt{2}}{2}$ or approximately 3.5 seconds.

99. *Verbal Model:* $\boxed{\text{Area}} = \frac{1}{2} \boxed{\text{Base}} \cdot \boxed{\text{Height}}$

Labels: Area $= 3000$ (square inches); Base $= b$ (inches); Height $= 1\frac{2}{3}(b)$ or $\frac{5}{3}b$ (inches)

Equation: $3000 = \frac{1}{2}b\left(\frac{5}{3}b\right)$

$3000 = \frac{5}{6}b^2$

$18{,}000 = 5b^2$

$3600 = b^2$

$\pm\sqrt{3600} = b$

$\pm 60 = b$

$b = 60$ and $\frac{5}{3}b = 100$

The base of the triangle is 60 inches and the height is 100 inches.

101. (a) $d = \sqrt{h^2 + 3^2}$

$d = \sqrt{h^2 + 9}$

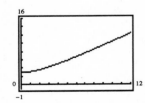

From the graph it appears that $h = 2.6$ when $d = 4$.

(b)

h	1	2	3	4	5	6	7
d	3.16	3.61	4.24	5	5.83	6.71	7.62

From the table it appears that h is approximately 2.5 when $d = 4$.

(c) $\sqrt{h^2 + 9} = d$

$\sqrt{h^2 + 9} = 4$

$\left(\sqrt{h^2 + 9}\right)^2 = 4^2$

$h^2 + 9 = 16$

$h^2 = 7$

$h = \sqrt{7}$ or $h \approx 2.646$

103. (a)

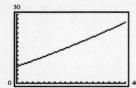

(b) $7.3 + 0.40t + 0.002t^2 = 20$

$$0.002t^2 + 0.40t - 12.7 = 0$$

$$t = \frac{-0.40 \pm \sqrt{0.40^2 - 4(0.002)(-12.7)}}{2(0.002)}$$

$$t = \frac{-0.40 \pm \sqrt{0.2616}}{0.004} \quad \text{(Discard value of } t \text{ that is not in domain.)}$$

$$t = \frac{-0.40 + \sqrt{0.2616}}{0.004} \implies t \approx 27.87$$

The percent of the population that graduated from college reached 20% during 1987.

(c) The percent of the population that graduated from college reached 20% during 1987.

105. *Verbal Model:* | Area of original lot | $+$ | Area of addition | $=$ | Area of the new lot |

     Labels: Area of original lot $= x(x)$ (square feet)

                 Area of addition $= x(20)$ (square feet)

                 Area of new lot $= 25,500$ (square feet)

     Equation: $x^2 + 20x = 25,500$

$$x^2 + 20x - 25,500 = 0$$

$$x = \frac{-20 \pm \sqrt{20^2 - 4(1)(-25,500)}}{2(1)}$$

$$x = \frac{-20 \pm \sqrt{102,400}}{2} \quad \text{(Discard negative value of } x.)$$

$$x = \frac{-20 + 320}{2} \implies x = 150 \quad \text{and} \quad x + 20 = 170$$

The dimensions of the new lot are 150 feet by 170 feet.

Chapter Test for Chapter 6

1. $x(x + 5) - 10(x + 5) = 0$

      $(x + 5)(x - 10) = 0$

             $x + 5 = 0 \implies x = -5$

           $x - 10 = 0 \implies x = 10$

2. $8x^2 - 21x - 9 = 0$

      $(8x + 3)(x - 3) = 0$

          $8x + 3 = 0 \implies x = -\frac{3}{8}$

           $x - 3 = 0 \implies x = 3$

3. $(x - 2)^2 = 50$

    $x - 2 = \pm\sqrt{50}$

    $x - 2 = \pm 5\sqrt{2}$

       $x = 2 \pm 5\sqrt{2}$

4. $(x + 3)^2 + 81 = 0$

    $(x + 3)^2 = -81$

     $x + 3 = \pm\sqrt{-81}$

     $x + 3 = \pm 9i$

        $x = -3 \pm 9i$

5. $x^2 - 3x + c$

$$x^2 - 3x + \left(-\frac{3}{2}\right)^2$$

$$x^2 - 3x + \frac{9}{4}$$

$$c = \frac{9}{4}$$

6. $2x^2 - 6x + 3 = 0$

$$2x^2 - 6x = -3$$

$$x^2 - 3x = -\frac{3}{2}$$

$$x^2 - 3x + \left(-\frac{3}{2}\right)^2 = -\frac{3}{2} + \frac{9}{4}$$

$$\left(x - \frac{3}{2}\right)^2 = \frac{3}{4}$$

$$x - \frac{3}{2} = \pm\frac{\sqrt{3}}{2}$$

$$x = \frac{3 \pm \sqrt{3}}{2}$$

7. $5x^2 - 12x + 10 = 0$

$$a = 5, \ b = -12, \ c = 10$$

$$b^2 - 4ac = (-12)^2 - 04(5)(10)$$

$$= 144 - 200$$

$$= -56$$

If the discriminant is a perfect square, there are two distinct rational solutions. If the discriminant is a positive nonperfect square, there are two distinct irrational solutions. If the discriminant is zero, there is one repeated rational solution. If the discriminant is negative, there are two distinct imaginary solutions.

8. $3x^2 - 8x + 3 = 0$

$$a = 3, \ b = -8, \ c = 3$$

$$x = \frac{-(-8) \pm \sqrt{(-8)^2 - 4(3)(3)}}{2(3)}$$

$$= \frac{8 \pm \sqrt{28}}{6}$$

$$= \frac{8 \pm 2\sqrt{7}}{6}$$

$$= \frac{4 \pm \sqrt{7}}{3}$$

9. $2y(y - 2) = 7$

$$2y^2 - 4y = 7$$

$$2y^2 - 4y - 7 = 0$$

$$a = 2, \ b = -4, \ c = -7$$

$$y = \frac{-(-4) \pm \sqrt{(-4)^2 - 4(2)(-7)}}{2(2)}$$

$$= \frac{4 \pm \sqrt{72}}{4}$$

$$= \frac{4 \pm 6\sqrt{2}}{4}$$

$$= \frac{2 \pm 3\sqrt{2}}{2}$$

10. $x^{2/3} - 6x^{1/3} + 8 = 0$ Let $u = x^{1/3}$

$$u^2 - 6u + 8 = 0$$

$$(u - 4)(u - 2) = 0$$

$$u - 4 = 0 \implies u = 4$$

$$u - 2 = 0 \implies u = 2$$

$$u = 4 \implies x^{1/3} = 4$$

$$x = 4^3$$

$$x = 64$$

$$u = 2 \implies x^{1/3} = 2$$

$$x = 2^3$$

$$x = 8$$

11. $y = -x^2 + 16$

$$y = -1(x - 0)^2 + 16$$

Vertex: $(0, 16)$

x-intercepts: Let $y = 0$

$$0 = -x^2 + 16$$

$$x^2 = 16$$

$$x = \pm 4$$

$(4, 0)$ and $(-4, 0)$

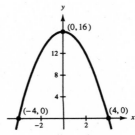

12. $y = x^2 - 2x - 15$

$y = x^2 - 2x + (-1)^2 - (-1)^2 - 15$

$y = (x^2 - 2x + 1) - 1 - 15$

$y = (x - 1)^2 - 16$

Vertex: $(1, -16)$

x-intercepts: Let $y = 0$

$$0 = x^2 - 2x - 15$$

$$0 = (x - 5)(x + 3)$$

$$x - 5 = 0 \implies x = 5$$

$$x + 3 = 0 \implies x = -3$$

$(5, 0)$ and $(-3, 0)$

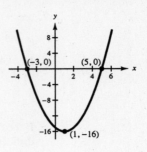

13. $2x(x - 3) < 0$

$2x(x - 3) = 0$

$$2x = 0 \implies x = 0$$

$$x - 3 = 0 \implies x = 3$$

Critical numbers: $0, 3$

Solution: $(0, 3)$

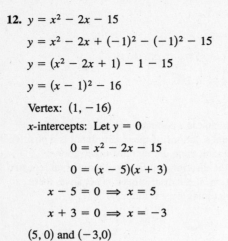

Test Interval	Representative x-value	Is inequality satisfied?
$(-\infty, 0)$	$x = -2$	$2(-2)(-2 - 3) \overset{?}{<} 0$ $20 \not< 0$
$(0, 3)$	$x = 1$	$2(1)(1 - 3) \overset{?}{<} 0$ $-4 < 0$
$(3, \infty)$	$x = 5$	$2(5)(5 - 3) \overset{?}{<} 0$ $20 \not< 0$

14. $x^2 - 4x \geq 12$

$x^2 - 4x - 12 \geq 0$

$x^2 - 4x - 12 = 0$

$(x - 6)(x + 2) = 0$

$$x - 6 = 0 \implies x = 6$$

$$x + 2 = 0 \implies x = -2$$

Critical numbers: $-2, 6$

Solution: $(-\infty, -2] \cup [6, \infty)$

Test Interval	Representative x-value	Is inequality satisfied?
$(-\infty, -2)$	$x = -4$	$(-4)^2 - 4(-4) - 12 \overset{?}{\geq} 0$ $20 \geq 0$
$(-2, 6)$	$x = 0$	$0^2 - 4(0) - 12 \overset{?}{\geq} 0$ $-12 \not\geq 0$
$(6, \infty)$	$x = 8$	$8^2 - 4(8) - 12 \overset{?}{\geq} 0$ $20 \geq 0$

15. $y = a(x - h)^2 + k$

$(h, k) = (3, -2) \implies y = a(x - 3)^2 - 2$

$(x, y) = (0, 4) \implies 4 = a(0 - 3)^2 - 2$

$$4 = 9a - 2$$

$$6 = 9a$$

$$\tfrac{2}{3} = a$$

$y = \tfrac{2}{3}(x - 3)^2 - 2$

$y = \tfrac{2}{3}(x^2 - 6x + 9) - 2$

$y = \tfrac{2}{3}x^2 - 4x + 6 - 2$

$y = \tfrac{2}{3}x^2 - 4x + 4$

16. *Verbal Model:* | Length | · | Width | = | Area |

    Labels: Length = l (feet); Width = $l - 8$ (feet); Area = 240 (square feet)

    Equation: $l(l - 8) = 240$

$$l^2 - 8l = 240$$

$$l^2 - 8l - 240 = 0$$

$$(l - 20)(l + 12) = 0$$

$$l - 20 = 0 \implies l = 20 \text{ and } l - 8 = 12$$

$$l + 12 = 0 \implies l = -12 \text{ (Discard this negative solution.)}$$

Thus, the length of the rectangle is 20 feet and the width is 12 feet.

17. $h = -16t^2 + 75$

$$35 = -16t^2 + 75$$

$$16t^2 - 40 = 0$$

$$16t^2 = 40$$

$$t^2 = \frac{40}{16}$$

$$t = \pm\sqrt{\frac{40}{16}}$$

$$t = \pm\frac{2\sqrt{10}}{4}$$

$$t = \pm\frac{\sqrt{10}}{2} \quad \text{(Choose the positive solution.)}$$

$$t = \frac{\sqrt{10}}{2} \approx 1.58$$

It will take the object approximately 1.58 seconds to fall to a height of 35 feet.

18. $A = \dfrac{2}{\pi}(100x - x^2), \quad 0 < x < 100$

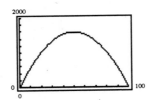

The value of A is maximum when the length $x = 50$ feet.

Cumulative Test for Chapters 1–6

1. $2x^2 + 5x - 7 = (2x + 7)(x - 1)$

2. $11x^2 + 6x - 5 = (11x - 5)(x + 1)$

3. $12x^3 - 27x = 3x(4x^2 - 9)$

$$= 3x[(2x)^2 - 3^2]$$

$$= 3x(2x + 3)(2x - 3)$$

4. $8x^3 + 125 = (2x)^3 + 5^3$

$$= (2x + 5)(4x^2 - 10x + 25)$$

5. $125 - 50x = 0$

$$-50x = -125$$

$$x = \frac{-125}{-50}$$

$$x = \frac{5}{2}$$

6. $t^2 - 8t = 0$

$$t(t - 8) = 0$$

$$t = 0$$

$$t - 8 = 0 \implies t = 8$$

7. $x^2(x + 2) - (x + 2) = 0$

$$(x + 2)(x^2 - 1) = 0$$

$$(x + 2)(x + 1)(x - 1) = 0$$

$$x + 2 = 0 \implies x = -2$$

$$x + 1 = 0 \implies x = -1$$

$$x - 1 = 0 \implies x = 1$$

8. $x(10 - x) = 25$

$$10x - x^2 = 25$$

$$0 = x^2 - 10x + 25$$

$$0 = (x - 5)^2$$

$$x - 5 = 0 \implies x = 5$$

9. $\dfrac{x + 3}{7} = \dfrac{x - 1}{4}$

$$4(x + 3) = 7(x - 1)$$

$$4x + 12 = 7x - 7$$

$$-3x + 12 = -7$$

$$-3x = -19$$

$$x = \frac{19}{3}$$

10. $\dfrac{x}{4} - \dfrac{x + 2}{6} = \dfrac{3}{2}$

$$12\left(\frac{x}{4} - \frac{x + 2}{6}\right) = 12\left(\frac{3}{2}\right)$$

$$3x - 2(x + 2) = 18$$

$$3x - 2x - 4 = 18$$

$$x - 4 = 18$$

$$x = 22$$

11. Point: $(4, -2)$

Slope: $m = \dfrac{5}{2}$

$$y - y_1 = m(x - x_1)$$

$$y - (-2) = \frac{5}{2}(x - 4)$$

$$y + 2 = \frac{5}{2}x - 10$$

$$y = \frac{5}{2}x - 12$$

12. $m = \dfrac{y_2 - y_1}{x_2 - x_1}$

$$m = \frac{2 - 8}{1 - (-5)}$$

$$= \frac{-6}{6}$$

$$= -1$$

Point: $(1, 2)$

Slope: $m = -1$

$$y - y_1 = m(x - x_1)$$

$$y - 2 = -1(x - 1)$$

$$y - 2 = -x + 1$$

$$y = -x + 3$$

13. No, the graph does not represent y as a function of x because a vertical line could cross the graph at more than one point.

14. Area of triangle $= \dfrac{1}{2}$(Base)(Height)

$$\text{Area} = \dfrac{1}{2}(5x)(2x + 9)$$

$$= \dfrac{5x}{2}(2x + 9) \text{ or } 5x^2 + \dfrac{45x}{2}$$

15. Increase in cost for next month $= 4\%(\$23{,}500)$

$$= \$940$$

$$\text{Interest penalty} = \$725$$

Based on this information, the customer should buy the car now instead of waiting another month.

16. $\begin{cases} 2x + 0.5y = 8 \implies -8x - 2y = -32 \\ 3x + 2y = 22 \implies \dfrac{3x + 2y = 22}{} \end{cases}$

$$\dfrac{\qquad\qquad}{-5x \qquad = -10}$$

$$x = 2 \text{ and } 3(2) + 2y = 22$$

$$2y = 16$$

$$y = 8$$

$(2, 8)$

17. *Verbal Model:* 0.12 $\boxed{\text{Liters of 12\% solution}}$ -0.00 $\boxed{\text{Liters of water}}$ $= 0.20$ $\boxed{\text{Liters of 20\% solution}}$

Labels: Liters of 12% solution $= 160$

Liters of water $= x$

Liters of 20% solution $= 160 - x$

Equation: $0.12(160) - (0)x = 0.20(160 - x)$

$$19.2 - 0 = 32 - 0.20x$$

$$19.2 = 32 - 0.20x$$

$$-12.8 = -0.20x$$

$$x = \dfrac{-12.8}{-0.20}$$

$$x = 64$$

So, 64 liters of water should be evaporated from the 12% solution.

18. $\begin{bmatrix} 3 & 7 \\ -2 & 6 \end{bmatrix} = 3(6) - (-2)(7)$

$$= 18 + 14$$

$$= 32$$

19. $\begin{vmatrix} 3 & -2 & 1 \\ 0 & 5 & 3 \\ 6 & 1 & 1 \end{vmatrix} = 3\begin{vmatrix} 5 & 3 \\ 1 & 1 \end{vmatrix} - 0 + 6\begin{vmatrix} -2 & 1 \\ 5 & 3 \end{vmatrix}$

$$= 3(5 - 3) - 0 + 6(-6 - 5)$$

$$= 3(2) + 6(-11)$$

$$= 6 - 66$$

$$= -60$$

20. (a) $\sqrt{75x^3y} = \sqrt{25x^2 \cdot 3xy}$

$\qquad\qquad = 5x\sqrt{3xy}$

(b) $\dfrac{40}{3 - \sqrt{5}} = \dfrac{40}{3 - \sqrt{5}} \cdot \dfrac{3 + \sqrt{5}}{3 + \sqrt{5}}$

$\qquad\quad = \dfrac{40(3 + \sqrt{5})}{3^2 - (\sqrt{5})^2}$

$\qquad\quad = \dfrac{40(3 + \sqrt{5})}{9 - 5}$

$\qquad\quad = \dfrac{\cancel{4}(10)(3 + \sqrt{5})}{\cancel{4}}$

$\qquad\quad = 10(3 + \sqrt{5}) \text{ or } 30 + 10\sqrt{5}$

21. (a) $(5 - 3i)(-2 - 4i) = -10 - 20i + 6i + 12i^2$

$\qquad\qquad\qquad\qquad = -10 - 14i + 12(-1)$

$\qquad\qquad\qquad\qquad = -22 - 14i$

(b) $(-3 - 5i)(-3 + 5i) = (-3)^2 - (5i)^2$

$\qquad\qquad\qquad\qquad = 9 - 25i^2$

$\qquad\qquad\qquad\qquad = 9 - 25(-1)$

$\qquad\qquad\qquad\qquad = 9 + 25$

$\qquad\qquad\qquad\qquad = 34$

22. $y = a(x - h)^2 + k$ and $(h, k) = (2, -1)$

$y = a(x - 2)^2 - 1$ and $(x, y) = (0, 0)$

$0 = a(0 - 2)^2 - 1$

$0 = a(4) - 1$

$1 = 4a$

$\frac{1}{4} = a \implies y = \frac{1}{4}(x - 2)^2 - 1$

$\qquad\qquad\quad = \frac{1}{4}(x^2 - 4x + 4) - 1$

$\qquad\qquad\quad = \frac{1}{4}x^2 - x + 1 - 1$

$\qquad\qquad\quad = \frac{1}{4}x^2 - x$

C H A P T E R 7
Rational Expressions and Rational Functions

CHAPTER 7
Rational Expressions and Rational Functions

Section 7.1   Simplifying Rational Expressions

Solutions to Odd-Numbered Exercises

1. The denominator is zero when $x - 8 = 0$ or $x = 8$.
So, the domain is all real values of x such that $x \neq 8$.

Domain: $(-\infty, 8) \cup (8, \infty)$

3. The denominator is zero when $x + 4 = 0$ or $x = -4$.
So, the domain is all real values of x such that $x \neq -4$.

Domain: $(-\infty, -4) \cup (-4, \infty)$

5. The denominator $4 \neq 0$ for all real values of x.

Domain: $(-\infty, \infty)$

7. The denominator, $t^2 - 16$ or $(t + 4)(t - 4)$, is zero when $t = -4$ or $t = 4$. So, the domain is all real values of t such that $t \neq -4$ and $t \neq 4$.

Domain: $(-\infty, -4) \cup (-4, 4) \cup (4, \infty)$

9. The denominator, $u^2 - 4u - 5$ or $(u - 5)(u + 1)$, is zero when $u = 5$ or $u = -1$. So, the domain is all real values of u such that $u \neq 5$ and $u \neq -1$.

Domain: $(-\infty, -1) \cup (-1, 5) \cup (5, \infty)$

11. The denominator $x^2 + 9 \neq 0$ for all real values of x.

Domain: $(-\infty, \infty)$

13. The denominator, $x^2 - 3x$ or $x(x - 3)$, is zero when $x = 0$ or $x = 3$. So, the domain is all real values of x such that $x \neq 0$ and $x \neq 3$.

Domain: $(-\infty, 0) \cup (0, 3) \cup (3, \infty)$

15. $\dfrac{32y^2}{12y} = \dfrac{(4y)(8y)}{(4y)(3)} = \dfrac{8y}{3}, \quad y \neq 0$

17. $\dfrac{18x^2y}{15xy^4} = \dfrac{6(3)(x)(x)(y)}{5(3)(x)(y)(y^3)}$

$= \dfrac{6x}{5y^3}, x \neq 0$

19. $\dfrac{x^2(x - 8)}{x(x - 8)} = \dfrac{x(x)(x - 8)}{x(x - 8)}$

$= x, x \neq 8, x \neq 0$

21. $\dfrac{6y^2 + 3y^3}{3y^2} = \dfrac{3y^2(2 + y)}{3y^2} = 2 + y, \quad y \neq 0$

23. $\dfrac{2x - 3}{4x - 6} = \dfrac{(2x - 3)(1)}{2(2x - 3)}$

$= \dfrac{1}{2}, x \neq \dfrac{3}{2}$

25. $\dfrac{81 - y^2}{2y - 18} = \dfrac{(9 - y)(9 + y)}{2(y - 9)}$

$= \dfrac{-1(y - 9)(9 + y)}{2(y - 9)}$

$= -\dfrac{9 + y}{2}, \quad y \neq 9$

27. $\dfrac{x^2 - 25z^2}{x + 5z} = \dfrac{(x + 5z)(x - 5z)}{1(x + 5z)}$

$= x - 5z, x \neq -5z$

29. $\dfrac{u^2 - 12u + 36}{u - 6} = \dfrac{(u - 6)(u - 6)}{1(u - 6)}$

$= u - 6, u \neq 6$

31. $\dfrac{x^2 - x - 20}{3x - 15} = \dfrac{(x - 5)(x + 4)}{3(x - 5)}$

$\qquad = \dfrac{x + 4}{3}, \quad x \neq 5$

33. $\dfrac{z^2 + 22z + 121}{3x + 33} = \dfrac{(z + 11)(z + 11)}{3(z + 11)}$

$\qquad = \dfrac{z + 11}{3}, \quad z \neq -11$

35. $\dfrac{y^3 - 4y}{y^2 + 4y - 12} = \dfrac{y(y^2 - 4)}{(y + 6)(y - 2)}$

$\qquad = \dfrac{y(y + 2)(y - 2)}{(y + 6)(y - 2)}$

$\qquad = \dfrac{y(y + 2)}{y + 6}, \quad y \neq 2$

37. $\dfrac{3 - x}{2x^2 - 3x - 9} = \dfrac{-1(x - 3)}{(2x + 3)(x - 3)}$

$\qquad = -\dfrac{1}{2x + 3}, \quad x \neq 3$

39. $\dfrac{x^2 + 3x - 40}{10 + 3x - x^2} = \dfrac{(x + 8)(x - 5)}{-1(x^2 - 3x - 10)}$

$\qquad = \dfrac{(x + 8)(x - 5)}{-1(x - 5)(x + 2)}$

$\qquad = -\dfrac{x + 8}{x + 2}, \quad x \neq 5$

41. $\dfrac{56z^2 - 3z - 20}{49z^2 - 16} = \dfrac{(8z - 5)(7z + 4)}{(7z + 4)(7z - 4)}$

$\qquad = \dfrac{8z - 5}{7z - 4}, \quad z \neq -\dfrac{4}{7}$

43. $\dfrac{4x^2y + x}{3x} = \dfrac{x(4xy + 1)}{x(3)}$

$\qquad = \dfrac{4xy + 1}{3}, \quad x \neq 0$

45. $\dfrac{5xy + 3x^2y^2}{xy^3} = \dfrac{(xy)(5 + 3xy)}{(xy)(y^2)}$

$\qquad = \dfrac{5 + 3xy}{y^2}, \quad x \neq 0$

47. $\dfrac{3m^2 - 12n^2}{m^2 + 4mn + 4n^2} = \dfrac{3(m^2 - 4n^2)}{(m + 2n)(m + 2n)}$

$\qquad = \dfrac{3(m + 2n)(m - 2n)}{(m + 2n)(m + 2n)}$

$\qquad = \dfrac{3(m - 2n)}{m + 2n}$

49. $\dfrac{x^3 + 27z^3}{x^2 + xz - 6z^2} = \dfrac{(x + 3z)(x^2 - 3xz + 9z^2)}{(x + 3z)(x - 2z)}$

$\qquad = \dfrac{x^2 - 3xz + 9z^2}{x - 2z}, \quad x \neq -3z$

51. $\dfrac{x^2 + 4xy}{x^2 - 16y^2} = \dfrac{x(x + 4y)}{(x + 4y)(x - 4y)}$

$\qquad = \dfrac{x}{x - 4y}, \quad x \neq -4y$

53. $\dfrac{mn + 3m - n^2 - 3n}{m^2 - n^2} = \dfrac{m(n + 3) - n(n + 3)}{(m + n)(m - n)}$

$\qquad = \dfrac{(n + 3)(m - n)}{(m + n)(m - n)}$

$\qquad = \dfrac{n + 3}{m + n}, \quad m \neq n$

55. $f(x) = \dfrac{4x}{x + 3}$

(a) $f(1) = \dfrac{4(1)}{1 + 3}$

$\qquad = \dfrac{4}{4} = 1$

(b) $f(-2) = \dfrac{4(-2)}{-2 + 3}$

$\qquad = \dfrac{-8}{1} = -8$

(c) $f(-3) = \dfrac{4(-3)}{-3 + 3}$

$\qquad = \dfrac{-12}{0}$ undefined

(d) $f(0) = \dfrac{4(0)}{0 + 3}$

$\qquad = \dfrac{0}{3} = 0$

$\dfrac{4x}{x + 3}$ is undefined for $x = -3$.

57. $h(s) = \dfrac{s^2}{s^2 - s - 2}$

(a) $h(10) = \dfrac{10^2}{10^2 - 10 - 2}$

$= \dfrac{100}{88} = \dfrac{25}{22}$

(b) $h(0) = \dfrac{0^2}{0^2 - 0 - 2}$

$= \dfrac{0}{-2} = 0$

(c) $h(-1) = \dfrac{(-1)^2}{(-1)^2 - (-1) - 2}$

$= \dfrac{1}{0}$ undefined

$\dfrac{s^2}{s^2 - s - 2}$ is undefined for $s = -1$.

(d) $h(2) = \dfrac{2^2}{2^2 - 2 - 2}$

$= \dfrac{4}{0}$ undefined

$\dfrac{s^2}{s^2 - s - 2}$ is undefined for $s = 2$.

59. The reduction is invalid; 5 is a term, but not a factor, of the numerator. You can divide out common factors, but not common terms.

61. The reduction is valid.

63. $\dfrac{x^{n+1} - 3x}{x} = \dfrac{\cancel{x}(x^n - 3)}{\cancel{x}(1)}$

$= x^n - 3, x \neq 0$

65. $\dfrac{x^{2n} - 4}{x^n + 2} = \dfrac{(\cancel{x^n + 2})(x^n - 2)}{1(\cancel{x^n + 2})}$

$= x^n - 2, x^n \neq -2$

67. $x = $ the number of units ordered

Domain $= \{1, 2, 3, 4, \ldots\}$, assuming a fractional number of units cannot be ordered.

69. $P = 2\left(x + \dfrac{500}{x}\right)$

The denominator x cannot be zero and x cannot be negative because x represents the length of a rectangle. Thus, the domain is $x > 0$.

Domain: $(0, \infty)$

71. Area of shaded portion $= x(x + 1)$

Total area of figure $= (x + 1)(x + 3)$

Ratio $= \dfrac{x\cancel{(x + 1)}}{\cancel{(x + 1)}(x + 3)} = \dfrac{x}{x + 3}, x > 0$

73. Area of shaded portion $= \dfrac{1}{2}(4x)(3x) = 6x^2$

Total area of figure $= \dfrac{1}{2}(8x)(6x) = 24x^2$

Ratio $= \dfrac{6x^2}{24x^2} = \dfrac{1\cancel{(6x^2)}}{4\cancel{(6x^2)}} = \dfrac{1}{4}, \quad x > 0$

75.

$x = $ depth of rectangular pool

$3x = $ width of rectangular pool

$3x + 6 = $ length of rectangular pool

Volume $= $ (length)(width)(depth)

$= (3x + 6)(3x)(x)$

$= (3x + 6)(3x^2)$

$= 9x^3 + 18x^2$

Ratio $= \dfrac{9\pi x^3 + 18\pi x^2}{9x^3 + 18x^2}$

$= \dfrac{\pi\cancel{(9x^3 + 18x^2)}}{1\cancel{(9x^3 + 18x^2)}} = \pi, \quad x > 0$

$2(3x) = 6x = $ diameter of circular pool

$3x = $ radius of circular pool

$x + 2 = $ depth of circular pool

Volume $= \pi(\text{radius})^2(\text{depth})$

$= \pi(3x)^2(x + 2)$

$= 9\pi x^2(x + 2)$

$= 9\pi x^3 + 18\pi x^2$

77. Average cost $= \dfrac{\text{Cost of Medicare}}{\text{Number of persons}}$

$= \dfrac{C}{P}$

$= \dfrac{111.5 + 13.91t \text{ billion}}{31.9 + 0.33t \text{ million}}$

$= \dfrac{(111.5 + 13.91t) \times 10^9}{(31.9 + 0.33t) \times 10^6}$

$= \dfrac{(111.5 + 13.91t) \times 10^3}{31.9 + 0.33t}$

$= \dfrac{1000(111.5 + 13.91t)}{31.9 + 0.33t}$

79. A rational expression is in reduced form if the numerator and denominator have no common factors other than ± 1.

81. You can divide out common factors from the numerator and denominator of a rational expression, but the error in this example is the dividing out of common terms.

83. $\dfrac{54}{62} = \dfrac{\cancel{2}(27)}{\cancel{2}(31)} = \dfrac{27}{31}$

85. $\dfrac{112}{200} = \dfrac{\cancel{8}(14)}{\cancel{8}(25)} = \dfrac{14}{25}$

87. $8x^2y - 24xy - 80y = 8y(x^2 - 3x - 10)$
$\qquad\qquad\qquad\qquad = 8y(x - 5)(x + 2)$

89. *Labels:*   Amount of 11% solution $= x$ (milliliters)

Amount of 6% solution $= y$ (milliliters)

Equations: $\begin{cases} x + \quad y = 600 \\ 0.11x + 0.06y = 0.08(600) \end{cases} \Rightarrow \begin{cases} -0.06x - 0.06y = -36 \\ \underline{\quad 0.11x + 0.06y = \quad 48} \\ \quad 0.05x \qquad\qquad = \quad 12 \\ \qquad x \qquad\qquad = \quad 240 \end{cases}$

$240 + y = 600$

$y = 360$

$(240, 360)$

So, 240 milliliters of the 11% solution and 360 milliliters of the 6% solution should be used.

Section 7.2   Multiplying and Dividing Rational Expressions

1. $16u^4 \cdot \dfrac{12}{8u^2} = \dfrac{16u^4}{1} \cdot \dfrac{12}{8u^2}$

$= \dfrac{16(12)u^4}{8u^2}$

$= \dfrac{\cancel{8}(2)(12)\cancel{(u^2)}(u^2)}{\cancel{8}\cancel{(u^2)}}$

$= 24u^2, \; u \neq 0$

3. $\dfrac{8s^3}{9s} \cdot \dfrac{6s^2}{32s} = \dfrac{8(6)s^5}{9(32)s^2}$

$= \dfrac{\cancel{8}\cancel{(3)}\cancel{(2)}\cancel{(s^2)}(s^3)}{\cancel{3}(3)\cancel{(8)}\cancel{(2)}(2)\cancel{(s^2)}}$

$= \dfrac{s^3}{6}, \; s \neq 0$

5. $\dfrac{-3x^4y}{7xy^4} \cdot \dfrac{8x^2y^2}{9y^3} = \dfrac{-3(8)x^6y^3}{7(9)xy^7}$

$= -\dfrac{\cancel{3}(8)\cancel{(x)}(x^5)\cancel{(y^3)}}{7\cancel{(3)}(3)\cancel{(x)}\cancel{(y^3)}(y^4)}$

$= -\dfrac{8x^5}{21y^4}, \; x \neq 0$

7. $\dfrac{1 - 3x}{4} \cdot \dfrac{46}{15 - 45x} = \dfrac{(1 - 3x)(46)}{4(15)(1 - 3x)}$

$= \dfrac{\cancel{(1 - 3x)}\cancel{(2)}(23)}{\cancel{2}(2)(15)\cancel{(1 - 3x)}}$

$= \dfrac{23}{30}, \; x \neq \dfrac{1}{3}$

9. $\dfrac{x + 25}{8} \cdot \dfrac{8}{x + 25} = \dfrac{(x + 25)(8)}{8(x + 25)}$

$= \dfrac{1\cancel{(x + 25)}\cancel{(8)}}{\cancel{8}\cancel{(x + 25)}}$

$= 1, \; x \neq -25$

11. $\dfrac{12 - r}{3} \cdot \dfrac{3}{r - 12} = \dfrac{(12 - r)(3)}{3(r - 12)}$

$\qquad\qquad = \dfrac{-1\cancel{(r - 12)}\cancel{(3)}}{\cancel{3}\cancel{(r - 12)}}$

$\qquad\qquad = -1,\ r \neq 12$

13. $\dfrac{6r}{r - 2} \cdot \dfrac{r^2 - 4}{33r^2} = \dfrac{6r(r + 2)(r - 2)}{(r - 2)(33r^2)}$

$\qquad\qquad = \dfrac{\cancel{3}(2)\cancel{(r)}(r + 2)\cancel{(r - 2)}}{\cancel{(r - 2)}\cancel{(3)}(11)\cancel{(r)}(r)}$

$\qquad\qquad = \dfrac{2(r + 2)}{11r},\ r \neq 2$

15. $\dfrac{(2x - 3)(x + 8)}{x^3} \cdot \dfrac{x}{3 - 2x} = \dfrac{(2x - 3)(x + 8)x}{x^3(3 - 2x)}$

$\qquad\qquad = \dfrac{\cancel{(2x - 3)}(x + 8)\cancel{(x)}}{\cancel{x}(x^2)(-1)\cancel{(2x - 3)}} = -\dfrac{x + 8}{x^2},\ x \neq \dfrac{3}{2}$

17. $\dfrac{3x - 15}{2x^2 - 50} \cdot \dfrac{2x^2 + 16x + 30}{6x + 9} = \dfrac{3(x - 5)(2)(x^2 + 8x + 15)}{2(x^2 - 25)(3)(2x + 3)}$

$\qquad\qquad = \dfrac{\cancel{3}\cancel{(x - 5)}\cancel{(2)}\cancel{(x + 5)}(x + 3)}{2\cancel{(x + 5)}\cancel{(x - 5)}\cancel{(3)}(2x + 3)}$

$\qquad\qquad = \dfrac{x + 3}{2x + 3},\ \ x \neq 5, x \neq -5$

19. $\dfrac{x^2 + 5x - 6}{x^2 + 4x} \cdot \dfrac{2x^2 + 4x - 16}{3x - 3} = \dfrac{(x + 6)(x - 1)(2)(x^2 + 2x - 8)}{x(x + 4)(3)(x - 1)}$

$\qquad\qquad = \dfrac{(x + 6)\cancel{(x - 1)}(2)\cancel{(x + 4)}(x - 2)}{x\cancel{(x + 4)}(3)\cancel{(x - 1)}}$

$\qquad\qquad = \dfrac{2(x + 6)(x - 2)}{3x},\ \ x \neq -4, x \neq 1$

21. $\dfrac{2t^2 - t - 15}{t + 2} \cdot \dfrac{t^2 - t - 6}{t^2 - 6t + 9} = \dfrac{(2t + 5)(t - 3)(t - 3)(t + 2)}{(t + 2)(t - 3)^2}$

$\qquad\qquad = \dfrac{(2t + 5)\cancel{(t - 3)}\cancel{(t - 3)}\cancel{(t + 2)}}{\cancel{(t + 2)}\cancel{(t - 3)}\cancel{(t - 3)}} = 2t + 5,\ t \neq 3, t \neq -2$

23. $\dfrac{x^2 + x}{2x + 3} \cdot \dfrac{3x^2 + 19x + 28}{x^2 + 5x + 4} = \dfrac{x(x + 1)(3x + 7)(x + 4)}{(2x + 3)(x + 4)(x + 1)}$

$\qquad\qquad = \dfrac{x(3x + 7)}{2x + 3},\ \ x \neq -4, x \neq -1$

25. $\dfrac{2x^2 - 2x - 12}{x^2 - 5x - 6} \cdot \dfrac{x^2 - 7x + 6}{3x^2 - 12x + 9} = \dfrac{2(x^2 - x - 6)(x - 6)(x - 1)}{(x - 6)(x + 1)(3)(x^2 - 4x + 3)}$

$\qquad\qquad = \dfrac{2\cancel{(x - 3)}(x + 2)\cancel{(x - 6)}\cancel{(x - 1)}}{\cancel{(x - 6)}(x + 1)(3)\cancel{(x - 3)}\cancel{(x - 1)}}$

$\qquad\qquad = \dfrac{2(x + 2)}{3(x + 1)},\ \ x \neq 6, x \neq 3, x \neq 1$

27. $(x^2 - 4y^2) \cdot \dfrac{xy}{(x - 2y)^2} = \dfrac{x^2 - 4y^2}{1} \cdot \dfrac{xy}{(x - 2y)^2}$

$\qquad\qquad = \dfrac{(x + 2y)(x - 2y)(xy)}{1(x - 2y)(x - 2y)} = \dfrac{(x + 2y)\cancel{(x - 2y)}(xy)}{\cancel{(x - 2y)}(x - 2y)} = \dfrac{xy(x + 2y)}{(x - 2y)}$

29. $(u - 2v)^2 \cdot \dfrac{u + 2v}{u - 2v} = \dfrac{(u - 2v)^2}{1} \cdot \dfrac{u + 2v}{u - 2v} = \dfrac{(u - 2v)^2(u + 2v)}{1(u - 2v)}$

$$= \dfrac{\cancel{(u - 2v)}(u - 2v)(u + 2v)}{\cancel{(u - 2v)}} = (u - 2v)(u + 2v), \; u \neq 2v$$

31. $\dfrac{x + 5}{x - 5} \cdot \dfrac{2x^2 - 9x - 5}{3x^2 + x - 2} \cdot \dfrac{x^2 - 1}{x^2 + 7x + 10} = \dfrac{(x + 5)(2x + 1)(x - 5)(x + 1)(x - 1)}{(x - 5)(3x - 2)(x + 1)(x + 5)(x + 2)}$

$$= \dfrac{\cancel{(x + 5)}(2x + 1)\cancel{(x - 5)}\cancel{(x + 1)}(x - 1)}{\cancel{(x - 5)}(3x - 2)\cancel{(x + 1)}\cancel{(x + 5)}(x + 2)}$$

$$= \dfrac{(2x + 1)(x - 1)}{(3x - 2)(x + 2)}, \; x \neq 5, \; x \neq -5, \; x \neq -1$$

33. $\dfrac{x^3 + 3x^2 - 4x - 12}{x^3 - 3x^2 - 4x + 12} \cdot \dfrac{x^2 - 9}{x} = \dfrac{x^2(x + 3) - 4(x + 3)}{x^2(x - 3) - 4(x - 3)} \cdot \dfrac{x^2 - 9}{x} = \dfrac{(x + 3)(x^2 - 4)}{(x - 3)(x^2 - 4)} \cdot \dfrac{(x + 3)(x - 3)}{x}$

$$= \dfrac{(x + 3)\cancel{(x^2 - 4)}(x + 3)\cancel{(x - 3)}}{\cancel{(x - 3)}\cancel{(x^2 - 4)}(x)} = \dfrac{(x + 3)^2}{x}, \; x \neq 3, \; x^2 \neq 4$$

35. $\dfrac{7xy^2}{10u^2v} \div \dfrac{21x^3}{45uv} = \dfrac{7xy^2}{10u^2v} \cdot \dfrac{45uv}{21x^3}$

$$= \dfrac{7(45)xy^2\,uv}{10(21)u^2\,vx^3} = \dfrac{7\cancel{(5)}(3)\cancel{(3)}\cancel{(x)}(y^2)\cancel{(u)}\cancel{(v)}}{\cancel{5}(2)\cancel{(7)}\cancel{(3)}\cancel{(u)}(u)\cancel{(v)}\cancel{(x)}(x^2)} = \dfrac{3y^2}{2ux^2}, \; v \neq 0$$

37. $\dfrac{3(a + b)}{4} \div \dfrac{(a + b)^2}{2} = \dfrac{3(a + b)}{4} \cdot \dfrac{2}{(a + b)^2}$

$$= \dfrac{3(a + b)(2)}{4(a + b)^2} = \dfrac{3\cancel{(a + b)}\cancel{(2)}}{2\cancel{(2)}\cancel{(a + b)}(a + b)} = \dfrac{3}{2(a + b)}$$

39. $\dfrac{(x^3y)^2}{(x + 2y)^2} \div \dfrac{x^2y}{(x + 2y)^3} = \dfrac{(x^3y)^2}{(x + 2y)^2} \cdot \dfrac{(x + 2y)^3}{x^2y}$

$$= \dfrac{x^6y^2(x + 2y)^3}{(x + 2y)^2(x^2y)} = \dfrac{\cancel{x^2}(x^4)(y)\cancel{(y)}\cancel{(x + 2y)^2}(x + 2y)}{\cancel{(x + 2y)^2}\cancel{(x^2)}\cancel{(y)}}$$

$$= x^4y(x + 2y), \; x \neq 0, \; y \neq 0, \; x \neq -2y$$

41. $\dfrac{\left(\dfrac{x^2}{12}\right)}{\left(\dfrac{5x}{18}\right)} = \dfrac{x^2}{12} \div \dfrac{5x}{18} = \dfrac{x^2}{12} \cdot \dfrac{18}{5x}$

$$= \dfrac{18x^2}{12(5)x} = \dfrac{\cancel{6}(3)\cancel{(x)}(x)}{\cancel{6}(2)(5)\cancel{(x)}}$$

$$= \dfrac{3x}{10}, \; x \neq 0$$

43. $\dfrac{\left(\dfrac{25x^2}{x - 5}\right)}{\left(\dfrac{10x}{5 - x}\right)} = \dfrac{25x^2}{x - 5} \div \dfrac{10x}{5 - x}$

$$= \dfrac{25x^2}{x - 5} \cdot \dfrac{5 - x}{10x} = \dfrac{25x^2(5 - x)}{(x - 5)(10x)}$$

$$= \dfrac{5\cancel{(5)}\cancel{(x)}(x)(-1)\cancel{(x - 5)}}{\cancel{(x - 5)}\cancel{(5)}(2)\cancel{(x)}}$$

$$= -\dfrac{5x}{2}, \; x \neq 0, \; x \neq 5$$

45. $\dfrac{16x^2 + 8x + 1}{3x^2 + 8x - 3} \div \dfrac{4x^2 - 3x - 1}{x^2 + 6x + 9} = \dfrac{16x^2 + 8x + 1}{3x^2 + 8x - 3} \cdot \dfrac{x^2 + 6x + 9}{4x^2 - 3x - 1}$

$$= \dfrac{(4x + 1)^2(x + 3)^2}{(3x - 1)(x + 3)(4x + 1)(x - 1)}$$

$$= \dfrac{\cancel{(4x + 1)}(4x + 1)\cancel{(x + 3)}(x + 3)}{(3x - 1)\cancel{(x + 3)}\cancel{(4x + 1)}(x - 1)}$$

$$= \dfrac{(4x + 1)(x + 3)}{(3x - 1)(x - 1)}, \ x \neq -3, \ x \neq -\dfrac{1}{4}$$

47. $\dfrac{y^3 - 8}{y^2 + y - 6} \div \dfrac{y^3 + 2y^2 + 4y}{y^2 + 3y} = \dfrac{y^3 - 8}{y^2 + y - 6} \cdot \dfrac{y^2 + 3y}{y^3 + 2y^2 + 4y}$

$$= \dfrac{\cancel{(y - 2)}\cancel{(y^2 + 2y + 4)}\cancel{(y)}\cancel{(y + 3)}}{\cancel{(y + 3)}\cancel{(y - 2)}\cancel{(y)}\cancel{(y^2 + 2y + 4)}}$$

$$= 1, \quad y \neq 0, \ y \neq -3, \ y \neq 2$$

49. $\dfrac{x(x + 3) - 2(x + 3)}{x^2 - 4} \div \dfrac{x}{x^2 + 4x + 4} = \dfrac{x(x + 3) - 2(x + 3)}{x^2 - 4} \cdot \dfrac{x^2 + 4x + 4}{x}$

$$= \dfrac{(x + 3)(x - 2)(x + 2)^2}{(x + 2)(x - 2)(x)}$$

$$= \dfrac{(x + 3)\cancel{(x - 2)}\cancel{(x + 2)}(x + 2)}{\cancel{(x + 2)}\cancel{(x - 2)}(x)}$$

$$= \dfrac{(x + 3)(x + 2)}{x}, \ x \neq -2, \ x \neq 2$$

51. $\dfrac{\left(\dfrac{x^2 - 4}{x - 7}\right)}{\left(\dfrac{3x + 6}{x^2 - 6x - 7}\right)} = \dfrac{x^2 - 4}{x - 7} \div \dfrac{3x + 6}{x^2 - 6x - 7}$

$$= \dfrac{x^2 - 4}{x - 7} \cdot \dfrac{x^2 - 6x - 7}{3x + 6}$$

$$= \dfrac{(x + 2)(x - 2)\cancel{(x - 7)}(x + 1)}{\cancel{(x - 7)}(3)(x + 2)}$$

$$= \dfrac{(x - 2)(x + 1)}{3}, \ x \neq -2, \ x \neq 7, \ x \neq -1$$

53. $\dfrac{\left(\dfrac{x^2 + 3x - 10}{x^2 + x - 6}\right)}{\left(\dfrac{x^2 - x - 30}{2x^2 - 15x + 18}\right)} = \dfrac{x^2 + 3x - 10}{x^2 + x - 6} \div \dfrac{x^2 - x - 30}{2x^2 - 15x + 18}$

$$= \dfrac{x^2 + 3x - 10}{x^2 + x - 6} \cdot \dfrac{2x^2 - 15x + 18}{x^2 - x - 30}$$

$$= \dfrac{(x + 5)\cancel{(x - 2)}(2x - 3)\cancel{(x - 6)}}{(x + 3)\cancel{(x - 2)}\cancel{(x - 6)}(x + 5)}$$

$$= \dfrac{2x - 3}{x + 3}, \ x \neq 2, \ x \neq 6, \ x \neq -5, \ x \neq \dfrac{3}{2}$$

55. $\left[\dfrac{x^2}{9} \cdot \dfrac{3(x + 4)}{x^2 + 2x}\right] \div \dfrac{x}{x + 2} = \dfrac{x(x)(3)(x + 4)}{3(3)(x)(x + 2)} \div \dfrac{x}{x + 2}$

$$= \dfrac{x(x + 4)}{3(x + 2)} \cdot \dfrac{x + 2}{x}$$

$$= \dfrac{\cancel{x}(x + 4)\cancel{(x + 2)}}{3\cancel{(x + 2)}\cancel{(x)}}$$

$$= \dfrac{x + 4}{3}, \ x \neq 0, \ x \neq -2$$

57. $\left[\dfrac{xy + y}{4x} \div (3x + 3)\right] \div \dfrac{y}{3x} = \dfrac{y(x + 1)}{4x} \cdot \dfrac{1}{3(x + 1)} \div \dfrac{y}{3x}$

$$= \dfrac{y\cancel{(x + 1)}(1)}{4(x)(3)\cancel{(x + 1)}} \cdot \dfrac{3x}{y}$$

$$= \dfrac{\cancel{y}\cancel{(3)}\cancel{(x)}}{4\cancel{(x)}\cancel{(3)}\cancel{(y)}}$$

$$= \dfrac{1}{4}, \ x \neq 0, \ y \neq 0, \ x \neq -1$$

59. $\dfrac{2x^2 + 5x - 25}{3x^2 + 5x + 2} \cdot \dfrac{3x^2 + 2x}{x + 5} \div \left(\dfrac{x}{x + 1}\right)^2 = \dfrac{2x^2 + 5x - 25}{3x^2 + 5x + 2} \cdot \dfrac{3x^2 + 2x}{x + 5} \cdot \dfrac{(x + 1)^2}{x^2}$

$$= \dfrac{(2x - 5)(x + 5)(x)(3x + 2)(x + 1)^2}{(3x + 2)(x + 1)(x + 5)(x^2)}$$

$$= \dfrac{(2x - 5)\cancel{(x + 5)}\cancel{(x)}\cancel{(3x + 2)}\cancel{(x + 1)}(x + 1)}{\cancel{(3x + 2)}\cancel{(x + 1)}\cancel{(x + 5)}\cancel{(x)}(x)}$$

$$= \dfrac{(2x - 5)(x + 1)}{x}, \; x \neq -1, \; x \neq -5, \; x \neq -\dfrac{2}{3}$$

61. $x^3 \cdot \dfrac{x^{2n} - 9}{x^{2n} + 4x^n + 3} \div \dfrac{x^{2n} - 2x^n - 3}{x} = \dfrac{x^3}{1} \cdot \dfrac{x^{2n} - 9}{x^{2n} + 4x^n + 3} \cdot \dfrac{x}{x^{2n} - 2x^n - 3}$

$$= \dfrac{x^3(x^n + 3)(x^n - 3)(x)}{1(x^n + 3)(x^n + 1)(x^n - 3)(x^n + 1)}$$

$$= \dfrac{x^4\cancel{(x^n + 3)}\cancel{(x^n - 3)}}{\cancel{(x^n + 3)}(x^n + 1)\cancel{(x^n - 3)}(x^n + 1)}$$

$$= \dfrac{x^4}{(x^n + 1)^2}, \; x^n \neq -3, \; x^n \neq 3, \; x \neq 0$$

63. $y_1 = \dfrac{3x + 2}{x} \cdot \dfrac{x^2}{9x^2 - 4}$

$$= \dfrac{\cancel{(3x + 2)}\cancel{(x)}(x)}{\cancel{x}\cancel{(3x + 2)}(3x - 2)}$$

$$= \dfrac{x}{3x - 2}, \; x \neq 0, \; x \neq -\dfrac{2}{3}$$

$$= y_2$$

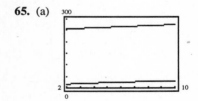

65. (a)

(b) Spending $= \dfrac{A(1000)}{B} = \dfrac{(-0.114t^2 + 3.09t + 10.5)(1000)}{2.47t + 250.7}$

(c)

Year, t	3	4	5	6	7	8
Amount per person	$72.62	$80.73	$87.82	$93.91	$99.05	$103.25

(d) Answers will vary.

67. To divide a rational expression by a polynomial, multiply the rational expression by the reciprocal of the polynomial.

69. In this example, the dividend (the first rational expression) was inverted instead of the divisor. To divide one rational expression by a second rational expression, you multiply the first rational expression by the reciprocal of the second expression.

71. $\dfrac{4}{9} \cdot \dfrac{15}{16} = \dfrac{\cancel{4}(\cancel{3})(5)}{3(\cancel{3})(\cancel{4})(4)}$

$= \dfrac{5}{12}$

73. $-\dfrac{8}{15} \cdot -\dfrac{4}{25} = \dfrac{8(4)}{(15)(25)}$

$= \dfrac{32}{375}$

75. $-\dfrac{4}{225} \div -\dfrac{2}{45} = -\dfrac{4}{225} \cdot -\dfrac{45}{2}$

$= \dfrac{2(\cancel{2})(\cancel{45})}{\cancel{45}(5)(\cancel{2})}$

$= \dfrac{2}{5}$

77. $3x^2 + 5x - 8 = (3x + 8)(x - 1)$

79. $9x^2 - 24x + 16 = (3x - 4)(3x - 4) = (3x - 4)^2$

81. *Verbal Model:* $\boxed{\text{Time of trip to island}} + \boxed{\text{Time of return trip}} = 7.5$

Label: Distance from harbor to island $= x$ (miles)

Equation: $\dfrac{x}{15} + \dfrac{x}{10} = 7.5$

$150\left(\dfrac{x}{15} + \dfrac{x}{10}\right) = 7.5(150)$

$10x + 15x = 1125$

$25x = 1125$

$x = 45$

The distance from the harbor to the island is 45 miles.

Section 7.3   Adding and Subtracting Rational Expressions

1. $\dfrac{5x}{8} + \dfrac{7x}{8} = \dfrac{5x + 7x}{8}$

$= \dfrac{12x}{8}$

$= \dfrac{\cancel{4}(3x)}{\cancel{4}(2)}$

$= \dfrac{3x}{2}$

3. $\dfrac{x}{9} - \dfrac{x + 2}{9} = \dfrac{x - (x + 2)}{9}$

$= \dfrac{x - x - 2}{9}$

$= -\dfrac{2}{9}$

5. $\dfrac{16 + z}{5z} - \dfrac{11 - z}{5z} = \dfrac{16 + z - (11 - z)}{5z}$

$= \dfrac{16 + z - 11 + z}{5z}$

$= \dfrac{5 + 2z}{5z}$

7. $\dfrac{3y}{3} - \dfrac{3y - 3}{3} - \dfrac{7}{3} = \dfrac{3y - (3y - 3) - 7}{3}$

$= \dfrac{3y - 3y + 3 - 7}{3}$

$= -\dfrac{4}{3}$

9. $\dfrac{2x - 1}{x(x - 3)} + \dfrac{1 - x}{x(x - 3)} = \dfrac{2x - 1 + 1 - x}{x(x - 3)}$

$= \dfrac{x}{x(x - 3)}$

$= \dfrac{\cancel{x}(1)}{\cancel{x}(x - 3)}$

$= \dfrac{1}{x - 3}, \ x \neq 0$

11. $\dfrac{3y - 22}{y - 6} - \dfrac{2y - 16}{y - 6} = \dfrac{3y - 22 - (2y - 16)}{y - 6}$

$= \dfrac{3y - 22 - 2y + 16}{y - 6}$

$= \dfrac{1\cancel{(y - 6)}}{\cancel{y - 6}}$

$= 1, \ y \neq 6$

13. $\dfrac{3x^2 - 4x + 6}{x^2 - 2x - 15} - \dfrac{2x^2 + 2x + 1}{x^2 - 2x - 15} = \dfrac{3x^2 - 4x + 6 - (2x^2 + 2x + 1)}{x^2 - 2x - 15}$

$$= \dfrac{3x^2 - 4x + 6 - 2x^2 - 2x - 1}{x^2 - 2x - 15}$$

$$= \dfrac{x^2 - 6x + 5}{x^2 - 2x - 15}$$

$$= \dfrac{(x - 5)(x - 1)}{(x - 5)(x + 3)}$$

$$= \dfrac{x - 1}{x + 3}, \quad x \neq 5$$

15. $\dfrac{x^2 - 3}{x + 3} + \dfrac{10x + 9}{x + 3} - \dfrac{2x - 9}{x + 3} = \dfrac{x^2 - 3 + 10x + 9 - (2x - 9)}{x + 3}$

$$= \dfrac{x^2 - 3 + 10x + 9 - 2x + 9}{x + 3}$$

$$= \dfrac{x^2 + 8x + 15}{x + 3}$$

$$= \dfrac{(x + 5)(x + 3)}{(x + 3)}$$

$$= x + 5, \quad x \neq -3$$

17. $5x^2 = 5(x^2)$

$20x^3 = 2^2(5)(x^3)$

Least common multiple: $2^2(5)(x^3) = 20x^3$

19. $15x^2 = 3(5)(x^2)$

$3(x + 5) = 3(x + 5)$

Least common multiple: $3(5)(x^2)(x + 5) = 15x^2(x + 5)$

21. $9y^3(y + 1)^2 = 3^2(y^3)(y + 1)^2$

$12y(y + 1) = 2^2(3)(y)(y + 1)$

Least common multiple: $2^2(3^2)(y^3)(y + 1)^2 = 36y^3(y + 1)^2$

23. $6(x^2 - 4) = 2(3)(x + 2)(x - 2)$

$2x(x + 2) = 2(x)(x + 2)$

Least common multiple:

$\quad 2(3)(x)(x + 2)(x - 2) = 6x(x + 2)(x - 2)$

25. $8t(t + 2) = 2^3 t(t + 2)$

$14t(t - 4) = 2(7)(t)(t - 4)$

Least common multiple:

$\quad 2^3(7)(t)(t + 2)(t - 4) = 56t(t + 2)(t - 4)$

27. $\dfrac{7}{3y} = \dfrac{7(x^2)}{3y(x^2)} = \dfrac{7x^2}{3y(x^2)}$

The missing factor is x^2.

29. $\dfrac{3u}{7v} = \dfrac{3u(u + 1)}{7v(u + 1)}$

The missing factor is $u + 1$.

31. $\dfrac{13x}{x - 2} = \dfrac{13x(x + 2)(-1)}{(x - 2)(x + 2)(-1)} = \dfrac{13x[-1(x + 2)]}{(x^2 - 4)(-1)}$

$$= \dfrac{13x[-(x + 2)]}{4 - x^2}$$

The missing factor is $-(x + 2)$ or $-x - 2$.

33. $3n - 12 = 3(n - 4)$

$6n^2 = 2(3)(n^2)$

Least common denominator: $2(3)(n^2)(n - 4) = 6n^2(n - 4)$

$$\frac{n + 8}{3(n - 4)} = \frac{(n + 8)(2n^2)}{3(n - 4)(2n^2)} = \frac{2n^2(n + 8)}{6n^2(n - 4)} \quad \text{and}$$

$$\frac{10}{6n^2} = \frac{10(n - 4)}{6n^2(n - 4)}$$

35. $x^2(x - 3)$

$x(x + 3)$

Least common denominator: $x^2(x - 3)(x + 3)$

$$\frac{2}{x^2(x - 3)} = \frac{2(x + 3)}{x^2(x - 3)(x + 3)} \quad \text{and}$$

$$\frac{5}{x(x + 3)} = \frac{5(x)(x - 3)}{x(x + 3)(x)(x - 3)}$$

$$= \frac{5x(x - 3)}{x^2(x + 3)(x - 3)}$$

37. $(s + 2)^2$

$s^3 + s^2 - 2s = s(s^2 + s - 2) = s(s + 2)(s - 1)$

Least common denominator $= s(s + 2)^2(s - 1)$.

$$\frac{8s}{(s + 2)^2} = \frac{8s(s)(s - 1)}{(s + 2)^2(s)(s - 1)} \quad \text{and} \quad \frac{3}{s(s + 2)(s - 1)} = \frac{3(s + 2)}{s(s + 2)(s - 1)(s + 2)}$$

$$= \frac{8s^2(s - 1)}{s(s + 2)^2(s - 1)} \qquad\qquad\qquad = \frac{3(s + 2)}{s(s + 2)^2(s - 1)}$$

39. $x^2 - 25 = (x + 5)(x - 5)$

$x^2 - 10x + 25 = (x - 5)^2$

Least common denominator: $(x + 5)(x - 5)^2$

$$\frac{x - 8}{(x + 5)(x - 5)} = \frac{(x - 8)(x - 5)}{(x + 5)(x - 5)(x - 5)} \quad \text{and} \quad \frac{9x}{(x - 5)^2} = \frac{9x(x + 5)}{(x - 5)^2(x + 5)}$$

$$= \frac{(x - 8)(x - 5)}{(x + 5)(x - 5)^2} \qquad\qquad\qquad = \frac{9x(x + 5)}{(x + 5)(x - 5)^2}$$

41. The least common denominator is a^2.

$$\frac{7}{a} + \frac{14}{a^2} = \frac{7(a)}{a(a)} + \frac{14}{a^2}$$

$$= \frac{7a}{a^2} + \frac{14}{a^2}$$

$$= \frac{7a + 14}{a^2} \text{ or } \frac{7(a + 2)}{a^2}$$

43. The least common denominator is $x - 8$.

$$\frac{3x}{x - 8} - \frac{6}{8 - x} = \frac{3x}{x - 8} - \frac{6(-1)}{(8 - x)(-1)}$$

$$= \frac{3x}{x - 8} - \frac{-6}{x - 8}$$

$$= \frac{3x - (-6)}{x - 8}$$

$$= \frac{3x + 6}{x - 8} \text{ or } \frac{3(x + 2)}{x - 8}$$

45. The least common denominator is $3x - 2$.

$$\frac{3x}{3x - 2} + \frac{2}{2 - 3x} = \frac{3x}{3x - 2} + \frac{2(-1)}{(2 - 3x)(-1)}$$

$$= \frac{3x}{3x - 2} + \frac{-2}{3x - 2}$$

$$= \frac{\cancel{3x - 2}}{\cancel{3x - 2}} = 1, \ x \neq \frac{2}{3}$$

47. $25 + \dfrac{10}{x + 4} = \dfrac{25}{1} + \dfrac{10}{x + 4}$

The least common denominator is $x + 4$.

$$25 + \frac{10}{x + 4} = \frac{25(x + 4)}{(x + 4)} + \frac{10}{x + 4}$$

$$= \frac{25(x + 4) + 10}{x + 4}$$

$$= \frac{25x + 100 + 10}{x + 4}$$

$$= \frac{25x + 110}{x + 4} \quad \text{or} \quad \frac{5(5x + 22)}{x + 4}$$

49. The least common denominator is $6x(x - 3)$.

$$-\frac{1}{6x} + \frac{1}{6(x - 3)} = \frac{-1(x - 3)}{6x(x - 3)} + \frac{1(x)}{6(x - 3)(x)}$$

$$= \frac{-(x - 3) + x}{6x(x - 3)}$$

$$= \frac{-x + 3 + x}{6x(x - 3)} = \frac{3}{6x(x - 3)}$$

$$= \frac{1(\cancel{3})}{2(\cancel{3})(x)(x - 3)}$$

$$= \frac{1}{2x(x - 3)}$$

51. The least common denominator is $(x + 3)(x - 2)$.

$$\frac{x}{x + 3} - \frac{5}{x - 2} = \frac{x(x - 2)}{(x + 3)(x - 2)} - \frac{5(x + 3)}{(x - 2)(x + 3)}$$

$$= \frac{x(x - 2) - 5(x + 3)}{(x + 3)(x - 2)}$$

$$= \frac{x^2 - 2x - 5x - 15}{(x + 3)(x - 2)}$$

$$= \frac{x^2 - 7x - 15}{(x + 3)(x - 2)}$$

53. The least common denominator is $(x - 5y)(x + 5y)$.

$$\frac{3}{x - 5y} + \frac{2}{x + 5y} = \frac{3(x + 5y)}{(x - 5y)(x + 5y)} + \frac{2(x - 5y)}{(x + 5y)(x - 5y)}$$

$$= \frac{3(x + 5y) + 2(x - 5y)}{(x - 5y)(x + 5y)}$$

$$= \frac{3x + 15y + 2x - 10y}{(x - 5y)(x + 5y)}$$

$$= \frac{5x + 5y}{(x - 5y)(x + 5y)} \quad \text{or}$$

$$= \frac{5(x + y)}{(x - 5y)(x + 5y)}$$

55. The least common denominator is $x^2(x^2 + 1)$.

$$\frac{4}{x^2} - \frac{4}{x^2 + 1} = \frac{4(x^2 + 1)}{x^2(x^2 + 1)} - \frac{4(x^2)}{(x^2 + 1)(x^2)}$$

$$= \frac{4(x^2 + 1) - 4x^2}{x^2(x^2 + 1)}$$

$$= \frac{4x^2 + 4 - 4x^2}{x^2(x^2 + 1)}$$

$$= \frac{4}{x^2(x^2 + 1)}$$

57. $\dfrac{x}{x^2 - 9} + \dfrac{3}{x(x - 3)} = \dfrac{x}{(x + 3)(x - 3)} + \dfrac{3}{x(x - 3)}$

The least common denominator is $x(x + 3)(x - 3)$.

$$\frac{x}{(x + 3)(x - 3)} + \frac{3}{x(x - 3)} = \frac{x(x)}{(x + 3)(x - 3)(x)} + \frac{3(x + 3)}{x(x - 3)(x + 3)}$$

$$= \frac{x^2 + 3(x + 3)}{x(x + 3)(x - 3)} = \frac{x^2 + 3x + 9}{x(x + 3)(x - 3)}$$

59. The least common denominator is $(x - 4)^2$.

$$\frac{4}{x - 4} + \frac{16}{(x - 4)^2} = \frac{4(x - 4)}{(x - 4)(x - 4)} + \frac{16}{(x - 4)^2}$$

$$= \frac{4(x - 4) + 16}{(x - 4)^2} = \frac{4x - 16 + 16}{(x - 4)^2}$$

$$= \frac{4x}{(x - 4)^2}$$

61. $\dfrac{y}{x^2 + xy} - \dfrac{x}{xy + y^2} = \dfrac{y}{x(x + y)} - \dfrac{x}{y(x + y)}$

The least common denominator is $xy(x + y)$.

$$\frac{y}{x(x + y)} - \frac{x}{y(x + y)} = \frac{y(y)}{x(x + y)(y)} - \frac{x(x)}{y(x + y)(x)}$$

$$= \frac{y^2 - x^2}{xy(x + y)} = \frac{(y + x)(y - x)}{xy(x + y)}$$

$$= \frac{y - x}{xy} \quad \text{or} \quad -\frac{x - y}{xy}, \ x \neq -y$$

63. $\dfrac{3u}{u^2 - 2uv + v^2} + \dfrac{2}{u - v} = \dfrac{3u}{(u - v)^2} + \dfrac{2}{u - v}$

The least common denominator is $(u - v)^2$.

$$\frac{3u}{(u - v)^2} + \frac{2}{u - v} = \frac{3u}{(u - v)^2} + \frac{2(u - v)}{(u - v)(u - v)}$$

$$= \frac{3u + 2(u - v)}{(u - v)^2} = \frac{3u + 2u - 2v}{(u - v)^2} = \frac{5u - 2v}{(u - v)^2}$$

65. $\dfrac{x}{x^2 - 9} - \dfrac{3x - 1}{x^2 + 7x + 12} = \dfrac{x}{(x + 3)(x - 3)} - \dfrac{3x - 1}{(x + 3)(x + 4)}$

The least common denominator is $(x + 3)(x - 3)(x + 4)$.

$$\frac{x}{(x + 3)(x - 3)} - \frac{3x - 1}{(x + 3)(x + 4)} = \frac{x(x + 4)}{(x + 3)(x - 3)(x + 4)} - \frac{(3x - 1)(x - 3)}{(x + 3)(x + 4)(x - 3)}$$

$$= \frac{x(x + 4) - (3x - 1)(x - 3)}{(x + 3)(x - 3)(x + 4)}$$

$$= \frac{x^2 + 4x - (3x^2 - 10x + 3)}{(x + 3)(x - 3)(x + 4)}$$

$$= \frac{x^2 + 4x - 3x^2 + 10x - 3}{(x + 3)(x - 3)(x + 4)}$$

$$= \frac{-2x^2 + 14x - 3}{(x + 3)(x - 3)(x + 4)}$$

67. The least common denominator is $x^2(x + 3)$.

$$\frac{4}{x} - \frac{2}{x^2} + \frac{4}{x + 3} = \frac{4(x)(x + 3)}{x(x)(x + 3)} - \frac{2(x + 3)}{x^2(x + 3)} + \frac{4(x^2)}{(x + 3)(x^2)}$$

$$= \frac{4x(x + 3) - 2(x + 3) + 4x^2}{x^2(x + 3)} = \frac{4x^2 + 12x - 2x - 6 + 4x^2}{x^2(x + 3)}$$

$$= \frac{8x^2 + 10x - 6}{x^2(x + 3)} \quad \text{or} \quad \frac{2(4x^2 + 5x - 3)}{x^2(x + 3)}$$

69. The least common denominator is xy.

$$\frac{1}{x} - \frac{3}{y} + \frac{3x - y}{xy} = \frac{1(y)}{x(y)} - \frac{3(x)}{y(x)} + \frac{3x - y}{xy} = \frac{y - 3x + 3x - y}{xy}$$

$$= \frac{0}{xy} = 0, \ x \neq 0, \ y \neq 0$$

71. $\dfrac{x}{x^2 + 15x + 50} + \dfrac{7}{2(x + 10)} - \dfrac{3}{2(x + 5)} = \dfrac{x}{(x + 10)(x + 5)} + \dfrac{7}{2(x + 10)} - \dfrac{3}{2(x + 5)}$

The least common denominator is $2(x + 10)(x + 5)$.

$$\dfrac{x}{(x + 10)(x + 5)} + \dfrac{7}{2(x + 10)} - \dfrac{3}{2(x + 5)} = \dfrac{x(2)}{(x + 10)(x + 5)(2)} + \dfrac{7(x + 5)}{2(x + 10)(x + 5)} - \dfrac{3(x + 10)}{2(x + 5)(x + 10)}$$

$$= \dfrac{2x + 7(x + 5) - 3(x + 10)}{2(x + 10)(x + 5)} = \dfrac{2x + 7x + 35 - 3x - 30}{2(x + 10)(x + 5)}$$

$$= \dfrac{6x + 5}{2(x + 10)(x + 5)}$$

73. $\dfrac{1}{x^2 + 7x + 12} + \dfrac{1}{x^2 - 9} + \dfrac{1}{x^2 - 16} = \dfrac{1}{(x + 3)(x + 4)} + \dfrac{1}{(x + 3)(x - 3)} + \dfrac{1}{(x + 4)(x - 4)}$

The least common denominator is $(x + 3)(x + 4)(x - 3)(x - 4)$.

$$\dfrac{1}{(x + 3)(x + 4)} + \dfrac{1}{(x + 3)(x - 3)} + \dfrac{1}{(x + 4)(x - 4)} = \dfrac{1(x - 3)(x - 4)}{(x + 3)(x + 4)(x - 3)(x - 4)} + \dfrac{1(x + 4)(x - 4)}{(x + 3)(x + 4)(x - 3)(x - 4)}$$

$$+ \dfrac{1(x + 3)(x - 3)}{(x + 3)(x + 4)(x - 3)(x - 4)}$$

$$= \dfrac{x^2 - 7x + 12 + x^2 - 16 + x^2 - 9}{(x + 3)(x + 4)(x - 3)(x - 4)}$$

$$= \dfrac{3x^2 - 7x - 13}{(x + 3)(x + 4)(x - 3)(x - 4)}$$

75. $\dfrac{4x^4}{x^3} - 2x = \dfrac{4x(x^3)}{x^3} - 2x$

$\qquad = 4x - 2x$

$\qquad = 2x, \quad x \neq 0$

77. $\dfrac{15x^3y}{10x^2} + \dfrac{3xy^2}{2y} = \dfrac{5(3)(x^2)(xy)}{5(2)(x^2)} + \dfrac{3x(y)(y)}{2(y)}$

$\qquad = \dfrac{3xy}{2} + \dfrac{3xy}{2}$

$\qquad = \dfrac{6xy}{2}$

$\qquad = \dfrac{2(3xy)}{2}$

$\qquad = 3xy, \quad x \neq 0, y \neq 0$

79. $\dfrac{\left(\dfrac{1}{2}\right)}{\left(3 + \dfrac{1}{x}\right)} = \dfrac{\left(\dfrac{1}{2}\right)}{\left(\dfrac{3x}{x} + \dfrac{1}{x}\right)} = \dfrac{\left(\dfrac{1}{2}\right)}{\left(\dfrac{3x + 1}{x}\right)} = \dfrac{1}{2} \cdot \dfrac{x}{3x + 1} = \dfrac{x}{2(3x + 1)}, x \neq 0$

or

$$\dfrac{\left(\dfrac{1}{2}\right)}{\left(3 + \dfrac{1}{x}\right)} = \dfrac{\left(\dfrac{1}{2}\right)(2x)}{\left(3 + \dfrac{1}{x}\right)(2x)} = \dfrac{x}{6x + 2}, x \neq 0$$

81. $\dfrac{\left(\dfrac{4}{x}+3\right)}{\left(\dfrac{4}{x}-3\right)} = \dfrac{\left(\dfrac{4}{x}+\dfrac{3x}{x}\right)}{\left(\dfrac{4}{x}-\dfrac{3x}{x}\right)} = \dfrac{\left(\dfrac{4+3x}{x}\right)}{\left(\dfrac{4-3x}{x}\right)} = \dfrac{4+3x}{x} \cdot \dfrac{x}{4-3x} = \dfrac{(4+3x)(x)}{x(4-3x)} = \dfrac{4+3x}{4-3x}, \ x \neq 0$

or

$\dfrac{\left(\dfrac{4}{x}+3\right)}{\left(\dfrac{4}{x}-3\right)} = \dfrac{\left(\dfrac{4}{x}+3\right) \cdot x}{\left(\dfrac{4}{x}-3\right) \cdot x} = \dfrac{4+3x}{4-3x}, \ x \neq 0$

83. $\dfrac{\left(16x - \dfrac{1}{x}\right)}{\left(\dfrac{1}{x}-4\right)} = \dfrac{\left(\dfrac{16x^2}{x} - \dfrac{1}{x}\right)}{\left(\dfrac{1}{x} - \dfrac{4x}{x}\right)} = \dfrac{\left(\dfrac{16x^2-1}{x}\right)}{\left(\dfrac{1-4x}{x}\right)}$

$= \dfrac{16x^2-1}{x} \cdot \dfrac{x}{1-4x} = \dfrac{(4x+1)(4x-1)(x)}{x(1-4x)}$

$= \dfrac{(4x+1)(4x-1)(x)}{x(4x-1)(-1)} = \dfrac{4x+1}{-1} = -4x-1, \ x \neq 0, \ x \neq \dfrac{1}{4}$

or

$\dfrac{\left(16x - \dfrac{1}{x}\right)}{\left(\dfrac{1}{x}-4\right)} = \dfrac{\left(16x - \dfrac{1}{x}\right)x}{\left(\dfrac{1}{x}-4\right)x} = \dfrac{16x^2-1}{1-4x} = \dfrac{(4x+1)(4x-1)}{-1(4x-1)} = -4x-1, \ x \neq 0, \ x \neq \dfrac{1}{4}$

85. $\dfrac{\left(3 + \dfrac{9}{x-3}\right)}{\left(4 + \dfrac{12}{x-3}\right)} = \dfrac{\left[\dfrac{3(x-3)}{x-3} + \dfrac{9}{x-3}\right]}{\left[\dfrac{4(x-3)}{x-3} + \dfrac{12}{x-3}\right]} = \dfrac{\left(\dfrac{3x-9+9}{x-3}\right)}{\left(\dfrac{4x-12+12}{x-3}\right)} = \dfrac{\left(\dfrac{3x}{x-3}\right)}{\left(\dfrac{4x}{x-3}\right)}$

$= \dfrac{3x}{x-3} \cdot \dfrac{x-3}{4x} = \dfrac{3(x)(x-3)}{(x-3)(4)(x)} = \dfrac{3}{4}, \ x \neq 0, \ x \neq 3$

or

$\dfrac{\left(3 + \dfrac{9}{x-3}\right)}{\left(4 + \dfrac{12}{x-3}\right)} = \dfrac{\left(3 + \dfrac{9}{x-3}\right) \cdot (x-3)}{\left(4 + \dfrac{12}{x-3}\right) \cdot (x-3)} = \dfrac{3(x-3)+9}{4(x-3)+12}$

$= \dfrac{3x-9+9}{4x-12+12} = \dfrac{3x}{4x} = \dfrac{3(x)}{4(x)} = \dfrac{3}{4}, \ x \neq 0, \ x \neq 3$

87. $\dfrac{\left(1 - \dfrac{1}{y^2}\right)}{\left(1 - \dfrac{4}{y} + \dfrac{3}{y^2}\right)} = \dfrac{\left(\dfrac{y^2}{y^2} - \dfrac{1}{y^2}\right)}{\left(\dfrac{y^2}{y^2} - \dfrac{4y}{y^2} + \dfrac{3}{y^2}\right)} = \dfrac{\left(\dfrac{y^2-1}{y^2}\right)}{\left(\dfrac{y^2-4y+3}{y^2}\right)}$

$= \dfrac{y^2-1}{y^2} \cdot \dfrac{y^2}{y^2-4y+3} = \dfrac{(y+1)(y-1)(y^2)}{y^2(y-3)(y-1)} = \dfrac{y+1}{y-3}, \ y \neq 0, y \neq 1$

or

$\dfrac{\left(1 - \dfrac{1}{y^2}\right)}{\left(1 - \dfrac{4}{y} + \dfrac{3}{y^2}\right)} = \dfrac{\left(1 - \dfrac{1}{y^2}\right) \cdot y^2}{\left(1 - \dfrac{4}{y} + \dfrac{3}{y^2}\right)y^2}$

$= \dfrac{y^2-1}{y^2-4y+3} = \dfrac{(y+1)(y-1)}{(y-1)(y-3)} = \dfrac{y+1}{y-3}, \ y \neq 1, \ y \neq 0$

89. $\dfrac{\left(\dfrac{y}{x} - \dfrac{x}{y}\right)}{\left(\dfrac{x+y}{xy}\right)} = \dfrac{\left(\dfrac{y^2}{xy} - \dfrac{x^2}{xy}\right)}{\left(\dfrac{x+y}{xy}\right)} = \dfrac{\left(\dfrac{y^2-x^2}{xy}\right)}{\left(\dfrac{x+y}{xy}\right)}$

$$= \dfrac{y^2 - x^2}{xy} \cdot \dfrac{xy}{x+y} = \dfrac{(y+x)(y-x)(xy)}{xy(x+y)} = y - x, x \neq 0, \ y \neq 0, \ x \neq -y$$

or

$$\dfrac{\left(\dfrac{y}{x} - \dfrac{x}{y}\right)}{\left(\dfrac{x+y}{xy}\right)} = \dfrac{\left(\dfrac{y}{x} - \dfrac{x}{y}\right)xy}{\left(\dfrac{x+y}{xy}\right)xy}$$

$$= \dfrac{y^2 - x^2}{x+y} = \dfrac{(y+x)(y-x)}{x+y} = y - x, x \neq -y, \ x \neq 0, \ y \neq 0$$

91. $\dfrac{\left(\dfrac{3}{x^2} + \dfrac{1}{x}\right)}{\left(2 - \dfrac{4}{5x}\right)} = \dfrac{\left(\dfrac{3}{x^2} + \dfrac{x}{x^2}\right)}{\left(\dfrac{10x}{5x} - \dfrac{4}{5x}\right)} = \dfrac{\left(\dfrac{3+x}{x^2}\right)}{\left(\dfrac{10x-4}{5x}\right)}$

$$= \dfrac{3+x}{x^2} \cdot \dfrac{5x}{2(5x-2)} = \dfrac{5x(x+3)}{x(x)(2)(5x-2)} = \dfrac{5(x+3)}{2x(5x-2)}$$

or

$$\dfrac{\left(\dfrac{3}{x^2} + \dfrac{1}{x}\right)}{\left(2 - \dfrac{4}{5x}\right)} = \dfrac{\left(\dfrac{3}{x^2} + \dfrac{1}{x}\right)5x^2}{\left(2 - \dfrac{4}{5x}\right)5x^2} = \dfrac{15 + 5x}{10x^2 - 4x} = \dfrac{5(3+x)}{2x(5x-2)}$$

93. $\dfrac{\left(\dfrac{x}{x-3} - \dfrac{2}{3}\right)}{\left(\dfrac{10}{3x} + \dfrac{x^2}{x-3}\right)} = \dfrac{\left(\dfrac{3x}{3(x-3)} - \dfrac{2(x-3)}{3(x-3)}\right)}{\left(\dfrac{10(x-3)}{3x(x-3)} + \dfrac{x^2(3x)}{3x(x-3)}\right)} = \dfrac{\left(\dfrac{3x - 2(x-3)}{3(x-3)}\right)}{\left(\dfrac{10(x-3) + x^2(3x)}{3x(x-3)}\right)}$

$$= \dfrac{\left(\dfrac{3x - 2x + 6}{3(x-3)}\right)}{\left(\dfrac{10x - 30 + 3x^3}{3x(x-3)}\right)}$$

$$= \dfrac{x+6}{3(x-3)} \cdot \dfrac{3x(x-3)}{3x^3 + 10x - 30}$$

$$= \dfrac{(x+6)(3)(x)(x-3)}{3(x-3)(3x^3 + 10x - 30)} = \dfrac{x(x+6)}{3x^3 + 10x - 30}, x \neq 0, \ x \neq 3$$

or

$$\dfrac{\left(\dfrac{x}{x-3} - \dfrac{2}{3}\right)}{\left(\dfrac{10}{3x} + \dfrac{x^2}{x-3}\right)} = \dfrac{\left(\dfrac{x}{x-3} - \dfrac{2}{3}\right)3x(x-3)}{\left(\dfrac{10}{3x} + \dfrac{x^2}{x-3}\right)3x(x-3)}$$

$$= \dfrac{3x^2 - 2x(x-3)}{10(x-3) + x^2(3x)} = \dfrac{3x^2 - 2x^2 + 6x}{10x - 30 + 3x^3} = \dfrac{x^2 + 6x}{3x^3 + 10x - 30} = \dfrac{x(x+6)}{3x^3 + 10x - 30}, x \neq 0, \ x \neq 3$$

95. $(u + v^{-2})^{-1} = \left(u + \dfrac{1}{v^2}\right)^{-1} = \dfrac{1}{u + \dfrac{1}{v^2}}$ or $(u + v^{-2})^{-1} = \left(\dfrac{u}{1} + \dfrac{1}{v^2}\right)^{-1} = \left(\dfrac{uv^2}{v^2} + \dfrac{1}{v^2}\right)^{-1}$

$$= \dfrac{1 \cdot v^2}{\left(u + \dfrac{1}{v^2}\right)v^2} = \dfrac{v^2}{uv^2 + 1}, v \neq 0 \qquad\qquad = \left(\dfrac{uv^2 + 1}{v^2}\right)^{-1} = \dfrac{v^2}{uv^2 + 1}, v \neq 0$$

97. $\dfrac{a + b}{ba^{-1} - ab^{-1}} = \dfrac{a + b}{\dfrac{b}{a} - \dfrac{a}{b}} = \dfrac{(a + b)ab}{\left(\dfrac{b}{a} - \dfrac{a}{b}\right)ab} = \dfrac{ab(a + b)}{b^2 - a^2} = \dfrac{ab\cancel{(a + b)}}{\cancel{(b + a)}(b - a)} = \dfrac{ab}{b - a}, a \neq 0, b \neq 0, a \neq -b$

or

$$\dfrac{a + b}{ba^{-1} - ab^{-1}} = \dfrac{a + b}{\left(\dfrac{b}{a} - \dfrac{a}{b}\right)} = \dfrac{a + b}{\left(\dfrac{b^2}{ab} - \dfrac{a^2}{ab}\right)} = \dfrac{a + b}{\left(\dfrac{b^2 - a^2}{ab}\right)} = \dfrac{a + b}{1} \cdot \dfrac{ab}{b^2 - a^2}$$

$$= \dfrac{\cancel{(a + b)}(ab)}{1\cancel{(b + a)}(b - a)} = \dfrac{ab}{b - a}, a \neq 0, b \neq 0, a \neq -b$$

99. $f(x) = \dfrac{1}{x}$

$$\dfrac{f(2 + h) - f(2)}{h} = \dfrac{\dfrac{1}{2 + h} - \dfrac{1}{2}}{h} = \dfrac{\dfrac{1(2)}{(2 + h)(2)} - \dfrac{1(2 + h)}{2(2 + h)}}{h}$$

$$= \dfrac{\left(\dfrac{2 - (2 + h)}{2(2 + h)}\right)}{h} = \dfrac{\left(\dfrac{-h}{2(2 + h)}\right)}{h}$$

$$= \dfrac{-h}{2(2 + h)} \div h = \dfrac{-h}{2(2 + h)} \cdot \dfrac{1}{h} = \dfrac{-1\cancel{(h)}}{2\cancel{(h)}(2 + h)} = -\dfrac{1}{2(2 + h)}$$

101. Note: The average of the two real numbers a and b is $(a + b)/2$.

$$\text{Average} = \dfrac{\left(\dfrac{x}{4} + \dfrac{x}{6}\right)}{2} = \dfrac{\left(\dfrac{6x}{24} + \dfrac{4x}{24}\right)}{2} = \dfrac{\left(\dfrac{10x}{24}\right)}{\left(\dfrac{2}{1}\right)} = \dfrac{10x}{24} \cdot \dfrac{1}{2} = \dfrac{10x}{48} = \dfrac{5\cancel{(2)}(x)}{24\cancel{(2)}} = \dfrac{5x}{24}$$

or

$$\text{Average} = \dfrac{\left(\dfrac{x}{4} + \dfrac{x}{6}\right)}{2} = \dfrac{\left(\dfrac{x}{4} + \dfrac{x}{6}\right) \cdot 24}{2 \cdot 24} = \dfrac{6x + 4x}{48} = \dfrac{10x}{48} = \dfrac{5\cancel{(2)}(x)}{24\cancel{(2)}} = \dfrac{5x}{24}$$

So, $5x/24$ is the average of $x/4$ and $x/6$.

103. Note: When the real number line between $x/6$ and $x/2$ is divided into four equal parts, the length of each part is $1/4$ of the distance between $x/6$ and $x/2$, or $1/4$ of the difference $(x/2) - (x/6)$.

$$\dfrac{\left(\dfrac{x}{2} - \dfrac{x}{6}\right)}{4} = \dfrac{\left(\dfrac{6x}{12} - \dfrac{2x}{12}\right)}{4} = \dfrac{\left(\dfrac{4x}{12}\right)}{\left(\dfrac{4}{1}\right)} = \dfrac{4x}{12} \cdot \dfrac{1}{4} = \dfrac{\cancel{4}(x)(1)}{12\cancel{(4)}} = \dfrac{x}{12}$$

So, each of the four equal parts is $x/12$ units long.

$$x_1 = \dfrac{x}{6} + \dfrac{x}{12} = \dfrac{2x}{12} + \dfrac{x}{12} = \dfrac{3x}{12} = \dfrac{x}{4}$$

or

$$\dfrac{\left(\dfrac{x}{2} - \dfrac{x}{6}\right)}{4} = \dfrac{\left(\dfrac{x}{2} - \dfrac{x}{6}\right) \cdot 12}{4 \cdot 12} = \dfrac{6x - 2x}{48} = \dfrac{4x}{48} = \dfrac{\cancel{4}(x)}{\cancel{4}(12)} = \dfrac{x}{12}$$

$$x_2 = x_1 + \dfrac{x}{12} = \dfrac{x}{4} + \dfrac{x}{12} = \dfrac{3x}{12} + \dfrac{x}{12} = \dfrac{4x}{12} = \dfrac{x}{3}$$

$$x_3 = x_2 + \dfrac{x}{12} = \dfrac{x}{3} + \dfrac{x}{12} = \dfrac{4x}{12} + \dfrac{x}{12} = \dfrac{5x}{12}$$

So, the three real numbers are $x/4$, $x/3$, and $5x/12$.

105. $\dfrac{t}{4} + \dfrac{t}{6} = \dfrac{t(6)}{4(6)} + \dfrac{t(4)}{6(4)} = \dfrac{6t}{24} + \dfrac{4t}{24} = \dfrac{6t + 4t}{24} = \dfrac{10t}{24} = \dfrac{5(2)(t)}{2(12)} = \dfrac{5t}{12}$

Thus, $5t/12$ of the task has been completed.

107. $\dfrac{1}{\left(\dfrac{1}{R_1} + \dfrac{1}{R_2}\right)} = \dfrac{1}{\left(\dfrac{R_2}{R_1 R_2} + \dfrac{R_1}{R_2 R_1}\right)} = \dfrac{1}{\left(\dfrac{R_2 + R_1}{R_1 R_2}\right)}$

$$= 1 \cdot \dfrac{R_1 R_2}{R_2 + R_1} = \dfrac{R_1 R_2}{R_2 + R_1}, R_1 \neq 0, R_2 \neq 0$$

or

$$\dfrac{1}{\left(\dfrac{1}{R_1} + \dfrac{1}{R_2}\right)} = \dfrac{1 \cdot R_1 R_2}{\left(\dfrac{1}{R_1} + \dfrac{1}{R_2}\right) R_1 R_2} = \dfrac{R_1 R_2}{R_2 + R_1}, R_1 \neq 0, R_2 \neq 0$$

109. $\begin{cases} A + B + C = 0 \\ \quad\; - B + C = 0 \\ -A \qquad\quad = 4 \end{cases} \Rightarrow A = -4 \Rightarrow -4 + B + C = 0$

$$B + C = 4$$

$\begin{cases} B + C = 4 \\ -B + C = 0 \end{cases}$

$$2C = 4 \Rightarrow C = 2 \text{ and } B + 2 = 4$$

$$B = 2$$

$(-4, 2, 2)$

To verify sum:

$$\dfrac{4}{x^3 - x} \stackrel{?}{=} \dfrac{-4}{x} + \dfrac{2}{x + 1} + \dfrac{2}{x - 1}$$

$$\stackrel{?}{=} \dfrac{-4(x + 1)(x - 1)}{x(x + 1)(x - 1)} + \dfrac{2(x)(x - 1)}{(x + 1)(x)(x - 1)} + \dfrac{2(x)(x + 1)}{(x - 1)(x)(x + 1)}$$

$$\stackrel{?}{=} \dfrac{-4(x^2 - 1) + 2x(x - 1) + 2x(x + 1)}{x(x + 1)(x - 1)}$$

$$\stackrel{?}{=} \dfrac{-4x^2 + 4 + 2x^2 - 2x + 2x^2 + 2x}{x(x^2 - 1)}$$

$$\dfrac{4}{x^3 - x} = \dfrac{4}{x^3 - x}$$

Yes, $\dfrac{4}{x^3 - x} = \dfrac{-4}{x} + \dfrac{2}{x + 1} + \dfrac{2}{x - 1}.$

111. Yes. Here is an example.

$$\dfrac{3}{2(x + 2)} + \dfrac{x}{x + 2}$$

The least common denominator is $2(x + 2)$, and this is the denominator of the first fraction.

113. $-6x(10 - 7x) = -60x + 42x^2$

115. $(11 - x)(11 + x) = 11^2 - x^2$

$$= 121 - x^2$$

117. $(x + 1)^2 = (x + 1)(x + 1)$

$$= x^2 + 2x + 1$$

119. $(x - 2)(x^2 + 2x + 4) = x^3 + 2x^2 + 4x - 2x^2 - 4x - 8$

$$= x^3 - 8$$

121. $P = (x + 3) + x + x + x + x + x + x + (2x + 3) + x + x + 2x$

$\quad = 12x + 6$

$A = x^2 + (x + 3)(3x) + x^2$

$\quad = x^2 + 3x^2 + 9x + x^2$

$\quad = 5x^2 + 9x$

123.

$$\frac{80 + 82 + 94 + 71 + x}{5} = 85$$

$$5\left(\frac{80 + 82 + 94 + 71 + x}{5}\right) = 85(5)$$

$$80 + 82 + 94 + 71 + x = 425$$

$$327 + x = 425$$

$$x = 98$$

The student needs a score of 98 on the next test to produce an average of 85.

Section 7.4   Dividing Polynomials

1. $\dfrac{10z^2 + 4z - 12}{4} = \dfrac{10z^2}{4} + \dfrac{4z}{4} - \dfrac{12}{4}$

$\qquad = \dfrac{5\cancel{(2)}z^2}{2\cancel{(2)}} + \dfrac{\cancel{4}(z)}{\cancel{4}} - \dfrac{\cancel{4}(3)}{\cancel{4}} = \dfrac{5z^2}{2} + z - 3$

3. $(7x^3 - 2x^2) \div x = \dfrac{7x^3 - 2x^2}{x} = \dfrac{7x^3}{x} - \dfrac{2x^2}{x}$

$\qquad = \dfrac{7\cancel{(x)}(x^2)}{\cancel{x}} - \dfrac{2\cancel{(x)}(x)}{\cancel{x}}$

$\qquad = 7x^2 - 2x, \; x \neq 0$

5. $\dfrac{50z^3 + 30z}{-5z} = \dfrac{50z^3}{-5z} + \dfrac{30z}{-5z}$

$\qquad = \dfrac{\cancel{5}(10)\cancel{(z)}(z^2)}{-1\cancel{(5)}\cancel{(z)}} + \dfrac{\cancel{5}(6)\cancel{(z)}}{-1\cancel{(5)}\cancel{(z)}}$

$\qquad = -10z^2 - 6, \; z \neq 0$

7. $\dfrac{8z^3 + 3z^2 - 2z}{2z} = \dfrac{8z^3}{2z} + \dfrac{3z^2}{2z} - \dfrac{2z}{2z}$

$\qquad = \dfrac{4\cancel{(2)}\cancel{(z)}(z^2)}{2\cancel{(z)}} + \dfrac{3z\cancel{(z)}}{2\cancel{(z)}} - \dfrac{1\cancel{(2z)}}{2\cancel{z}}$

$\qquad = 4z^2 + \dfrac{3z}{2} - 1, \; z \neq 0$

9. $\dfrac{m^4 + 2m^2 - 7}{-m} = \dfrac{m^4}{-m} + \dfrac{2m^2}{-m} - \dfrac{7}{-m}$

$\qquad = -\dfrac{\cancel{m}(m^3)}{\cancel{m}} - \dfrac{2m\cancel{(m)}}{\cancel{m}} + \dfrac{7}{m} = -m^3 - 2m + \dfrac{7}{m}$

11. $(5x^2y - 8xy + 7xy^2) \div 2xy = \dfrac{5x^2y - 8xy + 7xy^2}{2xy} = \dfrac{5x^2y}{2xy} - \dfrac{8xy}{2xy} + \dfrac{7xy^2}{2xy}$

$\qquad\qquad = \dfrac{5\cancel{(x)}(x)\cancel{(y)}}{2\cancel{(x)}\cancel{(y)}} - \dfrac{4\cancel{(2)}\cancel{(x)}\cancel{(y)}}{2\cancel{(x)}\cancel{(y)}} + \dfrac{7\cancel{(x)}\cancel{(y)}(y)}{2\cancel{(x)}\cancel{(y)}}$

$\qquad\qquad = \dfrac{5x}{2} - 4 + \dfrac{7y}{2}, \; x \neq 0, \; y \neq 0$

13.

$$
\begin{array}{r}
x - 5 \\
x - 3 \overline{) x^2 - 8x + 15} \\
\underline{x^2 - 3x} \\
-5x + 15 \\
\underline{-5x + 15} \\
0
\end{array}
$$

So, $\dfrac{x^2 - 8x + 15}{x - 3} = x - 5,\ x \neq 3$

or $\dfrac{x^2 - 8x + 15}{x - 3} = \dfrac{(x - 5)(x - 3)}{(x - 3)}$

$\qquad = \dfrac{(x - 5)\cancel{(x - 3)}}{\cancel{(x - 3)}} = x - 5,\ x \neq 3.$

15.

$$
\begin{array}{r}
x + 10 \\
x + 5 \overline{) x^2 + 15x + 50} \\
\underline{x^2 + 5x} \\
10x + 50 \\
\underline{10x + 50} \\
0
\end{array}
$$

So, $\dfrac{x^2 + 15x + 50}{x + 5} = x + 10,\ x \neq -5$ or

$\dfrac{x^2 + 15x + 50}{x + 5} = \dfrac{(x + 10)(x + 5)}{x + 5}$

$\qquad = \dfrac{(x + 10)\cancel{(x + 5)}}{\cancel{(x + 5)}}$

$\qquad = x + 10,\ x \neq -5.$

17.

$$
\begin{array}{r}
x + 7 \\
-x + 3 \overline{) -x^2 - 4x + 21} \\
\underline{-x^2 + 3x} \\
-7x + 21 \\
\underline{-7x + 21} \\
0
\end{array}
$$

So, $\dfrac{21 - 4x - x^2}{3 - x} = x + 7,\ x \neq 3$ or

$\dfrac{21 - 4x - x^2}{3 - x} = \dfrac{(7 + x)(3 - x)}{3 - x}$

$\qquad = \dfrac{(7 + x)\cancel{(3 - x)}}{\cancel{3 - x}}$

$\qquad = 7 + x,\ x \neq 3.$

19.

$$
\begin{array}{r}
y + 3 \\
2y + 1 \overline{) 2y^2 + 7y + 3} \\
\underline{2y^2 + y} \\
6y + 3 \\
\underline{6y + 3} \\
0
\end{array}
$$

So, $\dfrac{2y^2 + 7y + 3}{2y + 1} = y + 3,\ y \neq -\dfrac{1}{2}$ or

$\dfrac{2y^2 + 7y + 3}{2y + 1} = \dfrac{(2y + 1)(y + 3)}{2y + 1}$

$\qquad = \dfrac{\cancel{(2y + 1)}(y + 3)}{\cancel{(2y + 1)}}$

$\qquad = y + 3,\ y \neq -\dfrac{1}{2}.$

21.

$$
\begin{array}{r}
4x - 1 \\
4x + 1 \overline{) 16x^2 + 0x - 1} \\
\underline{16x^2 + 4x} \\
-4x - 1 \\
\underline{-4x - 1} \\
0
\end{array}
$$

So, $\dfrac{16x^2 - 1}{4x + 1} = 4x - 1,\ x \neq -\dfrac{1}{4}$ or

$\dfrac{16x^2 - 1}{4x + 1} = \dfrac{(4x - 1)(4x + 1)}{4x + 1}$

$\qquad = \dfrac{(4x - 1)\cancel{(4x + 1)}}{\cancel{4x + 1}}$

$\qquad = 4x - 1,\ x \neq -\dfrac{1}{4}.$

23.

$$
\begin{array}{r}
x^2 - 5x + 25 \\
x + 5 \overline{) x^3 + 0x^2 + 0x + 125} \\
\underline{x^3 + 5x^2} \\
-5x^2 + 0x \\
\underline{-5x^2 - 25x} \\
25x + 125 \\
\underline{25x + 125} \\
0
\end{array}
$$

So, $\dfrac{x^3 + 125}{x + 5} = x^2 - 5x + 25,\ x \neq -5$ or

$\dfrac{x^3 + 125}{x + 5} = \dfrac{(x + 5)(x^2 - 5x + 25)}{x + 5}$

$\qquad = \dfrac{\cancel{(x + 5)}(x^2 - 5x + 25)}{\cancel{(x + 5)}}$

$\qquad = x^2 - 5x + 25,\ x \neq -5.$

25.

$$
\begin{array}{r}
x^2 \qquad\; + 4 \\
x - 2 \overline{\smash{)}\; x^3 - 2x^2 + 4x - 8} \\
\underline{x^3 - 2x^2} \\
+4x - 8 \\
\underline{4x - 8} \\
0
\end{array}
$$

So, $\dfrac{x^3 - 2x^2 + 4x - 8}{x - 2} = x^2 + 4,\; x \neq 2$ or

$$\dfrac{x^3 - 2x^2 + 4x - 8}{x - 2} = \dfrac{x^2(x - 2) + 4(x - 2)}{x - 2}$$

$$= \dfrac{(x - 2)(x^2 + 4)}{x - 2}$$

$$= x^2 + 4,\; x \neq 2.$$

27.

$$
\begin{array}{r}
x^2 \qquad\; - 2 \\
x - 5 \overline{\smash{)}\; x^3 - 5x^2 - 2x + 10} \\
\underline{x^3 - 5x^2} \\
-2x + 10 \\
\underline{-2x + 10} \\
0
\end{array}
$$

So, $\dfrac{x^3 - 5x^2 - 2x + 10}{x - 5} = x^2 - 2,\;\; x \neq 5$ or

$$\dfrac{x^3 - 5x^2 - 2x + 10}{x - 5} = \dfrac{x^2(x - 5) - 2(x - 5)}{x - 5}$$

$$= \dfrac{(x - 5)(x^2 - 2)}{x - 5}$$

$$= x^2 - 2,\; x \neq 5.$$

29.

$$
\begin{array}{r}
2 + \dfrac{5}{x + 2} \\
x + 2 \overline{\smash{)}\; 2x + 9} \\
\underline{2x + 4} \\
5
\end{array}
$$

So, $\dfrac{2x + 9}{x + 2} = 2 + \dfrac{5}{x + 2}.$

31.

$$
\begin{array}{r}
x - 4 + \dfrac{32}{x + 4} \\
x + 4 \overline{\smash{)}\; x^2 + 0x + 16} \\
\underline{x^2 + 4x} \\
-4x + 16 \\
\underline{-4x - 16} \\
32
\end{array}
$$

So, $\dfrac{x^2 + 16}{x + 4} = x - 4 + \dfrac{32}{x + 4}.$

33.

$$
\begin{array}{r}
5x - 8 + \dfrac{19}{x + 2} \\
x + 2 \overline{\smash{)}\; 5x^2 + 2x + 3} \\
\underline{5x^2 + 10x} \\
-8x + 3 \\
\underline{-8x - 16} \\
19
\end{array}
$$

So, $\dfrac{5x^2 + 2x + 3}{x + 2} = 5x - 8 + \dfrac{19}{x + 2}.$

35.

$$
\begin{array}{r}
4x + 3 - \dfrac{11}{3x + 2} \\
3x + 2 \overline{\smash{)}\; 12x^2 + 17x - 5} \\
\underline{12x^2 + 8x} \\
9x - 5 \\
\underline{9x + 6} \\
-11
\end{array}
$$

So, $\dfrac{12x^2 + 17x - 5}{3x + 2} = 4x + 3 - \dfrac{11}{3x + 2}.$

37.

$$
\begin{array}{r}
5x^2 - 9x + 10 - \dfrac{10}{x + 3} \\
x + 3 \overline{\smash{)}\; 5x^3 + 6x^2 - 17x + 20} \\
\underline{5x^3 + 15x^2} \\
- 9x^2 - 17x \\
\underline{- 9x^2 - 27x} \\
10x + 20 \\
\underline{10x + 30} \\
-10
\end{array}
$$

So, $\dfrac{5x^3 + 6x^2 - 17x + 20}{x + 3} = 5x^2 - 9x + 10 - \dfrac{10}{x + 3}$

39.

$$
\begin{array}{r}
2x^2 + x + 4 + \dfrac{6}{x - 3} \\
x - 3 \overline{\smash{)}\; 2x^3 - 5x^2 + x - 6} \\
\underline{2x^3 - 6x^2} \\
x^2 + x \\
\underline{x^2 - 3x} \\
4x - 6 \\
\underline{4x - 12} \\
6
\end{array}
$$

So, $\dfrac{2x^3 - 5x^2 + x - 6}{x - 3} = 2x^2 + x + 4 + \dfrac{6}{x - 3}.$

41.

$$\begin{array}{r} x^5 + x^4 + x^3 + x^2 + x + 1 \\ x - 1 \overline{\smash{\big)}\, x^6 + 0x^5 + 0x^4 + 0x^3 + 0x^2 + 0x - 1} \end{array}$$

$\underline{x^6 - x^5}$

$x^5 + 0x^4$

$\underline{x^5 - x^4}$

$x^4 + 0x^3$

$\underline{x^4 - x^3}$

$x^3 + 0x^2$

$\underline{x^3 - x^2}$

$x^2 + 0x$

$\underline{x^2 - x}$

$x - 1$

$\underline{x - 1}$

0

So, $\dfrac{x^6 - 1}{x - 1} = x^5 + x^4 + x^3 + x^2 + x + 1$, $x \neq 1$.

43.

$$\begin{array}{r} x^2 + x + 1 + \dfrac{1}{x - 1} \\ x - 1 \overline{\smash{\big)}\, x^3 + 0x^2 + 0x + 0} \end{array}$$

$\underline{x^3 - x^2}$

$x^2 + 0x$

$\underline{x^2 - x}$

$x + 0$

$\underline{x - 1}$

1

So, $\dfrac{x^3}{x - 1} = x^2 + x + 1 + \dfrac{1}{x - 1}$.

45.

$$\begin{array}{r} x^3 - x + \dfrac{x}{x^2 + 1} \\ x^2 + 1 \overline{\smash{\big)}\, x^5 } \end{array}$$

$\underline{x^5 + x^3}$

$-x^3$

$\underline{-x^3 - x}$

x

So, $\dfrac{x^5}{x^2 + 1} = x^3 - x + \dfrac{x}{x^2 + 1}$.

47.

$$\begin{array}{r} x + 2 \\ x^2 + 2x + 3 \overline{\smash{\big)}\, x^3 + 4x^2 + 7x + 6} \end{array}$$

$\underline{x^3 + 2x^2 + 3x}$

$2x^2 + 4x + 6$

$\underline{2x^2 + 4x + 6}$

0

So, $\dfrac{x^3 + 4x^2 + 7x + 6}{x^2 + 2x + 3} = x + 2$.

49.

$$\begin{array}{r} x^2 - 2 + \dfrac{2x - 28}{3x^2 + x + 1} \\ 3x^2 + x + 1 \overline{\smash{\big)}\, 3x^4 + x^3 - 5x^2 + 0x - 30} \end{array}$$

$\underline{3x^4 + x^3 + x^2}$

$-6x^2 + 0x - 30$

$\underline{-6x^2 - 2x - 2}$

$2x - 28$

So, $\dfrac{3x^4 + x^3 - 5x^2 - 30}{3x^2 + x + 1} = x^2 - 2 + \dfrac{2x - 28}{3x^2 + x + 1}$

51.

$$\begin{array}{r} x + 4 + \dfrac{6x - 3}{x^2 + x - 4} \\ x^2 + x - 4 \overline{\smash{\big)}\, x^3 + 5x^2 + 6x - 19} \end{array}$$

$\underline{x^3 + x^2 - 4x}$

$4x^2 + 10x - 19$

$\underline{4x^2 + 4x - 16}$

$6x - 3$

So, $\dfrac{x^3 + 5x^2 + 6x - 19}{x^2 + x - 4} = x + 4 + \dfrac{6x - 3}{x^2 + x - 4}$

53.

$$\begin{array}{r} x^2 - 3x + 17 + \dfrac{-72x + 75}{x^2 + 3x - 5} \\ x^2 + 3x - 5 \overline{\smash{\big)}\, x^4 + 0x^3 + 3x^2 - 6x - 10} \end{array}$$

$\underline{x^4 + 3x^3 - 5x^2}$

$-3x^3 + 8x^2 - 6x$

$\underline{-3x^3 - 9x^2 + 15x}$

$17x^2 - 21x - 10$

$\underline{17x^2 + 51x - 85}$

$-72x + 75$

So, $\dfrac{x^4 + 3x^2 - 6x - 10}{x^2 + 3x - 5} = x^2 - 3x + 17 + \dfrac{-72x + 75}{x^2 + 3x - 5}$.

55.

$$\begin{array}{r} x^{2n} + x^n + 4 \\ x^n + 2 \overline{\smash{\big)}\, x^{3n} + 3x^{2n} + 6x^n + 8} \end{array}$$

$\underline{x^{3n} + 2x^{2n}}$

$x^{2n} + 6x^n$

$\underline{x^{2n} + 2x^n}$

$4x^n + 8$

$\underline{4x^n + 8}$

0

So, $\dfrac{x^{3n} + 3x^{2n} + 6x^n + 8}{x^n + 2} = x^{2n} + x^n + 4$, $x^n \neq -2$.

57.

$$
-4 \begin{array}{|rrrr} 1 & 3 & 0 & -1 \\ & -4 & 4 & -16 \\ \hline 1 & -1 & 4 & -17 \end{array}
$$

$$\frac{x^3 + 3x^2 - 1}{x + 4} = x^2 - x + 4 - \frac{17}{x + 4}$$

59.

$$
2 \begin{array}{|rrrrr} 1 & -4 & 0 & 1 & 10 \\ & 2 & -4 & -8 & -14 \\ \hline 1 & -2 & -4 & -7 & -4 \end{array}
$$

$$\frac{x^4 - 4x^3 + x + 10}{x - 2} = x^3 - 2x^2 - 4x - 7 - \frac{4}{x - 2}$$

61.

$$
-5 \begin{array}{|rrrr} 5 & 0 & 0 & 12 \\ & -25 & 125 & -625 \\ \hline 5 & -25 & 125 & -613 \end{array}
$$

$$\frac{5x^3 + 12}{x + 5} = 5x^2 - 25x + 125 - \frac{613}{x + 5}$$

63.

$$
-2 \begin{array}{|rr} 12 & -25 \\ & -24 \\ \hline 12 & -49 \end{array}
$$

$$\frac{12x - 25}{x + 2} = 12 - \frac{49}{x + 2}$$

65.

$$
4 \begin{array}{|rrrr} 5 & -6 & 0 & 8 \\ & 20 & 56 & 224 \\ \hline 5 & 14 & 56 & 232 \end{array}
$$

$$\frac{5x^3 - 6x^2 + 8}{x - 4} = 5x^2 + 14x + 56 + \frac{232}{x - 4}$$

67.

$$
6 \begin{array}{|rrrrr} 10 & -50 & 0 & 0 & -800 \\ & 60 & 60 & 360 & 2160 \\ \hline 10 & 10 & 60 & 360 & 1360 \end{array}
$$

$$\frac{10x^4 - 50x^3 - 800}{x - 6} = 10x^3 + 10x^2 + 60x + 360 + \frac{1360}{x - 6}$$

69.

$$
0.2 \begin{array}{|rrr} 0.1 & 0.8 & 1 \\ & 0.02 & 0.164 \\ \hline 0.1 & 0.82 & 1.164 \end{array}
$$

$$\frac{0.1x^2 + 0.8x + 1}{x - 0.2} = 0.1x + 0.82 + \frac{1.164}{x - 0.2}$$

71.

$$\frac{x^3 - 8x^2 + x + 42}{x - 3}$$

$$
3 \begin{array}{|rrrr} 1 & -8 & 1 & 42 \\ & 3 & -15 & -42 \\ \hline 1 & -5 & -14 & 0 \end{array}
$$

$$x^3 - 8x^2 + x + 42 = (x - 3)(x^2 - 5x - 14)$$
$$= (x - 3)(x - 7)(x + 2)$$

73.

$$\frac{3z^3 - 20z^2 + 36z - 16}{z - 4}$$

$$
4 \begin{array}{|rrrr} 3 & -20 & 36 & -16 \\ & 12 & -32 & 16 \\ \hline 3 & -8 & 4 & 0 \end{array}
$$

$$3z^3 - 20z^2 + 36z - 16 = (z - 4)(3z^2 - 8z + 4)$$
$$= (z - 4)(3z - 2)(z - 2)$$

75.

$$\frac{5t^3 - 27t^2 - 14t - 24}{t - 6}$$

$$
6 \begin{array}{|rrrr} 5 & -27 & -14 & -24 \\ & 30 & 18 & 24 \\ \hline 5 & 3 & 4 & 0 \end{array}
$$

$$5t^3 - 27t^2 - 14t - 24 = (t - 6)(5t^2 + 3t + 4)$$

77.

$$\frac{x^4 - 16}{x - 2}$$

$$
2 \begin{array}{|rrrrr} 1 & 0 & 0 & 0 & -16 \\ & 2 & 4 & 8 & 16 \\ \hline 1 & 2 & 4 & 8 & 0 \end{array}
$$

$$x^4 - 16 = (x - 2)(x^3 + 2x^2 + 4x + 8) \text{ or}$$
$$(x - 2)[x^2(x + 2) + 4(x + 2)] = (x - 2)(x + 2)(x^2 + 4)$$

79.

$$\frac{15x^2 - 2x - 8}{x - 4/5}$$

$$
\tfrac{4}{5} \begin{array}{|rrr} 15 & -2 & -8 \\ & 12 & 8 \\ \hline 15 & 10 & 0 \end{array}
$$

$$15x^2 - 2x - 8 = \left(x - \tfrac{4}{5}\right)(15x + 10) \text{ or}$$
$$5\left(x - \tfrac{4}{5}\right)(3x + 2) \text{ or } (5x - 4)(3x + 2)$$

81. $y_1 = \dfrac{x^3}{x^2 + 1}$

$\quad = x - \dfrac{x}{x^2 + 1}$ (See long division shown below.)

$\quad = y_2$

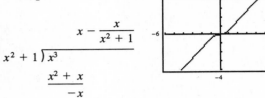

$$\begin{array}{r} x - \dfrac{x}{x^2 + 1} \\ x^2 + 1 \overline{\smash{)}\, x^3} \\ \underline{x^2 + x} \\ -x \end{array}$$

83. $\dfrac{\text{Polynomial}}{x - 6} = x^2 + x + 1 + \dfrac{-4}{x - 6}$

$\text{Polynomial} = (x - 6)(x^2 + x + 1) - 4$

$\qquad = x^3 + x^2 + x - 6x^2 - 6x - 6 - 4$

$\qquad = x^3 - 5x^2 - 5x - 10$

85.
$$\begin{array}{r|rrrr} 2 & 1 & 2 & -4 & c \\ & & 2 & 8 & 8 \\ \hline & 1 & 4 & 4 & (c + 8) \end{array}$$

The division comes out evenly when $c + 8 = 0$, or $c = -8$.

87.

$\qquad\qquad \text{Area} = (\text{Length})(\text{Width})$

$2x^3 + 3x^2 - 6x - 9 = (\text{Length})(2x + 3)$

$\dfrac{2x^3 + 3x^2 - 6x - 9}{2x + 3} = \text{Length}$

$$\begin{array}{r} x^2 \qquad\quad - 3 \\ 2x + 3 \overline{\smash{)}\, 2x^3 + 3x^2 - 6x - 9} \\ \underline{2x^3 + 3x^2} \\ -6x - 9 \\ \underline{-6x - 9} \\ 0 \end{array}$$

The length of the rectangle is $x^2 - 3$.

89.

$\qquad\qquad \text{Volume} = \dfrac{1}{2}(\text{Base})(\text{Height})(\text{Length})$

$x^3 + 18x^2 + 80x + 96 = \dfrac{1}{2}(x + 2)(h)(x + 12)$

$x^3 + 18x^2 + 80x + 96 = \dfrac{1}{2}h(x^2 + 14x + 24)$

$2x^3 + 36x^2 + 160x + 192 = h(x^2 + 14x + 24)$

$\dfrac{2x^3 + 36x^2 + 160x + 192}{x^2 + 14x + 24} = h$

$$\begin{array}{r} 2x + \quad 8 \\ x^2 + 14x + 24 \overline{\smash{)}\, 2x^3 + 36x^2 + 160x + 192} \\ \underline{2x^3 + 28x^2 + \quad 48x} \\ 8x^2 + 112x + 192 \\ \underline{8x^2 + 112x + 192} \\ 0 \end{array}$$

$h = 2x + 8$

The missing dimension is $2x + 8$ or $2(x + 4)$.

91. The x's cannot be divided out because x is not a factor of the numerator.

93. Yes. If $\dfrac{n(x)}{d(x)} = q(x)$, then $n(x) = d(x) \cdot q(x)$. So, both $d(x)$ and $q(x)$ are factors of $n(x)$.

95. $\dfrac{20t^2}{15t} = \dfrac{4(5)(t)(t)}{3(3)(t)}$

$\qquad = \dfrac{4t}{3}, \quad t \neq 0$

97. $\dfrac{8x^2y^3}{20xy^2} = \dfrac{2(4)(x)(x)(y^2)(y)}{4(5)(x)(y^2)}$

$\qquad = \dfrac{2xy}{5}, \quad x \neq 0, y \neq 0$

99. $3(2 - x) = 5x$

$\quad 6 - 3x = 5x$

$\qquad\quad 6 = 8x$

$\qquad\quad \dfrac{6}{8} = x$

$\qquad\quad \dfrac{3}{4} = x$

101. $8y^2 - 50 = 0$

$\qquad 8y^2 = 50$

$\qquad y^2 = \dfrac{50}{8}$

$\qquad y^2 = \dfrac{25}{4}$

$\qquad y = \pm\sqrt{\dfrac{25}{4}}$

$\qquad y = \pm\dfrac{5}{2}$

103. $x^2 + x - 42 = 0$

$\quad (x + 7)(x - 6) = 0$

$\qquad x + 7 = 0 \implies x = -7$

$\qquad x - 6 = 0 \implies x = 6$

105. $540t + 660(t - 1) = 1800$

$\quad 540t + 660t - 660 = 1800$

$\qquad 1200t - 660 = 1800$

$\qquad 1200t = 2460$

$\qquad t = 2.05$ hours or 2 hours, 3 minutes

$540(2.05) = 1107$ miles

Mid-Chapter Quiz for Chapter 7

1. $f(x) = \dfrac{x + 2}{x(x - 4)}$

The denominator, $x(x - 4)$ equals zero when $x = 0$ or when $x = 4$. So, the domain is the set of all real values of x such that $x \neq 0$, $x \neq 4$.

Domain: $(-\infty, 0) \cup (0, 4) \cup (4, \infty)$

2. $h(x) = \dfrac{x^2 - 9}{x^2 - x - 2}$

(a) $h(-3) = \dfrac{(-3)^2 - 9}{(-3)^2 - (-3) - 2}$

$\qquad = \dfrac{0}{10}$

$\qquad = 0$

(c) $h(-1) = \dfrac{(-1)^2 - 9}{(-1)^2 - (-1) - 2}$

$\qquad = \dfrac{-8}{0}$ undefined

$\dfrac{x^2 - 9}{x^2 - x - 2}$ is undefined for $x = -1$.

(b) $h(0) = \dfrac{0^2 - 9}{0^2 - 0 - 2}$

$\qquad = \dfrac{-9}{-2}$

$\qquad = \dfrac{9}{2}$

(d) $h(5) = \dfrac{5^2 - 9}{5^2 - 5 - 2}$

$\qquad = \dfrac{16}{18}$

$\qquad = \dfrac{8}{9}$

3. $\dfrac{9y^2}{6y} = \dfrac{3\cancel{(3)}\cancel{(y)}(y)}{2\cancel{(3)}\cancel{(y)}}$

$\quad = \dfrac{3y}{2}, \ y \neq 0$

4. $\dfrac{8u^3v^2}{36uv^3} = \dfrac{\cancel{4}(2)\cancel{(u)}(u^2)\cancel{(v^2)}}{\cancel{4}(9)\cancel{(u)}\cancel{(v^2)}(v)}$

$\quad = \dfrac{2u^2}{9v}, \ u \neq 0$

5. $\dfrac{4x^2 - 1}{x - 2x^2} = \dfrac{(2x + 1)(2x - 1)}{x(1 - 2x)}$

$\quad = \dfrac{(2x + 1)\cancel{(2x - 1)}}{x(-1)\cancel{(2x - 1)}}$

$\quad = -\dfrac{2x + 1}{x}, \ x \neq \dfrac{1}{2}$

6. $\dfrac{(z + 3)^2}{2z^2 + 5z - 3} = \dfrac{(z + 3)\cancel{(z + 3)}}{(2z - 1)\cancel{(z + 3)}}$

$\quad = \dfrac{z + 3}{2z - 1}, \ z \neq -3$

7. $\dfrac{7ab + 3a^2b^2}{a^2b} = \dfrac{\cancel{ab}(7 + 3ab)}{\cancel{ab}(a)}$

$\quad = \dfrac{7 + 3ab}{a}, \ b \neq 0$

8. $\dfrac{2mn^2 - n^3}{2m^2 + mn - n^2} = \dfrac{n^2\cancel{(2m - n)}}{\cancel{(2m - n)}(m + n)}$

$\quad = \dfrac{n^2}{m + n}, \ n \neq 2m$

9. $\dfrac{11t^2}{6} \cdot \dfrac{9}{33t} = \dfrac{\cancel{11}(t)\cancel{(t)}\cancel{(3)}(3)}{2\cancel{(3)}\cancel{(3)}\cancel{(11)}\cancel{(t)}}$

$\quad = \dfrac{t}{2}, \ t \neq 0$

10. $\dfrac{4}{3(x - 1)} \cdot \dfrac{2x}{(x^2 + 2x - 3)} = \dfrac{8x}{3(x - 1)(x + 3)(x - 1)}$

$\quad = \dfrac{8x}{3(x - 1)^2(x + 3)}$

11. $\dfrac{x - 1}{x^2 + 2x} \div \dfrac{1 - x^2}{x^2 + 5x + 6} = \dfrac{x - 1}{x^2 + 2x} \cdot \dfrac{x^2 + 5x + 6}{1 - x^2}$

$\quad = \dfrac{(x - 1)(x + 2)(x + 3)}{x(x + 2)(1 - x)(1 + x)}$

$\quad = \dfrac{\cancel{(x - 1)}\cancel{(x + 2)}(x + 3)}{x\cancel{(x + 2)}(-1)\cancel{(x - 1)}(x + 1)}$

$\quad = -\dfrac{x + 3}{x(x + 1)}, \quad x \neq -2, x \neq 1, x \neq -3$

12. $\dfrac{5u}{3(u + v)} \cdot \dfrac{2(u^2 - v^2)}{3v} \div \dfrac{25u^2}{18(u - v)} = \dfrac{5u(2)(u + v)(u - v)}{3(u + v)(3)(v)} \cdot \dfrac{18(u - v)}{25u^2}$

$\quad = \dfrac{\cancel{5}(2)\cancel{(3)}\cancel{(3)}(2)\cancel{(u)}\cancel{(u + v)}(u - v)^2}{\cancel{3}\cancel{(3)}\cancel{(5)}(5)(v)\cancel{(u)}(u)\cancel{(u + v)}}$

$\quad = \dfrac{4(u - v)^2}{5uv}, \ u \neq v, u \neq -v$

13. $\dfrac{\left(\dfrac{10}{x^2 + 2x}\right)}{\left(\dfrac{15}{x^2 + 3x + 2}\right)} = \dfrac{10}{x(x + 2)} \cdot \dfrac{(x + 2)(x + 1)}{15}$

$\quad = \dfrac{\cancel{5}(2)\cancel{(x + 2)}(x + 1)}{\cancel{5}(3)(x)\cancel{(x + 2)}}$

$\quad = \dfrac{2(x + 1)}{3x}, \ x \neq -2, x \neq -1$

14. $\dfrac{4x}{x + 5} - \dfrac{3x}{4} = \dfrac{4x(4)}{(x + 5)(4)} - \dfrac{3x(x + 5)}{4(x + 5)}$

$\quad = \dfrac{4x(4) - 3x(x + 5)}{4(x + 5)} = \dfrac{16x - 3x^2 - 15x}{4(x + 5)}$

$\quad = \dfrac{-3x^2 + x}{4(x + 5)}$

15. $\dfrac{x-y}{x^2-xy-6y^2} + \dfrac{3x-5y}{x^2+xy-2y^2} = \dfrac{x-y}{(x-3y)(x+2y)} + \dfrac{3x-5y}{(x+2y)(x-y)}$

$$= \dfrac{(x-y)(x-y)}{(x-3y)(x+2y)(x-y)} + \dfrac{(3x-5y)(x-3y)}{(x+2y)(x-y)(x-3y)}$$

$$= \dfrac{x^2-2xy+y^2+3x^2-14xy+15y^2}{(x-3y)(x+2y)(x-y)}$$

$$= \dfrac{4x^2-16xy+16y^2}{(x-3y)(x+2y)(x-y)}$$

$$= \dfrac{4(x^2-4xy+4y^2)}{(x-3y)(x+2y)(x-y)}$$

$$= \dfrac{4(x-2y)^2}{(x-3y)(x+2y)(x-y)}$$

16. $\dfrac{\left(1-\dfrac{2}{x}\right)}{\left(\dfrac{3}{x}-\dfrac{4}{5}\right)} = \dfrac{\left(1-\dfrac{2}{x}\right)5x}{\left(\dfrac{3}{x}-\dfrac{4}{5}\right)5x}$

$$= \dfrac{5x-10}{15-4x} \quad \text{or} \quad \dfrac{5(x-2)}{-4x+15}$$

$$\text{or} \quad -\dfrac{5(x-2)}{4x-15}, \quad x \neq 0$$

17.
$$
\begin{array}{r}
2x^2 - 4x + 3, \ x \neq \dfrac{2}{3} \\
3x-2\ \overline{)\ 6x^3 - 16x^2 + 17x - 6} \\
\underline{6x^3 - 4x^2} \\
-12x^2 + 17x \\
\underline{-12x^2 + 8x} \\
9x - 6 \\
\underline{9x - 6} \\
0
\end{array}
$$

18. $\dfrac{3x^3+4x^2-27x+20}{x+4}$

$$
\begin{array}{r|rrrr}
-4 & 3 & 4 & -27 & 20 \\
   &   & -12 & 32 & -20 \\
\hline
   & 3 & -8 & 5 & 0
\end{array}
$$

$$\dfrac{3x^3+4x^2-27x+20}{x+4} = 3x^2-8x+5, \quad x \neq -4$$

19. (a) Average cost per unit $= \dfrac{\text{Total costs}}{\text{Number of units}}$

$$= \dfrac{10.50x+6000}{x}$$

(b) Average cost per unit $= \dfrac{10.50x+6000}{x}, \ x=500$

Average cost per unit $= \dfrac{10.50(500)+6000}{500}$

$$= \dfrac{5250+6000}{500}$$

$$= \dfrac{11{,}250}{500}$$

$$= \$22.50$$

20.   Area of larger triangle $= \frac{1}{2}(x + 4)(x + 2)$

$$= \frac{(x + 4)(x + 2)}{2}$$

Area of smaller triangle $= \frac{1}{2}(x)\left(\frac{x(x + 2)}{x + 4}\right)$

$$= \frac{x^2(x + 2)}{2(x + 4)}$$

Area of shaded region $=$ Area of larger triangle $-$ Area of smaller triangle

$$= \frac{(x + 4)(x + 2)}{2} - \frac{x^2(x + 2)}{2(x + 4)}$$

$$= \frac{(x + 4)(x + 2)(x + 4)}{2(x + 4)} - \frac{x^2(x + 2)}{2(x + 4)}$$

$$= \frac{(x + 4)^2(x + 2) - x^2(x + 2)}{2(x + 4)}$$

$$= \frac{(x + 2)[(x + 4)^2 - x^2]}{2(x + 4)}$$

$$= \frac{(x + 2)(x^2 + 8x + 16 - x^2)}{2(x + 4)}$$

$$= \frac{(x + 2)(8x + 16)}{2(x + 4)}$$

$$= \frac{(x + 2)(8)(x + 2)}{2(x + 4)}$$

$$= \frac{\cancel{2}(4)(x + 2)^2}{\cancel{2}(x + 4)}$$

$$= \frac{4(x + 2)^2}{x + 4}$$

Ratio: $\dfrac{\text{Shaded Area}}{\text{Area of larger triangle}} = \dfrac{\left(\dfrac{4(x + 2)^2}{x + 4}\right)}{\left(\dfrac{(x + 2)(x + 4)}{2}\right)}$

$$= \frac{4(x + 2)(x + 2)}{x + 4} \cdot \frac{2}{(x + 2)(x + 4)}$$

$$= \frac{8(x + 2)\cancel{(x + 2)}}{(x + 4)^2\cancel{(x + 2)}}$$

$$= \frac{8(x + 2)}{(x + 4)^2}$$

Note: The restriction, $x \neq -2$, is not necessary because $x > 0$.

Section 7.5   Solving Rational Equations

1. (a) $x = 0$

$$\frac{0}{3} - \frac{0}{5} \overset{?}{=} \frac{4}{3}$$

$$0 - 0 \overset{?}{=} \frac{4}{3}$$

$$0 \neq \frac{4}{3}$$

No, 0 is *not* a solution.

(c) $x = \frac{1}{8}$

$$\frac{\left(\frac{1}{8}\right)}{3} - \frac{\left(\frac{1}{8}\right)}{5} \overset{?}{=} \frac{4}{3}$$

$$\frac{1}{24} - \frac{1}{40} \overset{?}{=} \frac{4}{3}$$

$$\frac{5}{120} - \frac{3}{120} \overset{?}{=} \frac{4}{3}$$

$$\frac{2}{120} \neq \frac{4}{3}$$

No, $\frac{1}{8}$ is *not* a solution.

(b) $x = -1$

$$\frac{-1}{3} - \frac{-1}{5} \overset{?}{=} \frac{4}{3}$$

$$\frac{-5}{15} - \frac{-3}{15} \overset{?}{=} \frac{4}{3}$$

$$-\frac{5}{15} + \frac{3}{15} \overset{?}{=} \frac{4}{3}$$

$$-\frac{2}{15} \neq \frac{4}{3}$$

No, -1 is *not* a solution.

(d) $x = 10$

$$\frac{10}{3} - \frac{10}{5} \overset{?}{=} \frac{4}{3}$$

$$\frac{10}{3} - 2 \overset{?}{=} \frac{4}{3}$$

$$\frac{10}{3} - \frac{6}{3} = \frac{4}{3}$$

Yes, 10 *is* a solution.

3. (a) $x = -1$

$$-\frac{1}{4} + \frac{3}{4(-1)} \overset{?}{=} 1$$

$$-\frac{1}{4} - \frac{3}{4} \neq 1$$

No, -1 is *not* a solution.

(c) $x = 3$

$$\frac{3}{4} + \frac{3}{4(3)} \overset{?}{=} 1$$

$$\frac{3}{4} + \frac{3}{12} \overset{?}{=} 1$$

$$\frac{3}{4} + \frac{1}{4} = 1$$

Yes, 3 *is* a solution.

(b) $x = 1$

$$\frac{1}{4} + \frac{3}{4(1)} \overset{?}{=} 1$$

$$\frac{1}{4} + \frac{3}{4} = 1$$

Yes, 1 *is* a solution.

(d) $x = \frac{1}{2}$

$$\frac{\left(\frac{1}{2}\right)}{4} + \frac{3}{4\left(\frac{1}{2}\right)} \overset{?}{=} 1$$

$$\frac{1}{8} + \frac{3}{2} \neq 1$$

No, $\frac{1}{2}$ is *not* a solution.

5. $\dfrac{x}{4} = \dfrac{3}{8}$ or $\dfrac{x}{4} = \dfrac{3}{8}$

$8x = 12$ $8\left(\dfrac{x}{4}\right) = 8\left(\dfrac{3}{8}\right)$

$x = \dfrac{12}{8}$ $2x = 3$

$x = \dfrac{3}{2}$ $x = \dfrac{3}{2}$

7. $6 - \dfrac{x}{4} = x - 9$

$4\left(6 - \dfrac{x}{4}\right) = 4(x - 9)$

$24 - x = 4x - 36$

$24 - 5x = -36$

$-5x = -60$

$x = 12$

9. $\dfrac{4t}{3} = 15 - \dfrac{t}{6}$

$6\left(\dfrac{4t}{3}\right) = 6\left(15 - \dfrac{t}{6}\right)$

$2(4t) = 90 - t$

$8t = 90 - t$

$9t = 90$

$t = \dfrac{90}{9}$

$t = 10$

11. $\dfrac{h + 2}{5} - \dfrac{h - 1}{9} = \dfrac{2}{3}$

$45\left(\dfrac{h + 2}{5} - \dfrac{h - 1}{9}\right) = 45\left(\dfrac{2}{3}\right)$

$9(h + 2) - 5(h - 1) = 30$

$9h + 18 - 5h + 5 = 30$

$4h + 23 = 30$

$4h = 7$

$h = \dfrac{7}{4}$

13. $\dfrac{9}{25 - y} = -\dfrac{1}{4}$

$36 = -1(25 - y) = -25 + y$

$61 = y$

or

$\dfrac{9}{25 - y} = -\dfrac{1}{4}$

$4(25 - y)\left(\dfrac{9}{25 - y}\right) = 4(25 - y)\left(-\dfrac{1}{4}\right)$

$36 = -(25 - y) = -25 + y$

$61 = y$

15. $5 - \dfrac{12}{a} = \dfrac{5}{3}$

$3a\left(5 - \dfrac{12}{a}\right) = 3a\left(\dfrac{5}{3}\right)$

$15a - 36 = 5a$

$-36 = -10a$

$\dfrac{-36}{-10} = a$

$\dfrac{18}{5} = a$

17. $\dfrac{12}{y + 5} + \dfrac{1}{2} = 2$

$2(y + 5)\left(\dfrac{12}{y + 5} + \dfrac{1}{2}\right) = 2(y + 5)2$

$24 + (y + 5) = 4(y + 5)$

$24 + y + 5 = 4y + 20$

$29 + y = 4y + 20$

$29 = 3y + 20$

$9 = 3y$

$\dfrac{9}{3} = y$

$3 = y$

19. $\dfrac{5}{x} = \dfrac{25}{3(x + 2)}$

$15(x + 2) = 25x$

$15x + 30 = 25x$

$30 = 10x$

$\dfrac{30}{10} = x$

$3 = x$

21. $\dfrac{8}{3x + 5} = \dfrac{1}{x + 2}$

$8(x + 2) = 3x + 5$

$8x + 16 = 3x + 5$

$5x + 16 = 5$

$5x = -11$

$x = -\dfrac{11}{5}$

23. $\dfrac{3}{x + 2} - \dfrac{1}{x} = \dfrac{1}{5x}$

$5x(x + 2)\left(\dfrac{3}{x + 2} - \dfrac{1}{x}\right) = 5x(x + 2)\left(\dfrac{1}{5x}\right)$

$5x(3) - 5(x + 2) = x + 2$

$15x - 5x - 10 = x + 2$

$10x - 10 = x + 2$

$9x - 10 = 2$

$9x = 12$

$x = \dfrac{12}{9} = \dfrac{4}{3}$

25.
$$\frac{1}{2} = \frac{18}{x^2}$$
$$x^2 = 36$$
$$x^2 - 36 = 0$$
$$(x + 6)(x - 6) = 0$$
$$x + 6 = 0 \implies x = -6$$
$$x - 6 = 0 \implies x = 6$$

27.
$$\frac{32}{t} = 2t$$
$$32 = 2t^2$$
$$0 = 2t^2 - 32$$
$$= 2(t^2 - 16) = 2(t + 4)(t - 4)$$
$$2 \neq 0$$
$$t + 4 = 0 \implies t = -4$$
$$t - 4 = 0 \implies t = 4$$

29.
$$x + 1 = \frac{72}{x}$$
$$x(x + 1) = x\left(\frac{72}{x}\right)$$
$$x^2 + x = 72$$
$$x^2 + x - 72 = 0$$
$$(x + 9)(x - 8) = 0$$
$$x + 9 = 0 \implies x = -9$$
$$x - 8 = 0 \implies x = 8$$

31.
$$x - \frac{24}{x} = 5$$
$$x\left(x - \frac{24}{x}\right) = 5(x)$$
$$x^2 - 24 = 5x$$
$$x^2 - 5x - 24 = 0$$
$$(x - 8)(x + 3) = 0$$
$$x - 8 = 0 \implies x = 8$$
$$x + 3 = 0 \implies x = -3$$

33.
$$\frac{10}{x(x - 2)} + \frac{4}{x} = \frac{5}{x - 2}$$
$$x(x - 2)\left[\frac{10}{x(x - 2)} + \frac{4}{x}\right] = x(x - 2)\left(\frac{5}{x - 2}\right)$$
$$10 + 4(x - 2) = 5x$$
$$10 + 4x - 8 = 5x$$
$$4x + 2 = 5x$$
$$2 = x \quad \text{Extraneous}$$

Substituting 2 for x in the original equation results in division by zero. Thus, $x = 2$ is an *extraneous* solution and the equation has *no* solution.

35.
$$\frac{10}{x + 3} + \frac{10}{3} = 6$$
$$3(x + 3)\left(\frac{10}{x + 3} + \frac{10}{3}\right) = 6(3)(x + 3)$$
$$3(10) + 10(x + 3) = 18(x + 3)$$
$$30 + 10x + 30 = 18x + 54$$
$$10x + 60 = 18x + 54$$
$$-8x + 60 = 54$$
$$-8x = -6$$
$$x = \frac{-6}{-8}$$
$$x = \frac{3}{4}$$

37.
$$\frac{2}{(x - 4)(x - 2)} = \frac{1}{x - 4} + \frac{2}{x - 2}$$
$$(x - 4)(x - 2)\left(\frac{2}{(x - 4)(x - 2)}\right) = \left(\frac{1}{x - 4} + \frac{2}{x - 2}\right)(x - 4)(x - 2)$$
$$2 = x - 2 + 2(x - 4)$$
$$2 = x - 2 + 2x - 8$$
$$2 = 3x - 10$$
$$12 = 3x$$
$$4 = x \quad \text{(Extraneous)}$$

Substituting 4 for x in the original equation results in division by zero. Thus, $x = 4$ in an *extraneous* solution and the equation has *no* solution.

39.
$$\frac{1}{x-5} + \frac{1}{x+5} = \frac{x+3}{x^2-25}$$

$$(x-5)(x+5)\left(\frac{1}{x-5} + \frac{1}{x+5}\right) = (x-5)(x+5)\left[\frac{x+3}{(x+5)(x-5)}\right]$$

$$x + 5 + x - 5 = x + 3$$

$$2x = x + 3$$

$$x = 3$$

41.
$$\frac{2}{x-10} - \frac{3}{x-2} = \frac{6}{x^2-12x+20}$$

$$(x-10)(x-2)\left(\frac{2}{x-10} - \frac{3}{x-2}\right) = (x-10)(x-2)\left[\frac{6}{(x-10)(x-2)}\right]$$

$$2(x-2) - 3(x-10) = 6$$

$$2x - 4 - 3x + 30 = 6$$

$$-x + 26 = 6$$

$$-x = -20$$

$$x = 20$$

43.
$$\frac{2(x+1)}{x^2-4x+3} + \frac{6x}{x-3} = \frac{3x}{x-1}$$

$$\frac{2(x+1)}{(x-3)(x-1)} + \frac{6x}{x-3} = \frac{3x}{x-1}$$

$$(x-3)(x-1)\left(\frac{2(x+1)}{(x-3)(x-1)} + \frac{6x}{x-3}\right) = (x-3)(x-1)\left(\frac{3x}{x-1}\right)$$

$$2(x+1) + 6x(x-1) = (x-3)(3x)$$

$$2x + 2 + 6x^2 - 6x = 3x^2 - 9x$$

$$6x^2 - 4x + 2 = 3x^2 - 9x$$

$$3x^2 + 5x + 2 = 0$$

$$(3x+2)(x+1) = 0$$

$$3x + 2 = 0 \Rightarrow 3x = -2 \Rightarrow x = -\frac{2}{3}$$

$$x + 1 = 0 \Rightarrow x = -1$$

45.
$$\frac{4}{2x+3} + \frac{17}{5(2x+3)} = 3$$

$$5(2x+3)\left[\frac{4}{2x+3} + \frac{17}{5(2x+3)}\right] = 5(2x+3)(3)$$

$$20 + 17 = 15(2x+3)$$

$$37 = 30x + 45$$

$$-8 = 30x$$

$$-\frac{8}{30} = x$$

$$-\frac{4}{15} = x$$

47.
$$1 = \frac{16}{y} - \frac{39}{y^2}$$

$$y^2(1) = y^2\left(\frac{16}{y} - \frac{39}{y^2}\right)$$

$$y^2 = 16y - 39$$

$$y^2 - 16y + 39 = 0$$

$$(y-13)(y-3) = 0$$

$$y - 13 = 0 \Rightarrow y = 13$$

$$y - 3 = 0 \Rightarrow y = 3$$

49.
$$\frac{1}{x-1} + \frac{3}{x+1} = 2$$

$$(x-1)(x+1)\left(\frac{1}{x-1} + \frac{3}{x+1}\right) = (x-1)(x+1)(2)$$

$$x + 1 + 3(x-1) = (x^2 - 1)(2)$$

$$x + 1 + 3x - 3 = 2x^2 - 2$$

$$4x - 2 = 2x^2 - 2$$

$$0 = 2x^2 - 4x = 2x(x-2)$$

$$2x = 0 \implies x = 0$$

$$x - 2 = 0 \implies x = 2$$

51.
$$\frac{x^2 - x}{x^4 + 4x^2 + 3} = \frac{1}{x^2 + 1} - \frac{1}{x^2 + 3}$$

$$\frac{x^2 - x}{(x^2 + 3)(x^2 + 1)} = \frac{1}{x^2 + 1} - \frac{1}{x^2 + 3}$$

$$(x^2 + 1)(x^2 + 3)\left[\frac{x^2 - x}{(x^2 + 3)(x^2 + 1)}\right] = (x^2 + 1)(x^2 + 3)\left(\frac{1}{x^2 + 1} - \frac{1}{x^2 + 3}\right)$$

$$x^2 - x = x^2 + 3 - (x^2 + 1)$$

$$x^2 - x = x^2 + 3 - x^2 - 1$$

$$x^2 - x = 2$$

$$x^2 - x - 2 = 0$$

$$(x - 2)(x + 1) = 0$$

$$x - 2 = 0 \implies x = 2$$

$$x + 1 = 0 \implies x = -1$$

53.
$$\frac{2x}{5} = \frac{x^2 - 5x}{5x}$$

$$2x(5x) = 5(x^2 - 5x)$$

$$10x^2 = 5x^2 - 25x$$

$$5x^2 + 25x = 0$$

$$5x(x + 5) = 0$$

$$5x = 0 \implies x = 0 \quad \text{(Extraneous)}$$

$$x + 5 = 0 \implies x = -5$$

Substituting 0 for x in the original equation results in division by zero. So, $x = 0$ is an *extraneous* solution and the only solution is $x = -5$.

55.
$$\frac{x}{2} = \frac{\left(2 - \dfrac{3}{x}\right)}{\left(1 - \dfrac{1}{x}\right)} = \frac{\left(2 - \dfrac{3}{x}\right)(x)}{\left(1 - \dfrac{1}{x}\right)(x)}$$

$$= \frac{2x - 3}{x - 1}$$

$$x(x - 1) = 2(2x - 3)$$

$$x^2 - x = 4x - 6$$

$$x^2 - 5x + 6 = 0$$

$$(x - 3)(x - 2) = 0$$

$$x - 3 = 0 \implies x = 3$$

$$x - 2 = 0 \implies x = 2$$

57. $$\frac{2(x + 7)}{x + 4} - 2 = \frac{2x + 20}{2x + 8}$$

$$2(x + 4)\left(\frac{2(x + 7)}{x + 4} - 2\right) = \left(\frac{2x + 20}{2(x + 4)}\right)(2)(x + 4)$$

$$4(x + 7) - 4(x + 4) = 2x + 20$$

$$4x + 28 - 4x - 16 = 2x + 20$$

$$12 = 2x + 20$$

$$-8 = 2x$$

$$-4 = x \quad \text{(Extraneous)}$$

Substituting -4 for x in the original equation results in division by zero. So, $x = -4$ is an *extraneous* solution and the equation has *no* solution.

59. *Verbal Model:* $\boxed{\text{Number}} + \boxed{\text{Reciprocal}} = \frac{65}{8}$

Labels: $\quad$ Number $= x$

$\qquad\qquad$ Reciprocal $= \dfrac{1}{x}$

Equation: $\qquad\qquad\qquad x + \dfrac{1}{x} = \dfrac{65}{8}$

$$8x\left(x + \frac{1}{x}\right) = 8x\left(\frac{65}{8}\right)$$

$$8x^2 + 8 = 65x$$

$$8x^2 - 65x + 8 = 0$$

$$(8x - 1)(x - 8) = 0$$

$$8x - 1 = 0 \implies x = \frac{1}{8}$$

$$x - 8 = 0 \implies x = 8$$

So, the number is $\frac{1}{8}$ or 8.

61. *Verbal Model:* $2\boxed{\text{Number}} + 3\boxed{\text{Reciprocal}} = \frac{97}{4}$

Labels: $\quad$ Number $= x$

$\qquad\qquad$ Reciprocal $= \dfrac{1}{x}$

Equations: $\qquad\qquad 2x + 3\left(\dfrac{1}{x}\right) = \dfrac{97}{4}$

$$2x + \frac{3}{x} = \frac{97}{4}$$

$$4x\left(2x + \frac{3}{x}\right) = 4x\left(\frac{97}{4}\right)$$

$$8x^2 + 12 = 97x$$

$$8x^2 - 97x + 12 = 0$$

$$(8x - 1)(x - 12) = 0$$

$$8x - 1 = 0 \implies x = \frac{1}{8}$$

$$x - 12 = 0 \implies x = 12$$

So, the number is $\frac{1}{8}$ or 12.

63. *Verbal Model:* $\boxed{\text{Time with tail wind}} = \boxed{\text{Time with head wind}}$

$\qquad\qquad\qquad \dfrac{\boxed{\text{Distance with tail wind}}}{\text{Rate with tail wind}} = \dfrac{\boxed{\text{Distance with head wind}}}{\text{Rate with head wind}}$

Labels: $\quad$ Speed of wind $= x$ (miles per hour)

$\qquad\qquad$ Tail wind: Distance $= 680$ (miles); Rate $= 300 + x$ (miles per hour)

$\qquad\qquad$ Head wind: Distance $= 520$ (miles); Rate $= 300 - x$ (miles per hour)

Equation: $\qquad\qquad \dfrac{680}{300 + x} = \dfrac{520}{300 - x}$

$$680(300 - x) = 520(300 + x)$$

$$204{,}000 - 680x = 156{,}000 + 520x$$

$$48{,}000 - 680x = 520x$$

—CONTINUED—

63. —CONTINUED—

$$48,000 = 1200x$$

$$\frac{48,000}{1200} = x$$

$$40 = x$$

So, the speed of the wind is 40 miles per hour.

65. *Verbal Model:*

$$\boxed{\text{Time at slower speed}} + \boxed{\text{Time at faster speed}} = 6$$

$$\boxed{\frac{\text{Distance at slower speed}}{\text{Slower speed}}} + \boxed{\frac{\text{Distance at faster speed}}{\text{Faster speed}}} = 6$$

Labels: Slower rate $= r$ (miles per hour); Distance $= 180$ (miles)

Faster rate $= t + 10$ (miles per hour); Distance $= 144$ (miles)

Equations:

$$\frac{180}{r} + \frac{144}{r + 10} = 6$$

$$r(r + 10)\left(\frac{180}{r} + \frac{144}{r + 10}\right) = 6r(r + 10)$$

$$180(r + 10) + 144r = 6r^2 + 60r$$

$$180r + 1800 + 144r = 6r^2 + 60r$$

$$324r + 1800 = 6r^2 + 60r$$

$$0 = 6r^2 - 264r - 1800$$

$$0 = 6(r^2 - 44r - 300)$$

$$0 = 6(r - 50)(r + 6)$$

$$6 \neq 0$$

$$r - 50 = 0 \implies r = 50$$

$$r + 6 = 0 \implies r = -6 \ (\text{Discard negative solution})$$

$$r = 50 \ \text{and} \ r + 10 = 60$$

The speeds are 50 miles per hour and 60 miles per hour.

67. *Verbal Model:*

$$\boxed{\text{Time for faster runner}} = \boxed{\text{Time for slower runner}}$$

$$\boxed{\frac{\text{Distance for faster runner}}{\text{Rate for faster runner}}} = \boxed{\frac{\text{Distance for slower runner}}{\text{Rate for slower runner}}}$$

Labels: Slower runner: Distance $= 4$ miles; Rate $= x$ (miles per hour)

Faster runner: Distance $= 5$ miles; Rate $= x + 2$ (miles per hour)

Equation:

$$\frac{5}{x + 2} = \frac{4}{x}$$

$$5x = 4(x + 2)$$

$$5x = 4x + 8$$

$$x = 8 \ \text{and} \ x + 2 = 10$$

The speeds of the runners are 8 miles per hour and 10 miles per hour.

69. *Verbal Model:*

| Future cost per person | = | Current cost per person | − | 4000 |

$$\frac{\text{Total cost}}{\text{Number of members in future}} = \frac{\text{Total cost}}{\text{Current number of members}} - \boxed{4000}$$

Labels: Current number of members $= x$ (persons)

Number of members in future $= x + 2$

Total cost $= 240{,}000$ (dollars)

Equation:

$$\frac{240{,}000}{x + 2} = \frac{240{,}000}{x} - 4000$$

$$x(x + 2)\left(\frac{240{,}000}{x + 2}\right) = \left(\frac{240{,}000}{x} - 4000\right)x(x + 2)$$

$$240{,}000x = 240{,}000(x + 2) - 4000x(x + 2)$$

$$240{,}000x = 240{,}000x + 480{,}000 - 4000x^2 - 8000x$$

$$4000x^2 + 8000x - 480{,}000 = 0$$

$$x^2 + 2x - 120 = 0$$

$$(x + 12)(x - 10) = 0$$

$$x + 12 = 0 \implies x = -12 \qquad \text{(Discard negative solution)}$$

$$x - 10 = 0 \implies x = 10$$

There are presently 10 members in the group.

71. *Verbal Model:*

| Rate for first person | + | Rate for second person | = | Rate together |

Equation:

$$\frac{1}{6} + \frac{1}{4} = \frac{1}{t}$$

$$24t\left(\frac{1}{6} + \frac{1}{4}\right) = 24t\left(\frac{1}{t}\right)$$

$$4t + 6t = 24$$

$$10t = 24$$

$$t = \frac{24}{10}$$

$$t = \frac{12}{5}$$

So, it will take $2\frac{2}{5}$ hours to complete the task if the two people work together.

73. *Verbal Model:*

| Rate for first electrician | + | Rate for second electrician | = | Rate together |

Equation:

$$\frac{1}{14} + \frac{1}{18} = \frac{1}{t}$$

$$126t\left(\frac{1}{14} + \frac{1}{18}\right) = 126t\left(\frac{1}{t}\right)$$

$$9t + 7t = 126$$

$$16t = 126$$

$$t = \frac{126}{16}$$

$$t = \frac{63}{8}$$

So, it will take the two electricians $7\frac{7}{8}$ hours working together.

75. *Verbal Model:* | Rate for one pipe | + | Rate for second pipe | = | Rate Together |

$$\frac{1}{t} + \frac{1}{(5/4)t} = \frac{1}{5}$$

$$\frac{1}{t} + \frac{4}{5t} = \frac{1}{5}$$

$$5t\left(\frac{1}{t} + \frac{4}{5t}\right) = 5t\left(\frac{1}{5}\right)$$

$$5 + 4 = t$$

$$9 = t \text{ and } \frac{5}{4}t = \frac{5}{4} \cdot 9 = \frac{45}{4} \text{ or } 11\frac{1}{4}$$

So, it would take $11\frac{1}{4}$ hours to fill the pool using only the pipe with the lower flow rate.

77.
$$\overline{C} = 1.50 + \frac{4200}{x}$$

$$2.90 = 1.50 + \frac{4200}{x}$$

$$1.40 = \frac{4200}{x}$$

$$1.40x = 4200$$

$$x = \frac{4200}{1.40}$$

$$x = 3000$$

So, 3000 units must be produced so that the average cost is $2.90.

79. As an equation is solved, a trial solution is sometimes found which does not satisfy the original equation. This is called an extraneous solution. You can identify an extraneous solution by checking solutions in the original equation.

81. $(4x - 7)(3x + 2) = 0$

$$4x - 7 = 0 \Longrightarrow 4x = 7 \Longrightarrow x = \frac{7}{4}$$

$$3x + 2 = 0 \Longrightarrow 3x = -2 \Longrightarrow x = -\frac{2}{3}$$

83. $7 - 3x > 4 - x$

$$7 - 2x > 4$$

$$-2x > -3$$

$$\frac{-2x}{-2} < \frac{-3}{-2}$$

$$x < \frac{3}{2}$$

85.
$$|x - 3| < 2$$

$$-2 < x - 3 < 2$$

$$1 < x < 5$$

87. *Label:* Amount invested at 8% = x (dollars)

$$0.055(2500) + 0.08x = 0.07(2500 + x)$$

$$137.5 + 0.08x = 175 + 0.07x$$

$$137.5 + 0.01x = 175$$

$$0.01x = 37.5$$

$$x = 3750$$

So, $3750 must be invested at 8%.

Section 7.6   Graphing Rational Functions

1. (a)

x	0	0.5	0.9	0.99	0.999
y	−4	−8	−40	−400	−4000

x	2	1.5	1.1	1.01	1.001
y	4	8	40	400	4000

x	2	5	10	100	1000
y	4	1	0.4444	0.0404	0.0040

(b) Vertical asymptote: $x = 1$

Horizontal asymptote: $y = 0$

(c) Domain: $(-\infty, 1) \cup (1, \infty)$

3. $f(x) = \dfrac{2}{x+1}$

Graph (d)

5. $f(x) = \dfrac{x-2}{x-1}$

Graph (b)

7. $f(x) = \dfrac{5}{x^2}$

Domain: $(-\infty, 0) \cup (0, \infty)$

Vertical asymptote: $x = 0$

Horizontal asymptote: $y = 0$

Because the denominator $x^2 = 0$ when $x = 0$, the domain is all real values of x except $x = 0$, and the vertical asymptote is $x = 0$. Because the degree of the numerator is less than the degree of the denominator, the horizontal asymptote is $y = 0$.

9. $f(x) = \dfrac{x}{x+8}$

Domain: $(-\infty, -8) \cup (-8, \infty)$

Vertical asymptote: $x = -8$

Horizontal asymptote: $y = 1$

Because the denominator $x + 8 = 0$ when $x = -8$, the domain is all real values of x except $x = -8$, and the vertical asymptote is $x = -8$. Because the degree of the numerator is equal to the degree of the denominator and the ratio of the leading coefficients is 1, the horizontal asymptote is $y = 1$.

11. $g(t) = \dfrac{3}{t(t-1)}$

Domain: $(-\infty, 0) \cup (0, 1) \cup (1, \infty)$

Vertical asymptotes: $t = 0$ and $t = 1$

Horizontal asymptote: $y = 0$

Because the denominator $t(t-1) = 0$ when $t = 0$ and $t = 1$, the domain is all real values of t except $t = 0$ and $t = 1$, and the vertical asymptotes are $t = 0$ and $t = 1$. Because the degree of the numerator is less than the degree of the denominator, the horizontal asymptote is $y = 0$.

13. $h(s) = \dfrac{2s^2}{s+3}$

Domain: $(-\infty, -3) \cup (-3, \infty)$

Vertical asymptotes: $s = -3$

Horizontal asymptotes: None

Because the denominator $s + 3 = 0$ when $s = -3$, the domain is all real values of s expect $s = -3$, and the vertical asymptote is $s = -3$. Because the degree of the numerator is greater than the degree of the denominator, there is no horizontal asymptote.

15. $y = \dfrac{3 - 5x}{1 - 3x}$

Domain: $\left(-\infty, \dfrac{1}{3}\right) \cup \left(\dfrac{1}{3}, \infty\right)$

Vertical asymptote: $x = \dfrac{1}{3}$

Horizontal asymptote: $y = \dfrac{5}{3}$

Because the denominator $1 - 3x = 0$ when $x = \frac{1}{3}$, the domain is all real values of x except $x = \frac{1}{3}$, and the vertical asymptote is $x = \frac{1}{3}$. Because the degree of the numerator is equal to the degree of the denominator and the ratio of the leading coefficients is $\frac{5}{3}$, the horizontal asymptote is $y = \frac{5}{3}$.

17. $y = \dfrac{5x^2}{x^2 - 1}$

Domain: $(-\infty, -1) \cup (-1, 1) \cup (1, \infty)$

Vertical asymptotes: $x = 1, x = -1$

Horizontal asymptote: $y = 5$

Because the denominator $x^2 - 1 = 0$ when $x = 1$ or $x = -1$, the domain is all real values of x except $x = 1$ and $x = -1$, and the vertical asymptotes are $x = 1$ and $x = -1$. Because the degree of the numerator is equal to the degree of the denominator and the ratio of the leading coefficients is 5, the horizontal asymptote is $y = 5$.

19. $f(x) = \dfrac{1}{x^2 + 2x - 3} = \dfrac{1}{(x + 3)(x - 1)}$

Domain: $(-\infty, -3) \cup (-3, 1) \cup (1, \infty)$

Vertical asymptotes: $x = -3, x = 1$

Horizontal asymptote: $y = 0$

Because the denominator $x^2 + 2x - 3 = 0$ when $x = -3$ or $x = 1$, the domain is all real values of x except $x = -3$ and $x = 1$, and the vertical asymptotes are $x = -3$ and $x = 1$. Because the degree of the numerator is less than the degree of the denominator, the horizontal asymptote is $y = 0$.

21. $f(x) = \dfrac{3}{x + 2}$

Graph (d)

23. $f(x) = \dfrac{3x^2}{x + 2}$

Graph (a)

25. $g(x) = \dfrac{5}{x}$

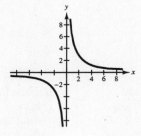

27. $f(x) = \dfrac{5}{x^2}$

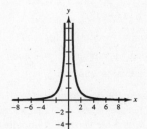

29. $f(x) = \dfrac{1}{x - 2}$

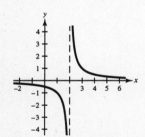

31. $g(x) = \dfrac{1}{2 - x}$

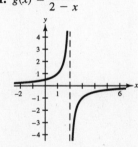

33. $y = \dfrac{2x + 4}{x}$

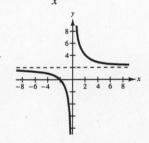

35. $y = \dfrac{3x}{x + 4}$

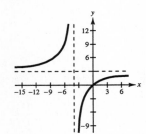

37. $y = \dfrac{2x^2}{x^2 + 1}$

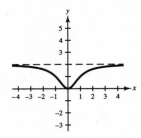

39. $g(t) = 3 - \dfrac{2}{t}$

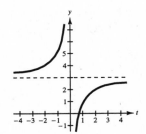

41. $y = \dfrac{4}{x^2 + 1}$

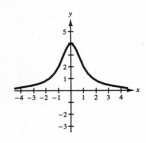

43. $y = -\dfrac{x}{x^2 - 4}$

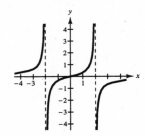

45. $h(x) = \dfrac{x - 3}{x - 1}$

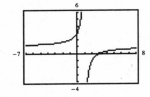

Domain: $(-\infty, 1) \cup (1, \infty)$
Horizontal asymptote: $y = 1$
Vertical asymptote: $x = 1$

47. $f(t) = \dfrac{6}{t^2 + 1}$

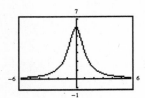

Domain: $(-\infty, \infty)$
Horizontal asymptote: $y = 0$
Vertical asymptote: None

49. $y = \dfrac{2(x^2 + 1)}{x^2}$

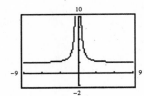

Domain: $(-\infty, 0) \cup (0, \infty)$
Horizontal asymptote: $y = 2$
Vertical asymptote: $x = 0$

51. $y = \dfrac{3}{x} + \dfrac{1}{x - 2}$

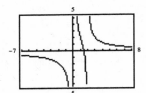

$$y = \frac{3}{x} + \frac{1}{x - 2} \implies y = \frac{4x - 6}{x(x - 2)}$$
Domain: $(-\infty, 0) \cup (0, 2) \cup (2, \infty)$
Horizontal asymptote: $y = 0$
Vertical asymptote: $x = 0, x = 2$

53. $g(x) = -\dfrac{1}{x}$

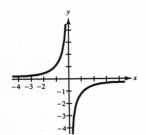

This transformation is a reflection in the x-axis.

55. $g(x) = \dfrac{1}{x - 2}$

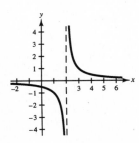

This transformation is a horizontal shift of two units to the right.

57. $g(x) = 2 + \dfrac{4}{x^2}$

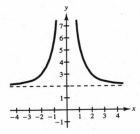

This transformation is a vertical shift two units upward.

59. $g(x) = -\dfrac{4}{(x-2)^2}$

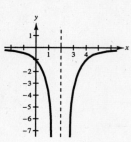

This transformation is a reflection in the x-axis and a horizontal shift of two units to the right.

61. $y = \dfrac{2x^2 + x}{x + 1}$

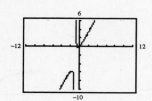

Domain: $(-\infty, -1) \cup (-1, \infty)$

Vertical asymptote: $x = -1$

Line: $y = 2x - 1$

63. (a) From the graph it appears that the x-intercept is $(-2, 0)$.

(b)
$$y = \frac{x+2}{x-2}, \quad y = 0$$

$$0 = \frac{x+2}{x-2}$$

$$(x-2)(0) = \left(\frac{x+2}{x-2}\right)(x-2)$$

$$0 = x + 2$$

$$-2 = x$$

$$(-2, 0)$$

65. (a) From the graph it appears that the x-intercepts are $(1, 0)$ and $(-1, 0)$.

(b)
$$y = x - \frac{1}{x}, \quad y = 0$$

$$0 = x - \frac{1}{x}$$

$$x(0) = \left(x - \frac{1}{x}\right)x$$

$$0 = x^2 - 1$$

$$1 = x^2$$

$$\pm\sqrt{1} = x$$

$$\pm 1 = x$$

$$(1, 0) \text{ and } (-1, 0)$$

67. (a) From the graph it appears that the x-intercept is $(1, 0)$.

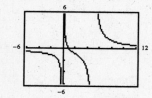

(b)
$$y = \frac{1}{x} + \frac{4}{x-5}, \quad y = 0$$

$$0 = \frac{1}{x} + \frac{4}{x-5}$$

$$x(x-5)(0) = \left(\frac{1}{x} + \frac{4}{x-5}\right)(x)(x-5)$$

$$0 = (x-5) + 4x$$

$$0 = 5x - 5$$

$$5 = 5x$$

$$1 = x$$

$$(1, 0)$$

69. (a) From the graph it appears that the x-intercept is $(4, 0)$.

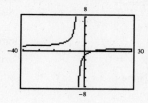

(b)
$$y = \frac{x-4}{x+5}, \quad y = 0$$

$$0 = \frac{x-4}{x+5}$$

$$(x+5)(0) = \left(\frac{x-4}{x+5}\right)(x+5)$$

$$0 = x - 4$$

$$4 = x$$

$$(4, 0)$$

71. (a) From the graph it appears that the x-intercepts are $(-3, 0)$ and $(2, 0)$.

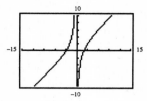

(b) $y = (x + 1) - \dfrac{6}{x}, \; y = 0$

$$0 = (x + 1) - \frac{6}{x}$$

$$x(0) = \left[(x + 1) - \frac{6}{x}\right]x$$

$$0 = x(x + 1) - 6$$

$$0 = x^2 + x - 6$$

$$0 = (x + 3)(x - 2)$$

$$x + 3 = 0 \implies x = -3$$

$$x - 2 = 0 \implies x = 2$$

$(-3, 0)$ and $(2, 0)$

73. (a) From the graph it appears that the x-intercepts are $(-3, 0)$ and $(4, 0)$.

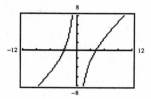

(b) $y = (x - 1) - \dfrac{12}{x}, \; y = 0$

$$0 = (x - 1) - \frac{12}{x}$$

$$x(0) = \left[(x - 1) - \frac{12}{x}\right]x$$

$$0 = x(x - 1) - 12$$

$$0 = x^2 - x - 12$$

$$0 = (x + 3)(x - 4)$$

$$x + 3 = 0 \implies x = -3$$

$$x - 4 = 0 \implies x = 4$$

$(-3, 0)$ and $(4, 0)$

75. (a) $C = 0.60x + 250$

(c) For $x = 1000$, $\overline{C} = \dfrac{0.60(1000) + 250}{1000} = \$0.85$

For $x = 10{,}000$, $\overline{C} = \dfrac{0.60(10{,}000) + 250}{10{,}000} = \$0.625$

(e)

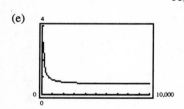

(b) $\overline{C} = \dfrac{0.60x + 250}{x}$

(d) The degree of the numerator is equal to the degree of the denominator, so the horizontal asymptote is $\overline{C} = \frac{0.60}{1}$, the ratio of the leading coefficients: $\overline{C} = 0.60$

The value of $\overline{C}$ will never be less than \$0.60.

77. (a) $C = \dfrac{2t}{4t^2 + 25}$

Because the degree of the numerator is less than the degree of the denominator, the horizontal asymptote of the function is $C = 0$. This indicates that the chemical is eliminated.

(b)

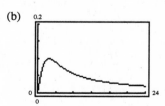

From the graph it appears that the concentration is greatest at approximately 2.5 hours.

79. (a) Area = (Length)(Width)

Length = $x \implies$ Area = x(Width)

$400 = x$(Width)

$\dfrac{400}{x} = $ Width

Perimeter P = 2(Length) + 2(Width)

$P = 2x + 2\left(\dfrac{400}{x}\right)$

$= 2\left(x + \dfrac{400}{x}\right)$

(b) The width x must be positive. Thus, the domain is $(0, \infty)$.

(c)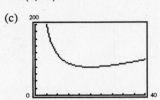

(d) From the graph it appears that the minimum perimeter occurs when the width x of the rectangle is 20 meters. The length is $400/x$ or 20 meters also.

81. An asymptote of a graph is a line to which the graph becomes arbitrarily close as $|x|$ or $|y|$ increases without bound.

83. No. There are separate portions of the graph on each side of the asymptote $x = 3$, and 3 is not in the domain of the function.

85. $g(x) = \dfrac{4 - 2x}{x - 2}$

$= \dfrac{-2(x - 2)}{x - 2}$

$= -2, \; x \neq 2$

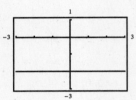

The fraction is not reduced to lowest terms.

87. $(2x - 15)^2 = (2x)^2 - 2(2x)(15) + 15^2$

$= 4x^2 - 60x + 225$

89. $[(x + 1) - y][(x + 1) + y] = (x + 1)^2 - y^2$

$= x^2 + 2x + 1 - y^2$

91.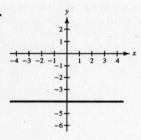

93.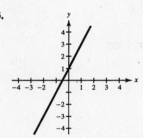

95.
$$\begin{array}{r} 4 - \dfrac{11}{x + 2} \\ x + 2 \enclose{longdiv}{4x - 3} \\ \underline{4x + 8} \\ -11 \end{array}$$

$\dfrac{4x - 3}{x + 2} = 4 - \dfrac{11}{x + 2}$

97. $\dfrac{1}{2}$(Base)(Height) = Area

$\dfrac{1}{2}(x)(x - 12) = 80$

$2\left[\dfrac{1}{2}(x)(x - 12)\right] = 2(80)$

$x(x - 12) = 160$

$x^2 - 12x = 160$

$x^2 - 12x - 160 = 0$

$(x - 20)(x + 8) = 0$

$x - 20 = 0 \implies x = 20 \quad$ and $\quad x - 12 = 8$

$x + 8 = 0 \implies x = -8 \qquad$ (Extraneous)

The base is 20 meters and the height is 8 meters.

Section 7.7   Rational Inequalities in One Variable

1. $\dfrac{2}{3-x} \le 0$

(a) $x = 3$

$\dfrac{2}{3-3} \overset{?}{\le} 0$

$\dfrac{2}{0}$ undefined

No, 3 is not a solution.

(b) $x = 4$

$\dfrac{2}{3-4} \overset{?}{\le} 0$

$-2 \le 0$

Yes, 4 is a solution.

(c) $x = -4$

$\dfrac{2}{3-(-4)} \overset{?}{\le} 0$

$\dfrac{2}{7} \not\le 0$

No, -4 is not a solution.

(d) $x = -\dfrac{1}{3}$

$\dfrac{2}{3-\left(-\frac{1}{3}\right)} \overset{?}{\le} 0$

$\dfrac{2}{\frac{10}{3}} \overset{?}{\le} 0$

$\dfrac{3}{5} \not\le 0$

No, $-\dfrac{1}{3}$ is not a solution.

3. $\dfrac{5}{x-3}$

$5 \ne 0$

$x - 3 = 0 \implies x = 3$

Critical number: 3

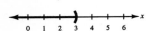

5. $\dfrac{2x}{x+5}$

$2x = 0 \implies x = 0$

$x + 5 = 0 \implies x = -5$

Critical numbers: $-5, 0$

7. $\dfrac{5}{x-3} > 0$

$5 \ne 0$

$x - 3 = 0 \implies x = 3$

Critical number: 3

The solution is $(3, \infty)$.

Test Interval	Representative x-value	Is inequality satisfied?
$(-\infty, 3)$	$x = 0$	$\dfrac{5}{0-3} > 0$ $-\dfrac{5}{3} \not> 0$
$(3, \infty)$	$x = 5$	$\dfrac{5}{5-3} > 0$ $\dfrac{5}{2} > 0$

9. $\dfrac{-5}{x-3} > 0$

$-5 \ne 0$

$x - 3 = 0 \implies x = 3$

Critical number: 3

Solution: $(-\infty, 3)$

Test Interval	Representative x-value	Is inequality satisfied?
$(-\infty, 3)$	$x = 0$	$\dfrac{-5}{0-3} \overset{?}{>} 0$ $\dfrac{5}{3} > 0$
$(3, \infty)$	$x = 5$	$\dfrac{-5}{5-3} \overset{?}{>} 0$ $-\dfrac{5}{2} \not> 0$

11. $\dfrac{4}{x-3} \le 0$

$4 \ne 0$

$x - 3 = 0 \implies x = 3$

Critical number: 3

The solution is $(-\infty, 3)$.

Test Interval	Representative x-value	Is inequality satisfied?
$(-\infty, 3)$	$x = 0$	$\dfrac{4}{0-3} \overset{?}{\le} 0$ $-\dfrac{4}{3} \le 0$
$(3, \infty)$	$x = 5$	$\dfrac{4}{5-3} \overset{?}{\le} 0$ $2 \not\le 0$

13. $\dfrac{x}{x-4} \le 0$

$x = 0$

$x - 4 = 0 \implies x = 4$

Critical numbers: 0, 4

The solution is $[0, 4)$.

Test Interval	Representative x-value	Is inequality satisfied?
$(-\infty, 0)$	$x = -2$	$\dfrac{-2}{-2-4} \overset{?}{\le} 0$ $\dfrac{1}{3} \le 0$
$(0, 4)$	$x = 2$	$\dfrac{2}{2-4} \overset{?}{\le} 0$ $-1 \le 0$
$(4, \infty)$	$x = 6$	$\dfrac{6}{6-4} \overset{?}{\le} 0$ $3 \not\le 0$

15. $\dfrac{y-3}{2y-11} \ge 0$

$y - 3 = 0 \implies y = 3$

$2y - 11 = 0 \implies y = \dfrac{11}{2}$

Critical numbers: $3, \dfrac{11}{2}$

The solution is $(-\infty, 3] \cup \left(\dfrac{11}{2}, \infty\right)$.

Test Interval	Representative y-value	Is inequality satisfied?
$(-\infty, 3)$	$y = 0$	$\dfrac{0-3}{2(0)-11} \overset{?}{\ge} 0$ $\dfrac{3}{11} \ge 0$
$\left(3, \dfrac{11}{2}\right)$	$y = 5$	$\dfrac{5-3}{2(5)-11} \overset{?}{\ge} 0$ $-2 \not\ge 0$
$\left(\dfrac{11}{2}, \infty\right)$	$y = 10$	$\dfrac{10-3}{2(10)-11} \overset{?}{\ge} 0$ $\dfrac{7}{9} \ge 0$

17. $\dfrac{3(u-3)}{u+1} < 0$

$3(u-3) = 0 \implies u = 3$

$u + 1 = 0 \implies u = -1$

Critical numbers: $3, -1$

The solution is $(-1, 3)$.

Test Interval	Representative x-value	Is inequality satisfied?
$(-\infty, -1)$	$u = -2$	$\dfrac{3(-2-3)}{-2+1} \overset{?}{<} 0$ $15 \not< 0$
$(-1, 3)$	$u = 0$	$\dfrac{3(0-3)}{0+1} \overset{?}{<} 0$ $-9 < 0$
$(3, \infty)$	$u = 5$	$\dfrac{3(5-3)}{5+1} \overset{?}{<} 0$ $1 \not< 0$

19. $\dfrac{6}{x-4} > 2$

$\dfrac{6}{x-4} - 2 > 0$

$\dfrac{6 - 2(x-4)}{x-4} > 0$

$\dfrac{6 - 2x + 8}{x-4} > 0$

$\dfrac{-2x + 14}{x-4} > 0$

$-2x + 14 = 0 \implies x = 7$

$x - 4 = 0 \implies x = 4$

Critical numbers: 4, 7

The solution is $(4, 7)$.

Test Interval	Representative x-value	Is inequality satisfied?
$(-\infty, 4)$	$x = 0$	$\dfrac{6}{0-4} \overset{?}{>} 0$ $-\dfrac{7}{2} \not> 0$
$(4, 7)$	$x = 5$	$\dfrac{6}{5-4} \overset{?}{>} 0$ $4 > 0$
$(7, \infty)$	$x = 10$	$\dfrac{6}{10-4} \overset{?}{>} 0$ $-1 \not> 0$

21. $\dfrac{x}{x-3} \le 2$

$\dfrac{x}{x-3} - 2 \le 0$

$\dfrac{x - 2(x-3)}{x-3} \le 0$

$\dfrac{x - 2x + 6}{x-3} \le 0$

$\dfrac{-x + 6}{x-3} \le 0$

$-x + 6 = 0 \implies x = 6$

$x - 3 = 0 \implies x = 3$

Critical numbers: 3, 6

The solution is $(-\infty, 3) \cup [6, \infty)$.

Test Interval	Representative x-value	Is inequality satisfied?
$(-\infty, 3)$	$x = 0$	$\dfrac{0}{0-3} - 2 \overset{?}{\le} 0$ $-2 \le 0$
$(3, 6)$	$x = 4$	$\dfrac{4}{4-3} - 2 \overset{?}{\le} 0$ $2 \not\le 0$
$(6, \infty)$	$x = 8$	$\dfrac{8}{8-3} - 2 \overset{?}{\le} 0$ $-\dfrac{2}{5} \le 0$

23. $\dfrac{x-5}{x^2-6x+9} \le 0$

$\dfrac{x-5}{(x-3)(x-3)} \le 0$

$x - 5 = 0 \implies x = 5$

$x - 3 = 0 \implies x = 3$

Critical numbers: 3, 5

The solution is $(-\infty, 3) \cup (3, 5]$.

Test Interval	Representative x-value	Is inequality satisfied?
$(-\infty, 3)$	$x = 0$	$\dfrac{0-5}{0^2 - 6(0) + 9} \overset{?}{\le} 0$ $-\dfrac{5}{9} \le 0$
$(3, 5)$	$x = 4$	$\dfrac{4-5}{4^2 - 6(4) + 9} \overset{?}{\le} 0$ $-1 \le 0$
$(5, \infty)$	$x = 6$	$\dfrac{6-5}{6^2 - 6(6) + 9} \overset{?}{\le} 0$ $\dfrac{1}{9} \not\le 0$

25.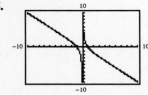

$\dfrac{1}{x} - x > 0$

The solution is $(-\infty, -1) \cup (0, 1)$.

27.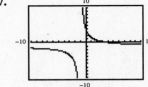

$\dfrac{x+6}{x+1} - 2 < 0$

The solution is $(-\infty, -1) \cup (4, \infty)$.

29.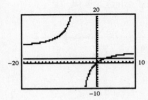

$\dfrac{6x}{x+5} < 2$

The solution is $\left(-5, \dfrac{5}{2}\right)$.

31.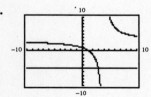

$\dfrac{3x-4}{x-4} < -5$

The solution is $(3, 4)$.

33.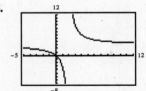

(a) $[0, 2)$

(b) $(2, 4]$

35.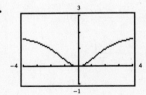

(a) $(-\infty, -2] \cup [2, \infty)$

(b) $(-\infty, \infty)$

37. *Verbal Model:* $\boxed{\text{Time traveling at first speed}}$ + $\boxed{\text{Time traveling at second speed}}$

Labels: First speed: rate $= r$ (miles per hour)

distance $= 144$ (miles)

time $= \dfrac{144}{r}$

Second speed: rate $= r + 10$ (miles per hour)

distance $= 260 - 144 = 116$ (miles)

time $= \dfrac{116}{(r+10)}$

—CONTINUED—

37. —CONTINUED—

Inequality:
$$\frac{144}{r} + \frac{116}{r+10} < 5$$

$$\frac{144(r+10) + 116r}{r(r+10)} < 5$$

$$\frac{260r + 1440}{r(r+10)} - 5 < 0$$

$$\frac{260r + 1440}{r(r+10)} - \frac{5r(r+10)}{r(r+10)} < 0$$

$$\frac{260r + 1440 - 5r^2 - 50r}{r(r+10)} < 0$$

$$\frac{-5r^2 + 210r + 1440}{r(r+10)} < 0$$

$$\frac{-5(r^2 - 42r - 288)}{r(r+10)} < 0$$

$$\frac{-5(r+6)(r-48)}{r(r+10)} < 0$$

$$r + 6 = 0 \implies r = -6 \qquad \text{(Discard negative solution.)}$$

$$r - 48 = 0 \implies r = 48$$

$$r(r+10) = 0 \implies r = 0 \text{ or } r = -10 \qquad \text{(Discard negative solution.)}$$

Both rates must be nonnegative and below the speed limit of 65 mph.

Critical numbers: 0 and 48

Test Interval	Representative x-value	Value of Expression	Conclusion
[0, 48)	10	$\dfrac{260(10) + 1440}{10(10+10)} - 5 = 15.2$	Positive
(48, 55]	50	$\dfrac{260(50) + 1440}{50(50+10)} - 5 \approx -0.19$	Negative

The solution for r is $(48, 55]$. Thus, $48 < r < 55$ and $58 < r + 10 < 65$.

39. (a)

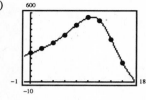

(b) From the graph, it appears that the temperature was at least $400°$ F for values of t in the interval $[5.7, 13.6]$.

41. The value of a rational expression can change sign only at the x values for which the rational expression is zero (x values for which the numerator is zero) and at the x values for which the rational expression is undefined (x values for which the denominator is zero). These x values are called the critical numbers, and they are used to determine the test intervals in solving rational inequalities.

43. $3x + 15 \geq 0$

$$3x \geq -15$$

$$x \geq -5$$

$$[-5, \infty)$$

45. $4(2 - 7x) > 10x - 1$

$8 - 28x > 10x - 1$

$8 - 38x > -1$

$-38x > -9$

$x < \frac{9}{38}$

$\left(-\infty, \frac{9}{38}\right)$

47. $|x - 6| < 3$

$-3 < x - 6 < 3$

$3 < \quad x < 9$

$(3, 9)$

49. $|x + 9| < 7$

$-7 < x + 9 < 7$

$-16 < x < -2$

$(-16, -2)$

51. $x^2 + 7x \geq 0$

$x(x + 7) \geq 0$

Critical numbers:

$x(x + 7) = 0$

$x = 0$

$x + 7 = 0 \implies x = -7$

Test: $x = 3 \qquad 3(3 + 7) \geq 0$

$x = -2 \qquad -2(-2 + 7) \not\geq 0$

$x = -10 \qquad -10(-10 + 7) \geq 0$

$(-\infty, -7] \cup [0, \infty)$

53. $2\pi \sqrt{\dfrac{L}{32}} = 0.8$

$\left(2\pi \sqrt{\dfrac{L}{32}}\right)^2 = (0.8)^2$

$4\pi^2 \left(\dfrac{L}{32}\right) = 0.64$

$\dfrac{\pi^2 L}{8} = 0.64$

$\pi^2 L = 5.12$

$L = \dfrac{5.12}{\pi^2}$

$L \approx 0.52$

The length of the pendulum is approximately 0.52 feet.

Review Exercises for Chapter 7

1. $f(x) = \dfrac{3x}{x - 8}$

The denominator is zero when $x - 8 = 0$ or $x = 8$.
So, the domain is all real values of x such that $x \neq 8$.

Domain $= (-\infty, 8) \cup (8, \infty)$.

3. $f(x) = \dfrac{x}{x^2 - 7x + 6}$

The denominator, $x^2 - 7x + 6$ or $(x - 6)(x - 1)$, is zero
when $x = 6$ or $x = 1$. So, the domain is all real values of
x such that $x \neq 6$ and $x \neq 1$.

Domain $= (-\infty, 1) \cup (1, 6) \cup (6, \infty)$.

5. $\dfrac{6x^4 y^2}{15xy^2} = \dfrac{3(2)(x)(x^3)(y^2)}{3(5)(x)(y^2)}$

$= \dfrac{2x^3}{5}, \; x \neq 0, \; y \neq 0$

7. $\dfrac{5b - 15}{30b - 120} = \dfrac{5(b - 3)}{30(b - 4)}$

$= \dfrac{5(b - 3)}{5(6)(b - 4)} = \dfrac{b - 3}{6(b - 4)}$

9. $\dfrac{9x - 9y}{y - x} = \dfrac{9(x - y)}{y - x}$

$= \dfrac{9(x - y)}{(-1)(x - y)} = -9, \; x \neq y$

11. $\dfrac{x^2 - 5x}{2x^2 - 50} = \dfrac{x(x - 5)}{2(x^2 - 25)}$

$= \dfrac{x(x - 5)}{2(x + 5)(x - 5)} = \dfrac{x}{2(x + 5)}, \; x \neq 5$

13. $\dfrac{7}{8} \cdot \dfrac{2x}{y} \cdot \dfrac{y^2}{14x^2} = \dfrac{14xy^2}{8(14)x^2 y}$

$= \dfrac{14(x)(y)(y)}{8(14)(x)(x)(y)} = \dfrac{y}{8x}, \; y \neq 0$

15. $\dfrac{60z}{z+6} \cdot \dfrac{z^2-36}{5} = \dfrac{60z(z+6)(z-6)}{(z+6)(5)}$

$\qquad = \dfrac{\cancel{5}(12)(z)\cancel{(z+6)}(z-6)}{\cancel{(z+6)}\cancel{(5)}}$

$\qquad = 12z(z-6),\ z \neq -6$

17. $25y^2 \div \dfrac{xy}{5} = \dfrac{25y^2}{1} \cdot \dfrac{5}{xy} = \dfrac{125y^2}{xy}$

$\qquad = \dfrac{125\cancel{(y)}(y)}{x\cancel{(y)}} = \dfrac{125y}{x},\ y \neq 0$

19. $\dfrac{x^2-7x}{x+1} \div \dfrac{x^2-14x+49}{x^2-1} = \dfrac{x^2-7x}{x+1} \cdot \dfrac{x^2-1}{x^2-14x+49} = \dfrac{x(x-7)(x+1)(x-1)}{(x+1)(x-7)^2}$

$\qquad = \dfrac{x\cancel{(x-7)}\cancel{(x+1)}(x-1)}{\cancel{(x+1)}\cancel{(x-7)}(x-7)} = \dfrac{x(x-1)}{x-7},\ x \neq -1,\ x \neq 1$

21. $\dfrac{\left(\dfrac{6}{x}\right)}{\left(\dfrac{2}{x^3}\right)} = \dfrac{6}{x} \cdot \dfrac{x^3}{2} = \dfrac{6x^3}{2x} = \dfrac{3\cancel{(2)}\cancel{(x)}(x^2)}{\cancel{2}\cancel{(x)}} = 3x^2,\ x \neq 0$ or $\quad \dfrac{\left(\dfrac{6}{x}\right)}{\left(\dfrac{2}{x^3}\right)} = \dfrac{\left(\dfrac{6}{x}\right) \cdot x^3}{\left(\dfrac{2}{x^3}\right) \cdot x^3} = \dfrac{6x^2}{2} = \dfrac{\cancel{2}(3)(x^2)}{\cancel{2}} = 3x^2,\ x \neq 0$

23. $\dfrac{\left(\dfrac{6x^2}{x^2+2x-35}\right)}{\left(\dfrac{x^3}{x^2-25}\right)} = \dfrac{6x^2}{x^2+2x-35} \cdot \dfrac{x^2-25}{x^3} = \dfrac{6x^2}{(x+7)(x-5)} \cdot \dfrac{(x+5)(x-5)}{x^3}$

$\qquad = \dfrac{6x^2(x+5)(x-5)}{(x+7)(x-5)(x^3)} = \dfrac{6\cancel{(x^2)}(x+5)\cancel{(x-5)}}{(x+7)\cancel{(x-5)}(x)\cancel{(x^2)}} = \dfrac{6(x+5)}{x(x+7)},\ x \neq 5,\ x \neq -5$

or

$\dfrac{\left(\dfrac{6x^2}{x^2+2x-35}\right)}{\left(\dfrac{x^3}{x^2-25}\right)} = \dfrac{\left[\dfrac{6x^2}{(x+7)(x-5)}\right](x+7)(x-5)(x+5)}{\left[\dfrac{x^3}{(x+5)(x-5)}\right](x+7)(x-5)(x+5)} = \dfrac{6x^2(x+5)}{x^3(x+7)}$

$\qquad = \dfrac{6\cancel{(x^2)}(x+5)}{\cancel{(x^2)}(x)(x+7)} = \dfrac{6(x+5)}{x(x+7)},\ x \neq 5,\ x \neq -5$

25. $\dfrac{4a}{9} - \dfrac{11a}{9} = \dfrac{4a-11a}{9} = -\dfrac{7a}{9}$

27. $\dfrac{x+7}{3x-1} - \dfrac{2x+3}{3x-1} = \dfrac{x+7-(2x+3)}{3x-1}$

$\qquad = \dfrac{x+7-2x-3}{3x-1}$

$\qquad = \dfrac{-x+4}{3x-1}$ or $\dfrac{4-x}{3x-1}$

29. $\dfrac{15}{16x} - \dfrac{5}{24x} - 1$

The least common denominator is $48x$.

$\dfrac{15}{16x} - \dfrac{5}{24x} - 1 = \dfrac{15(3)}{16x(3)} - \dfrac{5(2)}{24x(2)} - \dfrac{48x}{48x}$

$\qquad = \dfrac{45-10-48x}{48x} = \dfrac{-48x+35}{48x}$

31. $\dfrac{1}{x+5} + \dfrac{3}{x-12}$

The least common denominator is $(x+5)(x-12)$.

$\dfrac{1}{x+5} + \dfrac{3}{x-12} = \dfrac{1(x-12)}{(x+5)(x-12)} + \dfrac{3(x+5)}{(x-12)(x+5)}$

$\qquad = \dfrac{x-12+3x+15}{(x+5)(x-12)}$

$\qquad = \dfrac{4x+3}{(x+5)(x-12)}$

33. $5x + \dfrac{2}{x-3} - \dfrac{3}{x+2}$

The least common denominator is $(x-3)(x+2)$.

$$5x + \frac{2}{x-3} - \frac{3}{x+2} = \frac{5x(x-3)(x+2)}{1(x-3)(x+2)} + \frac{2(x+2)}{(x-3)(x+2)} - \frac{3(x-3)}{(x+2)(x-3)}$$

$$= \frac{5x(x^2 - x - 6) + 2(x+2) - 3(x-3)}{(x-3)(x+2)}$$

$$= \frac{5x^3 - 5x^2 - 30x + 2x + 4 - 3x + 9}{(x-3)(x+2)} = \frac{5x^3 - 5x^2 - 31x + 13}{(x-3)(x+2)}$$

35. $\dfrac{6}{x} - \dfrac{6x-1}{x^2+4}$

The least common denominator is $x(x^2+4)$.

$$\frac{6}{x} - \frac{6x-1}{x^2+4} = \frac{6(x^2+4)}{x(x^2+4)} - \frac{(6x-1)x}{(x^2+4)x} = \frac{6(x^2+4) - x(6x-1)}{x(x^2+4)}$$

$$= \frac{6x^2 + 24 - 6x^2 + x}{x(x^2+4)} = \frac{x+24}{x(x^2+4)}$$

37. $\dfrac{5}{x+3} - \dfrac{4x}{(x+3)^2} - \dfrac{1}{(x-3)}$

The least common denominator is $(x+3)^2(x-3)$.

$$\frac{5}{x+3} - \frac{4x}{(x+3)^2} - \frac{1}{x-3} = \frac{5(x+3)(x-3)}{(x+3)(x+3)(x-3)} - \frac{4x(x-3)}{(x+3)^2(x-3)} - \frac{1(x+3)^2}{(x-3)(x+3)^2}$$

$$= \frac{5(x^2-9) - 4x(x-3) - (x^2+6x+9)}{(x+3)^2(x-3)}$$

$$= \frac{5x^2 - 45 - 4x^2 + 12x - x^2 - 6x - 9}{(x+3)^2(x-3)}$$

$$= \frac{6x-54}{(x+3)^2(x-3)} \quad \text{or} \quad \frac{6(x-9)}{(x+3)^2(x-3)}$$

39. $\dfrac{3t}{\left(5 - \dfrac{2}{t}\right)} = \dfrac{\left(\dfrac{3t}{1}\right)}{\left(\dfrac{5t}{t} - \dfrac{2}{t}\right)} = \dfrac{\left(\dfrac{3t}{1}\right)}{\left(\dfrac{5t-2}{t}\right)} = \dfrac{3t}{1} \cdot \dfrac{t}{5t-2} = \dfrac{3t^2}{5t-2}, \; t \ne 0$

or $\quad \dfrac{3t}{\left(5 - \dfrac{2}{t}\right)} = \dfrac{(3t)t}{\left(5 - \dfrac{2}{t}\right) \cdot t} = \dfrac{3t^2}{5t-2}, \; t \ne 0$

41. $\dfrac{\left(\dfrac{1}{a^2-16} - \dfrac{1}{a}\right)}{\left(\dfrac{1}{a^2+4a} + 4\right)} = \dfrac{\left[\dfrac{1(a)}{(a^2-16)(a)} - \dfrac{1(a^2-16)}{a(a^2-16)}\right]}{\left[\dfrac{1}{a^2+4a} + \dfrac{4(a^2+4a)}{1(a^2+4a)}\right]} = \dfrac{\left(\dfrac{a-(a^2-16)}{a(a^2-16)}\right)}{\left(\dfrac{1+4(a^2+4a)}{a^2+4a}\right)}$

$$= \frac{a - a^2 + 16}{a(a^2-16)} \cdot \frac{a^2+4a}{1+4a^2+16a} = \frac{(-a^2+a+16)\cancel{(a)}\cancel{(a+4)}}{\cancel{a}\cancel{(a+4)}(a-4)(4a^2+16a+1)}$$

$$= \frac{-a^2+a+16}{(a-4)(4a^2+16a+1)}, \; a \ne 0, \; a \ne -4$$

—CONTINUED—

41. —CONTINUED—

or

$$\frac{\left(\dfrac{1}{a^2-16}-\dfrac{1}{a}\right)}{\left(\dfrac{1}{a^2+4a}+4\right)}=\frac{\left[\dfrac{1}{(a+4)(a-4)}-\dfrac{1}{a}\right]a(a+4)(a-4)}{\left[\dfrac{1}{a(a+4)}+4\right]a(a+4)(a-4)}=\frac{a-(a+4)(a-4)}{(a-4)+4(a)(a+4)(a-4)}$$

$$=\frac{a-(a^2-16)}{(a-4)[1+4a(a+4)]}=\frac{a-a^2+16}{(a-4)(1+4a^2+16a)}$$

$$=\frac{-a^2+a+16}{(a-4)(4a^2+16a+1)},\ a\neq-4,\ a\neq0$$

or $\dfrac{-a^2+a+16}{4a^3-63a-4}$

43. $(4x^3-x)\div2x=\dfrac{4x^3-x}{2x}=\dfrac{4x^3}{2x}-\dfrac{x}{2x}$

$$=\frac{2(2)(x)(x^2)}{2(x)}-\frac{1(x)}{2(x)}=2x^2-\frac{1}{2},\ x\neq0$$

45. $\dfrac{5a^6b^4-2a^4b^3+a^3b^2}{a^3b^2}=\dfrac{5a^6b^4}{a^3b^2}-\dfrac{2a^4b^3}{a^3b^2}+\dfrac{a^3b^2}{a^3b^2}$

$$=\frac{a^3b^2(5a^3b^2)}{a^3b^2}-\frac{a^3b^2(2ab)}{a^3b^2}+\frac{a^3b^2}{a^3b^2}$$

$$=5a^3b^2-2ab+1,\quad a\neq0,b\neq0$$

47.

$$\begin{array}{r}2x^2+\ \ x-1+\dfrac{1}{3x-1}\\[2pt]3x-1\overline{\smash{\big)}\,6x^3+\ x^2-4x+2}\\[2pt]\underline{6x^3-2x^2}\\[2pt]3x^2-4x\\[2pt]\underline{3x^2-\ x}\\[2pt]-3x+2\\[2pt]\underline{-3x+1}\\[2pt]1\end{array}$$

So, $\dfrac{6x^3+x^2-4x+2}{3x-1}=2x^2+x-1+\dfrac{1}{3x-1}.$

49.

$$\begin{array}{r}x^2-2\\[2pt]x^2-1\overline{\smash{\big)}\,x^4+0x^3-3x^2+0x+2}\\[2pt]\underline{x^4-x^2}\\[2pt]-2x^2+2\\[2pt]\underline{-2x+2}\\[2pt]0\end{array}$$

So, $\dfrac{x^4-3x^2+2}{x^2-1}=x^2-2,\ x\neq1,\ x\neq-1.$

51. $\dfrac{x^4-3x^2-25}{x-3}$

$$\begin{array}{r|rrrrr}3&1&0&-3&0&-25\\&&3&9&18&54\\\hline&1&3&6&18&29\end{array}$$

$$\frac{x^4-3x^2-25}{x-3}=x^3+3x^2+6x+18+\frac{29}{x-3}$$

53.

$$\begin{array}{r|rrrr}2&1&2&-5&-6\\&&2&8&6\\\hline&1&4&3&0\end{array}$$

$x^3+2x^2-5x-6=(x-2)(x^2+4x+3)$

$$=(x-2)(x+3)(x+1)$$

55. $\dfrac{3x}{8}=-15$

$3x=-120$

$x=-40$

57. $\dfrac{1}{3y - 4} = \dfrac{6}{4(y + 1)}$

$6(3y - 4) = 4(y + 1)$

$18y - 24 = 4y + 4$

$14y - 24 = 4$

$14y = 28$

$y = 2$

59. $r = 2 + \dfrac{24}{r}$

$r(r) = r\left(2 + \dfrac{24}{r}\right)$

$r^2 = 2r + 24$

$r^2 - 2r - 24 = 0$

$(r - 6)(r + 4) = 0$

$r - 6 = 0 \Rightarrow r = 6$

$r + 4 = 0 \Rightarrow r = -4$

61. $\dfrac{3}{y + 1} - \dfrac{8}{y} = 1$

$y(y + 1)\left(\dfrac{3}{y + 1} - \dfrac{8}{y}\right) = y(y + 1)1$

$3y - 8(y + 1) = y(y + 1)$

$3y - 8y - 8 = y^2 + y$

$-5y - 8 = y^2 + y$

$0 = y^2 + 6y + 8$

$0 = (y + 4)(y + 2)$

$y + 4 = 0 \Rightarrow y = -4$

$y + 2 = 0 \Rightarrow y = -2$

63. $\dfrac{12}{x^2 + x - 12} - \dfrac{1}{x - 3} = -1$

$(x + 4)(x - 3)\left[\dfrac{12}{(x + 4)(x - 3)} - \dfrac{1}{x - 3}\right] = -1(x + 4)(x - 3)$

$12 - (x + 4) = -1(x^2 + x - 12)$

$12 - x - 4 = -x^2 - x + 12$

$8 - x = -x^2 - x + 12$

$x^2 - 4 = 0$

$(x + 2)(x - 2) = 0$

$x + 2 = 0 \Rightarrow x = -2$

$x - 2 = 0 \Rightarrow x = 2$

65. $\dfrac{5}{x^2 - 4} - \dfrac{6}{x - 2} = -5$

$(x + 2)(x - 2)\left(\dfrac{5}{(x + 2)(x - 2)} - \dfrac{6}{x - 2}\right) = -5(x + 2)(x - 2)$

$5 - 6(x + 2) = -5(x^2 - 4)$

$5 - 6x - 12 = -5x^2 + 20$

$-6x - 7 = -5x^2 + 20$

$5x^2 - 6x - 27 = 0$

$(5x + 9)(x - 3) = 0$

$5x + 9 = 0 \Rightarrow x = -\dfrac{9}{5}$

$x - 3 = 0 \Rightarrow x = 3$

67. $f(x) = \dfrac{x - 3}{x + 1}$

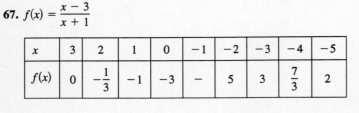

x	3	2	1	0	-1	-2	-3	-4	-5
$f(x)$	0	$-\dfrac{1}{3}$	-1	-3	$-$	5	3	$\dfrac{7}{3}$	2

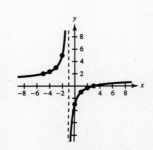

69. $f(x) = \dfrac{5}{x-6}$

Horizontal asymptote: $y = 0$

Vertical asymptote: $x = 6$

The degree of the numerator is less than the degree of the denominator, so the horizontal asymptote is $y = 0$.

The denominator is zero when $x = 6$, so the vertical asymptote is $x = 6$.

71. $f(x) = \dfrac{6x}{x-5}$

Horizontal asymptote: $y = 6$

Vertical asymptote: $x = 5$

The degree of the numerator is equal to the degree of the denominator, so the horizontal asymptote is $y = 6$, the ratio of the leading coefficients.

The denominator is zero when $x = 5$, so the vertical asymptote is $x = 5$.

73. $f(x) = \dfrac{x}{x+4}$

Domain: $(-\infty, -4) \cup (-4, \infty)$

Horizontal asymptote: $y = 1$

Vertical asymptote: $x = -4$

75. $g(x) = \dfrac{2x^2 - 2}{x^2 - 9}$

Domain: $(-\infty, -3) \cup (-3, 3) \cup (3, \infty)$

Horizontal asymptote: $y = 2$

Vertical asymptotes: $x = 3, x = -3$

77. $g(x) = \dfrac{2 + x}{1 - x}$

y-intercept: $(0, 2)$

x-intercept: $(-2, 0)$

Vertical asymptote: $x = 1$

Horizontal asymptote: $y = \dfrac{1}{-1} = -1$

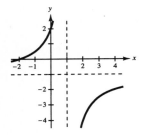

79. $f(x) = \dfrac{x}{x^2 + 1}$

y-intercept: $(0, 0)$

x-intercept: $(0, 0)$

Vertical asymptote: None

Horizontal asymptote: $y = 0$

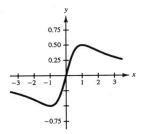

81. $P(x) = \dfrac{3x + 6}{x - 2}$

y-intercept: $(0, -3)$

x-intercept: $(-2, 0)$

Vertical asymptote: $x = 2$

Horizontal asymptote: $y = \dfrac{3}{1} = 3$

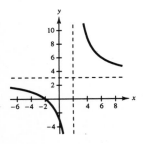

83. $h(x) = \dfrac{4}{(x - 1)^2}$

y-intercept: $(0, 4)$

x-intercept: None

Vertical asymptote: $x = 1$

Horizontal asymptote: $y = 0$

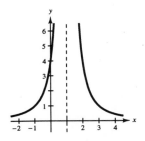

85. $f(x) = \dfrac{-5}{x^2}$

y-intercept: None

x-intercept: None

Vertical asymptote: $x = 0$

Horizontal asymptote: $y = 0$

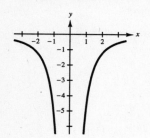

87. $y = \dfrac{x}{x^2 - 1}$

y-intercept: $(0, 0)$

x-intercept: $(0, 0)$

Vertical asymptotes: $x = 1, x = -1$

Horizontal asymptote: $y = 0$

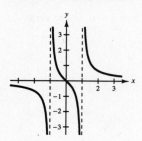

89. $\dfrac{x + 4}{x - 1}$

Critical numbers: $x + 4 = 0 \Rightarrow x = -4$

$\qquad\qquad\qquad x - 1 = 0 \Rightarrow x = 1$

Interval	Representative x-value	Value of expression	Conclusion
$(-\infty, -4)$	-10	$\dfrac{-10 + 4}{-10 - 1} = \dfrac{-6}{-11} = \dfrac{6}{11}$	Positive
$(-4, 1)$	0	$\dfrac{0 + 4}{0 - 1} = -4$	Negative
$(1, \infty)$	3	$\dfrac{3 + 4}{3 - 1} = \dfrac{7}{2}$	Positive

91. $\dfrac{x}{2x + 7}$

Critical numbers: $x = 0$

$\qquad\qquad 2x + 7 = 0 \Rightarrow x = -\dfrac{7}{2}$

Interval	Representative x-value	Value of expression	Conclusion
$\left(-\infty, -\dfrac{7}{2}\right)$	-5	$\dfrac{-5}{2(-5) + 7} = \dfrac{-5}{-3} = \dfrac{5}{3}$	Positive
$\left(-\dfrac{7}{2}, 0\right)$	-1	$\dfrac{-1}{2(-1) + 5} = \dfrac{-1}{3} = -\dfrac{1}{3}$	Negative
$(0, \infty)$	2	$\dfrac{2}{2(2) + 5} = \dfrac{2}{9}$	Positive

93. $\dfrac{x}{2x - 7} \geq 0$

$\qquad x = 0$

$\qquad 2x - 7 = 0 \Rightarrow x = \dfrac{7}{2}$

Critical numbers: $0, \dfrac{7}{2}$

Solution: $(-\infty, 0] \cup \left(\dfrac{7}{2}, \infty\right)$

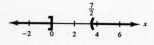

Test interval	Representative x-value	Is inequality satisfied?
$(-\infty, 0)$	$x = -2$	$\dfrac{-2}{2(-2) - 7} \overset{?}{\geq} 0$ $\dfrac{2}{11} \geq 0$
$\left(0, \dfrac{7}{2}\right)$	$x = 2$	$\dfrac{2}{2(2) - 7} \overset{?}{\geq} 0$ $\dfrac{-2}{3} \ngeq 0$
$\left(\dfrac{7}{2}, \infty\right)$	$x = 4$	$\dfrac{4}{2(4) - 7} \overset{?}{\geq} 0$ $4 \geq 0$

95.
$$\frac{x}{x+6} + 2 < 0$$

$$\frac{x}{x+6} + \frac{2(x+6)}{x+6} < 0$$

$$\frac{x + 2x + 12}{x+6} < 0$$

$$\frac{3x + 12}{x+6} < 0$$

$$3x + 12 = 0 \implies x = -4$$

$$x + 6 = 0 \implies x = -6$$

Critical numbers: $-4, -6$

The solution is $(-6, -4)$.

Test interval	Representative x-value	Is inequality satisfied?
$(-\infty, -6)$	$x = -8$	$\dfrac{-8}{-8+6} + 2 \overset{?}{<} 0$ $6 \not< 0$
$(-6, -4)$	$x = -5$	$\dfrac{-5}{-5+6} + 2 \overset{?}{<} 0$ $-3 < 0$
$(-4, \infty)$	$x = 0$	$\dfrac{0}{0+6} + 2 \overset{?}{<} 0$ $2 \not< 0$

97. *Verbal Model:* $\boxed{\text{Batting average}} = \boxed{\text{Total hits}} \div \boxed{\text{Total times at bat}}$

Labels: Batting average $= 0.400$

Additional consecutive hits $= x$

Current hits $= 45$

Total hits $= 45 + x$

Current times at bat $= 150$

Total times at bat $= 150 + x$

Equation:
$$0.400 = \frac{45 + x}{150 + x}$$

$$(150 + x)(0.400) = (150 + x)\left(\frac{45 + x}{150 + x}\right)$$

$$60 + 0.400x = 45 + x$$

$$15 + 0.400x = x$$

$$15 = 0.600x$$

$$\frac{15}{0.600} = x$$

$$25 = x$$

So, the player must hit safely for the next 25 times at bat.

99. *Verbal Model:* $\boxed{\begin{array}{c}\text{Supervisor's} \\ \text{rate}\end{array}} + \boxed{\begin{array}{c}\text{Your} \\ \text{rate}\end{array}} = \boxed{\begin{array}{c}\text{Rate} \\ \text{together}\end{array}}$

Labels: Supervisor: Time $= 12$ (minutes); Rate $= \dfrac{1}{12}$ (task per minute)

You: Time $= 15$ (minutes); Rate $= \dfrac{1}{15}$ (task per minute)

Together: Time $= t$ (minute); Rate $= \dfrac{1}{t}$ (task per minute)

—CONTINUED—

99. —CONTINUED—

Equation:
$$\frac{1}{12} + \frac{1}{15} = \frac{1}{t}$$

$$60t\left(\frac{1}{12} + \frac{1}{15}\right) = 60t\left(\frac{1}{t}\right)$$

$$5t + 4t = 60$$

$$9t = 60$$

$$t = \frac{60}{9} \text{ or } 6\frac{2}{3} \text{ minutes}$$

So, $6\frac{2}{3}$ minutes are required to complete the task together.

101. $\overline{C} = \dfrac{C}{x} = \dfrac{0.5x + 500}{x}, \quad x > 0$

Horizontal asymptote: $\overline{C} = \dfrac{0.5}{1} = \dfrac{1}{2}$

As x gets larger, the average cost approaches \$0.50.

103. $C = \dfrac{528p}{100 - p}, \quad 0 \le p < 100$

(a) $p = 25 \implies C = \dfrac{528(25)}{100 - 25} = \$176$ million

$p = 75 \implies C = \dfrac{528(75)}{100 - 75} = \$1584$ million

(b) No, $p \ne 100$ because 100 is not in the domain. If p were 100, the denominator would be zero, and division by zero is undefined. The line $p = 100$ is a vertical asymptote for the graph of the function.

(c) $C = \dfrac{528p}{100 - p}, \quad C < 1,500$

$$\frac{528p}{100 - p} < 1,500$$

$$\frac{528p}{100 - p} - 1,500 < 0$$

$$\frac{528p}{100 - p} - \frac{1,500(100 - p)}{100 - p} < 0$$

$$\frac{528p - 150,000 + 1,500p}{100 - p} < 0$$

$$\frac{2028p - 150,000}{100 - p} < 0$$

$$2028p - 150,000 = 0 \implies p \approx 74$$

$$100 - p = 0 \implies p = 100$$

Critical numbers: 74, 100

Domain: $[0, 100)$

Test interval	Representative x-value	Value of Polynomial	Conclusion
$[0, 74)$	50	$\dfrac{2028(50) - 150,000}{100 - 50} = \dfrac{-48,600}{50} = -972$	Negative
$(74, 100)$	80	$\dfrac{2028(80) - 150,000}{100 - 80} = \dfrac{12,240}{20} = 612$	Positive

The solution is $[0, 74)$.

Chapter Test for Chapter 7

1. $f(x) = \dfrac{3x}{x^2 - 25}$

The denominator, $x^2 - 25$ or $(x + 5)(x - 5)$, is zero when $x = -5$ or $x = 5$. So, the domain is all real values of x such that $x \neq -5$ and $x \neq 5$.

Domain $= (-\infty, -5) \cup (-5, 5) \cup (5, \infty)$.

2. $\dfrac{2 - x}{3x - 6} = \dfrac{2 - x}{3(x - 2)}$

$= \dfrac{-1\cancel{(x - 2)}}{3\cancel{(x - 2)}}$

$= -\dfrac{1}{3}, \ x \neq 2$

3. $\dfrac{y^2 + 8y + 16}{2(y - 2)} \cdot \dfrac{8y - 16}{(y + 4)^3} = \dfrac{(y + 4)^2(8)(y - 2)}{2(y - 2)(y + 4)^3}$

$= \dfrac{\cancel{(y + 4)^2}(4)(2)\cancel{(y - 2)}}{2\cancel{(y - 2)}\cancel{(y + 4)^2}(y + 4)}$

$= \dfrac{4}{y + 4}, \ y \neq 2$

4. $(4x^2 - 9) \cdot \dfrac{2x + 3}{2x^2 - x - 3} = \dfrac{(2x + 3)(2x - 3)(2x + 3)}{(2x - 3)(x + 1)}$

$= \dfrac{(2x + 3)^2\cancel{(2x - 3)}}{\cancel{(2x - 3)}(x + 1)}$

$= \dfrac{(2x + 3)^2}{x + 1}, \ x \neq \dfrac{3}{2}$

5. $\dfrac{x^2 + 4x}{2x^2 - 7x + 3} \div \dfrac{x^2 - 16}{x - 3} = \dfrac{x^2 + 4x}{2x^2 - 7x + 3} \cdot \dfrac{x - 3}{x^2 - 16}$

$= \dfrac{x\cancel{(x - 4)}\cancel{(x - 3)}}{(2x - 1)\cancel{(x - 3)}\cancel{(x + 4)}(x - 4)}$

$= \dfrac{x}{(2x - 1)(x - 4)}, \ x \neq 3, x \neq -4$

6. $\dfrac{\left(\dfrac{3x}{x + 2}\right)}{\left(\dfrac{12}{x^3 + 2x^2}\right)} = \dfrac{3x}{x + 2} \div \dfrac{12}{x^2(x + 2)} = \dfrac{3x}{x + 2} \cdot \dfrac{x^2(x + 2)}{12} = \dfrac{\cancel{3}(x^3)\cancel{(x + 2)}}{\cancel{3}(4)\cancel{(x + 2)}} = \dfrac{x^3}{4}, \ x \neq -2, x \neq 0$

7. $\dfrac{\left(9x - \dfrac{1}{x}\right)}{\left(\dfrac{1}{x} - 3\right)} = \dfrac{\left(9x - \dfrac{1}{x}\right) \cdot x}{\left(\dfrac{1}{x} - 3\right) \cdot x} = \dfrac{9x^2 - 1}{1 - 3x} = \dfrac{(3x + 1)\cancel{(3x - 1)}}{-1\cancel{(3x - 1)}}$

$= -(3x + 1) \quad \text{or} \quad -3x - 1, \ x \neq 0, \ x \neq \dfrac{1}{3}$

or

$\dfrac{\left(9x - \dfrac{1}{x}\right)}{\left(\dfrac{1}{x} - 3\right)} = \dfrac{\left(\dfrac{9x}{1} - \dfrac{1}{x}\right)}{\left(\dfrac{1}{x} - \dfrac{3}{1}\right)} = \dfrac{\left(\dfrac{9x^2}{x} - \dfrac{1}{x}\right)}{\left(\dfrac{1}{x} - \dfrac{3x}{x}\right)} = \dfrac{\left(\dfrac{9x^2 - 1}{x}\right)}{\left(\dfrac{1 - 3x}{x}\right)}$

$= \dfrac{(3x + 1)(3x - 1)}{x} \cdot \dfrac{x}{1 - 3x} = \dfrac{(3x + 1)\cancel{(3x - 1)}\cancel{(x)}}{\cancel{(x)}(-1)\cancel{(3x - 1)}}$

$= -(3x + 1) \quad \text{or} \quad -3x - 1, \ x \neq 0, \ x \neq \dfrac{1}{3}$

8. $2x + \dfrac{1 - 4x^2}{x + 1} = \dfrac{2x}{1} + \dfrac{1 - 4x^2}{x + 1}$

$\qquad\qquad = \dfrac{2x(x + 1)}{x + 1} + \dfrac{1 - 4x^2}{x + 1}$

$\qquad\qquad = \dfrac{2x(x + 1) + 1 - 4x^2}{x + 1}$

$\qquad\qquad = \dfrac{2x^2 + 2x + 1 - 4x^2}{x + 1}$

$\qquad\qquad = \dfrac{-2x^2 + 2x + 1}{x + 1}$ or $-\dfrac{2x^2 - 2x - 1}{x + 1}$

9. $\dfrac{5x}{x + 2} - \dfrac{2}{x^2 - x - 6} = \dfrac{5x}{x + 2} - \dfrac{2}{(x - 3)(x + 2)}$

$\qquad\qquad = \dfrac{5x(x - 3)}{(x + 2)(x - 3)} - \dfrac{2}{(x - 3)(x + 2)}$

$\qquad\qquad = \dfrac{5x(x - 3) - 2}{(x + 2)(x - 3)}$

$\qquad\qquad = \dfrac{5x^2 - 15x - 2}{(x + 2)(x - 3)}$

10. $\dfrac{3}{x} - \dfrac{5}{x^2} + \dfrac{2x}{x^2 + 2x + 1} = \dfrac{3}{x} - \dfrac{5}{x^2} + \dfrac{2x}{(x + 1)^2} = \dfrac{3x(x + 1)^2}{x(x)(x + 1)^2} - \dfrac{5(x + 1)^2}{x^2(x + 1)^2} + \dfrac{2x(x^2)}{(x + 1)^2(x^2)}$

$\qquad\qquad = \dfrac{3x(x^2 + 2x + 1) - 5(x^2 + 2x + 1) + 2x^3}{x^2(x + 1)^2}$

$\qquad\qquad = \dfrac{3x^3 + 6x^2 + 3x - 5x^2 - 10x - 5 + 2x^3}{x^2(x + 1)^2} = \dfrac{5x^3 + x^2 - 7x - 5}{x^2(x + 1)^2}$

11. $\dfrac{24a^7 + 42a^4 - 6a^3}{6a^2} = \dfrac{24a^7}{6a^2} + \dfrac{42a^4}{6a^2} - \dfrac{6a^3}{6a^2}$

$\qquad\qquad = \dfrac{\cancel{6}(4)(\cancel{a^2})(a^5)}{\cancel{6a^2}} + \dfrac{\cancel{6}(7)(\cancel{a^2})(a^2)}{\cancel{6a^2}} - \dfrac{\cancel{6a^2}(a)}{\cancel{6a^2}}$

$\qquad\qquad = 4a^5 + 7a^2 - a, \; a \neq 0$

12.

$$\begin{array}{r} t^2 + 3 + \dfrac{-6t + 6}{t^2 - 2} \text{ or } t^2 + 3 - \dfrac{6t - 6}{t^2 - 2} \end{array}$$

$$t^2 - 2 \, \overline{)\, t^4 + 0t^3 + t^2 - 6t + 0}$$

$$\underline{t^4 - 2t^2}$$

$$3t^2 - 6t$$

$$\underline{3t^2 - 6}$$

$$ - 6t + 6$$

Thus, $\dfrac{t^4 + t^2 - 6t}{t^2 - 2} = t^2 + 3 + \dfrac{-6t + 6}{t^2 - 2}$ or $t^2 + 3 - \dfrac{6t - 6}{t^2 - 2}$.

13. $\dfrac{2x^4 - 15x^2 - 7}{x - 3}$

$$\begin{array}{c|ccccc} 3 & 2 & 0 & -15 & 0 & -7 \\ & & 6 & 18 & 9 & 27 \\ \hline & 2 & 6 & 3 & 9 & 20 \end{array}$$

$\dfrac{2x^4 - 15x^2 - 7}{x - 3} = 2x^3 + 6x^2 + 3x + 9 + \dfrac{20}{x - 3}$

14. (a) $f(x) = \dfrac{3}{x-3}$

        y-intercept: $(0, -1)$

        x-intercept: None

        Vertical asymptote: $x = 3$

        Horizontal asymptote: $y = 0$

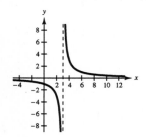

(b) $g(x) = \dfrac{3x}{x-3}$

        y-intercept: $(0, 0)$

        x-intercept: $(0, 0)$

        Vertical asymptote: $x = 3$

        Horizontal asymptote: $y = \dfrac{3}{1} = 3$

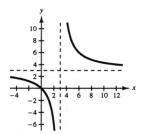

15.
$$\frac{2}{x+5} - \frac{3}{x+3} = \frac{1}{x}$$

$$x(x+5)(x+3)\left(\frac{2}{x+5} - \frac{3}{x+3}\right) = x(x+5)(x+3)\left(\frac{1}{x}\right)$$

$$2x(x+3) - 3x(x+5) = (x+5)(x+3)$$

$$2x^2 + 6x - 3x^2 - 15x = x^2 + 8x + 15$$

$$-x^2 - 9x = x^2 + 8x + 15$$

$$0 = 2x^2 + 17x + 15$$

$$0 = (2x + 15)(x + 1)$$

$$2x + 15 = 0 \implies x = -\frac{15}{2}$$

$$x + 1 = 0 \implies x = -1$$

16.
$$\frac{1}{x+1} + \frac{1}{x-1} = \frac{2}{x^2 - 1}$$

$$(x+1)(x-1)\left(\frac{1}{x+1} + \frac{1}{x-1}\right) = (x+1)(x-1)\left(\frac{2}{(x+1)(x-1)}\right)$$

$$x - 1 + x + 1 = 2$$

$$2x = 2$$

$$x = 1 \quad \text{(Extraneous)}$$

Substituting 1 for x in the original equation results in division by zero. So, 1 is an *extraneous* solution, and the equation has *no* solution.

17. $\dfrac{x-5}{x+8} \geq 0$

Critical numbers:

$x - 5 = 0 \implies x = 5$

$x + 8 = 0 \implies x = -8$

$(-\infty, -8) \cup [5, \infty)$

Test interval	Representative x-value	Is inequality satisfied?
$(-\infty, -8)$	$x = -10$	$\dfrac{-10-5}{-10+8} = \dfrac{-15}{-2} = \dfrac{15}{2} \overset{?}{\geq} 0$ $\dfrac{15}{2} \geq 0$
$(-8, 5)$	$x = 0$	$\dfrac{0-5}{0+8} = -\dfrac{5}{8} \overset{?}{\geq} 0$ $-\dfrac{5}{8} \not\geq 0$
$(5, \infty)$	$x = 6$	$\dfrac{6-5}{6+8} = \dfrac{1}{14} \overset{?}{\geq} 0$ $\dfrac{1}{14} \geq 0$

18. $\dfrac{3u+2}{u-3} \leq 2$

$\dfrac{3u+2}{u-3} - 2 \leq 0$

$\dfrac{3u+2}{u-3} - \dfrac{2(u-3)}{u-3} \leq 0$

$\dfrac{3u+2-2u+6}{u-3} \leq 0$

$\dfrac{u+8}{u-3} \leq 0$

$u + 8 = 0 \implies u = -8$

$u - 3 = 0 \implies u = 3$

Critical numbers: $-8, 3$

The solution is $[-8, 3)$.

Test interval	Representative u-value	Is inequality satisfied?
$(-\infty, -8)$	$u = -10$	$\dfrac{3(-10)+2}{-10-3} - 2 \overset{?}{\leq} 0$ $\dfrac{2}{13} \not\leq 0$
$(-8, 3)$	$u = 0$	$\dfrac{3(0)+2}{0-3} - 2 \overset{?}{\leq} 0$ $\dfrac{-8}{3} \leq 0$
$(3, \infty)$	$u = 5$	$\dfrac{3(5)+2}{5-3} - 2 \overset{?}{\leq} 0$ $\dfrac{13}{2} \not\leq 0$

19. *Verbal Model:* $\boxed{\begin{array}{c}\text{Rate of}\\\text{one painter}\end{array}} + \boxed{\begin{array}{c}\text{Rate of}\\\text{other painter}\end{array}} = \boxed{\begin{array}{c}\text{Rate}\\\text{together}\end{array}}$

Equation: $\dfrac{1}{t} + \dfrac{1}{\left(\dfrac{3}{2}t\right)} = \dfrac{1}{4}$

$\dfrac{1}{t} + \dfrac{2}{3t} = \dfrac{1}{4}$

$12t\left(\dfrac{1}{t} + \dfrac{2}{3t}\right) = 12t\left(\dfrac{1}{4}\right)$

$12 + 8 = 3t$

$20 = 3t$

$\dfrac{20}{3} = t$ and $\dfrac{3}{2}t = \dfrac{3}{2}\left(\dfrac{20}{3}\right) = 10$

So, the individual times are $\dfrac{20}{3}$ or $6\frac{2}{3}$ hours and 10 hours.

Cumulative Test for Chapters 1–7

1. $5(x + 2) - 4(2x - 3) = 5x + 10 - 8x + 12$

$\qquad\qquad\qquad\qquad = -3x + 22$

2. $0.12x + 0.05(2000 - 2x) = 0.12x + 100 - 0.1x$

$\qquad\qquad\qquad\qquad\qquad = 0.02x + 100$

3. $\quad f(x) = x^2 - 3$

$f(a + 2) = (a + 2)^2 - 3$

$\qquad\qquad = a^2 + 4a + 4 - 3$

$\qquad\qquad = a^2 + 4a + 1$

4. $\quad f(x) = \dfrac{3}{x + 5}$

$f(a + 2) = \dfrac{3}{(a + 2) + 5}$

$\qquad\quad = \dfrac{3}{a + 7}$

5. $\dfrac{-16x^2}{12x} = \dfrac{4(-4)(x)(x)}{4(3)(x)}$

$\qquad\quad = -\dfrac{4x}{3}, \ x \neq 0$

6. $\dfrac{6u^5v^{-3}}{7uv^3} = \dfrac{6u^5}{7uv^3v^3}$

$\qquad = \dfrac{6(u)(u^4)}{7(u)(v^6)}$

$\qquad = \dfrac{6u^4}{7v^6}, \ u \neq 0$

7. $\left(\sqrt{x} + 3\right)\left(\sqrt{x} - 3\right) = \left(\sqrt{x}\right)^2 - 3^2$

$\qquad\qquad\qquad\qquad = x - 9$

8. $\sqrt{u}\left(\sqrt{20} - \sqrt{5}\right) = \sqrt{20u} - \sqrt{5u}$

$\qquad\qquad\qquad = \sqrt{4 \cdot 5u} - \sqrt{5u}$

$\qquad\qquad\qquad = 2\sqrt{5u} - \sqrt{5u}$

$\qquad\qquad\qquad = \sqrt{5u}$

9. $\left(2\sqrt{t} + 3\right)^2 = \left(2\sqrt{t} + 3\right) \cdot \left(2\sqrt{t} + 3\right)$

$\qquad\qquad\quad = \left(2\sqrt{t}\right)^2 + 6\sqrt{t} + 6\sqrt{t} + 6\sqrt{t} + 9$

$\qquad\qquad\quad = 4t + 12\sqrt{t} + 9$

10. $\quad 5x - 2y = -25 \implies 15x - 6y = -75$

$-3x + 7y = 44 \implies -15x + 35y = 220$

$\qquad\qquad\qquad\qquad\quad 29y = 145$

$\qquad\qquad\qquad\qquad\qquad y = \dfrac{145}{29}$

$\qquad\qquad\qquad y = 5 \ \text{and} \ 5x - 2(5) = -25$

$\qquad\qquad\qquad\qquad\qquad\qquad 5x - 10 = -25$

$\qquad\qquad\qquad\qquad\qquad\qquad\qquad 5x = -15$

$\qquad\qquad\qquad\qquad\qquad\qquad\qquad x = -3$

$(-3, 5)$

11. $\quad 3x - 2y + z = 1$

$\quad x + 5y - 6z = 4$

$\ 4x - 3y + 2z = 2$

$D = \begin{vmatrix} 3 & -2 & 1 \\ 1 & 5 & -6 \\ 4 & -3 & 2 \end{vmatrix} = 3\begin{vmatrix} 5 & -6 \\ -3 & 2 \end{vmatrix} - 1\begin{vmatrix} -2 & 1 \\ -3 & 2 \end{vmatrix} + 4\begin{vmatrix} -2 & 1 \\ 5 & -6 \end{vmatrix}$

$\qquad\qquad\qquad = 3(10 - 18) - 1(-4 + 3) + 4(12 - 5)$

$\qquad\qquad\qquad = 3(-8) - 1(-1) + 4(7)$

$\qquad\qquad\qquad = -24 + 1 + 28 = 5$

—CONTINUED—

11. —CONTINUED—

$$D_x = \begin{vmatrix} 1 & -2 & 1 \\ 4 & 5 & -6 \\ 2 & -3 & 2 \end{vmatrix} = 1\begin{vmatrix} 5 & -6 \\ -3 & 2 \end{vmatrix} - 4\begin{vmatrix} -2 & 1 \\ -3 & 2 \end{vmatrix} + 2\begin{vmatrix} -2 & 1 \\ 5 & -6 \end{vmatrix}$$

$$= 1(10 - 18) - 4(-4 + 3) + 2(12 - 5)$$

$$= 1(-8) - 4(-1) + 2(7)$$

$$= -8 + 4 + 14 = 10$$

$$D_y = \begin{vmatrix} 3 & 1 & 1 \\ 1 & 4 & -6 \\ 4 & 2 & 2 \end{vmatrix} = 3\begin{vmatrix} 4 & -6 \\ 2 & 2 \end{vmatrix} - 1\begin{vmatrix} 1 & 1 \\ 2 & 2 \end{vmatrix} + 4\begin{vmatrix} 1 & 1 \\ 4 & -6 \end{vmatrix}$$

$$= 3(8 + 12) - 1(2 - 2) + 4(-6 - 4)$$

$$= 3(20) - 1(0) + 4(-10)$$

$$= 60 - 0 - 40 = 20$$

$$D_z = \begin{vmatrix} 3 & -2 & 1 \\ 1 & 5 & 4 \\ 4 & -3 & 2 \end{vmatrix} = 3\begin{vmatrix} 5 & 4 \\ -3 & 2 \end{vmatrix} - 1\begin{vmatrix} -2 & 1 \\ -3 & 2 \end{vmatrix} + 4\begin{vmatrix} -2 & 1 \\ 5 & 4 \end{vmatrix}$$

$$= 3(10 + 12) - 1(-4 + 3) + 4(-8 - 5)$$

$$= 3(22) - 1(-1) + 4(-13)$$

$$= 66 + 1 - 52 = 15$$

$$x = \frac{D_x}{D} = \frac{10}{5} = 2; \quad y = \frac{D_y}{D} = \frac{20}{5} = 4; \quad z = \frac{D_z}{D} = \frac{15}{5} = 3$$

$$(2, 4, 3)$$

12.
$$x + \frac{4}{x} = 4$$

$$x\left(x + \frac{4}{x}\right) = 4(x)$$

$$x^2 + 4 = 4x$$

$$x^2 - 4x + 4 = 0$$

$$(x - 2)(x - 2) = 0$$

$$x - 2 = 0 \implies x = 2$$

13.
$$\sqrt{x + 10} = x - 2$$

$$(\sqrt{x + 10})^2 = (x - 2)^2$$

$$x + 10 = x^2 - 4x + 4$$

$$0 = x^2 - 5x - 6$$

$$0 = (x - 6)(x + 1)$$

$$x - 6 = 0 \implies x = 6$$

$$x + 1 = 0 \implies x = -1$$
(Extraneous)

$$\text{Check: } \sqrt{6 + 10} \stackrel{?}{=} 6 - 2$$

$$\sqrt{16} = 4$$

$$\text{Check: } \sqrt{-1 + 10} \stackrel{?}{=} -1 - 2$$

$$\sqrt{9} \neq -3$$

14.
$$(x - 5)^2 + 50 = 0$$

$$(x - 5)^2 = -50$$

$$x - 5 = \pm\sqrt{-50}$$

$$x - 5 = \pm\sqrt{25(2)(-1)}$$

$$x - 5 = \pm 5\sqrt{2}i$$

$$x = 5 \pm 5\sqrt{2}i$$

15. The *x*-intercepts are approximately $(7.1, 0)$ and $(-1.1, 0)$.

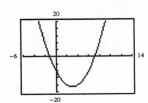

$$x^2 - 6x - 8 = 0$$

$$x = \frac{-(-6) \pm \sqrt{(-6)^2 - 4(1)(-8)}}{2(1)}$$

$$= \frac{6 \pm \sqrt{36 + 32}}{2}$$

$$= \frac{6 \pm \sqrt{68}}{2}$$

$$= \frac{6 \pm \sqrt{4 \cdot 17}}{2}$$

$$= \frac{6 \pm 2\sqrt{17}}{2}$$

$$= 3 \pm \sqrt{17}$$

$$x = 3 + \sqrt{17} \implies x \approx 7.1$$

$$x = 3 - \sqrt{17} \implies x \approx -1.1$$

The solutions of the equation are the first coordinates of the *x*-intercepts of the graph.

16. Vertex $(h, k) = (2, 3) \implies y = a(x - 2)^2 + 3$

Point $(x, y) = (0, 0) \implies 0 = a(0 - 2)^2 + 3$

$$0 = 4a + 3$$

$$-3 = 4a$$

$$-\frac{3}{4} = a$$

$$y = -\frac{3}{4}(x - 2)^2 + 3$$

$$y = -\frac{3}{4}(x^2 - 4x + 4) + 3$$

$$y = -\frac{3}{4}x^2 + 3x - 3 + 3$$

$$y = -\frac{3}{4}x^2 + 3x$$

17. $y = \dfrac{4}{x - 2}$

Domain: $x \neq 2$

Vertical asymptote: $x = 2$

Horizontal asymptote: $y = 0$

x-intercept: None

y-intercept: $(0, 2)$

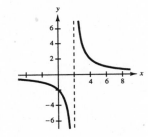

18. You can mow $\frac{1}{4}$ of the lawn in 1 hour. Your friend can mow $\frac{1}{5}$ of the lawn in 1 hour.

$$\frac{1}{4}t + \frac{1}{5}t = 1$$

$$20\left(\frac{1}{4}t + \frac{1}{5}t\right) = 20(1)$$

$$5t + 4t = 20$$

$$9t = 20$$

$$t = \frac{20}{9}$$

It will take $2\frac{2}{9}$ hours to mow the lawn together.

19. (Length)(Width)(Height) = Volume

$$(\text{Length})(h - 7)(h) = h^4 - 10h^3 + 21h^2$$

$$\text{Length} = \frac{h^4 - 10h^3 + 21h^2}{h(h - 7)}$$

$$= \frac{h^2(h^2 - 10h + 21)}{h(h - 7)}$$

$$= \frac{h(h)(h - 7)(h - 3)}{h(h - 7)}$$

$$= h(h - 3)$$

$$= h^2 - 3h$$

The length of the box is $h^2 - 3h$.

CHAPTER 8
More About Functions and Relations

CHAPTER 8
More About Functions and Relations

Section 8.1   Combinations of Functions

Solutions to Odd-Numbered Exercises

1. (a) $(f + g)(x) = 2x + x^2 = x^2 + 2x$ Domain = $\{x: x \text{ is a real number}\}$

   (b) $(f - g)(x) = 2x - x^2 = -x^2 + 2x$ Domain = $\{x: x \text{ is a real number}\}$

   (c) $(fg)(x) = 2x(x^2) = 2x^3$ Domain = $\{x: x \text{ is a real number}\}$

   (d) $(f/g) = \dfrac{2x}{x^2} = \dfrac{2}{x}$ Domain = $\{x: x \neq 0\}$

3. (a) $(f + g)(x) = 4x - 3 + x^2 - 9 = x^2 + 4x - 12$ Domain = $\{x: x \text{ is a real number}\}$

   (b) $(f - g)(x) = 4x - 3 - (x^2 - 9) = -x^2 + 4x + 6$ Domain = $\{x: x \text{ is a real number}\}$

   (c) $(fg)(x) = (4x - 3)(x^2 - 9)$

               $= 4x^3 - 3x^2 - 36x + 27$ Domain = $\{x: x \text{ is a real number}\}$

   (d) $(f/g)(x) = \dfrac{4x - 3}{x^2 - 9}$ Domain = $\{x: x \neq \pm 3\}$

5. (a) $(f + g)(x) = \dfrac{1}{x} + 5x \text{ or } \dfrac{5x^2 + 1}{x}$ Domain = $\{x: x \neq 0\}$

   (b) $(f - g)(x) = \dfrac{1}{x} - 5x \text{ or } \dfrac{-5x^2 + 1}{x}$ Domain = $\{x: x \neq 0\}$

   (c) $(fg)(x) = \dfrac{1}{x}(5x) = 5$ Domain = $\{x: x \neq 0\}$

   (d) $(f/g) = \dfrac{1/x}{5x} = \dfrac{1}{5x^2}$ Domain = $\{x: x \neq 0\}$

7. (a) $(f + g)(x) = \sqrt{x - 5} + \left(-\dfrac{3}{x}\right) = \dfrac{x\sqrt{x - 5}}{x} - \dfrac{3}{x} = \dfrac{x\sqrt{x - 5} - 3}{x}$ Domain = $\{x: x \geq 5\}$

   (b) $(f - g)(x) = \sqrt{x - 5} - \left(-\dfrac{3}{x}\right) = \dfrac{x\sqrt{x - 5}}{x} + \dfrac{3}{x} = \dfrac{x\sqrt{x - 5} + 3}{x}$ Domain = $\{x: x \geq 5\}$

   (c) $(fg)(x) = \sqrt{x - 5}\left(-\dfrac{3}{x}\right) = -\dfrac{3\sqrt{x - 5}}{x}$ Domain = $\{x: x \geq 5\}$

   (d) $(f/g)(x) = \dfrac{\sqrt{x - 5}}{\left(-\dfrac{3}{x}\right)} = \sqrt{x - 5}\left(-\dfrac{x}{3}\right) = -\dfrac{x\sqrt{x - 5}}{3}$ Domain = $\{x: x \geq 5\}$

9. (a) $(f + g)(x) = \sqrt{x + 4} + \sqrt{4 - x}$ Domain $= \{x: -4 \leq x \leq 4\}$

   (b) $(f - g)(x) = \sqrt{x + 4} - \sqrt{4 - x}$ Domain $= \{x: -4 \leq x \leq 4\}$

   (c) $(fg)(x) = \sqrt{x + 4} \cdot \sqrt{4 - x} = \sqrt{16 - x^2}$ Domain $= \{x: -4 \leq x \leq 4\}$

   (d) $(f/g) = \dfrac{\sqrt{x + 4}}{\sqrt{x - 4}}$ Domain $= \{x: -4 \leq x < 4\}$

11. (a) $(f + g)(x) = x^2 + 2x + 3$

   $(f + g)(2) = 2^2 + 2(2) + 3 = 11$

   or

   $(f + g)(2) = f(2) + g(2)$

   $= 2^2 + [2(2) + 3]$

   $= 4 + 7 = 11$

   (b) $(f - g)(x) = x^2 - (2x + 3)$

   $= x^2 - 2x - 3$

   $(f - g)(3) = 3^2 - 2(3) - 3 = 0$

   or

   $(f - g)(3) = f(3) - g(3)$

   $= 3^2 - [2(3) + 3]$

   $= 9 - 9 = 0$

13. (a) $(f + g)(x) = x^2 - 3x + 2 + 2x - 4$

   $= x^2 - x - 2$

   $(f + g)(-3) = (-3)^2 - (-3) - 2 = 10$

   or

   $(f + g)(-3) = f(-3) + g(-3)$

   $= (-3)^2 - 3(-3) + 2 + 2(-3) - 4$

   $= 9 + 9 + 2 - 6 - 4 = 10$

   (b) $(f - g)(x) = x^2 - 3x + 2 - (2x - 4)$

   $= x^2 - 3x + 2 - 2x + 4$

   $= x^2 - 5x + 6$

   $(f - g)(-1) = (-1)^2 - 5(-1) + 6 = 12$

   or

   $(f - g)(-1) = f(-1) + g(-1)$

   $= (-1)^2 - 3(-1) + 2 - [2(-1) - 4]$

   $= 1 + 3 + 2 - (-6)$

   $= 12$

15. (a) $(f + g)(x) = \sqrt{x} + x^2 + 1$

   $(f + g)(4) = \sqrt{4} + 4^2 + 1 = 19$

   or

   $(f + g)(4) = f(4) + g(4)$

   $= \sqrt{4} + (4^2 + 1)$

   $= 2 + 17 = 19$

   (b) $(f - g)(x) = \sqrt{x} - x^2 + 1$

   $= \sqrt{x} - x^2 - 1$

   $(f - g)(9) = \sqrt{9} - 9^2 - 1 = -79$

   or

   $(f - g)(9) = f(9) - g(9)$

   $= \sqrt{9} - (9^2 + 1)$

   $3 - 82 = -79$

17. (a) $(fg)(x) = |x|5 = 5|x|$

   $(fg)(-2) = 5|-2| = 10$

   or

   $(fg)(-2) = f(-2) \cdot g(-2)$

   $= |-2| \cdot 5 = 10$

   (b) $\left(\dfrac{f}{g}\right)(x) = \dfrac{|x|}{5}$

   $= \dfrac{|-2|}{5} = \dfrac{2}{5}$

   or

   $\left(\dfrac{f}{g}\right)(-2) = \dfrac{f(-2)}{g(-2)}$

   $= \dfrac{|-2|}{5} = \dfrac{2}{5}$

19. (a) $(fg)(x) = \frac{1}{x-4}(x) = \frac{x}{x-4}$

$(fg)(-2) = \frac{-2}{-2-4} = \frac{-2}{-6} = \frac{1}{3}$

or

$(fg)(-2) = f(-2)g(-2)$

$= \left(\frac{1}{-2-4}\right)(-2)$

$= \left(-\frac{1}{6}\right)(-2) = \frac{1}{3}$

(b) $\left(\frac{f}{g}\right)(x) = \frac{(1/(x-4))}{x} = \frac{1}{x(x-4)}$

$\left(\frac{f}{g}\right)(4) = \frac{1}{4(0)}$ Undefined

or

$\left(\frac{f}{g}\right)(4) = \frac{f(4)}{g(4)}$

Undefined, because $f(4)$ is undefined.

21. (a) $(f \circ g)(x) = f(g(x))$

$= f(x^2)$

$= x^2 - 3$

(b) $(g \circ f)(x) = g(f(x))$

$= g(x - 3)$

$= (x - 3)^2$ or $x^2 - 6x + 9$

(c) $(f \circ g)(4) = 4^2 - 3 = 16 - 3 = 13$

(d) $(g \circ f)(7) = (7 - 3)^2 = 4^2 = 16$

23. (a) $(f \circ g)(x) = f(g(x))$

$= f(3x)$

$= |3x - 3|$ or $3|x - 1|$

(b) $(g \circ f)(x) = g(f(x))$

$= g(|x - 3|)$

$= 3|x - 3|$

(c) $(f \circ g)(1) = |3(1) - 3| = |0| = 0$

(d) $(g \circ f)(2) = 3|2 - 3| = 3|-1| = 3$

25. (a) $(f \circ g)(x) = f(g(x)) = f(x + 5) = \sqrt{x + 5}$

(b) $(g \circ f)(x) = g(f(x)) = g(\sqrt{x}) = \sqrt{x} + 5$

(c) $(f \circ g)(4) = \sqrt{4 + 5} = \sqrt{9} = 3$

(d) $(g \circ f)(9) = \sqrt{9} + 5 = 3 + 5 = 8$

27. (a) $(f \circ g)(x) = f(g(x)) = f(\sqrt{x}) = \frac{1}{\sqrt{x} - 3}$

(b) $(g \circ f)(x) = g(f(x)) = g\left(\frac{1}{x - 3}\right)$

$= \sqrt{\frac{1}{x - 3}} = \frac{1}{\sqrt{x - 3}}$

(c) $(f \circ g)(49) = \frac{1}{\sqrt{49} - 3} = \frac{1}{7 - 3} = \frac{1}{4}$

(d) $(g \circ f)(12) = \frac{1}{\sqrt{12 - 3}} = \frac{1}{\sqrt{9}} = \frac{1}{3}$

29. (a) $(f \circ g)(x) = f(g(x)) = f(2x) = (2x)^2 + 1 = 4x^2 + 1$

Domain $= \{x : x \text{ is a real number}\}$

(b) $(g \circ f)(x) = g(f(x)) = g(x^2 + 1) = 2(x^2 + 1) = 2x^2 + 2$

Domain $= \{x : x \text{ is a real number}\}$

31. (a) $(f \circ g)(x) = f(g(x)) = f(x - 2) = \sqrt{x - 2}$

Domain $= \{x : x \geq 2\}$

(b) $(g \circ f)(x) = g(f(x)) = g(\sqrt{x}) = \sqrt{x} - 2$

Domain $= \{x : x \geq 0\}$

33. (a) $(f \circ g)(x) = f(g(x)) = f(x^2) = \frac{9}{x^2 + 9}$

Domain $= \{x : x \text{ is a real number}\}$

(b) $(g \circ f)(x) = g(f(x)) = g\left(\frac{9}{x + 9}\right)$

$= \left(\frac{9}{x + 9}\right)^2 = \frac{81}{(x + 9)^2}$

Domain $= \{x : x \neq -9\}$

35. $(f - g)(x) = x^2 - 3x - (5x + 3)$

$= x^2 - 3x - 5x - 3$

$= x^2 - 8x - 3$

$(f - g)(t) = t^2 - 8t - 3$

37. $f(x) = x^2 - 3x$, $g(x) = 5x + 3$

$$\left(\frac{f}{g}\right)(x) = \frac{x^2 - 3x}{5x + 3}$$

$$\left(\frac{f}{g}\right)(2) = \frac{2^2 - 3(2)}{5(2) + 3} = -\frac{2}{13}$$

39. $\dfrac{g(x + h) - g(x)}{h} = \dfrac{[5(x + h) + 3] - (5x + 3)}{h}$

$$= \frac{5x + 5h + 3 - 5x - 3}{h} = \frac{5h}{h} = 5$$

41. $(f \circ g)(-1) = f(g(-1))$

$$= f[5(-1) + 3]$$

$$= f(-2)$$

$$= (-2)^2 - 3(-2)$$

$$= 4 + 6 = 10$$

43. $(g \circ f)(y) = g(f(y))$

$$= g(y^2 - 3y)$$

$$= 5(y^2 - 3y) + 3$$

$$= 5y^2 - 15y + 3$$

45. $(f + g)(x) = x + 3 + x^2 - 2x - 15$

$$= x^2 - x - 12$$

$$(f + g)(t) = t^2 - t - 12$$

47. $(f/g)(x) = \dfrac{x + 3}{x^2 - 2x - 15}$

$$= \frac{x + 3}{(x + 3)(x - 5)}$$

$$= \frac{1}{x - 5}, x \neq -3$$

$$(f/g)(t) = \frac{1}{t - 5}, t \neq -3$$

49. $\dfrac{f(x + h) - f(x)}{h} = \dfrac{(x + h) + 3 - (x + 3)}{h}$

$$= \frac{x + h + 3 - x - 3}{h}$$

$$= \frac{h}{h}$$

$$= 1$$

51. $(f \circ g)(5) = f(g(5))$

$$= f[5^2 - 2(5) - 15]$$

$$= f(0)$$

$$= 0 + 3$$

$$= 3$$

53. $(g \circ f)(t) = g(f(t))$

$$= g(t + 3)$$

$$= (t + 3)^2 - 2(t + 3) - 15$$

$$= t^2 + 6t + 9 - 2t - 6 - 15$$

$$= t^2 + 4t - 12$$

55.

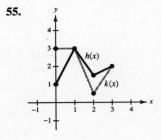

57.

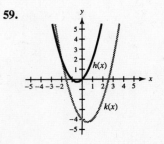

59.

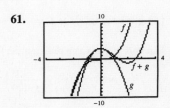

61.

The function $f(x)$ contributes more to the magnitude of the sum than $g(x)$ contributes.

63. (a) $f(1) = -1$ because $(1, -1)$ is an ordered pair of f.

   (b) $g(-1) = -2$ because $(-1, -2)$ is an ordered pair of g.

   (c) $(g \circ f)(1) = g(f(1)) = g(-1) = -2$

65. (a) $(f \circ g)(-3) = f(g(-3)) = f(1) = -1$

   (b) $(g \circ f)(-2) = g(f(-2)) = g(3) = 1$

67. (a) $f(3) = 10$ because $(3, 10)$ is an ordered pair of f.

   (b) $g(10) = 1$ because $(10, 1)$ is an ordered pair of g.

   (c) $(g \circ f)(3) = g(f(3)) = g(10) = 1$

69. (a) $(g \circ f)(4) = g(f(4)) = g(17) = 0$

   (b) $(f \circ g)(2) = f(g(2)) = f(3) = 10$

71. (a) $f(6) = -2$ because $(6, -2)$ is an ordered pair of f.

   (b) $g(6) = 4$ because $(6, 4)$ is an ordered pair of g.

   (c) $(f \circ g)(6) = f(g(6)) = f(4) = -2$

73. (a) $(f \circ g)(0) = f(g(0)) = f(7) = 10$

   (b) $(g \circ f)(5) = g(f(5)) = g(0) = 7$

75.

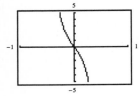

Domain $= \{x: x$ is a real number$\}$

77.

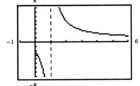

Domain $= \{x: 0 \le x < 1$ or $x > 1\}$

79. (a) $T(x) = (R + B)(x)$

$$= R(x) + B(x)$$
$$= \tfrac{3}{4}x + \tfrac{1}{15}x^2$$

   (b)

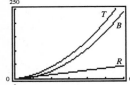

   (c) The function $B(x)$ contributes more to the magnitude of the sum at higher speeds.

81. (a) $f(g(x)) = f(0.02x) = 0.02x - 200{,}000$

   (b) $g(f(x)) = g(x - 200{,}000) = 0.02(x - 200{,}000)$

Thus, (b) or $g(f(x))$ represents the bonus, because it *first* subtracts \$200,000 from the total sales and *then* determines 2% of the difference.

83. $r(t) = 0.6t,\ A(r) = \pi r^2$

$$A(r(t)) = A(0.06t)$$
$$= \pi(0.6t)^2$$
$$= 0.36\pi t^2$$

85. The error is in the first line. Instead of $(f \circ g)(2)$, this is $(fg)(2)$

$$(f \circ g)(2) = f(g(2))$$
$$= f(2^3 + 1)$$
$$= f(9)$$
$$= 2(9) - 1$$
$$= 17$$

87. Domain $= \{x: x$ is a real number$\}$

89. Domain $= \{x: x \ne -3\}$

91.

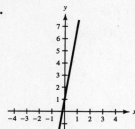

93.

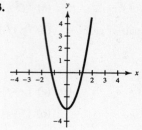

95.

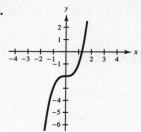

97. *Verbal Model:* | Length of shortest side | $+ 2$ | Length of each of longer sides | $=$ | Perimeter |

Labels: Length of shortest side $= x$ (centimeters)
             Length of each of longer sides $= 3x$ (centimeters)

Equation: $x + 2(3x) = 84$
                $x + 6x = 84$
                   $7x = 84$
                     $x = 12$ and $3x = 36$.

The length of the shortest side of the triangle is 12 centimeters and the other two sides are each 36 centimeters long.

Section 8.2   Inverse Functions

1. $f(x) = 5x \implies f^{-1}(x) = \dfrac{x}{5}$

$f(f^{-1}(x)) = f\left(\dfrac{x}{5}\right) = 5\left(\dfrac{x}{5}\right) = x$

$f^{-1}(f(x)) = f^{-1}(5x) = \dfrac{5x}{5} = x$

3. $f(x) = x + 10 \implies f^{-1}(x) = x - 10$

$f(f^{-1}(x)) = f(x - 10)$
$\qquad\qquad = (x - 10) + 10 = x$
$f^{-1}(f(x)) = f^{-1}(x + 10)$
$\qquad\qquad = (x + 10) - 10 = x$

5. $f(x) = \dfrac{1}{2}x \implies f^{-1}(x) = 2x$

$f(f^{-1}(x)) = f(2x) = \dfrac{1}{2}(2x) = x$

$f^{-1}(f(x)) = f^{-1}\left(\dfrac{1}{2}x\right) = 2\left(\dfrac{1}{2}x\right) = x$

7. $f(x) = x^7 \implies f^{-1}(x) = \sqrt[7]{x}$

$f(f^{-1}(x)) = f\left(\sqrt[7]{x}\right) = \left(\sqrt[7]{x}\right)^7 = x$
$f^{-1}(f(x)) = f^{-1}(x^7) = \sqrt[7]{x^7} = x$

9. $f(x) = \sqrt[3]{x} \implies f^{-1}(x) = x^3$

$f(f^{-1}(x)) = f(x^3) = \sqrt[3]{x^3} = x$
$f^{-1}(f(x)) = f^{-1}\left(\sqrt[3]{x}\right) = \left(\sqrt[3]{x}\right)^3 = x$

11. Domain and range of $f(x)$, $g(x)$:

$(-\infty, \infty)$

$f(g(x)) = f\left(\tfrac{1}{10}x\right) = 10\left(\tfrac{1}{10}\right)x = x$

$g(f(x)) = g(10x) = \tfrac{1}{10}(10x) = x$

So, f and g are inverses of each other.

13. Domain and range of $f(x)$, $g(x)$:

$(-\infty, \infty)$

$f(g(x)) = f(x - 15)$

$\qquad = (x - 15) + 15 = x$

$g(f(x)) = g(x + 15)$

$\qquad = (x + 15) - 15 = x$

So, f and g are inverses of each other.

15. Domain and range of $f(x)$, $g(x)$:

$(-\infty, \infty)$

$f(g(x)) = f\left(\dfrac{1 - x}{2}\right) = 1 - 2\left(\dfrac{1 - x}{2}\right)$

$\qquad = 1 - (1 - x) = x$

$g(f(x)) = g(1 - 2x) = \dfrac{1 - (1 - 2x)}{2}$

$\qquad = \dfrac{2x}{2} = x$

So, f and g are inverses of each other.

17. Domain and range of $f(x)$, $g(x)$:

$(-\infty, \infty)$

$f(g(x)) = f\left[\tfrac{1}{3}(2 - x)\right]$

$\qquad = 2 - 3\left(\tfrac{1}{3}\right)(2 - x)$

$\qquad = 2 - (2 - x) = x$

$g(f(x)) = g(2 - 3x) = \tfrac{1}{3}[2 - (2 - 3x)]$

$\qquad = \tfrac{1}{3}(3x) = x$

So, f and g are inverses of each other.

19. Domain and range of $f(x)$, $g(x)$:

$(-\infty, \infty)$

$f(g(x)) = f(x^3 - 1) = \sqrt[3]{(x^3 - 1) + 1}$

$\qquad = \sqrt[3]{x^3} = x$

$g(f(x)) = g\left(\sqrt[3]{x + 1}\right)$

$\qquad = \left(\sqrt[3]{x + 1}\right)^3 - 1$

$\qquad = (x + 1) - 1 = x$

So, f and g are inverses of each other.

21. Domain and range of $f(x)$, $g(x)$:

$(-\infty, 0) \cup (0, \infty)$

$f(g(x)) = f\left(\dfrac{1}{x}\right) = \dfrac{1}{(1/x)} = x$

$g(f(x)) = g\left(\dfrac{1}{x}\right) = \dfrac{1}{(1/x)} = x$

So, f and g are inverses of each other.

23. $f(x) = 8x$

$y = 8x$

$x = 8y$

$\dfrac{x}{8} = y$

$f^{-1}(x) = \dfrac{x}{8}$

25. $g(x) = x + 25$

$y = x + 25$

$x = y + 25$

$x - 25 = y$

$g^{-1}(x) = x - 25$

27. $g(x) = 3 - 4x$

$y = 3 - 4x$

$x = 3 - 4y$

$x - 3 = -4y$

$\dfrac{3 - x}{4} = y$

$g^{-1}(x) = \dfrac{3 - x}{4}$

29. $f(x) = 4x + 1$

$y = 4x + 1$

$x = 4y + 1$

$x - 1 = 4y$

$\dfrac{x - 1}{4} = y$

$f^{-1}(x) = \dfrac{x - 1}{4}$

31. $f(x) = -6x + 5$

$y = -6x + 5$

$x = -6y + 5$

$x - 5 = -6y$

$\dfrac{x - 5}{-6} = y$

$-\dfrac{x - 5}{6} = y$

$f^{-1}(x) = -\dfrac{x - 5}{6}$

33. $g(s) = 2s - 9$

$y = 2s - 9$

$s = 2y - 9$

$s + 9 = 2y$

$\dfrac{s + 9}{2} = y$

$g^{-1}(s) = \dfrac{s + 9}{2}$

35. $g(t) = \dfrac{1}{4}t + 2$

$y = \dfrac{1}{4}t + 2$

$t = \dfrac{1}{4}y + 2$

$t - 2 = \dfrac{1}{4}y$

$4(t - 2) = \left(\dfrac{1}{4}y\right)4$

$4t - 8 = y$

$g^{-1}(t) = 4t - 8$

37. $h(x) = \sqrt{x}$

$y = \sqrt{x}$

$x = \sqrt{y}$

$x^2 = y, \quad (x \geq 0)$

$h^{-1}(x) = x^2, \quad x \geq 0$

39. $f(x) = 2\sqrt{x + 1}$

$y = 2\sqrt{x + 1}$

$x = 2\sqrt{y + 1}$

$x^2 = 4(y + 1)$

$x^2 = 4y + 4$

$x^2 - 4 = 4y$

$\dfrac{x^2 - 4}{4} = y$

$f^{-1}(x) = \dfrac{x^2 - 4}{4}, \quad x \geq 0$

41. $h(t) = 1 + t^3$

$y = 1 + t^3$

$t = 1 + y^3$

$t - 1 = y^3$

$\sqrt[3]{t - 1} = y$

$h^{-1}(t) = \sqrt[3]{t - 1}$

43. $f(t) = t^3 - 1$

$y = t^3 - 1$

$t = y^3 - 1$

$t + 1 = y^3$

$\sqrt[3]{t + 1} = y$

$f^{-1}(t) = \sqrt[3]{t + 1}$

45. $g(s) = \dfrac{5}{s}$

$y = \dfrac{5}{s}$

$s = \dfrac{5}{y}$

$sy = 5$

$y = \dfrac{5}{s}$

$g^{-1}(s) = \dfrac{5}{s}$

47. $g(s) = \dfrac{-3s}{s + 4}$

$y = \dfrac{-3s}{s + 4}$

$s = \dfrac{-3y}{y + 4}$

$s(y + 4) = -3y$

$sy + 4s = -3y$

$4s = -sy - 3y$

$4s = -y(s + 3)$

$\dfrac{4s}{s + 3} = -y$

$\dfrac{-4s}{s + 3} = y$

$g^{-1}(s) = \dfrac{-4s}{s + 3}$

49. $h(x) = \dfrac{x - 5}{x}$

$y = \dfrac{x - 5}{x}$

$x = \dfrac{y - 5}{y}$

$xy = y - 5$

$xy - y = -5$

$y(x - 1) = -5$

$y = \dfrac{-5}{x - 1}$

$h^{-1}(s) = \dfrac{-5}{x - 1}$

51. Graph (b) is the graph of the inverse.

53. Graph (d) is the graph of the inverse.

55.

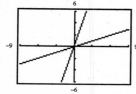

57.

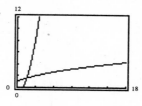

59.

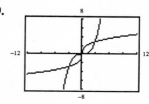

61.

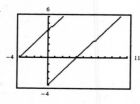

63.
$$f(x) = x^3 + 1$$
$$y = x^3 + 1$$
$$x = y^3 + 1$$
$$x - 1 = y^3$$
$$\sqrt[3]{x - 1} = y$$
$$f^{-1}(x) = \sqrt[3]{x - 1}$$

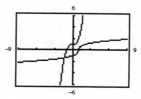

65.
$$f(x) = \sqrt{x^2 - 4}, \ x \geq 2$$
$$y = \sqrt{x^2 - 4}$$
$$x = \sqrt{y^2 - 4}$$
$$x^2 = y^2 - 4$$
$$x^2 + 4 = y^2$$
$$\sqrt{x^2 + 4} = y$$
$$f^{-1}(x) = \sqrt{x^2 + 4}, \ x \geq 0$$

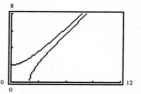

67. $f(x) = x^2 - 2$,

No.

It is possible to find a horizontal line that intersects the graph of f at more than one point. Thus, the function does *not* have an inverse.

69. No. It is possible to find a horizontal line that intersects the graph of f at more than one point. So, the function does *not* have an inverse.

71.

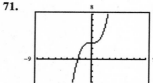

Yes. No horizontal line intersects the graph of f at more than one point. So, the function is one-to-one.

73.

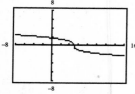

Yes. No horizontal line intersects the graph of f at more than one point. So, the function is one-to-one.

75.

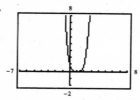

It is possible to find a horizontal line that intersects the graph of g at more than one point. So, the function is not one-to-one.

77.

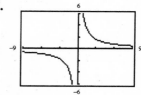

Yes. No horizontal line intersects the graph of h at more than one point. So, the function is one-to-one.

79.

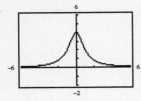

No. It is possible to find a horizontal line that intersects the graph f at more than one point. So, the function is not one-to-one.

81. $f(x) = x^4, x \geq 0$

$y = x^4$

$x = y^4$

$\sqrt[4]{x} = y$

$f^{-1}(x) = \sqrt[4]{x}$

Domain of $f^{-1} = \{x \mid x \geq 0\}$

83. $f(x) = (x - 2)^2, x \geq 2$

$y = (x - 2)^2$

$x = (y - 2)^2$

$\sqrt{x} = y - 2$

$\sqrt{x} + 2 = y$

$f^{-1}(x) = \sqrt{x} + 2$

Domain of $f^{-1} = \{x \mid x \geq 0\}$

85.

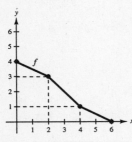

x	0	1	3	4
f^{-1}	6	4	2	0

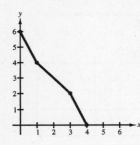

87.

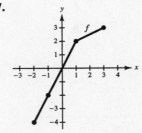

x	-4	-2	3	4
f^{-1}	-2	-1	2	0

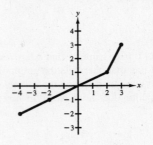

89. (a) $y = 9 + 0.65x$

$x = 9 + 0.65y$

$x - 9 = 0.65y$

$\dfrac{x - 9}{0.65} = y$ or

$y = \dfrac{20}{13}(x - 9)$

(b) x: hourly wage

y: number of units produced

(c) $y = \dfrac{x - 9}{0.65}$ and $x = 14.20$

$y = \dfrac{14.20 - 9}{0.65}$

$= \dfrac{5.20}{0.65} = 8$

So, 8 units are produced.

91. Yes.

The y-intercept of f would have coordinates $(0, a)$ for some real number a. If $(0, a)$ is on the graph of f, then $(a, 0)$ is on the graph of f^{-1}. The point $(a, 0)$ would be on the x-axis, so it is an x-intercept of f^{-1}.

93. Yes.

Interchanging the coordinates of the point $(2, 2)$ on the original function yields the same point $(2, 2)$ on the inverse function.

95. The polynomial can be any real number.

Domain $= \{x: x \text{ is a real number}\}$

97. The radicand must be greater than or equal to zero.

Domain $= \{x: -2 \leq x \leq 2\}$

99. *Verbal Model:* [Cost of first minute] $+$ [Number of additional minutes] $\cdot$ [Cost per additional minute] $=$ [Total cost of call]

Labels: Cost of first minute $= 0.95$ (dollars)

Cost per additional minute $= 0.35$ (dollars)

Number of additional minutes $= x$

Total cost of call $= 5.15$ (dollars)

Equation: $0.95 + 0.35x = 5.15$

$0.35x = 4.20$

$x = 12$

The number of additional minutes was 12, so the length of the call was $12 + 1$ or 13 minutes.

Section 8.3   Variation and Mathematical Models

1. $I = kV$

3. $u = kv^2$

5. $p = \dfrac{k}{d}$

7. $P = \dfrac{k}{\sqrt{1 + r}}$

9. $A = klw$

11. $P = \dfrac{k}{V}$

13. The area A of a triangle varies jointly as the base b of the triangle and the height h of the triangle (or A varies jointly as b and h).

15. The area A of a rectangle varies jointly as the length and the width.

17. The volume V of a right circular cylinder varies jointly as the square of the radius r of the cylinder and as the height h of the cylinder (or V varies jointly as the square of r and as h).

19. The average speed r varies directly as the distance d and inversely as the time t (or r varies directly as d and inversely as t).

$$\left[\text{Note: } r = \frac{d}{t} = \frac{(1)d}{t}\right]$$

21. $s = kt$

$20 = k(4)$

$\dfrac{20}{4} = k$

$5 = k \implies s = 5t$

23. $F = kx^2$

$500 = k(40)^2 = k(1600)$

$\dfrac{500}{1600} = k$

$\dfrac{5}{16} = k \implies F = \dfrac{5}{16}x^2$

25. $H = ku$

$100 = k(40)$

$\dfrac{100}{40} = k$

$\dfrac{5}{2} = k \implies H = \dfrac{5}{2}u$

27. $n = \dfrac{k}{m}$

$32 = \dfrac{k}{1.5}$

$1.5(32) = k$

$48 = k \implies n = \dfrac{48}{m}$

29. $g = \dfrac{k}{\sqrt{z}}$

$\dfrac{4}{5} = \dfrac{k}{\sqrt{25}}$

$\dfrac{4}{5} = \dfrac{k}{5}$

$5\left(\dfrac{4}{5}\right) = \left(\dfrac{k}{5}\right)5$

$4 = k \implies g = \dfrac{4}{\sqrt{z}}$

31. $L = \dfrac{k}{x^3}$

$18 = \dfrac{k}{3^3}$

$18 = \dfrac{k}{27}$

$27(18) = k$

$486 = k \implies L = \dfrac{486}{x^3}$

33. $P = kx^2$

$10 = k(5)^2$

$10 = k(25)$

$\dfrac{10}{25} = k$

$\dfrac{2}{5} = k \implies P = \dfrac{2}{5}x^2$

35. $F = kxy$

$500 = k(15)(8) = k(120)$

$\dfrac{500}{120} = k$

$\dfrac{25}{6} = k \implies F = \dfrac{25}{6}xy$

37. $c = ka\sqrt{b}$

$24 = k(9)\sqrt{4}$

$24 = k(18)$

$\dfrac{24}{18} = k$

$\dfrac{4}{3} = k \implies c = \dfrac{4}{3}a\sqrt{b}$

39. $d = \dfrac{kx^2}{r}$

$3000 = \dfrac{k(10)^2}{4}$

$12{,}000 = k(100)$

$\dfrac{12{,}000}{100} = k$

$120 = k \implies d = \dfrac{120x^2}{r}$

41. (a) $R = kx$

$3875 = k(500)$

$\frac{3875}{500} = k$

$7.75 = k \implies R = 7.75x$

$R = 7.75(635)$

$R = 4921.25$

Thus, the revenue is $4921.25.

(b) The constant of proportionality, $k = 7.75$ is the price per unit.

43. (a) $d = kF$

$3 = k(50)$

$\frac{3}{50} = k$

$0.06 = k \implies d = 0.06F$

$d = 0.06(20)$

$d = 1.2$

The force will stretch the spring 1.2 inches.

(b) $d = 0.06F$

$1.5 = 0.06F$

$\frac{1.5}{0.06} = F$

$25 = F$

The force is 25 pounds.

45. $d = kF$

$7 = k\left(10\frac{1}{2}\right)$

$7 = k\left(\frac{21}{2}\right)$

$14 = k(21)$

$\frac{14}{21} = k$

$\frac{2}{3} = k \implies d = \frac{2}{3}F$

$12 = \frac{2}{3}F$

$36 = 2F$

$\frac{36}{2} = F$

$18 = F$

The weight of the baby is 18 pounds.

47. $d = ks^2$

$75 = k(30)^2 = k(900)$

$\frac{75}{900} = k$

$\frac{1}{12} = k \implies d = \frac{1}{12}s^2$

$= \frac{1}{12}(50)^2$

$= \frac{1}{12}(2500)$

$= 208\frac{1}{3}$

The stopping distance is $208\frac{1}{3}$ feet.

49. $v = kt$

$96 = k(3)$

$32 = k$

The acceleration due to gravity is 32 ft/sec².

51. $P = kw^3$

$750 = k(25)^3 = k(15,625)$

$\frac{750}{15,625} = k$

$0.048 = k \implies P = 0.048w^3 = 0.048(40)^3$

$= 0.048(64,000) = 3072$

The turbine generates 3072 watts of power.

53. $x = \frac{k}{p}$

$800 = \frac{k}{5}$

$5(800) = k$

$4000 = k \implies x = \frac{4000}{p} = \frac{4000}{6} \approx 667$

At a price of $6, the demand would be approximately 667 units.

55. $M = kE$

$60 = k(360)$

$\frac{60}{360} = k$

$\frac{1}{6} = k \implies M = \frac{1}{6}E$

$54 = \frac{1}{6}E$

$6(54) = E$

$324 = E$

Tereshkova would have weighed 324 pounds with her equipment.

57. $p \stackrel{?}{=} kA$

9-inch pizza: $r = \dfrac{9}{2}$ and $A = \pi\left(\dfrac{9}{2}\right)^2 \approx 63.6$

$\quad 6.78 \approx k(63.6)$

$\quad \dfrac{6.78}{63.6} \approx k$

$\quad 0.107 \approx k$

12-inch pizza: $r = 6$ and $A = \pi(6)^2 \approx 113.1$

$\quad 9.78 \approx k(113.1)$

$\quad \dfrac{9.78}{113.1} \approx k$

$\quad 0.086 \approx k$

15-inch pizza: $r = \dfrac{15}{2}$ and $A = \pi\left(\dfrac{15}{2}\right)^2 \approx 176.7$

$\quad 12.18 \approx k(176.7)$

$\quad \dfrac{12.18}{176.7} \approx k$

$\quad 0.069 \approx k$

No, the price is *not* directly proportional to the surface area because the value of k is different for each pizza. The 15-inch pizza is the best buy.

59. $\quad A = kd^2$

$\quad 64 = k(20)^2$

$\quad 64 = 400k$

$\quad \dfrac{64}{100} = k$

$\quad \dfrac{4}{25} = k$

Using $k = \dfrac{4}{25}$,

$\quad 100 = \dfrac{4}{25}d^2$

$\quad \dfrac{25 \cdot 100}{4} = d^2$

$\quad \dfrac{2500}{4} = d^2$

$\quad \sqrt{\dfrac{2500}{4}} = d \quad$ Choose positive root.

$\quad \dfrac{50}{2} = d$

$\quad 25 = d$

So, the area of the projected picture on the screen is 100 square feet when the projector is 25 feet from the screen.

61. $\quad P = \dfrac{kp_1p_2}{d^2}$

$\quad 3840 = \dfrac{k(50,000)(64,000)}{80^2}$

$\quad 3840 = k(500,000)$

$\quad \dfrac{3840}{500,000} = k$

$\quad 0.00768 = k \implies P = \dfrac{0.00768p_1p_2}{d^2}$

$\quad P = \dfrac{0.00768(50,000)(144,000)}{160^2}$

$\quad P = 2160$

The airline could expect 2160 passengers per month between Alameda and Crystal Lake.

63. $\quad p = \dfrac{k}{t}$

$\quad 38 = \dfrac{k}{3}$

$\quad 3(38) = k$

$\quad 114 = k \implies p = \dfrac{114}{t}$

$\quad p = \dfrac{114}{6.5}$

$\quad p \approx 17.5$

Approximately 17.5% of the oil remained $6\frac{1}{2}$ years later. (This result can be verified graphically.)

65. If $y = kx$ with $k > 0$, then when one variable increases, the other variable also increases.

67. $y_1 = \dfrac{k}{x^2}$

$\quad y_2 = \dfrac{k}{(2x)^2}$

$\quad = \dfrac{k}{4x^2} = \dfrac{1}{4} \cdot \dfrac{k}{x^2} = \dfrac{1}{4}y_1$

When x is doubled, y is multiplied by 1/4.

69. (a) $(f + g)(x) = f(x) + g(x)$

$\qquad = 3x^2 + x + (-5x)$

$\qquad = 3x^2 - 4x$

(c) $(fg)(x) = f(x) \cdot g(x)$

$\qquad = (3x^2 + x)(-5x)$

$\qquad = -15x^3 - 5x^2$

(b) $(f - g)(x) = f(x) - g(x)$

$\qquad = 3x^2 + x - (-5x)$

$\qquad = 3x^2 + 6x$

(d) $(f/g)(x) = f(x)/g(x)$

$\qquad = \dfrac{3x^2 + x}{-5x}$

$\qquad = \dfrac{x(3x + 1)}{x(-5)}$

$\qquad = -\dfrac{3x + 1}{5}, \quad x \neq 0$

(e) $(f \circ g)(x) = f(g(x))$

$\qquad = f(-5x)$

$\qquad = 3(-5x)^2 + (-5x)$

$\qquad = 75x^2 - 5x$

(f) $(g \circ f)(x) = g(f(x))$

$\qquad = g(3x^2 + x)$

$\qquad = -5(3x^2 + x)$

$\qquad = -15x^2 - 5x$

71. (a) $(f + g)(x) = f(x) + g(x)$

$\qquad = \dfrac{1}{x + 4} + \dfrac{1}{6x}$

$\qquad = \dfrac{6x(1)}{6x(x + 4)} + \dfrac{1(x + 4)}{6x(x + 4)}$

$\qquad = \dfrac{7x + 4}{6x(x + 4)}$

(c) $(fg)(x) = f(x) \cdot g(x)$

$\qquad = \dfrac{1}{x + 4} \cdot \dfrac{1}{6x}$

$\qquad = \dfrac{1}{6x(x + 4)}$

(b) $(f - g)(x) = f(x) - g(x)$

$\qquad = \dfrac{1}{x + 4} - \dfrac{1}{6x}$

$\qquad = \dfrac{6x(1)}{6x(x + 4)} - \dfrac{1(x + 4)}{6x(x + 4)}$

$\qquad = \dfrac{5x - 4}{6x(x + 4)}$

(d) $(f/g)(x) = f(x)/g(x)$

$\qquad = \dfrac{1}{x + 4} \div \dfrac{1}{6x}$

$\qquad = \dfrac{1}{x + 4} \cdot \dfrac{6x}{1}$

$\qquad = \dfrac{6x}{x + 4}, \quad x \neq 0$

(e) $(f \circ g)(x) = f(g(x))$

$\qquad = f\left(\dfrac{1}{6}x\right)$

$\qquad = \dfrac{1}{\left(\dfrac{1}{6x} + 4\right)}$

$\qquad = \dfrac{1}{\left(\dfrac{1 + 24x}{6x}\right)}$

$\qquad = \dfrac{6x}{24x + 1}, \quad x \neq 0$

(f) $(g \circ f)(x) = g(f(x))$

$\qquad = g\left(\dfrac{1}{x + 4}\right)$

$\qquad = \dfrac{1}{6\left(\dfrac{1}{x + 4}\right)}$

$\qquad = \dfrac{1}{\left(\dfrac{6}{x + 4}\right)}$

$\qquad = \dfrac{x + 4}{6}, \quad x \neq -4$

73. *Verbal Model:* $\boxed{\begin{array}{c}\text{Variable}\\\text{cost}\end{array}} \cdot \boxed{\begin{array}{c}\text{Number of}\\\text{units}\end{array}} + \boxed{\text{Fixed costs}} = \boxed{\text{Total cost}}$

Label: Number of units $= x$

Equation: $5.75x + 12{,}000 = 17{,}060$

$5.74x = 5060$

$x = 880$

The number of games produced is 880.

Section 8.4   Polynomial Functions and Their Graphs

1. Degree 0

Constant function

Graph: Horizontal line

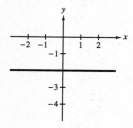

3. Degree 1

Linear function

Graph: Line with slope -3

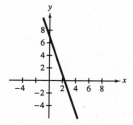

5. Degree 2

Quadratic function

Graph: Parabola

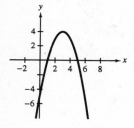

7. Degree 1

Linear function

Graph: Line with slope 2

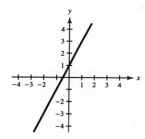

9. Degree: 5

Maximum numbers of turns: $5 - 1 = 4$

11. Degree: 2

Maximum number of turns: $2 - 1 = 1$

13. Degree: 6

Maximum number of turns: $6 - 1 = 5$

15. Horizontal translation 5 units to the right

17. $f(x) = x^5 - 2$, $g(x) = x^5$

Vertical translation 2 units downward

19. (a) $f(x) = (x - 2)^3$

Horizontal translation
2 units to the right

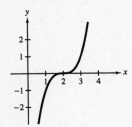

(b) $f(x) = x^3 - 2$

Vertical translation
2 units downward

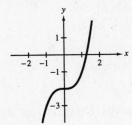

(c) $f(x) = -(x - 2)^3 - 2$

Reflection in x-axis, horizontal translation 2 units to
the right and vertical translation 2 units downward

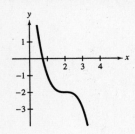

(d) $f(x) = -x^3$

Reflection in x-axis

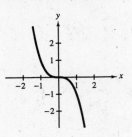

21. Leading coefficient, 2, *positive* $\Longrightarrow$ graph rises to right

Degree, 2, *even* (*same* behavior to left) $\Longrightarrow$ graph rises to left

23. Leading coefficient, $\frac{1}{3}$, *positive* $\Longrightarrow$ graph rises to right.

Degree, 3, *odd* (*opposite* behavior to left) $\Longrightarrow$ graph falls to left.

25. Leading coefficient, -5, *negative* $\Longrightarrow$ graph falls to right.

Degree, 3, *odd* (*opposite* behavior to left) $\Longrightarrow$ graph rises to left.

27. Leading coefficient, -1, *negative* $\Longrightarrow$ graph falls to the right.

Degree 4, *even*, (*same* behavior to the left) $\Longrightarrow$ graph falls to the left.

29. Leading coefficient, 6, *positive* $\Longrightarrow$ graph rises to the right.

Degree 4, *even*, (*same* behavior to the left) $\Longrightarrow$ graph rises to the left.

31. x-intercept:

$$f(x) = 0$$
$$x^2 - 25 = 0$$
$$(x + 5)(x - 5) = 0$$
$$x + 5 = 0 \Longrightarrow x = -5$$
$$x - 5 = 0 \Longrightarrow x = 5$$

y-intercept: $x = 0$

$$y = 0^2 - 25$$
$$y = -25$$

$(0, -25)$

$(-5, 0)$ and $(5, 0)$

33. x-intercept: $h(x) = 0$

$$x^2 - 6x + 9 = 0$$

$$(x - 3)^2 = 0$$

$$x - 3 = 0 \Rightarrow x = 3$$

$$(3, 0)$$

y-intercept: $x = 0$

$$y = 0^2 - 6(0) + 9$$

$$y = 9$$

$$(0, 9)$$

35. x-intercepts: $f(x) = 0$

$$x^2 + x - 2 = 0$$

$$(x + 2)(x - 1) = 0$$

$$x + 2 = 0 \Rightarrow x = -2$$

$$x - 1 = 0 \Rightarrow x = 1$$

$$(-2, 0) \text{ and } (1, 0)$$

y-intercept: $x = 0$

$$y = 0^2 + 0 - 2$$

$$y = -2$$

$$(0, -2)$$

37. x-intercepts: $f(x) = 0$

$$x^3 - 4x^2 + 4x = 0$$

$$x(x^2 - 4x + 4) = 0$$

$$x(x - 2)^2 = 0$$

$$x = 0$$

$$x - 2 = 0 \Rightarrow x = 2$$

$$(0, 0) \text{ and } (2, 0)$$

y-intercept: $x = 0$

$$y = 0^3 - 4(0)^2 + 4(0)$$

$$y = 0$$

$$(0, 0)$$

39. x-intercepts: $g(x) = 0$

$$\tfrac{1}{2}x^4 - \tfrac{1}{2} = 0$$

$$x^4 - 1 = 0$$

$$(x^2 + 1)(x^2 - 1) = 0$$

$$(x^2 + 1)(x + 1)(x - 1) = 0$$

$$x^2 + 1 = 0 \Rightarrow x^2 = -1$$

$$\Rightarrow x = \pm i$$

$$x + 1 = 0 \Rightarrow x = -1$$

$$x - 1 = 0 \Rightarrow x = 1$$

$$(-1, 0) \text{ and } (1, 0)$$

y-intercept: $x = 0$

$$y = \tfrac{1}{2}(0)^4 - \tfrac{1}{2}$$

$$y = -\tfrac{1}{2}$$

$$\left(0, -\tfrac{1}{2}\right)$$

41. x-intercepts: $f(x) = 0$

$$5x^4 + 15x^2 + 10 = 0$$

$$x^4 + 3x^2 + 2 = 0$$

$$(x^2 + 2)(x^2 + 1) = 0$$

$$x^2 + 2 = 0 \Rightarrow x^2 = -2$$

$$\Rightarrow x = \pm\sqrt{2}i$$

$$x^2 + 1 = 0 \Rightarrow x^2 = -1$$

$$\Rightarrow x = \pm i$$

No x-intercepts

y-intercept: $x = 0$

$$y = 5(0)^4 + 15(0)^2 + 10$$

$$y = 10$$

$$(0, 10)$$

43.

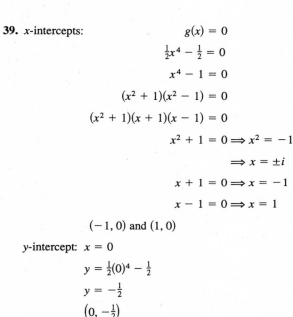

The x-intercepts are $(1, 0)$ and $(-1, 0)$.

45.

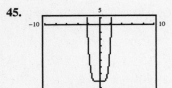

The *x*-intercepts are approximately
(2.236, 0) and (−2.236, 0).

Note: The exact values for these
intercepts are $(\sqrt{5}, 0)$ and
$(-\sqrt{5}, 0)$.

47.

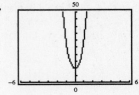

There are no *x*-intercepts.

49.

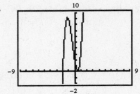

The *x*-intercepts are (−2, 0) and
$(\frac{1}{2}, 0)$.

51. Graph (e)   **53.** Graph (b)   **55.** Graph (a)   **57.** Graph (d)

59.

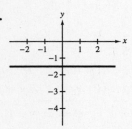

61.

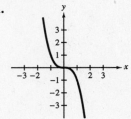

63.

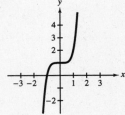

65.

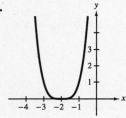

67.

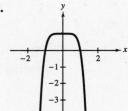

69.

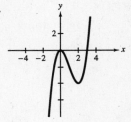

71. Degree: 3 ⟹ Maximum number
of turns: 3 − 1 = 2

Yes, the graph has the maximum
number of turns.

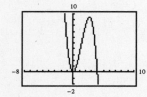

73. Degree: 4 ⟹ Maximum number
of turns: 4 − 1 = 3

Yes, the graph has the maximum
number of turns.

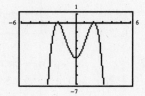

75. Degree: 5 ⟹ Maximum number
of turns: 5 − 1 = 4

No, the graph does not have the
maximum number of turns.

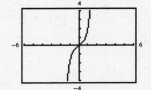

77. (a) Volume = (Length)(Width)(Height)

$$V = (12 - 2x)(12 - 2x)(x)$$

$$= 2(6 - x)(2)(6 - x)(x)$$

$$= 4x(6 - x)^2$$

(b) Domain: (0, 6)

(c)

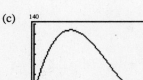

$V(x)$ is a maximum when $x = 2$.

79. To find the x-intercepts, set $f(x) = 0$ and solve for x. To find the y-intercept, set $x = 0$ and evaluate the function.

81. Not necessarily. A polynomial function of degree n can have from 0 to n x-intercepts. In other words, a polynomial function of degree n has at most n x-intercepts.

83. $6x^2 + 23x - 13 = (2x - 1)(3x + 13)$

85. $10x^4 + 22x^2 + 4x^2 = 2x^2(5x^2 + 11x + 2)$
$$= 2x^2(5x + 1)(x + 2)$$

87. $x^3 + 6x^2 - 2x - 12 = (x^3 + 6x^2) - (2x + 12)$
$$= x^2(x + 6) - 2(x + 6)$$
$$= (x + 6)(x^2 - 2)$$

89. $3x^2 + 5 = 0$
$$3x^2 = -5$$
$$x^2 = -\frac{5}{3}$$
$$x = \pm\sqrt{-\frac{5}{3}}$$
$$x = \pm\frac{\sqrt{5}\sqrt{-1}}{\sqrt{3}} \cdot \frac{\sqrt{3}}{\sqrt{3}}$$
$$x = \pm\frac{\sqrt{15}}{3}i$$

The solutions involve square roots of a negative number, and so the solutions are imaginary numbers. There are no real solutions.

91. $6x^2 + 3x - 5 = 0$
$$x = \frac{-3 \pm \sqrt{3^2 - 4(6)(-5)}}{2(6)}$$
$$x = \frac{-3 \pm \sqrt{129}}{12}$$

93. Perimeter = Length + 2 Wdth
$$320 = \text{Length} + 2w$$
$$320 - 2w = \text{Length}$$
Area = (Length)(Width)
$$= (320 - 2w)(w)$$
$$= 320w - 2w^2$$

The graph of the area function is a parabola opening downward. The maximum value of the function is found at the vertex of the parabola. The vertex can be determined by completing the square to write the function in standard form.

$$A = -2w^2 + 320w$$
$$= -2(w^2 - 160w)$$
$$= -2(w^2 - 160w + 6400) + 12{,}800$$
$$= -2(w - 80)^2 + 12{,}800$$

Vertex: $(80, 12{,}800)$

The first coordinate of the vertex is 80, so the width is 80 and the length is $320 - 2(80) = 160$. The dimensions of the lot are 80 feet by 160 feet.

Mid-Chapter Quiz for Chapter 8

1. $(f + g)(x) = 2x - 8 + x^2$
$= x^2 + 2x - 8$
Domain: $(-\infty, \infty)$

2. $(fg)(x) = (2x - 8)x^2$
$= 2x^3 - 8x^2$
Domain: $(-\infty, \infty)$

3. $(g - f)(x) = x^2 - (2x - 8)$
$= x^2 - 2x + 8$
Domain: $(-\infty, \infty)$

4. $\left(\dfrac{f}{g}\right)(x) = \dfrac{2x - 8}{x^2}$
Domain: $(-\infty, 0) \cup (0, \infty)$

5. $(f \circ g)(x) = f(g(x))$
$= f(x^2) = 2x^2 - 8$
Domain: $(-\infty, \infty)$

6. $(g \circ f)(x) = g(f(x))$
$= g(2x - 8) = (2x - 8)^2$
Domain: $(-\infty, \infty)$

7. $(f + g)(9) = f(9) + g(9)$
$= \sqrt{9} + [9^2 + 2(9)]$
$= 3 + 81 + 18$
$= 102$

8. $(f - g)(4) = f(4) - g(4)$
$= \sqrt{4} - [4^2 + 2(4)]$
$= 2 - 16 - 8$
$= -22$

9. $(fg)(1) = f(1)g(1)$
$= \sqrt{1}[1^2 + 2(1)]$
$= 1(3)$
$= 3$

10. $(f/g)(16) = f(16)/g(16)$
$= \dfrac{\sqrt{16}}{16^2 + 2(16)}$
$= \dfrac{4}{288}$
$= \dfrac{1}{72}$

11. $(f \circ g)(-5) = f(g(-5))$
$= f[(-5)^2 + 2(-5)]$
$= f(15)$
$= \sqrt{15}$

12. $(g \circ f)(7) = g(f(7))$
$= g(\sqrt{7})$
$= (\sqrt{7})^2 + 2\sqrt{7}$
$= 7 + 2\sqrt{7}$

13. $C(x) = 60x + 750$
$x(t) = 50t, 0 \leq t \leq 40$
$C(x(t)) = C(50t) = 60(50t) + 750$
$= 3000t + 750, 0 \leq t \leq 40$
$C(x(t))$ represents the cost of producing units for t hours, for $0 \leq t \leq 40$.

14. Domain and range of $f(x), g(x)$: $(-\infty, \infty)$
$(f \circ g)(x) = f(g(x)) = f\left[\frac{1}{15}(100 - x)\right]$
$= 100 - 15\left[\frac{1}{15}100 - x\right] = 100 - (100 - x) = x$
$(g \circ f)(x) = g(f(x)) = g(100 - 15x) = \frac{1}{15}[100 - (100 - 15x)] = \frac{1}{15}(15x) = x$

15.
$$f(x) = x^3 - 8$$
$$y = x^3 - 8$$
$$x = y^3 - 8$$
$$x + 8 = y^3$$
$$\sqrt[3]{x + 8} = y$$
$$f^{-1}(x) = \sqrt[3]{x + 8}$$

$$(f \circ f^{-1}) = f(f^{-1}(x))$$
$$= f(\sqrt[3]{x + 8})$$
$$= (\sqrt[3]{x + 8})^3 - 8$$
$$= x + 8 - 8$$
$$= x$$

$$(f^{-1} \circ f)(x) = f^{-1}(f(x))$$
$$= f^{-1}(x^3 - 8)$$
$$= \sqrt[3]{(x^3 - 8) + 8}$$
$$= \sqrt[3]{x^3}$$
$$= x$$

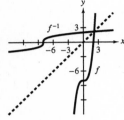

16.
$$z = kt$$
$$12 = k(4)$$
$$\frac{12}{4} = k$$
$$3 = k \implies z = 3t$$

17.
$$S = khr^2$$
$$120 = k(6)(2)^2$$
$$120 = k(24)$$
$$\frac{120}{24} = k$$
$$5 = k \implies S = 5hr^2$$

18.
$$N = \frac{kt^2}{s}$$
$$300 = \frac{k(10)^2}{5}$$
$$5(300) = k(10)^2$$
$$1500 = k(100)$$
$$\frac{1500}{100} = k$$
$$15 = k \implies N = \frac{15t^2}{s}$$

19.
$$m = kt$$
$$0.02 = k(12)$$
$$\frac{0.02}{12} = k$$
$$\frac{2}{1200} = k$$
$$\frac{1}{600} = k \implies m = \frac{1}{600}t$$
$$0.05 = \frac{1}{600}t$$
$$600(0.05) = t$$
$$30 = t$$

An explosion could occur 30 minutes after the leak began.

20. The graph rises to the right because the leading coefficient, 3, is positive.

21. (a) Transformation: Reflection in the x-axis.

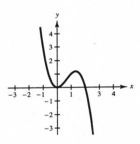

(b) Transformation: Horizontal shift of two units to the right.

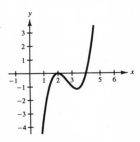

Section 8.5   Circles

1. Graphs (c)

3. Graph (a)

5. Graph (b)

7. $x^2 + y^2 = r^2$

$x^2 + y^2 = 5^2$

$x^2 + y^2 = 25$

9. $x^2 + y^2 = r^2$

$x^2 + y^2 = \left(\frac{2}{3}\right)^2$

$x^2 + y^2 = \frac{4}{9}$ or $9x^2 + 9y^2 = 4$

11. Radius = Distance between $(0, 0)$ and $(0, 8)$

$r = \sqrt{(0 - 0)^2 + (8 - 0)^2} = \sqrt{64} = 8$

$x^2 + y^2 = r^2$

$x^2 + y^2 = 8^2$

$x^2 + y^2 = 64$

13. Radius = Distance between $(0, 0)$ and $(5, 2)$

$r = \sqrt{(5 - 0)^2 + (2 - 0)^2} = \sqrt{29}$

$x^2 + y^2 = r^2$

$x^2 + y^2 = \left(\sqrt{29}\right)^2$

$x^2 + y^2 = 29$

15. Center (h, k): $(4, 3)$, Radius r: 10

$(x - h)^2 + (y - k)^2 = r^2$

$(x - 4)^2 + (y - 3)^2 = 10^2$

$(x - 4)^2 + (y - 3)^2 = 100$

17. Center (h, k): $(5, -3)$, Radius r: 9

$(x - h)^2 + (y - k)^2 = r^2$

$(x - 5)^2 + (y - (-3))^2 = 9^2$

$(x - 5)^2 + (y + 3)^2 = 81$

19. Radius $r = \sqrt{(-2 - 0)^2 + (1 - 1)^2}$

$= \sqrt{(-2)^2 + 0^2}$

$= \sqrt{4 + 0}$

$= \sqrt{4}$

$= 2$

Center (h, k): $(-2, 1)$, Radius r: 2

$(x - h)^2 + (y - k)^2 = r^2$

$(x - (-2))^2 + (y - 1)^2 = 2^2$

$(x + 2)^2 + (y - 1)^2 = 4$

21. Radius $r = \sqrt{(3 - 4)^2 + (2 - 6)^2}$

$= \sqrt{(-1)^2 + (-4)^2}$

$= \sqrt{1 + 16}$

$= \sqrt{17}$

Center (h, k): $(3, 2)$, Radius r: $\sqrt{17}$

$(x - h)^2 + (y - k)^2 = r^2$

$(x - 3)^2 + (y - 2)^2 = \left(\sqrt{17}\right)^2$

$(x - 3)^2 + (y - 2)^2 = 17$

23. $x^2 + y^2 = 16$

Center: $(0, 0)$

Radius: $\sqrt{16} = 4$

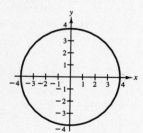

25. $x^2 + y^2 = 36$

Center: $(0, 0)$

Radius: $\sqrt{36} = 6$

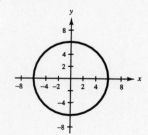

27. $4x^2 + 4y^2 = 1$

$x^2 + y^2 = \frac{1}{4}$

Center: $(0, 0)$

Radius: $\sqrt{\frac{1}{4}} = \frac{1}{2}$

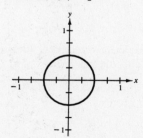

29. $25x^2 + 25y^2 - 144 = 0$

$$25x^2 + 25y^2 = 144$$

$$x^2 + y^2 = \frac{144}{25}$$

Center: $(0, 0)$

Radius: $\sqrt{\frac{144}{25}} = \frac{12}{5}$

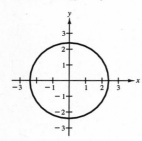

31. $(x + 1)^2 + (y - 5)^2 = 64$

Center: $(-1, 5)$

Radius: $\sqrt{64} = 8$

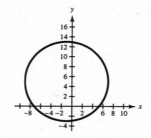

33. $(x - 2)^2 + (y - 3)^2 = 4$

Center: $(2, 3)$

Radius: $\sqrt{4} = 2$

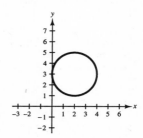

35. $\left(x + \frac{5}{2}\right)^2 + (y + 3)^2 = 9$

Center: $\left(-\frac{5}{2}, -3\right)$

Radius: $\sqrt{9} = 3$

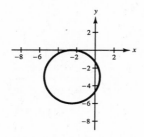

37.

$$x^2 + y^2 - 4x - 2y + 1 = 0$$

$$x^2 + 4x + y^2 - 2y = -1$$

$$(x^2 - 4x + 4) + (y^2 - 2y + 1) = -1 + 4 + 1$$

$$(x - 2)^2 + (y - 1)^2 = 4$$

Center: $(2, 1)$

Radius: $\sqrt{4} = 2$

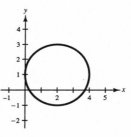

39.

$$x^2 + y^2 + 2x + 6y + 6 = 0$$

$$x^2 + 2x + y^2 + 6y = -6$$

$$(x^2 + 2x + 1) + (y^2 + 6y + 9) = -6 + 1 + 9$$

$$(x + 1)^2 + (y + 3)^2 = 4$$

Center: $(-1, -3)$

Radius: $\sqrt{4} = 2$

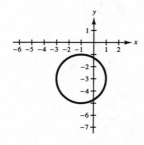

41.

$$x^2 + y^2 + 8x + 4y - 5 = 0$$

$$x^2 + 8x + y^2 + 4y = 5$$

$$(x^2 + 8x + 16) + (y^2 + 4y + 4) = 5 + 16 + 4$$

$$(x + 4)^2 + (y + 2)^2 = 25$$

Center: $(-4, -2)$

Radius: $\sqrt{25} = 5$

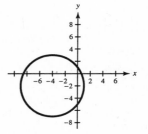

43. Radius $= 4000 + 500 = 4500$

$x^2 + y^2 = r^2$

$x^2 + y^2 = (4500)^2$

$x^2 + y^2 = 20{,}250{,}000$

45. Radius $r = 12$

$x^2 + y^2 = r^2$

$x^2 + y^2 = 12^2$

$x^2 + y^2 = 144$

Point (x, y) is on the circle $x^2 + y^2 = 144$, and the x-coordinate is 6.

$x^2 + y^2 = 144$

$6^2 + y^2 = 144$

$\quad y^2 = 144 - 36$

$\quad y^2 = 108$

$\quad y = \pm\sqrt{108}$ (Choose the positive value)

$\quad y = 6\sqrt{3}$

$\quad y \approx 10.4$

The height is approximately 10.4 feet.

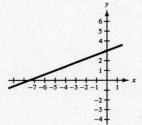

47. A circle is the set of all points (x, y) in a plane that are a given positive distance r from a fixed point (h, k) called the center.

The standard form of the equation of a circle centered at the origin is $x^2 + y^2 = r^2$.

49. $x^2 + 6x - 4 = 0$

$\quad x^2 + 6x = 4$

$x^2 + 6x + 9 = 4 + 9$

$\quad (x + 3)^2 = 13$

$\quad x + 3 = \pm\sqrt{13}$

$\quad x = -3 \pm \sqrt{13}$

51.

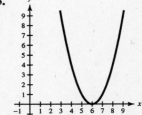

53.

55. *Verbal Model:* | Original cost per person | $-$ | New cost per person | $=$ | Difference in cost |

Label: New number of people $= x$

Original number of people $= x - 3$

Original cost per person $= \dfrac{288}{x - 3}$ (dollars)

New cost per person $= \dfrac{288}{x}$ (dollars)

Difference in cost $= 8$ (dollars)

Equation: $\dfrac{288}{x - 3} - \dfrac{288}{x} = 8$

$x(x - 3)\left(\dfrac{288}{x - 3} - \dfrac{288}{x} \right) = 8x(x - 3)$

$288x - 288(x - 3) = 8x^2 - 24x$

$864 = 8x^2 - 24x$

$0 = 8x^2 - 24x - 864$

$0 = x^2 - 3x - 108$

$0 = (x - 12)(x + 9)$

$x - 12 = 0 \implies x = 12$

$x + 9 = 0 \implies x = -9$

Discard the negative solution. The number of people going to the game is 12.

Section 8.6   Ellipses and Hyperbolas

1. Graph (a)

3. Graph (d)

5. Graph (e)

7. $\dfrac{x^2}{16} + \dfrac{y^2}{4} = 1$

$\dfrac{x^2}{4^2} + \dfrac{y^2}{2^2} = 1$

Vertices: $(-4, 0), (4, 0)$

Co-vertices: $(0, -2), (0, 2)$

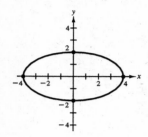

9. $\dfrac{x^2}{4} + \dfrac{y^2}{16} = 1$

$\dfrac{x^2}{2^2} + \dfrac{y^2}{4^2} = 1$

Vertices: $(0, -4), (0, 4)$

Co-vertices: $(-2, 0), (2, 0)$

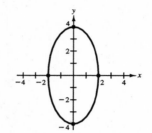

11. $\dfrac{x^2}{25/9} + \dfrac{y^2}{16/9} = 1$

$\dfrac{x^2}{(5/3)^2} + \dfrac{y^2}{(4/3)^2} = 1$

Vertices: $\left(-\frac{5}{3}, 0\right), \left(\frac{5}{3}, 0\right)$

Co-vertices: $\left(0, -\frac{4}{3}\right), \left(0, \frac{4}{3}\right)$

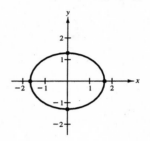

13. $\dfrac{9x^2}{4} + \dfrac{25y^2}{16} = 1$

$\dfrac{x^2}{4/9} + \dfrac{y^2}{16/25} = 1$

$\dfrac{x^2}{(2/3)^2} + \dfrac{y^2}{(4/5)^2} = 1$

Vertices: $(0, 4/5)$ and $(0, -4/5)$

Co-vertices: $(2/3, 0)$ and $(-2/3, 0)$

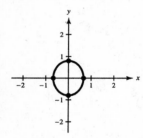

15. $16x^2 + 25y^2 - 9 = 0$

$16x^2 + 25y^2 = 0$

$\dfrac{16x^2}{9} + \dfrac{25y^2}{9} = 1$

$\dfrac{x^2}{9/16} + \dfrac{y^2}{9/25} = 1$

$\dfrac{x^2}{(3/4)^2} + \dfrac{y^2}{(3/5)^2} = 1$

Vertices: $(3/4, 0)$ and $(-3/4, 0)$

Co-vertices: $(0, 3/5)$ and $(0, -3/5)$

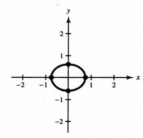

17. $4x^2 + y^2 - 4 = 0$

$$4x^2 + y^2 = 4$$

$$\frac{4x^3}{4} + \frac{y^2}{4} = \frac{4}{4}$$

$$\frac{x^2}{1} + \frac{y^2}{4} = 1$$

$$\frac{x^2}{1^2} + \frac{y^2}{2^2} = 1$$

Vertices: $(0, -2), (0, 2)$

Co-vertices: $(-1, 0), (1, 0)$

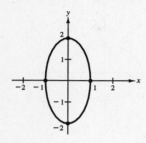

19. $x^2 + 2y^2 = 4$

$$\frac{x^2}{4} + \frac{2y^2}{4} = 1$$

$$\frac{x^2}{4} + \frac{y^2}{2} = 1 \implies a = 2$$

The vertices are $(2, 0)$ and $(-2, 0)$.

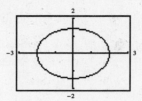

21. $3x^2 + y^2 - 12 = 0$

$$3x^2 + y^2 = 12$$

$$\frac{3x^2}{12} + \frac{y^2}{12} = 1$$

$$\frac{x^2}{4} + \frac{y^2}{12} = 1 \implies a = \sqrt{12} = 2\sqrt{3}$$

The vertices are $\left(0, 2\sqrt{3}\right)$ and $\left(0, -2\sqrt{3}\right)$

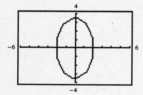

23. Vertices: $(-4, 0), (4, 0) \implies a = 4$

Major axis horizontal

Co-vertices: $(0, -3), (0, 3) \implies b = 3$

Minor axis vertical

$$\frac{x^2}{a^2} + \frac{y^2}{b^2} = 1$$

$$\frac{x^2}{4^2} + \frac{y^2}{3^2} = 1$$

$$\frac{x^2}{16} + \frac{y^2}{9} = 1$$

25. Vertices: $(-2, 0), (2, 0) \implies a = 2$

Major axis horizontal

Co-vertices: $(0, -1), (0, 1) \implies b = 1$

Minor axis vertical

$$\frac{x^2}{a^2} + \frac{y^2}{b^2} = 1$$

$$\frac{x^2}{2^2} + \frac{y^2}{1^2} = 1$$

$$\frac{x^2}{4} + \frac{y^2}{1} = 1$$

27. Vertices: $(0, -4), (0, 4) \implies a = 4$

Major axis vertical

Co-vertices: $(-3, 0), (3, 0) \implies b = 3$

Minor axis horizontal

$$\frac{x^2}{b^2} + \frac{y^2}{a^2} = 1$$

$$\frac{x^2}{3^2} + \frac{y^2}{4^2} = 1$$

$$\frac{x^2}{9} + \frac{y^2}{16} = 1$$

29. Vertices: $(0, -2), (0, 2) \implies a = 2$

Major axis vertical

Co-vertices: $(-1, 0), (1, 0) \implies b = 1$

Minor axis horizontal

$$\frac{x^2}{b^2} + \frac{y^2}{a^2} = 1$$

$$\frac{x^2}{1^2} + \frac{y^2}{2^2} = 1$$

$$\frac{x^2}{1} + \frac{y^2}{4} = 1$$

31. Major axis vertical length $10 \implies a = \frac{10}{2} = 5$

Minor axis horizontal with length $6 \implies b = \frac{6}{2} = 3$

$$\frac{x^2}{b^2} + \frac{y^2}{a^2} = 1$$

$$\frac{x^2}{3^2} + \frac{y^2}{5^2} = 1$$

$$\frac{x^2}{9} + \frac{y^2}{25} = 1$$

33. Major axis horizontal with length $20 \implies a = \frac{20}{2} = 10$

Minor axis vertical with length $12 \implies b = \frac{12}{2} = 6$

$$\frac{x^2}{a^2} + \frac{y^2}{b^2} = 1$$

$$\frac{x^2}{10^2} + \frac{y^2}{6^2} = 1$$

$$\frac{x^2}{100} + \frac{y^2}{36} = 1$$

35. $\dfrac{(x + 5)^2}{16} + y^2 = 1$

Center: $(-5, 0)$

$-5 - 4 = -9$ and $-5 + 4 = -1 \implies$ Vertices: $(-9, 0)$ and $(-1, 0)$

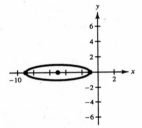

37. $\dfrac{(x - 1)^2}{9} + \dfrac{(y - 5)^2}{25} = 1$

Center: $(1, 5)$

$5 - 5 = 0$ and $5 + 5 = 10 \implies$ Vertices: $(1, 0)$ and $(1, 10)$

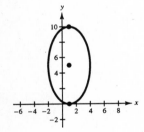

39.
$$9x^2 + 4y^2 + 36x - 24 + 36 = 0$$
$$9x^2 + 36x + 4y^2 - 24y = -36$$
$$9(x^2 + 4x) + 4(y^2 - 6y) = -36$$
$$9(x^2 + 4x + 4) + 4(y^2 - 6y + 9) = -36 + 9(4) + 4(9)$$
$$9(x + 2)^2 + 4(y - 3)^2 = 36$$
$$\frac{9(x + 2)^2}{36} + \frac{4(y - 3)^2}{36} = \frac{36}{36}$$
$$\frac{(x + 2)^2}{4} + \frac{(y - 3)^2}{9} = 1$$

Center: $(-2, 3)$

$3 - 3 = 0$ and $3 + 3 = 6 \implies$ Vertices: $(-2, 0)$ and $(-2, 6)$

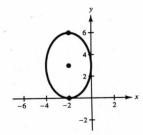

41. Center: $(0, 0)$, major axis vertical

Vertices: $(0, 2)$ and $(0, -2) \Longrightarrow a = 2$

Co-vertices: $(1, 0)$ and $(-1, 0) \Longrightarrow b = 1$

Equation: $\dfrac{x^2}{1} + \dfrac{y^2}{4} = 1$

43. Center: $(4, 0)$, major axis vertical

Vertices: $(4, 4)$ and $(4, -4) \Longrightarrow a = 4$

Co-vertices: $(1, 0)$ and $(7, 0) \Longrightarrow b = |4 - 1| = 3$

Equation: $\dfrac{(x - 4)^2}{9} + \dfrac{y^2}{16} = 1$

45. Graph (c)

47. Graph (a)

49. Graph (b)

51. $x^2 - y^2 = 9$

$\dfrac{x^2}{9} - \dfrac{y^2}{9} = \dfrac{9}{9}$

$\dfrac{x^2}{3^2} - \dfrac{y^2}{3^2} = 1 \Longrightarrow a = 3, b = 3$

Vertices: $(-3, 0), (3, 0)$

Asymptotes: $y = -\frac{3}{3}x$ or $y = -x$, $y = \frac{3}{3}x$ or $y = x$

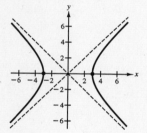

53. $y^2 - x^2 = 9$

$\dfrac{y^2}{9} - \dfrac{x^2}{9} = \dfrac{9}{9}$

$\dfrac{y^2}{3^2} - \dfrac{x^2}{3^2} = 1 \Longrightarrow a = 3, b = 3$

Vertices: $(0, -3), (0, 3)$

Asymptotes: $y = -\frac{3}{3}x$ or $y = -x$, $y = \frac{3}{3}x$ or $y = x$

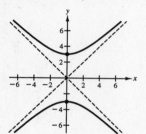

55. $\dfrac{x^2}{9} - \dfrac{y^2}{25} = 1$

$\dfrac{x^2}{3^2} - \dfrac{y^2}{5^2} = 1 \Longrightarrow a = 3, b = 5$

Vertices: $(-3, 0), (3, 0)$

Asymptotes: $y = -\frac{5}{3}x$, $y = \frac{5}{3}x$

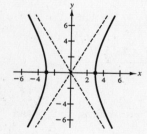

57. $\dfrac{y^2}{9} - \dfrac{x^2}{25} = 1$

$\dfrac{y^2}{3^2} - \dfrac{x^2}{5^2} = 1 \Longrightarrow a = 3, b = 5$

Vertices: $(0, -3), (0, 3)$

Asymptotes: $y = -\frac{3}{5}x$, $y = \frac{3}{5}x$

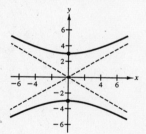

59. $\dfrac{x^2}{1} - \dfrac{y^2}{(9/4)} = 1$

$\dfrac{x^2}{1^2} - \dfrac{y^2}{(3/2)^2} = 1 \Rightarrow a = 1, b = \dfrac{3}{2}$

Vertices: $(-1, 0), (1, 0)$

Asymptotes: $y = \dfrac{-3/2}{1}x$ or $y = -\dfrac{3}{2}x,\ y = \dfrac{3/2}{1}x$

$\qquad\qquad$ or $y = \dfrac{3}{2}x$

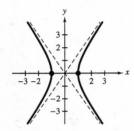

61. $4y^2 - x^2 + 16 = 0$

$\qquad 4y^2 - x^2 = -16$

$\qquad x^2 - 4y^2 = 16$

$\qquad \dfrac{x^2}{16} - \dfrac{4y^2}{16} = \dfrac{16}{16}$

$\qquad \dfrac{x^2}{16} - \dfrac{y^2}{4} = 1$

$\qquad \dfrac{x^2}{4^2} - \dfrac{y^2}{2^2} = 1 \Rightarrow a = 4, b = 2$

Vertices: $(-4, 0), (4, 0)$

Asymptotes: $y = -\frac{2}{4}x$ or $y = -\frac{1}{2}x,\ y = \frac{2}{4}x$ or $y = \frac{1}{2}x$

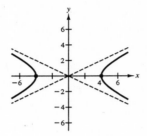

63. Vertices: $(-4, 0), (4, 0) \Rightarrow a = 4,$

Transverse axis is horizontal

Asymptotes: $y = 2x,$

$\qquad y = -2x \Rightarrow \dfrac{b}{a} = 2 \Rightarrow \dfrac{b}{4} = 2 \Rightarrow b = 8$

$\dfrac{x^2}{a^2} - \dfrac{y^2}{b^2} = 1$

$\dfrac{x^2}{4^2} - \dfrac{y^2}{8^2} = 1$

$\dfrac{x^2}{16} - \dfrac{y^2}{64} = 1$

65. Vertices: $(0, -4), (0, 4) \Rightarrow a = 4,$

Transverse axis is vertical

Asymptotes: $y = \dfrac{1}{2}x,$

$\qquad y = -\dfrac{1}{2}x \Rightarrow \dfrac{a}{b} = \dfrac{1}{2} \Rightarrow \dfrac{4}{b} = \dfrac{1}{2} \Rightarrow b = 8$

$\dfrac{y^2}{a^2} - \dfrac{x^2}{b^2} = 1$

$\dfrac{y^2}{4^2} - \dfrac{x^2}{8^2} = 1$

$\dfrac{y^2}{16} - \dfrac{x^2}{64} = 1$

67. Vertices: $(-9, 0), (9, 0) \Rightarrow a = 9,$

Transverse axis is horizontal

Asymptotes: $y = \dfrac{2}{3}x,$

$\qquad y = -\dfrac{2}{3}x \Rightarrow \dfrac{b}{a} = \dfrac{2}{3} \Rightarrow \dfrac{b}{9} = \dfrac{2}{3} \Rightarrow b = 6$

$\dfrac{x^2}{a^2} - \dfrac{y^2}{b^2} = 1$

$\dfrac{x^2}{9^2} - \dfrac{y^2}{6^2} = 1$

$\dfrac{x^2}{81} - \dfrac{y^2}{36} = 1$

69. Vertices: $(0, -1)$ and $(0, 1) \Rightarrow a = 1,$

Transverse axis is vertical

Asymptotes: $y = -2x$ and

$\qquad y = 2x \Rightarrow \dfrac{a}{b} = 2 \Rightarrow \dfrac{1}{b} = 2 \Rightarrow b = \dfrac{1}{2}$

$\dfrac{y^2}{a^2} - \dfrac{x^2}{b^2} = 1$

$\dfrac{y^2}{1^2} - \dfrac{x^2}{(1/2)^2} = 1$

$\dfrac{y^2}{1} - \dfrac{x^2}{1/4} = 1$

71. $(y + 4)^2 - (x - 3)^2 = 25$

$$\frac{(y + 4)^2}{25} - \frac{(x - 3)^2}{25} = 1$$

Center: $(3, -4)$

Vertices: $(3, -9)$ and $(3, 1)$

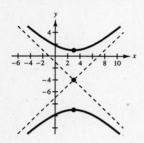

73. $\dfrac{(x - 1)^2}{4} - \dfrac{(y + 2)^2}{1} = 1$

Center: $(1, -2)$

Vertices: $(3, -2)$ and $(-1, -2)$

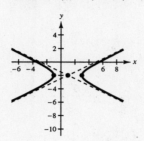

75.
$$9x^2 - y^2 - 36x - 6y + 18 = 0$$
$$9x^2 - 36x - y^2 - 6y = -18$$
$$9(x^2 - 4x) - (y^2 + 6y) = -18$$
$$9(x^2 - 4x + 4) - (y^2 + 6y + 9) = -18 + 9(4) - 1(9)$$
$$9(x - 2)^2 - (y + 3)^2 = 9$$
$$\frac{9(x - 2)^2}{9} - \frac{(y + 3)^2}{9} = \frac{9}{9}$$
$$\frac{(x - 2)^2}{1} - \frac{(y + 3)^2}{9} = 1$$

Center: $(2, -3)$

Vertices: $(3, -3)$ and $(1, -3)$

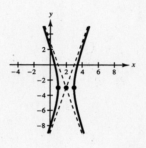

77. Center: $(0, 0)$, Transverse axis is vertical

Vertices: $(0, 3)$ and $(0, -3) \Rightarrow a = 3$

$$\frac{y^2}{3^2} - \frac{x^2}{b^2} = 1$$

$(-2, 5)$ on graph $\Rightarrow \dfrac{5^2}{3^2} - \dfrac{(-2)^2}{b^2} = 1$

$$\frac{25}{9} - \frac{4}{b^2} = 1$$

$$25b^2 - 36 = 9b^2$$

$$16b^2 = 36$$

$$b^2 = \frac{36}{16} = \frac{9}{4}$$

$$\frac{y^2}{9} - \frac{x^2}{9/4} = 1$$

79. Center: $(3, 2)$, Transverse axis is horizontal

Vertices: $(1, 2)$ and $(5, 2) \Rightarrow a = 2$

$$\frac{(x - 3)^2}{2^2} - \frac{(y - 2)^2}{b^2} = 1$$

$(0, 0)$ on graph $\Rightarrow \dfrac{(0 - 3)^2}{2^2} - \dfrac{(0 - 2)^2}{b^2} = 1$

$$\frac{9}{4} - \frac{4}{b^2} = 1$$

$$9b^2 - 16 = 4b^2$$

$$5b^2 = 16$$

$$b^2 = \frac{16}{5}$$

$$\frac{(x - 3)^2}{4} - \frac{(y - 2)^2}{16/5} = 1$$

81. $\dfrac{x^2}{324} + \dfrac{y^2}{196} = 1$

$\dfrac{x^2}{18^2} + \dfrac{y^2}{14^2} = 1$

Vertices: $(18, 0)$ and $(-18, 0)$

Co-vertices: $(0, 14)$ and $(0, -14)$

The distance between the vertices is the longest distance across the pool. This longest distance is 36 feet.

The distance between the co-vertices is the shortest distance across the pool. This shortest distance is 28 feet.

83. The maximum width is 170 yards, so each co-vertex is 85 yards from the center. The maximum length is 200 yards, so each vertex is 100 yards from the center. The center is $(0, 85)$.

$\dfrac{x^2}{85^2} + \dfrac{(y - 85)^2}{100^2} = 1$

$\dfrac{x^2}{7225} + \dfrac{(y - 85)^2}{10{,}000} = 1$

85. A circle is an ellipse in which the major axis and the minor axis have the same length. Both circles and ellipses have foci; however, in a circle, both foci are at the same point, the center of the circle.

87. The difference of the distances between each point on the hyperbola and the two foci is a positive constant.

89. This is the equation of the left half of the hyperbola.

91. $\begin{aligned} d &= \sqrt{(x_2 - x_1)^2 + (y_2 - y_1)^2} \\ &= \sqrt{(-1 - 5)^2 + (4 - 2)^2} \\ &= \sqrt{(-6)^2 + 2^2} \\ &= \sqrt{36 + 4} \\ &= \sqrt{40} \\ &= \sqrt{4(10)} \\ &= 2\sqrt{10} \end{aligned}$

93.

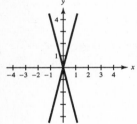

95.

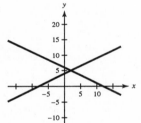

97. $\begin{aligned} c^2 &= a^2 - b^2 \\ c^2 &= 25^2 - 7^2 \\ c^2 &= 625 - 49 \\ c^2 &= 576 \\ c &= \sqrt{576} \\ c &= 24 \end{aligned}$

99. $\begin{aligned} c^2 &= a^2 - b^2 \\ 12^2 &= a^2 - 5^2 \\ 144 &= a^2 - 25 \\ 169 &= a^2 \\ \sqrt{169} &= a \\ 13 &= a \end{aligned}$

101. *Verbal Model:* $\boxed{\begin{array}{c}\text{Distance of}\\\text{first rider}\end{array}}$ + $\boxed{\begin{array}{c}\text{Distance of}\\\text{second rider}\end{array}}$ = 55

Labels: Time = t (hours)

Distance of first rider = $10t$ (miles)

Distance of second rider = $12t$ (miles)

Equation: $10t + 12t = 55$

$$22t = 55$$

$$t = \frac{55}{22}$$

$$t = \frac{5}{2}$$

The two riders will be 55 miles apart in 2.5 hours.

Section 8.7   Parabolas

1. Graph (b)

3. Graph (e)

5. Graph (f)

7. $x^2 = 4py$

$y = \frac{1}{2}x^2 \Longrightarrow x^2 = 2y$

$4p = 2 \Longrightarrow p = \frac{1}{2}$

Vertex: $(0, 0)$

Focus: $\left(0, \frac{1}{2}\right)$

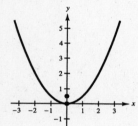

9. $y^2 = 4px$

$y^2 = -6x$

$4p = -6 \Longrightarrow p = -\frac{3}{2}$

Vertex: $(0, 0)$

Focus: $\left(-\frac{3}{2}, 0\right)$

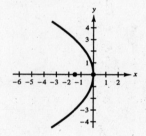

11. $x^2 = 4py$

$x^2 + 8y = 0$

$x^2 = -8y$

$4p = -8 \Longrightarrow p = -2$

Vertex: $(0, 0)$

Focus: $(0, -2)$

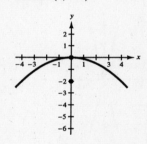

13. $(x - h)^2 = 4p(y - k)$

$(x - 1)^2 + 8(y + 2) = 0$

$(x - 1)^2 = -8(y + 2)$

$4p = -8 \Longrightarrow$

$p = -2$

Vertex: $(1, -2)$

Focus: $(1, -4)$

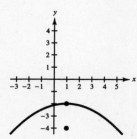

15. $(y - k)^2 = 4p(x - h)$

$\left(y + \frac{1}{2}\right)^2 = 2(x - 5)$

$4p = 2 \Longrightarrow p = \frac{1}{2}$

Vertex: $\left(5, -\frac{1}{2}\right)$

Focus: $\left(\frac{11}{2}, -\frac{1}{2}\right)$

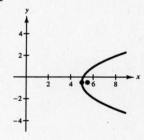

17. $(x - h)^2 = 4p(y - k)$

$$y = \tfrac{1}{4}(x^2 - 2x + 5)$$

$$4y = (x^2 - 2x) + 5$$

$$4y - 5 = (x^2 - 2x)$$

$$4y - 5 + 1 = (x^2 - 2x + 1)$$

$$4y - 4 = (x - 1)^2$$

$$4(y - 1) = (x - 1)^2$$

$$4p = 4 \Longrightarrow p = 1$$

Vertex: $(1, 1)$

Focus: $(1, 2)$

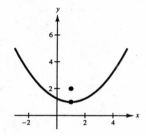

19. $(y - k)^2 = 4p(x - h)$

$$y^2 + 6y + 8x + 25 = 0$$

$$y^2 + 6y = -8x - 25$$

$$y^2 + 6y + 9 = -8x - 25 + 9$$

$$(y + 3)^2 = -8x - 16$$

$$(y + 3)^2 = -8(x + 2)$$

$$4p = -8x \Longrightarrow p = -2$$

Vertex: $(-2, -3)$

Focus: $(-4, -3)$

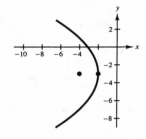

21. $(x - h)^2 = 4p(y - k)$

$$y = -\tfrac{1}{6}(x^2 + 4x - 2)$$

$$-6y = (x^2 + 4x) - 2$$

$$-6y + 2 = (x^2 + 4x)$$

$$-6y + 2 + 4 = (x^2 + 4x + 4)$$

$$-6y + 6 = (x + 2)^2$$

$$-6(y - 1) = (x + 2)^2$$

$$4p = -6 \Longrightarrow p = -\tfrac{3}{2}$$

Vertex: $(-2, 1)$

Focus: $\left(-2, -\tfrac{1}{2}\right)$

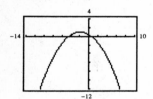

23. $(y - k)^2 = 4p(x - h)$

$$y^2 + x + y = 0$$

$$y^2 + y = -x$$

$$y^2 + y + \tfrac{1}{4} = -x + \tfrac{1}{4}$$

$$\left(y + \tfrac{1}{2}\right)^2 = -1\left(x - \tfrac{1}{4}\right)$$

$$4p = -1 \Longrightarrow p = -\tfrac{1}{4}$$

Vertex: $\left(\tfrac{1}{4}, -\tfrac{1}{2}\right)$

Focus: $\left(0, -\tfrac{1}{2}\right)$

Note: To graph this parabola, solve for y:

$$y = -\tfrac{1}{2} + \sqrt{-x + \tfrac{1}{4}} \text{ and } y = -\tfrac{1}{2} - \sqrt{-x + \tfrac{1}{4}}$$

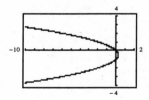

25. The graph is the top half of the parabola.

$$y^2 = -4x$$

$$y = \sqrt{-4x}$$

$$y = 2\sqrt{-x}$$

27. The graph is the bottom half of the parabola.

$$y^2 = 16x$$

$$y = -\sqrt{16x}$$

$$y = -4\sqrt{x}$$

29. Vertical axis $\Longrightarrow x^2 = 4py$

$(3, 6)$ on graph $\Longrightarrow 3^2 = 4p(6)$

$$9 = 24p$$

$$p = \tfrac{9}{24}$$

$$p = \tfrac{3}{8} \text{ and } 4p = \tfrac{3}{2}$$

$$x^2 = \tfrac{3}{2}y \text{ or } y = \tfrac{2}{3}x^2$$

31. Vertex: $(0, 0)$

Focus: $\left(0, -\tfrac{3}{2}\right) \Longrightarrow$ Vertical axis; $x^2 = 4py$

$$p = -\tfrac{3}{2} \Longrightarrow 4p = -6$$

$$x^2 = -6y$$

33. Vertex: $(0, 0)$

Focus: $(-2, 0) \Longrightarrow$ Horizontal axis; $y^2 = 4px$

$\quad p = -2 \Longrightarrow 4p = -8$

$y^2 = -8x$

35. Vertex: $(0, 0)$

Focus: $(0, 1) \Longrightarrow$ Vertical axis, $x^2 = 4py$

$\quad p = 1 \Longrightarrow 4p = 4$

$x^2 = 4y$

37. Vertex: $(0, 0)$

Focus: $(4, 0) \Longrightarrow$ Horizontal axis; $y^2 = 4px$

$\quad p = 4 \Longrightarrow 4p = 16$

$y^2 = 16x$

39. Horizontal axis $\Longrightarrow y^2 = 4px$

$\quad (4, 6)$ on graph $\Longrightarrow 6^2 = 4p(4)$

$\qquad\qquad\qquad 36 = 16p$

$\qquad\qquad\qquad p = \frac{36}{16}$

$\qquad\qquad\qquad p = \frac{9}{4}$ and $4p = 9$

$\qquad\qquad\qquad y^2 = 9x$

41.   Vertical axis $\Longrightarrow (x - h)^2 = 4p(y - k)$

Vertex: $(3, 1) \Longrightarrow (x - 3)^2 = 4p(y - 1)$

$(4, 0)$ on graph $\Longrightarrow (4 - 3)^2 = 4p(0 - 1)$

$\qquad\qquad\qquad 1^2 = 4p(-1)$

$\qquad\qquad\qquad 1 = -4p$

$\qquad\qquad\qquad -\frac{1}{4} = p$ and $4p = -1$

$(x - 3)^2 = -(y - 1)$

43.   Horizontal axis $\Longrightarrow (y - k)^2 = 4p(x - h)$

Vertex: $(-2, 0) \Longrightarrow (y - 0)^2 = 4p(x + 2)$

$\qquad\qquad\qquad y^2 = 4p(x + 2)$

$(0, 2)$ on graph $\Longrightarrow 2^2 = 4p(0 + 2)$

$\qquad\qquad\qquad 4 = 4p(2)$

$\qquad\qquad\qquad 4 = 8p$

$\qquad\qquad\qquad \frac{1}{2} = p$ and $4p = 2$

$y^2 = 2(x + 2)$

45. Vertex: $(3, 2)$ and Focus: $(1, 2)$

The focus is two units to the left of the vertex.

$p = -2$ and $4p = -8$

Horizontal axis $\Longrightarrow (y - k)^2 = 4p(x - h)$

$\qquad\qquad\qquad (y - 2)^2 = -8(x - 3)$

47. Vertex: $(0, 4)$ and Focus: $(0, 6)$

The focus is two units above the vertex.

$p = 2$ and $4p = 8$

Vertical axis $\Longrightarrow (x - h)^2 = 4p(y - k)$

$\qquad\qquad\qquad (x - 0)^2 = 8(y - 4)$

$\qquad\qquad\qquad x^2 = 8(y - 4)$

49. Horizontal axis $\Longrightarrow (y - k)^2 = 4p(x - h)$

Vertex: $(0, 2) \Longrightarrow (y - 2)^2 = 4p(x - 0)$

$\qquad\qquad\qquad (y - 2)^2 = 4px$

$(1, 3)$ on graph $\Longrightarrow (3 - 2)^2 = 4p(1)$

$\qquad\qquad\qquad 1^2 = 4p$

$\qquad\qquad\qquad 1 = 4p$

$\qquad\qquad\qquad \frac{1}{4} = p$ and $4p = 1$

$(y - 2)^2 = 1(x)$

$(y - 2)^2 = x$

51. (a) Vertex $(0, 0)$ and Vertical axis $\Rightarrow x^2 = 4py$

$(60, 20)$ on graph $\Rightarrow 60^2 = 4p(20)$

$$3600 = 80p$$

$$p = \frac{3600}{80}$$

$$p = 45 \text{ and } 4p = 180$$

$$x^2 = 180y \quad \text{or} \quad y = \frac{1}{180}x^2$$

(b) $x = 0 \Rightarrow y = \dfrac{0^2}{180} = \dfrac{0}{180} = 0$

$x = 20 \Rightarrow y = \dfrac{20^2}{180} = \dfrac{400}{180} = \dfrac{20}{9}$

$x = 40 \Rightarrow y = \dfrac{40^2}{180} = \dfrac{1600}{180} = \dfrac{80}{9}$

$x = 60 \Rightarrow y = \dfrac{60^2}{180} = \dfrac{3600}{180} = 20$

x	0	20	40	60
$y = \frac{1}{180}x^2$	0	$\frac{20}{9}$	$\frac{80}{9}$	20

53. $R = 375x - \frac{3}{2}x^2$

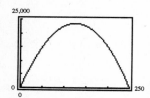

The maximum revenue occurs when $x = 125$.

55. All points on the parabola are equidistant from the directrix and the focus.

57. No. If the graph intersected the graph of the directrix, there would exist points on the parabola nearer to the directrix than the focus.

59. $(x + 6)^2 - 5 = x^2 + 12x + 36 - 5$

$= x^2 + 12x + 31$

61. $12 - (x - 8)^2 = 12 - (x^2 - 16x + 64)$

$= 12 - x^2 + 16x - 64$

$= -x^2 + 16x - 52$

63. $x^2 - 4x + 1 = (x^2 - 4x + 4) - 4 + 1$

$= (x - 2)^2 - 3$

65. $-x^2 + 6x + 5 = -(x^2 - 6x) + 5$

$= -(x^2 - 6x + 9) + 9 + 5$

$= -(x - 3)^2 + 14$

67. $y - y_1 = m(x - x_1)$

$y - 5 = \frac{5}{8}[x - (-2)]$

$y - 5 = \frac{5}{8}(x + 2)$

$y - 5 = \frac{5}{8}x + \frac{5}{4}$

$y = \frac{5}{8}x + \frac{5}{4} + 5$

$y = \frac{5}{8}x + \frac{25}{4}$

69. $\dfrac{90 + 74 + 82 + 90 + 2x}{6} = 85$

$$\dfrac{336 + 2x}{6} = 85$$

$$336 + 2x = 510$$

$$2x = 174$$

$$x = 87$$

The student needs to score 87 on the final exam.

Section 8.8   Nonlinear Systems of Equations

1.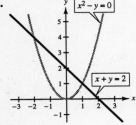

The solutions are $(-2, 4)$ and $(1, 1)$.

3.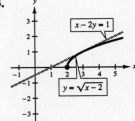

The solution is $(3, 1)$.

5.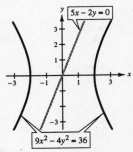

The system has no real solution.

7.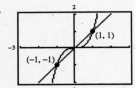

The solutions are $(-1, 1)$, $(0, 0)$, and $(1, 1)$.

9.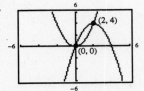

The solutions are $(0, 0)$ and $(2, 4)$.

11.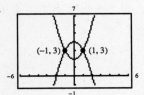

The solutions are $(-1, 3)$ and $(1, 3)$.

13.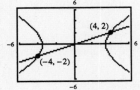

The solutions are $(4, 2)$ and $(-4, -2)$.

15.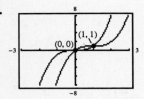

The solutions are $(0, 0)$ and $(1, 1)$.

17. $\begin{cases} y = 2x^2 \\ y = -2x + 12 \end{cases}$

$$2x^2 = -2x + 12$$
$$2x^2 + 2x - 12 = 0$$
$$x^2 + x - 6 = 0$$
$$(x + 3)(x - 2) = 0$$
$$x + 3 = 0 \Longrightarrow x = -3$$
$$x - 2 = 0 \Longrightarrow x = 2$$

$x = -3 \Longrightarrow -2(-3) + 12 = 18 \qquad (-3, 18)$

$x = 2 \Longrightarrow -2(2) + 12 = 8 \qquad\qquad (2, 8)$

19. $\begin{cases} x^2 + y = 9 \\ x - y = -3 \Longrightarrow y = x + 3 \end{cases}$

$$x^2 + (x + 3) = 9$$
$$x^2 + x - 6 = 0$$
$$(x + 3)(x - 2) = 0$$
$$x + 3 = 0 \Longrightarrow x = -3$$
$$x - 2 = 0 \Longrightarrow x = 2$$

$x = -3 \Longrightarrow y = -3 + 3 = 0 \qquad (-3, 0)$

$x = 2 \Longrightarrow y = 2 + 3 = 5 \qquad\quad (2, 5)$

21. $\begin{cases} x^2 + 2y = 6 \\ x - y = -4 \Longrightarrow x = y - 4 \end{cases}$

$$(y - 4)^2 + 2y = -6$$
$$y^2 - 8y + 16 + 2y = -6$$
$$y^2 - 6y + 22 = 0$$
$$y = \frac{-(-6) \pm \sqrt{36 - 4(1)(22)}}{2(1)}$$
$$= \frac{6 \pm \sqrt{-52}}{2} \qquad \text{Not real}$$

No real solution

23. $\begin{cases} x^2 + y^2 = 25 \\ 2x - y = -5 \Longrightarrow y = 2x + 5 \end{cases}$

$$x^2 + (2x + 5)^2 = 25$$
$$x^2 + 4x^2 + 20x + 25 = 25$$
$$5x^2 + 20x = 0$$
$$5x(x + 4) = 0$$
$$5x = 0 \Longrightarrow x = 0$$
$$x + 4 = 0 \Longrightarrow x = -4$$

$x = 0 \Longrightarrow y = 2(0) + 5 = 5 \qquad\qquad (0, 5)$

$x = -4 \Longrightarrow y = 2(-4) + 5 = -3 \qquad (-4, -3)$

25. $\begin{cases} x^2 + y^2 = 64 \\ -3x + y = 8 \Longrightarrow y = 3x + 8 \end{cases}$

$$x^2 + (3x + 8)^2 = 64$$
$$x^2 + 9x^2 + 48x + 64 = 64$$
$$10x^2 + 48x = 0$$
$$2x(5x + 24) = 0$$
$$2x = 0 \Longrightarrow x = 0$$
$$5x + 24 = 0 \Longrightarrow 5x = -24 \Longrightarrow x = -\tfrac{24}{5}$$

$x = 0 \Longrightarrow y = 3(0) + 8 = 8 \qquad\qquad\qquad (0, 8)$

$x = -\tfrac{24}{5} \Longrightarrow y = 3\left(-\tfrac{24}{5}\right) + 8 = -\tfrac{32}{5} \qquad \left(-\tfrac{24}{5}, -\tfrac{32}{5}\right)$

27. $\begin{cases} 4x + y^2 = 2 \\ 2x - y = -11 \Longrightarrow -y = -2x - 11 \Longrightarrow y = 2x + 11 \end{cases}$

$$4x + (2x + 11)^2 = 2$$
$$4x + 4x^2 + 44x + 121 = 2$$
$$4x^2 + 48x + 119 = 0$$
$$(2x + 7)(2x + 17) = 0$$

$2x + 7 = 0 \Longrightarrow 2x = -7 \Longrightarrow x = -\tfrac{7}{2} \quad \text{and} \quad y = 2\left(-\tfrac{7}{2}\right)$

$2x + 17 = 0 \Longrightarrow x = -\tfrac{17}{2} \quad \text{and} \quad y = 2\left(-\tfrac{17}{2}\right) + 11 = -6 + 11 = 4$

$\left(-\tfrac{7}{2}, 4\right) \quad \text{and} \quad \left(-\tfrac{17}{2}, -6\right)$

29. $\begin{cases} y = \sqrt{4 - x} \\ x + 3y = \qquad 6 \implies x = 6 - 3y \end{cases}$

$$y = \sqrt{4 - (6 - 3y)} = \sqrt{3y - 2}$$

$$y^2 = (\sqrt{3y - 2})^2 = 3y - 2$$

$$y^2 - 3y + 2 = 0$$

$$(y - 2)(y - 1) = 0$$

$$y - 2 = 0 \implies y = 2$$

$$y - 1 = 0 \implies y = 1$$

$$y = 2 \implies x = 6 - 3(2) = 0 \qquad (0, 2)$$

$$y = 1 \implies x = 6 - 3(1) = 3 \qquad (3, 1)$$

31. $\begin{cases} 16x^2 + 9y^2 = 144 \\ 4x + 3y = \quad 12 \implies y = \dfrac{12 - 4x}{3} \end{cases}$

$$16x^2 + 9\left(\frac{12 - 4x}{3}\right)^2 = 144$$

$$16x^2 + 144 - 96x + 16x^2 = 144$$

$$32x^2 - 96x = 0$$

$$32x(x - 3) = 0$$

$$32x = 0 \implies x = 0$$

$$x - 3 = 0 \implies x = 3$$

$$x = 0 \implies y = \frac{12 - 4(0)}{3} = 4 \qquad (0, 4)$$

$$x = 3 \implies y = \frac{12 - 4(3)}{3} = 0 \qquad (3, 0)$$

33. $\begin{cases} x^2 - y^2 = 9 \\ x^2 + y^2 = 1 \implies x^2 = 1 - y^2 \end{cases}$

$$(1 - y^2) - y^2 = 9$$

$$1 - 2y^2 = 9$$

$$-2y^2 = 8$$

$$y^2 = -4$$

$$y = \pm\sqrt{-4} \qquad \text{Not real}$$

No real solution

35. $\begin{cases} x^2 + 2y = 1 \implies -x^2 - 2y = -1 \\ x^2 + y^2 = 4 \implies \underline{\quad x^2 + y^2 = \quad 4} \end{cases}$

$$\qquad\qquad\qquad\qquad y^2 - 2y = 3$$

$$y^2 - 2y - 3 = 0$$

$$(y - 3)(y + 1) = 0$$

$$y - 3 = 0 \implies y = 3$$

$$y + 1 = 0 \implies y = -1$$

$$\begin{array}{ll} x^2 + 2(3) = 1 & x^2 + 2(-1) = 1 \\ x^2 + 6 = 1 & x^2 - 2 = 1 \\ x^2 = -5 & x^2 = 3 \\ x = \pm\sqrt{-5} \ \text{Not real} & x = \pm\sqrt{3} \end{array}$$

$$(\sqrt{3}, -1) \text{ and } (-\sqrt{3}, -1)$$

37. $\begin{cases} -x + y^2 = \quad 10 \\ \underline{\quad x^2 - y^2 = -8} \\ \quad x^2 - x = 2 \end{cases}$

$$x^2 - x - 2 = 0$$

$$(x - 2)(x + 1) = 0$$

$$x - 2 = 0 \implies x = 2$$

$$x + 1 = 0 \implies x = -1$$

$$\begin{array}{ll} -2 + y^2 = 10 & -(-1) + y^2 = 10 \\ y^2 = 12 & y^2 = 9 \\ y = \pm\sqrt{12} & y = \pm\sqrt{9} \\ \quad = \pm 2\sqrt{3} & \quad = \pm 3 \end{array}$$

$$(2, 2\sqrt{3}), (2, -2\sqrt{3}) \qquad (-1, 3), (-1, -3)$$

39. $\begin{cases} x^2 + y^2 = 7 \\ \underline{x^2 - y^2 = 1} \\ \quad 2x^2 = 8 \end{cases}$

$$\begin{array}{ll} 2x^2 = 8 & x^2 + y^2 = 7 \\ x^2 = 4 & 4 + y^2 = 7 \\ x = \pm\sqrt{4} & y^2 = 3 \\ x = \pm 2 & y = \pm\sqrt{3} \end{array}$$

$$(2, \sqrt{3}), (2, -\sqrt{3}), (-2, \sqrt{3}), (-2, -\sqrt{3})$$

41.
$$\begin{cases} \dfrac{x^2}{4} + y^2 = 1 \implies \dfrac{x^2}{4} + y^2 = 1 \\ x^2 + \dfrac{y^2}{4} = 1 \implies -4x^2 - y^2 = -4 \end{cases}$$

$$-\dfrac{15}{4}x^2 = -3 \qquad\qquad x^2 + \dfrac{y^2}{4} = 1$$

$$15x^2 = 12 \qquad\qquad \dfrac{4}{5} + \dfrac{y^2}{4} = 1$$

$$x^2 = \dfrac{12}{15} \qquad\qquad \dfrac{y^2}{4} = \dfrac{1}{5}$$

$$= \dfrac{4}{5} \qquad\qquad y^2 = \dfrac{4}{5}$$

$$x = \pm\sqrt{\dfrac{4}{5}} = \pm\dfrac{2\sqrt{5}}{5} \qquad\qquad y = \pm\sqrt{\dfrac{4}{5}} = \pm\dfrac{2\sqrt{5}}{5}$$

$$\left(\dfrac{2\sqrt{5}}{5}, \dfrac{2\sqrt{5}}{5}\right), \left(\dfrac{2\sqrt{5}}{5}, -\dfrac{2\sqrt{5}}{5}\right), \left(-\dfrac{2\sqrt{5}}{5}, \dfrac{2\sqrt{5}}{5}\right), \left(-\dfrac{2\sqrt{5}}{5}, -\dfrac{2\sqrt{5}}{5}\right)$$

43.
$$\begin{cases} y^2 - x^2 = 10 \implies -x^2 + y^2 = 10 \\ x^2 + y^2 = 16 \implies x^2 + y^2 = 16 \end{cases}$$

$$2y^2 = 26 \qquad\qquad x^2 + \left(\sqrt{13}\right)^2 = 16$$

$$y^2 = 13 \qquad\qquad x^2 + 13 = 16$$

$$y = \pm\sqrt{13} \qquad\qquad x^2 = 3$$

$$x = \pm\sqrt{3}$$

$$\left(\pm\sqrt{3}, \pm\sqrt{13}\right)$$

45. The slope of the line through $(0, 10)$ and $(5, 0)$ is $m = \dfrac{0 - 10}{5 - 0} = \dfrac{-10}{5} = -2$. The equation of the line with slope $m = -2$ and y-intercept $(0, 10)$ is $y = -2x + 10$.

$$\begin{cases} y = -2x + 10 \\ \dfrac{x^2}{9} - \dfrac{y^2}{16} = 1 \end{cases}$$

By substitution,

$$\dfrac{x^2}{9} - \dfrac{(-2x + 10)^2}{16} = 1$$

$$\dfrac{x^2}{9} - \dfrac{4x^2 - 40x + 100}{16} = 1$$

$$16x^2 - 9(4x^2 - 40x + 100) = 1(9)(16)$$

$$16x^2 - 36x^2 + 360x - 900 = 144$$

$$-20x^2 + 360x - 1044 = 0$$

$$\dfrac{-20x^2}{-4} + \dfrac{360}{-4} + \dfrac{-1044}{-4} = 0$$

$$5x^2 + 90x + 261 = 0$$

By the quadratic formula: $x = \dfrac{90 \pm \sqrt{(-90)^2 - 4(5)(261)}}{2(5)}$

$$x = \dfrac{90 \pm \sqrt{2880}}{10}$$

$x \approx 14.367$ and $y \approx -2(14.367) + 10 \approx -18.734$

$x \approx 3.633$ and $y \approx -2(3.633) + 10 \approx 2.733$

Discarding the solution in the fourth quadrant, $(3.633, 2.733)$ is the point on the mirror at which light from the point $(0, 10)$ will reflect to the focus.

47. "Equation of Clark Street" line through $(-2, -1)$ and $(5, 0)$

$$m = \frac{0 - (-1)}{5 - (-2)} = \frac{1}{7}$$

$$y - 0 = \frac{1}{7}(x - 5)$$

$$y = \frac{1}{7}x - \frac{5}{7}$$

Equation of circle: $x^2 + y^2 = 1^2$

$$x^2 + y^2 = 1$$

$$\begin{cases} x^2 + y^2 = 1 \\ \qquad y = \frac{1}{7}x - \frac{5}{7} \end{cases}$$

$$x^2 + \left(\frac{1}{7}x - \frac{5}{7}\right)^2 = 1$$

$$x^2 + \frac{1}{49}x^2 - \frac{10}{49}x + \frac{25}{49} = 1$$

$$49x^2 + x^2 - 10x + 25 = 49$$

$$50x^2 - 10x - 24 = 0$$

$$(5x - 4)(5x + 3) = 0$$

$$5x - 4 = 0 \Longrightarrow x = \frac{4}{5}$$

$$5x + 3 = 0 \Longrightarrow x = -\frac{3}{5}$$

$$y = \frac{1}{7}\left(\frac{4}{5}\right) - \frac{5}{7}$$

$$= \frac{4}{35} - \frac{25}{35} = -\frac{21}{35} = -\frac{3}{5}$$

$$y = \frac{1}{7}\left(-\frac{3}{5}\right) - \frac{5}{7}$$

$$= -\frac{3}{35} - \frac{25}{35} = -\frac{28}{35} = -\frac{4}{5}$$

$$\left(\frac{4}{5}, -\frac{3}{5}\right)\left(-\frac{3}{5}, -\frac{4}{5}\right)$$

Residents who live between $\left(\frac{4}{5}, -\frac{3}{5}\right)$ and $\left(-\frac{3}{5}, -\frac{4}{5}\right)$ on Clark Street are *not* eligible to ride the school bus.

49. Answers will vary.

1. Solve one of the equations for one variable in terms of the other.

2. Substitute the expression found in Step 1 into the other equation to obtain an equation of one variable.

3. Solve the equation obtained in Step 2.

4. Back-substitution the solution from Step 3 into the expression obtained in Step 1 to find the value of the other variable.

5. Check the solution to see that it satisfies each of the original equations.

51. $\begin{cases} 3x + 5y = 9 \implies 9x + 15y = 27 \\ 2x - 3y = -13 \implies 10x - 15y = -65 \end{cases}$

$$19x = -38$$

$$x = -2$$

$3(-2) + 5y = 9$

$-6 + 5y = 9$

$5y = 15$

$y = 3 \qquad (-2, 3)$

53. $4600(0.068) + x(0.09) = (4600 + x)(0.08)$

$$312.80 + 0.09x = 368 + 0.08x$$

$$312.80 + 0.01x = 368$$

$$0.01x = 55.20$$

$$x = 5520$$

An additional $5520 must be invested.

Review Exercises for Chapter 8

1. (a) $(f + g)(x) = f(x) + g(x) = x + 2 + x - 2 = 2x$ Domain $= \{x: x \text{ is a real number}\}$

(b) $(f - g)(x) = f(x) - g(x) = x + 2 - (x - 2) = x + 2 - x + 2 = 4$ Domain $= \{x: x \text{ is a real number}\}$

(c) $(fg)(x) = f(x)g(x) = (x + 2)(x - 2) = x^2 - 4$ Domain $= \{x: x \text{ is a real number}\}$

(d) $(f/g)(x) = f(x)/g(x) = \dfrac{x + 2}{x - 2}$ Domain $= \{x: x \neq 2\}$

3. $f(x) = \dfrac{2}{x - 1}, \; g(x) = x$

(a) $(f + g)(x) = \dfrac{2}{x - 1} + x = \dfrac{2 + x^2 - x}{x - 1}$ Domain $= \{x: x \neq 1\}$

(b) $(f - g)(x) = \dfrac{2}{x - 1} - x = \dfrac{2 - x^2 + x}{x - 1}$ Domain $= \{x: x \neq 1\}$

(c) $(fg)(x) = \left(\dfrac{2}{x - 1}\right)(x) = \dfrac{2x}{x - 1}$ Domain $= \{x: x \neq 1\}$

(d) $\left(\dfrac{f}{g}\right)(x) = \dfrac{\left(\dfrac{2}{x - 1}\right)}{x} = \dfrac{2}{x(x - 1)}$ Domain $= \{x: x \neq 0, 1\}$

5. $f(x) = x^2, \; g(x) = 4x - 5$

(a) $(f + g)(x) = x^2 + 4x - 5$ or   (a) $(f + g)(-5) = f(-5) + g(-5)$

$(f + g)(-5) = (-5)^2 + 4(-5) - 5$ $ = (-5)^2 + [4(-5) - 5]$

$ = 25 - 20 - 5 = 0$ $ = 25 + (-25) = 0$

(b) $(f - g)(x) = x^2 - (4x - 5)$ or   (b) $(f - g)(0) = f(0) - g(0)$

$ = x^2 - 4x + 5$ $ = 0^2 - [4(0) - 5]$

$(f - g)(0) = 0^2 - 4(0) + 5 = 5$ $ = 0 - (-5) = 5$

(c) $(fg)(x) = x^2(4x - 5)$ or   (c) $(fg)(2) = f(2)g(2)$

$ = 4x^3 - 5x^2$ $ = 2^2[4(2) - 5]$

$(fg)(2) = 4(2)^3 - 5(2)^2$ $ = 4(3) = 12$

$ = 32 - 20 = 12$

(d) $\left(\dfrac{f}{g}\right)(x) = \dfrac{x^2}{4x - 5}$ or   (d) $\left(\dfrac{f}{g}\right)(1) = \dfrac{f(1)}{g(1)}$

$\left(\dfrac{f}{g}\right)(1) = \dfrac{1^2}{4(1) - 5} = \dfrac{1}{-1} = -1$ $\phantom{\left(\dfrac{f}{g}\right)(1) } = \dfrac{1^2}{4(1) - 5} = \dfrac{1}{-1} = -1$

7. $f(x) = \frac{2}{3}\sqrt{x}$, $g(x) = -x^2$

(a) $(f + g)(x) = \frac{2}{3}\sqrt{x} + (-x^2)$ or   (a) $(f + g)(1) = f(1) + g(1)$

$$(f + g)(1) = \frac{2}{3}\sqrt{1} + (-1^2)$$
$$= \frac{2}{3}\sqrt{1} + (-1^2)$$
$$= \frac{2}{3} - 1^2 = \frac{2}{3} - 1 = -\frac{1}{3}$$
$$= \frac{2}{3} + (-1) = -\frac{1}{3}$$

(b) $(f - g)(x) = \frac{2}{3}\sqrt{x} - (-x^2)$ or   (b) $(f - g)(9) = f(9) - g(9)$

$$= \frac{2}{3}\sqrt{x} + x^2$$
$$= \frac{2}{3}\sqrt{9} - (-9^2)$$
$$(f - g)(9) = \frac{2}{3}\sqrt{9} + 9^2 = 2 + 81 = 83$$
$$= 2 - (-81) = 83$$

(c) $(fg)(x) = \left(\frac{2}{3}\sqrt{x}\right)(-x^2)$ or   (c) $(fg)\left(\frac{1}{4}\right) = f\left(\frac{1}{4}\right)g\left(\frac{1}{4}\right)$

$$= -\frac{2}{3}x^2\sqrt{x} = -\frac{2}{3}x^{5/2}$$
$$= \left(\frac{2}{3}\sqrt{\frac{1}{4}}\right)\left[-\left(\frac{1}{4}\right)^2\right]$$
$$(fg)\left(\frac{1}{4}\right) = -\frac{2}{3}\left(\frac{1}{4}\right)^{5/2}$$
$$= \left[\frac{2}{3}\left(\frac{1}{2}\right)\right]\left(-\frac{1}{16}\right)$$
$$= -\frac{2}{3}\left(\frac{1}{32}\right) = -\frac{1}{48}$$
$$= \frac{1}{3}\left(-\frac{1}{16}\right) = -\frac{1}{48}$$

(d) $\left(\frac{f}{g}\right)(x) = \frac{(2/3)\sqrt{x}}{-x^2}$ or   (d) $\left(\frac{f}{g}\right)(2) = \frac{f(2)}{g(2)} = \frac{(2/3)\sqrt{2}}{-2^2}$

$$\left(\frac{f}{g}\right)(2) = \frac{(2/3)\sqrt{2}}{-2^2} = \frac{2\sqrt{2}}{3(-4)} = \frac{2\sqrt{2}}{-12} = -\frac{\sqrt{2}}{6}$$
$$= \frac{(2/3)\sqrt{2}}{-4} = -\frac{2}{12}\sqrt{2} = -\frac{\sqrt{2}}{6}$$

9. $f(x) = x + 2$, $g(x) = x^2$

(a) $(f \circ g)(x) = f(g(x))$
$$= f(x^2) = x^2 + 2$$

(b) $(g \circ f)(x) = g(f(x))$
$$= g(x + 2) = (x + 2)^2$$

(c) $(f \circ g)(2) = 2^2 + 2$
$$= 4 + 2 = 6$$

(d) $(g \circ f)(-1) = (-1 + 2)^2$
$$= 1^2 = 1$$

11. $f(x) = \sqrt{x + 1}$, $g(x) = x^2 - 1$

(a) $(f \circ g)(x) = f(g(x))$
$$= f(x^2 - 1)$$
$$= \sqrt{(x^2 - 1) + 1}$$
$$= \sqrt{x^2} = |x|$$

(b) $(g \circ f)(x) = g(f(x))$
$$= g\left(\sqrt{x + 1}\right)$$
$$= \left(\sqrt{x + 1}\right)^2 - 1$$
$$= x + 1 - 1 = x, \; x \geq -1$$

(c) $(f \circ g)(5) = |5| = 5$

(d) $(g \circ f)(-1) = -1$

13. $f(x) = \sqrt{x - 4}$, $g(x) = 2x$

(a) $(f \circ g)(x) = f(g(x))$
$$= f(2x) = \sqrt{2x - 4}$$

Domain: $[2, \infty)$

(b) $(g \circ f)(x) = g(f(x))$
$$= g\left(\sqrt{x - 4}\right) = 2\sqrt{x - 4}$$

Domain: $[4, \infty)$

15. $f(x) = \frac{1}{6}x \implies f^{-1}(x) = 6x$

17. $f(x) = x^3 \implies f^{-1}(x) = \sqrt[3]{x}$

19. $f(x) = \frac{1}{4}x$

$y = \frac{1}{4}x$

$x = \frac{1}{4}y$

$4x = y$

$f^{-1}(x) = 4x$

21. $h(x) = \sqrt{x}$

$y = \sqrt{x}$

$x = \sqrt{y}$

$x^2 = y \quad (x \geq 0)$

$h^{-1}(x) = x^2, \; x \geq 0$

23. $f(t) = |t + 3|$ No inverse

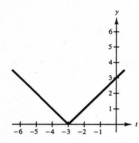

It is possible to find a horizontal line that intersects the graph of f at more than one point. Thus, the function is not one-to-one and it does *not* have an inverse.

25. Restrict $f(x) = 2(x - 4)^2$ to the domain $[4, \infty)$.

$y = 2(x - 4)^2, \; x \geq 4$

$x = 2(y - 4)^2, \; y \geq 4$

$\frac{x}{2} = (y - 4)^2, \; y \geq 4$

$\sqrt{\frac{x}{2}} = y - 4$

$\sqrt{\frac{x}{2}} + 4 = y$

$f^{-1}(x) = \sqrt{\frac{x}{2}} + 4 \; \text{ or } \; f^{-1}(x) = \frac{\sqrt{2x}}{2} + 4$

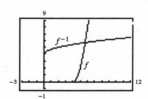

27.

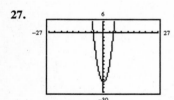

It is possible to find a horizontal line that intersects the graph of f at more than one point. Thus, the function is not one-to-one and it does *not* have an inverse.

29.

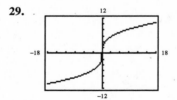

No horizontal line intersects the graph of f at more than one point. Thus, the function is one-to-one and it does have an inverse.

31. $y = k\sqrt[3]{x}$

$12 = k\sqrt[3]{8}$

$12 = k(2)$

$\frac{12}{2} = k$

$6 = k \implies y = 6\sqrt[3]{x}$

33. $T = krs^2$

$5000 = k(0.09)(1000)^2$

$5000 = k(90,000)$

$\frac{5000}{90,000} = k$

$\frac{5}{90} = k$

$\frac{1}{18} = k \implies T = \frac{1}{18}rs^2$

35. (a)

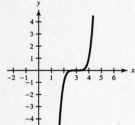

(b)

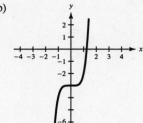

(c)

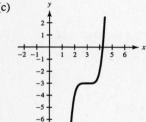

(d)

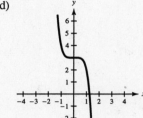

37. $f(x) = -(x - 2)^3$

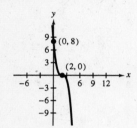

x-intercept:	$f(x) = 0$	y-intercept: $x = 0$
	$-(x - 2)^3 = 0$	$y = -(0 - 2)^3$
	$(x - 2)^3 = 0$	$y = -(-2)^3$
	$x - 2 = 0 \implies x = 2$	$y = 8$
	$(2, 0)$	$(0, 8)$

39. $g(x) = x^4 - x^3 - 2x^2$

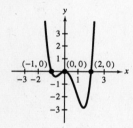

x-intercept:	$g(x) = 0$	y-intercept: $x = 0$
	$x^4 - x^3 - 2x^2 = 0$	$y = 0^4 - 0^3 - 2(0)^2$
	$x^2(x^2 - x - 2) = 0$	$y = 0$
	$x^2(x - 2)(x + 1) = 0$	$(0, 0)$
	$x^2 = 0 \implies x = 0$	
	$x - 2 = 0 \implies x = 2$	
	$x + 1 = 0 \implies x = -1$	
	$(0, 0)$, $(2, 0)$, and $(-1, 0)$	

41. $f(x) = x(x + 3)^2$

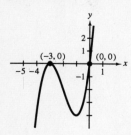

x-intercept:	$f(x) = 0$	y-intercept: $x = 0$
	$x(x + 3)^2 = 0$	$y = 0(0 + 3)^2$
	$x = 0$	$y = 0(9)$
	$x + 3 = 0 \implies x = -3$	$y = 0$
	$(0, 0)$ and $(-3, 0)$	$(0, 0)$

43. $f(x) = -x^2 + 6x + 9$

Leading coefficient, -1, *negative* $\Rightarrow$ graph falls to right

Degree, 2, *even* (*same* behavior to left) $\Rightarrow$ graph falls to left

45. $g(x) = \frac{3}{4}(x^4 + 3x^2 + 2)$

Leading coefficient, $\frac{3}{4}$, *positive* $\Rightarrow$ graph rises to right

Degree, 4, *even* (*same* behavior to left) $\Rightarrow$ graph rises to left

47. This conic is a hyperbola.   **49.** This conic is a circle.

51. This conic is a circle.   **53.** This conic is a parabola.

55. $x^2 + y^2 = r^2$

Radius: $r = 12$

$x^2 + y^2 = 12^2$

$x^2 + y^2 = 144$

57. $(x - h)^2 + (y - k)^2 = r^2$

Center: $(0, 0)$, Passes through $(-1, 3)$

Radius: $r = \sqrt{(0 - (-1))^2 + (0 - 3)^2}$

$= \sqrt{1^2 + (-3)^2}$

$= \sqrt{10}$

$(x - 0)^2 + (y - 0)^2 = \left(\sqrt{10}\right)^2$

$x^2 + y^2 = 10$

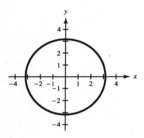

59. $x^2 + y^2 = 64$

$x^2 + y^2 = 8^2$

Circle

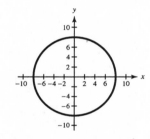

61. Vertices: $(0, -5), (0, 5) \Rightarrow$ Major axis vertical

Co–vertices: $(-2, 0), (2, 0) \Rightarrow$ Minor axis horizontal

$\dfrac{x^2}{b^2} + \dfrac{y^2}{a^2} = 1$

$\dfrac{x^2}{2^2} + \dfrac{y^2}{5^2} = 1$

$\dfrac{x^2}{4} + \dfrac{y^2}{25} = 1$

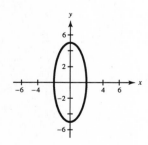

63. Center: $(5, 3)$

Vertices: $(10, 3), (0, 3) \Rightarrow$ Major axis horizontal, $a = 5$

Co–vertices: $(5, 0), (5, 6) \Rightarrow$ Minor axis vertical, $b = 3$

$\dfrac{(x - h)^2}{a^2} + \dfrac{(y - k)^2}{b^2} = 1$

$\dfrac{(x - 5)^2}{5^2} + \dfrac{(y - 3)^2}{3^2} = 1$

$\dfrac{(x - 5)^2}{25} + \dfrac{(y - 3)^2}{9} = 1$

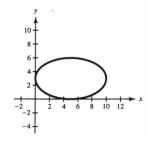

65. $x^2 + 4y^2 = 64$

$$\frac{x^2}{64} + \frac{4y^2}{64} = \frac{64}{64}$$

$$\frac{x^2}{64} + \frac{y^2}{16} = 1$$

$$\frac{x^2}{8^2} + \frac{y^2}{4^2} = 1$$

Center: $(0, 0)$

Vertices: $(-8, 0), (8, 0)$

Co-vertices: $(0, 4), (0, -4)$

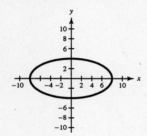

67. $9x^2 + 4y^2 - 18x + 16y - 299 = 0$

$$(9x^2 - 18x) + (4y^2 + 16y) = 299$$

$$9(x^2 - 2x) + 4(y^2 + 4y) = 299$$

$$9(x^2 - 2x + 1) + 4(y^2 + 4y + 4) = 299 + 9(1) + 4(4)$$

$$9(x - 1)^2 + 4(y + 2)^2 = 324$$

$$\frac{9(x - 1)^2}{324} + \frac{4(y + 2)^2}{324} = \frac{324}{324}$$

$$\frac{(x - 1)^2}{36} + \frac{(y + 2)^2}{81} = 1$$

$$\frac{(x - 1)^2}{6^2} + \frac{(y + 2)^2}{9^2} = 1$$

Center: $(1, -2)$

Vertices: $(1, 7)$ and $(1, -11)$

Co-vertices: $(-5, -2)$ and $(7, -2)$

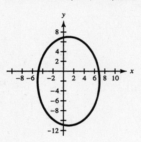

69. Vertices: $(6, 0)$ and $(-6, 0) \implies a = 6$, Transverse axis is horizontal

Asymptotes: $y = -\frac{1}{3}x$ and $y = \frac{1}{3}x \implies \frac{b}{a} = \frac{1}{3} \implies \frac{b}{6} = \frac{1}{3} \implies b = 2$

$$\frac{x^2}{a^2} - \frac{y^2}{b^2} = 1$$

$$\frac{x^2}{6^2} - \frac{y^2}{2^2} = 1$$

$$\frac{x^2}{36} - \frac{y^2}{4} = 1$$

71. Vertices: $(1, 3)$ and $(-1, 3) \implies a = 1$, Transverse axis is horizontal

Asymptotes: $y = -2x + 3$ and $y = 2x + 3 \implies \frac{b}{a} = 2 \implies \frac{b}{1} = \frac{2}{1} \implies b = 2$

$$\frac{(x - h)^2}{a^2} - \frac{(y - k)^2}{b^2} = 1$$

$$\frac{(x - 0)^2}{1^2} - \frac{(y - 3)^2}{2^2} = 1$$

$$\frac{x^2}{1} - \frac{(y - 3)^2}{4} = 1$$

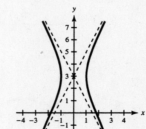

73. $\dfrac{x^2}{25} - \dfrac{y^2}{4} = 1$

$\dfrac{x^2}{5^2} - \dfrac{y^2}{2^2} = 1$

Center: $(0, 0)$

Vertices: $(5, 0)$ and $(-5, 0)$

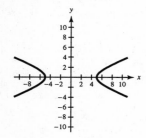

75.
$$y = x^2 - 4x + 2$$
$$x^2 - 4x + 2 = y$$
$$x^2 - 4x = y - 2$$
$$x^2 - 4x + 4 = y - 2 + 4$$
$$(x - 2)^2 = y + 2$$
$$(x - 2)^2 = 4\left(\tfrac{1}{4}\right)(y + 2)$$

Vertex: $(2, -2)$

Focus: $\left(2, -\tfrac{7}{4}\right)$

77. Vertex: $(0, 0)$

Focus: $(-2, 0) \implies$ Horizontal axis; $y^2 = 4px$

$p = -2 \implies 4p = -8$

$y^2 = -8x$

79. Vertical axis $\implies (x - h)^2 = 4p(y - k)$

Vertex: $(-6, 4) \implies (x + 6)^2 = 4p(y - 4)$

Focus: $(-6, -1) \implies p = -1 - 4 = -5$

$4p = -20$

$(x + 6)^2 = -20(y - 4)$

81.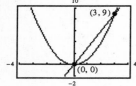

The solutions are $(0, 0)$ and $(3, 9)$.

83.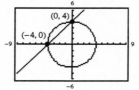

The solutions are $(-4, 0)$ and $(0, 4)$.

85. $\begin{cases} y = 5x^2 \\ y = -15x - 10 \end{cases}$

$5x^2 = -15x - 10$

$5x^2 + 15x + 10 = 0$

$x^2 + 3x + 2 = 0$

$(x + 2)(x + 1) = 0$

$x + 2 = 0 \implies x = -2$ and $y = 5(-2)^2 = 20$

$x + 1 = 0 \implies x = -1$ and $y = 5(-1)^2 = 5$

$(-2, 20)$ and $(-1, 5)$

87. $\begin{cases} x^2 + y^2 = 1 \\ x + y = -1 \implies y = -1 - x \end{cases}$

$x^2 + (-1 - x)^2 = 1$

$x^2 + 1 + 2x + x^2 = 1$

$2x^2 + 2x = 0$

$2x(x + 1) = 0$

$2x = 0 \implies x = 0$ and $y = -1 - 0 = -1$

$x + 1 = 0 \implies x = -1$ and $y = -1 - (-1) = 0$

$(0, -1), (-1, 0)$

89. $\begin{cases} \dfrac{x^2}{16} + \dfrac{y^2}{4} = 1 \quad \Rightarrow x^2 + 4y^2 = 16 \\ \quad\quad\quad y = x + 2 \Rightarrow \end{cases}$

$\quad\Rightarrow x^2 + 4y^2 = 16$

$y^2 = x^2 + 4x + 4 \Rightarrow \dfrac{4x^2 - 4y^2 = -16x - 16}{5x^2 \quad\quad\quad = -16x}$

$5x^2 + 16x = 0$

$x(5x + 16) = 0$

$\quad x = 0 \Rightarrow y = 0 + 2 = 2$

$5x + 16 = 0 \Rightarrow x = -\dfrac{16}{5} \quad \text{and} \quad y = -\dfrac{16}{5} + 2 = -\dfrac{6}{5}$

$(0, 2), \left(-\dfrac{16}{5}, -\dfrac{6}{5}\right)$

91. $\begin{cases} \dfrac{x^2}{25} + \dfrac{y^2}{9} = 1 \Rightarrow 9x^2 + 25y^2 = 225 \\ \dfrac{x^2}{25} - \dfrac{y^2}{9} = 1 \Rightarrow \dfrac{9x^2 - 25y^2 = 225}{18x^2 \quad\quad = 450} \end{cases}$

$\quad\quad\quad\quad\quad\quad x^2 = \dfrac{450}{18}$

$\quad\quad\quad\quad\quad\quad x^2 = 25$

$\quad\quad\quad\quad\quad\quad x = \pm\sqrt{25}$

$\quad\quad\quad\quad\quad\quad x = \pm 5$

$\dfrac{x^2}{25} + \dfrac{y^2}{9} = 1$

$\dfrac{25}{25} + \dfrac{y^2}{9} = 1$

$1 + \dfrac{y^2}{9} = 1$

$\dfrac{y^2}{9} = 0$

$y^2 = 0$

$y = 0$

$(5, 0), (-5, 0)$

93. $P = kw^3$

(a) $1000 = k(20)^3$

$1000 = k(8000)$

$\dfrac{1000}{8000} = k$

$\dfrac{1}{8} = k$

(b) $P = \dfrac{1}{8} w^3$

$P = \dfrac{1}{8}(25)^3$

$P = \dfrac{1}{8}(15{,}625)$

$P = 1953.125$

The output is 1953.125 kilowatts of power.

95. $\quad x = \dfrac{k}{\sqrt{p}}$

$1000 = \dfrac{k}{\sqrt{25}}$

$1000 = \dfrac{k}{5}$

$5(1000) = k$

$5000 = k \Rightarrow x = \dfrac{5000}{\sqrt{p}}$

$\quad\quad\quad\quad\quad x = \dfrac{5000}{\sqrt{28}}$

$\quad\quad\quad\quad\quad x \approx 945$

The demand would be approximately 945 units.

97. Radius $r = 4000 + 1000$

$r = 5000$

$x^2 + y^2 = r^2$

$x^2 + y^2 = (5000)^2$

$x^2 + y^2 = 25{,}000{,}000$

99. (a)

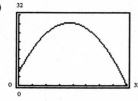

(c) The maximum height of the ball occurs at the vertex $(15, 28.5)$, so the maximum height is 28.5 feet.

(d) The ball strikes the ground where the height y is zero. This occurs at approximately $(31.9, 0)$. The ball strikes the ground approximately at a distance of approximately 31.9 feet from the child.

(b) When $x = 0$, $y = 6$. The ball is 6 feet high when it leaves the child's hand.

Chapter Test for Chapter 8

1. $(f - g)(x) = f(x) - g(x)$

$$= \frac{1}{2}x - (x^2 - 1)$$

$$= \frac{1}{2}x - x^2 + 1$$

$$= -x^2 + \frac{1}{2}x + 1$$

2. $(fg)(x) = f(x)g(x)$

$$= \frac{1}{2}x(x^2 - 1)$$

$$= \frac{1}{2}x^3 - \frac{1}{2}x$$

3. $(f/g)(x) = f(x)/g(x)$

$$= \frac{\frac{1}{2}x}{x^2 - 1}$$

$$= \frac{x}{2(x^2 - 1)}$$

4. $(f \circ g)(x) = f(g(x))$

$$= f(x^2 - 1)$$

$$= \frac{1}{2}(x^2 - 1)$$

5. $f(x) = \sqrt{25 - x}$, $g(x) = x^2$

$$(f \circ g)(x) = f(g(x))$$

$$= f(x^2) = \sqrt{25 - x^2}$$

Domain $= \{x: -5 \le x \le 5\}$

6.

$$f(x) = \tfrac{1}{2}x - 1$$

$$y = \tfrac{1}{2}x - 1$$

$$x = \tfrac{1}{2}y - 1$$

$$x + 1 = \tfrac{1}{2}y$$

$$2(x + 1) = y$$

$$2x + 2 = y$$

$$f^{-1}(x) = 2x + 2$$

$$f(x) = \tfrac{1}{2}x - 1, f^{-1}(x) = 2x + 2$$

$$f(f^{-1}(x)) = f(2x + 2)$$

$$= \tfrac{1}{2}(2x + 2) - 1$$

$$= (x + 1) - 1$$

$$= x$$

$$f^{-1}(f(x)) = f^{-1}\left(\tfrac{1}{2}x - 1\right)$$

$$= 2\left(\tfrac{1}{2}x - 1\right) + 2$$

$$= x - 2 + 2$$

$$= x$$

7. $S = \dfrac{kx^2}{y}$

8.

$$v = k\sqrt{u}$$

$$\frac{3}{2} = k\sqrt{36}$$

$$\frac{3}{2} = k(6)$$

$$\frac{3/2}{6} = k$$

$$\frac{1}{4} = k \Rightarrow v = \frac{1}{4}\sqrt{u}$$

9. $f(x) = -2x^3 + 3x^2 - 4$

Leading coefficient, -2, *negative* $\Rightarrow$ graph falls to right.

Degree, 3, *odd* (*opposite* behavior to left) $\Rightarrow$ graph rises to left.

10. $f(x) = 1 - (x - 2)^3 = -(x - 2)^3 + 1$

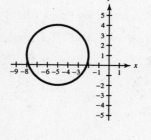

x-intercept:

$f(x) = 0$

$-(x - 2)^3 + 1 = 0$

$-(x - 2)^3 = -1$

$(x - 2)^3 = 1$

$x - 2 = \sqrt[3]{1}$

$x - 2 = 1$ $(0, 9)$

$x = 3$

$(3, 0)$

y-intercept: $x = 0$

$y = -(0 - 2)^3 + 1$

$y = -(-2)^3 + 1$

$y = 8 + 1$

$y = 9$

11. Circle

Center: $(0, 0)$

Radius: 5

$x^2 + y^2 = 5^2$

$x^2 + y^2 = 25$

12. Vertices: $(0, -10), (0, 10) \Longrightarrow a = 10$, Major axis is
vertical

Co-vertices: $(-3, 0), (3, 0) \Longrightarrow b = 3$, Minor axis is
horizontal

$$\frac{x^2}{a^2} + \frac{y^2}{b^2} = 1$$

$$\frac{x^2}{3^2} + \frac{y^2}{10^2} = 1$$

$$\frac{x^2}{9} + \frac{y^2}{100} = 1$$

13. Vertical: $(-3, 0), (3, 0) \Longrightarrow a = 3$, Transverse axis is
horizontal

Asymptotes: $y = \dfrac{1}{2}x$,

$$y = -\frac{1}{2}x \Longrightarrow \frac{b}{a} = \frac{1}{2} \Longrightarrow \frac{b}{3} = \frac{1}{2} \Longrightarrow b = \frac{3}{2}$$

$$\frac{x^2}{a^2} - \frac{y^2}{b^2} = 1$$

$$\frac{x^2}{3^2} - \frac{y^2}{(3/2)^2} = 1$$

$$\frac{x^2}{9} - \frac{y^2}{9/4} = 1$$

14. Parabola with horizontal axis

$(y - k)^2 = 4p(x - h)$

Vertex: $(-2, 1) \Longrightarrow (y - 1)^2 = 4p(x + 2)$

$(6, 9)$ on graph $\Longrightarrow (9 - 1)^2 = 4p(6 + 2)$

$8^2 = 4p(8)$

$64 = 32p$

$2 = p$ and $4p = 8$

$(y - 1)^2 = 8(x + 2)$

15. (a) $(x + 5)^2 + (y - 1)^2 = 9$

$(x + 5)^2 + (y - 1)^2 = 3^2$

$(x - h)^2 + (y - k)^2 = r^2$ Circle

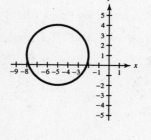

(b) $\dfrac{x^2}{9} + \dfrac{y^2}{16} = 1$

$$\frac{x^2}{3^2} + \frac{y^2}{4^2} = 1$$

$$\frac{x^2}{b^2} + \frac{y^2}{a^2} = 1 \quad \text{(Ellipse)}$$

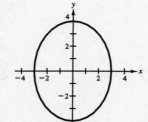

(c) $\dfrac{x^2}{9} - \dfrac{y^2}{16} = 1$

$$\frac{x^2}{3^2} - \frac{y^2}{4^2} = 1$$

$$\frac{x^2}{a^2} - \frac{y^2}{b^2} = 1 \quad \text{(Hyperbola)}$$

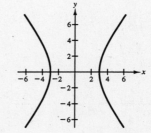

16. (a) $\begin{cases} y = \frac{1}{2}x^2 \\ y = -4x + 6 \end{cases}$

$$-4x + 6 = \frac{1}{2}x^2$$

$$-\frac{1}{2}x^2 - 4x + 6 = 0$$

$$x^2 + 8x - 12 = 0$$

$$x = \frac{-b \pm \sqrt{b^2 - 4ac}}{2a}$$

$$= \frac{-8 \pm \sqrt{8^2 - 4(1)(-12)}}{2(1)}$$

$$= \frac{-8 \pm \sqrt{64 + 48}}{2}$$

$$= \frac{-8 \pm \sqrt{112}}{2} = \frac{-8 \pm 4\sqrt{7}}{2}$$

$$= -4 \pm 2\sqrt{7}$$

$$x = -4 + 2\sqrt{7} \implies y = -4\left(-4 + 2\sqrt{7}\right) + 6$$
$$= 16 - 8\sqrt{7} + 6$$
$$= 22 - 8\sqrt{7}$$

$$x = -4 - 2\sqrt{7} \implies y = -4\left(-4 - 2\sqrt{7}\right) + 6$$
$$= 16 + 8\sqrt{7} + 6$$
$$= 22 + 8\sqrt{7}$$

$$\left(-4 - 2\sqrt{7}, 22 + 8\sqrt{7}\right), \left(-4 + 2\sqrt{7}, 22 - 8\sqrt{7}\right)$$

(b) $\begin{cases} x^2 + y^2 = 100 \\ x + y = 14 \implies y = 14 - x \end{cases}$

$$x^2 + (14 - x)^2 = 100$$

$$x^2 + 196 - 28x + x^2 = 100$$

$$2x^2 - 28x + 96 = 0$$

$$x^2 - 14x + 48 = 0$$

$$(x - 6)(x - 8) = 0$$

$$x - 6 = 0 \implies x = 6$$
$$\text{and}$$
$$14 - x = 14 - 6 = 8$$

$$x - 8 = 0 \implies x = 8$$
$$\text{and}$$
$$14 - x = 14 - 8 = 6$$

$$(6, 8), (8, 6)$$

17. $\qquad P = \dfrac{k}{V}$

$$1 = \frac{k}{180}$$

$$180(1) = k$$

$$180 = k \implies \quad P = \frac{180}{V}$$

$$0.75 = \frac{180}{V}$$

$$0.75V = 180$$

$$V = \frac{180}{0.75}$$

$$V = 240$$

The volume is 240 cubic meters.

Cumulative Test for Chapters 1–8

1. $-(-3x^2)^3(2x^4) = -(-27x^6)(2x^4) = 54x^{10}$

2. $-\dfrac{(2u^2v)^2}{-3uv^2} = -\dfrac{4u^4v^2}{-3uv^2} = \dfrac{-4u^3(u)(v^2)}{-3(u)(v^2)} = \dfrac{4u^3}{3}, u \neq 0, v \neq 0$

3. $\dfrac{5}{\sqrt{12} - 2} = \dfrac{5(\sqrt{12} + 2)}{(\sqrt{12} - 2)(\sqrt{12} + 2)}$

$= \dfrac{5\sqrt{4 \cdot 3} + 10}{(\sqrt{12})^2 - 2^2}$

$= \dfrac{10\sqrt{3} + 10}{12 - 4}$

$= \dfrac{10(\sqrt{3} + 1)}{8}$

$= \dfrac{5\cancel{(2)}(\sqrt{3} + 1)}{4\cancel{(2)}}$

$= \dfrac{5(\sqrt{3} + 1)}{4}$

4. $\sqrt{\dfrac{14xz^5}{6xz}} = \sqrt{\dfrac{7z^4}{3}}$

$= \sqrt{\dfrac{z^4(7)}{3}} \cdot \dfrac{\sqrt{3}}{\sqrt{3}}$

$= \dfrac{z^2\sqrt{21}}{3}, \quad x \neq 0, \quad z \neq 0$

5. $\dfrac{6x - 5}{x - 3} - \dfrac{3x - 8}{x - 3} = \dfrac{6x - 5 - (3x - 8)}{x - 3}$

$= \dfrac{6x - 5 - 3x + 8}{x - 3}$

$= \dfrac{3x + 3}{x - 3}$

$= \dfrac{3(x + 1)}{x - 3}$

6. $\dfrac{x}{x - 5} + \dfrac{7x}{x + 3} = \dfrac{x(x + 3)}{(x - 5)(x + 3)} + \dfrac{7x(x - 5)}{(x + 3)(x - 5)}$

$= \dfrac{x^2 + 3x + 7x^2 - 35x}{(x + 3)(x - 5)}$

$= \dfrac{8x^2 - 32x}{(x + 3)(x - 5)}$

$= \dfrac{8x(x - 4)}{(x + 3)(x - 5)}$

7. $3{,}770{,}000{,}000 = 3.77 \times 10^9$

8. $0.00000000026 = 2.6 \times 10^{-10}$

9. $(8, 1)$ and $(3, 6)$

$d = \sqrt{(3 - 8)^2 + (6 - 1)^2}$

$= \sqrt{(-5)^2 + 5^2}$

$= \sqrt{25 + 25}$

$= \sqrt{50}$

$= \sqrt{25 \cdot 2}$

$= 5\sqrt{2}$

10. $(-6, 7)$ and $(6, -2)$

$d = \sqrt{(6 + 6)^2 + (-2 - 7)^2}$

$= \sqrt{(12)^2 + (-9)^2}$

$= \sqrt{144 + 81}$

$= \sqrt{225}$

$= 15$

11. $(-1, -2)$ and $(3, 6)$

$y - y_1 = m(x - x_1)$

Point: $(-1, -2)$

Slope: $m = \dfrac{6 - (-2)}{3 - (-1)} = \dfrac{8}{4} = 2$

$y + 2 = 2(x + 1)$

$y + 2 = 2x + 2$

$y = 2x$

12. $(1, 5)$ and $(6, 0)$

$y - y_1 = m(x - x_1)$

Point: $(6, 0)$

Slope: $m = \dfrac{0 - 5}{6 - 1} = \dfrac{-5}{5} = -1$

$y - 0 = -1(x - 6)$

$y = -x + 6$

13. Vertex: $(1, 3) \implies y = a(x - 1)^2 + 3$

Point: $(0, 5) \implies 5 = a(0 - 1)^2 + 3$

$5 = a(1) + 3$

$2 = a$

$y = 2(x - 1)^2 + 3$

$y = 2(x^2 - 2x + 1) + 3$

$y = 2x^2 - 4x + 2 + 3$

$y = 2x^2 - 4x + 5$

14. (a) Function

Every element in A is matched with exactly one element in B.

(b) Function

Every element in A is matched with exactly one element in B.

(c) Not a function

The element 1 in A is matched with two elements in B.

(b) Function

Every element in A is matched with exactly one element in B.

15. $3x^2 - 21x = 3x(x - 7)$

16. $y^3 + 6y^2 + 9y = y(y^2 + 6y + 9)$
$$= y(y + 3)(y + 3)$$
$$= y(y + 3)^2$$

17. $16x^4 - 1 = (4x^2)^2 - 1^2$
$$= (4x^2 - 1)(4x^2 + 1)$$
$$= [(2x)^2 - 1^2](4x^2 + 1)$$
$$= (2x + 1)(2x - 1)(4x^2 + 1)$$

18. $5x(2x - 5) - (2x - 5)^2 = (2x - 5)[5x - (2x - 5)]$
$$= (2x - 5)(5x - 2x + 5)$$
$$= (2x - 5)(3x + 5)$$

19. $\dfrac{\left(\dfrac{9}{x}\right)}{\left(\dfrac{6}{x} + 2\right)} = \dfrac{\left(\dfrac{9}{x}\right)(x)}{\left(\dfrac{6}{x} + 2\right)(x)}$

$$= \dfrac{9}{6 + 2x}, \ x \neq 0$$

20. $\dfrac{\left(1 + \dfrac{2}{x}\right)}{\left(x - \dfrac{4}{x}\right)} = \dfrac{\left(1 + \dfrac{2}{x}\right)(x)}{\left(x - \dfrac{4}{x}\right)(x)} = \dfrac{x + 2}{x^2 - 4}$

$$= \dfrac{1\cancel{(x + 2)}}{\cancel{(x + 2)}(x - 2)}$$

$$= \dfrac{1}{x - 2}, \ x \neq 0, \ x \neq -2$$

21. (a) $f(7) = 7^2 - 11 = 49 - 11 = 38$

(b) $f(-5) = (-5)^2 - 11 = 25 - 11 = 14$

(c) $f(3h) = (3h)^2 - 11 = 9h^2 - 11$

(d) $f(w + 3) = (w + 3)^2 - 11 = w^2 + 6w + 9 - 11 = w^2 + 6w - 2$

22. $y = 3 - \frac{1}{2}x$

$y = -\frac{1}{2}x + 3$

Line

Slope $m = -\frac{1}{2}$

y-intercept: $(0, 3)$

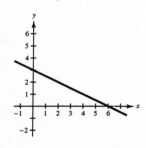

23. $(x - 5)^2 + (y + 2)^2 = 25$

Circle

Center: $(5, -2)$

Radius: $r = 5$

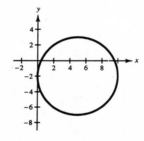

24. $\dfrac{x^2}{9} - \dfrac{y^2}{25} = 1$

Hyperbola

Horizontal transverse axis

Vertices: $(3, 0)$ and $(-3, 0)$

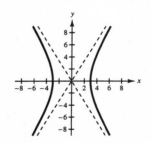

25. (a) The graph can be obtained by shifting the original graph 3 units downward.

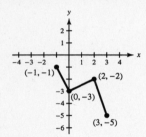

(b) The graph can be obtained by reflecting the original graph in the *x*-axis.

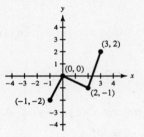

(c) The graph can be obtained by shifting the original graph 1 unit to the left.

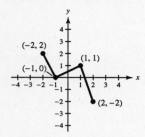

(d) The graph can be obtained by shifting the original graph 2 units to the right.

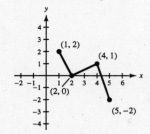

(e) The graph can be obtained by reflecting the original graph about the *y*-axis.

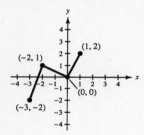

(f) The graph can be obtained by shifting the original graph 3 units upward.

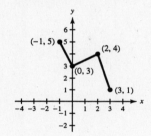

26. Center: $(3, 0)$

Vertices: $(6, 0), (0, 0) \Rightarrow$ Major axis horizontal, $a = 3$

Co-vertices: $(3, 2)(3, -2) \Rightarrow$ Minor axis vertical, $b = 2$

$$\frac{(x - h)^2}{a^2} + \frac{(y - k)^2}{b^2} = 1$$

$$\frac{(x - 3)^2}{3^2} + \frac{(y - 0)^2}{2^2} = 1$$

$$\frac{(x - 3)^2}{9} + \frac{y^2}{4} = 1$$

27. $x^2 + 3x - 2 = 0$

$$x = \frac{-3 \pm \sqrt{3^2 - 4(1)(-2)}}{2(1)}$$

$$= \frac{-3 \pm \sqrt{9 + 8}}{2}$$

$$= \frac{-3 \pm \sqrt{17}}{2}$$

28. $|2x + 14| = 60$

$2x + 14 = -60$ and $2x + 14 = 60$

$2x = -74$ $2x = 46$

$x = -37$ $x = 23$

29.

$$x = \sqrt{12x - 35}$$

$$x^2 = \left(\sqrt{12x - 35}\right)^2$$

$$x^2 = 12x - 35$$

$$x^2 - 12x + 35 = 0$$

$$(x - 7)(x - 5) = 0$$

$$x - 7 = 0 \Rightarrow x = 7$$

$$x - 5 = 0 \Rightarrow x = 5$$

30.
$$\frac{5x}{x-3} - 7 = \frac{15}{x-3}$$

$$(x-3)\left(\frac{5x}{x-3} - 7\right) = \left(\frac{15}{x-3}\right)(x-3)$$

$$5x - 7(x-3) = 15$$

$$5x - 7x + 21 = 15$$

$$-2x + 21 = 15$$

$$-2x = -6$$

$$x = 3 \quad \text{(Extraneous)}$$

The equation has no solution.

31.
$$6x^2 - 4 \le 5x$$

$$6x^2 - 5x - 4 \le 0$$

Critcal numbers: $\quad 6x^2 - 5x - 4 = 0$

$$(3x-4)(2x+1) = 0$$

$$3x - 4 = 0 \implies 3x = 4 \implies x = \tfrac{4}{3}$$

$$2x + 1 = 0 \implies 2x = -1 \implies x = -\tfrac{1}{2}$$

Test Interval	Representative x-value	Is inequality satisfied?
$\left(-\infty, -\tfrac{1}{2}\right]$	-3	$6(-3)^2 - 4 \overset{?}{\le} 5(-3)$ No, $50 \not\le -15$
$\left[-\tfrac{1}{2}, \tfrac{4}{3}\right]$	1	$6(1)^2 - 4 \overset{?}{\le} 5(1)$ Yes, $2 \le 5$
$\left[\tfrac{4}{3}, \infty\right)$	3	$6(3)^2 - 4 \overset{?}{\le} 5(3)$ No, $50 \not\le 15$

$$-\tfrac{1}{2} \le x \le \tfrac{4}{3}$$

32. $\quad |2x - 9| < 7$

$$-7 < 2x - 9 < 7$$

$$2 < 2x < 16$$

$$1 < x < 8$$

33. $\begin{cases} 6x + 7y = -4 \implies \\ 2x + 5y = 4 \implies \end{cases}$ $\begin{aligned} 6x + 7y &= -4 \\ -6x - 15y &= -12 \\ \hline -8y &= -16 \implies y = 2 \end{aligned}$

$$6x + 7(2) = -4$$

$$6x + 14 = -4$$

$$6x = -18$$

$$x = -3$$

$$(-3, 2)$$

34. $\begin{cases} y = x^2 - x - 1 \\ 3x - y = 4 \end{cases} \Longrightarrow 3x - (x^2 - x - 1) = 4$

$$3x - x^2 + x + 1 = 4$$

$$-x^2 + 4x - 3 = 0$$

$$x^2 - 4x + 3 = 0$$

$$(x - 3)(x - 1) = 0$$

$$x - 3 = 0 \Longrightarrow x = 3 \quad \text{and} \quad y = 3^2 - 3 - 1 = 5$$

$$x - 1 = 0 \Longrightarrow x = 1 \quad \text{and} \quad y = 1^2 - 1 - 1 = -1$$

$(3, 5), (1, -1)$

35. $\begin{vmatrix} 5 & -3 & -1 \\ -2 & 1 & -1 \\ 1 & 0 & 2 \end{vmatrix} = 5\begin{vmatrix} 1 & -1 \\ 0 & 2 \end{vmatrix} - (-2)\begin{vmatrix} -3 & -1 \\ 0 & 2 \end{vmatrix} + 1\begin{vmatrix} -3 & -1 \\ 1 & -1 \end{vmatrix}$

$$= 5(2) + 2(-6) + 1(4)$$

$$= 10 - 12 + 4$$

$$= 2$$

36. Cost $= 1.10 + 0.45x$, $x =$ number of additional minutes

$$1.10 + 0.45x \leq 11$$

$$0.45x \leq 9.90$$

$$x \leq \frac{9.90}{0.45}$$

$$x \leq 22$$

The maximum number of additional minutes is 22. With the first minute, the phone call can last up to 23 minutes.

Interval time for call: $0 < t \leq 23$

37. Cost $= 5.35x + 30{,}000$

Revenue $= 11.60x$

Break $-$ even: Revenue $=$ Cost

$$11.60x = 5.35x + 30{,}000$$

$$6.25x = 30{,}000$$

$$x = \frac{30{,}000}{6.25}$$

$$x = 4800$$

38. Two consecutive odd integers: x and $x + 2$

$$x + (x + 2) = x(x + 2) - 47$$

$$2x + 2 = x^2 + 2x - 47$$

$$0 = x^2 - 49$$

$$0 = (x + 7)(x - 7)$$

$$x + 7 = 0 \Longrightarrow x = -7 \quad \text{(Discard negative answer)}$$

$$x - 7 = 0 \Longrightarrow x = 7 \quad \text{and } x + 2 = 9$$

The two consecutive integers are 7 and 9.

39. $L = \dfrac{kD^4}{H^2}$

$$6 = \frac{k(2)^4}{10^2}$$

$$6 = \frac{16k}{100}$$

$$600 = 16k$$

$$\frac{600}{16} = k$$

$$\frac{75}{2} = k \Longrightarrow L = \frac{75D^4}{2H^2}$$

$$L = \frac{75(3)^4}{2(14)^2}$$

$$L \approx 15.5$$

CHAPTER 9
Exponential and Logarithmic
Functions and Equations

CHAPTER 9
Exponential and Logarithmic Functions and Equations

Section 9.1   Exponential Functions and Their Graphs

Solutions to Odd-Numbered Exercises

1. $4^{\sqrt{3}} \approx 11.036$

3. $e^{1/3} \approx 1.396$

5. $\dfrac{4e^3}{12e^2} = \dfrac{\cancel{4}(e^2)e}{3(\cancel{4})(\cancel{e^2})}$

$= \dfrac{e}{3}$

≈ 0.906

7. $(-8e^3)^{1/3} = (-8)^{1/3}e^{3 \cdot (1/3)} = -2e \approx -5.437$

9. (a) $f(-2) = 3^{-2} = \dfrac{1}{9}$

(b) $f(0) = 3^0 = 1$

(c) $f(1) = 3^1 = 3$

11. $g(x) = 5^x$

(a) $g(-1) = 5^{-1} = \dfrac{1}{5}$

(b) $g(1) = 5^1 = 5$

(c) $g(3) = 5^3 = 125$

13. (a) $f(0) = 500\left(\dfrac{1}{2}\right)^0 = 500(1) = 500$

(b) $f(1) = 500\left(\dfrac{1}{2}\right)^1 = 500\left(\dfrac{1}{2}\right) = 250$

(c) $f(\pi) = 500\left(\dfrac{1}{2}\right)^\pi \approx 56.657$

15. (a) $f(0) = 1000(1.05)^{(2)(0)}$

$= 1000(1.05)^0$

$= 1000(1) = 1000$

(b) $f(5) = 1000(1.05)^{(2)(5)}$

$= 1000(1.05)^{10} \approx 1628.895$

(c) $f(10) = 1000(1.05)^{(2)(10)}$

$= 1000(1.05)^{20} \approx 2653.298$

17. (a) $g(1) = \dfrac{500}{(1.02)^{5(1)}}$

$= \dfrac{500}{(1.02)^5}$

≈ 452.865

(b) $g(8) = \dfrac{500}{(1.02)^{5(8)}}$

$= \dfrac{500}{(1.02)^{40}}$

≈ 226.445

(c) $g(25) = \dfrac{500}{(1.02)^{5(25)}}$

$= \dfrac{500}{(1.02)^{125}}$

≈ 42.068

19. (a) $f(-1) = 6e^{3(-1)}$ (b) $f(0) = 6e^{3(0)}$

$= 6e^{-3}$ $= 6e^0$

$= \dfrac{6}{e^3} \approx 0.299$ $= 6 \cdot 1 = 6$

21. (a) $g(-4) = 10e^{-0.5(-4)}$

$= 10e^2 \approx 73.891$

(b) $g(4) = 10e^{-0.5(4)}$

$= 10e^{-2} \approx 1.353$

23. (a) $A(0) = \dfrac{42}{1 + e^{-1.2(0)}}$

$= \dfrac{42}{1 + e^0}$

$= \dfrac{42}{1 + 1}$

$= \dfrac{42}{2}$

$= 21$

(b) $A(20) = \dfrac{42}{1 + e^{-1.2(20)}}$

$= \dfrac{42}{1 + e^{-24}}$

≈ 42

25. Graph (b)

27. Graph (e)

29. Graph (f)

31. Graph (h)

33. Vertical shift two units downward

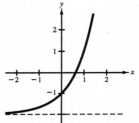

35. Horizontal shift five units to the left

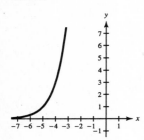

37. Reflection in the y-axis

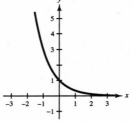

39. Horizontal shift four units to the left and reflection in the y-axis

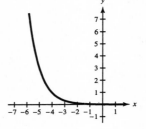

41. Vertical shift 10 units upward and reflection in the x-axis

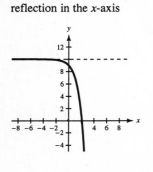

43. Vertical shift four units downward and horizontal shift two units to the left

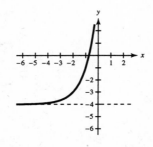

45.

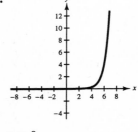

$y = 0$

47.

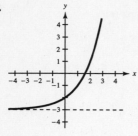

$y = -3$

49.

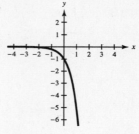

$y = 0$

51.

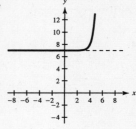

$y = 7$

53.

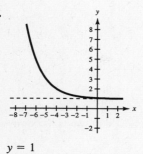

$y = 1$

55.

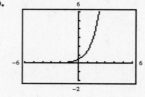

57.

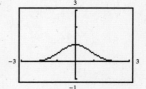

59.

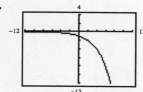

61.

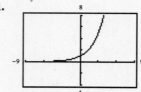

63.

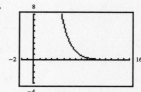

65.

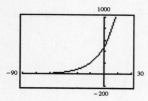

67.

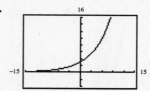

69.

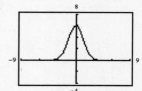

71.

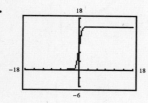

73. $A = P\left(1 + \dfrac{r}{n}\right)^{nt}$

n	1	4	12	365	Continuous
A	\$466.10	\$487.54	\$492.68	\$495.22	\$495.30

$$A = 100\left(1 + \frac{0.08}{1}\right)^{1(20)} = \$466.10 \qquad A = 100\left(1 + \frac{0.08}{4}\right)^{4(20)} = \$487.54$$

$$A = 100\left(1 + \frac{0.08}{12}\right)^{12(20)} = \$492.68 \qquad A = 100\left(1 + \frac{0.08}{365}\right)^{365(20)} = \$495.22 \qquad A = Pe^{rt} = 100e^{0.08(20)} = \$495.30$$

75. $A = P\left(1 + \dfrac{r}{n}\right)^{nt}$

n	1	4	12	365	Continuous
P	\$2541.75	\$2498.00	\$2487.98	\$2483.09	\$2482.93

$$5000 = P\left(1 + \frac{0.07}{1}\right)^{1(10)} \qquad\qquad 5000 = P\left(1 + \frac{0.07}{4}\right)^{4(10)}$$

$$\frac{5000}{(1.07)^{10}} = P \qquad\qquad\qquad \frac{5000}{(1.0175)^{40}} = P$$

$$\$2541.75 = P \qquad\qquad\qquad \$2498.00 = P$$

$$5000 = P\left(1 + \frac{0.07}{12}\right)^{12(10)} \qquad\qquad 5000 = P\left(1 + \frac{0.07}{365}\right)^{365(10)}$$

$$\frac{5000}{(1.0058\overline{3})^{120}} = P \qquad\qquad\qquad \frac{5000}{(1.0001918)^{3650}} = P$$

$$\$2487.98 = P \qquad\qquad\qquad \$2483.09 = P$$

$$A = Pe^{rt}$$

$$5000 = Pe^{0.07(10)}$$

$$\frac{5000}{e^{0.7}} = P$$

$$\$2482.93 = P$$

77. (a)
$$A = P\left(1 + \frac{r}{n}\right)^{nt}$$

$$1{,}000{,}000 = P\left(1 + \frac{0.105}{4}\right)^{4(20)}$$

$$1{,}000{,}000 = P\left(1 + \frac{0.105}{4}\right)^{80}$$

$$\frac{1{,}000{,}000}{\left(1 + \frac{0.105}{4}\right)^{80}} = P$$

$$\$125{,}819.05 = P$$

(b)
$$A = P\left(1 + \frac{r}{n}\right)^{nt}$$

$$1{,}000{,}000 = P\left(1 + \frac{0.105}{12}\right)^{12(20)}$$

$$1{,}000{,}000 = P\left(1 + \frac{0.105}{12}\right)^{240}$$

$$\frac{1{,}000{,}000}{\left(1 + \frac{0.105}{12}\right)^{240}} = P$$

$$\$123{,}580.10 = P$$

(c)
$$A = P\left(1 + \frac{r}{n}\right)^{nt}$$

$$1{,}000{,}000 = P\left(1 + \frac{0.105}{365}\right)^{365(20)}$$

$$1{,}000{,}000 = P\left(1 + \frac{0.105}{365}\right)^{7300}$$

$$\frac{1{,}000{,}000}{\left(1 + \frac{0.105}{365}\right)^{7300}} = P$$

$$\$122{,}493.42 = P$$

(d)
$$A = Pe^{rt}$$

$$1{,}000{,}000 = Pe^{0.105(20)}$$

$$1{,}000{,}000 = Pe^{2.1}$$

$$\frac{1{,}000{,}000}{e^{2.1}} = P$$

$$\$122{,}456.43 = P$$

79. $V(t) = 64{,}000(2)^{t/15}$

   (a) $t = 5$

       $V(5) = 64{,}000(2)^{5/15}$

            $= 64{,}000(2)^{1/3}$

            $\approx 80{,}634.95$

   (b) $t = 20$

       $V(20) = 64{,}000(2)^{20/15}$

             $= 64{,}000(2)^{4/3}$

             $\approx 161{,}269.89$

The approximate value of the property five years after the date of purchase is $80,634.95; after 20 years, the approximate value is $161,269.89.

81. $v(t) = 16{,}000\left(\frac{3}{4}\right)^{t}$

   $v(2) = 16{,}000\left(\frac{3}{4}\right)^{2} = 16{,}000\left(\frac{9}{16}\right) = 9000$

The value of the car after two years is approximately $9000.

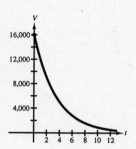

83. $y = 16\left(\frac{1}{2}\right)^{80/30}$

     $= 16\left(\frac{1}{2}\right)^{8/3}$

      ≈ 2.52 grams

85. (a)

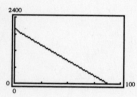

   (b) $t = 0 \implies h = 1950 + 50e^{-0.4433(0)} - 22(0) = 2000$ feet

       $t = 25 \implies h = 1950 + 50e^{-0.4433(25)} - 22(25) = 1400$ feet

       $t = 50 \implies h = 1950 + 50e^{-0.4433(50)} - 22(50) = 850$ feet

       $t = 75 \implies h = 1950 + 50e^{-0.4433(75)} - 22(75) = 300$ feet

   (c) The parachutist will reach the ground at 88.6 seconds.

87. (a) and (b)

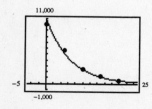

(c)

h	0	5	10	15	20
P	10,332	5583	2376	1240	517
P(Model)	10,958	5176	2445	1155	546

   (d) When h is 8, P is approximately 3300 kilograms per square meter.

   (e) When P is 2000, h is approximately 11.3 kilometers.

89. No. $e \neq \dfrac{271{,}801}{99{,}990}$ because e is an irrational number.

91. $f(x) = 1^{x}$ is a constant function.

93. (a) $f(x) = 2x$ (linear function)

(b) $f(x) = 2x^2$ (quadratic function)

(c) $f(x) = 2^x$ (exponential function)

(d) $f(x) = 2^{-x}$ (exponential function)

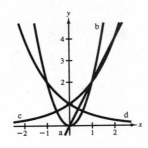

95. $5x^4 \cdot x^3 = 5x^{4+3} = 5x^7$

97. $\left(-\dfrac{4}{x^2}\right)^2 = \dfrac{4^2}{(x^2)^2} = \dfrac{16}{x^4}$

99. $x > 6 - y$

$x + y > 6$

$y > -x + 6$

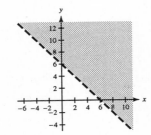

101.

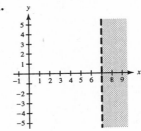

103. $5x - 2y < 5$

$-2y < -5x + 5$

$y > \frac{5}{2}x - \frac{5}{2}$

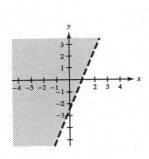

105. $a^2 + b^2 = c^2$ (Pythagorean Theorem)

$60^2 + 30^2 = c^2$

$3600 + 900 = c^2$

$4500 = c^2$

$\sqrt{4500} = c$

$30\sqrt{5} = c$

$67.08 \approx c$

The length of each diagonal should be $30\sqrt{5}$ or approximately 67.08 feet.

Section 9.2   Logarithmic Functions and Their Graphs

1. $\log_2 8 = 3 \implies 2^3 = 8$

3. $\log_4 256 = 4 \implies 4^4 = 256$

5. $\log_4 \frac{1}{16} = -2 \implies 4^{-2} = \frac{1}{16}$

7. $\log_{36} 6 = \frac{1}{2} \implies 36^{1/2} = 6$

9. $7^2 = 49 \implies \log_7 49 = 2$

11. $3^{-2} = \frac{1}{9} \implies \log_3 \frac{1}{9} = -2$

13. $5^4 = 625 \implies \log_5 625 = 4$

15. $8^{2/3} = 4 \implies \log_8 4 = \frac{2}{3}$

17. $25^{-1/2} = \frac{1}{5} \implies \log_{25} \frac{1}{5} = -\frac{1}{2}$

19. The power to which 2 must be raised to obtain 8 is 3.

$\log_2 8 = 3$

21. The power to which 10 must be raised to obtain 10 is 1.

$\log_{10} 10 = 1$

23. The power to which 2 must be raised to obtain $\frac{1}{4}$ is -2.

$\log_2 \frac{1}{4} = -2$

25. The power to which 3 must be raised to obtain $\frac{1}{81}$ is -4.

$\log_3 \frac{1}{81} = -4$

27. The power to which 4 must be raised to obtain $\frac{1}{64}$ is -3.

$\log_4 \frac{1}{64} = -3$

29. There is *no power* to which 2 can be raised to obtain -3. (For any x, $2^x > 0$.)

$$\log_2(-3) \text{ undefined}$$

31. The power to which 10 must be raised to obtain 1000 is 3.

$$\log_{10} 1000 = 3$$

33. The power to which 4 must be raised to obtain 1 is 0.

$$\log_4 1 = 0$$

35. The power to which 9 must be raised to obtain 3 is $\frac{1}{2}$. $\left(\textbf{Note:} 9^{1/2} = \sqrt{9} = 3\right)$

$$\log_9 3 = \frac{1}{2}$$

37. There is *no power* to which 4 can be raised to obtain -4.

$$\log_4(-4) \text{ undefined}$$

39. The power to which 16 must be raised to obtain 4 is $\frac{1}{2}$. $\left(\textbf{Note:} 16^{1/2} = \sqrt{16} = 4\right)$

41. The power to which 9 must be raised to obtain 9^4 is 4.

$$\log_9 9^4 = 4$$

43. $\log_{10} \dfrac{\sqrt{3}}{2} \approx -0.0625$

45. $\log_{10} 0.85 \approx -0.0706$

47. $\ln 0.75 \approx -0.2877$

49. $\ln\left(\sqrt{2} + 4\right) \approx 1.6890$

51. Graph (e)

53. Graph (d)

55. Graph (a)

57. Vertical shift 5 units upward

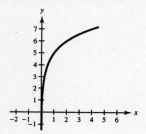

59. Horizontal shift 7 units to the right

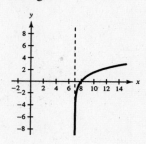

61. Reflection in the y-axis

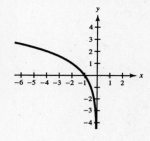

63. Horizontal shift 2 units to the right and vertical shift 5 units upward

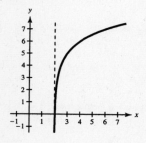

65. Horizontal shift 4 units to the left and vertical shift 8 units downward

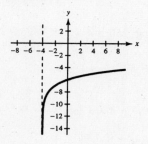

67. Domain: $(3, \infty)$

Vertical asymptote: $x = 3$

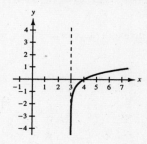

69. Domain: $(0, \infty)$

Vertical asymptote: $x = 0$

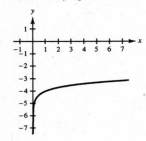

71. Domain: $(0, \infty)$

Vertical asymptote: $x = 0$

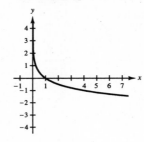

73. Domain: $(1, \infty)$

Vertical asymptote: $x = 1$

75. Domain: $(-2, \infty)$

Vertical asymptote: $x = -2$

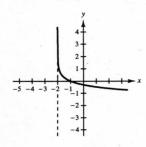

77. Graph (b)

79. Graph (d)

81. Graph (f)

83. Vertical shift 7 units upward

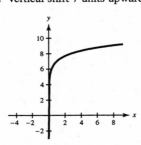

Domain: $(0, \infty)$

Vertical asymptote: $x = 0$

85. Horizontal shift 2 units to the right

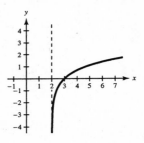

Domain: $(2, \infty)$

Vertical asymptote: $x = 2$

87. Reflection in the x-axis

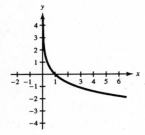

Domain: $(0, \infty)$

Vertical asymptote: $x = 0$

89. Reflection in the y-axis and vertical shift 3 units upward

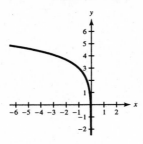

Domain: $(-\infty, 0)$

Vertical asymptote: $x = 0$

91. Horizontal shift 1 unit to the left and vertical shift 5 units downward

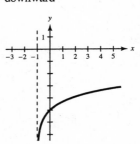

Domain: $(-1, \infty)$

Vertical asymptote: $x = -1$

93. Domain: $(0, \infty)$

Vertical asymptote: $x = 0$

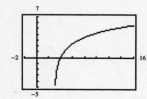

95. Domain: $(-\infty, 0)$

Vertical asymptote: $x = 0$

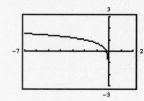

97. Domain: $(4, \infty)$

Vertical asymptote: $t = 4$

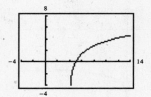

99. Domain: $(0, \infty)$

Vertical asymptote: $t = 0$

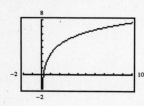

101. $\log_8 132 = \dfrac{\log_{10} 132}{\log_{10} 8} \approx 2.3481$

$\log_8 132 = \dfrac{\ln 132}{\ln 8} \approx 2.3481$

103. $\log_3 7 = \dfrac{\log_{10} 7}{\log_{10} 3} \approx 1.7712$

$\log_3 7 = \dfrac{\ln 7}{\ln 3} \approx 1.7712$

105. $\log_2 0.72 = \dfrac{\log_{10} 0.72}{\log_{10} 2} \approx -0.4739$

$\log_2 0.72 = \dfrac{\ln 0.72}{\ln 2} \approx -0.4739$

107. $\log_{15} 1250 = \dfrac{\log_{10} 1250}{\log_{10} 15} \approx 2.6332$

$\log_{15} 1250 = \dfrac{\ln 1250}{\ln 15} \approx 2.6332$

109. $\log_{1/2} 4 = \dfrac{\log_{10} 4}{\log_{10}(1/2)} = -2$

$\log_{1/2} 4 = \dfrac{\ln 4}{\ln(1/2)} = -2$

111. $\log_4 \sqrt{42} = \dfrac{\log_{10} \sqrt{42}}{\log_{10} 4} \approx 1.3481$

$\log_4 \sqrt{42} = \dfrac{\ln \sqrt{42}}{\ln 4} \approx 1.3481$

113. $h = 116 \log_{10}(a + 40) - 176$ and $a = 55$

$= 116 \log_{10}(55 + 40) - 176 = 116 \log_{10} 95 - 176 \approx 116(1.9777) - 176 \approx 53.4$

The shoulder height of the elk is approximately 53.4 inches.

115. $t = \dfrac{\ln 2}{r}$

$r = 7\% \implies t = \dfrac{\ln 2}{0.07} \approx 9.9$ yrs

$r = 8\% \implies t = \dfrac{\ln 2}{0.08} \approx 8.7$ yrs

$r = 9\% \implies t = \dfrac{\ln 2}{0.09} \approx 7.7$ yrs

$r = 10\% \implies t = \dfrac{\ln 2}{0.10} \approx 6.9$ yrs

$r = 11\% \implies t = \dfrac{\ln 2}{0.11} \approx 6.3$ yrs

$r = 12\% \implies t = \dfrac{\ln 2}{0.12} \approx 5.8$ yrs

r	0.07	0.08	0.09	0.10	0.11	0.12
t	9.9	8.7	7.7	6.9	6.3	5.8

117. $f(x) = 2^x$

$g(x) = \log_2 x$

These two functions are inverses of each other.

119. $\log_a a = 1$ because $a^1 = a$

121. $(a^{-3}b^4)^{-2} = a^{-3(-2)}b^{4(-2)}$

$= a^6 b^{-8}$

$= \dfrac{a^6}{b^8}$

123. $\left(\dfrac{a^2}{3b^{-1}}\right)^{-3} = \dfrac{a^{2(-3)}}{3^{-3}(b^{-1})^{(-3)}}$

$= \dfrac{a^{-6}}{27^{-1}b^3}$

$= \dfrac{27}{a^6 b^3}$

125. $y = -(x - 2)^2 + 1$

$a = -1 \implies$ parabola opens downward

Vertex: $(2, 1)$

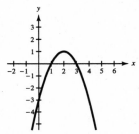

127. *Verbal Model:* $\boxed{\begin{array}{c}\text{Original cost} \\ \text{per person}\end{array}} - \boxed{\begin{array}{c}\text{Lower cost} \\ \text{per person}\end{array}} = 6250$

Labels: Number of people presently in group $= n$

Original cost per person $= \dfrac{250{,}000}{n}$ (dollars)

Lower cost per person $= \dfrac{250{,}000}{n + 3}$ (dollars)

Equation: $\dfrac{250{,}000}{n} - \dfrac{250{,}000}{n + 3} = 6250$

$n(n + 2)\left(\dfrac{250{,}000}{n} - \dfrac{250{,}000}{n + 3}\right) = 6250\, n(n + 2)$

$250{,}000(n + 2) - 250{,}000n = 6250n^2 + 12500n$

$250{,}000n + 500{,}000 - 250{,}000n = 6250n^2 + 12500n$

$500{,}000 = 6250n^2 + 12500n$

$0 = 6250n^2 + 12500n - 500{,}000$

$0 = n^2 + 2n - 80$

$0 = (n + 10)(n - 8)$

$n + 10 = 0 \implies n = -10$ (Discard negative solution.)

$n - 8 = 0 \implies n = 8$

There are presently 8 people in the group.

Section 9.3   Properties of Logarithms

1. $\log_5 5^2 = 2 \log_5 5$

$= 2(1) = 2$

3. $\log_{18} 9 + \log_{18} 2 = \log_{18}(9 \cdot 2)$

$= \log_{18} 18$

$= 1$

5. $\log_4 8 + \log_4 2 = \log_4 (8 \cdot 2)$

$= \log_4 16$

$= \log_4 4^2$

$= 2 \log_4 4$

$= 2(1)$

$= 2$

7. $\log_4 8 - \log_4 2 = \log_4 \left(\frac{8}{2}\right)$

$= \log_4 4$

$= 1$

9. $\log_6 72 - \log_6 2 = \log_6 \left(\frac{72}{2}\right)$

$= \log_6 36$

$= \log_6 6^2$

$= 2 \log_6 6$

$= 2(1)$

$= 2$

11. $\log_3 \frac{2}{3} + \log_3 \frac{1}{2} = \log_3 \left(\frac{2}{3} \cdot \frac{1}{2}\right)$

$= \log_3 \frac{1}{3}$

$= \log_3 3^{-1}$

$= (-1) \log_3 3$

$= (-1)(1)$

$= -1$

13. $\ln 1 = 0$

15. $\log_3 9 = \log_3 3^2$

$= 2 \log_3 3$

$= 2(1)$

$= 2$

17. $\log_2 \frac{1}{8} = \log_2 1 - \log_2 8$

$= 0 - 3 = -3$

19. $\ln e^5 - \ln e^2 = 5 \ln e - 2 \ln e$ or $\ln e^5 - \ln e^2 = \ln \frac{e^5}{e^2}$

$= 5(1) - 2(1)$ $= \ln e^3$

$= 3$ $= 3 \ln e$

$= 3(1)$

$= 3$

21. $\ln e^4 = 4 \ln e = 4(1) = 4$

23. $\ln \frac{e^3}{e^2} = \ln e^3 - \ln e^2$ or $\ln \frac{e^3}{e^2} = \ln e^{3-2}$

$= 3 \ln e - 2 \ln e$ $= \ln e^1$

$= 3 - 2$ $= 1 \ln e$

$= 1$ $= 1(1) = 1$

25. $\ln (e^2 \cdot e^4) = \ln e^2 + \ln e^4$ or $\ln (e^2 \cdot e^4) = \ln (e^{2+4})$

$= 2 \ln e + 4 \ln e$ $= \ln e^6$

$= 2 + 4$ $= 6 \ln e$

$= 6$ $= 6(1) = 6$

27. $\log_3 11x = \log_3 11 + \log_3 x$

29. $\ln y^3 = 3 \ln y$

31. $\log_2 \frac{z}{17} = \log_2 z - \log_2 17$

33. $\log_5 x^{-2} = -2 \log_5 x$

35. $\log_5 \sqrt[3]{x} = \log_5 x^{1/3}$

$= \frac{1}{3} \log_5 x$

37. $\log_3 \sqrt[3]{x+1} = \log_3 (x+1)^{1/3}$

$= \frac{1}{3} \log_3 (x+1)$

39. $\log_5 \frac{1}{\sqrt[3]{x}} = \log_5 x^{-1/3}$

$= -\frac{1}{3} \log_5 x$

41. $\ln 3x^2 y = \ln 3 + \ln x^2 + \ln y$

$= \ln 3 + 2 \ln x + \ln y$

43. $\ln\sqrt{x(x + 2)} = \ln[x(x + 2)]^{1/2}$

$\qquad = \dfrac{1}{2}\ln[x(x + 2)]$

$\qquad = \dfrac{1}{2}[\ln x + \ln(x + 2)]$

$\qquad \text{or } \dfrac{1}{2}\ln x + \dfrac{1}{2}\ln(x + 2)$

45. $\ln\dfrac{t + 4}{15} = \ln(t + 4) - \ln 15$

47. $\log_2\dfrac{x^2}{x - 3} = \log_2 x^2 - \log_2(x - 3)$

$\qquad = 2\log_2 x - \log_2(x - 3)$

49. $\log_3\sqrt{\dfrac{4x}{y^2}} = \log_3\left(\dfrac{4x}{y^2}\right)^{1/2}$

$\qquad = \dfrac{1}{2}\log_3\dfrac{4x}{y^2}$

$\qquad = \dfrac{1}{2}(\log_3(4x) - \log_3 y^2)$

$\qquad = \dfrac{1}{2}(\log_3 4 + \log_3 x - 2\log_3 y)$

$\qquad \text{or } \dfrac{1}{2}\log_3 4 + \dfrac{1}{2}\log_3 x - \log_3 y$

51. $\ln\sqrt[3]{x(x + 5)} = \ln[x(x + 5)]^{1/3}$

$\qquad = \dfrac{1}{3}\ln[x(x + 5)]$

$\qquad = \dfrac{1}{3}[\ln x + \ln(x + 5)]$

$\qquad \text{or } \dfrac{1}{3}\ln x + \dfrac{1}{3}\ln(x + 5)$

53. $\ln\left(\dfrac{x + 1}{x - 1}\right)^2 = 2\ln\left(\dfrac{x + 1}{x - 1}\right)$

$\qquad = 2[\ln(x + 1) - \ln(x - 1)]$

$\qquad \text{or } 2\ln(x + 1) - 2\ln(x - 1)]$

55. $\log_4[x^6(x - 7)^2] = \log_4 x^6 + \log_4(x - 7)^2$

$\qquad = 6\log_4 x + 2\log_4(x - 7)$

57. $\ln\dfrac{a^3(b - 4)}{c^2} = \ln[a^3(b - 4)] - \ln c^2$

$\qquad = \ln a^3 + \ln(b - 4) - \ln c^2$

$\qquad = 3\ln a + \ln(b - 4) - 2\ln c$

59. $\log_{10}\dfrac{(4x)^3}{x - 7} = \log_{10}(4x)^3 - \log_{10}(x - 7)$

$\qquad = 3\log_{10}(4x) - \log_{10}(x - 7)$

$\qquad = 3(\log_{10} 4 + \log_{10} x) - \log_{10}(x - 7)$

$\qquad \text{or } 3\log_{10} 4 + 3\log_{10} x - \log_{10}(x - 7)$

61. $\ln\dfrac{x\sqrt[3]{y}}{(wz)^4} = \ln(x\sqrt[3]{y}) - \ln(wz)^4$

$\qquad = \ln x + \ln y^{1/3} - 4\ln(wz)$

$\qquad = \ln x + \dfrac{1}{3}\ln y - 4(\ln w + \ln z)$

$\qquad \text{or } \ln x + \dfrac{1}{3}\ln y - 4\ln w - 4\ln z$

63. $\log_5[(xy)^2(x + 3)^4] = \log_5(xy)^2 + \log_5(x + 3)^4$

$\qquad = 2\log_5(xy) + 4\log_5(x + 3)$

$\qquad = 2(\log_5 x + \log_5 y) + 4\log_5(x + 3)$

$\qquad \text{or } 2\log_5 x + 2\log_5 y + 4\log_5(x + 3)$

65. $\log_2 3 + \log_2 x = \log_2 3x$

67. $\log_{10} 4 - \log_{10} x = \log_{10} \dfrac{4}{x}$

69. $4 \ln b = \ln b^4, \ b > 0$

71. $-2 \log_5 2x = \log_5 (2x)^{-2}, \ x > 0$

     or $\log_5 \dfrac{1}{4x^2}, \ x > 0$

73. $\dfrac{1}{3} \ln(2x + 1) = \ln(2x + 1)^{1/3}$

     $= \ln \sqrt[3]{2x + 1}$

75. $\log_3 2 + \dfrac{1}{2} \log_3 y = \log_3 2 + \log_3 y^{1/2}$

     $= \log_3 2 + \log_3 \sqrt{y}$

     $= \log_3 2\sqrt{y}$

77. $2 \ln x + 3 \ln y - \ln z = \ln x^2 + \ln y^3 - \ln z, \ x > 0$

     $= \ln x^2 y^3 - \ln z, \ z > 0$

     $= \ln \dfrac{x^2 y^3}{z}, \ x > 0, y > 0, z > 0$

79. $4(\ln x + \ln y) = 4 \ln xy = \ln(xy)^4 \ \text{or} \ \ln x^4 y^4, \ x > 0, y > 0$

81. $\dfrac{1}{2}(\ln 8 + \ln 2x) = \dfrac{1}{2}(\ln 8(2x))$

     $= \dfrac{1}{2}(\ln 16x)$

     $= \ln(16x)^{1/2}$

     $= \ln \sqrt{16x}$

     $= \ln 4\sqrt{x}$

83. $\log_4(x + 8) - 3 \log_4 x = \log_4(x + 8) - \log_4 x^3$

     $= \log_4 \dfrac{x + 8}{x^3}, \ x > 0$

85. $\dfrac{1}{2} \log_5 x + \log_5(x - 3) = \log_5 x^{1/2} + \log_5(x - 3)$

     $= \log_5 \left[\sqrt{x}(x - 3) \right]$

87. $5 \log_6(c + d) - \log_6(m - n) = \log_6(c + d)^5 - \log_6(m - n)$

     $= \log_6 \dfrac{(c + d)^5}{m - n}, \ c > -d, m > n$

89. $\dfrac{1}{5}(3 \log_2 x - \log_2 y) = \dfrac{1}{5}(\log_2 x^3 - \log_2 y)$

     $= \dfrac{1}{5} \log_2 \dfrac{x^3}{y}$

     $= \log_2 \sqrt[5]{\dfrac{x^3}{y}}, \ x > 0, y > 0$

91. $\dfrac{1}{5} \log_6(x - 3) - 2 \log_6 x - 3 \log_6(x + 1) = \dfrac{1}{5} \log_6(x - 3) - \left[2 \log_6 x + 3 \log_6(x + 1) \right]$

     $= \log_6(x - 3)^{1/5} - \left[\log_6 x^2 + \log_6(x + 1)^3 \right]$

     $= \log_6 \sqrt[5]{x - 3} - \log_6 [x^2(x + 1)^3]$

     $= \log_6 \dfrac{\sqrt[5]{x - 3}}{x^2(x + 1)^3}, \ x > 3$

93. $\log_4 4 = \log_4 2^2$ or $\log_4 4 = \log_4(2 \cdot 2)$

$\qquad\quad = 2 \log_4 2$ $\qquad\quad = \log_4 2 + \log_4 2$

$\qquad\quad \approx 2(0.5000) = 1$ $\qquad\quad \approx 0.5000 + 0.5000 = 1$

95. $\log_4 6 = \log_4(3 \cdot 2)$

$\qquad = \log_4 3 + \log_4 2$

$\qquad \approx 0.7925 + 0.5000$

$\qquad = 1.2925$

97. $\log_4 \dfrac{3}{2} = \log_4 3 - \log_4 2$

$\qquad\quad \approx 0.7925 - 0.5000$

$\qquad\quad = 0.2925$

99. $\log_4 \sqrt{2} = \log_4 2^{1/2}$

$\qquad\quad = \dfrac{1}{2} \log_4 2$

$\qquad\quad \approx \dfrac{1}{2}(0.5000)$

$\qquad\quad = 0.2500$

101. $\log_4(3 \cdot 2^4) = \log_4 3 + \log_4 2^4$

$\qquad\qquad = \log_4 3 + 4 \log_4 2$

$\qquad\qquad \approx 0.7925 + 4(0.5000)$

$\qquad\qquad = 2.7925$

103. $\log_4 3^0 = \log_4 1 = 0$

105. $\log_4 \sqrt[3]{\dfrac{1}{9}} = \log_4 \left(\dfrac{1}{9}\right)^{1/3}$

$\qquad\qquad = \dfrac{1}{3} \log_4 \dfrac{1}{9}$

$\qquad\qquad = \dfrac{1}{3} \log_4 3^{-2}$

$\qquad\qquad = \dfrac{1}{3}(-2 \log_4 3)$

$\qquad\qquad = -\dfrac{2}{3} \log_4 3$

$\qquad\qquad \approx -\dfrac{2}{3}(0.7925)$

$\qquad\qquad \approx -0.5283$

107. $\log_{10} 9 = \log_{10} 3^2$

$\qquad\quad = 2 \log_{10} 3$

$\qquad\quad \approx 2(0.477)$

$\qquad\quad = 0.954$

109. $\log_{10} 36 = \log_{10}(3 \cdot 12)$

$\qquad\qquad = \log_{10} 3 + \log_3 12$

$\qquad\qquad \approx 0.477 + 1.079$

$\qquad\qquad = 1.556$

111. $\log_{10} \sqrt[3]{36} = \log_{10} 36^{1/3}$

$\qquad\qquad = \dfrac{1}{3} \log_{10} 36$

$\qquad\qquad = \dfrac{1}{3} \log_{10}(3 \cdot 12)$

$\qquad\qquad = \dfrac{1}{3}(\log_{10} 3 + \log_{10} 12)$

$\qquad\qquad \approx \dfrac{1}{3}(0.477 + 1.079)$

$\qquad\qquad = \dfrac{1}{3}(1.556)$

$\qquad\qquad = 0.5187$

113. $\log_4 \dfrac{4}{x} = \log_4 4 - \log_4 x$

$\qquad\quad = 1 - \log_4 x$

115. $\log_2(5 \cdot 2^3) = \log_2 5 + \log_2 2^3$

$\qquad\qquad = \log_2 5 + 3 \log_2 2$

$\qquad\qquad = \log_2 5 + 3(1)$

$\qquad\qquad = \log_2 5 + 3$

117. $\log_5 \sqrt{50} = \log_5 50^{1/2}$

$\qquad\quad = \dfrac{1}{2} \log_5(25 \cdot 2)$

$\qquad\quad = \dfrac{1}{2}(\log_5 25 + \log_5 2)$

$\qquad\quad = \dfrac{1}{2}(2 + \log_5 2)$

$\qquad\quad = \dfrac{1}{2}(2) + \dfrac{1}{2} \log_5 2$

$\qquad\quad = 1 + \dfrac{1}{2} \log_5 2$

119. $\ln 3e^2 = \ln 3 + \ln e^2$

$= \ln 3 + 2 \ln e$

$= \ln 3 + 2(1)$

$= \ln 3 + 2$

121. $\ln \dfrac{12}{e^3} = \ln 12 - \ln e^3$

$= \ln 12 - 3 \ln e$

$= \ln 12 - 3(1)$

$= \ln 12 - 3$

123. $\log_{10} 50 + \log_{10} 4 = \log_{10}(50 \cdot 4)$

$= \log_{10} 200$

$= \log_{10}(10^2 \cdot 2)$

$= \log_{10} 10^2 + \log_{10} 2$

$= 2 \log_{10} 10 + \log_{10} 2$

$= 2(1) + \log_{10} 2$

$= 2 + \log_{10} 2$

125. (a) $f(t) = 80 - \log_{10}(t + 1)^{12}, \ 0 \le t \le 12$

$= 80 - 12 \log_{10}(t + 1), \ 0 \le t \le 12$

$f(2) = 80 - 12 \log_{10}(2 + 1)$

$f(2) = 80 - 12 \log_{10} 3$

$f(2) \approx 74.27$

$f(8) = 80 - 12 \log_{10}(8 + 1)$

$f(8) = 80 - 12 \log_{10} 9$

$f(8) \approx 68.55$

(b)

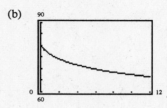

127. $E = 1.4(\log_{10} C_2 - \log_{10} C_1)$

$E = 1.4 \log_{10} \dfrac{C_2}{C_1}$ or $E = \log_{10}\left(\dfrac{C_2}{C_1}\right)^{1.4}$

129. $kt = \ln \dfrac{T - S}{T_0 - S} = \ln \dfrac{84.8 - 70}{98.6 - 70} \approx -0.7$

$kt = -0.7 \implies k \approx \dfrac{-0.7}{t}$

$k(t + 1) = \ln \dfrac{T - S}{T_0 - S} = \ln \dfrac{83.3 - 70}{98.6 - 70} \approx -0.8$

$k(t + 1) \approx -0.8$ and $k \approx \dfrac{-0.7}{t} \implies \dfrac{-0.7}{t}(t + 1) \approx -0.8$

$-0.7 - \dfrac{0.7}{t} \approx -0.8$

$-\dfrac{0.7}{t} \approx -0.1$

$-0.7 \approx -0.1t$

$\dfrac{-0.7}{-0.1} \approx t$

$7 \approx t$

The first temperature reading took place 7 hours after death, so the death occurred at about 1:30 A.M.

131. $\log_2 8x = \log_2 8 + \log_2 x = 3 + \log_2 x$

True

133. $\log_3(u + v) \neq \log_3 u + \log_3 v$

False

135. $\log_5 x^2 = 2 \log_5 x = \log_5 x + \log_5 x$

True

137. $\ln(xy) \neq \ln x \cdot \ln y$

False

For example, $\ln(2.3) \overset{?}{=} \ln 2 \cdot \ln 3$

$$1.7918 \neq 0.7615$$

$$\ln(xy) = \ln x + \ln y$$

139. $f(x) = \ln x$

$f(0) \neq 0$

False. Zero is not in the domain of $f(x)$.

141. $f(x) = \ln x$

$f(2x) = \ln 2 + \ln x$

True

143. Yes. $f(ax) = \log_a(ax)$

$$= \log_a a + \log_a x$$

$$= 1 + f(x)$$

145. $\log_7 49 = 2 \implies 7^2 = 49$

147. $\ln e^{4x} = 4x \implies e^{4x} = e^{4x}$

149.
$$\begin{aligned} 25\sqrt{3x} - 3\sqrt{12x} &= 25\sqrt{3x} - 3\sqrt{4(3x)} \\ &= 25\sqrt{3x} - 3(2)\sqrt{3x} \\ &= 25\sqrt{3x} - 6\sqrt{3x} \\ &= 19\sqrt{3x} \end{aligned}$$

151.
$$\begin{aligned} \sqrt{u}\left(\sqrt{20} - \sqrt{5}\right) &= \sqrt{20u} - \sqrt{5u} \\ &= \sqrt{4(5u)} - \sqrt{5u} \\ &= 2\sqrt{5u} - \sqrt{5u} \\ &= \sqrt{5u} \end{aligned}$$

153.
$$\begin{aligned} \frac{50x}{\sqrt{2}} &= \frac{50x}{\sqrt{2}} \cdot \frac{\sqrt{2}}{\sqrt{2}} \\ &= \frac{50x\sqrt{2}}{2} \\ &= 25x\sqrt{2} \end{aligned}$$

155.
$$\begin{aligned} 26.76 &= 30 - \sqrt{0.5(x - 1)} \\ -3.24 &= -\sqrt{0.5(x - 1)} \\ 3.24 &= \sqrt{0.5(x - 1)} \\ (3.24)^2 &= 0.5(x - 1) \\ 10.4976 &= 0.5x - 0.5 \\ 10.9976 &= 0.5x \\ 21.9952 &= x \\ x &\approx 22 \text{ units} \end{aligned}$$

The demand is approximately 22 units.

Mid-Chapter Quiz for Chapter 9

1. $f(x) = \left(\dfrac{4}{3}\right)^x$

(a) $f(2) = \left(\dfrac{4}{3}\right)^2 = \dfrac{16}{9}$

(b) $f(0) = \left(\dfrac{4}{3}\right)^0 = 1$

(c) $f(-1) = \left(\dfrac{4}{3}\right)^{-1} = \dfrac{3}{4}$

(d) $f(1.5) = \left(\dfrac{4}{3}\right)^{1.5} = \left(\dfrac{4}{3}\right)^{3/2} = \sqrt{\dfrac{64}{27}} = \dfrac{8}{3\sqrt{3}} = \dfrac{8\sqrt{3}}{9}$

or $\left(\dfrac{4}{3}\right)^{1.5} \approx 1.54$

2. $g(x) = 2^{-0.5x}$

Domain: $(-\infty, \infty)$

Range: $(0, \infty)$

3. Domain: $(-\infty, \infty)$

Horizontal asymptote: $y = 0$

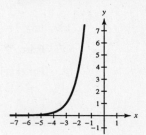

4. Domain: $(-\infty, \infty)$

Horizontal asymptote: $y = 1$

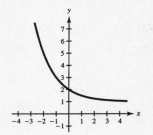

5. Domain: $(-\infty, \infty)$

Horizontal asymptote: $y = 0$

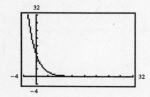

6. Domain: $(-\infty, \infty)$

Horizontal asymptote: $y = 0$

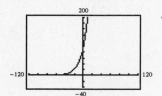

7.

n	1	4	12	365	Continuous
A	\$3185.89	\$3314.90	\$3345.61	\$3360.75	\$3361.27

$$A = 750\left(1 + \frac{0.075}{n}\right)^{(n)(20)}$$

For $n = 1$: $A = 750\left(1 + \frac{0.075}{1}\right)^{(1)(20)} = \$3185.89$ For $n = 4$: $A = 750\left(1 + \frac{0.075}{4}\right)^{(4)(20)} = \$3314.90$

For $n = 12$: $A = 750\left(1 + \frac{0.075}{12}\right)^{(12)(20)} = \$3345.61$ For $n = 365$: $A = 750\left(1 + \frac{0.075}{365}\right)^{(365)(20)} = \$3360.75$

Continuous compounding: $A = 750e^{0.075(20)} = \$3361.27$

8. $y = 14\left(\frac{1}{2}\right)^{t/40}, t \geq 0$

$y = 14\left(\frac{1}{2}\right)^{125/40}$

$y \approx 1.6$ grams

9. $\log_4\left(\frac{1}{16}\right) = -2 \implies 4^{-2} = \frac{1}{16}$

10. $3^4 = 8 \implies \log_3 81 = 4$

11. $\text{Log}_5 125 = 3$ because 3 is the power to which 5 must be raised to obtain 125.

12. Domain: $(-4, \infty)$

Horizontal asymptote: $t = -4$

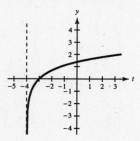

13. Domain: $(0, \infty)$

Horizontal asymptote: $x = 0$

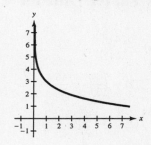

14. This is the graph of $f(x) = \log_5(x - 2) + 1$.
Therefore $h = 2$ and $k = 1$.

15. $\log_6 450 = \dfrac{\ln 450}{\ln 6}$

≈ 3.4096

16. $\ln\left(\dfrac{6x^2}{\sqrt{x^2 + 1}}\right) = \ln(6x^2) - \ln\sqrt{x^2 + 1}$

$= \ln 6 + \ln x^2 - \ln(x^2 + 1)^{1/2}$

$= \ln 6 + 2\ln x - \dfrac{1}{2}\ln(x^2 + 1)$

17. $2(\ln x - 3\ln y) = 2(\ln x - \ln y^3)$

$= 2\ln\dfrac{x}{y^3}$

$= \ln\left(\dfrac{x}{y^3}\right)^2$

Section 9.4   Solving Exponential and Logarithmic Equations

1. $3^{2x-5} = 27$

(a) $x = 1$: $3^{2(1)-5} \overset{?}{=} 27$

$3^{-3} \neq 27$

No, 1 is not a solution.

(b) $x = 4$: $3^{2(4)-5} \overset{?}{=} 27$

$3^3 = 27$

Yes, 4 is a solution.

3. $e^{x+5} = 45$

(a) $x = -5 + \ln 45$: $e^{-5+\ln 45+5} \overset{?}{=} 45$

$e^{\ln 45} = 45$

Yes, $-5 + \ln 45$ is a solution.

(b) $x = -5 + e^{45}$: $e^{-5+e^{45}+5} \overset{?}{=} 45$

$e^{e^{45}} \neq 45$

No, $-5 + e^{45}$ is not a solution.

5. $\log_9(6x) = \dfrac{3}{2}$

(a) $x = 27$: $\log_9(6 \cdot 27) \overset{?}{=} \dfrac{3}{2}$

$\log_9(162) \neq \dfrac{3}{2}$

No, 27 is not a solution.

(b) $x = \dfrac{9}{2}$: $\log_9\left(6 \cdot \dfrac{9}{2}\right) \overset{?}{=} \dfrac{3}{2}$

$\log_9(27) = \dfrac{3}{2}$

Yes, $\dfrac{9}{2}$ is a solution.

7. $2^x = 2^5$

$x = 5$

9. $3^{x+4} = 3^{12}$

$x + 4 = 12$

$x = 8$

11. $3^{x-1} = 3^7$

$x - 1 = 7$

$x = 8$

13. $4^{x-1} = 16 = 4^2$

$x - 1 = 2$

$x = 3$

15. $2^{x+2} = \dfrac{1}{16}$

$2^{x+2} = 2^{-4}$

$x + 2 = -4$

$x = -6$

17. $5^x = \dfrac{1}{125} = 5^{-3}$

$x = -3$

19. $\log_2(x + 3) = \log_2 7$

$x + 3 = 7$

$x = 4$

21. $\log_4(x - 4) = \log_4 12$

$x - 4 = 12$

$x = 16$

23. $\ln 5x = \ln 22$

$5x = 22$

$x = \dfrac{22}{5}$

25. $\ln(2x - 3) = \ln 15$

$2x - 3 = 15$

$2x = 18$

$x = 9$

27. $\log_3 x = 4$

$x = 3^4 = 81$

29. $\log_{10} 2x = 6$

$2x = 10^6$

$= 1,000,000$

$x = 500,000$

31. $\ln e^{2x-1} = 2x - 1$

33. $10^{\log_{10} 2x} = 2x, \; x > 0$

35. $2^x = 45$

$x = \log_2 45$

$= \dfrac{\log_{10} 45}{\log_{10} 2}$

≈ 5.49

37. $10^{2y} = 52$

$2y = \log_{10} 52 \approx 1.72$

$y \approx 0.86$

39. $4^x = 8$

$x = \log_4 8$

$= \dfrac{\ln 8}{\ln 4}$

$= 1.5$

41. $3^{x+4} = 6$

$x + 4 = \log_3 6$

$x = -4 + \log_3 6$

$x \approx -2.37$

43. $2 \cdot 10^{x+6} = 500$

$10^{x+6} = 250$

$x + 6 = \log_{10} 250$

$x = -6 + \log_{10} 250$

$x \approx -3.60$

45. $4 \cdot 7^{x-2} = 428$

$7^{x-2} = 107$

$x - 2 = \log_7 107$

$x = 2 + \log_7 107$

$x \approx 4.40$

47. $2^{x-1} - 6 = 1$

$2^{x-1} = 7$

$x - 1 = \log_2 7$

$x = 1 + \log_2 7$

$x \approx 3.81$

49. $\frac{1}{5}(4^{x+2}) = 300$

$4^{x+2} = 1500$

$x + 2 = \log_4 1500$

$x = -2 + \log_4 1500$

≈ 3.28

51. $5(2)^{3x} - 4 = 13$

$5(2)^{3x} = 17$

$2^{3x} = \frac{17}{5}$

$2^{3x} = 3.4$

$3x = \log_2 3.4$

$x = \frac{1}{3} \log_2 3.4$

$x \approx 0.59$

53. $3e^x = 42$

$e^x = 14$

$x = \ln 14$

$x \approx 2.64$

55. $\frac{1}{4} e^x = 5$

$e^x = 20$

$x = \ln 20$

$x \approx 3.00$

57. $2e^{3x} = 80$

$e^{3x} = 40$

$3x = \ln 40$

$x = \dfrac{\ln 40}{3}$

$x \approx 1.23$

59. $4 + e^{2x} = 150$

$e^{2x} = 146$

$2x = \ln 146$

$x = \dfrac{\ln 146}{2} \approx 2.49$

61. $8 - 12e^{-x} = 7$

$-12e^{-x} = -1$

$12e^{-x} = 1$

$e^{-x} = \dfrac{1}{12}$

$-x = \ln\left(\dfrac{1}{12}\right)$

$x = -\ln\left(\dfrac{1}{12}\right)$

$x \approx 2.48$

63. $23 - 5e^{x+1} = 3$

$-5e^{x+1} = -20$

$e^{x+1} = \dfrac{-20}{-5} = 4$

$x + 1 = \ln 4$

$x = -1 + \ln 4$

≈ 0.39

65. $300e^{x/2} = 9000$

$e^{x/2} = 30$

$\dfrac{x}{2} = \ln 30$

$x = 2 \ln 30$

≈ 6.80

67. $6000e^{-2t} = 1200$

$e^{-2t} = 0.2$

$-2t = \ln 0.2$

$t = \dfrac{\ln 0.2}{-2}$

≈ 0.80

69. $250(1.04)^x = 1000$

$(1.04)^x = 4$

$x = \log_{1.04} 4$

≈ 35.35

71. $\dfrac{1600}{(1.1)^x} = 200$

$1600 = 200(1.1)^x$

$\dfrac{1600}{200} = (1.1)^x$

$8 = 1.1^x$

$x = \log_{1.1} 8$

≈ 21.82

73. $4(1 + e^{x/3}) = 84$

$1 + e^{x/3} = \dfrac{84}{4} = 21$

$e^{x/3} = 20$

$\dfrac{x}{3} = \ln 20$

$x = 3 \ln 20$

≈ 8.99

75. $\log_{10} x = 0$

$x = 10^0$

$= 1$

77. $\log_{10} 4x = \dfrac{3}{2}$

$4x = 10^{3/2}$

$x = \dfrac{10^{3/2}}{4} \approx 7.91$

79. $\log_2 x = 4.5$

$x = 2^{4.5}$

$x \approx 22.63$

81. $4 \log_3 x = 28$

$\log_3 x = 7$

$x = 3^7$

$x = 2187$

83. $2 \log_{10}(x + 5) = 15$

$\log_{10}(x + 5) = \dfrac{15}{2} = 7.5$

$x + 5 = 10^{7.5}$

$x = -5 + 10^{7.5}$

$\approx 31,622,771.60$

85. $16 \ln x = 30$

$\ln x = \dfrac{30}{16}$

$\ln x = \dfrac{15}{8}$

$x = e^{15/8}$

$x \approx 6.52$

87. $\ln 2x = 3$

$2x = e^3$

$x = \dfrac{e^3}{2}$

$x \approx 10.04$

89. $1 - 2 \ln x = -4$

$-2 \ln x = -5$

$2 \ln x = 5$

$\ln x = \dfrac{5}{2}$

$x = e^{5/2}$

$x \approx 12.18$

91. $\dfrac{2}{3} \ln(x + 1) = -1$

$2 \ln(x + 1) = -3$

$\ln(x + 1) = -\dfrac{3}{2}$

$x + 1 = e^{-3/2}$

$x = -1 + e^{-3/2} \approx -0.78$

93. $\ln x^2 = 6$

$x^2 = e^6$

$x = \pm \sqrt{e^6} \approx \pm 20.09$

95. $\log_4 x + \log_4 5 = 2$

$\log_4 5x = 2$

$5x = 4^2$

$5x = 16$

$x = \dfrac{16}{5}$

$x = 3.2$

97. $\log_6(x + 8) + \log_6 3 = 2$

$\log_6 3(x + 8) = 2$

$3(x + 8) = 6^2$

$3x + 24 = 36$

$3x = 12$

$x = 4$

99. $\log_{10} x + \log_{10}(x - 3) = 1$

$\log_{10} x(x - 3) = 1$

$x(x - 3) = 10^1$

$x^2 - 3x = 10$

$x^2 - 3x - 10 = 0$

$(x - 5)(x + 2) = 0$

$x - 5 = 0 \implies x = 5$

$x + 2 = 0 \implies x = -2$ (Extraneous)

(**Note:** $\log_{10}(-2)$ undefined)

101. $\log_5(x - 3) - \log_5 x = 1$

$\log_5 \dfrac{x + 3}{x} = 1$

$\dfrac{x + 3}{x} = 5^1$

$x + 3 = 5x$

$3 = 4x$

$\dfrac{3}{4} = x$ or $x = 0.75$

103. $\log_6(x - 5) + \log_6 x = 2$

$\log_6 x(x - 5) = 2$

$x(x - 5) = 6^2$

$x^2 - 5x = 36$

$x^2 - 5x - 36 = 0$

$(x - 9)(x + 4) = 0$

$x - 9 = 0 \implies x = 9$

$x + 4 = 0 \implies x = -4$ (Extraneous)

(**Note:** $\log_6(-4)$ undefined)

105. $\log_2(x - 1) + \log_2(x + 3) = 3$

$\log_2[(x - 1)(x + 3)] = 3$

$(x - 1)(x + 3) = 2^3$

$x^2 + 2x - 3 = 8$

$x^2 + 2x - 11 = 0$

$x = \dfrac{-2 \pm \sqrt{2^2 - 4(1)(-11)}}{2(1)}$

$x = \dfrac{-2 \pm \sqrt{48}}{2}$

$x = \dfrac{-2 \pm 4\sqrt{3}}{2}$

$x = -1 \pm 2\sqrt{3}$

$x = -1 + 2\sqrt{3}$ (Choose positive answer.)

$x \approx 2.46$

107. $\log_4 3x + \log_4(x - 2) = \dfrac{1}{2}$

$$\log_4[3x(x - 2)] = \dfrac{1}{2}$$

$$3x(x - 2) = 4^{1/2}$$

$$3x^2 - 6x = 2$$

$$3x^2 - 6x - 2 = 0$$

$$x = \dfrac{-(-6) \pm \sqrt{(-6)^2 - 4(3)(-2)}}{2(3)}$$

$$x = \dfrac{6 \pm \sqrt{60}}{6}$$

$$x = \dfrac{6 \pm 2\sqrt{15}}{6}$$

$$x = \dfrac{3 \pm \sqrt{15}}{3}$$

$$x = \dfrac{3 + \sqrt{15}}{3} \qquad \text{(Choose positive answer.)}$$

$$x \approx 2.29$$

109. $\log_3 2x + \log_3(x - 1) = \log_3 4 = 1$

$$\log_3 \dfrac{2x(x - 1)}{4} = 1$$

$$\dfrac{2x(x - 1)}{4} = 3^1$$

$$\dfrac{2x^2 - 2x}{4} = 3$$

$$2x^2 - 2x = 12$$

$$2x^2 - 2x - 12 = 0$$

$$x^2 - x - 6 = 0$$

$$(x - 3)(x + 2) = 0$$

$$x - 3 = 0 \implies x = 3$$

$$x + 2 = 0 \implies x = -2 \quad \text{(Extraneous)}$$

111. $e^x = 2$

Graph $y = e^x$ and $y = 2$.

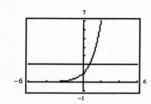

$x \approx 0.69$

113. $2 \ln(x + 3) = 3$

Graph $y = 2 \ln(x + 3)$ and $y = 3$.

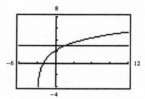

$x \approx 1.48$

115. $5000 = 2500e^{0.09t}$

$$\frac{5000}{2500} = e^{0.09t}$$

$$2 = e^{0.09t}$$

$$0.09t = \ln 2$$

$$t = \frac{\ln 2}{0.09} \approx 7.70$$

Approximately 7.7 years are required for the investment to double in value.

117.
$$B = 10 \log_{10}\left(\frac{I}{10^{-16}}\right)$$

$$75 = 10 \log_{10}\left(\frac{I}{10^{-16}}\right)$$

$$\frac{75}{10} = \log_{10}\left(\frac{I}{10^{-16}}\right)$$

$$7.5 = \log_{10}\left(\frac{I}{10^{-16}}\right)$$

$$10^{7.5} = \frac{I}{10^{-16}}$$

$$10^{-16}(10^{7.5}) = I$$

$$10^{-8.5} = I$$

The intensity is approximately $10^{-8.5}$ watts per centimeter squared.

119.
$$F = 200e^{-0.5\pi\theta/180}$$

$$80 = 200e^{-0.5\pi\theta/180}$$

$$\frac{80}{200} = e^{-0.5\pi\theta/180}$$

$$0.4 = e^{-0.5\pi\theta/180}$$

$$\ln 0.4 = \frac{-0.5\pi\theta}{180}$$

$$180 \ln 0.4 = -0.5\pi\theta$$

$$\frac{180 \ln 0.4}{-0.5\pi} = \theta$$

$$105 \approx \theta$$

The smallest value of θ is approximately $105°$.

121.
$$T = 15.7 - 2.48 \ln m$$

$$2.5 = 15.7 - 2.48 \ln m$$

$$-13.2 = -2.48 \ln m$$

$$13.2 = 2.48 \ln m$$

$$\frac{13.2}{2.48} = \ln m$$

$$m = e^{13.2/2.48}$$

$$m \approx 205$$

Approximately 205 muon decays were recorded.

123. $\quad 2^{x-1} = 32 \qquad 2^{x-1} = 30$

$\qquad 2^{x-1} = 2^5 \qquad x - 1 = \log_2 30$

$\qquad x - 1 = 5 \qquad x = 1 + \log_2 30$

$\qquad\quad x = 6 \qquad\quad x \approx 5.907$

The first equation does not require logarithms for its solution because both sides of the equation can be written as powers of 2.

125. $\log_a(uv) = \log_a u + \log_a v$

$$\log_a\left(\frac{u}{v}\right) = \log_a u - \log_a v$$

$$\log_a u^n = n \log_a u$$

127.
$$\frac{2}{3}x + \frac{2}{3} = 4x - 6$$

$$3\left(\frac{2}{3}x + \frac{2}{3}\right) = 3(4x - 6)$$

$$2x + 2 = 12x - 18$$

$$-10x + 2 = -18$$

$$-10x = -20$$

$$x = \frac{-20}{-10}$$

$$x = 2$$

129. $\dfrac{5}{2x} - \dfrac{4}{x} = 3$

$2x\left(\dfrac{5}{2x} - \dfrac{4}{x}\right) = 2x(3)$

$5 - 8 = 6x$

$-3 = 6x$

$\dfrac{-3}{6} = x$

$-\dfrac{1}{2} = x$

131. $|x - 4| = 3$

$x - 4 = -3 \quad \text{or} \quad x - 4 = 3$

$x = 1 \qquad\qquad x = 7$

133. *Labels:*   Amount of 30% solution $= x$ (liters)

Amount of 80% solution $= y$ (liters)

Equations: $\begin{cases} x + \quad y = 100 \\ 0.30x + 0.80y = .50(100) \end{cases} \Rightarrow \begin{cases} -0.30x - 0.30y = -30 \\ \underline{0.30x + 0.80y = 50} \end{cases}$

$0.50y = 20$

$y = \dfrac{20}{0.50}$

$y = 40$

$x + 40 = 100$

$x = 60$

So, 60 liters of the 30% alcohol solution and 40 liters of the 80% alcohol solution are required.

Section 9.5   Exponential and Logarithmic Applications

1. $A = P\left(1 + \dfrac{r}{n}\right)^{nt}$

$1004.83 = 500\left(1 + \dfrac{r}{12}\right)^{12(10)}$

$\dfrac{1004.83}{500} = \left(1 + \dfrac{r}{12}\right)^{120}$

$2.00966 = \left(1 + \dfrac{r}{12}\right)^{120}$

$2.00966^{1/120} = 1 + \dfrac{r}{12}$

$1.00583 \approx 1 + \dfrac{r}{12}$

$0.00583 \approx \dfrac{r}{12}$

$12(0.00583) \approx r$

$0.07 \approx r \text{ or } r \approx 7\%$

3. $A = P\left(1 + \dfrac{r}{n}\right)^{nt}$

$36{,}581 = 1000\left(1 + \dfrac{r}{365}\right)^{365(40)}$

$\dfrac{36{,}581}{1000} = \left(1 + \dfrac{r}{365}\right)^{14{,}600}$

$36.581 = \left(1 + \dfrac{r}{365}\right)^{14{,}600}$

$36.581^{(1/14{,}600)} = 1 + \dfrac{r}{365}$

$1.000247 \approx 1 + \dfrac{r}{365}$

$0.000247 \approx \dfrac{r}{365}$

$365(0.000247) \approx r$

$0.090 \approx r \text{ or } r \approx 9\%$

5.
$$A = Pe^{rt}$$
$$8267.38 = 750e^{r(30)}$$
$$\frac{8267.38}{750} = e^{30r}$$
$$11.0232 \approx e^{30r}$$
$$30r \approx \ln 11.0232$$
$$r \approx \frac{\ln 11.0232}{30}$$
$$r \approx 0.0800 \text{ or } r \approx 8\%$$

7.
$$A = P\left(1 + \frac{r}{n}\right)^{nt}$$
$$22,405.68 = 5000\left(1 + \frac{r}{365}\right)^{365(25)}$$
$$\frac{22,405.68}{5000} = \left(1 + \frac{r}{365}\right)^{9125}$$
$$\left(\frac{22,405.68}{5000}\right)^{1/9125} = 1 + \frac{r}{365}$$
$$1.000164384 \approx 1 + \frac{r}{365}$$
$$0.000164384 = \frac{r}{365}$$
$$365(0.000164384) = r$$
$$0.06 \approx r \text{ or } r \approx 6\%$$

9.
$$A = Pe^{rt}$$
$$24,666.97 = 1500e^{r(40)}$$
$$\frac{24,667.97}{1500} = e^{40r}$$
$$40r \approx \ln\left(\frac{24,667.97}{1500}\right)$$
$$r = \frac{1}{40}\ln\left(\frac{24,667.97}{1500}\right)$$
$$r \approx 0.07 \text{ or } r \approx 7\%$$

11.
$$A = P\left(1 + \frac{r}{n}\right)^{nt}$$
$$12,000 = 6000\left(1 + \frac{0.08}{4}\right)^{4t}$$
$$\frac{12,000}{6000} = (1.02)^{4t}$$
$$2 = (1.02)^{4t}$$
$$\log_{1.02} 2 = 4t$$
$$\frac{\log_{1.02} 2}{4} = t$$
$$8.75 \text{ years} \approx t$$

13.
$$A = P\left(1 + \frac{r}{n}\right)^{nt}$$
$$4000 = 2000\left(1 + \frac{0.105}{365}\right)^{365t}$$
$$\frac{4000}{2000} = \left(1 + \frac{0.105}{365}\right)^{365t}$$
$$2 = \left(1 + \frac{0.105}{365}\right)^{365t}$$
$$\approx (1.0002877)^{365t}$$
$$\log_{1.0002877} 2 \approx 365t$$
$$\frac{\log_{1.0002877} 2}{365} \approx t$$
$$6.60 \text{ years} \approx t$$

15.
$$A = Pe^{rt}$$
$$3000 = 1500e^{0.075t}$$
$$\frac{3000}{1500} = e^{0.075t}$$
$$2 = e^{0.075t}$$
$$\ln 2 = 0.075t$$
$$\frac{\ln 2}{0.075} = t$$
$$9.24 \text{ years} \approx t$$

17.
$$A = P\left(1 + \frac{r}{n}\right)^{nt}$$
$$600 = 300\left(1 + \frac{0.05}{1}\right)^{1t}$$
$$2 = (1.05)^{t}$$
$$t = \log_{1.05} 2$$
$$t \approx 14.21 \text{ years}$$

19. $A = P\left(1 + \dfrac{r}{n}\right)^{nt}$

$12{,}000 = 6000\left(1 + \dfrac{0.07}{4}\right)^{4t}$

$2 = \left(1 + \dfrac{0.07}{4}\right)^{4t}$

$2 = (1.0175)^{4t}$

$4t = \log_{1.0175} 2$

$t = \dfrac{1}{4}\log_{1.0175} 2$

$t \approx 9.99$ years

21. $A = Pe^{rt}$ or $A = P\left(1 + \dfrac{r}{n}\right)^{nt}$

$1587.75 \stackrel{?}{=} 750e^{0.075(10)}$

$1587.75 \stackrel{?}{=} 750e^{0.75}$

$1587.75 = 1587.75$

Continuous compounding

23. $A = Pe^{rt}$ or

$141.48 \stackrel{?}{=} 100e^{0.07(5)}$

$141.48 \stackrel{?}{=} 100e^{0.35}$

$141.48 \neq 141.90$

$A = P\left(1 + \dfrac{r}{n}\right)^{nt}$

$141.48 = 100\left(1 + \dfrac{0.07}{n}\right)^{n(5)}$

$n = 1:\ 141.48 \stackrel{?}{=} 100\left(1 + \dfrac{0.07}{1}\right)^{1(5)}$

$141.48 \neq 140.26$

$n = 4:\ 141.48 \stackrel{?}{=} \left(1 + \dfrac{0.07}{4}\right)^{4(5)}$

$141.48 \stackrel{?}{=} 100(1.0175)^{20}$

$141.48 = 141.48$

Quarterly compounding

25. $A = Pe^{rt}$

$= 1000e^{0.08(1)}$

$= 1083.29$

Interest $= 1083.29 - 1000$

$= 83.29$

Effective yield $= \dfrac{83.29}{1000}$

$= 0.08329$

$\approx 8.33\%$

27. $A = P\left(1 + \dfrac{r}{n}\right)^{nt}$

$= 1000\left(1 + \dfrac{0.07}{12}\right)^{12(1)}$

≈ 1072.29

Interest $= 1072.29 - 1000$

$= 72.29$

Effective yield $= \dfrac{72.29}{1000}$

$= 0.07229$

$\approx 7.23\%$

29. $A = P\left(1 + \dfrac{r}{n}\right)^{nt}$

$= 1000\left(1 + \dfrac{0.06}{4}\right)^{4(1)}$

≈ 1061.36

Interest $= 1061.36 - 1000$

$= 61.36$

Effective yield $= \dfrac{61.36}{1000}$

$= 0.06136$

$= 6.136\%$

31. $A = P\left(1 + \dfrac{r}{n}\right)^{nt}$

$ = 1000\left(1 + \dfrac{0.08}{12}\right)^{12(1)}$

$ \approx 1083.00$

Interest $= 1083 - 1000$

$ = 83$

Effective yield $= \dfrac{83}{1000}$

$ = 0.083$

$ = 8.3\%$

33. $A = Pe^{rt}$

$ = 1000e^{0.075(1)}$

$ \approx 1077.88$

Interest $= 1077.88 - 1000$

$ = 77.88$

Effective yield $= \dfrac{77.88}{1000}$

$ \approx 0.0779$

$ \approx 7.79\%$

35. No. The equation could be written as

$2P = P\left(1 + \dfrac{r}{n}\right)^{nt}$ or $2P = Pe^{rt}$.

As a first step, you could divide both sides of the equation by P and then solve the resulting equation.

37. $ A = Pe^{rt}$

$10{,}000 = Pe^{0.09(20)} = Pe^{1.8}$

$\dfrac{10{,}000}{e^{1.8}} = P$

$\$1652.99 \approx P$

39. $ A = P\left(1 + \dfrac{r}{n}\right)^{nt}$

$750 = P\left(1 + \dfrac{0.06}{365}\right)^{365(3)}$

$ = P\left(1 + \dfrac{0.06}{365}\right)^{1095}$

$\dfrac{750}{\left(1 + \dfrac{0.06}{365}\right)^{1095}} = P$

$\$626.46 \approx P$

41. $ A = P\left(1 + \dfrac{r}{n}\right)^{nt}$

$25{,}000 = P\left(1 + \dfrac{0.07}{12}\right)^{12(30)}$

$\phantom{25{,}000} = P\left(1 + \dfrac{0.07}{12}\right)^{360}$

$\dfrac{25{,}000}{\left(1 + \dfrac{0.07}{12}\right)^{360}} = P$

$\$3080.15 \approx P$

43. $ A = P\left(1 + \dfrac{r}{n}\right)^{nt}$

$1000 = P\left(1 + \dfrac{0.05}{365}\right)^{365(1)}$

$ = P\left(1 + \dfrac{0.05}{365}\right)^{365}$

$\dfrac{1000}{\left(1 + \dfrac{0.05}{365}\right)^{365}} = P$

$\$951.23 \approx P$

45. $ A = Pe^{rt}$

$500{,}000 = Pe^{0.08(25)}$

$500{,}000 = Pe^{2}$

$\dfrac{500{,}000}{e^{2}} = P$

$\$67{,}667.64 \approx P$

47. $A = \dfrac{P(e^{rt} - 1)}{e^{r/12} - 1}$

$ = \dfrac{30(e^{0.08(10)} - 1)}{e^{0.08/12} - 1} \approx \$5496.57$

49. $A = \dfrac{P(e^{rt} - 1)}{e^{r/12} - 1} = \dfrac{50(e^{0.10(40)} - 1)}{e^{0.10/12} - 1} \approx \$320{,}250.81$

51. $(0, 3)$: $3 = Ce^{k(0)}$

$3 = Ce^0$

$3 = C(1)$

$3 = C \implies y = 3e^{kt}$

$(2, 8)$: $8 = 3e^{k(2)}$

$\frac{8}{3} = e^{2k}$

$\ln \frac{8}{3} = 2k$

$\frac{1}{2} \ln \frac{8}{3} = k$

$0.4904 \approx k$

53. $(0, 400)$:

$400 = Ce^{k(0)}$

$400 = Ce^0$

$400 = C(1)$

$400 = C \implies y = 400e^{kt}$

$(3, 200)$:

$200 = 400e^{k(3)}$

$\frac{200}{400} = e^{3k}$

$\frac{1}{2} = e^{3k}$

$\ln \frac{1}{2} = 3k$

$\frac{1}{3} \ln \frac{1}{2} = k$

$-0.2310 \approx k$

55. $y = Ce^{kt}$

$(0, 13.1) \implies 13.1 = Ce^{k(0)}$

$13.1 = C(1)$

$13.1 = C \implies y = 13.1e^{kt}$

$(15, 14.1) \implies 14.1 = 13.1e^{k(15)}$

$\frac{14.1}{13.1} = e^{15k}$

$15k = \ln\left(\frac{14.1}{13.1}\right)$

$k = \frac{1}{15} \ln\left(\frac{14.1}{13.1}\right)$

$k \approx 0.0049$

$y = 13.1e^{0.0049t}$

For 2020, $t = 20$: $y = 13.1e^{0.0049(20)}$

$y \approx 14.4$ million

57. $y = Ce^{kt}$

$(0, 12.9) \implies 12.9 = Ce^{k(0)}$

$12.9 = C(1)$

$12.9 = C \implies y = 12.9e^{kt}$

$(15, 14.6) \implies 14.6 = 12.9e^{k(15)}$

$\frac{14.6}{12.9} = e^{15k}$

$15k = \ln\left(\frac{14.6}{12.9}\right)$

$k = \frac{1}{15} \ln\left(\frac{14.6}{12.9}\right)$

$k \approx 0.0083$

$y = 12.9e^{0.0083t}$

For 2020, $t = 20$: $y = 12.9e^{0.0083(20)}$

$y \approx 15.2$ million

59. $y = Ce^{kt}$

$(0, 18.1) \implies 18.1 = Ce^{k(0)}$

$18.1 = C(1)$

$18.1 = C \implies y = 18.1e^{kt}$

$(15, 26.1) \implies 26.1 = 18.1e^{k(15)}$

$\frac{26.1}{18.1} = e^{15k}$

$15k = \ln\left(\frac{26.1}{18.1}\right)$

$k = \frac{1}{15} \ln\left(\frac{26.1}{18.1}\right)$

$k \approx 0.0244$

$y = 18.1e^{0.0244t}$

For 2020, $t = 20$: $y = 18.1e^{0.0244(20)}$

$y \approx 29.5$ million

61. $y = Ce^{kt}$

$(0, 18.1) \implies 18.1 = Ce^{k(0)}$

$18.1 = C(1)$

$18.1 = C \implies y = 18.1e^{kt}$

$(15, 19.2) \implies 19.2 = 18.1e^{k(15)}$

$\frac{19.2}{18.1} = e^{15k}$

$15k = \ln\left(\frac{19.2}{18.1}\right)$

$k = \frac{1}{15} \ln\left(\frac{19.2}{18.1}\right)$

$k \approx 0.0039$

$y = 18.1e^{0.0039t}$

For 2020, $t = 20$: $y = 18.1e^{0.0039(20)}$

$y \approx 19.6$ million

63. The value of k is larger in Exercise 59 because the population of Bombay is increasing faster than the population of Shanghai.

65.

t (years)	0	1620	1000
Ce^{kt} (grams)	$Ce^{k(0)} = 5$	$Ce^{k(1620)} = 2.5$	$Ce^{k(1000)} = ?$

$(0, 5): 5 = Ce^{k(0)}$ $(1620, 2.5): 2.5 = 5e^{k(1620)}$

$\qquad\quad 5 = Ce^0 \qquad\qquad\qquad\qquad\quad \frac{2.5}{5} = e^{1620k}$

$\qquad\quad 5 = C(1) \qquad\qquad\qquad\qquad\quad \frac{1}{2} = e^{1620k}$

$\qquad\quad 5 = C \Rightarrow y = 5e^{kt} \qquad\qquad \ln\frac{1}{2} = 1620k$

$\qquad\qquad\qquad\qquad\qquad\qquad\qquad\quad \frac{1}{1620}\ln\frac{1}{2} = k$

$\qquad\qquad\qquad\qquad\qquad -0.00043 \approx k \Rightarrow y = 5e^{-0.00043t}$

$\qquad\qquad\qquad (1000, y): y = 5e^{-0.00043(1000)}$

$\qquad\qquad\qquad\qquad\qquad\qquad y = 5e^{-0.43}$

$\qquad\qquad\qquad\qquad\qquad\qquad y \approx 3.3 \text{ grams}$

67.

t (years)	0	5730	1000
Ce^{kt} (grams)	$Ce^{k(0)} = 5$	$Ce^{k(5730)} = 2.5$	$Ce^{k(1000)} = ?$

$(0, 5): 5 = Ce^{k(0)} \qquad\quad (5730, 2.5): 2.5 = 5e^{k(5730)}$

$\qquad\quad 5 = C(1) \qquad\qquad\qquad\qquad \frac{2.5}{5} = e^{5730k}$

$\qquad\quad 5 = C$

$\qquad\qquad\qquad\qquad\qquad\qquad\qquad \frac{1}{2} = e^{5730k}$

$\qquad\qquad\qquad\qquad\qquad\qquad 5730k = \ln\frac{1}{2}$

$\qquad\qquad\qquad\qquad\qquad\qquad\qquad k = \frac{\ln 0.5}{5730}$

$\qquad\qquad\qquad\qquad\qquad k \approx -0.000121 \Rightarrow y = 5e^{-0.00121t}$

$\qquad\qquad (1000, y): y = 5e^{-0.000121(1000)}$

$\qquad\qquad\qquad\qquad\qquad\quad = 5e^{-0.121}$

$\qquad\qquad\qquad\qquad\qquad\quad = 4.43 \text{ grams}$

69. $y = Ce^{kt}$

t (years)	0	13	10
Ce^{kt} (grams)	$Ce^{k(0)} = 100$	$Ce^{k(13)} = 50$	$Ce^{k(10)} = ?$

$(0, 100)$: $100 = Ce^{k(0)}$

$\qquad 100 = Ce^0$

$\qquad 100 = C(1)$

$\qquad 100 = C \implies y = 100e^{kt}$

$(13, 50)$: $50 = 100e^{k(13)}$

$\qquad 0.5 = e^{13k}$

$\qquad \ln 0.5 = 13k$

$\qquad \dfrac{\ln 0.5}{13} = k$

$-0.05332 \approx k \implies y = 100e^{-0.05332t}$

$\qquad y = 100e^{-0.05332(10)}$

$\qquad y = 100e^{-0.5332}$

$\qquad y \approx 58.7$

Approximately 58.7 grams will remain after 10 years.

71.

t (years)	0	5730	?
Ce^{kt} (grams)	$Ce^{k(0)} = C$	$Ce^{k(5730)} = 0.5C$	$Ce^{k(?)} = 0.15/C$

$(5730, 0.5C)$: $0.5C = Ce^{k(5730)}$

$\qquad \dfrac{0.5C}{C} = e^{5730k}$

$\qquad \dfrac{1}{2} = e^{5730k}$

$\qquad \ln \dfrac{1}{2} = 5730k$

$\qquad \dfrac{1}{5730} \ln \dfrac{1}{2} = k$

$-0.000121 \approx k \implies y = Ce^{-0.000121t}$

$(t, 0.15C)$: $0.15C = Ce^{-0.000121t}$

$\qquad \dfrac{0.15C}{C} = e^{-0.000121t}$

$\qquad 0.15 = e^{0.000121t}$

$\qquad \ln 0.15 = 0.000121t$

$\qquad \dfrac{\ln 0.15}{-0.000121} = t$

$\qquad 15{,}700 \text{ years} \approx t$

The tree burned approximately 15,700 years ago. (Answers may vary slightly depending on rounding.)

73. $y = 32{,}000(0.8)^x$

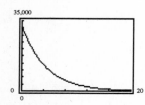

(a) After one year, the value of the truck is approximately $25,6000.

(b) The value of the truck will be approximately $16,000 in about 3.1 years.

75. 20 years:
$A = \$17{,}729.42$ (from graph)

Amount deposited:
$30(12)(20) = \$7200$

Interest:
$17{,}729.42 - 7200 = \$10{,}529.42$

77. $R = \log_{10} I \implies I = 10^R$

Alaska: $I = 10^{8.4}$

San Fernando Valley: $I = 10^{6.6}$

$\qquad \text{Ratio} = \dfrac{10^{8.4}}{10^{6.6}}$

$\qquad = 10^{8.4 - 6.6}$

$\qquad = 10^{1.8}$

$\qquad \approx 63$

The earthquake in Alaska was 63 times as great as the earthquake in the San Fernando Valley.

79. $R = \log_{10} I \implies I = 10^R$

Mexico City: $I = 10^{8.1}$

Nepal: $I = 10^{6.5}$

$$\text{Ratio} = \frac{10^{8.1}}{10^{6.5}}$$

$$= 10^{8.1 - 6.5}$$

$$= 10^{1.6}$$

$$\approx 40$$

The earthquake in Mexico City was 40 times as great as the earthquake in Nepal.

81. $p(t) = \dfrac{5000}{1 + 4e^{-t/6}}$

(a)

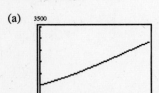

(b) $t = 0$: $p(0) = \dfrac{5000}{1 + 4e^{-0/6}}$

$$= \frac{5000}{1 + 4e^0}$$

$$= \frac{5000}{1 + 4}$$

$$= \frac{5000}{5}$$

$$= 1000$$

The original population was 1000.

(c) $t = 9$: $p(9) = \dfrac{5000}{1 + 4e^{-9/6}}$

$$= \frac{5000}{1 + 4e^{-3/2}}$$

$$\approx 2642$$

After 9 years, the population is approximately 2642.

(d)

$$2000 = \frac{5000}{1 + 4e^{-t/6}}$$

$$2000(1 + 4e^{-t/6}) = 5000$$

$$1 + 4e^{-t/6} = \frac{5000}{2000}$$

$$1 + 4e^{-t/6} = 2.5$$

$$4e^{-t/6} = 1.5$$

$$e^{-t/6} = \frac{1.5}{4}$$

$$e^{-t/6} = 0.375$$

$$-t/6 = \ln 0.375$$

$$t = -6 \ln 0.375$$

$$t \approx 5.9$$

The population will be 2000 after approximately 5.9 years.

83. (a) $2.5 = 10(1 - e^{k(5)})$

$$\frac{2.5}{10} = (1 - e^{5k})$$

$$0.25 = 1 - e^{5k}$$

$$-0.75 = -e^{5k}$$

$$0.75 = e^{5k}$$

$$\ln 0.75 = 5k$$

$$\frac{\ln 0.75}{5} = k$$

$$-0.0575 \approx k \implies S = 10(1 - e^{-0.0575x})$$

(b) $S = 10(1 - e^{-0.0575(7)})$

$S = 10(1 - e^{-0.4025})$

$S \approx 3.3$ (thousands)

If advertising expenditures are raised to $700, sales should be approximately 3300 units. (Answers may vary slightly depending on rounding.)

85. If $y = Ce^{kt}$ is a model of exponential decay, then $k < 0$.

87.
$$R = \log_{10} I_1 \implies I_1 = 10^R$$
$$R + 1 = \log_{10} I_2 \implies I_2 = 10^{R+1}$$
$$= 10 \cdot 10^R$$
$$= 10 I_1$$

If the reading on the Richter scale is increased by 1, the intensity of the earthquake is increased by a factor of 10.

89. $\log_{10} 10{,}000 = \log_{10} 10^4 = 4$

91. $\log_3 \dfrac{1}{81} = \log_3 3^{-4} = -4$

93.
$$\begin{cases} x - y = 0 \\ x + 2y = 9 \end{cases} \implies \begin{cases} -x + y = 0 \\ \underline{x + 2y = 9} \end{cases}$$
$$3y = 9$$
$$y = 3$$
$$x - 3 = 0$$
$$x = 3$$
$$(3, 3)$$

95.
$$\begin{cases} y = x^2 \\ -3x + 2y = 2 \end{cases}$$
$$-3x + 2(x^2) = 2$$
$$2x^2 - 3x - 2 = 0$$
$$(2x + 1)(x - 2) = 0$$
$$2x + 1 = 0 \implies 2x = -1 \implies x = -\frac{1}{2}$$
$$x - 2 = 0 \implies x = 2$$
$$x = -\frac{1}{2} \implies y = \left(-\frac{1}{2}\right)^2 = \frac{1}{4}$$
$$x = 2 \implies y = (2)^2 = 4$$
$$\left(-\frac{1}{2}, \frac{1}{4}\right) \quad \text{and} \quad (2, 4)$$

97.
$$\begin{cases} x - y = -1 \\ x + 2y - 2z = 3 \\ 3x - y + 2z = 3 \end{cases}$$
$$\begin{cases} x - y = -1 \\ 3y - 2z = 4 \\ 2y + 2z = 6 \end{cases}$$
$$\begin{cases} x - y = -1 \\ 3y - 2z = 4 \\ y + z = 3 \end{cases}$$
$$\begin{cases} x - y = -1 \\ y + z = 3 \\ 3y - 2z = 4 \end{cases}$$
$$\begin{cases} x - y = -1 \\ y + z = 3 \\ -5z = -5 \end{cases}$$
$$\begin{cases} x - y = -1 \\ y + z = 3 \\ z = 1 \end{cases}$$
$$y + 1 = 3 \qquad x - 2 = -1$$
$$y = 2 \qquad x = 1$$
$$(1, 2, 1)$$

99. *Verbal Model:* $\boxed{\begin{array}{c}\text{Distance of}\\ \text{first car}\end{array}} + \boxed{\begin{array}{c}\text{Distance of}\\ \text{second car}\end{array}} = 690$

Labels: Speed of first car $= x$ (miles per hour)

Speed of second car $= x + 15$ (miles per hour)

Distance of first car $= 6x$ (miles)

Distance of second car $= 6(x + 15)$ (miles)

Equation: $6x + 6(x + 15) = 690$

$$6x + 6x + 90 = 690$$

$$12x + 90 = 690$$

$$12x = 600$$

$$x = 50$$

$$x + 15 = 65$$

The speeds of the two cars are 50 kilometers per hour and 65 kilometers per hour.

Review Exercises for Chapter 9

1. $f(x) = 2^x$

(a) $f(-3) = 2^{-3} = \dfrac{1}{2^3} = \dfrac{1}{8}$

(b) $f(1) = 2^1 = 2$

(c) $f(2) = 2^2 = 4$

3. $f(x) = 4 + 3^x$

(a) $f(1) = 4 + 3^1 = 4 + 3 = 7$

(b) $f(-2) = 4 + 3^{-2} = 4 + \dfrac{1}{9} = \dfrac{37}{9}$

(c) $f(10) = 4 + 3^{10} = 4 + 59{,}049 = 59{,}053$

5.

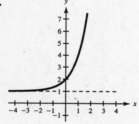

Horizontal asymptote: $y = 1$

7.

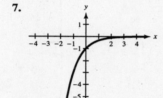

Horizontal asymptote: $y = 0$

9.

Horizontal asymptote: $y = 0$

11.

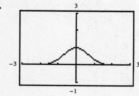

13. $g(t) = e^{-t/3}$

(a) $g(-3) = e^{-(-3)/3} = e \approx 2.718$

(b) $g(\pi) = e^{-\pi/3} \approx 0.351$

(c) $g(6) = e^{-6/3} = e^{-2} \approx 0.135$

15. $h(s) = 1 - e^{0.2s}$

(a) $h(0) = 1 - e^{0.2(0)}$

$= 1 - e^0$

$= 1 - 1 = 0$

(b) $h(2) = 1 - e^{0.2(2)}$

$= 1 - e^{0.4}$

≈ -0.492

(c) $h(\sqrt{10}) = 1 - e^{0.2\sqrt{10}}$

$= -0.882$

17.

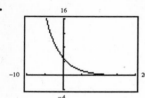

19.

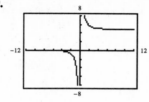

21. $\ln e = 1 \implies e^1 = e$

23. $\log_5\left(\frac{1}{25}\right) = -2 \implies 5^{-2} = \frac{1}{25}$

25. $4^3 = 64 \implies \log_4 64 = 3$

27. $25^{3/2} = 125 \implies \log_{25} 125 = \frac{3}{2}$

29. $\log_{10} 1000$

The power to which 10 must be raised to obtain 1000 is 3.

$$\log_{10} 1000 = 3$$

31. $\log_3 \frac{1}{9}$

The power to which 3 must be raised to obtain $\frac{1}{9}$ is -2.

$$\log_3 \frac{1}{9} = -2$$

33. $\log_2 64$

The power to which 2 must be raised to obtain 64 is 6.

$$\log_2 64 = 6$$

35. $\log_{10} \frac{1}{1000}$

The power to which 10 must be raised to obtain $\frac{1}{1000}$ is -3.

$$\log_{10} \frac{1}{1000} = -3$$

37. $\log_3 27$

The power to which 3 must be raised to obtain 27 is 3.

$$\log_3 27 = 3$$

39. $\log_{10} 0.01$

The power to which 10 must be raised to obtain 0.01 is -2.

$$\log_{10} 0.01 = -2$$

41. $\log_4 \frac{1}{64}$

The power to which 4 must be raised to obtain $\frac{1}{64}$ is -3.

$$\log_4 \frac{1}{64} = -3$$

43. $\log_2 \sqrt{2} = \log_2 2^{1/2}$

The power to which 2 must be raised to obtain $\sqrt{2}$ is $\frac{1}{2}$.

$$\log_2 \sqrt{2} = \frac{1}{2}$$

45.

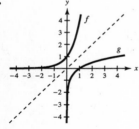

47. $f(x) = -2 + \log_3 x$

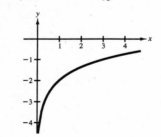

49.

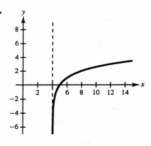

51. $\ln e^3 = 3$

53. $\ln\left(\frac{5}{4}\right) \approx 0.223$

55.

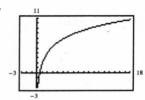

57. $\log_4 9 = \dfrac{\log_{10} 9}{\log_{10} 4} \approx 1.585$

$\log_4 9 = \dfrac{\ln 9}{\ln 4} \approx 1.585$

59. $\log_{12} 200 = \dfrac{\log_{10} 200}{\log_{10} 12} \approx 2.132$

$\log_{12} 200 = \dfrac{\ln 200}{\ln 12} \approx 2.132$

61. $\log_5 18 = \log_5(3^2 \cdot 2)$

$= \log_5 3^2 + \log_5 2$

$= 2 \log_5 3 + \log_5 2$

$\approx 2(0.6826) + (0.4307)$

≈ 1.7959

63. $\log_5\left(\frac{1}{2}\right) = \log_5 1 - \log_5 2$

$\approx 0 - (0.4307)$

≈ -0.4307

65. $\log_5 (12)^{2/3} = \frac{2}{3} \log_5 12$

$= \frac{2}{3} \log_5 (2^2 \cdot 3)$

$= \frac{2}{3}(\log_5 2^2 + \log_5 3)$

$= \frac{2}{3}(2 \log_5 2 + \log_5 3)$

$= \frac{4}{3} \log_5 2 + \frac{2}{3} \log_5 3$

$\approx \frac{4}{3}(0.4307) + \frac{2}{3}(0.6826)$

≈ 1.0293

67. $\log_4 6x^4 = \log_4 6 + \log_4 x^4$

$= \log_4 6 + 4 \log_4 x$

69. $\log_5 \sqrt{x + 2} = \log_5 (x + 2)^{1/2}$

$= \frac{1}{2} \log_5 (x + 2)$

71. $\ln \frac{x + 2}{x - 2} = \ln(x + 2) - \ln(x - 2)$

73. $\ln\left[\sqrt{2x}\,(x + 3)^5\right] = \ln \sqrt{2x} + \ln(x + 3)^5$

$= \ln(2x)^{1/2} + \ln(x + 3)^5$

$= \frac{1}{2} \ln(2x) + 5 \ln(x + 3)$

$= \frac{1}{2}(\ln 2 + \ln x) + 5 \ln(x + 3)$

or $\frac{1}{2} \ln 2 + \frac{1}{2} \ln x + 5 \ln(x + 3)$

75. $\log_4 x - \log_4 10 = \log_4 \frac{x}{10}$

77. $\log_8 16x + \log_8 2x^2 = \log_8[16x(2x^2)]$

$= \log_8(32x^3)$

79. $-2(\ln 2x - \ln 3) = -2 \ln \frac{2x}{3}$

$= \ln\left(\frac{2x}{3}\right)^{-2}, \ x > 0$

or $\ln \frac{9}{4x^2}, \ x > 0$

81. $3 \ln x + 4 \ln y + \ln z = \ln x^3 + \ln y^4 + \ln z$

$= \ln(x^3 y^4 z)$

83. $4[\log_2 k - \log_2(k - t)] = 4 \log_2\left(\frac{k}{k - t}\right)$

$= \log_2\left(\frac{k}{k - t}\right)^4$

85. $\log_2 4x \neq 2 \log_2 x$

False

Note: $\log_2 4x = \log_2 4 + \log_2 x$

$= 2 + \log_2 x$

87. $\log_{10} 10^{2x} = 2x \log_{10} 10$

$= 2x(1) = 2x$

True

89. $\log_4 \frac{16}{x} = \log_4 16 - \log_4 x = 2 - \log_4 x$

True

91. $2^{x+1} = 64$

$2^{x+1} = 2^6$

$x + 1 = 6$

$x = 5$

93. $4^{x-3} = \frac{1}{16}$

$4^{x-3} = 4^{-2}$

$x - 3 = -2$

$x = 1$

95. $\log_3 5x = \log_3 85$

$5x = 85$

$x = 17$

97. $\ln(x + 4) = \ln 17$

$x + 4 = 17$

$x = 13$

99. $3^x = 500$

$x = \log_3 500$

$x \approx 5.66$

101. $2e^{x/2} = 45$

$e^{x/2} = \frac{45}{2}$

$e^{x/2} = 22.5$

$\frac{x}{2} = \ln 22.5$

$x = 2 \ln 22.5$

$x \approx 6.23$

103. $\dfrac{500}{(1.05)^x} = 100$

$500 = 100(1.05)^x$

$\dfrac{500}{100} = (1.05)^x$

$5 = 1.05^x$

$x = \log_{1.05} 5$

$x \approx 32.99$

105. $\log_3 x = 5$

$x = 3^5$

$x = 243$

107. $\log_5(x - 10) = 2$

$x - 10 = 5^2$

$x - 10 = 25$

$x = 35$

109. $\log_{10} 2x = 1.5$

$2x = 10^{1.5}$

$x = \dfrac{10^{1.5}}{2}$

$x \approx 15.81$

111. $\ln x = 7.25$

$x = e^{7.25}$

$x \approx 1408.10$

113. $\log_2 2x = -0.65$

$2x = 2^{-0.65}$

$x = \dfrac{2^{-0.65}}{2}$

$x \approx 0.32$

115. $\log_2 x + \log_2 3 = 3$

$\log_2(3x) = 3$

$3x = 2^3$

$3x = 8$

$x = \frac{8}{3}$

or $x \approx 2.67$

117. $\log_3 (x + 2) - \log_3 x = 3$

$\log_3 \dfrac{x + 2}{x} = 3$

$\dfrac{x + 2}{x} = 3^3$

$\dfrac{x + 2}{x} = 27$

$x + 2 = 27x$

$2 = 26x$

$\dfrac{2}{26} = x$

$\dfrac{1}{13} = x$

$0.08 \approx x$

119. $A = P\left(1 + \dfrac{r}{n}\right)^{nt}$

n	1	4	12	365	Continuous
A	\$3806.13	\$4009.59	\$4058.25	\$4082.26	\$4083.08

$A = 500\left(1 + \dfrac{0.07}{1}\right)^{1(30)}$

≈ 3806.13

$A = 500\left(1 + \dfrac{0.07}{4}\right)^{4(30)}$

≈ 4009.59

$A = 500\left(1 + \dfrac{0.07}{12}\right)^{12(30)}$

≈ 4058.25

$A = 500\left(1 + \dfrac{0.07}{365}\right)^{365(30)}$

≈ 4082.26

$A = Pe^{rt}$

$= 500e^{0.07(30)}$

≈ 4083.08

121. $A = P\left(1 + \dfrac{r}{n}\right)^{nt}$

n	1	4	12	365	Continuous
P	\$2301.55	\$2103.50	\$2059.87	\$2038.82	\$2038.11

$50{,}000 = P\left(1 + \dfrac{0.08}{1}\right)^{1(40)}$

$\dfrac{50{,}000}{(1.08)^{40}} = P$

$2301.55 \approx P$

$50{,}000 = P\left(1 + \dfrac{0.08}{4}\right)^{4(40)}$

$\dfrac{50{,}000}{(1.02)^{160}} = P$

$2103.50 \approx P$

$50{,}000 = P\left(1 + \dfrac{0.08}{12}\right)^{12(40)}$

$\dfrac{50{,}000}{\left(1 + \dfrac{0.08}{12}\right)^{480}} = P$

$2059.87 \approx P$

$50{,}000 = P\left(1 + \dfrac{0.08}{365}\right)^{365(40)}$

$\dfrac{50{,}000}{\left(1 + \dfrac{0.08}{365}\right)^{14{,}600}} = P$

$2038.82 \approx P$

$A = Pe^{rt}$

$50{,}000 = Pe^{0.08(40)}$

$\dfrac{50{,}000}{e^{3.2}} = P$

$2038.11 \approx P$

123. $A = P\left(1 + \dfrac{r}{n}\right)^{nt}$

$= 1000\left(1 + \dfrac{0.085}{12}\right)^{12(1)}$

≈ 1088.39

Interest $= 1088.39 - 1000$

$= 88.39$

Effective yield $= \dfrac{88.39}{1000}$

≈ 0.0884

$\approx 8.84\%$

125. $y = Ce^{kt}$

t (years)	0	223,000	100,000
Ce^{kt} (grams)	$Ce^{k(0)} = 5$	$Ce^{k(223{,}000)} = 2.5$	$Ce^{k(100{,}000)} = ?$

$(0, 5): 5 = Ce^{k(0)}$

$5 = Ce^0$

$5 = C(1)$

$5 = C \implies y = 5e^{kt}$

$(223{,}000, 2.5): 2.5 = 5e^{k(223{,}000)}$

$0.5 = e^{223{,}000k}$

$\ln 0.5 = 223{,}000k$

$\dfrac{\ln 0.5}{223{,}000} = k$

$-0.000003108 \approx k \implies y = 5e^{-0.000003108t}$

$y = 5e^{-0.000003108(100{,}000)}$

$y = 5e^{-0.3108}$

$y \approx 3.7$ grams

Approximately 3.7 grams will remain after 100,000 years.

127.
$$p = 25 - 0.4e^{0.02x}$$
$$16.97 = 25 - 0.4e^{0.02x}$$
$$-8.03 = -0.4e^{0.02x}$$
$$8.03 = 0.4e^{0.02x}$$
$$\frac{8.03}{0.4} = e^{0.02x}$$
$$20.075 = e^{0.02x}$$
$$\ln 20.075 = 0.02x$$
$$\frac{\ln 20.075}{0.02} = x$$
$$150 \approx x$$

The demand is approximately 150 units.

129. (a) $P = \dfrac{500}{1 + 4e^{-0.36t}}$

$$= \frac{500}{1 + 4e^{-0.36(5)}}$$

$$= \frac{500}{1 + 4e^{-1.8}}$$

$$\approx \frac{500}{1 + 4(0.1653)}$$

$$= \frac{500}{1.6612}$$

$$\approx 300.987$$

After 5 years, the population is approximately 301 deer.

(b)
$$P = \frac{500}{1 + 4e^{-0.36t}}$$

$$250 = \frac{500}{1 + 4e^{-0.36t}}$$

$$1 + 4e^{-0.36t} = \frac{500}{250}$$

$$1 + 4e^{-0.36t} = 2$$

$$4e^{-0.36t} = 1$$

$$e^{-0.36t} = 0.25$$

$$\ln e^{-0.36t} = \ln 0.25$$

$$-0.36t = \ln 0.25$$

$$t = \frac{\ln 0.25}{-0.36} \approx 3.85$$

The deer population will be 250 in approximately 3.85 years.

(c)
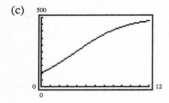

(d) There is a horizontal asymptote at $y = 0$ because the population cannot be less than zero. There is also a horizontal asymptote at $P = 500$ because the deer population will level off at 500.

Chapter Test for Chapter 9

1. $f(t) = 54\left(\dfrac{2}{3}\right)^t$

$f(-1) = 54\left(\dfrac{2}{3}\right)^{-1} = 54\left(\dfrac{3}{2}\right) = 81$

$f(0) = 54\left(\dfrac{2}{3}\right)^0 = 54(1) = 54$

$f\left(\dfrac{1}{2}\right) = 54\left(\dfrac{2}{3}\right)^{1/2} = 54\sqrt{\dfrac{2}{3}} \approx 44.09$

$f(2) = 54\left(\dfrac{2}{3}\right)^2 = 54\left(\dfrac{4}{9}\right) = 24$

2.

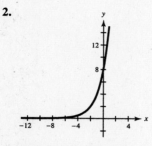

Horizontal asymptote: $y = 0$

3.

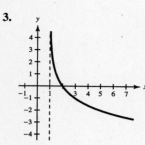

Domain: $(1, \infty)$

Vertical asymptote: $x = 1$

4. $\log_5 125 = 3 \implies 5^3 = 125$

5. $4^{-2} = \dfrac{1}{16} \implies \log_4 \dfrac{1}{16} = -2$

6. $\log_8 2$

The power to which 8 must be raised to obtain 2 is $\frac{1}{3}$.

$\left(8^{1/3} = \sqrt[3]{8} = 2\right)$

$\log_8 2 = \frac{1}{3}$

7. $f(x) = \log_5 x$ and $g(x) = 5^x$

f is the inverse function of g.

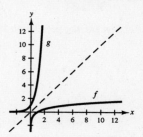

8. $\log_4\left(5x^2/\sqrt{y}\right) = \log_4 5x^2 - \log_4 \sqrt{y}$

$\qquad = \log_4 5 + \log_4 x^2 - \log_4 y^{1/2}$

$\qquad = \log_4 5 + 2\log_4 x - \dfrac{1}{2}\log_4 y$

9. $8\ln a + \ln b - 3\ln c = \ln a^8 + \ln b - \ln c^3$

$\qquad = \ln(a^8 b) - \ln c^3$

$\qquad = \ln\left(\dfrac{a^8 b}{c^3}\right)$

10. $\log_4 x = 3$

$\quad x = 4^3$

$\quad x = 64$

11. $10^{3y} = 832$

$3y = \log_{10} 832$

$y = \dfrac{\log_{10} 832}{3}$

$y \approx 0.97$

12. $400e^{0.08t} = 1200$

$e^{0.08t} = \dfrac{1200}{400}$

$e^{0.08t} = 3$

$0.08t = \ln 3$

$t = \dfrac{\ln 3}{0.08}$

$t \approx 13.73$

13. $\dfrac{1.06^x - 1}{0.06} = 2.56$

$1.06^x - 1 = 2.56(0.06)$

$1.06^x - 1 = 0.1536$

$1.06^x = 1.1536$

$x = \log_{1.06} 1.1536$

$x \approx 2.452$

14. $\dfrac{1 - 1.0045^{-x}}{0.0045} = 92.57$

$1 - 1.0045^{-x} = 92.57(0.0045)$

$1 - 1.0045^{-x} = 0.416565$

$-1.0045^{-x} = -0.583435$

$1.0045^{-x} = 0.583435$

$-x = \log_{1.0045} 0.583435$

$x = -\log_{1.0045} 0.583435$

$x \approx 120.007$

15. $3 \ln(2x - 3) = 10$

$\ln(2x - 3) = \dfrac{10}{3}$

$2x - 3 = e^{10/3}$

$2x = 3 + e^{10/3}$

$x = \dfrac{3 + e^{10/3}}{2}$

≈ 15.516

16. $\log_2 x + \log_2(x + 4) = 5$

$\log_2[x(x + 4)] = 5$

$x(x + 4) = 2^5$

$x^2 + 4x = 32$

$x^2 + 4x - 32 = 0$

$(x - 4)(x + 8) = 0$

$x - 4 = 0 \Longrightarrow x = 4$

$x + 8 = 0 \Longrightarrow x = -8$ (Extraneous)

(Note: $\log_2(-8)$ undefined.**)**

17. $2 \log_{10} x - \log_{10} 9 = 2$

$\log_{10} x^2 - \log_{10} 9 = 2$

$\log_{10}\left(\dfrac{x^2}{9}\right) = 2$

$\dfrac{x^2}{9} = 10^2$

$x^2 = 900$

$x = \pm\sqrt{900}$

$x = 30$

Discard negative answer; $\log_{10}(-30)$ undefined.

18. (a) $A = P\left(1 + \dfrac{r}{n}\right)^{nt}$

$A = 2000\left(1 + \dfrac{0.07}{4}\right)^{4(20)}$

$A = 2000\left(1 + \dfrac{0.07}{4}\right)^{80}$

$A \approx 8012.78$

The balance in the account compounded quarterly is $8012.78.

(b) $A = Pe^{rt}$

$A = 2000e^{0.07(20)}$

$A \approx 8110.40$

The balance in the account compounded continuously is $8110.40.

19. $A = P\left(1 + \dfrac{r}{n}\right)^{nt}$

$100{,}000 = P\left(1 + \dfrac{0.09}{4}\right)^{4(25)}$

$100{,}000 = P(1.0225)^{100}$

$\dfrac{100{,}000}{(1.0225)^{100}} = P$

$\$10{,}806.08 \approx P$

The principal is $10,806.08.

20. $A = Pe^{rt}$

$1006.88 = 500e^{r(10)}$

$\dfrac{1006.88}{500} = e^{10r}$

$2.01376 = e^{10r}$

$\ln 2.01376 = 10r$

$\dfrac{\ln 2.01376}{10} = r$

$0.0700 \approx r$

$7\% \approx r$

The annual interest rate is 7%.

21

t (years)	0	1	3
Ce^{kt} (dollars)	$Ce^{k(0)} = 18{,}000$	$Ce^{k(1)} = 14{,}000$	$Ce^{k(3)} = ?$

$y = Ce^{kt}$

$(0, 18{,}000)$: $18{,}000 = Ce^{k(0)}$

$\qquad\qquad 18{,}000 = Ce^0$

$\qquad\qquad 18{,}000 = C(1)$

$\qquad\qquad 18{,}000 = C \implies y = 18{,}000e^{kt}$

$(1, 14{,}000)$: $14{,}000 = 18{,}000e^{k(1)}$

$\qquad\qquad \dfrac{14{,}000}{18{,}000} = e^k$

$\qquad\qquad \dfrac{7}{9} = e^k$

$\qquad\qquad \ln\dfrac{7}{9} = k$

$\qquad -0.25131 \approx k \implies y = 18{,}000e^{-0.25131(t)}$

$(3, y)$: $y = 18{,}000e^{-0.25131(3)}$

$\qquad\quad y = 18{,}000e^{-0.75393}$

$\qquad\quad y \approx \$8469$

The value of the car after 3 years is approximately \$8469. (Answers may vary slightly depending upon rounding.)

22. $p(t) = \dfrac{2400}{1 + 3e^{-t/4}}$

(a) $p(4) = \dfrac{2400}{1 + 3e^{-4/4}}$

$\qquad\quad = \dfrac{2400}{1 + 3e^{-1}}$

$\qquad\quad \approx 1141$

The population after 4 years is approximately 1141.

(b) $\qquad 1200 = \dfrac{2400}{1 + 3e^{-t/4}}$

$\qquad 1200(1 + 3e^{-t/4}) = 2400$

$\qquad\qquad 1 + 3e^{-t/4} = \dfrac{2400}{1200}$

$\qquad\qquad 1 + 3e^{-t/4} = 2$

$\qquad\qquad 3e^{-t/4} = 1$

$\qquad\qquad e^{-t/4} = \dfrac{1}{3}$

$\qquad\qquad -\dfrac{t}{4} = \ln\dfrac{1}{3}$

$\qquad\qquad t = -4\ln\dfrac{1}{3}$

$\qquad\qquad t \approx 4.4$

The population will be 1200 in approximately 4.4 years

Cumulative Test for Chapters 1–9

1. $4 - \frac{1}{2}x = 6$

$2\left(4 - \frac{1}{2}x\right) = 2(6)$

$8 - x = 12$

$-x = 4$

$x = -4$

2. $12(3 - x) = 5 - 7(2x + 1)$

$36 - 12x = 5 - 14x - 7$

$36 - 12x = -14x - 2$

$36 + 2x = -2$

$2x = -38$

$x = -19$

3. $x^2 - 10 = 0$

$x^2 = 10$

$x = \pm\sqrt{10}$

4. $-2 \leq 1 - 2x \leq 2$

$-3 \leq -2x \quad \leq 1$

$\dfrac{-3}{-2} \geq \dfrac{-2x}{-2} \quad \geq \dfrac{1}{-2}$

$\dfrac{3}{2} \geq \quad x \quad \geq -\dfrac{1}{2}$ or $-\dfrac{1}{2} \leq x \leq \dfrac{3}{2}$

5. $|x + 7| \geq 2$

$x + 7 \leq -2 \quad$ or $\quad x + 7 \geq 2$

$x \leq -9 \quad$ or $\quad x \geq -5$

6. $x^2 + 3x - 10 < 0$

Critical numbers: $\quad x^2 + 3x - 10 = 0$

$(x + 5)(x - 2) = 0$

$x + 5 = 0 \implies x = -5$

$x - 2 = 0 \implies x = 2$

The critical numbers are -5 and 2.

The solution is $(-5, 2)$ or $-5 < x < 2$.

Interval	Representative x-value	Is inequality satisfied?
$(-\infty, -5)$	$x = -6$	$(-6)^2 + 3(-6) - 10 \stackrel{?}{<} 0$ $8 \not< 0$
$(-5, 2)$	$x = 0$	$0^2 + 3(0) - 8 \stackrel{?}{<} 0$ $-8 < 0$
$(2, \infty)$	$x = 5$	$5^2 + 3(5) - 8 \stackrel{?}{<} 0$ $32 \not< 0$

7. $(x^2 \cdot x^3)^4 = (x^5)^4$

$= x^{20}$

8. $\left(\dfrac{3x^2}{2y}\right)^{-2} = \dfrac{3^{-2}x^{-4}}{2^{-2}y^{-2}}$

$= \dfrac{2^2 y^2}{3^2 x^4}$

$= \dfrac{4y^2}{9x^4}$

9. $(9x^8)^{3/2} = 9^{3/2}x^{8(3/2)}$

$= \left(\sqrt{9}\right)^3 x^{12}$

$= 27x^{12}$

10. $\dfrac{\left(\dfrac{4}{x^2 - 9} + \dfrac{2}{x - 2}\right)}{\left(\dfrac{1}{x + 3} + \dfrac{1}{x - 3}\right)} = \dfrac{\left(\dfrac{4(x - 2) + 2(x^2 - 9)}{(x + 3)(x - 3)(x - 2)}\right)}{\left(\dfrac{1(x - 3) + 1(x + 3)}{(x - 3)(x + 3)}\right)}$

$= \dfrac{\left(\dfrac{4x - 8 + 2x^2 - 18}{(x + 3)(x - 3)(x - 2)}\right)}{\left(\dfrac{x - 3 + x + 3}{(x - 3)(x + 3)}\right)}$

$= \dfrac{\left(\dfrac{2x^2 + 4x - 26}{(x + 3)(x - 3)(x - 2)}\right)}{\left(\dfrac{2x}{(x - 3)(x + 3)}\right)}$

$= \dfrac{2x^2 + 4x - 26}{(x + 3)(x - 3)(x - 2)} \div \dfrac{2x}{(x - 3)(x + 3)}$

$= \dfrac{2(x^2 + 2x - 13)}{(x + 3)(x - 3)(x - 2)} \cdot \dfrac{(x - 3)(x + 3)}{2x}$

$= \dfrac{2(x^2 + 2x - 13)\cancel{(x - 3)}\cancel{(x + 3)}}{\cancel{(x + 3)}\cancel{(x - 3)}(x - 2)\cancel{(2)}(x)}$

$= \dfrac{x^2 + 2x - 13}{x(x - 2)}, \; x \neq 3, x \neq -3$

11. $\dfrac{\left(\dfrac{1}{x+1}+\dfrac{1}{2}\right)}{\left(\dfrac{3}{2x^2+4x+2}\right)} = \dfrac{\left(\dfrac{1(2)+(x+1)}{(x+1)(2)}\right)}{\left(\dfrac{3}{2(x^2+2x+1)}\right)}$

$= \dfrac{\left(\dfrac{2+x+1}{(x+1)(2)}\right)}{\left(\dfrac{3}{2(x+1)(x+1)}\right)}$

$= \dfrac{\left(\dfrac{x+3}{(x+1)(2)}\right)}{\left(\dfrac{3}{2(x+1)(x+1)}\right)}$

$= \dfrac{x+3}{(x+1)(2)} \div \dfrac{3}{2(x+1)(x+1)}$

$= \dfrac{x+3}{(x+1)(2)} \cdot \dfrac{2(x+1)(x+1)}{3}$

$= \dfrac{(x+3)(2)(x+1)(x+1)}{(x+1)(2)(3)}$

$= \dfrac{(x+3)(x+1)}{3}, \; x \neq -1$

12.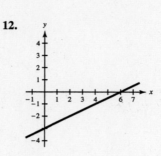

13.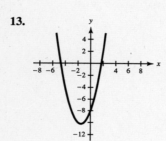

14.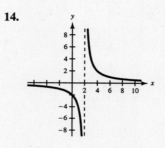

15.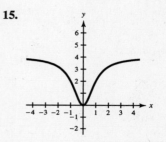

16. $\sqrt{x}-x+12=0$ Let $u=\sqrt{x}$

$u-u^2+12=0$

$-u^2+u+12=0$

$u^2-u-12=0$

$(u-4)(u+3)=0$

$u-4=0 \Rightarrow u=4$

$u+3=0 \Rightarrow u=-3$

$u=4 \Rightarrow \sqrt{x}=4 \Rightarrow x=16$

$u=-3 \Rightarrow \sqrt{x}=-3$ (Extraneous)

17. $\sqrt{5-x}+10=11$

$\sqrt{5-x}=1$

$\left(\sqrt{5-x}\right)^2=1^2$

$5-x=1$

$-x=-4$

$x=4$

18.
$$\frac{1}{x} + \frac{4}{10 - x} = 1$$

$$x(10 - x)\left(\frac{1}{x} + \frac{4}{10 - x}\right) = 1(x)(10 - x)$$

$$10 - x + 4x = x(10 - x)$$

$$3x + 10 = 10x - x^2$$

$$x^2 - 7x + 10 = 0$$

$$(x - 5)(x - 2) = 0$$

$$x - 5 = 0 \Rightarrow x = 5$$

$$x - 2 = 0 \Rightarrow x = 2$$

19.
$$\frac{x - 3}{x} + 1 = \frac{x - 4}{x - 6}$$

$$x(x - 6)\left(\frac{x - 3}{x} + 1\right) = \left(\frac{x - 4}{x - 6}\right)x(x - 6)$$

$$(x - 6)(x - 3) + x(x - 6) = (x - 4)x$$

$$x^2 - 9x + 18 + x^2 - 6x = x^2 - 4x$$

$$2x^2 - 15x + 18 = x^2 - 4x$$

$$x^2 - 11x + 18 = 0$$

$$(x - 9)(x - 2) = 0$$

$$x - 9 = 0 \Rightarrow x = 9$$

$$x - 2 = 0 \Rightarrow x = 2$$

20. Perimeter $= 4(12 - 2 \cdot 4) + 4\sqrt{4^2 + 4^2}$

$$= 4(4) + 4\sqrt{32}$$

$$= 16 + 16\sqrt{2}$$

$$\approx 38.63 \text{ inches}$$

21. This transformation is a horizontal shift of 4 units to the left.

22. This transformation is a reflection in the x-axis.

23. $\sqrt{-36}\sqrt{-3} = 6i\left(\sqrt{3}\,i\right)$

$$= 6\sqrt{3}\,i^2$$

$$= -6\sqrt{3}$$

24. $(7 + 4i) - (6 - 2i) = 7 + 4i - 6 + 2i$

$$= 1 + 6i$$

25. $(2 + 9i)(5 - i) = 10 - 2i + 45i - 9i^2$

$$= 10 + 43i - 9(-1)$$

$$= 10 + 43i + 9$$

$$= 19 + 43i$$

26. $\dfrac{1 + 2i}{2 + 3i} = \dfrac{1 + 2i}{2 + 3i} \cdot \dfrac{2 - 3i}{2 - 3i}$

$$= \frac{2 - 3i + 4i - 6i^2}{4 - 6i + 6i - 9i^2}$$

$$= \frac{8 + i}{13}$$

$$= \frac{8}{13} + \frac{1}{13}i$$

27. $\begin{cases} 4x - 3y + 2z = -2 \\ -2x + y + z = 1 \\ x - 2y - 6z = -12 \end{cases}$

$\begin{cases} x - 2y - 6z = -12 \\ -2x + y + z = 1 \\ 4x - 3y + 2z = -2 \end{cases}$ Interchange Eqn. 1 and Eqn. 3

$\begin{cases} x - 2y - 6z = -12 \\ -3y - 11z = -23 \\ 4x - 3y + 2z = -2 \end{cases}$ 2Eqn. 1 + Eqn. 2

$\begin{cases} x - 2y - 6z = -12 \\ -3y - 11z = -23 \\ 5y + 26z = 46 \end{cases}$ −4Eqn. 1 + Eqn. 3

$\begin{cases} x - 2y - 6z = -12 \\ -3y - 11z = -23 \\ 23z = 23 \end{cases}$ 5Eqn. 2 + 3Eqn. 3

$\begin{cases} x - 2y - 6z = -12 \\ -3y - 11z = -23 \\ z = 1 \end{cases}$ $\frac{1}{23}$Eqn. 3

$z = 1 \implies -3y - 11(1) = -23$

$-3y - 11 = -23$

$-3y = -12$

$y = 4$

$y = 4, z = 1 \implies x - 2(4) - 6(1) = -12$

$x - 14 = -12$

$x = 2$

Solution: $(2, 4, 1)$

28. $\begin{cases} 2x - y = 4 \\ 3x + y = -5 \end{cases}$

$5x = -1$

$x = -\frac{1}{5}$

$3\left(-\frac{1}{5}\right) + y = -5$

$-\frac{3}{5} + y = -5$

$y = -\frac{25}{5} + \frac{3}{5}$

$y = -\frac{22}{5}$

Solution: $\left(-\frac{1}{5}, -\frac{22}{5}\right)$

29. $\begin{vmatrix} 10 & 25 \\ 6 & -5 \end{vmatrix} = 10(-5) - 6(25)$

$= -50 - 150$

$= -200$

30. $\begin{vmatrix} 4 & 3 & 5 \\ 3 & 2 & -2 \\ 5 & -2 & 0 \end{vmatrix} = 4\begin{vmatrix} 2 & -2 \\ -2 & 0 \end{vmatrix} - 3\begin{vmatrix} 3 & 5 \\ -2 & 0 \end{vmatrix} + 5\begin{vmatrix} 3 & 5 \\ 2 & -2 \end{vmatrix}$

$= 4(0 - 4) - 3(0 + 10) + 5(-6 - 10)$

$= 4(-4) - 3(10) + 5(-16)$

$= -16 - 30 - 80$

$= -126$

31. $\begin{vmatrix} x & y & 1 \\ 3 & 5 & 1 \\ 1 & 2 & 1 \end{vmatrix} = 0$

$x\begin{vmatrix} 5 & 1 \\ 2 & 1 \end{vmatrix} - 3\begin{vmatrix} y & 1 \\ 2 & 1 \end{vmatrix} + 1\begin{vmatrix} y & 1 \\ 5 & 1 \end{vmatrix} = 0$

$x(3) - 3(y - 2) + 1(y - 5) = 0$

$3x - 3y + 6 + y - 5 = 0$

$3x - 2y + 1 = 0$

32. $f(x) = x - 3$

$g(x) = 4x + 1$

(a) $(f + g)(x) = f(x) + g(x)$

$= x - 3 + 4x + 1$

$= 5x - 2$

(b) $(f - g)(x) = f(x) - g(x)$

$= x - 3 - (4x + 1)$

$= x - 3 - 4x - 1$

$= -3x - 4$

(c) $(fg)(x) = f(x) \cdot g(x)$

$= (x - 3)(4x + 1)$

$= 4x^2 + x - 12x - 3$

$= 4x^2 - 11x - 3$

(d) $(f/g)(x) = \dfrac{f(x)}{g(x)}$

$= \dfrac{x - 3}{4x + 1}$

Domain of $(f/g)(x)$: $\left(-\infty, -\frac{1}{4}\right) \cup \left(-\frac{1}{4}, \infty\right)$

33. $f(x) = \sqrt{x - 1}$

$g(x) = x^2 + 1$

(a) $(f + g)(x) = f(x) + g(x)$

$= \sqrt{x - 1} + x^2 + 1$

(b) $(f - g)(x) = f(x) - g(x)$

$= \sqrt{x - 1} - (x^2 + 1)$

$= \sqrt{x - 1} - x^2 - 1$

(c) $(fg)(x) = f(x) \cdot g(x)$

$= \sqrt{x - 1}\,(x^2 + 1)$

(d) $(f/g)(x) = \dfrac{f(x)}{g(x)}$

$= \dfrac{\sqrt{x - 1}}{x^2 + 1}$

Domain of $(f/g)(x)$: $[1, \infty)$

34. $f(x) = 2x^2$

$g(x) = \sqrt{x + 6}$

(a) $(f \circ g)(x) = f\left(\sqrt{x + 6}\right)$

$= 2\left(\sqrt{x + 6}\right)^2$

$= 2(x + 6)$

$= 2x + 12$

(b) $(g \circ f)(x) = g(2x^2)$

$= \sqrt{2x^2 + 6}$

35. $f(x) = x - 2$

$g(x) = |x|$

(a) $(f \circ g)(x) = f(|x|)$

$= |x| - 2$

(b) $(g \circ f)(x) = g(x - 2)$

$= |x - 2|$

36. $h(x) = 5x - 2$

This linear function is a one-to-one function so it has an inverse.

$y = 5x - 2$

$x = 5y - 2$

$x + 2 = 5y$

$\dfrac{x + 2}{5} = y$

$h^{-1}(x) = \dfrac{x + 2}{5}$

37. $d = ks^2$

$50 = k \cdot 25^2$

$50 = k \cdot 625$

$\dfrac{50}{625} = k$

$\dfrac{2}{25} = k \implies d = \dfrac{2}{25}s^2$

$d = \dfrac{2}{25}(40)^2$

$d = \dfrac{2}{25}(1600)$

$d = 128 \text{ feet}$

38.

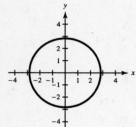

39.

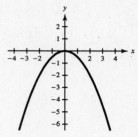

40.

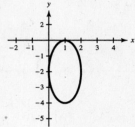

41.

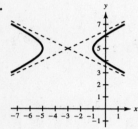

42. $\log_4 64 = 3 \implies 4^3 = 64$

43. $\log_3\left(\frac{1}{81}\right) = -4 \implies 3^{-4} = \frac{1}{81}$

44. $\ln 1 = 0 \implies e^0 = 1$

45. $3(\log_2 x + \log_2 y) - \log_2 z = 3\log_2 xy - \log_2 z$
$$= \log_2(xy)^3 - \log_2 z$$
$$= \log_2 \frac{(xy)^3}{z} \quad \text{or} \quad \log_2 \frac{x^3 y^3}{z}$$

46. $\ln \dfrac{5x}{(x+1)^2} = \ln 5x - \ln(x+1)^2$
$$= \ln 5 + \ln x - 2\ln(x+1)$$

47. (a) $\log_3 x = -2$
$$x = 3^{-2}$$
$$x = \frac{1}{9}$$

    (b) $4\ln x = 10$
$$\ln x = \frac{10}{4}$$
$$\ln x = \frac{5}{2}$$
$$x = e^{5/2}$$
$$x \approx 12.182$$

    (c) $3(1 + e^{2x}) = 20$
$$1 + e^{2x} = \frac{20}{3}$$
$$e^{2x} = \frac{17}{3}$$
$$2x = \ln \frac{17}{3}$$
$$x = \frac{1}{2} \ln \frac{17}{3}$$
$$x \approx 0.867$$

48. $0.30x + 0.60(20 - x) = 0.40(20)$
$$0.30x + 12 - 0.60x = 8$$
$$-0.30x + 12 = 8$$
$$-0.30x = -4$$
$$x = \frac{-4}{-0.30}$$
$$x = \frac{40}{3} \text{ and } 20 - x = \frac{60}{3} - \frac{40}{3}$$
$$20 - x = \frac{20}{3}$$

Thus, $13\frac{1}{3}$ gallons of the 30% solution and $6\frac{2}{3}$ gallons of the 60% solution are used.

49. $\quad V(t) = 22{,}000(0.8)^t$
$$15{,}000 = 22{,}000(0.8)^t$$
$$\frac{15{,}000}{22{,}000} = (0.8)^t$$
$$t = \log_{0.8}\left(\frac{15{,}000}{22{,}000}\right)$$
$$t \approx 1.7$$

The value of the car is \$15,000 after approximately 1.7 years.

C H A P T E R   1 0
Sequences, Series and the Binomial Theorem

CHAPTER 10
Sequences, Series, and the Binomial Theorem

Section 10.1   Sequences and Series

Solutions to Odd-Numbered Exercises

1. $a_n = 2n$

$a_1 = 2(1) = 2$

$a_2 = 2(2) = 4$

$a_3 = 2(3) = 6$

$a_4 = 2(4) = 8$

$a_5 = 2(5) = 10$

3. $a_n = (-1)^n 2n$

$a_1 = (-1)^1 \cdot 2(1) = -2$

$a_2 = (-1)^2 \cdot 2(2) = 4$

$a_3 = (-1)^3 \cdot 2(3) = -6$

$a_4 = (-1)^4 \cdot 2(4) = 8$

$a_5 = (-1)^5 \cdot 2(5) = -10$

5. $a_n = \left(\frac{1}{2}\right)^n$

$a_1 = \left(\frac{1}{2}\right)^1 = \frac{1}{2}$

$a_2 = \left(\frac{1}{2}\right)^2 = \frac{1}{4}$

$a_3 = \left(\frac{1}{2}\right)^3 = \frac{1}{8}$

$a_4 = \left(\frac{1}{2}\right)^4 = \frac{1}{16}$

$a_5 = \left(\frac{1}{2}\right)^5 = \frac{1}{32}$

7. $a_n = \left(-\frac{1}{2}\right)^{n+1}$

$a_1 = \left(-\frac{1}{2}\right)^{1+1} = \frac{1}{4}$

$a_2 = \left(-\frac{1}{2}\right)^{2+1} = -\frac{1}{8}$

$a_3 = \left(-\frac{1}{2}\right)^{3+1} = \frac{1}{16}$

$a_4 = \left(-\frac{1}{2}\right)^{4+1} = -\frac{1}{32}$

$a_5 = \left(-\frac{1}{2}\right)^{5+1} = \frac{1}{64}$

9. $a_n = (-0.2)^{n-1}$

$a_1 = (-0.2)^{1-1} = (-0.2)^0 = 1$

$a_2 = (-0.2)^{2-1} = (-0.2)^1 = -0.2$

$a_3 = (-0.2)^{3-1} = (-0.2)^2 = 0.04$

$a_4 = (-0.2)^{4-1} = (-0.2)^3 = -0.008$

$a_5 = (-0.2)^{5-1} = (-0.2)^4 = 0.0016$

11. $a_n = \dfrac{1}{n+1}$

$a_1 = \dfrac{1}{1+1} = \dfrac{1}{2}$

$a_2 = \dfrac{1}{2+1} = \dfrac{1}{3}$

$a_3 = \dfrac{1}{3+1} = \dfrac{1}{4}$

$a_4 = \dfrac{1}{4+1} = \dfrac{1}{5}$

$a_5 = \dfrac{1}{5+1} = \dfrac{1}{6}$

13. $a_n = \dfrac{2n}{3n+2}$

$a_1 = \dfrac{2(1)}{3(1)+2} = \dfrac{2}{5}$

$a_2 = \dfrac{2(2)}{3(2)+2} = \dfrac{4}{8} = \dfrac{1}{2}$

$a_3 = \dfrac{2(3)}{3(3)+2} = \dfrac{6}{11}$

$a_4 = \dfrac{2(4)}{3(4)+2} = \dfrac{8}{14} = \dfrac{4}{7}$

$a_5 = \dfrac{2(5)}{3(5)+2} = \dfrac{10}{17}$

15. $a_n = \dfrac{(-1)^n}{n^2}$

$a_1 = \dfrac{(-1)^1}{1^2} = -1$

$a_2 = \dfrac{(-1)^2}{2^2} = \dfrac{1}{4}$

$a_3 = \dfrac{(-1)^3}{3^2} = \dfrac{-1}{9}$

$a_4 = \dfrac{(-1)^4}{4^2} = \dfrac{1}{16}$

$a_5 = \dfrac{(-1)^5}{5^2} = \dfrac{-1}{25}$

17. $a_n = \dfrac{2+(-1)^{n+1}}{n}$

$a_1 = \dfrac{2+(-1)^{1+1}}{1} = 3$

$a_2 = \dfrac{2+(-1)^{2+1}}{2} = \dfrac{1}{2}$

$a_3 = \dfrac{2+(-1)^{3+1}}{3} = 1$

$a_4 = \dfrac{2+(-1)^{4+1}}{4} = \dfrac{1}{4}$

$a_5 = \dfrac{2+(-1)^{5+1}}{5} = \dfrac{3}{5}$

19. $a_n = 5 - \dfrac{1}{2^n}$

$a_1 = 5 - \dfrac{1}{2^1} = \dfrac{9}{2}$

$a_2 = 5 - \dfrac{1}{2^2} = \dfrac{19}{4}$

$a_3 = 5 - \dfrac{1}{2^3} = \dfrac{39}{8}$

$a_4 = 5 - \dfrac{1}{2^4} = \dfrac{79}{16}$

$a_5 = 5 - \dfrac{1}{2^5} = \dfrac{159}{32}$

21. $a_n = \dfrac{2^n}{n!}$

$a_1 = \dfrac{2^1}{1!} = \dfrac{2}{1} = 2$

$a_2 = \dfrac{2^2}{2!} = \dfrac{4}{2} = 2$

$a_3 = \dfrac{2^3}{3!} = \dfrac{8}{6} = \dfrac{4}{3}$

$a_4 = \dfrac{2^4}{4!} = \dfrac{16}{24} = \dfrac{2}{3}$

$a_5 = \dfrac{2^5}{5!} = \dfrac{32}{120} = \dfrac{4}{15}$

23. $a_n = 2 + (-2)^n$

$a_1 = 2 + (-2)^1 = 0$

$a_2 = 2 + (-2)^2 = 6$

$a_3 = 2 + (-2)^3 = -6$

$a_4 = 2 + (-2)^4 = 18$

$a_5 = 2 + (-2)^5 = -30$

25. $a_n = \dfrac{5 + (-1)^{n+1}}{2n^2}$

$a_1 = \dfrac{5 + (-1)^{1+1}}{2(1)^2} = \dfrac{6}{2} = 3$

$a_2 = \dfrac{5 + (-1)^{2+1}}{2(2)^2} = \dfrac{4}{8} = \dfrac{1}{2}$

$a_3 = \dfrac{5 + (-1)^{3+1}}{2(3)^2} = \dfrac{6}{18} = \dfrac{1}{3}$

$a_4 = \dfrac{5 + (-1)^{4+1}}{2(4)^2} = \dfrac{4}{32} = \dfrac{1}{8}$

$a_5 = \dfrac{5 + (-1)^{5+1}}{2(5)^2} = \dfrac{6}{50} = \dfrac{3}{25}$

27. $a_1 = 1,\ a_{n+1} = a_n + 4$

$a_1 = 1$

$a_2 = 1 + 4 = 5$

$a_3 = 5 + 4 = 9$

$a_4 = 9 + 4 = 13$

$a_5 = 13 + 4 = 17$

$a_6 = 17 + 4 = 21$

29. $a_1 = 2,\ a_{n+1} = 2a_n + 1$

$a_1 = 2$

$a_2 = 2(2) + 1 = 5$

$a_3 = 2(5) + 1 = 11$

$a_4 = 2(11) + 1 = 23$

$a_5 = 2(23) + 1 = 47$

$a_6 = 2(47) + 1 = 95$

31. $a_1 = 4,\ a_{n+1} = \dfrac{1}{a_n} + 1$

$a_1 = 4$

$a_2 = \dfrac{1}{4} + 1 = \dfrac{1}{4} + \dfrac{4}{4} = \dfrac{5}{4}$

$a_3 = \dfrac{1}{5/4} + 1 = \dfrac{4}{5} + \dfrac{5}{5} = \dfrac{9}{5}$

$a_4 = \dfrac{1}{(9/5)} + 1 = \dfrac{5}{9} + \dfrac{9}{9} = \dfrac{14}{9}$

$a_5 = \dfrac{1}{(14/9)} + 1 = \dfrac{9}{14} + \dfrac{14}{14} = \dfrac{23}{14}$

$a_6 = \dfrac{1}{(23/14)} + 1 = \dfrac{14}{23} + \dfrac{23}{23} = \dfrac{37}{23}$

33. $a_1 = 30,\ a_{n+1} = \dfrac{1}{2}a_n + 5$

$a_1 = 30$

$a_2 = \dfrac{1}{2}(30) + 5 = 15 + 5 = 20$

$a_3 = \dfrac{1}{2}(20) + 5 = 10 + 5 = 15$

$a_4 = \dfrac{1}{2}(15) + 5 = \dfrac{15}{2} + \dfrac{10}{2} = \dfrac{25}{2}$

$a_5 = \dfrac{1}{2}\left(\dfrac{25}{2}\right) + 5 = \dfrac{25}{4} + \dfrac{20}{4} = \dfrac{45}{4}$

$a_6 = \dfrac{1}{2}\left(\dfrac{45}{4}\right) + 5 = \dfrac{45}{8} + \dfrac{40}{8} = \dfrac{85}{8}$

35. $a_1 = -2,\ a_2 = 0,\ a_{n+2} = a_{n+1} - 2a_n$

$a_1 = -2$

$a_2 = 0$

$a_3 = 0 - 2(-2) = 4$

$a_4 = 4 - 2(0) = 4$

$a_5 = 4 - 2(4) = -4$

$a_6 = -4 - 2(4) = -12$

37. $a_n = (-1)^n(5n - 3)$

$a_{15} = (-1)^{15}[5(15) - 3]$

$\quad\;\; = -(75 - 3)$

$\quad\;\; = -72$

39. $a_n = \dfrac{n}{5 + \sqrt{n}}$

$a_{25} = \dfrac{25}{5 + \sqrt{25}}$

$\quad\;\; = \dfrac{25}{10}$

$\quad\;\; = \dfrac{5}{2}$

41. $a_n = \dfrac{n!}{(n - 2)!}$

$a_8 = \dfrac{8!}{(8 - 2)!}$

$\quad\; = \dfrac{8!}{6!}$

$\quad\; = \dfrac{8 \cdot 7 \cdot 6!}{6!}$

$\quad\; = 56$

43. $\dfrac{5!}{4!} = \dfrac{5 \cdot 4 \cdot 3 \cdot 2 \cdot 1}{4 \cdot 3 \cdot 2 \cdot 1} = 5$

45. $\dfrac{10!}{12!} = \dfrac{10!}{12 \cdot 11 \cdot 10!} = \dfrac{1}{132}$

47. $\dfrac{25!}{27!} = \dfrac{25!}{27 \cdot 26 \cdot 25!}$

$\quad\;\; = \dfrac{1}{27 \cdot 26} = \dfrac{1}{702}$

49. $\dfrac{n!}{(n + 1)!} = \dfrac{n!}{(n + 1)n!} = \dfrac{1}{n + 1}$

51. $\dfrac{(n + 1)!}{(n - 1)!} = \dfrac{(n + 1) \cdot (n) \cdot (n - 1)!}{(n - 1)!}$

$\quad\;\; = \dfrac{(n + 1)(n)}{1} = n^2 + n$

53. $\dfrac{(2n)!}{(2n - 1)!} = \dfrac{2n(2n - 1)!}{(2n - 1)!} = 2n$

55. $1, -1, 1, -1, 1, \ldots$

$a_n = (-1)^{n+1}$

(There are many correct answers.)

57. $1, 4, 9, 16, 25, \ldots$

$a_n = n^2$

(There are many correct answers.)

59. $4, 2, \dfrac{4}{3}, 1, \dfrac{4}{5}, \ldots$

$a_n = \dfrac{4}{n}$

(There are many correct answers.)

61. $1, 3, 5, 7, 9, \ldots$

$a_n = 2n - 1$

(There are many correct answers.)

63. $\displaystyle\sum_{k=1}^{6} 3k = 3(1) + 3(2) + 3(3) + 3(4) + 3(5) + 3(6) = 3 + 6 + 9 + 12 + 15 + 18 = 63$

65. $\displaystyle\sum_{i=0}^{6} (2i + 5) = [2(0) + 5] + [2(1) + 5] + [2(2) + 5] + [2(3) + 5] + [2(4) + 5] + [2(5) + 5] + [2(6) + 5]$

$\quad\quad\quad\quad\;\; = 5 + 7 + 9 + 11 + 13 + 15 + 17 = 77$

67. $\displaystyle\sum_{i=0}^{4} (3i + 2) = [3(0) + 2] + [3(1) + 2] + [3(2) + 2] + [3(3) + 2] + [3(4) + 2]$

$\quad\quad\quad\quad\;\; = 2 + 5 + 8 + 11 + 14$

$\quad\quad\quad\quad\;\; = 40$

69. $\displaystyle\sum_{i=0}^{5} (i^2 + 3) = (0^2 + 3) + (1^2 + 3) + (2^2 + 3) + (3^2 + 3) + (4^2 + 3) + (5^2 + 3)$

$\quad\quad\quad\quad\;\; = 3 + 4 + 7 + 12 + 19 + 28$

$\quad\quad\quad\quad\;\; = 73$

71. $\displaystyle\sum_{i=0}^{4} 2^i = 2^0 + 2^1 + 2^2 + 2^3 + 2^4$

$\qquad = 1 + 2 + 4 + 8 + 16$

$\qquad = 31$

73. $\displaystyle\sum_{j=1}^{5} \frac{(-1)^{j+1}}{j} = \frac{(-1)^{1+1}}{1} + \frac{(-1)^{2+1}}{2} + \frac{(-1)^{3+1}}{3} + \frac{(-1)^{4+1}}{4} + \frac{(-1)^{5+1}}{5} = 1 + \frac{-1}{2} + \frac{1}{3} + \frac{-1}{4} + \frac{1}{5} = \frac{47}{60}$

75. $\displaystyle\sum_{m=2}^{6} \frac{2m}{2(m-1)} = \frac{2(2)}{2(2-1)} + \frac{2(3)}{2(3-1)} + \frac{2(4)}{2(4-1)} + \frac{2(5)}{2(5-1)} + \frac{2(6)}{2(6-1)}$

$\qquad = \frac{4}{2(1)} + \frac{6}{2(2)} + \frac{8}{2(3)} + \frac{10}{2(4)} + \frac{12}{2(5)}$

$\qquad = \frac{4}{2} + \frac{6}{4} + \frac{8}{6} + \frac{10}{8} + \frac{12}{10}$

$\qquad = 2 + \frac{3}{2} + \frac{4}{3} + \frac{5}{4} + \frac{6}{5}$

$\qquad = \frac{120}{60} + \frac{90}{60} + \frac{80}{60} + \frac{75}{60} + \frac{72}{60}$

$\qquad = \frac{437}{60}$

77. $\displaystyle\sum_{k=1}^{6} (-8) = (-8) + (-8) + (-8) + (-8) + (-8) + (-8) = -48$

79. $\displaystyle\sum_{i=1}^{8} \left(\frac{1}{i} - \frac{1}{i+1} \right) = \left[\frac{1}{1} - \frac{1}{1+1} \right] + \left[\frac{1}{2} - \frac{1}{2+1} \right] + \left[\frac{1}{3} - \frac{1}{3+1} \right] + \left[\frac{1}{4} - \frac{1}{4+1} \right] + \left[\frac{1}{5} - \frac{1}{5+1} \right]$

$\qquad + \left[\frac{1}{6} - \frac{1}{6+1} \right] + \left[\frac{1}{7} - \frac{1}{7+1} \right] + \left[\frac{1}{8} - \frac{1}{8+1} \right]$

$\qquad = \frac{1}{2} + \frac{1}{6} + \frac{1}{12} + \frac{1}{20} + \frac{1}{30} + \frac{1}{42} + \frac{1}{56} + \frac{1}{72}$

$\qquad = \frac{1260 + 420 + 210 + 126 + 84 + 60 + 45 + 35}{2520} = \frac{2240}{2520} = \frac{8}{9}$

or

$\displaystyle\sum_{i=1}^{8} \left(\frac{1}{i} - \frac{1}{i+1} \right) = \left(1 - \frac{1}{2} \right) + \left(\frac{1}{2} - \frac{1}{3} \right) + \left(\frac{1}{3} - \frac{1}{4} \right) + \left(\frac{1}{4} - \frac{1}{5} \right) + \left(\frac{1}{5} - \frac{1}{6} \right) + \left(\frac{1}{6} - \frac{1}{7} \right) + \left(\frac{1}{7} - \frac{1}{8} \right) + \left(\frac{1}{8} - \frac{1}{9} \right)$

$\qquad = 1 - \frac{1}{9} = \frac{8}{9}$

81. $\displaystyle\sum_{n=0}^{5} \left(-\frac{1}{3} \right)^n = \left(-\frac{1}{3} \right)^0 + \left(-\frac{1}{3} \right)^1 + \left(-\frac{1}{3} \right)^2 + \left(-\frac{1}{3} \right)^3 + \left(-\frac{1}{3} \right)^4 + \left(-\frac{1}{3} \right)^5$

$\qquad = 1 + \left(-\frac{1}{3} \right) + \frac{1}{9} + \left(-\frac{1}{27} \right) + \frac{1}{81} + \left(-\frac{1}{243} \right) = \frac{243 - 81 + 27 - 9 + 3 - 1}{243} = \frac{182}{243}$

83. $\displaystyle\sum_{n=1}^{6} n(n+1) = 112$ $\qquad\qquad\qquad\qquad$ **85.** $\displaystyle\sum_{j=2}^{6} (j! - j) = 852$

87. $\displaystyle\sum_{k=1}^{6} \ln k \approx 6.5793$

Note: $\displaystyle\sum_{k=1}^{6} \ln k = \ln 1 + \ln 2 + \ln 3 + \ln 4 + \ln 5 + \ln 6 = \ln(1 \cdot 2 \cdot 3 \cdot 4 \cdot 5 \cdot 6) = \ln 720 \approx 6.5793$

89. $1 + 2 + 3 + 4 + 5 = \displaystyle\sum_{k=1}^{5} k$

91. $2 + 4 + 6 + 8 + 10 = \displaystyle\sum_{k=1}^{5} 2k$

93. $\dfrac{1}{2(1)} + \dfrac{1}{2(2)} + \dfrac{1}{2(3)} + \dfrac{1}{2(4)} + \cdots + \dfrac{1}{2(10)} = \displaystyle\sum_{k=1}^{10} \dfrac{1}{2k}$

95. $\dfrac{1}{1^2} + \dfrac{1}{2^2} + \dfrac{1}{3^2} + \dfrac{1}{4^2} + \cdots + \dfrac{1}{20^2} = \displaystyle\sum_{k=1}^{20} \dfrac{1}{k^2}$

97. $\dfrac{1}{3^0} - \dfrac{1}{3^1} + \dfrac{1}{3^2} - \dfrac{1}{3^3} + \cdots - \dfrac{1}{3^9} = \displaystyle\sum_{k=0}^{9} \dfrac{1}{(-3)^k}$

or $\displaystyle\sum_{k=0}^{9} \left(-\dfrac{1}{3}\right)^k$

99. $\dfrac{4}{1 + 3} + \dfrac{4}{2 + 3} + \dfrac{4}{3 + 3} + \cdots + \dfrac{4}{20 + 3} = \displaystyle\sum_{k=1}^{20} \dfrac{4}{k + 3}$

101. $\dfrac{1}{2} + \dfrac{2}{3} + \dfrac{3}{4} + \dfrac{4}{5} + \dfrac{5}{6} + \cdots + \dfrac{11}{12} = \displaystyle\sum_{k-1}^{11} \dfrac{k}{k + 1}$

103. $\dfrac{2}{4} + \dfrac{4}{5} + \dfrac{6}{6} + \dfrac{8}{7} + \cdots + \dfrac{40}{23} = \displaystyle\sum_{k=1}^{20} \dfrac{2k}{k + 3}$

105. $1 + 1 + 2 + 6 + 24 + 120 + 720 = \displaystyle\sum_{k=0}^{6} k!$

107. Graph (c)

109. Graph (b)

111.
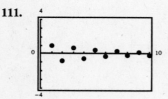

113.
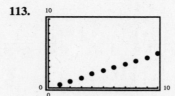

115.
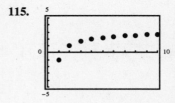

117. $\bar{x} = \dfrac{3 + 7 + 2 + 1 + 5}{5} = 3.6$

119. $\bar{x} = \dfrac{0.5 + 0.8 + 1.1 + 0.8 + 0.7 + 0.7 + 1.0}{7} = 0.8$

121. (a) $A_N = 500(1 + 0.07)^N$

$A_1 = 500(1 + 0.07)^1 = \$535$

$A_2 = 500(1 + 0.07)^2 = \$572.45$

$A_3 = 500(1 + 0.07)^3 \approx \$612.52$

$A_4 = 500(1 + 0.07)^4 \approx \$655.40$

$A_5 = 500(1 + 0.07)^5 \approx \$701.28$

$A_6 = 500(1 + 0.07)^6 \approx \$750.37$

$A_7 = 500(1 + 0.07)^7 \approx \$802.89$

$A_8 = 500(1 + 0.07)^8 \approx \$859.09$

(b) $A_{40} = 500(1 + 0.07)^{40} \approx \$7487.23$

(c)
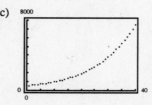

(d) Yes. Investment earning compound interest increases at an increasing rate.

123. $a_n = \dfrac{180(n - 2)}{n}, n \ge 3$

$a_5 = \dfrac{180(5 - 2)}{5} = \dfrac{180(3)}{5} = 108°$

$a_6 = \dfrac{180(6 - 2)}{6} = \dfrac{180(4)}{6} = 120°$

At any point where a pentagon and two hexagons meet, the sum of the degrees of the three angles is

$a_5 + 2(a_6) = 108° + 2(120°) = 348°.$

Because this sum is less than 360°, there are gaps between the hexagons.

125. The odd terms are negative.

$(-1)^n = -1$ for odd values of n.

127. $\displaystyle\sum_{k=1}^{4} 3k = 3\sum_{k=1}^{4} k$

True.

$3 + 6 + 9 + 12 = 3(1 + 2 + 3 + 4)$

129. $\dfrac{1}{2} \cdot \dfrac{2}{3} \cdot \dfrac{3}{4} = \dfrac{1 \cdot \cancel{2} \cdot \cancel{3}}{\cancel{2} \cdot \cancel{3} \cdot 4} = \dfrac{1}{4}$

131. $\dfrac{1}{n} + \dfrac{1}{n + 1} = \dfrac{1(n + 1)}{n(n + 1)} + \dfrac{1(n)}{(n + 1)(n)}$

$= \dfrac{n + 1 + n}{n(n + 1)}$

$= \dfrac{2n + 1}{n(n + 1)}$

133. $(x + 10)^{-2} = \dfrac{1}{(x + 10)^2}$

$= \dfrac{1}{x^2 + 20x + 100}$

135. $(a^2)^{-4} = a^{-8} = \dfrac{1}{a^8}$

137. $\sqrt{128x^3} = \sqrt{64x^2(2x)} = 8x\sqrt{2x}$

139.

$22{,}000(0.8)^t = 15{,}000$

$(0.8)^t = \dfrac{15{,}000}{22{,}000}$

$(0.8)^t = \dfrac{15}{22}$

$t = \log_{0.8}\left(\dfrac{15}{22}\right)$

$t \approx 1.72$ years

Section 10.2 Arithmetic Sequences

1. $d = 3$

3. $d = -6$

5. $d = 3$

7. $d = -4$

9. $d = \frac{2}{3}$

11. $d = -\frac{5}{4}$

13. The sequence is arithmetic.

$d = 2$

15. The sequence is arithmetic.

$d = -16$

17. The sequence is arithmetic.

$d = -57$

19. The sequence is *not* arithmetic.

21. The sequence is arithmetic.

$d = 0.8$

23. The sequence is arithmetic.

$d = \frac{3}{2}$

25. The sequence is *not* arithmetic.

Note: $\frac{2}{3} - \frac{1}{3} \neq \frac{4}{3} - \frac{2}{3}$

27. The sequence is *not* arithmetic.

Note: $\ln 8 - \ln 4 \overset{?}{=} \ln 12 - \ln 8$

$\ln \frac{8}{4} \overset{?}{=} \ln \frac{12}{8} \Rightarrow \ln \frac{8}{4} \overset{?}{=} \ln \frac{12}{8}$

$\ln 2 \neq \ln \frac{3}{2}$

29. $a_1 = 3(1) + 4 = 7$

$a_2 = 3(2) + 4 = 10$

$a_3 = 3(3) + 4 = 13$

$a_4 = 3(4) + 4 = 16$

$a_5 = 3(5) + 4 = 19$

31. $a_1 = -2(1) + 8 = 6$

$a_2 = -2(2) + 8 = 4$

$a_3 = -2(3) + 8 = 2$

$a_4 = -2(4) + 8 = 0$

$a_5 = -2(5) + 8 = -2$

33. $a_1 = \frac{5}{2}(1) - 1 = \frac{3}{2}$

$a_2 = \frac{5}{2}(2) - 1 = 4$

$a_3 = \frac{5}{2}(3) - 1 = \frac{13}{2}$

$a_4 = \frac{5}{2}(4) - 1 = 9$

$a_5 = \frac{5}{2}(5) - 1 = \frac{23}{2}$

35. $a_n = \frac{3}{5}n + 1$

$a_1 = \frac{3}{5}(1) + 1 = \frac{8}{5}$

$a_2 = \frac{3}{5}(2) + 1 = \frac{11}{5}$

$a_3 = \frac{3}{5}(3) + 1 = \frac{14}{5}$

$a_4 = \frac{3}{5}(4) + 1 = \frac{17}{5}$

$a_5 = \frac{3}{5}(5) + 1 = 4$

37. $a_n = 3(n + 3) - 1$

$a_1 = 3(1 + 3) - 1 = 11$

$a_2 = 3(2 + 3) - 1 = 14$

$a_3 = 3(3 + 3) - 1 = 17$

$a_4 = 3(4 + 3) - 1 = 20$

$a_5 = 3(5 + 3) - 1 = 23$

39. $a_1 = -\frac{1}{4}(1 - 1) + 4 = 4$

$a_2 = -\frac{1}{4}(2 - 1) + 4 = \frac{15}{4}$

$a_3 = -\frac{1}{4}(3 - 1) + 4 = \frac{7}{2}$

$a_4 = -\frac{1}{4}(4 - 1) + 4 = \frac{13}{4}$

$a_5 = -\frac{1}{4}(5 - 1) + 4 = 3$

41. $a_1 = 25$

$a_2 = a_{1+1} = a_1 + 3 = 25 + 3 = 28$

$a_3 = a_{2+1} = a_2 + 3 = 28 + 3 = 31$

$a_4 = a_{3+1} = a_3 + 3 = 31 + 3 = 34$

$a_5 = a_{4+1} = a_4 + 3 = 34 + 3 = 37$

43. $a_1 = 9$

$a_2 = a_{1+1} = a_1 - 3 = 9 - 3 = 6$

$a_3 = a_{2+1} = a_2 - 3 = 6 - 3 = 3$

$a_4 = a_{3+1} = a_3 - 3 = 3 - 3 = 0$

$a_5 = a_{4+1} = a_4 - 3 = 0 - 3 = -3$

45. $a_1 = -10$

$a_2 = a_{1+1} = a_1 + 6 = -10 + 6 = -4$

$a_3 = a_{2+1} = a_2 + 6 = -4 + 6 = 2$

$a_4 = a_{3+1} = a_3 + 6 = 2 + 6 = 8$

$a_5 = a_{4+1} = a_4 + 6 = 8 + 6 = 14$

47. $a_1 = 100, a_{k+1} = a_k - 20$

$a_1 = 100$

$a_2 = a_{1+1} = a_1 - 20 = 100 - 20 = 80$

$a_3 = a_{2+1} = a_2 - 20 = 80 - 20 = 60$

$a_4 = a_{3+1} = a_3 - 20 = 60 - 20 = 40$

$a_5 = a_{4+1} = a_4 - 20 = 40 - 20 = 20$

49. Graph (b)

51. Graph (e)

53. Graph (c)

55.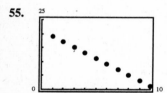

57.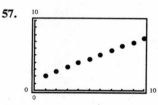

59.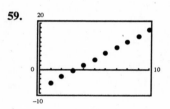

61. $a_n = a_1 + (n-1)d$

$a_n = 3 + (n-1)\frac{1}{2}$

$a_n = 3 + \frac{1}{2}n - \frac{1}{2}$

$a_n = \frac{1}{2}n + \frac{5}{2}$

63. $a_n = a_1 + (n-1)d$

$a_n = 1000 + (n-1)(-25)$

$a_n = 1000 - 25n + 25$

$a_n = -25n + 1025$

65. $a_1 = 3, d = \frac{3}{2}$

$a_n = a_1 + (n-1)d$

$a_n = 3 + (n-1)\frac{3}{2}$

$a_n = 3 + \frac{3}{2}n - \frac{3}{2}$

$a_n = \frac{3}{2}n + \frac{3}{2}$

67. $a_3 = 20, d = -4$

$a_n = a_1 + (n-1)d$

$a_3 = a_1 + (3-1)(-4)$

$20 = a_1 + 2(-4)$

$20 = a_1 - 8$

$28 = a_1$

$a_n = 28 + (n-1)(-4)$

$a_n = 28 - 4n + 4$

$a_n = -4n + 32$

69. $a_1 = 5, a_5 = 15$

$a_n = a_1 + (n-1)d$

$15 = 5 + (5-1)d$

$15 = 5 + 4d$

$10 = 4d$

$\frac{10}{4} = d$

$\frac{5}{2} = d$

$a_n = 5 + (n-1)\frac{5}{2}$

$a_n = 5 + \frac{5}{2}n - \frac{5}{2}$

$a_n = \frac{5}{2}n + \frac{5}{2}$

71. $a_3 = 16, a_4 = 20$

$d = a_4 - a_3 = 20 - 16 = 4$

$a_n = a_1 + (n-1)d$

$a_4 = a_1 + (4-1)(4)$

$20 = a_1 + 3(4)$

$20 = a_1 + 12$

$8 = a_1$

$a_n = 8 + (n-1)(4)$

$a_n = 8 + 4n - 4$

$a_n = 4n + 4$

73. $a_n = a_1 + (n-1)d$

$a_3 = a_1 + (3-1)d$

$30 = 50 + 2d$

$-20 = 2d$

$-10 = d$

$a_n = a_1 + (n-1)d$

$a_n = 50 + (n-1)(-10)$

$a_n = 50 - 10n + 10$

$a_n = -10n + 60$

75. $a_2 = 10, a_6 = 8$

$a_n = a_1 + (n-1)d$

$a_6 = a_1 + (6-1)d \Longrightarrow 8 = a_1 + 5d \Longrightarrow \quad 8 = a_1 + 5d$

$a_2 = a_1 + (2-1)d \Longrightarrow 10 = a_1 + d \Longrightarrow \underline{-10 = -a_1 - d}$

$$-2 = 4d$$

$$-\frac{1}{2} = d$$

$a_2 = a_1 + 1d$

$10 = a_1 - \frac{1}{2}$

$\frac{21}{2} = a_1$

$a_n = a_1 + (n-1)d$

$a_n = \frac{21}{2} + (n-1)\left(-\frac{1}{2}\right)$

$a_n = \frac{21}{2} - \frac{1}{2}n + \frac{1}{2}$

$a_n = -\frac{1}{2}n + 11$

77. $a_1 = 0.35, a_2 = 0.30$

$a_n = a_1 + (n - 1)d,$

$a_2 = a_1 + (2 - 1)d$

$0.30 = 0.35 + d,$

$-0.05 = d$

$a_n = 0.35 + (n - 1)(-0.05)$

$a_n = 0.35 - 0.05n + 0.05,$

$a_n = -0.05n + 0.40$

79. $\displaystyle\sum_{k=1}^{20} k = 1 + 2 + 3 + \cdots + 20$

Sum of *arithmetic* sequence: $\dfrac{n}{2}(a_1 + a_n)$

$n = 20, a_1 = 1, a_n = 20$

$\displaystyle\sum_{k=1}^{20} k = \dfrac{20}{2}(1 + 20) = 10(21) = 210$

81. $\displaystyle\sum_{k=1}^{10} 5k = 5 + 10 + 15 + \cdots + 50$

Sum of *arithmetic* sequence: $\dfrac{n}{2}(a_1 + a_n)$

$n = 10, a_1 = 5(1) = 5, a_n = 5(10) = 50$

$\displaystyle\sum_{k=10}^{10} 5k = \dfrac{10}{2}(5 + 50)$

$= 5(55)$

$= 275$

83. $\displaystyle\sum_{k=1}^{50} (k + 3) = 4 + 5 + 6 + \cdots + 53$

Sum of *arithmetic* sequence: $\dfrac{n}{2}(a_1 + a_n)$

$n = 50, a_1 = 1 + 3 = 4, a_n = 50 + 3 = 53$

$\displaystyle\sum_{k=1}^{50} (k + 3) = \dfrac{50}{2}(4 + 53)$

$= 25(57)$

$= 1425$

85. $\displaystyle\sum_{n=1}^{500} \dfrac{n}{2} = \dfrac{1}{2} + \dfrac{2}{2} + \dfrac{3}{2} + \cdots + \dfrac{500}{2}$

Sum of *arithmetic* sequence: $\dfrac{n}{2}(a_1 + a_n)$

$n = 500, \ a_1 = \dfrac{1}{2}, a_n = \dfrac{500}{2}$

$\displaystyle\sum_{n=1}^{500} \dfrac{n}{2} = \dfrac{500}{2}\left(\dfrac{1}{2} + \dfrac{500}{2}\right)$

$= 250\left(\dfrac{501}{2}\right) = 62{,}625$

87. $\displaystyle\sum_{n=1}^{30} \left(\dfrac{1}{3}n - 4\right) = -\dfrac{11}{3} - \dfrac{10}{3} - 3 - \cdots + 6$

Sum of *arithmetic* sequence: $\dfrac{n}{2}(a_1 + a_n)$

$n = 30, \ a_1 = \dfrac{1}{3}(1) - 4 = -\dfrac{11}{3}, a_n = \dfrac{1}{3}(30) - 4 = 6$

$\displaystyle\sum_{n=1}^{30} \left(\dfrac{1}{3}n - 4\right) = \dfrac{30}{2}\left(-\dfrac{11}{3} + 6\right)$

$= 15\left(\dfrac{7}{3}\right) = 35$

89. $\displaystyle\sum_{j=1}^{20} \left(750 - \dfrac{1}{2}j\right) = 14{,}895$

91. $\displaystyle\sum_{n=1}^{40} (1000 - 25n) = 19{,}500$

93. $\displaystyle\sum_{n=1}^{50} (2.15n + 5.4) = 3011.25$

95. Arithmetic sequence with $d = 7$

$a_n = a_1 + (n - 1)d$

$= 5 + (n - 1)7$

$= 5 + 7n - 7$

$= 7n - 2$

Sum: $\dfrac{n}{2}(a_1 + a_n)$

$n = 12, a_1 = 5, a_{12} = 7(12) - 2 = 82$

$\displaystyle\sum_{n=1}^{12} (7n - 2) = \dfrac{12}{2}(5 + 82) = 6(87) = 522$

97. Arithmetic sequence with $d = 6$

$a_n = a_1 + (n - 1)d$

$a_n = 2 + (n - 1)6 = 2 + 6n - 6 = 6n - 4$

Sum: $\dfrac{n}{2}(a_1 + a_n)$

$n = 25, a_1 = 2, a_{25} = 6(25) - 4 = 146$

$\displaystyle\sum_{n=1}^{25} (6n - 4) = \dfrac{25}{2}(2 + 146) = \dfrac{25}{2}(148) = 1850$

99. Arithmetic sequence with $d = -25$

$$a_n = a_1 + (n - 1)d$$
$$= 200 + (n - 1)(-25)$$
$$= 200 - 25n + 25$$
$$= -25n + 225$$

Sum: $\frac{n}{2}(a_1 + a_n)$

$n = 8, a_1 = 200, a_8 = -25(8) + 225 = 25$

$$\sum_{n=1}^{8}(-25n + 225) = \frac{8}{2}(200 + 25) = 4(225) = 900$$

101. Arithmetic sequence with $d = 12$

$$a_n = a_1(n - 1)d$$
$$= -50 + (n - 1)(12)$$
$$= 12n - 62$$

Sum: $\frac{n}{2}(a_1 + a_n)$

$n = 50, a_1 = -50, a_{50} = 12(50) - 62 = 538$

$$\sum_{n=1}^{50}(12n - 62) = \frac{50}{2}(-50 + 538)$$
$$= 25(488)$$
$$= 12,200$$

103. Arithmetic sequence with $d = 3.5$

$$a_n = a_1 + (n - 1)d = 1 + (n - 1)(3.5) = 1 + 3.5n - 3.5 = 3.5n - 2.5$$

Sum: $\frac{n}{2}(a_1 + a_n)$

$n = 12, a_1 = 1, a_{12} = 3.5(12) - 2.5 = 39.5$

$$\sum_{n=1}^{12}(3.5n - 2.5) = \frac{12}{2}(1 + 39.5) = 6(40.5) = 243$$

105. $0.5, 0.9, 1.3, 1.7, \ldots, n = 10$

Arithmetic sequence with $d = 0.4$

$$a_n = a_1 + (n - 1)d$$
$$a_n = 0.5 + (n - 1)(0.4)$$
$$= 0.5 + 0.4n - 0.4 = 0.4n + 0.1$$

Sum: $\frac{n}{2}(a_1 + a_n)$

$n = 10, a_1 = 0.5, a_{10} = 0.4(10) + 0.1 = 4.1$

$$\sum_{n=1}^{10}(0.4n + 0.1) = \frac{10}{2}(0.5 + 4.1) = 5(4.6) = 23$$

107. $1 + 2 + 3 + \cdots + 75 = \sum_{k=1}^{75} k$

Sum: $\frac{n}{2}(a_1 + a_n)$

$n = 75, a_1 = 1, a_{75} = 75$

$$\sum_{k=1}^{75} k = \frac{75}{2}(1 + 75) = \frac{75}{2}(76) = 2850$$

109. $2 + 4 + 6 + 8 + \cdots, n = 50$

Sum of arithmetic sequence with $d = 2$

$$a_n = a_1 + (n - 1)d = 2 + (n - 1)2$$
$$= 2 + 2n - 2 = 2n$$

Sum: $\frac{n}{2}(a_1 + a_n)$

$n = 50, a_1 = 2, a_{50} = 2(50) = 100$

$$\sum_{n=1}^{50} 2n = \frac{50}{2}(2 + 100) = 25(102) = 2550$$

111. $36,000 + 38,000 + 40,000 + 42,000 + 44,000 + 46,000$

Sum of arithmetic sequence: $\frac{n}{2}(a_1 + a_n)$

$n = 6, a_1 = 36,000, a_6 = 46,000$

Total salary $= \frac{6}{2}(36,000 + 46,000)$

$$= 3(82,000) = 246,000$$

The total salary for the first six years will be \$246,000.

113. $20 + 21 + 22 + \cdots, n = 20$

Arithmetic sequence with $d = 1$

$$a_n = a_1 + (n - 1)d = 20 + (n - 1)(1) = 20 + n - 1 = n + 19$$

Sum: $\dfrac{n}{2}(a_1 + a_n)$

$n = 20, \; a_1 = 20, \; a_{20} = 20 + 19 = 39$

$$\sum_{n=1}^{20} (n + 19) = \frac{20}{2}(20 + 39) = 10(59) = 590$$

There are 590 seats on the main floor.

$$\frac{15,000}{590} \approx 25.424$$

You should charge approximately $25.43 per ticket to obtain $15,000. (**Note:** When the ticket price of 15,000/590 is rounded to the nearest cent, the result is $25.42. However, this ticket price will not quite generate $15,000, so the price should be set at $25.43.)

115. During each one-hour period, the clock strikes the hour, n, and it strikes *once* at 15, 30, and 45 minutes after the hour. Thus, it strikes $n + 3$ times per hour.

In a one twelve-hour period, the clock strikes $\displaystyle\sum_{n=1}^{12} (n + 3)$ times.

Sum of arithmetic sequence: $\dfrac{n}{2}(a_1 + a_n)$

$n = 12$

$a_1 = (1 + 3) = 4$

$a_{12} = 12 + 3 = 15$

$$\sum_{n=1}^{12} (n + 3) = \frac{12}{2}(4 + 15) = 6(19) = 114$$

The clock strikes 114 times in a 12-hour period.

117. $93 + 89 + \cdots, n = 2 + 6 = 8$

Sum of arithmetic sequence with $d = -4$

$$a_n = a_1 + (n - 1)d$$
$$= 93 + (n - 1)(-4)$$
$$= 93 - 4n + 4 = -4n + 97$$

Sum: $\dfrac{n}{2}(a_1 + a_n)$

$n = 8, \; a_1 = 93, \; a_8 = -4(8) + 97 = 65$

$$\sum_{n=1}^{8} (-4n + 97) = \frac{8}{2}(93 + 65) = 4(158) = 632$$

In 8 trips around the field 632 bales of hay will be made.

119. $16 + 48 + 80 + \cdots, n = 8$

Sum of arithmetic sequence with $d = 32$

$$a_n = a_1 + (n - 1)d$$
$$= 16 + (n - 1)(32)$$
$$= 16 + 32n - 32$$
$$= 32n - 16$$

Sum: $\dfrac{n}{2}(a_1 + a_n)$

$n = 8, \; a_1 = 16, \; a_8 = 32(8) - 16 = 240$

$$\sum_{n=1}^{8} (32n - 16) = \frac{8}{2}(16 + 240) = 4(256) = 1024$$

The object will fall a total distance of 1024 feet.

121. (a) $1 + 3 = 4$

$1 + 3 + 5 = 9$

$1 + 3 + 5 + 7 = 16$

$1 + 3 + 5 + 7 + 9 = 25$

$1 + 3 + 5 + 7 + 9 + 11 = 36$

The sums are 4, 9, 16, 25, and 36.

(b) Conjecture: All of these sums are perfect squares. The next perfect square is 49.

Check: $1 + 3 + 5 + 7 + 9 + 11 + 13 = 49$

(c) $1, 3, 5, 7, 9, \ldots$

Arithmetic sequence with $a_1 = 1$ and $d = 2$

$a_n = a_1 + (n - 1)d$

$a_n = 1 + (n - 1)(2)$

$a_n = 1 + 2n - 2$

$a_n = 2n - 1$

Sum: $\frac{n}{2}(a_1 + a_n)$

$$\sum_{k=1}^{n} (2k - 1) = \frac{n}{2}[1 + (2n - 1)] = \frac{n}{2}(2n) = n^2$$

123. A recursion formula gives the relationship between the terms a_{n+1} and a_n.

125. $a_2 = 12, a_3 = 15 \implies d = 15 - 12 = 3$

$a_1 + 3 = a_2$

$a_1 + 3 = 12$

$a_1 = 9$

127. $10\left(\dfrac{7 + 15}{2}\right) = 10\left(\dfrac{22}{2}\right)$

$= 10(11)$

$= 110$

129. $\frac{3}{2}[4(-3) + 5(4)] = \frac{3}{2}(-12 + 20) = \frac{3}{2}(8) = 12$

131. $f(x) = x^3 - 2x$

Domain: $(-\infty, \infty)$

133. $h(x) = \sqrt{16 - x^2}$

Domain: $[-4, 4]$

135. $g(t) = \ln(t - 2)$

Domain: $(2, \infty)$

137. $A = P\left(1 + \dfrac{r}{n}\right)^{nt}$

$= 10,000\left(1 + \dfrac{0.075}{365}\right)^{365(15)}$

$\approx \$30,798.61$

Mid-Chapter Quiz for Chapter 10

1. $a_n = 32\left(\dfrac{1}{4}\right)^{n-1}$

$a_1 = 32\left(\dfrac{1}{4}\right)^{1-1} = 32\left(\dfrac{1}{4}\right)^0 = 32(1) = 32$

$a_2 = 32\left(\dfrac{1}{4}\right)^{2-1} = 32\left(\dfrac{1}{4}\right)^1 = 32\left(\dfrac{1}{4}\right) = 8$

$a_3 = 32\left(\dfrac{1}{4}\right)^{3-1} = 32\left(\dfrac{1}{4}\right)^2 = 32\left(\dfrac{1}{16}\right) = 2$

$a_4 = 32\left(\dfrac{1}{4}\right)^{4-1} = 32\left(\dfrac{1}{4}\right)^3 = 32\left(\dfrac{1}{64}\right) = \dfrac{1}{2}$

$a_5 = 32\left(\dfrac{1}{4}\right)^{5-1} = 32\left(\dfrac{1}{4}\right)^4 = 32\left(\dfrac{1}{256}\right) = \dfrac{1}{8}$

2. $a_n = \dfrac{(-3)^n n}{n + 4}$

$a_1 = \dfrac{(-3)^1 1}{1 + 4} = \dfrac{(-3)}{5} = -\dfrac{3}{5}$

$a_2 = \dfrac{(-3)^2 2}{2 + 4} = \dfrac{(9)2}{6} = \dfrac{18}{6} = 3$

$a_3 = \dfrac{(-3)^3 3}{3 + 4} = \dfrac{(-27)3}{7} = -\dfrac{81}{7}$

$a_4 = \dfrac{(-3)^4 4}{4 + 4} = \dfrac{(81)4}{8} = \dfrac{324}{8} = \dfrac{81}{2}$

$a_5 = \dfrac{(-3)^5 5}{5 + 4} = \dfrac{(-243)5}{9} = -\dfrac{1215}{9} = -135$

3. $a_n = \dfrac{n!}{(n+1)!}$

$a_1 = \dfrac{1!}{(1+1)!} = \dfrac{1}{2!} = \dfrac{1}{2}$

$a_2 = \dfrac{2!}{(2+1)!} = \dfrac{2}{3!} = \dfrac{2}{6} = \dfrac{1}{3}$

$a_3 = \dfrac{3!}{(3+1)!} = \dfrac{6}{4!} = \dfrac{6}{24} = \dfrac{1}{4}$

$a_4 = \dfrac{4!}{(4+1)!} = \dfrac{24}{5!} = \dfrac{24}{120} = \dfrac{1}{5}$

$a_5 = \dfrac{5!}{(5+1)!} = \dfrac{120}{6!} = \dfrac{120}{720} = \dfrac{1}{6}$

4. $a_n = \dfrac{n}{(n+1)^2}$

$a_1 = \dfrac{1}{(1+1)^2} = \dfrac{1}{2^2} = \dfrac{1}{4}$

$a_2 = \dfrac{2}{(2+1)^3} = \dfrac{2}{9}$

$a_3 = \dfrac{3}{(3+1)^2} = \dfrac{3}{16}$

$a_4 = \dfrac{4}{(4+1)^2} = \dfrac{4}{25}$

$a_5 = \dfrac{5}{(5+1)^2} = \dfrac{5}{36}$

5. $a_1 = 4,\ a_{n+1} = 3a_n - 2$

$a_1 = 4$

$a_2 = 3(4) - 2 = 10$

$a_3 = 3(10) - 2 = 28$

$a_4 = 3(28) - 2 = 82$

$a_5 = 3(82) - 2 = 244$

$a_6 = 3(244) - 2 = 730$

6. $a_1 = 1,\ a_2 = 3,\ a_{n+2} = \dfrac{3a_n}{a_{n+1}}$

$a_1 = 1$

$a_2 = 3$

$a_3 = \dfrac{3(1)}{3} = 1$

$a_4 = \dfrac{3(3)}{1} = 9$

$a_5 = \dfrac{3(1)}{9} = \dfrac{1}{3}$

$a_6 = \dfrac{3(9)}{1/3} = 81$

7. $1, 4, 7, 10, 13, \ldots$

$a_n = 3n - 2$

(There are many correct answers.)

8. $0, 3, 8, 15, 24, \ldots$

$a_n = n^2 - 1$

(There are many correct answers.)

9. $\dfrac{2}{3}, \dfrac{3}{4}, \dfrac{4}{5}, \dfrac{5}{6}, \dfrac{6}{7}, \ldots$

$a_n = \dfrac{n+1}{n+2}$

(There are many correct answers.)

10. $\dfrac{1}{2}, -\dfrac{1}{4}, \dfrac{1}{8}, -\dfrac{1}{16}, \ldots$

$a_n = \dfrac{(-1)^{n+1}}{2^n}$

(There are many correct answers.)

11. $\displaystyle\sum_{k=1}^{4} 10k^3 = 10(1)^3 + 10(2)^3 + 10(3)^3 + 10(4)^3$

$= 10 + 80 + 270 + 640$

$= 1000$

12. $\displaystyle\sum_{i=1}^{10} (i^2 + 4) = (1^2 + 4) + (2^2 + 4) + (3^2 + 4) + (4^2 + 4) + (5^2 + 4) + (6^2 + 4) + (7^2 + 4) + (8^2 + 4)$

$\qquad + (9^2 + 4) + (10^2 + 4)$

$= 5 + 8 + 13 + 20 + 29 + 40 + 53 + 68 + 85 + 104$

$= 425$

13. $\displaystyle\sum_{j=1}^{5} \frac{60}{j+1} = \frac{60}{1+1} + \frac{60}{2+1} + \frac{60}{3+1} + \frac{60}{4+1} + \frac{60}{5+1}$

$\qquad = \dfrac{60}{2} + \dfrac{60}{3} + \dfrac{60}{4} + \dfrac{60}{5} + \dfrac{60}{6}$

$\qquad = 30 + 20 + 15 + 12 + 10$

$\qquad = 87$

14. $\dfrac{2}{3(1)} + \dfrac{2}{3(2)} + \dfrac{2}{3(3)} + \cdots + \dfrac{2}{3(20)} = \displaystyle\sum_{k=1}^{20} \frac{2}{3k}$

15. $\dfrac{1}{1^3} - \dfrac{1}{2^3} + \dfrac{1}{3^3} - \cdots + \dfrac{1}{25^3} = \displaystyle\sum_{k=1}^{25} \frac{(-1)^{k+1}}{k^3}$

16. $a_n = a_1 + (n-1)d$

$\quad a_n = 5 + (n-1)7$

$\qquad = 5 + 7n - 7$

$\qquad = 7n - 2$

17. $a_1 = 20,\ a_4 = 11$

$\quad a_n = a_1 + (n-1)d$

$\quad a_4 = a_1 + (4-1)d$

$\quad 11 = 20 + 3d$

$\ -9 = 3d$

$\ -3 = d$

$\quad a_n = 20 + (n-1)(-3) = 20 - 3n + 3 = -3n + 23$

18. $\displaystyle\sum_{i=1}^{n} a_i = \frac{n}{2}[a_1 + a_n]$

$\displaystyle\sum_{i=1}^{50} (3i + 5) = 8 + 11 + 14 + \cdots + 155$

$\qquad = \dfrac{50}{2}[8 + 155]$

$\qquad = 25(163)$

$\qquad = 4075$

19. $\displaystyle\sum_{i=1}^{n} a_i = \frac{n}{2}[a_1 + a_n]$

$\displaystyle\sum_{j=1}^{300} \frac{j}{5} = \frac{1}{5} + \frac{2}{5} + \frac{3}{5} + \cdots + \frac{300}{5}$

$\qquad = \dfrac{300}{2}\left[\dfrac{1}{5} + \dfrac{300}{5}\right]$

$\qquad = 150\left(\dfrac{301}{5}\right)$

$\qquad = 9030$

20. $\displaystyle\sum_{i=1}^{n} a_i = \frac{n}{2}(a_1 + a_n)$

$\displaystyle\sum_{i=1}^{60} (0.2i + 3) = 3.2 + 3.4 + 3.6 + \cdots + 15$

$\qquad = \dfrac{60}{2}(3.2 + 15)$

$\qquad = 30(18.2)$

$\qquad = 546$

21. $\displaystyle\sum_{i=1}^{n} a_i = \frac{n}{2}(a_1 + a_n)$

$\displaystyle\sum_{i=1}^{50} \left(\frac{1}{3}i - 4\right) = -\frac{11}{3} + \left(-\frac{10}{3}\right) + (-3) + \cdots + \frac{38}{3}$

$\qquad = \dfrac{50}{2}\left(-\dfrac{11}{3} + \dfrac{38}{3}\right)$

$\qquad = 25(9)$

$\qquad = 225$

22. $25.75, 23.50, 21.25, 18.75, \ldots$

$\quad a_1 = 23.75$ and $d = -2.25$

$\quad a_n = a_1 + (n-1)d$

$\quad a_{10} = 25.75 + (10-1)(-2.25)$

$\qquad = 25.75 + 9(-2.25)$

$\qquad = 25.75 - 20.25$

$\qquad = 5.5°$

The temperature decreased by 5.5 degrees.

Section 10.3   Geometric Sequences and Series

1. $2, 6, 18, 54, \ldots$
$r = 3$

3. $1, -3, 9, -27, \ldots$
$r = -3$

5. $192, -48, 12, -3, \ldots$
$r = -\frac{1}{4}$

7. $1, -\frac{3}{2}, \frac{9}{4}, -\frac{27}{8}, \ldots$
$r = -\frac{3}{2}$

9. $9, 6, 4, \frac{8}{3}, \ldots$
$r = \frac{2}{3}$

11. $1, \pi, \pi^2, \pi^3, \ldots$
$r = \pi$

13. $1.1, (1.1)^2, (1.1)^3, (1.1)^4, \ldots$
$r = 1.1$

15. $64, 32, 16, 8, \ldots$
The sequence is geometric.
$r = \frac{1}{2}$

17. $10, 15, 20, 25, \ldots$
The sequence is *not* geometric.
$\frac{15}{10} \neq \frac{20}{15}$

19. $5, 10, 20, 40, \ldots$
The sequence is geometric.
$r = 2$

21. $1, 8, 27, 64, 125, \ldots$
The sequence is *not* geometric.
$\frac{8}{1} \neq \frac{27}{8}$

23. The sequence is geometric.
$r = -\frac{2}{3}$

25. $\frac{1}{4}, \frac{2}{5}, \frac{3}{6}, \frac{4}{7}, \ldots$
The sequence is *not* geometric.
$\frac{2/5}{1/4} \neq \frac{3/6}{2/5}$

27. The sequence is geometric.
$r = (1 + 0.02)$ or 1.02

29. $a_n = a_1 r^{n-1} = 4(2)^{n-1}$
$a_1 = 4(2)^{1-1} = 4$
$a_2 = 4(2)^{2-1} = 8$
$a_3 = 4(2)^{3-1} = 16$
$a_4 = 4(2)^{4-1} = 32$
$a_5 = 4(2)^{5-1} = 64$

31. $a_n = a_1 r^{n-1} = 5(-2)^{n-1}$
$a_1 = 5(-2)^{1-1} = 5$
$a_2 = 5(-2)^{2-1} = -10$
$a_3 = 5(-2)^{3-1} = 20$
$a_4 = 5(-2)^{4-1} = -40$
$a_5 = 5(-2)^{5-1} = 80$

33. $a_n = a_1 r^{n-1} = 6\left(\frac{1}{3}\right)^{n-1}$
$a_1 = 6\left(\frac{1}{3}\right)^{1-1} = 6$
$a_2 = 6\left(\frac{1}{3}\right)^{2-1} = 2$
$a_3 = 6\left(\frac{1}{3}\right)^{3-1} = \frac{2}{3}$
$a_4 = 6\left(\frac{1}{3}\right)^{4-1} = \frac{2}{9}$
$a_5 = 6\left(\frac{1}{3}\right)^{5-1} = \frac{2}{27}$

35. $a_n = a_1 r^{n-1} = 1\left(-\frac{1}{2}\right)^{n-1}$
$a_1 = 1\left(-\frac{1}{2}\right)^{1-1} = 1$
$a_2 = 1\left(-\frac{1}{2}\right)^{2-1} = -\frac{1}{2}$
$a_3 = 1\left(-\frac{1}{2}\right)^{3-1} = \frac{1}{4}$
$a_4 = 1\left(-\frac{1}{2}\right)^{4-1} = -\frac{1}{8}$
$a_5 = 1\left(-\frac{1}{2}\right)^{5-1} = \frac{1}{16}$

37. $a_1 = 4, r = -\frac{1}{2}$
$a_n = a_1 r^{n-1}$
$a_n = 4\left(-\frac{1}{2}\right)^{n-1}$
$a_1 = 4\left(-\frac{1}{2}\right)^{1-1} = 4$
$a_2 = 4\left(-\frac{1}{2}\right)^{2-1} = -2$
$a_3 = 4\left(-\frac{1}{2}\right)^{3-1} = 1$
$a_4 = 4\left(-\frac{1}{2}\right)^{4-1} = -\frac{1}{2}$
$a_5 = 4\left(-\frac{1}{2}\right)^{5-1} = \frac{1}{4}$

39. $a_n = a_1 r^{n-1} = 1000(1.01)^{n-1}$
$a_1 = 1000(1.01)^{1-1} = 1000$
$a_2 = 1000(1.01)^{2-1} = 1010$
$a_3 = 1000(1.01)^{3-1} = 1020.1$
$a_4 = 1000(1.01)^{4-1} = 1030.301$
$a_5 = 1000(1.01)^{5-1} = 1040.60401$

41. $a_1 = 4000, \ r = \dfrac{1}{1.01}$

$a_n = a_1 r^{n-1}$

$a_n = 4000\left(\dfrac{1}{1.01}\right)^{n-1}$

$a_1 = 4000\left(\dfrac{1}{1.01}\right)^{1-1} = 4000$

$a_2 = 4000\left(\dfrac{1}{1.01}\right)^{2-1} \approx 3960.396$

$a_3 = 4000\left(\dfrac{1}{1.01}\right)^{3-1} \approx 3921.184$

$a_4 = 4000\left(\dfrac{1}{1.01}\right)^{4-1} \approx 3882.361$

$a_5 = 4000\left(\dfrac{1}{1.01}\right)^{5-1} \approx 3843.921$

43. $a_1 = 10, r = \dfrac{3}{5}, \ a_n = a_1 r^{n-1}$

$a_n = 10\left(\dfrac{3}{5}\right)^{n-1}$

$a_1 = 10\left(\dfrac{3}{5}\right)^{1-1} = 10$

$a_2 = 10\left(\dfrac{3}{5}\right)^{2-1} = 6$

$a_3 = 10\left(\dfrac{3}{5}\right)^{3-1} = \dfrac{18}{5}$

$a_4 = 10\left(\dfrac{3}{5}\right)^{4-1} = \dfrac{54}{25}$

$a_5 = 10\left(\dfrac{3}{5}\right)^{5-1} = \dfrac{162}{125}$

45. $a_n = a_1 r^{n-1}$

$a_8 = 3(-3)^{8-1}$

$= 3(-3)^7$

$= -6561$

47. $a_n = a_1 r^{n-1}$

$a_{10} = 120\left(-\dfrac{1}{3}\right)^{10-1}$

$= 120\left(-\dfrac{1}{3}\right)^9$

$= -\dfrac{120}{19,683}$

$= -\dfrac{40}{6621}$

≈ -0.006

49. $a_n = a_1 r^{n-1}$

$a_{10} = 6\left(\dfrac{1}{2}\right)^{10-1}$

$= 6\left(\dfrac{1}{2}\right)^9$

$= \dfrac{6}{512}$

$= \dfrac{3}{256}$

≈ 0.012

51. $a_n = a_1 r^{n-1}$

$a_{10} = 3\left(\sqrt{2}\right)^{10-1}$

$= 3(2^{1/2})^9$

$= 3(2)^{9/2}$

$= 3(2)^4 2^{1/2}$

$= 3(16)\sqrt{2}$

$= 48\sqrt{2}$

≈ 67.882

53. $a_n = a_1 r^{n-1}$

$a_{12} = 200(1.2)^{12-1} = 200(1.2)^{11} \approx 1486.02$

55. $r = \dfrac{3}{4}$

$a_n = a_1 r^{n-1}$

$a_5 = 4\left(\dfrac{3}{4}\right)^{5-1}$

$= 4\left(\dfrac{3}{4}\right)^4$

$= \dfrac{4(81)}{256}$

$= \dfrac{81}{64}$

≈ 1.266

57. $a_3 = a_1 r^{3-1}$

$\dfrac{9}{4} = 1(r)^2$

$\pm\sqrt{\dfrac{9}{4}} = \pm\dfrac{3}{2} = r$

$a_n = a_1 r^{n-1}$

$a_6 = 1\left(\pm\dfrac{3}{2}\right)^{6-1}$

$= \left(\pm\dfrac{3}{2}\right)^5$

$= \pm\dfrac{243}{32}$

59. $a_3 = 6, \ a_5 = \dfrac{8}{3}$

$\dfrac{a_5}{a_3} = \dfrac{a_1 r^4}{a_1 r^2}$

$= r^2$

$= \dfrac{8/3}{6}$

$= \dfrac{8}{18} = \dfrac{4}{9}$

$r^2 = \dfrac{4}{9} \Rightarrow r = \pm\dfrac{2}{3}$

$a_6 = a_5 r$

$= \dfrac{8}{3} \cdot \left(\pm\dfrac{2}{3}\right)$

$= \pm\dfrac{16}{9}$

61. $a_n = a_1 r^{n-1} = 2(3)^{n-1}$

63. $a_n = a_1 r^{n-1} = 1(2)^{n-1} = 2^{n-1}$

65. $a_n = a_1 r^{n-1} = 4\left(-\frac{1}{2}\right)^{n-1}$

67. $r = \frac{2}{8} = \frac{1}{4}$

$a_n = a_1 r^{n-1} = 8\left(\frac{1}{4}\right)^{n-1}$

69. $r = \frac{8}{12} = \frac{2}{3}$

$a_n = a_1 r^{n-1}$

$a_n = 12\left(\frac{2}{3}\right)^{n-1}$

71. $r = \frac{12}{4} = 3$

$a_n = a_1 r^{n-1}$

$a_n = 4(3)^{n-1}$

73. $4, -6, 9, -\frac{27}{2}, \ldots$

$a_1 = 4, \ r = \frac{-6}{4} = -\frac{3}{2}$

$a_n = a_1 r^{n-1}$

$a_n = 4\left(-\frac{3}{2}\right)^{n-1}$

75. Graph (b)

77. Graph (a)

79.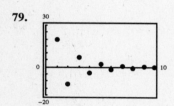

81.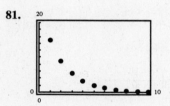

83. $\displaystyle\sum_{i=1}^{10} 2^{i-1} = 2^0 + 2^1 + 2^2 + \cdots + 2^9$

Sum of geometric sequence: $a_1\left(\dfrac{r^n - 1}{r - 1}\right)$

$a_1 = 1, \ r = 2, \ n = 10$

$\displaystyle\sum_{i=1}^{10} 2^{i-1} = 1\left(\frac{2^{10} - 1}{2 - 1}\right) = \left(\frac{1024 - 1}{1}\right) = 1023$

85. $\displaystyle\sum_{i=1}^{12} 3\left(\frac{3}{2}\right)^{i-1} = 3\left(\frac{3}{2}\right)^0 + 3\left(\frac{3}{2}\right) + 3\left(\frac{3}{2}\right)^2 + \cdots + 3\left(\frac{3}{2}\right)^{11}$

Sum of geometric sequence: $a_1\left(\dfrac{r^n - 1}{r - 1}\right)$

$a_1 = 3, \ r = \frac{3}{2}, \ n = 12$

$\displaystyle\sum_{i=1}^{12} 3\left(\frac{3}{2}\right)^{i-1} = 3\left[\frac{(3/2)^{12} - 1}{(3/2) - 1}\right] = 3\frac{(531{,}441/4096) - 1}{1/2} = 3\left(\frac{527{,}345}{4096}\right)(2) = \frac{1{,}582{,}035}{2048} \approx 772{,}478$

or $\displaystyle\sum_{i=1}^{12} 3\left(\frac{3}{2}\right)^{i-1} = 3\left(\frac{1.5^{12} - 1}{1.5 - 1}\right) \approx \frac{3(128.7463)}{0.5} \approx 772.478$

87. $\displaystyle\sum_{i=1}^{15} 3\left(-\frac{1}{3}\right)^{i-1} = 3\left(-\frac{1}{3}\right)^0 + 3\left(-\frac{1}{3}\right)^1 + 3\left(-\frac{1}{3}\right)^2 + \cdots + 3\left(-\frac{1}{3}\right)^{14}$

Sum of geometric sequence: $a_1\left(\dfrac{r^n - 1}{r - 1}\right)$

$a_1 = 3, \ r = -\frac{1}{3}, \ n = 15$

$\displaystyle\sum_{i=1}^{15} 3\left(-\frac{1}{2}\right)^{i-1} = 3\left[\frac{(-1/3)^{15} - 1}{(-1/3) - 1}\right] = 3\left[\frac{(-1/3)^{15} - 1}{-4/3}\right] = 3\left[\left(-\frac{1}{3}\right)^{15} - 1\right]\left(-\frac{3}{4}\right) \approx 2.250$

89. $\displaystyle\sum_{i=1}^{8} 6(0.1)^{i-1} = 6(0.1)^0 + 6(0.1)^1 + 6(0.1)^2 + \cdots + 6(0.1)^8$

Sum of geometric sequence: $a_1 \dfrac{(r^n - 1)}{r - 1}$

$a_1 = 6, r = 0.1, n = 8$

$$\sum_{i=1}^{8} 6(0.1)^{i-1} = 6\left(\frac{(0.1)^8 - 1}{0.1 - 1}\right) = \frac{6(-0.999999999)}{-0.9} = 6.6666666 \approx 6.67$$

91. $\displaystyle\sum_{i=1}^{10} 15(0.3)^{i-1} = 15(0.3)^0 + 15(0.3)^1 + 15(0.3)^2 + \cdots + 15(0.3)^9$

Sum of geometric sequence: $a_1 \dfrac{(r^n - 1)}{r - 1}$

$a_1 = 15, \; r = 0.3, \; n = 10$

$$\sum_{i=1}^{10} 15(0.3)^{i-1} = 15\left(\frac{(0.3)^{10} - 1}{0.3 - 1}\right) \approx 21.43$$

93. $\displaystyle\sum_{i=1}^{30} 100(0.75)^{i-1} \approx 399.93$

95. $\displaystyle\sum_{i=1}^{20} 100(1.1)^{i} = 6300.25$

97. Geometric sequence with $r = -3$

$a_n = 1(-3)^{n-1}$

Sum: $a_1\left(\dfrac{r^n - 1}{r - 1}\right)$

$$\sum_{k=1}^{10} (-3)^{k-1} = 1\left[\frac{(-3)^{10} - 1}{-3 - 1}\right]$$

$$= \frac{59{,}049 - 1}{-4}$$

$$= \frac{59{,}048}{-4}$$

$$= -14{,}762$$

99. 5, 10, 20, 40, 80, . . .

Geometric sequence with $r = 2$

$a_n = 5(2)^{n-1}$

Sum: $a_1\left(\dfrac{r^n - 1}{r - 1}\right)$

$$\sum_{k=1}^{9} 5(2)^{k-1} = 5\left(\frac{2^9 - 1}{2 - 1}\right)$$

$$= 5\left(\frac{511}{1}\right)$$

$$= 2555$$

101. Geometric sequence with $r = \dfrac{1}{2}$

$a_n = 8\left(\dfrac{1}{2}\right)^{n-1}$

Sum: $a_1\left(\dfrac{r^n - 1}{r - 1}\right)$

$$\sum_{k=1}^{15} 8\left(\frac{1}{2}\right)^{k-1} = 8\left[\frac{(1/2)^{15} - 1}{(1/2) - 1}\right]$$

$$= 8\left[\frac{(1/2)^{15} - 1}{-1/2}\right]$$

$$= 8\left[\left(\frac{1}{2}\right)^{15} - 1\right](-2)$$

$$\approx 16.000$$

103. 4, 12, 36, 108, . . . , $n = 8$

Geometric sequence with $r = 3$

$a_n = 4(3)^{n-1}$

Sum: $a_1\left(\dfrac{r^n - 1}{r - 1}\right)$

$$\sum_{k=1}^{8} 4(3)^{n-1} = 4\left(\frac{3^8 - 1}{3 - 1}\right)$$

$$= \frac{4(6560)}{2}$$

$$= 13{,}120$$

105. $60, -15, \dfrac{15}{4}, -\dfrac{15}{16}, \ldots, n = 12$

Geometric sequence with $r = -\dfrac{1}{4}$

$a_n = 60\left(-\dfrac{1}{4}\right)^{n-1}$

Sum: $a_1\left(\dfrac{r^n - 1}{r - 1}\right)$

$\displaystyle\sum_{k=1}^{12}\left[60\left(-\dfrac{1}{4}\right)^{n-1}\right] = 60\left(\dfrac{(-0.25)^{12} - 1}{-0.25 - 1}\right)$

≈ 48.00

107. Geometric sequence with $r = 1.06$

$a_n = 30(1.06)^{n-1}$

Sum: $a_1\left(\dfrac{r^n - 1}{r - 1}\right)$

$\displaystyle\sum_{k=1}^{20} 30(1.06)^{k-1} = 30\left[\dfrac{(1.06)^{20} - 1}{1.06 - 1}\right]$

$= 30\left[\dfrac{(1.06)^{20} - 1}{0.06}\right]$

≈ 1103.568

109. $500, 500(1.04), 500(1.04)^2, \ldots, n = 18$

Geometric sequence with $r = 1.04$

$a_n = 500(1.04)^{n-1}$

Sum: $a_1\left(\dfrac{r^n - 1}{r - 1}\right)$

$\displaystyle\sum_{k=1}^{18} 500(1.04)^{n-1} = 500\left(\dfrac{(1.04)^{18} - 1}{1.04 - 1}\right)$

$\approx 12{,}822.71$

111. $\displaystyle\sum_{i=1}^{\infty} a_i = \dfrac{a_1}{1 - r}$

This series is geometric with $a_1 = \left(\dfrac{1}{2}\right)^0 = 1$ and $r = \dfrac{1}{2}$.

$\displaystyle\sum_{n=0}^{\infty}\left(\dfrac{1}{2}\right)^n = \dfrac{1}{1 - (1/2)}$

$= \dfrac{1}{1/2}$

$= 2$

113. $\displaystyle\sum_{i=1}^{\infty} a_i = \dfrac{a_1}{1 - r}$

This series is geometric with $a_1 = \left(-\dfrac{1}{2}\right)^0 = 1$ and $r = -\dfrac{1}{2}$.

$\displaystyle\sum_{n=0}^{\infty}\left(-\dfrac{1}{2}\right)^n = \dfrac{1}{1 - (-1/2)}$

$= \dfrac{1}{3/2}$

$= \dfrac{2}{3}$

115. $\displaystyle\sum_{i=1}^{\infty} a_i = \dfrac{a_1}{1 - r}$

This series is geometric with $a_1 = 2\left(-\dfrac{2}{3}\right)^0 = 2$ and $r = -\dfrac{2}{3}$.

$\displaystyle\sum_{i=0}^{\infty} 2\left(-\dfrac{2}{3}\right)^i = \dfrac{2}{1 - (-2/3)}$

$= \dfrac{2}{5/3}$

$= \dfrac{6}{5}$

117. $\displaystyle\sum_{i=1}^{\infty} a_i = \dfrac{a_1}{1 - r}$

This series is geometric with $a_1 = 8$ and $r = \dfrac{6}{8} = \dfrac{3}{4}$.

$\displaystyle\sum_{i=0}^{\infty} 8\left(\dfrac{3}{4}\right)^i = \dfrac{8}{1 - (3/4)}$

$= \dfrac{8}{1/4}$

$= 32$

119. (a) $a_n = 250{,}000(0.75)^n$

(b) $a_5 = 250{,}000(0.75)^5 \approx \$59{,}326.17$

The depreciated value at the end of 5 years is approximately $59,326.17.

(c) The machine depreciated the most during the first year.

121. $a_n = 30,000(1.05)^{n-1}$

$$\sum_{k=1}^{40} 30,000(1.05)^{k-1} = 30,000\left[\frac{(1.05)^{40} - 1}{1.05 - 1}\right]$$

$$= 30,000\left[\frac{(1.05)^{40} - 1}{0.05}\right]$$

$$\approx 3,623,993$$

The total salary over the 40-year period would be approximately \$3,623,993.

123. $\sum_{i=1}^{n} a_1 r^{i-1} = a_1\left(\frac{r^n - 1}{r - 1}\right)$

$a_1 = 100\left(1 + \frac{0.09}{12}\right),\ r = \left(1 + \frac{0.09}{12}\right)$

$$\sum_{k=1}^{120} 100\left(1 + \frac{0.09}{12}\right)^k = 100\left(1 + \frac{0.10}{12}\right)\left[\frac{\left(1 + \frac{0.10}{12}\right)^{120} - 1}{\left(1 + \frac{0.10}{12}\right) - 1}\right]$$

$$= \$19,496.56$$

125. $\sum_{i=1}^{n} a_1 r^{i-1} = a_1\left(\frac{r^n - 1}{r - 1}\right)$

$a_1 = 30\left(1 + \frac{0.08}{12}\right),\ r = \left(1 + \frac{0.08}{12}\right)$

$$\sum_{k=1}^{480} 30\left(1 + \frac{0.08}{12}\right)^k = 30\left(1 + \frac{0.08}{12}\right)\left[\frac{\left(1 + \frac{0.08}{12}\right)^{480} - 1}{\left(1 + \frac{0.08}{12}\right) - 1}\right]$$

$$= \$105,428.44$$

127. $\sum_{i=1}^{n} a_1 r^{i-1} = a_1\left(\frac{r^n - 1}{r - 1}\right)$

$a_1 = 75\left(1 + \frac{0.06}{12}\right),\ r = \left(1 + \frac{0.06}{12}\right)$

$$\sum_{k=1}^{360} 75\left(1 + \frac{0.06}{12}\right)^k = 75\left(1 + \frac{0.06}{12}\right)\left[\frac{\left(1 + \frac{0.06}{12}\right)^{360} - 1}{\left(1 + \frac{0.06}{12}\right) - 1}\right]$$

$$= \$75,715.32$$

129. $a_1 = 0.01,\ r = 2$

$a_n = a_1 r^{n-1} = (0.01)2^{k-1}$

(a) $\sum_{k=1}^{29} (0.01)2^{k-1} = 0.01\left[\frac{2^{29} - 1}{2 - 1}\right] = 0.01(2^{29} - 1) = 5,368,709.11$

The total income for 29 days would be \$5,368,709.11.

(b) $\sum_{k=1}^{30} (0.01)2^{k-1} = 0.01\left[\frac{2^{30} - 1}{2 - 1}\right] = 0.01(2^{30} - 1) = \$10,737,418.23$

The total income for 30 days would be \$10,737,418.23.

131. (a) $a_n = P(0.999)^n$

(b) $a_6 = P(0.999)^{365} \approx P(0.694)$

Approximately 69.4% of the initial power is available.

(c)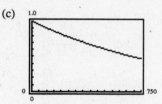

The power supply needs to be changed after 693 days.

133. $100 + 2(100)(0.75) + \cdots + 2(100)(0.75)^{10} = 100 + \sum_{k=1}^{10} 2(100)(0.75)^k$

$$= 100 + 2(100)(0.75)\left(\frac{(0.75)^{10} - 1}{0.75 - 1}\right) \approx 666.21$$

135. Answers will vary.

$$a_n = 100(-0.01)^n \quad \text{or} \quad a_n = \left(-\tfrac{2}{3}\right)^{n-1}$$

137. The nth partial sum of a sequence is the sum of the first n terms of the sequence.

139. $\dfrac{n!}{n(n+1)!} = \dfrac{\cancel{n!}}{n(n+1)\cancel{n!}}$

$$= \frac{1}{n(n+1)} \quad \text{or} \quad \frac{1}{n^2 + n}$$

141. $\dfrac{3}{2}y + 11 < 20$

$$2\left(\frac{3}{2}y + 11\right) < 40$$

$$3y + 22 < 40$$

$$3y < 18$$

$$y < 6$$

143. $\qquad 2x - \dfrac{5}{x} > 3$

$$2x - \frac{5}{x} - 3 > 0$$

$$\frac{2x^2 - 5 - 3x}{x} > 0$$

$$\frac{2x^2 - 3x - 5}{x} > 0$$

Critical numbers:

$$2x^2 - 3x - 5 = 0$$

$$(2x - 5)(x + 1) = 0$$

$$2x - 5 = 0 \implies 2x = 5 \implies x = \frac{5}{2}$$

$$x + 1 = 0 \implies x = -1$$

$$x = 0$$

Tests:

4: $\dfrac{2(4)^2 - 3(4) - 5}{4} \overset{?}{>} 0$ Yes

2: $\dfrac{2(2)^2 - 3(2) - 5}{2} \overset{?}{>} 0$ No

$-\tfrac{1}{2}$: $\dfrac{2\left(\frac{1}{2}\right)^2 - 3\left(\frac{1}{2}\right) - 5}{-\frac{1}{2}} \overset{?}{>} 0$ Yes

-3: $\dfrac{2(-3)^2 - 3(-3) - 5}{-3} \overset{?}{>} 0$ No

$$-1 < x < 0 \quad \text{or} \quad x > \frac{5}{2}$$

145. $a^2 + b^2 + c^2$ (Pythagorean Theorem)

$$25^2 + 40^2 = c^2$$
$$625 + 1600 = c^2$$
$$2225 = c^2$$
$$\sqrt{2225} = c$$
$$5\sqrt{89} = c$$
$$47.17 \approx c$$

The length of the diagonal should be approximately 47.17 feet.

Section 10.4   The Binomial Theorem

1. $_6C_4 = \dfrac{6!}{2!4!} = \dfrac{6 \cdot 5 \cdot 4 \cdot 3}{4 \cdot 3 \cdot 2 \cdot 1} = 15$

Note: $_6C_4 = {_6}C_2 = \dfrac{6 \cdot 5}{2 \cdot 1} = 15$

3. $_{10}C_5 = \dfrac{10!}{5!5!} = \dfrac{10 \cdot 9 \cdot 8 \cdot 7 \cdot 6}{5 \cdot 4 \cdot 3 \cdot 2 \cdot 1} = 252$

5. $_{20}C_{20} = \dfrac{20!}{0!20!} = 1$ (**Note:** $0! = 1$)

7. $_{18}C_{18} = \dfrac{18!}{0! \cdot 18!} = \dfrac{18!}{1 \cdot 18!} = 1$

9. $_{50}C_{48} = {_{50}}C_2 = \dfrac{50!}{48!2!} = \dfrac{50 \cdot 49}{2 \cdot 1} = 1225$

11. $_{25}C_4 = \dfrac{25!}{21!4!} = \dfrac{25 \cdot 24 \cdot 23 \cdot 22}{4 \cdot 3 \cdot 2 \cdot 1} = 12{,}650$

13. $_{52}C_5 = 2{,}598{,}960$

15. $_{200}C_{195} = 2{,}535{,}650{,}040$

17. $_{15}C_3 = \dfrac{15!}{12!3!} = \dfrac{15 \cdot 14 \cdot 13 \cdot \cancel{12!}}{\cancel{12!}3 \cdot 2 \cdot 1} = 455$

$_{15}C_{12} = \dfrac{15!}{3!12!} = \dfrac{15 \cdot 14 \cdot 13 \cdot \cancel{12!}}{3 \cdot 2 \cdot 1 \cdot \cancel{12!}} = 455$

19. $_{25}C_5 = \dfrac{25!}{20!5!} = \dfrac{25 \cdot 24 \cdot 23 \cdot 22 \cdot 21 \cdot \cancel{20!}}{\cancel{20!}5 \cdot 4 \cdot 3 \cdot 2 \cdot 1} = 53{,}130$

$_{25}C_{20} = \dfrac{25!}{5!20!} = \dfrac{25 \cdot 24 \cdot 23 \cdot 22 \cdot 21 \cdot \cancel{20!}}{5 \cdot 4 \cdot 3 \cdot 2 \cdot 1 \cdot \cancel{20!}} = 53{,}130$

21. $_5C_2 = \dfrac{5!}{3! \cdot 2!} = \dfrac{5 \cdot 4 \cdot \cancel{3!}}{\cancel{3!}2 \cdot 1} = 10$

$_5C_3 = \dfrac{5!}{2! \cdot 3!} = \dfrac{5 \cdot 4 \cdot \cancel{3!}}{2 \cdot 1 \cdot \cancel{3!}} = 10$

23. $_{12}C_5 = \dfrac{12!}{7! \cdot 5!} = \dfrac{12 \cdot 11 \cdot 10 \cdot 9 \cdot 8 \cdot \cancel{7!}}{\cancel{7!}5 \cdot 4 \cdot 3 \cdot 2 \cdot 1} = 792$

$_{12}C_7 = \dfrac{12!}{5! \cdot 7!} = \dfrac{12 \cdot 11 \cdot 10 \cdot 9 \cdot 8 \cdot \cancel{7!}}{5 \cdot 4 \cdot 3 \cdot 2 \cdot 1 \cdot \cancel{7!}} = 792$

25. $_{10}C_0 = \dfrac{10!}{10! \cdot 0!} = \dfrac{10!}{10! \cdot 1} = 1$

$_{10}C_{10} = \dfrac{10!}{0! \cdot 10!} = \dfrac{10!}{1 \cdot 10!} = 1$

27. $_6C_2$ Sixth row:

	1	6	**15**	20	15	6	1	
	$_6C_0$	$_6C_1$	$_6C_2$	$_6C_3$	$_6C_4$	$_6C_5$	$_6C_6$	

29. $_7C_3$ Seventh row:

	1	7	21	**35**	35	21	7	1
	$_7C_0$	$_7C_1$	$_7C_2$	$_7C_3$	$_7C_4$	$_7C_5$	$_7C_6$	$_7C_7$

31. $_8C_4$ Eighth row:

1	8	28	56	70	56	28	8	1
$_8C_0$	$_8C_1$	$_8C_2$	$_8C_3$	$_8C_4$	$_8C_5$	$_8C_6$	$_8C_7$	$_8C_8$

33. $(x + 3)^6 = x^6 + 6x^5(3) + {_6C_2}x^4(3)^2 + {_6C_3}x^3(3)^3 + {_6C_4}x^2(3)^4 + 6x(3)^5 + (3)^6$

$(x + 3)^6 = x^6 + 6x^5(3) + 15x^4(3)^2 + 20x^3(3)^3 + 15x^2(3)^4 + 6x(3)^5 + (3)^6$

$\qquad = x^6 + 18x^5 + 135x^4 + 540x^3 + 1215x^2 + 1458x + 729$

35. $(x + 1)^5 = x^5 = 5x^4(1) + {_5C_2}x^3(1)^2 + {_5C_3}x^2(1)^3 + 5x(1)^4 + 1^5$

$(x + 1)^5 = x^5 + 5x^4(1) + 10x^3(1)^2 + 10x^2(1)^3 + 5x(1)^4 + (1)^5$

$\qquad = x^5 + 5x^4 + 10x^3 + 10x^2 + 5x + 1$

37. $(x - 4)^6 = x^6 + 6x^5(-4) + {_6C_2}x^4(-4)^2 + {_6C_3}x^3(-4)^3 + {_6C_4}x^2(-4)^4 + 6x(-4)^5 + (-4)^6$

$\qquad = x^6 - 24x^5 + 15x^4(16) + 20x^3(-64) + 15x^2(256) + 6x(-1024) + 4096$

$\qquad = x^6 - 24x^5 + 240x^4 - 1280x^3 + 3840x^2 - 6144x + 4096$

39. $(x + y)^8 = x^8 + 8x^7y + {_8C_2}x^6y^2 + {_8C_3}x^5y^3 + {_8C_4}x^4y^4 + {_8C_5}x^3y^5 + {_8C_6}x^2y^6 + 8xy^7 + y^8$

$\qquad = x^8 + 8x^7y + 28x^6y^2 + 56x^5y^3 + 70x^4y^4 + 56x^3y^5 + 28x^2y^6 + 8xy^7 + y^8$

41. $(u - 2v)^3 = u^3 + 3u^2(-2v) + 3u(-2v)^2 + (-2v)^3$

$(u - 2v)^3 = u^3 - 3u^2(2v) + 3u(2v)^2 - (2v)^3$

$\qquad = u^3 - 6u^2v + 12uv^2 - 8v^3$

43. $(2x - 1)^4$ Fourth row: 1 4 6 4 1

$(2x - 1)^4 = (2x)^4 + 4(2x)^3(-1) + 6(2x)^2(-1)^2 + 4(2x)(-1)^3 + (-1)^4$

$\qquad = 16x^4 - 4(8x^3) + 6(4x^2) - 4(2x) + 1$

$\qquad = 16x^4 - 32x^3 + 24x^2 - 8x + 1$

45. $(2y + z)^6$ Sixth row: 1 6 15 20 15 6 1

$(2y + z)^6 = (2y)^6 + 6(2y)^5z + 15(2y)^4z^2 + 20(2y)^3z^3 + 15(2y)^2z^4 + 6(2y)z^5 + z^6$

$\qquad = 64y^6 + 6(32y^5)z + 15(16y^4)z^2 + 20(8y^3)z^3 + 15(4y^2)z^4 + 6(2y)z^5 + z^6$

$\qquad = 64y^6 + 192y^5z + 240y^4z^2 + 160y^3z^3 + 60y^2z^4 + 12yz^5 + z^6$

47. $(3a + 2b)^4 = (3a)^4 + 4(3a)^3(2b) + {_4C_2}(3a)^2(2b)^2 + 4(3a)(2b)^3 + (2b)^4$

$(3a + 2b)^4 = (3a)^4 + 4(3a)^3(2b) + 6(3a)^2(2b)^2 + 4(3a)(2b)^3 + (2b)^4$

$\qquad = 81a^4 + 4(27a^3)(2b) + 6(9a^2)(4b^2) + 4(3a)(8b^3) + 16b^4$

$\qquad = 81a^4 + 216a^3b + 216a^2b^2 + 96ab^3 + 16b^4$

49. 4th term: $_{10}C_3x^{10-3}y^3 = 120x^7y^3$

51. 6th term: $_9C_5a^{9-5}(4b)^5 = 126a^4(1024b^5)$

$\qquad\qquad\qquad\qquad = 129{,}024a^4b^5$

53. 3rd term: $_5C_2x^{5-2}(-6y)^2 = 10x^3(36y^2)$

$\qquad\qquad\qquad\qquad = 360x^3y^2$

55. 10th term: $_{12}C_9(3a)^{12-9}(-b)^9 = 220(27a^3)(-b^9)$

$\qquad\qquad\qquad\qquad\qquad = -5940a^3b^9$

57. 8th term: $_9C_7(4x)^{9-7}(3y)^7 = 36(16x^2)(2187y^7)$

$$= 1,259,712\, x^2y^7$$

59. 9th term: $_{12}C_8(10x)^{12-8}(-3y)^8 = 495(10,000x^4)(6561y^8)$

$$= 32,476,950,000x^4y^8$$

61. $_{10}C_3 = \dfrac{10 \cdot 9 \cdot 8}{3 \cdot 2 \cdot 1} = 120$

$_{10}C_3 x^7(1)^3 = 120x^7$

Coefficient: 120

63. $_{15}C_{11} = {}_{15}C_4$

$$= \dfrac{15 \cdot 14 \cdot 13 \cdot 12}{4 \cdot 3 \cdot 2 \cdot 1}$$

$$= 1365$$

$_{15}C_{11}x^4(-y)^{11} = -1365x^4y^{11}$

Coefficient: -1365

65. $_{12}C_9 = {}_{12}C_3 = \dfrac{12 \cdot 11 \cdot 10}{3 \cdot 2 \cdot 1}$

$$= 220$$

$_{12}C_9(2x)^3y^9 = 220(8x^3)y^9$

$$= 1760x^3y^9$$

Coefficient: 1760

67. $_4C_2 = \dfrac{4!}{2! \cdot 2!} = 6$

$_4C_2(x^2)^2(-3)^2 = 6x^4(9) = 54x^4$

Coefficient: 54

69. $(1.02)^8 = (1 + 0.02)^8$

$$= 1^8 + 8(1)^7(0.02) + 28(1)^6(0.02)^2 + 56(1)^5(0.02)^3 + 70(1)^4(0.02)^4 + \cdots$$

$$= 1 + 0.16 + 0.0112 + 0.000448 + 0.0000112 + \cdots$$

$$\approx 1.172$$

71. $(2.99)^{12} = (3 - 0.01)^{12}$

$$= 3^{12} - 12(3)^{11}(0.01) + 66(3)^{10}(0.01)^2 - 220(3)^9(0.01)^3 + 495(3)^8(0.01)^4 - 792(3)^7(0.01)^5 + \cdots$$

$$= 531{,}441 - 21257.64 + 389.7234 - 4.33026 + 0.03247695 - 0.0001732104 + \cdots$$

$$\approx 510{,}568.785$$

73. The graph of g is a transformation of the graph of f—a horizontal shift of two units to the right.

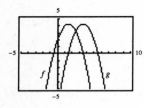

$f(x) = -x^2 + 3x + 2$

$g(x) = f(x - 2) = -(x - 2)^2 + 3(x - 2) + 2$

$\qquad = -(x^2 - 4x + 4) + 3x - 6 + 2$

$\qquad = -x^2 + 4x - 4 + 3x - 6 + 2$

$\qquad = -x^2 + 7x - 8$

75. $\left(\frac{1}{2} + \frac{1}{2}\right)^5 = \left(\frac{1}{2}\right)^5 + 5\left(\frac{1}{2}\right)^4\left(\frac{1}{2}\right) + 10\left(\frac{1}{2}\right)^3\left(\frac{1}{2}\right)^2 + 10\left(\frac{1}{2}\right)^2\left(\frac{1}{2}\right)^3 + 5\left(\frac{1}{2}\right)\left(\frac{1}{2}\right)^4 + \left(\frac{1}{2}\right)^5$

$\qquad = \frac{1}{32} + 5\left(\frac{1}{16}\right)\left(\frac{1}{2}\right) + 10\left(\frac{1}{8}\right)\left(\frac{1}{4}\right) + 10\left(\frac{1}{4}\right)\left(\frac{1}{8}\right) + 5\left(\frac{1}{2}\right)\left(\frac{1}{16}\right) + \frac{1}{32}$

$\qquad = \frac{1}{32} + \frac{5}{32} + \frac{10}{32} + \frac{10}{32} + \frac{5}{32} + \frac{1}{32}$

Note that $\left(\frac{1}{2} + \frac{1}{2}\right) = 1$, and $\left(\frac{1}{2} + \frac{1}{2}\right)^5 = 1^5 = 1$. The expanded version, $\frac{1}{32} + \frac{5}{32} + \frac{10}{32} + \frac{10}{32} + \frac{5}{32} + \frac{1}{32}$ also simplifies to $\frac{32}{32}$ or 1.

77. $\left(\frac{1}{4} + \frac{3}{4}\right)^4 = \left(\frac{1}{4}\right)^4 + 4\left(\frac{1}{4}\right)^3\left(\frac{3}{4}\right) + 6\left(\frac{1}{4}\right)^2\left(\frac{3}{4}\right)^2 + 4\left(\frac{1}{4}\right)\left(\frac{3}{4}\right)^3 + \left(\frac{3}{4}\right)^4$

$\qquad = \frac{1}{256} + 4\left(\frac{1}{64}\right)\left(\frac{3}{4}\right) + 6\left(\frac{1}{16}\right)\left(\frac{9}{16}\right) + 4\left(\frac{1}{4}\right)\left(\frac{27}{64}\right) + \frac{81}{256}$

$\qquad = \frac{1}{256} + \frac{12}{256} + \frac{54}{256} + \frac{108}{256} + \frac{81}{256}$

Note that $\left(\frac{1}{4} + \frac{3}{4}\right) = 1$, and $\left(\frac{1}{4} + \frac{3}{4}\right)^4 = 1^4 = 1$. The expanded version, $\frac{1}{256} + \frac{12}{256} + \frac{54}{256} + \frac{108}{256} + \frac{81}{256}$, also simplifies to $\frac{256}{256}$ or 1.

79. $(0.7 + 0.3)^3 = (0.7)^3 + 3(0.7)^2(0.3) + 3(0.7)(0.3)^2 + (0.3)^3$

$\qquad\qquad = 0.343 + 0.441 + 0.189 + 0.027$

Note that $(0.7 + 0.3) = 1$ and $(0.7 + 0.3)^3 = 1^3 = 1$. The expanded version also simplifies to 1.

81. In the shaded portion, the successive differences between consecutive entries increase by 1.

$3 - 1 = 2$

$6 - 3 = 3$

$10 - 6 = 4$

$15 - 10 = 5$

83. The terms of the expansion of $(x - y)^n$ alternate in sign.

85. The first and last numbers in each row are 1. The other numbers in the row are formed by adding the two numbers immediately above the number.

87. $\begin{bmatrix} 3 & 7 \\ -2 & 6 \end{bmatrix} = 3(6) - (-2)(7)$

$\qquad\qquad = 18 - (-14)$

$\qquad\qquad = 32$

89. $\begin{bmatrix} 4 & 3 & 5 \\ 3 & 2 & -2 \\ 5 & -2 & 0 \end{bmatrix} = 4\begin{vmatrix} 2 & -2 \\ -2 & 0 \end{vmatrix} - 3\begin{vmatrix} 3 & 5 \\ -2 & 0 \end{vmatrix} + 5\begin{vmatrix} 3 & 5 \\ 2 & -2 \end{vmatrix}$

$\qquad\qquad = 4(-4) - 3(10) + 5(-16)$

$\qquad\qquad = -16 - 30 - 80$

$\qquad\qquad = -126$

91. $\log_3 \frac{1}{81} = x$

$\qquad 3^x = \frac{1}{81}$

$\qquad 3^x = 3^{-4}$

$\qquad x = -4$

93. $e^{x/2} = 8$

$\qquad \frac{x}{2} = \ln 8$

$\qquad x = 2\ln 8$

$\qquad x \approx 4.16$

95. $\ln(x + 3) = 10$

$\qquad x + 3 = e^{10}$

$\qquad x = -3 + e^{10}$

$\qquad x \approx 22{,}023.47$

97. $y = ax^2 + bx + c$

$(0, 2):$ $2 = a(0)^2 + b(0) + c$ $\Rightarrow$ $c = 2$

$(10, 8):$ $8 = a(10)^2 + b(10) + c$ $\Rightarrow$ $100a + 10b + c = 8$

$(20, 0):$ $0 = a(20)^2 + b(20) + c$ $\Rightarrow$ $400a + 20b + c = 0$

$$D = \begin{vmatrix} 0 & 0 & 1 \\ 100 & 10 & 1 \\ 400 & 20 & 1 \end{vmatrix} = 0 - 100\begin{vmatrix} 0 & 1 \\ 20 & 1 \end{vmatrix} + 400\begin{vmatrix} 0 & 1 \\ 10 & 1 \end{vmatrix}$$

$$= 0 - 100(-20) + 400(-10)$$

$$= 0 + 2000 - 4000$$

$$= -2000$$

$$D_x = \begin{vmatrix} 2 & 0 & 1 \\ 8 & 10 & 1 \\ 0 & 20 & 1 \end{vmatrix} = 2\begin{vmatrix} 10 & 1 \\ 20 & 1 \end{vmatrix} - 8\begin{vmatrix} 0 & 1 \\ 20 & 1 \end{vmatrix} + 0$$

$$= 2(-10) - 8(-20) + 0$$

$$= -20 + 160$$

$$= 140$$

$$D_y = \begin{vmatrix} 0 & 2 & 1 \\ 100 & 8 & 1 \\ 400 & 0 & 1 \end{vmatrix} = 0 - 100\begin{vmatrix} 2 & 1 \\ 0 & 1 \end{vmatrix} + 400\begin{vmatrix} 2 & 1 \\ 8 & 1 \end{vmatrix}$$

$$= 0 - 100(2) + 400(-6)$$

$$= 0 - 200 - 2400$$

$$= -2600$$

$$D_z = \begin{vmatrix} 0 & 0 & 2 \\ 100 & 10 & 8 \\ 400 & 20 & 0 \end{vmatrix} = 0 - 100\begin{vmatrix} 0 & 2 \\ 20 & 0 \end{vmatrix} + 400\begin{vmatrix} 0 & 2 \\ 10 & 8 \end{vmatrix}$$

$$= 0 - 100(-40) + 400(-20)$$

$$= 0 + 4000 - 8000$$

$$= -4000$$

$$x = \frac{D_x}{D} = \frac{140}{-2000} = -0.07$$

$$y = \frac{D_y}{D} = \frac{-2600}{-2000} = 1.3$$

$$z = \frac{D_z}{D} = \frac{-4000}{-2000} = 2$$

$$y = -0.07x^2 + 1.3x + 2$$

Review Exercises for Chapter 10

1. $a_n = 3n + 5$

$a_1 = 3(1) + 5 = 8$

$a_2 = 3(2) + 5 = 11$

$a_3 = 3(3) + 5 = 14$

$a_4 = 3(4) + 5 = 17$

$a_5 = 3(5) + 5 = 20$

3. $a_n = \dfrac{1}{2^n} + \dfrac{1}{2}$

$a_1 = \dfrac{1}{2^1} + \dfrac{1}{2} = 1$

$a_2 = \dfrac{1}{2^2} + \dfrac{1}{2} = \dfrac{1}{4} + \dfrac{2}{4} = \dfrac{3}{4}$

$a_3 = \dfrac{1}{2^3} + \dfrac{1}{2} = \dfrac{1}{8} + \dfrac{4}{8} = \dfrac{5}{8}$

$a_4 = \dfrac{1}{2^4} + \dfrac{1}{2} = \dfrac{1}{16} + \dfrac{8}{16} = \dfrac{9}{16}$

$a_5 = \dfrac{1}{2^5} + \dfrac{1}{2} = \dfrac{1}{32} + \dfrac{16}{32} = \dfrac{17}{32}$

5. $a_1 = 1,\ a_{n+1} = 5a_n - 2$

$a_1 = 1$

$a_2 = 5(1) - 2 = 3$

$a_3 = 5(3) - 2 = 13$

$a_4 = 5(13) - 2 = 63$

$a_5 = 5(63) - 2 = 313$

$a_6 = 5(313) - 2 = 1563$

7. $a_1 = 2,\ a_2 = 3,$
$a_{n+2} = a_{n+1} - a_n$

$a_1 = 2$

$a_2 = 3$

$a_3 = 3 - 2 = 1$

$a_4 = 1 - 3 = -2$

$a_5 = -2 - 1 = -3$

$a_6 = -3 - (-2) = -1$

9. $a_n = (-1)^n(n + 4)$

$a_{10} = (-1)^{10}(10 + 4)$

$\quad = 14$

11.

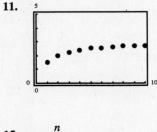

13. $a_n = (n + 1)!$

$a_1 = (1 + 1)! = 2! = 2 \cdot 1 = 2$

$a_2 = (2 + 1)! = 3! = 3 \cdot 2 \cdot 1 = 6$

$a_3 = (3 + 1)! = 4! = 4 \cdot 3 \cdot 2 \cdot 1 = 24$

$a_4 = (4 + 1)! = 5! = 5 \cdot 4 \cdot 3 \cdot 2 \cdot 1 = 120$

$a_5 = (5 + 1)! = 6! = 6 \cdot 5 \cdot 4 \cdot 3 \cdot 2 \cdot 1 = 720$

15. $a_n = \dfrac{n}{n!}$

$a_1 = \dfrac{1}{1!} = \dfrac{1}{1} = 1$

$a_2 = \dfrac{2}{2!} = \dfrac{2}{2} = 1$

$a_3 = \dfrac{3}{3!} = \dfrac{3}{6} = \dfrac{1}{2}$

$a_4 = \dfrac{4}{4!} = \dfrac{4}{24} = \dfrac{1}{6}$

$a_5 = \dfrac{5}{5!} = \dfrac{5}{120} = \dfrac{1}{24}$

17. $1, 3, 5, 7, 9, \ldots$

$a_n = 2n - 1$

19. $\dfrac{1}{4}, \dfrac{2}{9}, \dfrac{3}{16}, \dfrac{4}{25}, \dfrac{5}{36}, \cdots$

$a_n = \dfrac{n}{(n + 1)^2}$

21. $\displaystyle\sum_{k=1}^{4} 7 = 7 + 7 + 7 + 7 = 28$

23. $\displaystyle\sum_{n=1}^{4}\left(\dfrac{1}{n} - \dfrac{1}{n + 1}\right) = \left(\dfrac{1}{1} - \dfrac{1}{1 + 1}\right) + \left(\dfrac{1}{2} - \dfrac{1}{2 + 1}\right) + \left(\dfrac{1}{3} - \dfrac{1}{3 + 1}\right) + \left(\dfrac{1}{4} - \dfrac{1}{4 + 1}\right)$

$\qquad = 1 - \dfrac{1}{2} + \dfrac{1}{2} - \dfrac{1}{3} + \dfrac{1}{3} - \dfrac{1}{4} + \dfrac{1}{4} - \dfrac{1}{5}$

$\qquad = 1 - \dfrac{1}{5} = \dfrac{4}{5}$

25. $[5(1) - 3] + [5(2) - 3] + [5(3) - 3] + [5(4) - 3] = \displaystyle\sum_{k=1}^{4}(5k - 3)$

27. $\dfrac{1}{3(1)} + \dfrac{1}{3(2)} + \dfrac{1}{3(3)} + \dfrac{1}{3(4)} + \dfrac{1}{3(5)} + \dfrac{1}{3(6)} = \displaystyle\sum_{k=1}^{6}\dfrac{1}{3k}$

29. $a_n = 132 - 5n$

$a_1 = 132 - 5(1) = 127$

$a_2 = 132 - 5(2) = 122$

$a_3 = 132 - 5(3) = 117$

$a_4 = 132 - 5(4) = 112$

$a_5 = 132 - 5(5) = 107$

31. $a_n = \frac{3}{4}n + \frac{1}{2}$

$a_1 = \frac{3}{4}(1) + \frac{1}{2} = \frac{5}{4}$

$a_2 = \frac{3}{4}(2) + \frac{1}{2} = 2$

$a_3 = \frac{3}{4}(3) + \frac{1}{2} = \frac{11}{4}$

$a_4 = \frac{3}{4}(4) + \frac{1}{2} = \frac{7}{2}$

$a_5 = \frac{3}{4}(5) + \frac{1}{2} = \frac{17}{4}$

33. $a_1 = 5, \ a_{k+1} = a_k + 3$

$a_1 = 5$

$a_2 = 5 + 3 = 8$

$a_3 = 8 + 3 = 11$

$a_4 = 11 + 3 = 14$

$a_5 = 14 + 3 = 17$

35. $a_1 = 80, \ a_{k+1} = a_k - \frac{5}{2}$

$a_1 = 80$

$a_2 = 80 - \frac{5}{2} = \frac{160 - 5}{2} = \frac{155}{2}$

$a_3 = \frac{155}{2} - \frac{5}{2} = \frac{150}{2} = 75$

$a_4 = 75 - \frac{5}{2} = \frac{150 - 5}{2} = \frac{145}{2}$

$a_5 = \frac{145}{2} - \frac{5}{2} = \frac{140}{2} = 70$

37. $a_1 = 10, \ d = 4$

$a_n = a_1 + (n - 1)d$

$a_n = 10 + (n - 1)4$

$a_n = 10 + 4n - 4$

$a_n = 4n + 6$

39. $a_1 = 1000, \ a_2 = 950$

$d = a_2 - a_1$

$\quad = 950 - 1000 = -50$

$a_n = a_1 + (n - 1)d$

$a_n = 1000 + (n - 1)(-50)$

$a_n = 1000 - 50n + 50$

$a_n = -50n + 1050$

41. $\displaystyle\sum_{k=1}^{12}(7k - 5) = 2 + 9 + 16 + \cdots + 79$

Sum of arithmetic sequence: $\frac{n}{2}(a_1 + a_n)$

$n = 12, \ a_1 = 7(1) - 5 = 2, \ a_{12} = 7(12) - 5 = 79$

$\displaystyle\sum_{k=1}^{12}(7k - 5) = \frac{12}{2}(2 + 79) = 6(81) = 486$

43. $\displaystyle\sum_{j=1}^{100}\frac{j}{4} = \frac{1}{4} + \frac{2}{4} + \frac{3}{4} + \cdots + \frac{100}{4}$

Sum of arithmetic sequence: $\frac{n}{2}(a_1 + a_n)$

$n = 100, \ a_1 = \frac{1}{4}, \ a_{100} = \frac{100}{4} = 25$

$\displaystyle\sum_{j=1}^{100}\frac{j}{4} = \frac{100}{2}\left(\frac{1}{4} + 25\right) = 50\left(\frac{101}{4}\right) = \frac{2525}{2}$ or 1262.5

45. $\displaystyle\sum_{i=1}^{60}(1.25i + 4) = 2527.5$

47. $a_n = a_1 r^{n-1}$

$a_n = 10(3)^{n-1}$

$a_1 = 10(3)^{1-1} = 10$

$a_2 = 10(3)^{2-1} = 30$

$a_3 = 10(3)^{3-1} = 90$

$a_4 = 10(3)^{4-1} = 270$

$a_5 = 10(3)^{5-1} = 810$

49. $a_n = a_1 r^{n-1}$

$a_n = 100\left(-\frac{1}{2}\right)^{n-1}$

$a_1 = 100\left(-\frac{1}{2}\right)^{1-1} = 100$

$a_2 = 100\left(-\frac{1}{2}\right)^{2-1} = -50$

$a_3 = 100\left(-\frac{1}{2}\right)^{3-1} = 25$

$a_4 = 100\left(-\frac{1}{2}\right)^{4-1} = -12.5$

$a_5 = 100\left(-\frac{1}{2}\right)^{5-1} = 6.25$

51. $a_1 = 3$, $a_{k+1} = 2a_k$

$a_1 = 3$

$a_2 = 2(3) = 6$

$a_3 = 2(6) = 12$

$a_4 = 2(12) = 24$

$a_5 = 2(24) = 48$

53. $a_1 = 1$, $r = -\frac{2}{3}$

$a_n = a_1 r^{n-1}$

$a_n = 1\left(-\frac{2}{3}\right)^{n-1}$

$a_n = \left(-\frac{2}{3}\right)^{n-1}$

55. $a_1 = 24$, $a_2 = 48$

$r = \frac{48}{24} = 2$

$a_n = a_1 r^{n-1}$

$a_n = 24(2)^{n-1}$

57. $a_1 = 12$, $a_4 = -\frac{3}{2}$

$a_4 = a_1 r^3$

$-\frac{3}{2} = 12r^3$

$-\frac{3}{24} = r^3$

$-\frac{1}{8} = r^3$

$\sqrt[3]{-\frac{1}{8}} = r$

$-\frac{1}{2} = r$

$a_n = a_1 r^{n-1}$

$a_n = 12\left(-\frac{1}{2}\right)^{n-1}$

59. $\displaystyle\sum_{n=1}^{12} 2^n = 2^1 + 2^2 + 2^3 + \cdots + 2^{12}$

Sum of geometric sequence: $a_1\left(\dfrac{r^n - 1}{r - 1}\right)$

$a_1 = 2$, $r = 2$, $n = 12$

$$\sum_{n=1}^{12} 2^n = 2\left(\frac{2^{12} - 1}{2 - 1}\right) = 2(2^{12} - 1) = 8190$$

61. $\displaystyle\sum_{n=1}^{8} 5\left(-\frac{3}{4}\right) = 5\left(-\frac{3}{4}\right)^1 + 5\left(-\frac{3}{4}\right)^2 + 5\left(-\frac{3}{4}\right)^3 + \cdots + 5\left(-\frac{3}{4}\right)^8$

Sum of geometric sequence: $a_1\left(\dfrac{r^n - 1}{r - 1}\right)$

$a_1 = 5\left(-\frac{3}{4}\right)$, $r = -\frac{3}{4}$, $n = 8$

$$\sum_{k=1}^{8} 5\left(-\frac{3}{4}\right)^k = 5\left(-\frac{3}{4}\right)\left[\frac{\left(-\frac{3}{4}\right)^8 - 1}{-\frac{3}{4} - 1}\right] = 5\left(-\frac{3}{4}\right)\left[\frac{\left(-\frac{3}{4}\right)^8 - 1}{-\frac{7}{4}}\right] = 5\left(-\frac{3}{4}\right)\left[\left(-\frac{3}{4}\right)^8 - 1\right]\left(-\frac{4}{7}\right) \approx -1.928$$

63. $\displaystyle\sum_{i=1}^{8} (1.25)^{i-1} = (1.25)^0 + (1.25)^1 + (1.25)^2 + \cdots + (1.25)^7$

Sum of geometric sequence: $a_1\left(\dfrac{r^n - 1}{r - 1}\right)$

$a_1 = 1$, $r = 1.25$, $n = 8$

$$\sum_{i=1}^{8} (1.25)^{i-1} = 1\left[\frac{(1.25)^8 - 1}{1.25 - 1}\right] = \left[\frac{(1.25)^8 - 1}{0.25}\right] \approx 19.842$$

65. $\displaystyle\sum_{n=1}^{120} 500(1.01)^n = 500(1.01)^1 + 500(1.01)^2 + 500(1.01)^3 + \cdots + 500(1.01)^{120}$

Sum of geometric sequence: $a_1\left(\dfrac{r^n - 1}{r - 1}\right)$

$a_1 = 500(1.01)$, $r = 1.01$, $n = 120$

$$\sum_{n=1}^{120} 500(1.01)^n = 500(1.01)\left[\frac{(1.01)^{120} - 1}{1.01 - 1}\right] = 500(1.01)\left[\frac{(1.01)^{120} - 1}{0.01}\right] \approx 116{,}169.54$$

67. $\displaystyle\sum_{k=1}^{50} 50(1.2)^{k-1} = 2{,}274{,}859.538$

69. $\displaystyle\sum_{i=0}^{\infty} a_1 r^i = \dfrac{a_1}{1-r}$

$a_1 = 1$ and $r = \dfrac{7}{8}$

$\displaystyle\sum_{i=1}^{\infty} \left(\dfrac{7}{8}\right)^{i-1} = \dfrac{7/8}{1-(7/8)}$

$\qquad = \dfrac{1}{1/8}$

$\qquad = 8$

71. $\displaystyle\sum_{i=0}^{\infty} a_1 r^{i-1} = \dfrac{a_1}{1-r}$

$a_1 = 4$ and $r = \dfrac{2}{3}$

$\displaystyle\sum_{k=1}^{\infty} 4\left(\dfrac{2}{3}\right)^{k-1} = \dfrac{4}{1-(2/3)}$

$\qquad = \dfrac{4}{1/3}$

$\qquad = 12$

73. $\displaystyle {}_8C_3 = \dfrac{8!}{5! \cdot 3!}$

$\qquad = \dfrac{8 \cdot 7 \cdot 6 \cdot 5!}{5! \cdot 3 \cdot 2 \cdot 1} = 56$

75. $\displaystyle {}_{12}C_0 = \dfrac{12!}{12! \cdot 0!}$

$\qquad = \dfrac{12!}{12! \cdot 1} = 1$

77. $\displaystyle {}_{40}C_4 = 91{,}390$

79. $\displaystyle {}_{25}C_6 = 177{,}100$

81. Fourth row: 1   4   6   4   1

$(x+9)^4 = x^4 + 4x^3(9) + 6x^2(9)^2 + 4x(9)^3 + 9^4$

$\qquad = x^4 + 36x^3 + 486x^2 + 2916x + 6561$

83. Third row: 1   3   3   1

$(5-2x)^3 = 5^3 + 3(5)^2(-2x) + 3(5)(-2x)^2 + (-2x)^3$

$\qquad = 125 - 150x + 60x^2 - 8x^3$

85. $(x+1)^{10} = x^{10} + 10x^9(1) + {}_{10}C_2 x^8(1)^2 + {}_{10}C_3 x^7(1)^3 + {}_{10}C_4 x^6(1)^4 + {}_{10}C_5 x^5(1)^5 + {}_{10}C_6 x^4(1)^6 + {}_{10}C_7 x^3(1)^7 + {}_{10}C_8 x^2(1)^8 +$

$\qquad 10x(1)^9 + (1)^{10}$

$(x+1)^{10} = x^{10} + 10x^9(1) + 45x^8(1)^2 + 120x^7(1)^3 + 210x^6(1)^4 + 252x^5(1)^5 + 210x^4(1)^6 + 120x^3(1)^7 + 45x^2(1)^8 +$

$\qquad 10x(1)^9 + (1)^{10}$

$\qquad = x^{10} + 10x^9 + 45x^8 + 120x^7 + 210x^6 + 252x^5 + 210x^4 + 120x^3 + 45x^2 + 10x + 1$

87. $(y-2)^6 = y^6 - 6y^5(2) + {}_6C_2 y^4(2)^2 - {}_6C_3 y^3(2)^3 + {}_6C_4 y^2(2)^4 - 6y(2)^5 + (2)^6$

$(y-2)^6 = y^6 - 6y^5(2) + 15y^4(2)^2 - 20y^3(2)^3 + 15y^2(2)^4 - 6y(2)^5 + (2)^6$

$\qquad = y^6 - 12y^5 + 15y^4(4) - 20y^3(8) + 15y^2(16) - 6y(32) + 64$

$\qquad = y^6 - 12y^5 + 60y^4 - 160y^3 + 240y^2 - 192y + 64$

89. $\left(\dfrac{1}{2} - x\right)^8 = \left(\dfrac{1}{2}\right)^8 - 8\left(\dfrac{1}{2}\right)^7 x + {}_8C_2\left(\dfrac{1}{2}\right)^6 x^2 - {}_8C_3\left(\dfrac{1}{2}\right)^5 x^3 + {}_8C_4\left(\dfrac{1}{2}\right)^4 x^4 - {}_8C_5\left(\dfrac{1}{2}\right)^3 x^5 + {}_8C_6\left(\dfrac{1}{2}\right)^2 x^6 - 8\left(\dfrac{1}{2}\right)x^7 + x^8$

$\left(\dfrac{1}{2} - x\right)^8 = \left(\dfrac{1}{2}\right)^8 - 8\left(\dfrac{1}{2}\right)^7 x + 28\left(\dfrac{1}{2}\right)^6 x^2 - 56\left(\dfrac{1}{2}\right)^5 x^3 + 70\left(\dfrac{1}{2}\right)^4 x^4 - 56\left(\dfrac{1}{2}\right)^3 x^5 + 28\left(\dfrac{1}{2}\right)^2 x^6 - 8\left(\dfrac{1}{2}\right)x^7 + x^8$

$\qquad = \dfrac{1}{256} - \dfrac{1}{16}x + \dfrac{7}{16}x^2 - \dfrac{7}{4}x^3 + \dfrac{35}{8}x^4 - 7x^5 + 7x^6 - 4x^7 + x^8$

91. $(10x + 3y)^4 = (10x)^4 + 4(10x)^3(3y) + {}_4C_2(10x)^2(3y)^2 + 4(10x)(3y)^3 + (3y)^4$

$\qquad = 10{,}000x^4 + 12{,}000x^3y + 5400x^2y^2 + 1080xy^3 + 81y^4$

93. $(u^2 + v^3)^9 = (u^2)^9 + 9(u^2)^8(v^3) + {}_9C_2(u^2)^7(v^3)^2 + {}_9C_3(u^2)^6(v^3)^3 + {}_9C_4(u^2)^5(v^3)^4 + {}_9C_5(u^2)^4(v^3)^5 + {}_9C_6(u^2)^3(v^3)^6 +$

$\qquad {}_9C_7(u^2)^2(v^3)^7 + 9(u^2)(v^3)^8 + (v^3)^9$

$\qquad = u^{18} + 9u^{16}v^3 + 36u^{14}v^6 + 84u^{12}v^9 + 126u^{10}v^{12} + 126u^8v^{15} + 84u^6v^{18} + 36u^4v^{21} + 9u^2v^{24} + v^{27}$

95. The term involving x^5 in the expansion of $(x - 3)^{10}$:

$$_{10}C_5 x^5 (-3)^5 = 252x^5(-243) = -61,236x^5$$

Coefficient: $-61,236$

97. The term involving ab^5 in the expansion of $(a + b)^6$

$$_6C_5 ab^5 = 6ab^5$$

Coefficient: 6

99. Fourth term: $_5C_3(2x)^2(-3y)^3 = 10(4x^2)(-27y^3)$

$$= -1080x^2y^3$$

101. Seventh term: $_9C_6(10x)^3(-3y)^6 = 84(1000x^3)(729y^6)$

$$= 61,236,000x^3y^6$$

103. $A_n = 5000\left(1 + \dfrac{0.08}{4}\right)^n$

(a) $A_1 = 5000\left(1 + \frac{0.08}{4}\right)^1 = \$5100.00$

$A_2 = 5000\left(1 + \frac{0.08}{4}\right)^2 = \$5202.00$

$A_3 = 5000\left(1 + \frac{0.08}{4}\right)^3 \approx \$5306.04$

$A_4 = 5000\left(1 + \frac{0.08}{4}\right)^4 \approx \$5412.16$

$A_5 = 5000\left(1 + \frac{0.08}{4}\right)^5 \approx \$5520.40$

$A_6 = 5000\left(1 + \frac{0.08}{4}\right)^6 \approx \$5630.81$

$A_7 = 5000\left(1 + \frac{0.08}{4}\right)^7 \approx \$5743.43$

$A_8 = 5000\left(1 + \frac{0.08}{4}\right)^8 \approx \$5858.30$

(b) $A_{40} = 5000\left(1 + \frac{0.08}{4}\right)^{40} = \$11,040.20$

(c)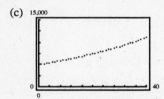

105. $225 + 226 + 227 + \cdots + 300$

Sum of arithmetic sequence with $d = 1$

$a_1 = 225, \ d = 1$

$a_n = a_1 + (n - 1)d$

$a_n = 225 + (n - 1)1$

$a_n = 225 + n - 1$

$a_n = n + 224$

$300 = n + 224$

$76 = n$

Sum: $\dfrac{n}{2}(a_1 + a_n)$

$n = 76, \ a_1 = 225, \ a_{76} = 300$

$$\sum_{n=1}^{76}(n + 224) = \frac{76}{2}(225 + 300) = 38(525) = 19,950$$

107. $4.9, 14.7, 24.5, \ldots$

$a_1 = 4.9, \ d = 9.8$

$a_n = a_1 + (n - 1)d$

$a_n = 4.9 + (n - 1)(9.8)$

$\quad = 4.9 + 9.8n - 9.8$

$\quad = 9.8n - 4.9$

$a_{10} = 9.8(10) - 4.9$

$\quad = 93.1$

Sum: $\dfrac{n}{2}(a_1 + a_n)$

$$\sum_{k=1}^{10}(9.8n - 4.9) = \frac{10}{2}(4.9 + 93.1)$$

$$= 5(98)$$

$$= 490 \text{ meters}$$

109. (a) $a_n = 85,000(1.012)^n$

(b) $a_{50} = 85,000(1.012)^{50} \approx 154,328$

The estimated population is approximately 154,328.

111. $\sum\limits_{i=1}^{n} a_1 r^{i-1} = a_1\left(\dfrac{r^n - 1}{r - 1}\right)$

$a_1 = 50\left(1 + \dfrac{0.07}{12}\right)^1$, $r = \left(1 + \dfrac{0.07}{12}\right)$

$\sum\limits_{k=1}^{240} 50\left(1 + \dfrac{0.07}{12}\right)^k = 50\left(1 + \dfrac{0.07}{12}\right)\left[\dfrac{\left(1 + \dfrac{0.07}{12}\right)^{240} - 1}{\left(1 + \dfrac{0.07}{12}\right) - 1}\right]$

$= \$26{,}198.27$

113. $\sum\limits_{i=1}^{n} a_1 r^{i-1} = a_1\left(\dfrac{r^n - 1}{r - 1}\right)$

$a_1 = 100\left(1 + \dfrac{0.10}{12}\right)^1$, $r = \left(1 + \dfrac{0.10}{12}\right)$

$\sum\limits_{k=1}^{480} 100\left(1 + \dfrac{0.10}{12}\right)^k = 100\left(1 + \dfrac{0.10}{12}\right)\left[\dfrac{\left(1 + \dfrac{0.10}{12}\right)^{480} - 1}{\left(1 + \dfrac{0.10}{12}\right) - 1}\right]$

$= \$637{,}678.02$

Chapter Test for Chapter 10

1. $a_n = \left(-\dfrac{2}{3}\right)^{n-1}$

$a_1 = \left(-\dfrac{2}{3}\right)^{1-1} = 1$

$a_2 = \left(-\dfrac{2}{3}\right)^{2-1} = -\dfrac{2}{3}$

$a_3 = \left(-\dfrac{2}{3}\right)^{3-1} = \dfrac{4}{9}$

$a_4 = \left(-\dfrac{2}{3}\right)^{4-1} = -\dfrac{8}{27}$

$a_5 = \left(-\dfrac{2}{3}\right)^{5-1} = \dfrac{16}{81}$

2. $a_1 = 76, a_{k+1} = \dfrac{a_k}{2} + 6$

$a_1 = 76$

$a_2 = \dfrac{76}{2} + 6 = 44$

$a_3 = \dfrac{44}{2} + 6 = 28$

$a_4 = \dfrac{28}{2} + 6 = 20$

$a_5 = \dfrac{20}{2} + 6 = 16$

3. $\dfrac{1}{2 \cdot 3}, \dfrac{1}{3 \cdot 4}, \dfrac{1}{4 \cdot 5}, \dfrac{1}{5 \cdot 6}, \ldots$

$a_n = \dfrac{1}{(n + 1)(n + 2)}$

(There are many correct answers.)

4. $\sum\limits_{n=1}^{5} \dfrac{n!}{7} = \dfrac{1!}{7} + \dfrac{2!}{7} + \dfrac{3!}{7} + \dfrac{4!}{7} + \dfrac{5!}{7}$

$= \dfrac{1}{7} + \dfrac{2}{7} + \dfrac{6}{7} + \dfrac{24}{7} + \dfrac{120}{7}$

$= \dfrac{153}{7}$

5. $\dfrac{2}{3(1) + 1} + \dfrac{2}{3(2) + 1} + \cdots + \dfrac{2}{3(12) + 1} = \sum\limits_{k=1}^{12} \dfrac{2}{3k + 1}$

6. $a_1 = 12, d = 4$

$a_n = 12 + (n - 1)4$

$a_n = 12 + 4n - 4$

$a_n = 4n + 8$

$a_1 = 4(1) + 8 = 12$

$a_2 = 4(2) + 8 = 16$

$a_3 = 4(3) + 8 = 20$

$a_4 = 4(4) + 8 = 24$

$a_5 = 4(5) + 8 = 28$

7. $a_n = 5000, d = -100$

$a_n = a_1 + (n - 1)d$

$a_n = 5000 + (n - 1)(-100)$

$a_n = 5000 - 100n + 100$

$a_n = -100n + 5100$

8. $2, -3, \dfrac{9}{2}, -\dfrac{27}{4}, \dfrac{81}{8}, \ldots$

$r = \dfrac{a_2}{a_1} = -\dfrac{3}{2}$

Note: $\dfrac{9/2}{-3} = \dfrac{-27/4}{9/2}$

$= \dfrac{81/8}{-27/4}$

$= -\dfrac{3}{2}$

9. $a_1 = 4$, $r = \frac{1}{2}$

$a_n = a_1 r^{n-1}$

$a_n = 4\left(\frac{1}{2}\right)^{n-1}$

10. $\displaystyle\sum_{j=0}^{4} (3j + 1) = [3(0) + 1] + [3(1) + 1] + [3(2) + 1] + [3(3) + 1] + [3(4) + 1] = 1 + 4 + 7 + 10 + 13 = 35$

11. $\displaystyle\sum_{n=1}^{10} 3\left(\frac{1}{2}\right)^n$

Sum of a geometric sequence with $a_1 = \frac{3}{2}$, $n = 10$, $r = \frac{1}{2}$

$$\sum_{n=1}^{10} 3\left(\frac{1}{2}\right)^n = \frac{\frac{3}{2}\left[\left(\frac{1}{2}\right)^{10} - 1\right]}{\left(\frac{1}{2}\right) - 1} = \frac{3069}{1024}$$

or

$$\sum_{n=1}^{10} 3\left(\frac{1}{2}\right)^n = 3\left(\frac{1}{2}\right)^1 + 3\left(\frac{1}{2}\right)^2 + 3\left(\frac{1}{2}\right)^3 + 3\left(\frac{1}{2}\right)^4 + 3\left(\frac{1}{2}\right)^5 + 3\left(\frac{1}{2}\right)^6 + 3\left(\frac{1}{2}\right)^7 + 3\left(\frac{1}{2}\right)^8 + 3\left(\frac{1}{2}\right)^9 + 3\left(\frac{1}{2}\right)^{10}$$

$$= \frac{3}{2} + \frac{3}{4} + \frac{3}{8} + \frac{3}{16} + \frac{3}{32} + \frac{3}{64} + \frac{3}{128} + \frac{3}{256} + \frac{3}{512} + \frac{3}{1024}$$

$$= \frac{3069}{1024}$$

12. $\displaystyle\sum_{i=0}^{\infty} a_1 r^i = \frac{a_1}{1 - r}$

$a_1 = 4, r = \dfrac{2}{3}$

$\displaystyle\sum_{i=1}^{\infty} 4\left(\frac{2}{3}\right)^{i-1} = \frac{4}{1 - (2/3)}$

$$= \frac{4}{1/3}$$

$$= 12$$

13. $3 + 6 + 9 + 12 + \cdots$, $n = 50$

Sum of arithmetic sequence with $d = 3$

$a_n = a_1 + (n - 1)d$

$a_n = 3 + (n - 1)3$

$a_n = 3 + 3n - 3$

$a_n = 3n$

Sum: $\dfrac{n}{2}(a_1 + a_n)$

$n = 50$, $a_1 = 3(1) = 3$, $a_{50} = 3(50) = 150$

$\displaystyle\sum_{n=1}^{50} 3n = \frac{50}{2}(3 + 150) = 25(153) = 3825$

14. Total $= 50\left(1 + \dfrac{0.08}{12}\right)^1 + 50\left(1 + \dfrac{0.08}{12}\right)^2 + 50\left(1 + \dfrac{0.08}{12}\right)^3 + \cdots + 50\left(1 + \dfrac{0.08}{12}\right)^{300}$

$$= 50\left(1 + \frac{0.08}{12}\right)\frac{\left(1 + \dfrac{0.08}{12}\right)^{300} - 1}{\left(1 + \dfrac{0.08}{12}\right) - 1} = \$47,868.33$$

15. $_{20}C_3 = \dfrac{20!}{(20 - 3)!3!} = \dfrac{20 \cdot 19 \cdot 18}{3 \cdot 2 \cdot 1} = 1140$

16. $(x - 2)^5$ 5th row coefficients: $1, 5, 10, 10, 5, 1$

$(x - 2)^5 = x^5 - 5x^4(2) + 10x^3(2)^2 - 10x^2(2)^3 + 5x(2)^4 - 2^5$

$= x^5 - 10x^4 + 10x^3(4) - 10x^2(8) + 5x(16) - 32$

$= x^5 - 10x^4 + 40x^3 - 80x^2 + 80x - 32$

17. $(x + y)^8$, x^3y^5

$${}_8C_5x^3y^5 = {}_8C_3x^3y^5$$

$$= \frac{8 \cdot 7 \cdot 6}{3 \cdot 2 \cdot 1}x^3y^5 = 56x^3y^5$$

Coefficient: 56

18. $\displaystyle\sum_{r=1}^{n} a_1r^{i-1} = a_1\left(\frac{r^n - 1}{r - 1}\right)$

$a_1 = 35,000, r = 1.05$

$$\sum_{k=1}^{15} 35,000(1.05)^{k-1} = 35,000\left(\frac{1.05^{15} - 1}{1.05 - 1}\right)$$

$$= \$755,249.73$$

Cumulative Test for Chapters 1–10

1. $(-3x^2y^3)^2 \cdot (4xy^2) = (9x^4y^6)(4xy^2) = 36x^5y^8$

2. $\frac{3}{8}x - \frac{1}{12}x + 8 = \frac{9}{24}x - \frac{2}{24}x + 8 = \frac{7}{24}x + 8$

3. $\frac{64r^2s^4}{16rs^2} = 4rs^2$

4. $\left(\frac{3x}{4y^3}\right)^2 = \frac{(3x)^2}{(4y^3)^2} = \frac{9x^2}{16y^6}$

5. $\frac{8}{\sqrt{10}} = \frac{8\sqrt{10}}{\sqrt{10} \cdot \sqrt{10}} = \frac{8\sqrt{10}}{10} = \frac{4\sqrt{10}}{5}$

6. $\log_4 64 = 3$

7. $5x - 20x^2 = 5x(1 - 4x)$

8. $64 - (x - 6)^2 = 8^2 - (x - 6)^2$

$$= [8 + (x - 6)][8 - (x - 6)]$$

$$= (2 + x)(14 - x)$$

9. $15x^2 - 16x - 15 = (5x + 3)(3x - 5)$

10. $8x^3 + 1 = (2x)^3 + 1^3 = (2x + 1)(4x^2 - 2x + 1)$

11. $y = 2x - 3$

12. $y = -\frac{3}{4}x + 2$

13. $9x^2 + 4y^2 = 36$

$$\frac{9x^2}{36} + \frac{4y^2}{36} = \frac{36}{36}$$

$$\frac{x^2}{4} + \frac{y^2}{9} = 1$$

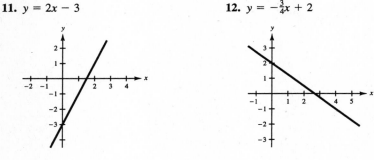

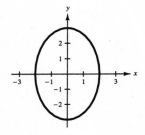

14. $(x - 2)y = 5$

$$y = \frac{5}{x - 2}$$

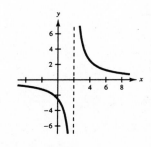

15. $\left(\frac{3}{2}, 8\right)$, $\left(\frac{11}{2}, \frac{5}{2}\right)$

Point: $\left(\frac{3}{2}, 8\right)$

Slope $m = \dfrac{\frac{5}{2} - 8}{\frac{11}{2} - \frac{3}{2}} = \dfrac{-\frac{11}{2}}{4} = -\frac{11}{8}$

$$y - y_1 = m(x - x_1)$$

$$y - 8 = -\frac{11}{8}\left(x - \frac{3}{2}\right)$$

$$y = -\frac{11}{8}x + \frac{33}{16} + 8$$

$$y = -\frac{11}{8}x + \frac{161}{16}$$

16. $h(x) = \sqrt{16 - x^2}$

Domain: $[-4, 4]$

17. $g(x) = \dfrac{x}{x + 10}$

$g(c - 6) = \dfrac{c - 6}{(c - 6) + 10} = \dfrac{c - 6}{c + 4}$

18. $N = \dfrac{k}{t + 1}$

$300 = \dfrac{k}{0 + 1}$

$300 = \dfrac{k}{1} = k$

$N = \dfrac{300}{5 + 1} = \dfrac{300}{6} = 50$

19. Vertex: $(2, 2) \implies y = a(x - 2)^2 + 2$

Point: $(0, 4) \implies 4 = a(0 - 2)^2 + 2$

$4 = a(4) + 2$

$2 = 4a$

$\tfrac{1}{2} = a$

Parabola: $y = \tfrac{1}{2}(x - 2)^2 + 2$ or $(x - 2)^2 = 2(y - 2)$

20. $P = 6000 + 20{,}000(3)^{-0.2t}$

(a) $13{,}000 = 6000 + 20{,}000(3)^{-0.2t}$

$7000 = 20{,}000(3)^{-0.2t}$

$0.35 = 3^{-0.2t}$

$-0.2t = \log_3 0.35$

$t = \dfrac{\log_3 0.35}{-0.2}$

$t \approx 4.8$ years

The profit reached \$13,000 during the fourth year.

(b)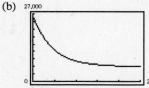

21. $H = 62 + 35 \log_{10}(x - 4), x = 10$

$H = 62 + 35 \log_{10}(10 - 4)$

$H = 62 + 35 \log_{10} 6$

$H \approx 89.24 \implies 4'6'' = 89.2\%$ (Adult height)

54 inches $= 89.2\%$ (Adult height)

$\dfrac{54}{0.892} =$ Adult height

60.5 inches $\approx$ Adult height

5 feet $\approx$ Adult height

The girl can expect to be approximately 5 feet tall as an adult.

22. $(z - 3)^4 = z^4 + {}_4C_1 z^3(-3) + {}_4C_2 z^2(-3)^2 + {}_4C_3 z(-3)^3 + (-3)^4$

$(z - 3)^4 = z^4 + 4z^3(-3) + 6z^2(-3)^2 + 4z(-3)^3 + (-3)^4$

$= z^4 - 12z^3 + 54z^2 - 108z + 81$